Greenhouse Effect, Sea Level and Drought

NATO ASI Series

Advanced Science Institutes Series

A Series presenting the results of activities sponsored by the NATO Science Committee, which aims at the dissemination of advanced scientific and technological knowledge, with a view to strengthening links between scientific communities.

The Series is published by an international board of publishers in conjunction with the NATO Scientific Affairs Division

A Life Sciences	Plenum Publishing Corporation
B Physics	London and New York
C Mathematical	Kluwer Academic Publishers
and Physical Sciences	Dordrecht, Boston and London
D Behavioural and Social Sciences	
E Applied Sciences	
F Computer and Systems Sciences	Springer-Verlag
G Ecological Sciences	Berlin, Heidelberg, New York, London,
H Cell Biology	Paris and Tokyo

Series C: Mathematical and Physical Sciences - Vol. 325

Greenhouse Effect, Sea Level and Drought

edited by

Roland Paepe
Earth Technology Institute,
Vrije Universiteit Brussel and Belgian Geological Survey,
Brussels, Belgium

Rhodes W. Fairbridge
NASA, GSFC Institute for Space Studies,
New York, NY, U.S.A.

and

Saskia Jelgersma
Geological Survey of the Netherlands,
Haarlem, The Netherlands

Technical editor:
Marinus A. Pool
Earth Technology Institute,
Vrije Universiteit Brussel,
Brussels, Belgium

Kluwer Academic Publishers

Dordrecht / Boston / London

Published in cooperation with NATO Scientific Affairs Division

Proceedings of the NATO Advanced Research Workshop on
Geohydrological Management of Sea Level and Mitigation of Drought
Fuerteventura, Canary Islands, Spain
March 1–7, 1989

Library of Congress Cataloging-in-Publication Data

NATO Advanced Research Workshop on Geohydrological Management of Sea
    Level and Mitigation of Drought (1989 : Fuerteventura, Canary
    Islands)
    Greenhouse effect, sea level, and drought : proceedings of the
    NATO Advanced Research Workshop on Geohydrological Management of Sea
    Level and Mitigation of Drought, Fuerteventura, Canary Islands
    (Spain), March 1–7, 1989 / edited by Roland Paepe, Rhodes W.
    Fairbridge, and Saskia Jelgersma.
        p.   cm. -- (NATO ASI series. Series C., Mathematical and
    physical sciences ; no. 325)
    "Published in cooperation with NATO Scientific Affairs Division."
    Includes index.

    ISBN-13: 978-94-010-6801-7   e-ISBN-13: 978-94-009-0701-0
    DOI: 10.1007/978-94-009-0701-0

    1. Sea level--Congresses. 2. Greenhouse effect, Atmospheric-
    -Congresses. 3. Droughts--Congresses.   I. Paepe, Roland.
    II. Fairbridge, Rhodes Whitmore, 1914-   . III. Jelgersma, Saskia,
    1929-   . IV. Title. V. Series.
    GC89.N37 1989
    551.46--dc20   90-48531

ISBN-13: 978-94-010-6801-7

Published by Kluwer Academic Publishers,
P.O. Box 17, 3300 AA Dordrecht, The Netherlands.

Kluwer Academic Publishers incorporates the publishing programmes of
D. Reidel, Martinus Nijhoff, Dr W. Junk and MTP Press.

Sold and distributed in the U.S.A. and Canada
by Kluwer Academic Publishers,
101 Philip Drive, Norwell, MA 02061, U.S.A.

In all other countries, sold and distributed
by Kluwer Academic Publishers Group,
P.O. Box 322, 3300 AH Dordrecht, The Netherlands.

Printed on acid-free paper

TABLE OF CONTENTS

PREFACE

Shortly after the creation of the Vrije Universiteit Brussel (Free University Brussels) in 1970, currently labelled as VUB, a Department of Quaternary Geology was installed within the Faculty of Science in 1974. At the beginning it dealt mainly with the study of periglacial loess deposits of the Pleistocene Glacial Period in Central Belgium and with coastal deposits in relation to sea level rise during the warm Holocene period covering the last 10,000 years, in which the dawn of civilization took place step by step.

Today the same research teams widen their scope of interest: they are presently studying the loess plateau in the People's Republic of China and the world-wide problems associated with sea level rise, coastal erosion being one of the most devastating natural hazards. More and more emphasis is put on problems concerning environmental engineering and those dealing with global change. Since 1975 UNESCO sponsored a number of symposia of the International Union for Quaternary Research (INQUA), whose secretariat was located on the VUB Campus grounds from 1973 to 1982. In 1981 the Applied Geology Department of the Faculty of Applied Sciences was created.

The NATO-Advanced Research Workshop (ARW), organized in Fuerteventura (Canary Islands, Spain) in March 1989 was a climax of this series of Global Change gatherings. As Rector of the VUB, I am satisfied that the VUB, through its Earth Technology Institute, could cooperate with NATO and the National Science Foundations of both USA and Belgium in cosponsoring such an initiative.

Prof. Dr. S. Loccufier, Rector of the Vrije Universiteit Brussel.

INTRODUCTION

It was again one of Rhodes W. Fairbridge's brilliant ideas to organize a meeting on "Greenhouse Effects, Sea Level Change and Mitigation of Drought" in Europe's southernmost territory, namely the Canary Islands (Spain) and moreover, on one of its most desertic islands, Fuerteventura which literally means a "strong venture", or was it "all the best for the challenge". What's in a name? It reminds me of "fuerte ventoso" very windy in the English translation. Certainly the Island's well known periodical strong winds control, since long, the future of the island as a hazardous venture for which some "good luck" may help to survive!

"Strong winds", or even stronger winds in the future are indeed an indication of dramatically changing climatic conditions or at least of an increasing aridity precisely in that part of the world just offshore the Sahara and more specifically the Sahel.

However, the theme was sufficiently attractive to motivate more then fifty top scientists, not only from NATO countries but also from Australia and Africa as well as from the USSR and the People's Republic of China, to bring their worldwide knowledge together in order to answer one of the most challenging and controversial questions: is there a warming of the Earth causing a sea level rise on one side and a steady growing desertification on the other?

Fairbridge formulated this apparently contradictional statement rather concisely into one sentence: *How to reduce sea level rise and mitigate drought in the Sahel and other similar regions*. A question which at first seemed easy to answer became more and more difficult during the one week meeting at Fuerteventura, the first week of March 1989.

During a four-days session climatologists and geologists exchanged ideas on the causes of the Greenhouse Effect and discussed in particular the general opinion that the warming of the climate during the last century is a direct result of industrial pollution and in particular of the burning of fossil fuels. It is, and becomes more and more controversial whether this trend in the Earth's climatic evolution is naturally biassed and if so, how to disentangle its cyclicity from man-induced temperature rise. Once again the lack of proxy data was held responsible for the weakness of modelling of the past and future climatic conditions. Particular stress was put upon the lack of continental geo-data on the long term or last three million years, the middle term or the last interglacial/glacial cycle, the short term or holocene record of say the last 10,000 years, and even only of historical times.

Other questions regarding the direct consequences of global warming (such as desertification processes and sea level rise) have been discussed in the same spirit of controversial possible explanations. It clearly came to the fore that, whereas desertification was generally recognized as a global ongoing process, sea level rise was not. Indeed, it was stated that in many places of the world even a drop in sea level is observed. Historical desertifications have been detected by studying a number of archaeological sites in Greece and Israel where droughts seem to respond to a 1000 year cycle, namely: during the 8th century BC, the Late Roman Period (2nd century AD), the 12th century AD, and finally the present drought.

The importance of the effect of neotectonic, and sea floor movements was generally recognized although little, if any, specific studies have shown the direct relationship between such movements and the atmospheric CO_2-level.

From the technical papers on case studies and management in the various fields of greenhouse effects and sea level changes it became evident that the control of e.g. sea level rise is not economically feasible for the next decades. Other, more efficient measures might be worked out in the meantime to mitigate drought in badly affected parts of the globe such as the Chad basin.

The session reports, the panel discussions as well as the recommendations were reviewed and finally endorsed during a one-morning plenary session. The entire results are reproduced here as an annex at the end of this volume. Aspects of the feasibility, rainfall distribution, general climate, deserts and droughts, environmental changes and hazards, continental water, planning techniques in coastal areas are highlighted as well as engineering solutions and their cost-benefit ratios.

Even though bringing up far more questions about the Greenhouse Effect, sea level changes and the mitigation of droughts than solutions, this publication should be considered as a fair contribution of the State of the Art of all topics concerned and an encouragement for the development of further studies on these matters in the future

Acknowledgments

The preparation of the NATO-Advanced Research Workshop (ARW) was initiated by the President R.W. Fairbridge (NASA - Goddard Space Center, New York) and the Director R. Paepe (Vrije Universiteit Brussel - VUB; Belgian Geological Survey). At a later stage they received strong support from S. Jelgersma (Geological Survey of the Netherlands), H. Faure (Laboratoire de Géologie du Quaternaire, Marseille) and C. de Meyer (Harbour and Engineering Consultants - HAECON Ltd, Belgium). This mixed group of earth scientists and engineering experts constituted the organising committee of the symposium.

The local committee in Fuerteventura was composed of Emilio Custodio (Politechnic University of Catalonia, Barcelona), J. Jimenez (Municipality of Las Palmas), J.A. Nunez (Water Resources Service, Las Palmas), L. Puga (Water Resources Service, Tenerife) and J.J. Braojos (Water Resources Service, Tenerife). Together with J. Meco (University of La Laguna, Las Palmas) E. Custodio has organized a most interesting and agreeable field trip to all interesting geological and hydrological points of the islands.

To all Spanish colleagues, our honourable hosts in the deepest South of Europe, our heartfelt thanks for their never ending endeavours to make the symposium a success and offer a most comfortable stay for all foreign participants at the beaches of Fuerteventura.

The financial support of the NATO-ARW Programme, US National Science Foundation (NSF), the Belgian National Fund for Scientific Research, the Earth Technology Institute (ETI) of the VUB (Brussels), and the Belgian Geological Survey (BGD-SGB) is greatly acknowledged.

We feel greatly indebted to Dra. E. van Overloop and I. Demolin from ETI-VUB for their great help at the ARW Secretariat to the Organising Committee before, during and after the symposium as well as warm thanks to Mrs. M. Bucquoie and Mrs. E. Mariolakos for helping with both the secretarial and the session work.

Finally, we are most grateful to Drs. M.A. Pool (ETI-VUB) for all his efforts together with A. van der Veen (Leiden, The Netherlands) in taking up the considerable work of the technical production of this book. At the same time we like to express our gratitude to B. Denness (Bureau of Applied Sciences, Isle of Wight) for taking up the thankless task of reviewing the session papers.

R. Paepe, Director of the NATO-ARW.

It is a pity that some of the participants, who produced a quite interesting contribution to the workshop, did not send in their relevant paper for publication in the proceedings, despite tedious efforts from the editors (R. Paepe).

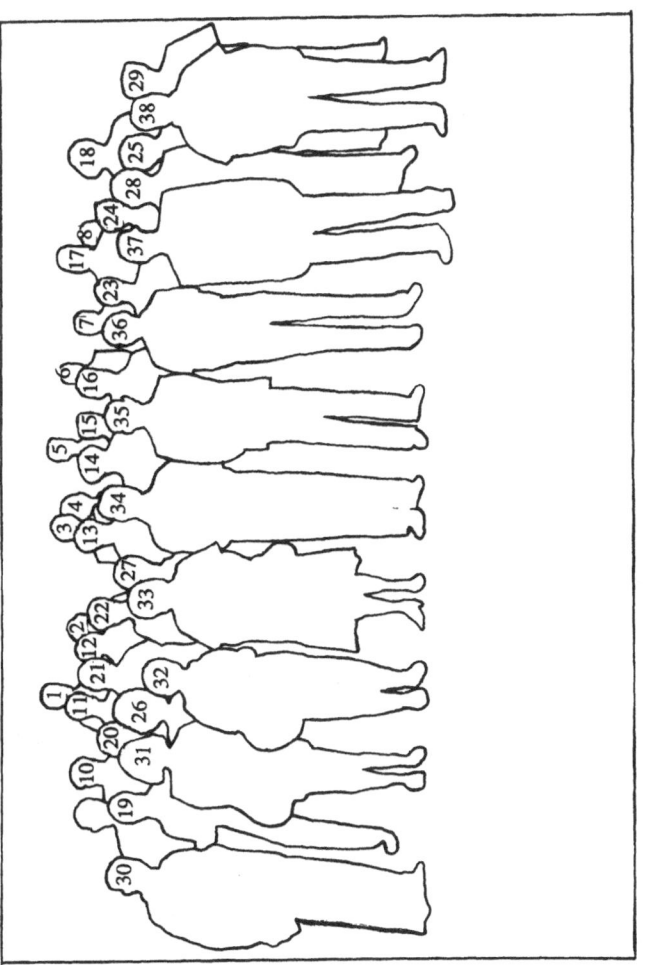

1. WELLS.G.L.
2. SHARP.J.M. Jr.
3. POOL.M.A.
4. CUSTODIO.E.
5. TARDY.Y.
6. KAPLIN.P.
7. KLIGE.R.
8. LLAMAS.M.R.
9. KOGBE.C.A.
10. PATFOORT.G.A.
11. ROMIN.E.
12. JELGERSMA.S.
13. ROEBERT.A.J.

14. WEDEL.P.
15. RAMPINO.M.
16. BLUMENTHAL.M.B.
17. PIRAZZOLI.P.A.
18. MOGUEDET.G.
19. GOLDSMITH.V.
20. DEMOLIN.I.
21. ISSAR.A.
22. PETIT-MAIRE.N.
23. GUIRAUD.R.
24. OLLIER.C.
25. DENNESS.B.
26. Mrs.MARIOLAKOS.I.

27. HOWARD.K.
28. MARIOLAKOS.I.
29. LIU TUNGSHENG
30. FAIRBRIDGE.R.W.
31. VAN OVERLOOP.E.
32. ZAZO.C.
33. BAETEMAN.C.
34. DE BOODT.M.
35. CAMFIELD.S.A.
36. BERGER.A.
37. PAEPE.R.
38. RUTTER.N.W.

LIST OF AUTHORS AND PARTICIPANTS

(a) Director:

Prof.Dr. PAEPE R.
Vrije Universiteit Brussel
Earth Technology Institute
Pleinlaan 2
1050 BRUSSELS (Belgium)
or
Belgian Geological Survey
Jennerstraat 13
1040 BRUSSELS (Belgium)

(b) Key speakers:

Prof.Dr. BERGER A.
Université Catholique de Louvain
Institut d'Astronomie et
d'Astrophysique G. Lemaître
Chemin du Cyclotron 2
1348 LOUVAIN-LA-NEUVE (Belgium)

Prof.Dr.Ir. DE BOODT M.
Rijksuniversiteit Gent
Laboratorium voor Bodemfysika
Coupure Links 653
9000 GENT (Belgium)

Dr.Ir. DE MEYER C.
HAECON S.A.
Deinsesteenweg 110
9810 GENT (Belgium)

Prof.Dr. JELGERSMA S.
Geological Survey of the Netherlands
P.O.box 157
2000 AD HAARLEM (The Netherlands)

Ir. ROMIJN E.
International Association of
Hydrogeologists
c/o Mariënbergweg 1
6862 ZL OOSTERBEEK (The Netherlands)

Prof. Dr. AYALA-GARCEDO F.J.
Instituto Técnológico GeoMinero de España
Rios Rosas 23
28003 MADRID (Spain)

Prof. Dr. FAIRBRIDGE R.W.
Columbia University and
NASA - Goddard Institute for Space Studies
2880 Broadway
NEW YORK, NY 10025 (USA)

Prof. SHARP J. M., Jr
Department of Geological Sciences
University of Texas at Austin
TX 78713-7909 AUSTIN (USA)

(c) Other participants:

Prof. OLLIER C.
University of New England
Geography Department
2351 ARMIDALE (Australia)

Prof.Dr. BAETEMAN C.
Belgian Geological Survey
Jennerstraat 13
1040 BRUSSELS (Belgium)

Prof. PATFOORT G.
Vrije Universiteit Brussel
Afdeling Burgerlijke Bouwkunde
Pleinlaan 2
1050 BRUSSELS (Belgium)

Prof. Dr. DIENG Babacar
Ecole Inter-Etats de l'Equipement Rural
B.P. 7023
OUAGADOUGOU (Burkina-Faso)

Prof. HOWARD K.
Groundwater Research Group
University of Toronto
1265 Military Trail
SCARBOROUGH, ONTARIO MICIA4 (Canada)

Prof. RUTTER N.W.
University of Alberta
Department of Geology
1-26 Earth Sciences Building
T6G 2E3 EDMONTON, ALBERTA (Canada)

Prof. LIU Tungsheng
Academia Sinica - Institute of Geology
P.O. BOX 634
100011 BEIJING (China)

Prof. GUIRAUD R.
Laboratoire de Géologie
Faculté des Sciences
Rue Louis Pasteur 33
84000 AVIGNON (France)

Dr. KOGBE C.A.
Rockview International
Tour Onyx - Rue Vandrezanne 10
75644 PARIS Cedex 13 (France)

Prof. LE HOUÉROU H.N.
CNRS - Centre d'Ecologie
Fonctionnelle et Evolutive
Route de Mende - BP 5051
34003 MONTPELLIER Cedex (France)

Prof. LEROUX M.
Université Jean Moulin
CNRS UA 260 - PICG 252
B.P. 0638
69239 LYON (France)

Dr. MOGUEDET G.
Université d'Angers
Faculté des Sciences
Laboratoire de Géologie
Boulevard Lavoisier - Belle-Beille
49045 ANGERS Cedex (France)

Dr. PETIT-MAIRE N.
Laboratoire de Géologie du Quaternaire
CNRS-Luminy Case 907
13288 MARSEILLE Cedex 9 (France)

Dr. PIRAZZOLI P.A.
(CNRS-Intergeo)
Place Aridstide Bruant 1
92190 Meudon-Bellevue (France)

Prof. TARDY Y.
Université Louis Pasteur-Inst. Géologie
Rue Blessig 1
67084 STRASBOURG (France)
or Institut Francais de Recherche Scientifique
pour le Développement en Coopération
Centre ORSTOM
B. P. 2528
BAMAKO (Mali)

Prof. MARIOLAKOS I.
University of Athens - Department of Geology
Panepistimioupolis
Zographou
15771 ATHENS (Greece)

Prof. ISSAR A.
J. Blaustein Institute for Desert Research
Ben Gurion University of the Negev
Sede Boqer Campus
84993 BEER SHEVA (Israel)

Drs POOL M.A.
Vrije Universiteit Brussel
Earth Technology Institute
Pleinlaan 2
1050 BRUSSELS (Belgium)

Ir ROEBERT A.J.
Gemeentewaterleidingen
Condensatorweg 54
1014 AX AMSTERDAM (The Netherlands)

Dr. CUSTODIO E.
International Groundwater Hydrology Course
Polytechnic University of Catalonia
Jordi Girona Salgado 31
08034 BARCELONA (Spain)

Prof. LLAMAS M.R.
Department of Geodynamics
Complutense University
28040 MADRID (Spain)

Dr. ZAZO C.
Universidad Complutense
Dpto. Geodinamica
28040 MADRID (Spain)

Dr. WEDEL P.
Chalmers University of Technology
Department of Geology
S-41296 GÖTEBORG (Sweden)

Dr. DENNESS B.
Bureau of Applied Sciences
Wydcombe Manor, Whitwell
ISLE OF WIGHT, PO38 2NY (England)

Dr. BLUMENTHAL M.B.
Lamont-Doherty Geological Observatory of
Columbia University
PALISADES, NY 10964 (USA)

Dr. CAMFIELD F.E.
Coastal Engineering Research Center
US Army Engineers Waterways Experiment
Station
P.O.box 631
VICKSBURG, MISSISSIPI 39181-0631 (USA)

Dr. GOLDSMITH V.
Department of Geology and Geography
Hunter College
(City University of New York)
695 Park Avenue
NEW YORK, NY 10021 (USA)

Dr. G. KUKLA
Lamont-Doherty Geological Observatory of
Columbia University
PALISADES, NY 10964 (USA)

Dr. RAMPINO M.
Earth Systems Group
Department of Applied Science
New York University
NEW YORK, NY 10003 (USA)

Dr. WELLS G.
Lunar and Planetary Institute
3303 NASA Road 1
77058-4399 HOUSTON, TEXAS (USA)

Prof. KAPLIN P.
State University of Moscow
Department of Geography
119899 MOSCOW (USSR)

Dr. KLIGE R.
USSR-Academy of Sciences
Water Problems Institute
Sadovo-Chernogriazkaya ul. 13/3
103064 MOSCOW (USSR)

Prof.Dr. ILUNGA LUTUMBA
Institut Supérieur Pédagogique
B.P. 854
BUKAVU (Zaire)

Dr. DA CUNHA L.
NATO - Scientific Affairs Division
1110 Brussels (Belgium)

DEMARÉE G.R.
Royal Meteorological Institute of Belgium
Ringlaan 3
B-1180 BRUSSELS (Belgium)

Dra. VAN OVERLOOP E.
Vrije Universiteit Brussel
Earth Technology Institute
Pleinlaan 2
1050 BRUSSELS (Belgium)

PART I: GREENHOUSE EFFECTS

THE GREENHOUSE EFFECT, STRATOSPHERIC OZONE, MARINE PRODUCTIVITY, AND GLOBAL HYDROLOGY: FEEDBACKS IN THE GLOBAL CLIMATE SYSTEM

MICHAEL R. RAMPINO
Earth Systems group
Department of Applied Science
New York University
New York, NY 10003
and
NASA, Goddard Space Flight Center
Institute for Space Studies
New York, NY 10025

ROBERT ETKINS
NOAA
National Climate Program Office
Rockville, MD 20852

ABSTRACT: The problems of greenhouse warming, ozone depletion and changes in global hydrology are inter-related through a number of feedback mechanisms. Recognized feedbacks act through two important loops involving changes in atmospheric chemistry, UV-B radiation at the earth's surface, and marine productivity. Reliable predictions of future changes in global and regional climate require an understanding of the direction, magnitude and time constants of the various climatic and biochemical feedback processes. Changes in the amount and distribution of precipitation can also have important feedback effects on soil moisture, vegetation, cloudiness, ground and cloud albedo, ocean salinity (through changes in evaporation and runoff), bottom water formation and productivity. Model studies incorporating these and other feedbacks should be useful in determining the sensitivity of the climate system and anthropogenic changes.

1. Introduction

The increasing amounts of trace gases in the atmosphere are contributing to two important global-scale environmental problems - an enhanced greenhouse warming and the depletion of stratospheric ozone (National Research Council 1989). Recent studies of various aspects of the global ocean/atmospheric system suggest that the two may be interrelated through a number of feedback mechanisms (Figure 1). Furthermore, changes in global temperatures and atmospheric composition are expected to cause major changes in global hydrological conditions that will affect natural ecosystems, agriculture and water supply.

R. Paepe et al. (eds.), Greenhouse Effect, Sea Level and Drought, 3–20.

1.1. THE GREENHOUSE EFFECT

The primary greenhouse gases in the earth's atmosphere are water vapour, CO_2, and ozone, along with methane, chlorofluorocarbons (CFC's) and nitrous oxide. These gases are rather inefficient absorbers of solar radiation, and thus allow much of the incoming solar energy to pass through the atmosphere to heat the earth's surface. They are, however, good absorbers of the long-wave (IR) radiation that is re-emitted from the surface. (Some molecules are much better IR absorbers than others - one molecule of CFC 12, for example, is 10,000 times more effective than a molecule of CO_2). A proportion of outgoing energy absorbed by these gases is reradiated back toward the planet producing an additional warming of the troposphere and earth's surface. This is the "greenhouse effect".

A simple energy-balance calculation of the earth shows the importance of the global greenhouse effect. The Earth-atmosphere system is heated by solar radiation at a mean rate of $S_o(1-a)/4$ where S_o is the solar constant, a is the fraction of radiation reflected by the earth and atmosphere, and the factor 4 comes from the spherical geometry of the earth. The input must be balanced by emission of long-wave radiation back to space, where the rate is given by σT_e^4, where σ is the Stefan-Boltzmann constant, and T_e is the effective radiation temperature of the system. Assuming the current albedo of 0.3, the energy balance model gives a global average temperature of 255 K (-18°C), in the absence of an atmosphere (Mitchell 1989). This is well below freezing, and when one considers that the young sun was significantly dimmer, it means that the earth should have suffered total and permanent glaciation early in its history, although we know that this did not happen (the so-called "faint early sun paradox"). The reason is most likely the existence of a CO_2 (and water vapour?) mediated greenhouse warming.

The current greenhouse warming is estimated to be 33°C. The greenhouse gases in the earth's atmosphere thus raise the global mean surface temperature from well below the freezing point of water to a comfortable level (15°C). The presence of greenhouse gases in the earth's atmosphere has therefore been essential for planetary habitability. Although the amount of CO_2 and other greenhouse gases in the atmosphere have varied significantly over geologic timescales, primarily because of changes in the balance between volcanic emissions and the uptake of CO_2 by rock weathering reactions and/or organic activity, the mean global temperature has probably not varied greatly during the last 3.5 billion years or more. The presence of equable temperatures has enabled life to evolve on the earth, and life has probably contributed to the maintenance of global mean temperatures within limits (Schwartzman and Volk, 1989).

In the past century, however, human activities have directly and indirectly increased the amount of greenhouse gases in the atmosphere, mostly CO_2, chlorofluorocarbons, N_2O and methane, leading to an increased radiative heating of the surface by about 2 Wm^{-2}, and a possible global temperature increase of about 0.5°C since 1900 AD (Hansen and Lebedeff, 1988; Jones et al., 1988). Carbon dioxide concentrations in the atmosphere are currently increasing by 1.5 ppm per year, mainly from the burning of fossil fuels. This accounts for only about 58% of the total CO_2 released by fossil fuels, and uptake by the oceans probably accounts for the remainder. The biosphere may now be a net source of atmospheric CO_2 as a result of deforestation, but this is still uncertain.

Current global climate simulations predict a warming of from 1.5 to 4.5°C for a doubling of atmospheric CO_2 (Schlesinger and Mitchell, 1985). The more complex GCMs, with water vapour and cloud feedbacks, give results in the upper part of that range (Hansen et al., 1988; Mitchell 1989). At projected rates of increases in greenhouse gases , this doubling of atmospheric CO_2 is expected to take place around the year 2050. A 1°C increase in atmospheric temperature should be associated with a 6% increase in water vapour - this increase in water vapour enhances the greenhouse heating of the troposphere in a positive feedback effect (Mitchell, 1989).

1.2. STRATOSPHERIC OZONE DEPLETION

Ozone (O_3) is created in the stratosphere by intense interaction of molecular oxygen with solar ultraviolet radiation. The two principal reactions are: $O_2 + hv \Rightarrow O + O$ and $O + O_2 + M \Rightarrow O_3 + M$, where M is any atmospheric molecule. The presence of ozone in the stratosphere, mostly between 15 and 30 km altitude, affects the temperature structure of the atmosphere (ozone is a greenhouse gas) and shields the earth's surface from biologically harmful ultraviolet radiation (ozone absorbs virtually all solar ultraviolet radiation in wavelengths from 240 to 290 nm, which is lethal to most cells)(Wayne, 1985).

Ozone is destroyed in other reactions in the stratosphere, for example, photodissociation: $O_3 + hv \Rightarrow O_2 + O$ (at wavelengths <7100 Å), and reactions with NO_x and OH. Ozone undergoes natural fluctuations on daily, seasonal and solar-cycle timescales (for detailed ozone chemistry, see Wayne, 1985). The addition of CFC's to the atmosphere has introduced free chlorine molecules into the stratosphere, which are important catalysts in ozone destruction reactions (Wuebbles et al., 1989). Reduction of ozone in the Antarctic springtime stratosphere has been observed, and linked to heterogeneous reactions involving CFC-derived chlorine compounds in polar stratospheric clouds (see below)(Toon at al., 1986; McElroy et al., 1986). A similar effect may now be taking place in the Arctic stratosphere (Hofmann et al., 1989).

Ozone reductions at mid-latitudes since 1983 have also been reported (Heath, 1988), with a significant loss during winter months (Rowland, 1989). After the eruption of El Chichon in April 1982 and the intense El Nino of 1982-1983, decreases in total ozone of 6.5% in mid-latitudes and 8% in the north polar regions were observed. The ozone content of the atmosphere from 65°N to 65°S declined by about 2.5% after these events (Heath, 1988). The magnitude of the depletion is a matter of some debate as a result of possible instrumental calibration and drift problems (Rodgers, 1988).

2. Feedback in the Climate System

The greenhouse and ozone problems are not entirely independent, since small aspects of the major earth systems are linked. Lashof (1989) has provided estimates of the sensitivity of the global mean temperature to a range of internal climate feedbacks, including changes in water vapour, clouds and sea-ice albedo, and biogeochemical feedbacks, such as changes in methane emission, ocean CO_2 uptake, and vegetation albedo. The input, I. is defined as the global average equilibrium temperature change due to direct radiative effects of a particular perturbation, and the output, O, is the equilibrium temperature change when the feedback process (or processes) is included. Lashof (1989) made some useful preliminary estimates of the gain, g, and amplification, f, of these climatic and biogeochemical feedbacks:

$$g = \frac{O-I}{O} \qquad f = \frac{O}{I} = \frac{1}{1-g}$$

In this case, negative feedback is represented by $(g < 0)$, positive feedback by $(g > 0)$, and an unstable system by $(g > 1)$. A useful consequence of defining gain in this way is that the gain for a system with several feedbacks is simply the sum of the gains of each feedback loop. The total gain of the global greenhouse warming is the sum of the gains for each individual process, and is a fundamental determinant of the sensitivity of the climate system to perturbations - natural and anthropogenic. Lashof (1989) estimates the gain from biogeochemical feedbacks as 0.08 to 0.44 compared with 0.17 to 0.77 for internal climate feedbacks. This shows that biogeochemical feedbacks have the potential to increase the climate associated with any given forcing.

Feedbacks to the CO_2 greenhouse problem in climate models include changes in atmospheric water vapour, lapse rate, cloud cover and altitude, cloud optical depth, an surface albedo (Schlesinger, 1985). Biogeochemical feedbacks include release of methane hydrates, increased flux of carbon dioxide and methane from soil organic matter to the atmosphere due to higher rates of microbial activity, changes in ocean circulation and mixing, which affects ocean productivity, increased flux of carbon dioxide from the atmosphere to the biosphere as a result of CO_2 fertilization, and potential effects of ecological reorganizations which can cause changes in surface albedo, terrestrial carbon storage, and changes in biological pumping of carbon from the ocean surface to deeper waters.

For example, the most direct feedback is on ocean-carbon chemistry. As the ocean warms, the solubility of CO_2 decreases and the carbonate equilibrium shifts toward carbonic acid; these effects combine to increase pCO_2 in the ocean by 4 to 5 %/°C for a fixed total carbon content. Lashof has also investigated the coupling between climate change and ocean CO_2 uptake. Performing the calculation with and without coupling temperature and carbonate chemistry, he finds, for example, that the coupling produces a gain of 0.008 based on the temperature change to be realised by the year AD 2100.

2.1. MAJOR FEEDBACKS BETWEEN GREENHOUSE GASES AND STRATOSPHERIC OZONE

Figure 1 shows a schematic diagram of the recognized possible interactions between the greenhouse effect and stratospheric ozone depletion. Trace gas releases (natural and anthropogenic) affect both the global temperature and the ozone layer, so that for the temperature, $\Delta T_{surf} = f(CO_{2,source}, CFC_{source}, CH_{4,source},$ etc.), and for ozone, $\Delta_{ozone} = f(CFC_{source}, CO_{2,source}, CH_{4,source},$ etc.). These equations should also include parameterizations for all relevant feedback processes, so that $\Delta_{ozone} = \Delta_{ozone}(\Delta T_{surf})$, and $\Delta T_{surf} = \Delta T_{surf}(\Delta_{ozone})$ through various feedback loops.

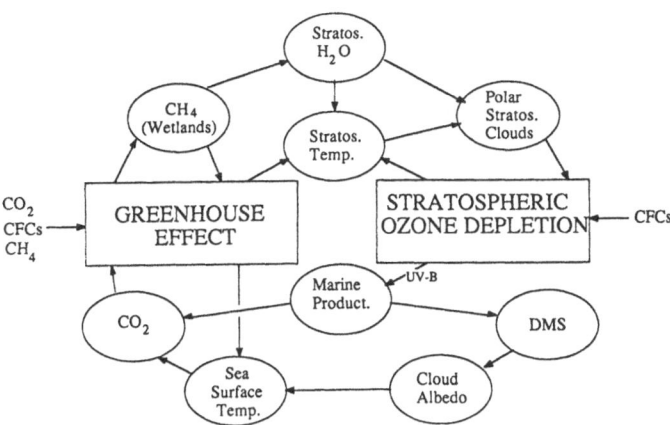

Figure 1. Schematic diagram showing the recognized possible interactions and biogeochemical feedback processes involving the greenhouse effect of increasing trace gases and depletion of stratospheric ozone.

At least two major feedback loops exist, an atmospheric chemistry loop and a marine productivity loop (Figure 1). The recognized and proposed interactions and feedbacks are summarized below. They include: (a) Increasing greenhouse effect and climatic warming from increased in CO_2 and other trace gases; (b) Increasing methane release from wetlands, hydrates, etc. as a result of climatic warming; (c) Increase in stratospheric water vapour as a result of methane increase; (d) Increase in stratospheric clouds and increased reduction of ozone as a result of heterogeneous reactions; (e) Increased UV-B radiation at the ground as a result of decreased ozone. The possible effects of increase UV-B radiation on primary productivity; (f) Effects of changes in primary productivity on atmospheric CO_2; (g) Decrease in marine productivity leads to decreased DMS production, decreased marine cloud albedo, and climate warming feedback. These feedback processes can influence, and be influenced by, global hydrology as described below.

2.1.1. *Increasing greenhouse effect from increases in carbon dioxide, CFC's, methane, and tropospheric ozone.* The increase in atmospheric carbon dioxide over the last 100 years is calculated to produce an equilibrium global surface warming of about 0.5°C, which is amplified by 50% by CH_4, $CFCl_3$(F11), CF_2Cl_2(F12), and tropospheric ozone (Ramanathan et al., 1985). GCM studies by Hansen et al. (1988) have shown that different scenarios of CO_2 emission rates will produce surface warming in the range of 0.6 to 1.5°C over the next 40 years (including the other trace gas feedbacks). Other climatic feedbacks to be taken into account include changes in atmospheric water vapour, cloudiness, and surface albedo. The GISS/GCM results, with additional feedbacks, gives a total warming of about

4.5°C for a doubling of the atmospheric PCO_2 (Hansen et al., 1988). Temperature increases should show and amplification at high latitudes (Hansen et al., 1988), and this may already have been detected in the southern hemisphere (Karoly, 1987). One feedback mechanism, warming of the ocean surface waters, however, will produce only a relatively small release of ocean CO_2 to the atmosphere, about 1.5 ppm per degree C (Macintyre, 1978). Radiation balance effects of a surface warming should cause a stratospheric cooling, with possible effects on ozone as discussed below.

2.1.2. Increasing methane release from wetlands, hydrates and other sources may be induced by greenhouse warming. Methane is currently increasing in the atmosphere by 0.016 ppmv per year (Blake and Rowland, 1988). Present distributions of methane in the global atmosphere shows highest values in the northern hemisphere high latitudes (Figure 2), most likely representing a source in high-latitude bogs and wetlands (Matthews and Fung, 1987). This release of methane should act as a positive feedback on greenhouse warming. Study of the Vostok ice core suggests an increase in atmospheric methane from 0.34 to 0.62 ppmv from the penultimate ice age to the following interglacial, from about 160 to 120 kyr BP. In that case, the contribution of methane (including chemical feedback) to the global climatic warming would have been about 25% of that due to CO_2 (Raynaud et al., 1988). Raynaud et al. (1988) suggest that the CH_4 increase could have come largely from freshwater wetlands, where increasing temperature may lead to an exponential response of methane emission rate. Based on the CH_4 emission-rate response calculated by Hameed and Cess (1983), a temperature change of 4.5°C would cause an interglacial source strength of methane higher by a factor of 1.7 over that during glacial periods. With this in mind, it is important to note that atmospheric methane has apparently increased from 0.7 to its present value of 1.68 ppmv over the last 300 years (Stauffer et al., 1985).

Analysis of temperature profiles measured in permafrost in northernmost Alaska indicate a surface warming of 2 to 4°C in the last few decades to a century (Lachenbruch and Marshall, 1986). This warming of permafrost areas may be leading to increased release of methane from soil bacterial activity and possibly from gas hydrates (Bell, 1982; Kvenvolden, 1988). The amount of methane stored in gas hydrates in permafrost and continental margin sediments is probably large, but the exact amount is rather poorly known (from 10^3 to 10^6 Gt). The amount of methane that would be released as a result of global warming is unknown. Taking the area of sea floor at depths of from 200 to 1,000 metres as the region potentially subject to hydrates becoming unstable, Revelle (1983) estimates that for a 1° temperature at the water-sediment interface, about 3×10^{14} g y^{-1} would be released. Better estimates could be made on the basis of the temperature dependence of wetland methane release, and the vulnerability of the various gas hydrate reservoirs to projected temperature increases.

Atmospheric CO from anthropogenic sources has also been increasing by about 0.8 to 1.4% per year on average (Khalil and Rasmussen, 1988); this could lead to a decrease in atmospheric OH⁻ radicals, which would result in a decreased sink for methane, and hence higher atmospheric methane values. Because of increased Co and methane, present tropospheric OH⁻ abundances may be about 20% less than several hundred years ago (Khalil

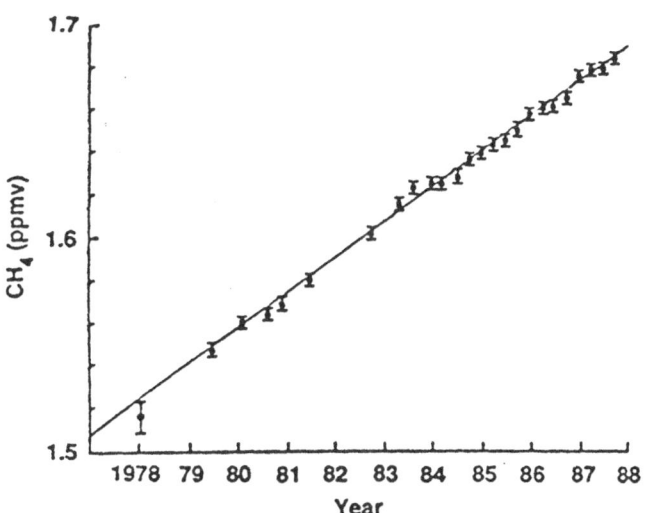

Figure 2. Worldwide average tropospheric CH_4 mixing ratio versus time from January 1978 through September 1987. Least squares fitting of all points except the first gives a slope of 0.0165 ppmv per year (after Blake and Rowland 1988)

and Rasmussen, 1985).

2.1.3. Increased methane in the atmosphere leads to an increase in stratospheric water. Evaluation of atmospheric infra-red spectra has confirmed methane increases of about 1% per year from 1951 to 1981 (Rinsland et al., 1985) and more recently at 0.7% per year. Recently, Blake and Rowland (1988) reported an 11% increase in tropospheric methane over the last 9 years, from 1.52 ppmv in 1978 to 1.68 ppmv in 1987 for a yearly increment of 0.016 ppmv. This should have been accompanied by an increase in stratospheric water vapour (by the reaction $CH_4 + OH \Rightarrow CH_3 + H_2O$) from 5 to 6.4 ppmv, or as much as 28% since the 1940's, and 45% over the past few centuries. This could have increased the mass of precipitable water available for the formation of the polar stratospheric clouds (PSC's) observed over Antarctica during the Austral Spring since the late 1970's. Formation of these clouds requires stratospheric temperatures of 160 to 190 K. Increased water vapour in the stratosphere due to the reduction of methane should cause local radiative cooling, favouring the growth of PSC's. Increased trace gases are also expected to cause a cooling of the stratosphere of up to 6°C at altitudes of 35 to 40 km by the year 2030, and this could lead to an increase in areas of regions cold enough for the formation of polar stratospheric clouds (Ramanathan et al., 1985)(Figure 3).

2.1.4. Increase in polar stratospheric clouds that provide a substrate for heterogeneous reactions with CFC's, and increase the rate of ozone destruction. Significant reductions

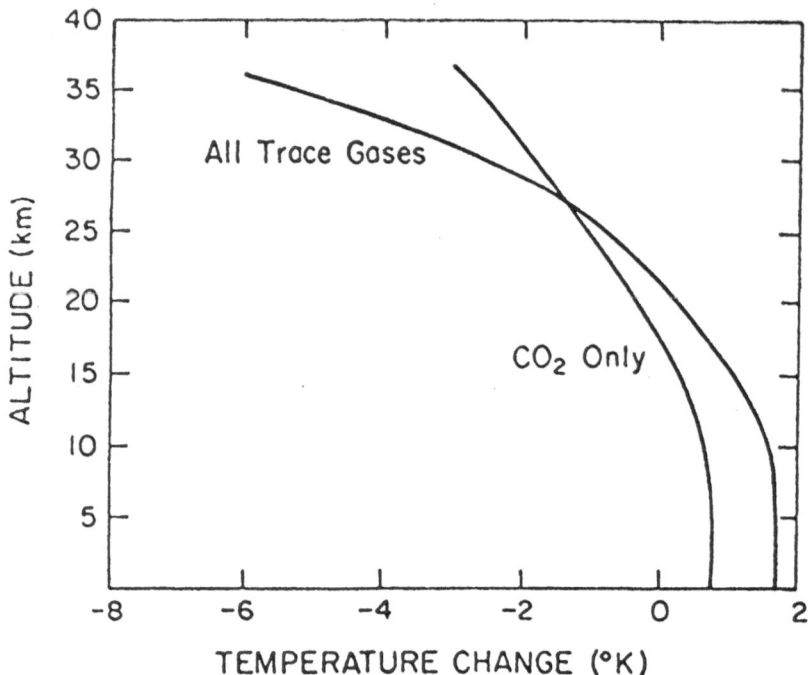

Figure 3. Vertical profiles of projected temperature changes from 1980 to 2030 due to CO_2 and all trace gases (Ramanathan et al., 1985).

in stratospheric ozone have been observed in the Antarctic "ozone hole" region (Farman et al., 1985; Stolarski et al., 1986; Hofmann et al., 1987), and increases in observed OClO and changes in ozone in the wintertime stratosphere suggest that significant heterogeneous chemistry takes place in the Arctic as it does in the Antarctic. (Mount et al., 1988), with possible ozone depletion (Hofmann et al., 1989). As mentioned above, increased methane in the global atmosphere leads to an increase of precipitable water in the stratosphere. Nitric acid in the form of mono- and tri-hydrates ($HNO_3.H_2O$ and $HNO_3.3H_2O$) can now condense at the cold temperatures of the winter polar stratosphere. Loss of vapour phase NO_x allows halogen-based chemistry to destroy ozone (Toon et al., 1986). These heterogeneous chemical reactions in PSC's (Solomon et al., 1988; Toon et al., 1986; McElroy et al., 1986) may have led to formation of the ozone hole in the cold Antarctic polar regions of the stratosphere (Molina et al., 1987), with a total reduction in ozone 30 to 50% in the Antarctic Spring over the past decade, or about 0.5% of total global ozone,

Another factor of possible importance is the response of the height and temperature of the tropical tropopause to greenhouse warming. Tropical tropopause temperatures control the water vapour content of the lower stratosphere (Bruhl and Crutzen, 1988). Changes in these temperatures will introduce a further feedback on ozone depletions (Liu et al.,

1976). The region near the tropical tropopause is radiatively heated by all anthropogenic trace gases, thus tropopause temperature will rise, allowing more water vapour to enter the stratosphere. This would lead to additional ozone loss. On the other hand, the warming of the lower troposphere, especially over the oceans, may lead to increased moist convection which tends to raise and cool the tropical tropopause (Reid and Gage, 1981). Bruhl and Grutzen 1988) have shown that for steady state conditions with doubled CO_2 and methane, and increased CFC's, that the tropopause temperatures could drop by as much as 4 °C and the tropopause height could increase by 900 m. Under these conditions the calculated surface temperature increased by about 3 °C. The expansion of the tropical troposphere could cause an additional decrease in total ozone by about 3% in the tropics. On the other hand, water vapour would increase to 5 ppm and temperatures by 2.1 K in the same case if the pressure level of the tropopause were fixed (Bruhl and Crutzen, 1988).

2.1.5. *The effect of ozone depletion on marine primary productivity.* UV-B can penetrate to ecologically significant depths in natural waters, and there is strong evidence that UV-B adversely affects phytoplankton productivity. Frederick and Snell (1988) have shown that even present day reduction of Antarctic ozone levels implies a greatly extended period of summerlike UV radiation that might adversely affect life. Using a simple model with recent experimental data (Smith and Baker, 1982) have attempted to quantify these effects.

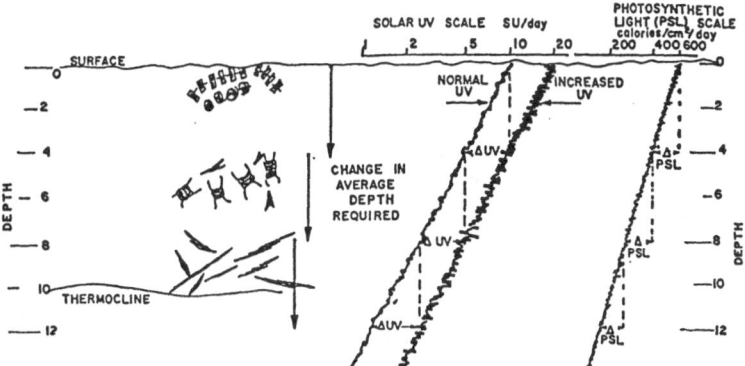

Figure 4. A model illustrating the loss of photosynthetic light whenever organism are forced to move deeper in the water column to avoid increased UV-B radiation, as would be the case with ozone depletion (Calkins and Blakefield 1986).

12

Whenever surface dwelling phytoplankton are forced to move deeper in the water column to avoid increased UV-B, there is a loss of some photosynthetic light, and hence a decrease in primary productivity (Calkins and Blakefield, 1986)(Figure 4). Eilers and Peeters (1988) recently developed a dynamic model of the relationship between light intensity and the rate of photosynthesis of phytoplankton. Smith and Baker (1982) examined the potential magnitude of photosynthesis reduction for the whole euphotic zone. Their calculations predict that with a 25% reduction in ozone thickness, primary productivity would decline 9%, whereas a 16% reduction in ozone is calculated to cause a 5% decrease in primary productivity (Kelly, 1986; Calkins, 1982).

2.1.6. *The effect of primary productivity on atmospheric carbon dioxide.* Annually, marine plankton produce 30-80 Gt of organic carbon, some 10-30% if which enters the deep-ocean waters where organic matter is oxidized - the so called *new production.* Less than 1% is ultimately buried in a long term steady state with the phosphate nutrient influx from weathering of continental rocks. In contrast, new production depends on light availability, recycling of nutrients and plankton growth/mortality rates, where the latter may be adversely affected by increased UV-B levels. A reduction of new production flux would cause an increase in atmospheric CO_2, and complete cessation of productivity could increase atmospheric CO_2 some 300 to 350 ppm. Plankton also produce calcium carbonate, yielding roughly 7 Gt of $CACO_3$ annually, 5 Gt of which is buried in the bottom sediments. Pumping of calcium ions from the surface to deep waters causes a decrease in surface water alkalinity.

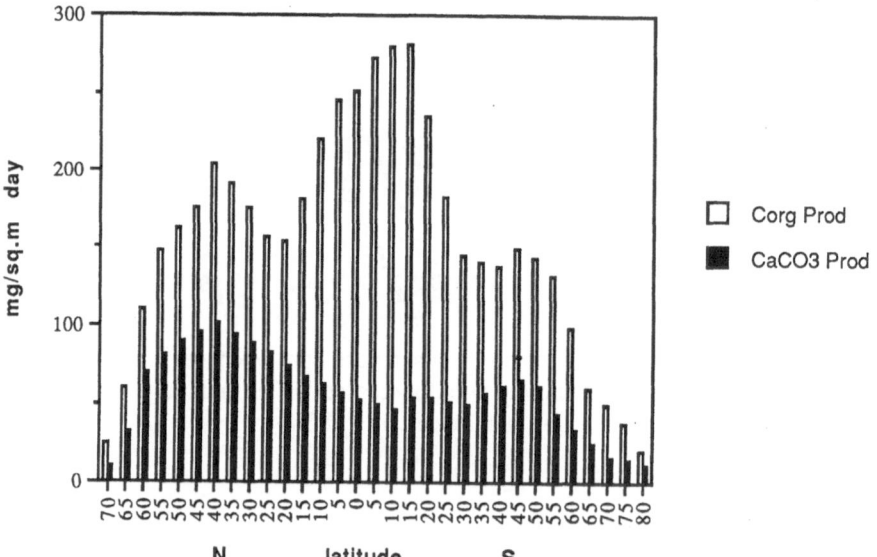

Figure 5. Zonally averaged values of initial production of organic carbon and calcium carbonate (shaded area) in the open ocean (after Lapenis, 1988).

A cessation of this $CACO_3$-pump would lead to an increase in surface water alkalinity and a decrease in atmospheric CO_2 of about 80 to 100 ppm. Thus, the net effect of a cessation of ocean productivity should be an increase in atmospheric CO_2 by about 200 to 300 ppm (Lapenis, 1988).

Figure 5 shows the latitudinal distribution of productivity (P) of organic carbon and calcium carbonate (Lapenis, 1988). It can be seen that the ratio varies significantly with latitude. The difference in this so called "Broecker parameter",

$$Br = \frac{P_{Corg}}{P_{CACO_3}},$$

between high latitude and tropical surface oceans can be used in box models to determine the changes in productivity in tropical surface waters that may be related to a repartitioning of nutrients as a result of reduced high-latitude productivity (Lapenis et al., 1989). In the multi-box ocean model of Lapenis (1988), for example, Br_H is greater in high latitude surface oceans of area S_H than Br_T in tropical surface oceans of area S_T. atmospheric carbon dioxide mixes "rapidly" with latitude such that atmospheric PCO_2 levels are much more uniform than those of the surface ocean. To the extent that the marine biosphere provides biological fixation of CO_2 (which sinks to deeper levels as either organic detritus or carbonate shells depending on the local productivity and the Broecker number) the net atmosphere/ocean carbon dioxide gas exchange can be considered to be driven by oceanic pCO_2 associated with a global mean Broecker parameter

$$
 = \frac{S_H Br_H + S_T Br_T}{S_H S_T}.$$

The result is that increased productivity within specific latitudinal bands can result in either increases or decreases in atmospheric carbon dioxide (Lapenis et al., 1989). These latitude-dependent effects make it possible to relate the latitudinal pattern of ozone depletion to changes in the latitudinal distribution of productivity of organic carbon and calcium carbonate. Such biologically-mediated changes in atmospheric carbon dioxide levels can also be caused by changes in circulation patterns leading to changes in nutrient supply to the surface biota.

Viecelli (1984) has estimated that a 1% decrease in global marine biosphere productivity could result in an increase in CO_2 of 0.5 to 2.5% depending on rate of vertical mixing. The increase of CO_2 would lead to increased greenhouse warming, thus providing a positive feedback to the warming already in progress, while at the same time cooling the atmosphere. The additional surface warming might cause an increase in the release of methane from wetlands and gas hydrates, producing a positive feedback on methane-related water in the stratosphere, formation of polar stratospheric clouds, and further reduction in atmospheric ozone. Hence, a reduction of primary productivity can lead to a chain of feedbacks that eventually causes further decreases in primary productivity, a net positive feedback.

2.1.7. A decrease in marine productivity could cause a decrease in DMS production, which leads to reduced cloud albedo, and climatic warming. Dimethyl sulphide (DMS), produced by marine phytoplankton, may be the major precursor of cloud condensation nuclei in the marine atmosphere, and hence acts as a control of cloud albedo and global temperatures (Andreae et al., 1985; Charlson et al., 1987; Bates et al, 1987; Rampino and Volk, 1988). Saigne and Legrand (1987) measured methanesulphonic acid (a product of DMS) in Antarctic ice, and found that concentrations during the last glacial period were 2 to 5 times higher than today. The resulting 20% to 46% increase in non-seasalt sulphate content in the Antarctic atmosphere during full glacial (compared with interglacial) conditions is consistent with higher DMS emissions from increased glacial marine productivity. Similar spectral features with Milankovitch periodicities, and correlation of carbon dioxide and non-seasalt sulphate in the ice core suggest a possible global climatic significance for these changes, and a possible link between CO_2 and DMS emissions.

Assuming that the average global population of cloud condensation nuclei (CCN) is proportional to $NSS-SO_4$ in the Antarctic atmosphere, a global radiative cooling of up to 1°C would result from the increase in DMS during glacial times, reinforcing the negative CO_2 effect (-0.6°C) and that from total insolation changes (-0.2°C, Legrand et al., 1988) so providing another positive feedback to global cooling. Schwartz (1988) has recently suggested that DMS contributes only about 20% of the total global atmospheric sulphur, compared with 50% from anthropogenic sources. Since most of the anthropogenic sulphur is emitted in the Northern Hemisphere, atmospheric sulphate should increased there in the last century, leading to increased CCN and cloud albedo, and a cooling relative to the Southern Hemisphere. Hemispheric temperature dat presented by Schwartz do not show such differences, but it must be realised, that other factors affecting cloudiness and temperature may be masking the effects of anthropogenic sulphur in the Northern Hemisphere. Wigley (1989) has suggested that changes in cloud albedo caused by SO_2 from industrial pollution might have led to cooling that has offset the temperature changes resulting from the greenhouse effect. Savoie and Prospero (1989) provide evidence that most of the non-seasalt sulphate particles in the atmosphere over the Pacific Ocean are derived from marine sources.

3. Links with Global Hydrology

Global precipitation patterns are intimately involved in the climate system and carbon cycle, and must be taken into account in studies of feedbacks in the greenhouse/depletion problems. Some of the potentially important feedbacks involving global hydrology include: 1) changes in availability if cloud condensation nuclei related to changes in plankton-produced DMS, and in SO_4 from pollution sources. This could change cloudiness and affect global albedo, precipitation, soil moisture, vegetation patterns; 2) increase in precipitation would lead to increase in soil moisture, leading to waterlogged soils in some areas with increased production of soil CH_4, and hence enhanced greenhouse warming; 3) changes in natural vegetation and agriculture resulting from variations in rainfall that can change soil moisture, regional climate, and the uptake and release of CO_2 from soils and vegetation; 4) changes in continental runoff, ocean evaporation and salinity in sensitive areas, affecting the production of ocean bottom waters, upwelling and organic productivity, which can affect atmospheric CO_2 concentrations. Changes in salinity in high-latitude oceans would affect the production of North Atlantic Deep Water and Antarctic bottom water and thus change the pole-to - equator, and interhemispheric heat transport.

A number of studies have shown that greenhouse warming will be accompanied by significant alterations in the global patterns of precipitation. Two basic approaches to forecasting precipitation changes have been taken: 1) GCM model simulations with increased atmospheric trace gases, and 2) Use of palaeoclimatic analogs of the climate warming expected from greenhouse effect.

Climate simulations for doubled atmospheric CO_2 using various GCM's give somewhat different results. For four current GCM's, the annual percentage change in precipitation for the entire globe varies from $+11.9\%$ to $+17.9\%$, and for the Northern Hemisphere from 11.6% to 15.3%, with greater percentage changes seen at high latitudes (Grotch, 1988). Model studies show an increase high latitude precipitation, especially in winter, with generally little change in the subtropics. There is some disagreement among GCM's on regional changes in precipitation. The GISS and NCAR models produce enhanced soil moisture in the northern continents in winter and summer, whereas other models give drier conditions in the northern mid-latitudes in summer (Mitchell, 1989).

In palaeoclimatic analog studies, the roughly $1°C$ warming predicted for about the year 2000 is compared with the Holocene Climatic Optimum of 5,000 to 6,000 years BP, the $2.5°C$ warming projected by the year 2025 is compared with the Eemian Interglacial Maximum (about 125,000 yr BP) and the Pliocene Climatic Optimum of 3 to 4 Myr BP is used an analog for the $4°C$ warming projected with CO_2 doubling by about 2050 (Budyko and Izrael, 1987; Velichko, 1987). For example, during the Holocene climatic optimum, precipitation in the central regions of the Sahara was greater than present by 200 to 300 mm/yr. In the Mediterranean region (including the Middle East) precipitation was greater by about 50 mm/yr. Reduction in precipitation of 25 to 50 mm/yr is estimated to have occurred in Central Russia, and up to 100 mm/yr reduction in precipitation in the central and eastern United States, and along the Pacific Coast from California to Alaska (Velichko, 1989). In the peak interglacial scenario, palaeoclimatic data are more limited, but increases in precipitation are projected for high northern latitudes (up to 250 mm/yr in Northern Europe).

4. Model Studies of Feedback in the Climate System

Once the feedbacks are identified, model studies can be very useful in investigating the role of the feedbacks in the climate system (Schlesinger, 1989; Lashof, 1989). Lashof (1989) considered the sensitivity of a single variable (the global mean surface temperature, $<T>$) to greenhouse gas forcing, including the influence of a number of positive and negative feedbacks assumed to depend only on $<T>$. However, this represents a zeroth-order steady-state analysis, and as such may mask nonlinear effects associated with the spatially varying climatic, dynamic and compositional response to the atmosphere-ocean-biology system. An objective of future studies is a "first-order" climatic response analysis including latitude-dependent effects. Such a first-order analysis would, in the climatic context, predict the effect on equator-to-pole temperature gradient, of ice-albedo feedbacks, changes in equator-to-pole heat flux (Hoffert et al., 1983), and other feedbacks.

Similarly, assessing the feedbacks from circulation changes on the structure and biology of the oceans, and atmospheric transport and radiation feedbacks on the structure of the atmosphere, requires consideration of (at least) horizontal variability. The composition of the ocean and in particular its biological productivity at the surface is known to vary markedly with latitude, where distinct bands of high and low productivity are readily identified in satellite imagery. Atmospheric composition is also a function of latitude (as well as altitude), particularly as regards the "thickness" of the ozone column and the amount of UV-B radiation penetrating to the surface. Thus, potential effects from biota from such changes, and possible feedbacks on the carbon cycle, must be addressed in a latitudinally-resolved context.

Future studies utilizing 2-D ocean models, and 2-D atmospheric climate and photochemistry models are planned in collaboration with Lawrence Livermore National Laboratories, and with the State Hydrological Institute and Main Geophysical Observatory in Leningrad, USSR. In these experiments we will attempt to determine the sign, magnitude and time constants of climatic and biogeochemical feedbacks among the various aspects of the global climate system. The goal is an integrated systems model incorporating the important feedback processes.

5. Acknowledgments

We thank K. Caldeira, A. Hecht, M. Hoffert, A. Lapenis, E. Rosanov, T. Volk, and D. Wuebbles for helpful discussions and contributions to this ongoing study. MRR thanks the Center for Global Habitability at Columbia University and its director Ruth Levenson for partial support.

6. References

Andreae, M.O., Ferek, R.J., Bermond, F., Byrd, K.P., Engstrom, R.T., Hardin, S., Houmere, P.D., LeMarrec, F., & Raemdonck, H. (1985) "Dimethyl sulphide in the marine atmosphere.", Jour. Geophys. Res. 90, pp. 12891-12900

Bates, T.S., Charlson, R.J., & Gammon, R.H. (1987) "Evidence for the climatic role of marine biogenic sulphur.", in *Nature*, 329, pp. 319-321

Bell, P. (1982) "Methane hydrate and the carbon dioxide question.", in *Carbon Dioxide Review*, 1982, New York, Oxford University Press

Blake, D.R., & Rowland, F.S. (1988) "Continuing world-wide increase in tropospheric methane, 1978-1987.", in *Science*, 239, pp. 1129-1131

Bruhl, C. & Crutzen, P.J. (1988) "Scenarios of possible changes in atmospheric temperatures and ozone concentrations due to man's activities, estimated with a one-dimensional coupled photochemical climate model.", Clim. Dynam. 2, pp. 173-203

Budyko, M.I. & Izrael, Ya. A. (eds)(1987) "Anthropogenic Climatic Changes.", Leningrad, Gidrometeoizdat

Calkins, J. (1982) "Modelling light loss versus UV-B increase for organisms which control their vertical position in the water column.", in Calkins, J. (ed), *The Role of Solar Ultraviolet Radiation in Marine Ecosystems*, New York, Plenum Press, pp. 539-542

Calkins, J. & Blakefield, M. (1986) :An estimate of the role of current levels of solar ultraviolet radiation in aquatic ecosystems.", in *Effects of Changes in Stratospheric Ozone on Global Climate*, UNEP/EPA Report, pp. 211-235

Charlson, R.J., Lovelock, J.E., Andreae, M.O., & Warren, S.G. (1987) "Oceanic phytoplankton, atmospheric sulphur, cloud albedo and climate.", in *Nature* 326, pp. 655-661

Eilers, P.H.C. & Peeters, J.C.H. (1988) "A model for the relationship between light intensity and photosynthesis in phytoplankton.", Eco. Model. 42, pp. 199-215

Farman, J.C., Gardiner, B.G., & Shanklin, J.D. (1985) "Large losses of total ozone in Antarctica reveal seasonal ClO_x/NO_x interaction.", in *Nature* 315, pp. 207-210

Frederick, J.E., & Snell, H.E. (1988) "Ultraviolet radiation levels during the Antarctic Spring.", in *Science*, 241, pp. 439-440

Grotch,S. (1988) "Regional intercomparisons of general circulation model predictions and historical climate data.", Technical Report TR041, U.S. Dept. of Energy, Washington D.C.

Hameed, S. & Cess, R. (1983) "Impact of a global warming on biosphere sources of methane and its climatic consequences.", Tellus, 35B, pp.1-7

Hansen, J.E., Fung, I., Lacis, A., Lebedeff, S., Ruedy, R., & Russell, G. (1988) "Global climate changes as forecast by the Goddard Institute for Space Studies three-dimensional model.", in Jour. Geophys. Res., 93, pp. 9341-9364

Hansen, J.E., & Lebedeff, S. (1988) "Global surface air temperatures: Update through 1987.", in Geophys. Res. Lett., 15, pp. 323-326

Heath, D.F. (1988) "Non-seasonal changes in total column ozone from satellite observations, 1970-86.", in *Nature* 332, pp. 219-227

Hoffert, M.I., Flannery, B.P., Callegari, A.J., Hsieh, C.T., & Wiscomb, W. (1983)

"Evaporation-limited tropical ocean temperatures as a constraint on climate sensitivity.", J. Atmos. Sci. 40, pp. 1659-1668

Hofmann, D.J., Harder, J.W., Rolf, S.R., & Rosen, J.M. (1987) "Balloon-borne observations of the development and vertical structure of the Antarctic zone hole in 1986.", in *Nature* 326, pp. 59-62

Hofmann, D.J., Deshler, T.L., Aimedieu, P., Matthews, W.A., Johnston, P.V., Kondo, Y., Sheldon, W.R., Byrne, G.J., & Benbrook, J.R. (1989) "Stratospheric clouds and ozone depletion in the Arctic during January 1989.", in *Nature* 340, pp. 117-121

Jones, P.D., Wigley, T.M.L., Folland, C.K., Parker, D.E., Angell, J.K., Lebedeff, S., & Hansen, J.E. (1988) "Evidence for global warming in the past decade.", in *Nature* 338, p. 790

Karoly, D.J. (1987) "Southern hemisphere temperature trends: A possible greenhouse gas effect?", in Geophys. Res. Lett. 14, pp. 1139-1141

Kelly, J. (1986) "How might enhanced levels of solar ultraviolet UV-B radiation affect marine ecosystems?", in *Effects of Changes in Stratospheric Ozone and Global Climate*, UNEP/EPA, pp. 237-258

Khalil, M.A.K., & Rasmussen, R.A. (1985) "Causes of increasing atmospheric methane: Depletion of hydroxyl radicals and the rise of emissions.", in Atmos. Environ. 19, pp. 397-407

Khalil, M.A.K., & Rasmussen, R.A. (1988) "Carbon monoxide in the Earth's atmosphere: Indications of a global increase.", in *Nature* 332, pp. 242-245

Kvenvolden, K.A. (1988) "Methane hydrates and global climate.", in Global Geochem. Cycles 2, pp. 221-229

Lachenbruch, A.H., & Marshall, B.V. (1986) "Changing climate: Geothermal evidence from permafrost in the Alaskan Arctic.", in *Science* 234, pp. 689-696

Lapenis, A.G. (1988) "Biodynamic mechanism of changes in atmospheric CO_2 concentration.", Akad. Nauk. SSSR. 6, pp. 794-799

Lapenis, A.G., Rosanov, E.V., & Caldeira, K.G. (1989) "Effect of ocean circulation on the ratio between organic carbon and calcium carbonate primary productivity.", Global Biogeochem. Cycles (submitted)

Lashof, D.A. (1989) "The dynamic greenhouse: Feedback processes that may influence future concentrations of atmospheric trace gases.", Climatic Change 14, pp. 213-242

Legrand, M.R., Delmas, R.J., & Charlson, R.J. (1988) "Climate forcing implications from Wostok ice-core sulphate data.", in *Nature* 334, pp. 418-420

Liu, S.C., Donahue, T.M., Cicerone, R.J., & Chameides, W.L. (1976) "Effect of water vapour on the destruction of ozone in the stratosphere perturbed by ClO_x or NO_x pollutants.", in Jour. Geophys. Res. 81, pp. 3111-3118

Macintyre, F. (1978) "On the temperature coefficient of PCO_2 in seawater.", Clim. Change 1, pp. 349-354

Matthews, E., Fung, I. (1987) "Methane emission from natural wetlands; Global distribution, area, and environmental characteristics of sources.", Global Biogeochem. Cycles 1, pp. 61-86

McElroy, M.B., Salawitch, R.J., & Wofsy, S.C. (1986) "Antarctic O_3: Chemical mechanisms for the spring decrease.", Geophys, Res. Lett. 13, pp. 1296-1299

Mitchell, J.F.B. (1989) "The "Greenhouse" effect and climate change.", Rev. of Geophys. 27, pp. 115-139

Molina, M.J., Tso, T., Molina, L.T., & Wang, F.C. (1987) "Antarctic stratospheric chemistry of chlorine nitrate, hydrogen chlorine, and ice: Release of active chlorine.", in *Science* 238, pp. 1253-1257

Mount, G.H., Solomon, S., Sanders, R.W., Jakoubek, R.O., & Schmeltekopf, A.L. (1988) "Observations of stratospheric NO_2 and O_3 at Thule, Greenland.", in *Science* 242, pp. 555-558

National Research Council (1989) "Ozone depletion, greenhouse gases, and climate change.", Washington D.C., National Academy Press

Ramanathan, V., Cicerone, R.J., Singh, H.B., & Kiehl, J.T. (1985) "Trace gas trends and their potential role in climate change.", Jour. Geophys. Res. 90, pp. 5547-5566

Rampino, M.R., & Volk, T. (1988) "Mass extinctions, atmospheric sulphur and climatic warming at the K/T boundary.", in *Nature* 332, pp. 63-65

Raynaud, D., Chappellaz, J., Barnola, J.M., Korotkevitch, Y.S., & Lorius, C. (1988) "Climatic and CH_4 cycle implications of glacial-interglacial CH_4 change in the Vostok ice core.", in *Nature* 333, pp. 655-657

Reid, G.C., & Cage, K.S. (1981) "on the annual variation in the height of the tropical tropopause.", J. Atmos. Sci. 38, pp. 1928-1938

Revelle, R. (1983) "Methane hydrates in continental slope sediments and increasing atmospheric carbon dioxide.", in Changing Climate, Washington D.C., National Academy Press

Rinsland, C.P., Levine, J.S., & Miles, T. (1985) "Concentration of methane in the troposphere deduced from 1951 infrared solar spectra.", in *Nature* 318, pp. 245-249

Rodgers, C. (1988) "Global ozone trends reassessed.", in *Nature* 332, p. 201

Rowland, F.S. (1989) "The role of halocarbons in stratospheric ozone depletion.", in *Ozone Depletion, Greenhouse Gases, and Climate Change*, Washington D.C., National Academy Press

Saigne, C., & Legrand, M. (1987) "Measurements of methanesulphonic acid in Antarctic ice." in *Nature* 330, pp. 240-242

Savoie, D.L., & Prospero, J.M. (1989) "Comparison of oceanic and continental sources of non-sea-salt sulphate over the Pacific Ocean.", in *Nature* 339, pp. 685-689

Schlesinger, M.E. (1985) "Analysis of results from energy balance and radiative-convective models.", in *Projecting the Climatic Effects of Increasing Carbon Dioxide*, Washington, D.C., U.S. Dept. of Energy, pp. 280-319

Schlesinger, M.E. (1989) "Quantitative analysis of feedbacks in climate model simulations.", in Berger, A., Dickinson, R.E. and Kidson, J.W. (eds.) *Understanding Climate Change*, Geophysical Monograph 52, Washington, D.C., American Geophysical Union, pp. 177-187

Schwartz, S.E. (1988) "Are global cloud albedo and climate controlled by marine phytoplankton?", in *Nature* 336, pp. 441-445

Schwartzman, D.W., & Volk, T. (1989) "Biotic enhancement of weathering and the habitability of the Earth.", in *Nature* (in press)

Smith, R.C., & Baker, K.S. (1982) "Assessment of the influence of enhanced UV-B on marine primary productivity.", in Calkins, J. (ed.), *The Role of Solar Ultraviolet Radiation in Marine Ecosystems*, New York, Plenum Press, pp. 509-537

Solomon, S., Mount, G.H., Sanders, R.W., Jakoubek, R.O., & Schmeltekopf, A.L. (1988) "Observations of nighttime abundance of Oclo in the winter stratosphere above Thule, Greenland.", in *Science* 242, pp. 550-555

Stauffer, B., Fischer, G., Neftel, A., & Oeschger, H. (1985) "Increases in atmospheric methane recorded in Antarctic ice core.", in *Science* 229, pp. 1386-1388

Stolarski, R.S., Krueger, A.J., Schoeberl, M.R., McPeters, R.D., Newman, P.A., & Alpert, J.C. (1986) "Nimbus 7 satellite measurements of the springtime Antarctic ozone decrease.", in *Nature* 322, pp. 808-811

Toon, O.B., Hamill, P., Turco, R.P., & Pinto, J. (1986) "Condensation of HNO_3 and Hcl in the winter polar stratosphere.", Geophys. Res. Lett. 13, pp. 1284-1287

Velichko, A.A. (1989) "Palaeoclimatic reconstructions as a forecast element.", preprint

Viecelli, J.A. (1984) "The atmospheric carbon dioxide response to oceanic primary productivity fluctuations.", Clim. Change 6, pp. 153-166

Wayne, R.P. (1985) "Chemistry of atmospheres.", Oxford, Clarendon Press

Wigley, T.L. (1989) "Possible climate change due to SO_2-derived cloud condensation nuclei.", in *Nature* 339, pp. 365-367

Wuebbles, D.J., Grant, K.E., Connell, P.S., & Penner, J.E. (1989) "The role of atmospheric chemistry in climate change.", Jour. Air Pollution Control 39, pp. 22-28

EFFECTS OF WEST AFRICAN AIR HUMIDITY ON ATLANTIC SEA SURFACE TEMPERATURE

M. BENNO BLUMENTHAL
Lamont-Doherty Geological Observatory
of Columbia University
Palisades, NY 07641

ABSTRACT: Simple models of both the atmospheric and oceanic sides of the ocean/atmosphere boundary are used to model the air specific humidity over the tropical Atlantic. It is shown that most of the climatological patterns of specific humidity can be reproduced with these fairly simple models.

These models can then be used to predict the effects of changes in West African surface specific humidity on the sea surface temperature (SST) patterns of the tropical Atlantic. They suggest that small changes in specific humidity (1 g/kg) could result in significant SST changes (1 °C) over the area which the land humidity influences. These SST changes in turn could effect West African climate, the Sahel in particular.

1. Introduction

A number of studies indicate that Atlantic sea surface temperature (SST) patterns are correlated with drought patterns in Western Africa, particularly the Sahel (Lamb, 1978; Lamb, 1983; Folland et al., 1986). The first Lamb study compares patterns of wind, surface pressure, and SST from a set of drought years to the patterns found during relatively wet years. The second looks at air humidity over the ocean, again comparing wet and dry years. The studies together conclude that there are distinct patterns over the ocean correlated with drought in the Sahel, and that the variations in moisture content over the Gulf of Guinea are not responsible for drought in the Sahel: in fact, the air humidity over the Gulf of Guinea tends to be higher during drought years as compared to both non-drought years and climatology (Lamb, 1983). The controlling factor instead is the meridional position of the Intertropical convergence zone (ITCZ): when the ITCZ is in its normal relatively northern position, moist air from the Gulf of Guinea is blown over the Sahel, leading to normal (relatively high) rainfall. When the ITCZ shifts slightly southward (200-300 km), however, air in the Sahel is supplied from the waters west of Western Africa, making that air relatively

21

R. Paepe et al. (eds.), Greenhouse Effect, Sea Level and Drought, 21–40.
© 1990 *Kluwer Academic Publishers.*

dry. This mechanism has also been observed in runs of atmospheric global circulation models (GCMs): Charney et al. (1975) used a GCM forced with different patterns of North African vegetation to investigate the effects of different surface conditions on Sahel rainfall. They observed that a dry Sahel tended to push the ITCZ southward, maintaining the air's dryness. This feedback is one possible mechanism for maintaining long periods of drought in the Sahel.

This paper explores an interaction between the ocean and atmosphere that could be part of another feedback mechanism governing drought in the Sahel. The first part of the mechanism is essentially an ocean dynamics problem: the effects of West African dry air on oceanic SST patterns. A simple dynamical model for moisture in the lower atmosphere shows that climatological patterns of air humidity can be understood as resulting from a combination of the patterns of SST and the air humidity over certain land areas at certain times. This ability to reproduce the climatology suggests that the mechanisms contained in the model have enough of the essential physics to have some predictive power. This air humidity model, coupled with a simple model for SST, lets us explore the effects on SST of changes in specific humidity over Western Africa. This mechanism suggests that an increase in specific humidity would increase SST off the coast of Western Africa, decreasing the severity of droughts when wind conditions are such that northwesterly winds determine the oceanic contribution to the moisture over Africa. More importantly, these changes in SST pattern could change the meridional position of the ITCZ, potentially a powerful feedback mechanism for controlling rainfall conditions over the Sahel.

The paper is divided into four more sections. The next section discusses the physical model for the air specific humidity q_a and the model implementation. The third section presents a diagnostic calculation to verify the model physics against ship-determined climatology, concluding that the essential physics has indeed been captured. The fourth section discusses the relevant dynamics of SST, laying the groundwork for the final section, which discusses the effects of changes in land air humidity on SST.

2.   Surface Humidity Model

This model for air specific humidity q_a is designed to model the behaviour of q_a at temporal and spacial scales for which air humidity is important to the ocean: i.e. scales of order 100 km and weeks to months. In fact, the original motivation for the creation of this model was to improve the behaviour of an ocean model: in order to reproduce the observed SST structure, the model required heat fluxes out of the ocean surface layer that were significantly different from measurements even when the considerable uncertainty of the measurements were taken into account (Blumenthal and Cane, 1989; henceforth BC). It was suggested there, given the structure of the surface heat flux errors, that the air humidity model was inadequate, particularly because the effects of dry air blowing from land to sea were missing. This model, then, is designed to include those horizontal advection effects as they appear on oceanic temporal and spacial scales.

Figure 1, which is copied from Malkus (1962) but originally appeared in Bunker et al.

(1949), shows an aircraft sounding made in clear air over the Caribbean Sea. It illustrates several important about the vertical structure of the atmosphere that are relevant here (the

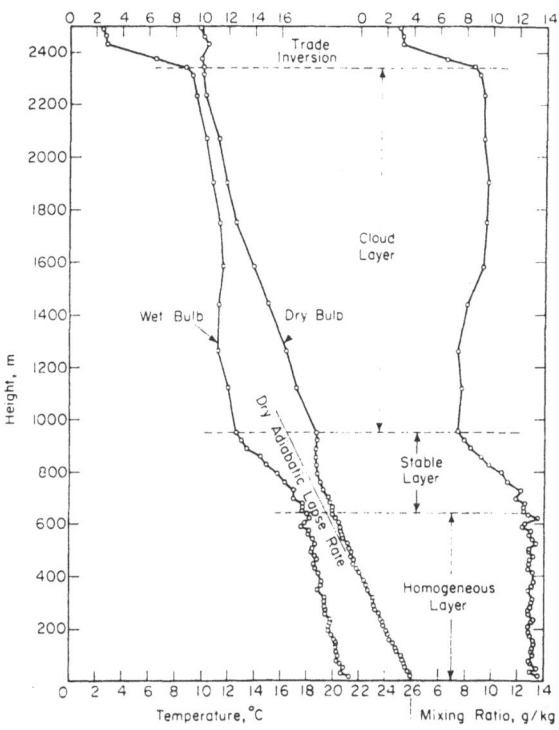

Figure 1. Typical vertical temperature and moisture sounding of the western Atlantic trade-wind region. Sounding made in clear space between cloud groups. From Malkus, 1962, fig 52 (originally from Bunker et al., 1949, fig 58).

points are all discussed in detail in Bunker et al., 1949). Near the surface, there is a homogeneous layer, here 600 m in height, where the mixing ratio (almost equivalent to the specific humidity) is independent of height. This homogeneous layer is capped by a stable layer where there is relatively little vertical motion. Acceleration measurements made in the aircraft that took the soundings indicate that in the homogeneous layer

"the friction force of the air flowing over the water surface is the source of the turbulent motion, rather than the penetration or development of convection in the homogeneous layer."

(Bunker et al., 1949). This is important for the construction of our model because it implies that it is appropriate to base the dynamics of the homogeneous layer primarily on what occurs within it and in the ocean: detailed information about what occurs above it is not as important. Furthermore, because of the stable layer which caps the homogeneous layer, we do not need to model vertical velocity there and the consequent advection of water vapour through the column top. Were we to model the cloud layer, on the other hand, convection would be quite important, and information about the atmosphere above 1000 m would be required.

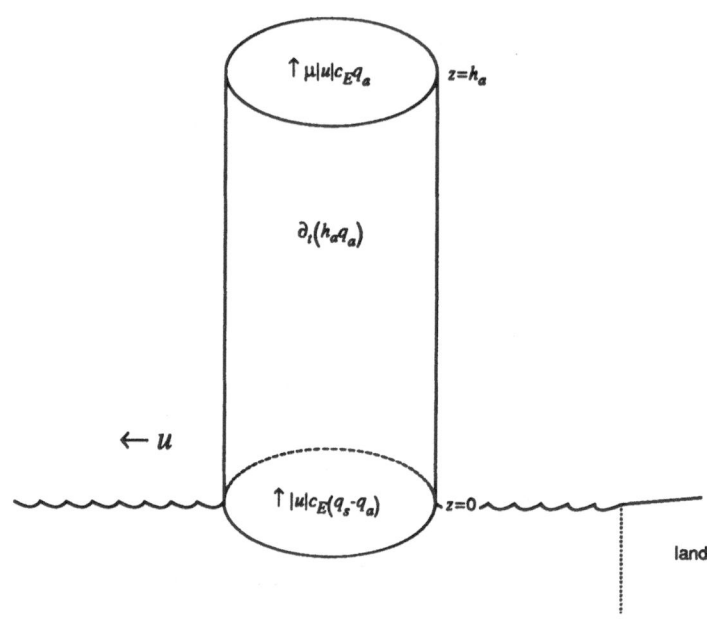

Figure 2. Schematic diagram of Q_e model.

Our starting point will then be to determine the fluxes of water vapour into a well-mixed column of air that sits directly above the sea. The situation is illustrated schematically in Figure 2. The air column is advected from land to water at which point the moisture flux from the ocean to the air causes the specific humidity of the air to equilibrate towards $q_e = R_H q_s(T_s)$, a function only of SST. The time rate of change of the water content of the air column of height h_a being activated by a wind of speed $|u|$ is a result of the difference between the flux through the bottom of the column and the flux through the top. As a matter of convenience we write water content in terms of the air specific humidity q_a: for the accuracy and scales of this model differences between the dynamics and thermodynamics of specific humidity and the corresponding dynamics of water content are not important. The balance equation following an air column is then

$$\frac{D(h_a q_a)}{Dt} = |u|c_E(q_s - q_a) - \mu|u|c_E q_a \tag{1}$$

where $|u|$ is the wind speed, $c_E = 1.3 \times 10^{-3}$ is the exchange coefficient, q_s is the saturated specific humidity evaluated at the sea surface temperature, $D/Dt = \delta_t + u\delta_x + v\delta_y$ represents the time derivate following an air column, and μ is a parameter that represents the relative weakness of the process at the top of the air column as compared to the bottom (this will be explained further below).

The formula for the flux through the bottom of the air column is the standard one, familiar through the standard bulk formula for latent heat flux (Weare et al., 1980). Here the wind speed is modified slightly: in order to include the flux due to the high frequency wind variability that is filtered out of monthly wind data, a minimum wind speed (4 m/s) is imposed> The formula for determining $q_s(T_s)$ is also a standard one (Weare et al., 1980), namely

$$q_s(T) - \frac{1580848985}{P} \times 10^{-2354/(T+273.15)} \qquad (2)$$

where T is in °C. For the purposes of this paper the pressure P is always taken to be one atmosphere.

The formula for the flux through the top is chosen to have similar structure to the flux through the bottom: the wind speed $|u|$ is again used as a measure of the strength of the turbulent motions, and the gradient of specific humidity (here chosen so that the specific humidity difference $q_a - q_{above}$ is proportional to q_a) is also linearly related to the flux. The coefficient μ thus represents the net effect of two factors: both the relation of specific humidity and adjustments in the exchange coefficient.

This choice of formula for the flux through the top means that (1) can be rewritten in terms of a somewhat more useful parameter R_H, namely

$$\frac{D(h_a q_a)}{Dt} - \frac{|u|c_E}{R_H}(R_H q_s - q_a) \qquad (3)$$

where $R_H = 1/(1 + \mu)$. The solution to this equation is an exponential decay such that, in the limit of long time and unchanging SST (and thus unchanging q_s), q_a approaches an equilibrium value $q_e = R_H q_s$. Thus R_H is more or less the equilibrium relative humidity (they differ by the factor $q_s(T_a)/q_s(T_s)$ which is close to unity), and climatological relative humidity is order 80% (R_H is estimated from data in the next section). The e-folding time for q_a to reach equilibrium is $R_H h_a/|u|c_E$, which for typical values ($|u| = 5$ m/s, $h_a = 600$ m) is roughly 1 day, much shorter than the week to month time scale that we are interested in modelling.

If we now consider the specific humidity field $q_a(x,y,t)$ rather than the q_a of a particular air column, we can further simplify the equations. Because the equilibrium time scale is shorter than the time scale of interest, the $\delta_t q_a$ term can be neglected, leaving us with a purely spatial equation to be solved,

$$(1 + \frac{R_H h_a}{|u| c_E} u \delta_x + \frac{R_H h_a}{|u| c_E} v \delta_y) q_a - R_H q_s \qquad (4)$$

The solutions to (4) will give the spatial structure of q_a for every time. Physically the air columns advected by the wind renders the process as a spacial adjustment scale. Furthermore, because the advection speed and the strength of the flux are proportional to the wind speed (as long as $|u| > 4$ m/s), for sufficiently strong winds the model is dependent only on wind direction, not wind speed (for low wind speeds the advection terms tend towards zero; physically this corresponds to air columns equilibrating in place). Consequently, the spacial scale is only weakly dependent on wind speed, simplifying the spacial structure.

We have parameterized a process that is rapid compared to the monthly timescale of interest: given an wind field $u(x,y) = (u,v)$, an SST field $T_s(x,y)$ (from which we can calculate the saturated specific humidity $q_a(T_s)$), and suitable boundary conditions, we have a prediction for the air specific humidity $q_a(x,y)$. The model has two parameters: R_H, which is the ration of the equilibrium to saturated specific humidity q_e/q_s, and h_a, the height of the homogenous layer in the air column. The next section checks whether the model is consistent with some of the available data, and determines values for the two model parameters.

3. Climatology of Surface Humidity

3.1. DIAGNOSTIC CALCULATION

In order to check whether the physics included in the q_a model are consistent with the available data, a diagnostic calculation using a q_a data set was made. Esbensen and Kushnir (1981) have made available a global monthly climatology of a number of atmospheric and oceanic parameters on a 5° by 4° grid. The available data include the sea surface temperature T_s and air specific humidity q_a. Hellerman and Rosenstein (1983) provide a similarly global data set of wind stress, from which we have calculated wind speed and direction. A slight rearrangement of (4) gives

$$q_a R_H - [\frac{1}{c_E |u|} (u \nabla_x q_a + v \nabla_y q_a)] R - q_a \qquad (5)$$

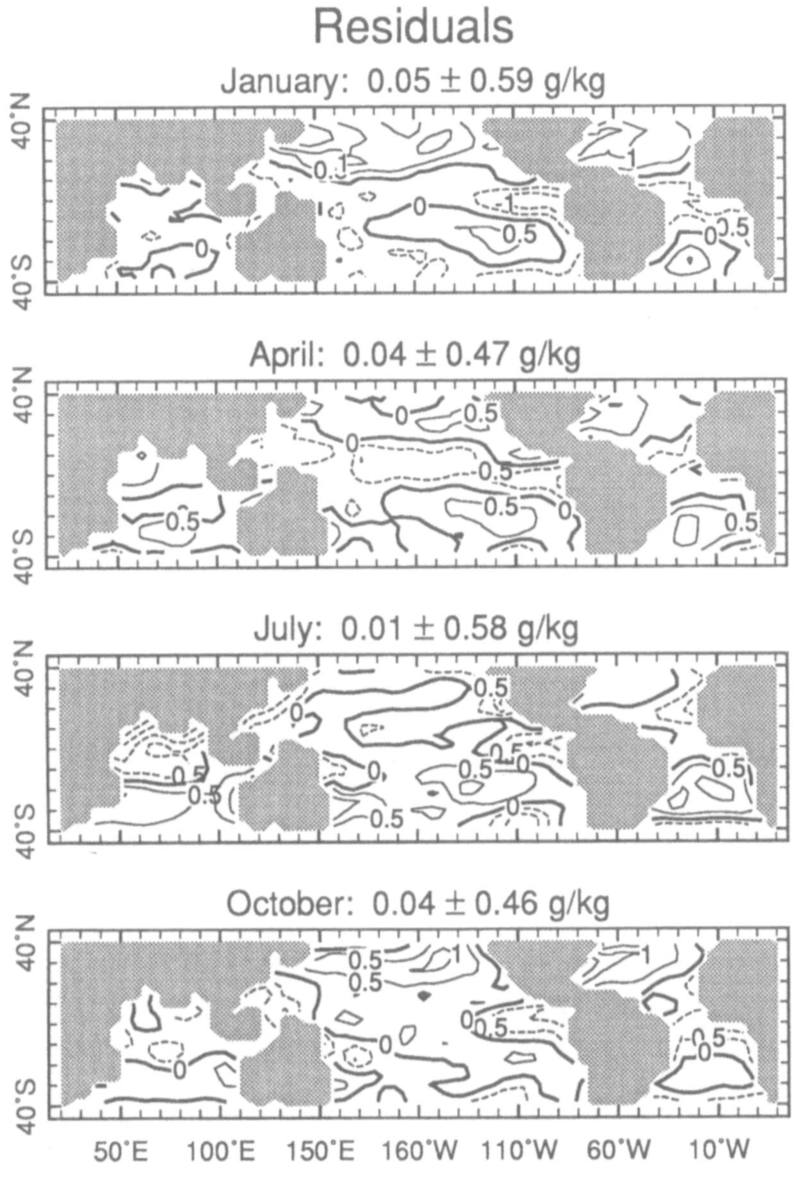

Figure 3. Residuals from diagnostic calculation.

28

January data comparison

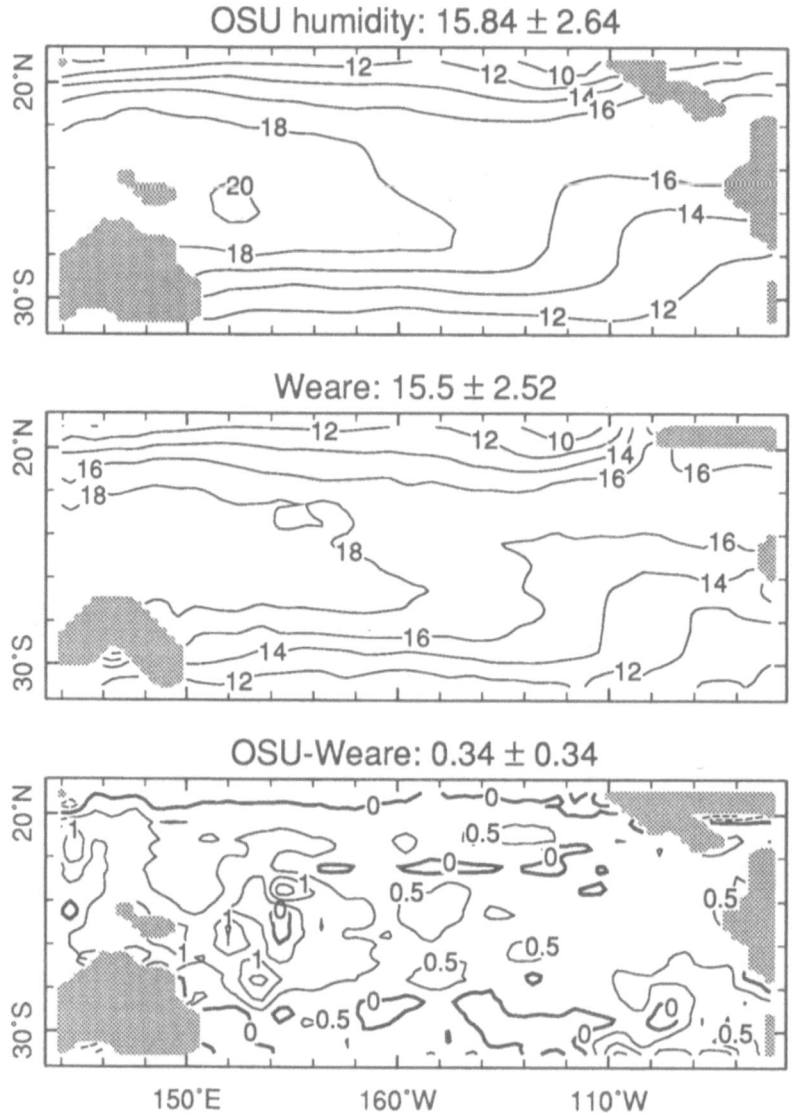

Figure 4. Comparison of q_a from Esbensen and Kushnir, 1981, with Weare et al., 1980.

where $R = h_a R_H r$, r being a factor to compensate for the coarse grid of the available data ($r \approx 0.4$, see appendix A), and ∇_x and ∇_y are centered finite difference approximations to the partial derivative operators. This relation exists at each grid point to the domain, giving us an overdetermined system for calculating R_H and R. In matrix notation this becomes

$$A \begin{bmatrix} R_H \\ R \end{bmatrix} - q_a \qquad (6)$$

which can be easily inverted if the errors are relatively homogeneous and both parameters well resolved

$$\begin{bmatrix} R_H \\ R \end{bmatrix} - (A^T A)^{-1} A^T q_a \qquad (7)$$

The results of this calculation using all 12 month of data between 40° S and 40° N is R_H = 0.77, R = 170 m, which corresponds to a scale height of h_a = 600 m, quite consistent with the Bunker et al., 1949 results reproduced in Figure 1. The residuals from the fit are plotted in Figure 3: the mean difference between the prediction and data for each month is between -0.02 to 0.04, the rms between 0.47 to 0.58 g/kg, both fairly reasonable given the uncertainties in the data. As an illustration of the difficulties with the data, consider Figure 4 which compares two data products over the tropical Pacific: Esbensen and Kushnir (marked 'OSU' in the Figure), and Weare et al., 1980. The mean difference between the two over the Pacific varies monthly from 0.34-0.39, the rms from 0.32 to 0.43, giving a total rms of order 0.5. Thus the data product differences are the same size as the model-data differences.

When estimating more than one parameter, there is a question of whether all the parameters have significant estimates. Plotted in Figure 5 are the residuals when R is arbitrarily set to zero: The mean error is now 0.26 and the rms also slightly higher, demonstrating that the estimate of R is indeed significantly different from zero.

3.2. PREDICTIVE CALCULATION

Given that the diagnostic calculation has verified that the model physics are sufficient for modelling the climatological data on the scales of interest, we would like to use the model to predict q_a given the wind and SST fields. This was done with a pair of implicit one-dimensional inversions of an upwind differencing scheme, namely (4) rewritten as

$$(1 + A\delta_x)(1 + B\delta_y) q_a - q_e \qquad (8)$$

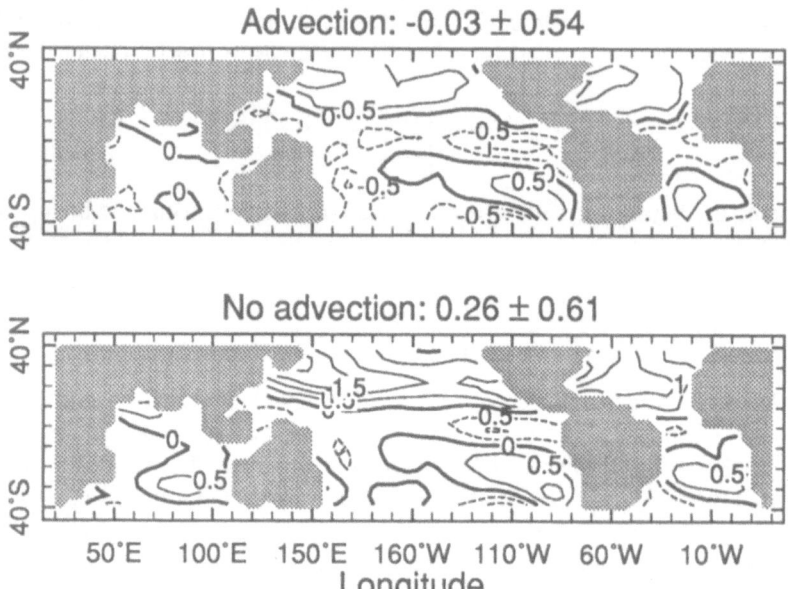

Figure 5. Effects on residuals of omitting the advection.

where $A = Hu/(c_E|u|)$ and $B = Hv/(c_E|u|)$. Inverted this becomes

$$q_a - (1 + B\delta_y)^{-1}(1 + A\delta_x)^{-1}q_e \qquad (9)$$

An upwind differencing scheme was used as a finite difference approximation to the spacial derivatives, e.g.

$$\delta_x q - \begin{array}{ll} (q_{i+1} - q_i)/\Delta_x & u < 0 \\ (q_i - q_{i-1})/\Delta_x & u > 0 \end{array} \qquad (10)$$

Upwind differencing insures that the matrix to be inverted is diagonally dominant, which implies that the inversion is stable. Upwind differencing also means that only upwind boundary conditions enter the domain (as in the continuous system).

In order to calculate q_a from (9), in addition to the wind and SST fields we require upwind boundary values for q_a, i.e. the values that are advected into the domain. Since we did not have a source of measured q_a over land, specifying values at all coasts turned out to be difficult, and we instead chose to extend the calculation to predict values over land as well. This requires forcing values q_e) over land as well as water: here the land forcing values were chosen to roughly match climatology (values from 8 to 12 g/kg, drier in winter). Boundary values at the edges of the model domain were fixed to match the equilibrium value. Consequently differences between prediction and data are expected near the boundaries of the domain.

Figure 6 shows both the predicted field q_a and the forcing field q_e for a July climatology. Note that there are only slight discontinuities of q_e at the coasts, indicating that the chosen land values are not too inappropriate. The Figure also shows a comparison of the model prediction with the Esbensen and Kushnir calculation from data: agreement is very good, with almost no mean error and an rms of ± 0.61, only slightly higher than the diagnostic calculation.

4. SST Model

The SST model is that used in Seager et al., 1988 (henceforth SZC) and BC; the following discussion is taken from the latter. The SST is determined from a forced advective equation. The temperature is assumed to be uniform (well-mixed) in the top layer, so that the SST
a) q_a prediction b) q_e forcing values (values over land were chosen to give reasonable results) c) difference between q_a prediction and data.

reflects the temperature throughout the layer. This temperature is determined from the net balance of horizontal advection, upwelling, and surface heat flux, depicted schematically in Figure 7:

$$\delta(h_{mix}T_s) + u_T\delta_y(h_{mix}T_s) + v_T\delta_y(h_{mix}T_s) + \gamma w(T_s - T_d) - Q + \kappa(\delta_{xx} + \delta_{yy})(h_{mix}T_s) \tag{11}$$

32

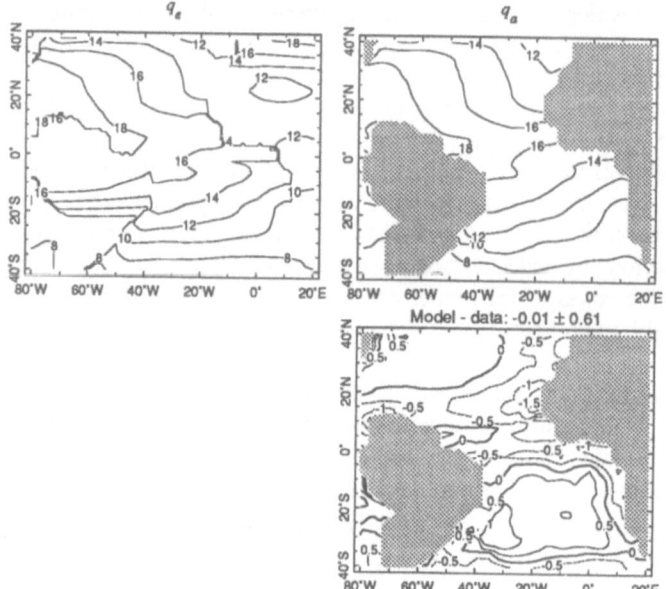

Figure 6. Comparison of q_a models with data in the Atlantic:

where T_d is a subsurface temperature determined by the dynamics below the mixed layer. The upwelling term is usually written as $w(T_s - T_e)$, where T_e is the temperature of water entrained into the mixed layer. The two temperatures are related by

$$T_e - (1 - \gamma)T_s + \gamma T_d \tag{12}$$

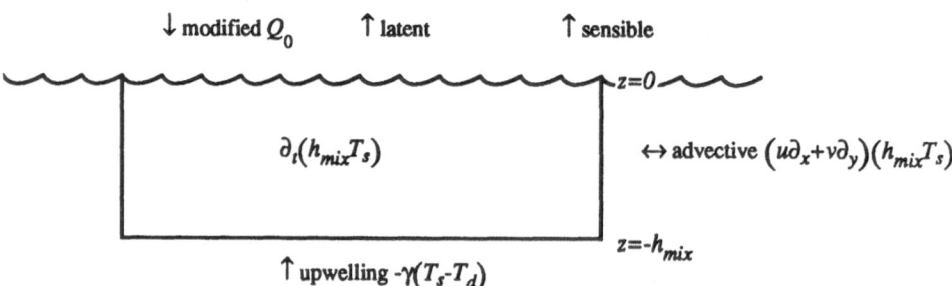

Figure 7. Schematic diagram of SST model: The changes in SST are determined by the net difference between the solar flux modulated by albedo and cloudiness effects, latent heat flux, sensible heat flux, horizontal advection, and upwelling.

The more important equation for the purpose of determining the effects of air humidity changes on SST is that for the surface heat flux Q. The surface heat flux is the sum of the solar flux, latent heat flux, sensible heat flux, and back radiation. The clear sky solar flux Q_0 is determined from a formula that accounts for latitude and time of year. At the ocean's surface it is reduced by the effects of surface albedo (assumed 0.06 over the ocean), measured cloud over C, and the absorption and reflection of the atmosphere, which depends on solar angle α (cf Weare et al., 1980). The sensible heat flux and back radiation are modelled together as being proportional to the difference between the sea surface and temperature T_s and reference temperature T_0. In the tropics this term is small relative to the solar and latent heat fluxes (Weare et al., 1980), having a variability of less than 20 W/m^2 over the tropical Pacific, so an imprecise parameterization of this term does not significantly affect the results. The latent heat flux is computed from the standard bulk formula: it involves wind speed $|\mathbf{u}|$, sea surface temperature T_s, and the air specific humidity. The net surface heat flux is thus given by

$$Q - (0.94)Q_0(1 - a_c C + a_\alpha \alpha) - \rho_a C_E L|u|(q_s(T_s) - q_a) - a_{sst}(T_s - T_0). \qquad (13)$$

As discussed in the section on the air humidity model, $|\mathbf{u}|$ is not allowed to fall below 4 m/s in order to compensate for the loss of variability in using monthly winds. There are two important implications of this equation for determining the effects of q_a changes

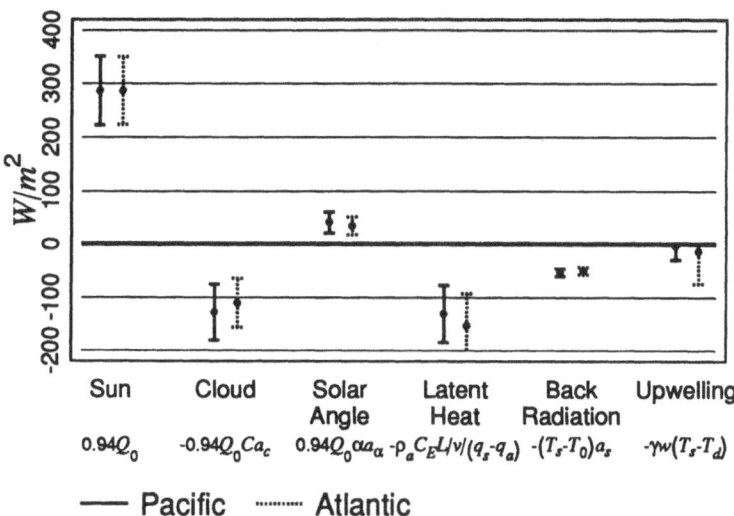

Figure 8. Relative size of SST heat flux terms averaged from 20° S to 20° N.

34

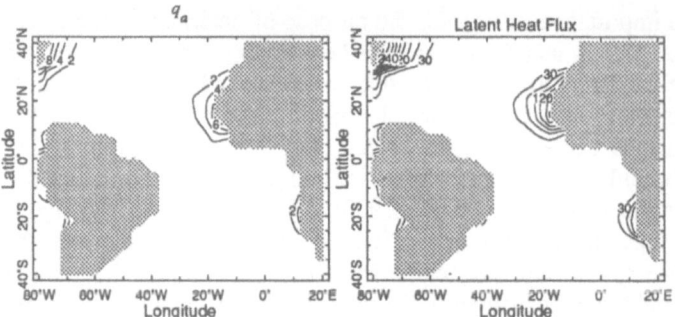

Figure 9. Immediate effect of a unit change in land specific humidity.

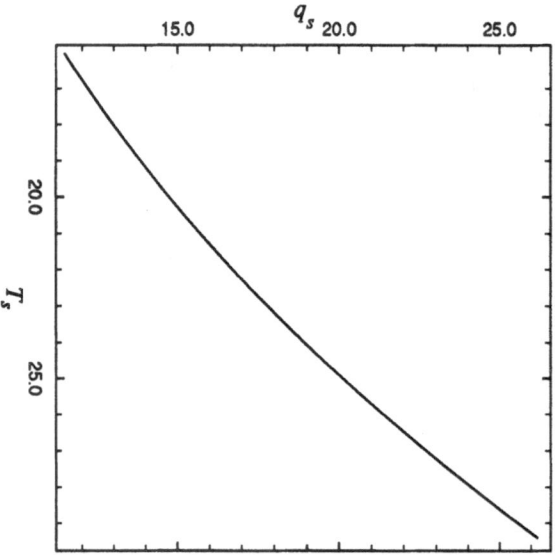

Figure 10. Saturated specific humidity as a function of temperature.

on SST. The first is that the latent heat is a relatively important term for the heat flux balance: Figure 8 shows that on average latent heat and cloudiness essentially balance the solar input in the tropical ocean. The second is that this relatively important term is quite sensitive to small changes in air humidity: since the latent heat flux depends on the difference between saturated specific humidity at SST $q_s(T_s)$ and the current air humidity q_a, a 10% change in air humidity of measuring air humidity (just over 1 g/kg) could result in a 50% change in latent heat flux. This shows both the difficulty of measuring air humidity to sufficient accuracy for heat flux purposes, and the possibility of a measurable oceanic reaction to changes in air humidity.

5. SST Changes from q_a Changes

If the air humidity q_a being advected from West Africa were to decrease, other things being the same, the latent heat flux from the ocean to the atmosphere would increase. This is illustrated in Figure 9, which shows the immediate response to a unit change in the air humidity over land given January winds. There are three points of influence: off the coast of South Africa, North America, and West Africa. The change in air humidity is manifest as a change in the latent heat flux. Off the coast of West Africa, for example, a decrease in land humidity of 5 g/kg would result in a decrease in latent flux of as much as 60 W/m².

This change in latent flux would tend to decrease the SSt as well. Since a heat flux of 50 W/m² would raise the temperature of 35 m of water 1°C in a month, we would see a significant response to a 5 g/kg change on the order of a month. A decrease in SST would decrease the cooling effect of both the latent heat and the upwelling. In areas where advection is unimportant, the latent heat flux term is the only term dependent on the SST,, and the SST stops decreasing when the latent heat flux is the same as it was before the q_a decrease. This is the point where the change in saturated specific humidity q_s matches the change in q_a. Figure 10 plots q_s against temperature, revealing that its formulation (2) notwithstanding, q_s is a relatively linear function over the range over the range of interest: an SST change of 1°C results in a change in the specific humidity of 1 g/kg. This implies that if the 5 g/kg air humidity change were entirely balanced by latent heat flux adjustments the SST would change by 5°C.

If the change in cooling due to upwelling is important, when the surface temperature T_s decreases it becomes closer to the deep temperature T_d and upwelling becomes less effective. Thus a smaller change in T_s is required to equilibrate than in the no-advection case. In the absence of an ocean model we can parameterize this process by saying that a fixed fraction λ of the change is borne by latent heat, the remainder by other processes. This parameterization allows us to rewrite the air humidity equation (4) for the change in air humidity Δq_a,

$$(u\delta_x + v\delta_y)\Delta q_a - \frac{|u|c_E}{R_H h_a}(R_H \Delta q_s - \Delta q_a) \tag{14}$$

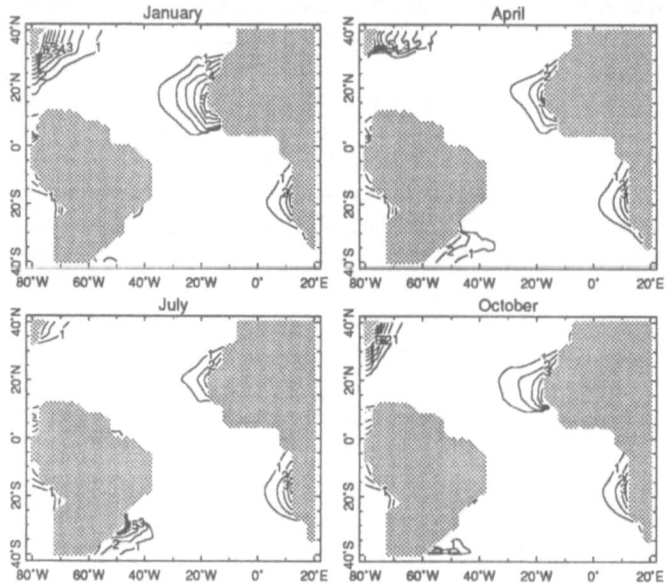

Figure 11. Patterns resulting from a unit specific humidity change over land assuming some SST change due to the change in response in q_a 10^{-1} g/kg or 0.06C.

$$- \frac{|u|c_E}{R_H h_a}(R_H \lambda - 1)\Delta q_a \tag{15}$$

In light of the relative size of the heat flux terms (Fig. 8), an appropriate λ is order 70% ($R_H\lambda = 0.67$), leaving most of the change in SST balanced by the latent heat flux. This results in patterns of air humidity changes that are given in Figure 11. These are also the patterns of SST, suggesting that a 5 g/kg change in specific humidity in West Africa would result in SST change such that the peak value is 4°C, with a change of 1°C over a fairly large area. This would indeed be a significant change in SST. Doing this calculation with a full ocean model would allow λ to vary spacially; this would not change the nature of the patterns.

6. Summary

This paper uses a simple advection model for air humidity to construct a numerical scheme for calculating a specific humidity field q_a from wind and sea surface temperature field. The model is designed for time scales of weeks to months; on that time scale equilibration is rapid, thus the q_a field at any particular time is a function only of the wind and SST at that time. A diagnostic calculation shows that the model dynamics are consistent with available data to approximately the accuracy of the data. In a predictive calculation, the air humidity model was shown to be able to reproduce the essential features of the tropical Atlantic humidity field.

Coupled to the air humidity model is a similarly simple model for the behavior of SST. The models together allow us to start to answer the question of what might happen if there were a significant change in average specific humidity over parts of Africa. The results show that there would be significant SST changes (as large as 4°C) if the specific humidity over Western Africa were to change by 5 g/kg. These calculations are done assuming no change to either the wind or the oceanic velocity field. If the changes in SST were indeed that large, the wind field would quite likely change, leading to changes in the oceanic circulation as well. Changes in the wind field, while beyond the scope of the simple models discussed here, could be modelled to give a more complete picture of the resulting changes, including any possible feedback mechanism where the changes modelled here lead to changes in air humidity over Africa, completing the cycle.

7. Acknowledgments

I would like to thank Dr. Stephen Zebiak and Dr. Yves Tourre for their helpful discussion and good advice, and Dr. Zebiak and Dr. Mark Cane for critically reading an earlier version of the manuscript. This work was supported by NSF Grant (OCE 86-08386)

8. References

Blumenthal, M.B., & Cane, M.A. (1989) "Accounting for parameter uncertainties in model verification: an illustration with tropical sea surface temperature", J. Phys. Oceanogr., (in press)

Bunker, A., Haurwitz, B., Malkus, J.S., & Stommel, H. (1949) "Vertical distribution of temperature and humidity over the Caribbean sea", in *Papers in Phys. Oceanogr. and Met.*, Mass. Inst. Tech. and Woods Hole Oceanogr. Inst., 11(1), pp. 1-82

Charney, J.G., Stone, P.H., & Quirk, W.J. (1975) "Drought in the Sahara: A biogeophysical feedback mechanism", Science, 187, pp. 434-435

Esbensen, S.K., & Kushnir, V. (1981) *The Heat Budget of the Global Ocean: An Atlas based on Estimates from Surface Marine Observations.*, Climate Research Institute Rep. 29, Oregon State University, 27 p., 188 Figs.

Folland, C.E., Parker, D.E., Ward, M.N., & Colman, A.W. (1986) *Sahel Rainfall, Northern Hemisphere Circulation Anomalies and Worldwide Sea Temperature Changes*, (unpublished)

Helleman, S., & Rosenstein, M. (1983) "Normal monthly wind stress over the world ocean with error estimates", J. of Phys. Oceanogr., 13, pp. 1093-1104

Lamb, P.J. (1978) "Large scale tropical Atlantic surface circulation patterns associated with Subsaharan weather anomalies", Tellus, 30, pp. 240-251

Lamb, P.J. (1983) "West African water vapor variations between recent contrasting Subsaharan rainy seasons", Tellus, 35A, pp. 198-212

Lough, J.M. (1986) "Tropical Atlantic sea surface temperature and rainfall variations in Subsaharan Africa", Mon. Wea. Rev., 114, pp. 561-570

Malkus, J.S. (1962) *The Sea: Ideas and Observations on Progress in the Study of the Seas, 1: Physical Oceanography*, Chapter 4, Hill, M.N. (ed.), Wiley Interscience, New York, pp. 88-294

Seager, R., Zebiak, S.E., & Cane, M.A. (1988) "A model of the tropical Pacific sea surface temperature climatology", J. Geophys. Res., 93, pp. 1265-1280

Weare, B.C., Strub, P.T., & Samuel, M.D. (1980) *Marine Climate Atlas of the Tropical Pacific Ocean.*, Dept. of Land, Air and Water Resources, University of California, Davis, 147 p.

Weare, B.C., Strub, P.T., & Samuel, M.D. (1981) "Annual mean surface heat fluxes in the tropical Pacific Ocean", J. Phys. Oceanogr., 11, pp. 705-717

Appendix A Using averaged values to compute q_a gradients

In using smoothed data sets to compute gradients in the diagnostic calculation of section 3, the gradients are distorted. To estimate (and compensate for) this effect, consider the distortion introduced in a simple test case: one spacial coordinate (x) which increases downwind and a constant equilibrium specific humidity q_e so that

$$\partial_x q = \gamma (q_e - q) \tag{A1}$$

where $\gamma = c_E/(R_H h_a)$. Suppose we are estimating gradients on a grid with separation Δx from data that has been box averaged with a box size of Δ_a. We would like to calculate the gradient at position x. We want the gradient to be representative of the entire interval Δx, so we calculate the average gradient,

$$\overline{\partial_x q} = \frac{1}{\Delta x} \int_{x-\Delta x/2}^{x+\Delta x/2} \partial_x q(x') \, dx' = \frac{q(x + \Delta x/2) - q(x - \Delta x/2)}{\Delta x} \tag{A2}$$

We can now use the solution to (A1) and explicitly include the exponential dependence, namely

$$q(x') - q_E = (q(x) - q_E) e^{-\gamma(x'-x)} \tag{A3}$$

where we have chosen to write the solution $q(x')$ at any point x' in terms of $q(x)$ the value in the center of the interval. We thus get an expression for the average gradient

$$\overline{\partial_x q} = -(q(x) - q_E) \frac{\sinh \gamma \Delta x/2}{\Delta x/2} \tag{A4}$$

On the other hand, because averaged values are being used in computing the gradients, what actually gets computed is

$$\frac{\overline{q}(x + \Delta x/2) - \overline{q}(x - \Delta x/2)}{\Delta x} \tag{A5}$$

where the average in $\overline{q}$ is computed over a width Δ_a. Again we can use the solution to relate these average values to $q(x)$ and γ,

$$\overline{q}(x \pm \Delta x/2) = q_e + \frac{1}{\Delta_a} \int_{-\Delta_a/2}^{\Delta_a/2} [q(x \pm \Delta x/2) - q_e] e^{-\gamma x'} \, dx' \tag{A6}$$

$$= q_e + (q(x) - q_e) e^{\mp \gamma \Delta x/2} \frac{\sinh \gamma \Delta_a/2}{\Delta_a/2} \tag{A7}$$

so that the expression (A5) becomes

$$\frac{\overline{q}(x + \Delta x/2) - \overline{q}(x - \Delta x/2)}{\Delta x} = -(q(x) - q_E) \frac{\sinh \gamma \Delta x/2}{\Delta x/2} \frac{\sinh \gamma \Delta_a/2}{\gamma \Delta_a/2} \tag{A8}$$

Therefore, the ratio r which converts what is calculated (A5) to what is required (A4) is

$$r(h_a) = \frac{\gamma \Delta_a/2}{\sinh \gamma \Delta_a/2}. \tag{A9}$$

The Esbensen and Kushnir data set uses a weighed averaging scheme with a radius of 1100 km; the scheme effectively surpresses wavelengths under 1600 km (15°) (Esbensen and Kushnir 1981).

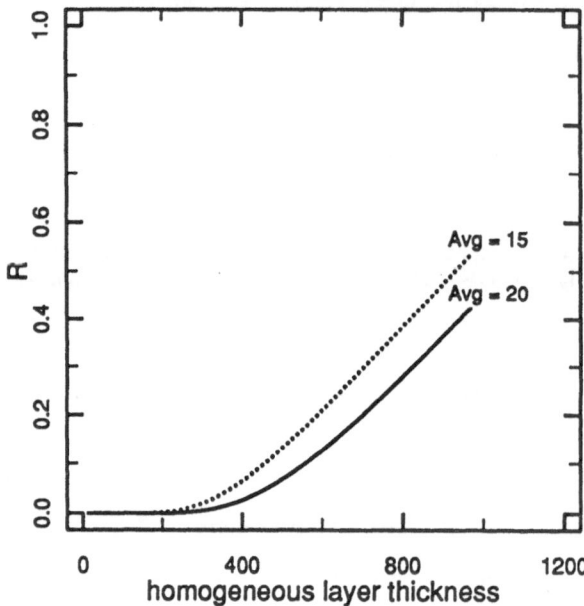

Figure 12: Coarse grid effects: relation of homogeneous layer thickness h_a to inversion parameter R.

There is additional smoothing in data poor regions, so that an appropriate box size would be 15°-20°. Figure 12 shows the inversion parameter $R = h_a R_H r$ as a function of scale height h_a for averaging scales Δ_a of both 15° and 20°. The curves indicate that the observed value $R = 170$ m corresponds to a scale height $h_a = 600$ m. This means the correction factor $r \approx .4$ is an important effect.

DETERMINING NATURAL AND MANMADE CLIMATE CHANGE: HISTORICAL REVIEW AND IMPLICATIONS FOR THE 1900'S AND BEYOND

BRUCE DENNESS
Bureau of Applied Sciences
Wydcombe Manor
Whitwell
Isle of Wight
PO38 2NY England

Abstract: Global climate is forever changing over every timescale. During the past century the earth has warmed up by more than 0.5 degrees centigrade. This is largely in keeping with predictions for the greenhouse effect. Unfortunately, there have been periods of cooling during that time which must be explained by some other mechanism, i.e. short-term natural temperature changes. These natural changes confuse the interpretation of the steady greenhouse increase and have resulted in the reluctance of politicians and planners to take decisions for remedial action to accommodate expected increases in sea level and changes in the geography of drought.

Here recent models to represent natural climate change are reviewed with particular emphasis on a deterministic model which describes the variation of global temperature. The output of that model is then combined with a steadily increasing greenhouse temperature to give a composite natura and manmade global temperature which very closely matches measured changes over the past century.

The variation of local, regional and global precipitation described by the deterministic model is illustrated over various historical periods with examples from North and South America, Africa and the world as a whole. Implications for regional and global economy are discussed and comment is made on the gross global depression that the model forecasts for the 1990's - whether or not the greenhouse effect has a substantial influence. Sea level changes are similarly reviewed and related to the model both recently and historically. Attention is drawn to accelerations of atmospheric CO_2 concentration in about 1870 and 1950 according to the historical record and these are related directly to simultaneous acceleration in world Population growth.

The combination of the natural and manmade deterministic models forecasts a global temperature rise of almost 1 degree centigrade by the year 2000. Political and socio-economic implications of misinterpreting this rise as solely due to the greenhouse effect are addressed. Finally a range of practical means of containing use of fossil fuels within levels that need not exacerbate the greenhouse effect are discussed with due regard to economic considerations and secondary spin-off problems and their solution.

R. Paepe et al. (eds.), Greenhouse Effect, Sea Level and Drought, 41–74.
© 1990 *Kluwer Academic Publishers.*

1. Introduction

Climate has changed on all timescales in different ways around the world since the earliest geological records. Over the past couple of centuries since the birth of the industrial revolution an anthropogenic component has been added to these natural changes. Observation and measurement of climate change, particularly temperature and precipitation, over the past century has made us increasingly aware of the scale and rate of change over even relatively short periods but has not preciously enabled the accurate differentiation of natural and manmade climate change. This problem is addressed here with a view to enabling planners and politicians to respond to a more precisely defined problem than has hitherto been the case. The background to this study encompasses climate change itself and the development of modelling techniques to simulate climate change.

1.1. GENERAL CLIMATE CHANGE

Global climate is forever changing on every timescale. This is reflected in regional and local variations of any of a vast range of climate-related parameters. Over the past century alone the Earth has warmed up by more than 0.5 degrees centigrade: this can be compared with an increase of about 6 degrees centigrade at the end of the last glaciation from about 17,000 to 12,000 years before present (BP) when ice sheets retreated from Eurasia and North America to raise eustatic sea level by about 100 metres. The geological past has seen even greater natural changes but it is the more recent historical past that is better documented and generally thought to be more relevant to interpreting trends and forecasting future climate.

1.2. DIFFERENTIATION NATURAL AND MANMADE CLIMATE CHANGES

Natural climate changes are generally thought to be largely controlled by astronomical phenomena concerned with the Earth's aspect to and orbit around the Sun. The thesis developed here does not challenge that relationship but could accommodate a more fundamental link than simply cause-and-effect. It also seeks to extend the range of frequencies of climate change from those appropriate to the recognition of a link between ice ages and astronomical phenomena to those impacting on recent history and present and future planning.

Manmade climate changes are generally considered to stem from discharges of so-called greenhouse gases into the atmosphere largely as a result of increasing industrialisation and deforestation. To date perhaps the main culprit has been carbon dioxide (CO_2) released from the burning of fossil fuels and no longer able to be as substantially taken up by the decreasing forests. Lesser gaseous elements further aggravate the greenhouse effect, thus termed because the whole family of greenhouse gases enter the upper atmosphere where they behave like the glass in a greenhouse by allowing all the Sun's energy in while reflecting back to Earth much of the infra-red part of the spectrum that would otherwise pass out again through the atmosphere.

The temperature rise thought to result from the greenhouse effect has been calculated from the ledger of CO_2 input to the atmosphere theoretically to lead to a steadily increasing global temperature of about 0.6 degrees centigrade over the past century (Hansen et al., 1981; Denness, 1984 a). The measured response of recorded global temperature over the same period has been more confusing: the first half of the period saw the expected rise but since the 1930's the global temperature has fluctuated about a stable mean. This might at first sight appear to challenge the reality of the greenhouse effect. However, a model of natural global temperature change is extant that, coupled with conventional interpretations of the greenhouse effect, leads to a closer approximation to the observed signal than is otherwise available. The model is totally deterministic and has been extensively reported by Denness (1981, 1982, 1983 a, b & c, 1984 a, b, c & d, 1985 a & b, 1986 a & b, 1987 and 1989 a & b) both with respect to its derivation and the physical and socio-economic implications of its forecasts.

2. Climate Forecasting Models

Forecasters seem to be divided philosophically into two camps: those who believe that natural phenomena change according to a recognisable though possibly complex pattern and are, therefore, deterministic, and those who believe such changes to be stochastic. The former group seeks forecasting models, the ultimate goal of which is to predict events at a given time in the future with certainty - any probability statement being only a measure of ignorance regarding the baseline facts on which the model is calibrated. The latter's models are statistically inclined and necessarily attach a probability to each forecast - even if baseline facts are absolutely known - on the basis of the recurrence frequency of a given event in the past.

The writer is a determinist. a position he arrived at from the need to forecast future climatic events on the strength of only short-term (perhaps 10 or 20 years) baseline input data at sea and in many parts of the developing world. Such data commonly reflect only the climatic "standstill" since the 1940's so that a probabilistic method of forecasting could never predict an unusual event such as the exit from the Little Ice Age in the middle of the previous century - yet such events happen.

2.1. GENERAL CIRCULATION MODELS (GCMS)

GCMs perform as their title suggest: they describe the general circulation pattern of the global atmosphere subject to the input of a wide range of known, inferred or conjectural data concerning the constitution and other physical and chemical properties of the atmosphere and the oceans, and the ocean/air and land/air interface. By so doing they permit interpretations of regional variations of temperature and precipitation in response to various input data suites. It is generally conceded that, matched against current atmospheric conditions, GCMs are better at forecasting temperature than precipitation.

The success of the models can be only as good as the input data and the validity of the relationships used to describe the multitude of interactions and feedbacks (servo-mechanisms) in the total system. Should any of these, say a density-related parameter having an influence on atmospheric pressure, change in a way not reflected by the input data the consequent forecasts will go awry. Overall it would appear that GCMs have much to offer, particularly, with respect to forecasting in areas for which there is little or no observed or measured information, subject to reservations about the validity of input data to simulate future conditions.

2.2. DETERMINISTIC MODELS

Deterministic climate modelling began with the recognition by Croll (1875) that there appears to have been a relationship between the recurrence of widespread glaciation and the variation of certain special planetary characteristics. Milankovitch (1938) consolidated and extended Croll's ideas with the recognition of a more precise connection between planetary deportment and climate over recent geological time. This so-called astronomical theory has firmly entrenched itself in the climatological literature as a cause-and-effect doctrine.

More recently Hansen et al. (1981) have embellished the astronomical principle for the shorter term appertaining to the present century and near-future forecasting by considering the possible influence of the Sun, volcanic activity and so on. Constructive reviews of such modelling were given by Imbrie & Imbrie (1979) for the long term and Liss and Crane (1983) for the short term with particular attention to the greenhouse effect.

The most recent entry to the deterministic forecasting field appears to be the writer's own model which he has, perhaps pretentiously, called a geophysical model to allow for its further application outside the immediate area of climate forecasting. In concert with a forecast temperature rise resulting from the known input of atmospheric pollutants to the greenhouse effect the latter has been tested with interesting results in forecast since 1980.

2.2.1. *Comparison of Selected Models.* Various historical and modern deterministic models are compared here to illustrate the direction of development and the recent urgency to incorporate the greenhouse element properly differentiated from natural changes.

2.2.1.1. Astronomical Models. Croll may have had little idea that his observation of relationships between astronomical periodicities within the solar system and the recurrence of ice ages, only themselves recognised a few years earlier by Carpentier and Agazziz (Imbrie & Imbrie, 1979), would provide the basis for deterministic climate modelling of the future. With Croll's experience behind him and the mind of a mathematician and engineer Milutin Milankovitch was in a better position to realise that his association of glacial recurrences with planetary changes are more or less consistent and therefore forecastable. Milankovitch's observations must rank as one of the cornerstones of modern climatology, albeit that they refer to low frequency climate variation of little direct interest to contemporary engineering or planning.

Hansen appears to have spanned the astronomical gap between the long timescale of ice age recurrence and the short timescale of modern planning. In addition to astronomical phenomena such as sunspots (the variation of which appears to have an association with earthly climate change but for which no concrete explanation has yet been convincingly demonstrated). Hansen's model also embraces a consideration of nearer geophysical phenomena such as changes in the intensity of volcanity. Nevertheless, published forecasts using Hansen's empirically deterministic model do not appear to allow for dramatic changes in the near future. Several other deterministic models proliferating in the literature seem be equally enthusiastic in their representation of past events but also to point on average to a uniform future.

2.2.1.2. Geophysical Model. Among the more recent newcomers to the deterministic scene is the writer's own forecasting model. The origin of this model was described by Denness (1981) who pointed to the separation of the multi-frequency components of variation in a 7 million year oxygen isotope time series from Shackleton and Cita (1979) as the starting point for its derivation from published climate time series. The initial components of the subsequently much more extensive compound sine series from that work are shown in Figure 1 with periods of about 4.8, 2.4 and 1.2 million years - ever doubling frequency and amplitude simultaneously reducing by 0.81. Through a series of other publications Denness (up. cit.) extended that sine series by matching geological timescales and beyond and to the short-term interpretation of less then a year.

The governing equation of the geophysical model, which describes global temperature variation on all timescales as well as other geophysical parameters, is

$$G(t) = \sum_{n=N(T)}^{\infty} A(T) a^{n-1} . \sin\{b^{-n} . \pi (\frac{t}{T})\}$$

which is zero-registered at time T_0 and in which:
G(t) is a time-based climate, index, e.g. global temperature,
A(T) is the amplitude of a reference periodicity T,
N(T) is the reference integer for periodicity T,
a, b are absolute constants, here taken as 0.81 and 0.5 respectively, n is an integer, i.e., the reference number of a particular sine component and t is time in years.

The geophysical model can be used in forecasting mode with t negative from the end of 1980 when the model was completed. Of course, it represents only the variation of natural global temperature with no anthropogenic influence

2.2.1.3. Greenhouse Effect Models. As noted earlier the greenhouse effect describes that (rising) component of overall global temperature change that is due to the pollution of the atmosphere by manmade gases that allow all elements of the Sun's energy in to the Earth but do not allow all of the infra-red part of the spectrum to be reflected out again. The main pollutant is CO_2 and it has been calculated by Moore et al. (1981) as a mean among other estimates that about three gigatonnes of carbon in the form of CO_2 enters the atmosphere each year. In turn this has been shown by, for example, Hansen et al.(1981) and Denness

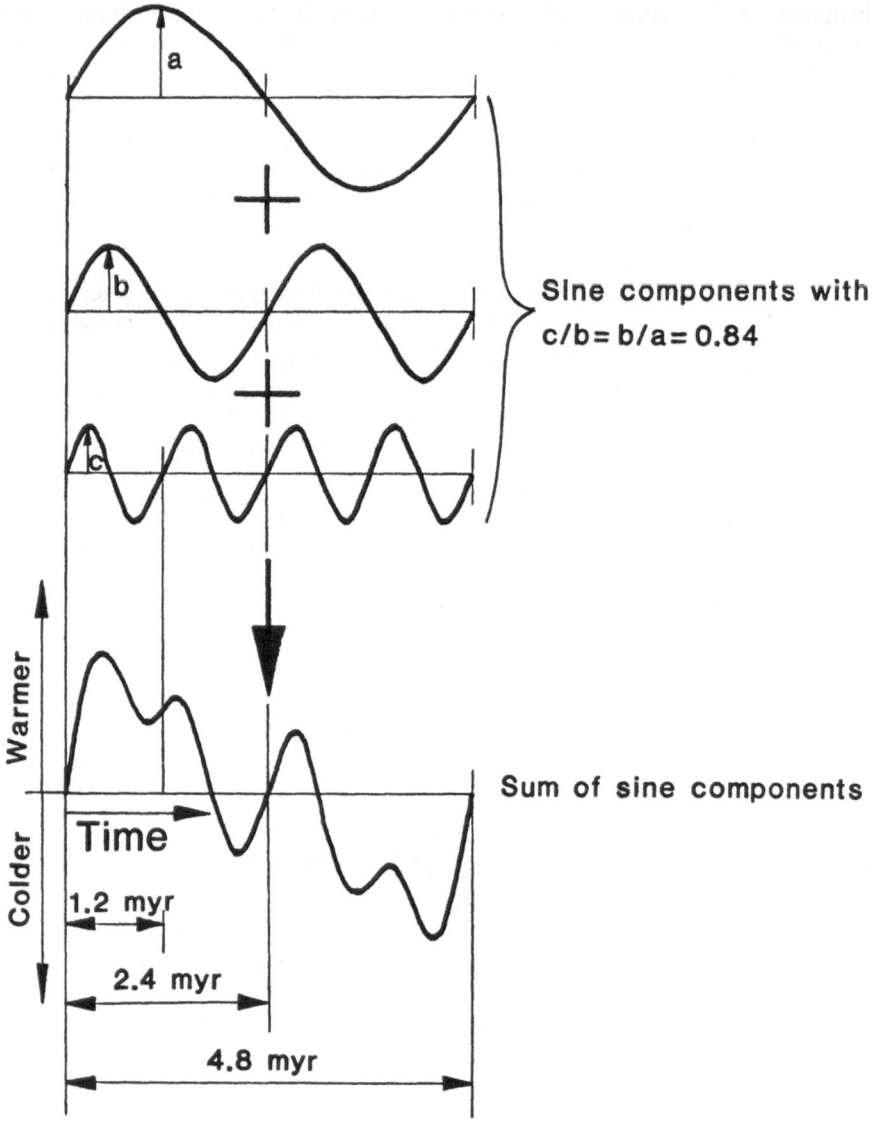

Figure 1. Combination of sinusoidal components.

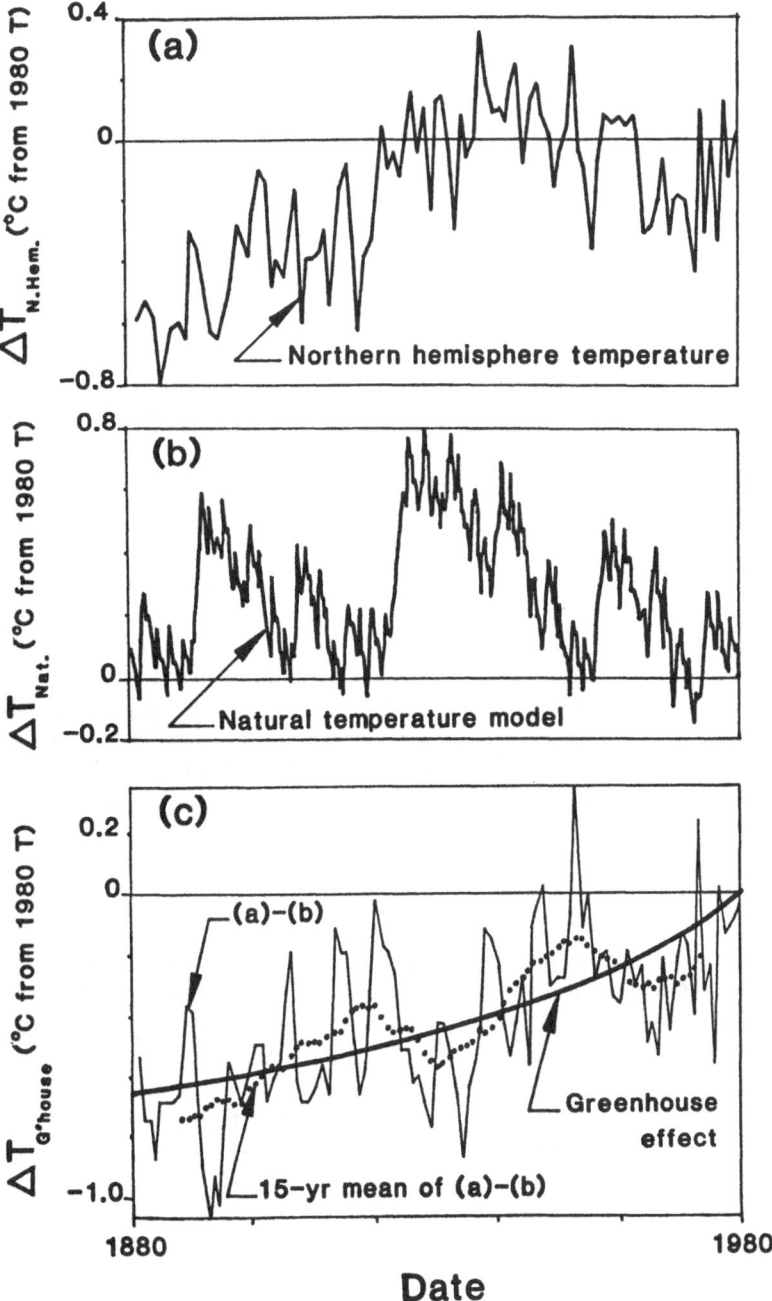

Figure 2. Separation of natural and manmade temperature change.

(1984 a) theoretically to have resulted in a rise of about 0.6 degrees centigrade over the last 100 years; doubling atmospheric CO_2 is commonly estimated to raise the global temperature by about 3 degrees centigrade, e.g. Wetherald and Manabe (1981)

2.2.1.4. Natural and Manmade Global Temperature Models. Measurement of global temperature since 1881 has been accomplished by several workers (e.g., Jones and Wigley, 1980) with the result that a fairly steadily increasing trend until about the end of the 1930's is seen to be followed by fluctuations about a steady mean up to recent years. This is superficially at odds with the steady rise throughout required by the greenhouse effect. The differences between the postulated greenhouse component and the observed temperature is thought to be due to natural temperature change, as indicated for example by Hansen, which often tends to be treated as "noise" on the main signal.

Better recognition of the form variation of the natural component would allow more confidence to be placed in the greenhouse calculations. The geophysical model described above permits this. Figure 2, based upon data presented by Denness (1984 a), illustrates how the subtraction of the temperature represented by that model from the observed global temperature leads to a residual temperature that has generally increased over the past century much in keeping with the trend commonly postulated for the greenhouse effect as shown. It is suggested that this is an encouraging demonstration for proponents of the greenhouse hypothesis.

2.2.2. *Testing the Geophysical Model.* The ability of the geophysical model to match climate-related rime series has been tested against several hundred published data suites over geological, historical and instrumental timescales in hindcast. The present theme is concerned with water variation: therefore, additional examples of hindcast matching to precipitation time series that have recently come to the writer's attention are presented here by way of further demonstration of the model's potential.

2.2.2.1. Colombia, South America. Figure 3 portrays the variation of annual rainfall at the cities of Bogotá and Popayán situated in very different geographical regions of Colombia. Both show virtually the same cyclic fluctuations with a period of little under 17 years, one of the components of the geophysical model, precisely in step with the model in such a way as to indicate a reduction of rainfall in response to natural global warming (hindcast by the model) as previously calibrated by Klein (1982).

2.2.2.2. Northern Africa. In view of the current theme figure 4, rearranged from Denness (1985 a), is presented to illustrate the matching in hindcast of the geophysical model with a series of historical time series from Lamb (1977) relating to drought and flood in a region still notorious for such events today. The model is presented here in its basic form, i.e. not subject to moving average smoothing; this illustrates the greater complexity of interpreting its unadulterated output but still permits attention to be drawn to general peaks and troughs. It should be noted that periods of relatively high global temperature correspond essentially to drought in the west and flood in the east of the region, whereas the precipitation response

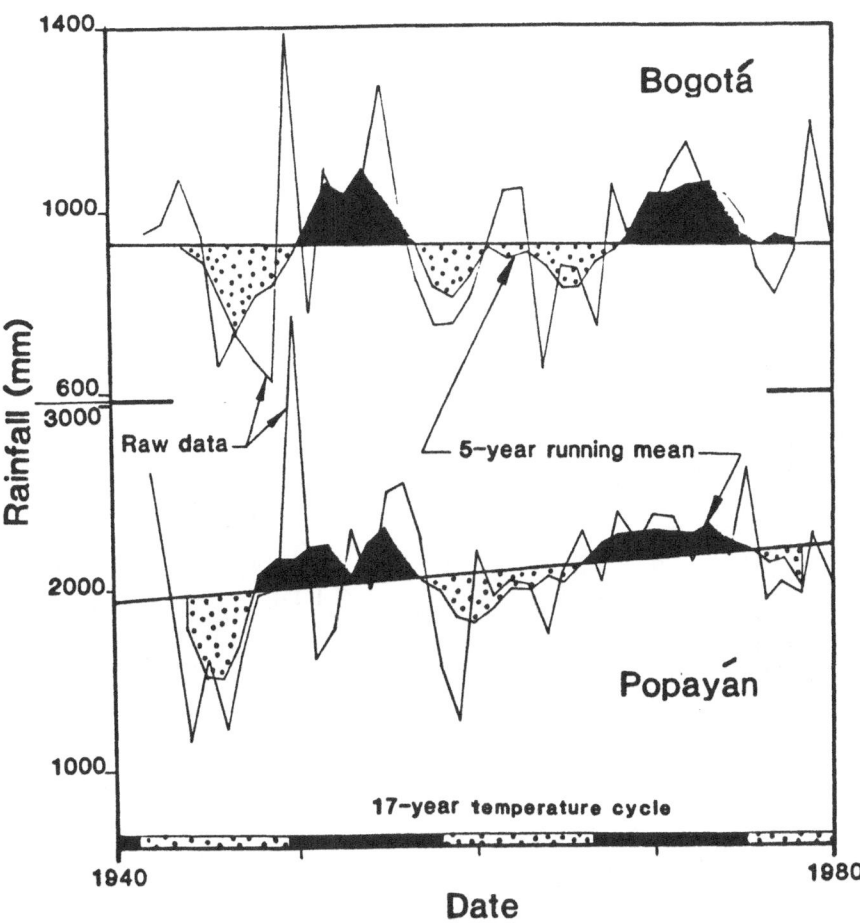

Figure 3. Colombian rainfall cycles.

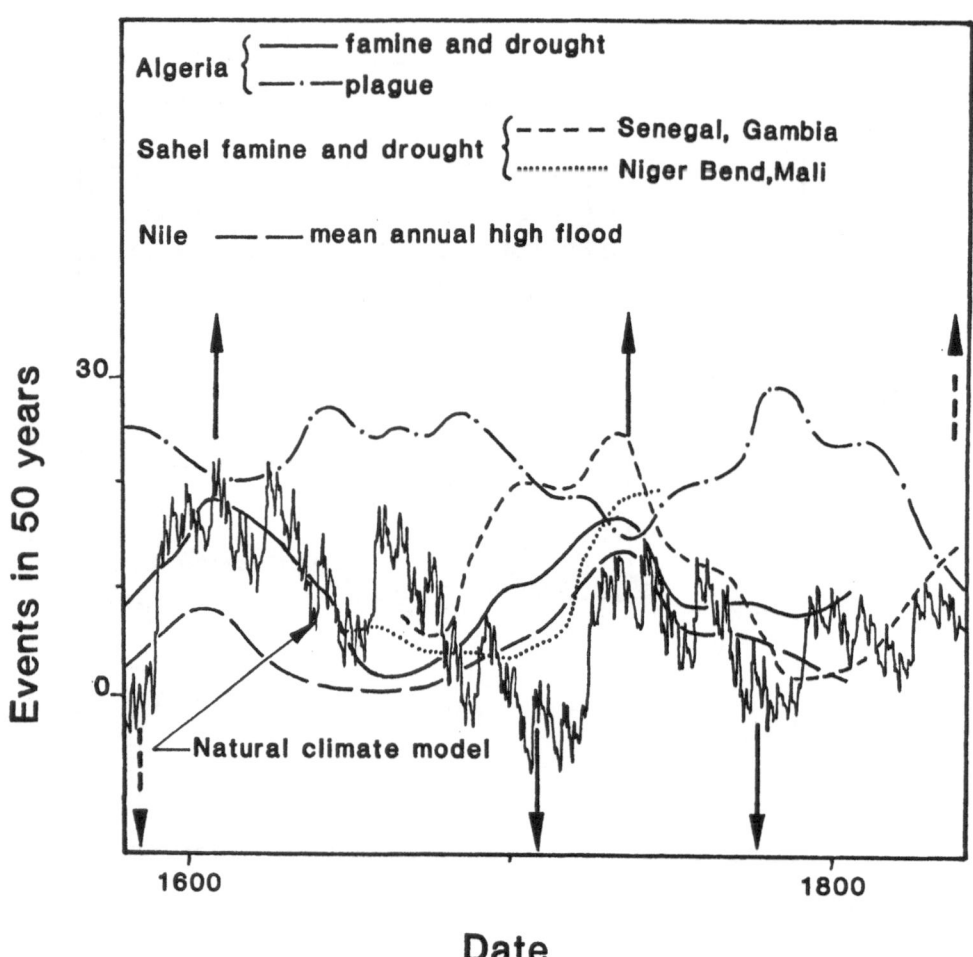

Figure 4. Historical drought in northern Africa.

is reversed for relatively low global temperature.

3. Climate and Man

Man is by nature deeply concerned by economic trends. Here the impact of drought and sea level rise on those trends and other distant historical events is demonstrated with regard to both natural and manmade climate changes.

3.1. RELATION OF DROUGHT AND ECONOMY TO GEOPHYSICAL MODEL

Here we are concerned with the natural component of climate change - that part that happens with or without a greenhouse effect. In order to explore the regional variation of precipitation (especially drought) in response to northern hemisphere temperature change it is appropriate to consider Figure 5, taken from Wigley et al. (1980). Here it is seen that as global temperature increases so some parts of the world become wetter and others drier. This influences agricultural productivity; generally the drier land becomes the less productive it is. This results either directly in an economic machine to agricultural irrigation and/or artificial fertilisers as described in greater depth by Denness (1983 c and 1984 d).

Among the earliest examples reported by the writer of a significant economic response to drought introduced by global temperature change is the expansion and withdrawal of agriculture in Roman Tripolitania (approximately modern Tunisia) during the period 300 BC - AD 300 (Burns and Denness, 1985). The progress of economic trends in England and France for a similar period after AD 1300 was followed by a review of the economic progress of the whole world since the early 1800's by Denness (1984 d). Many other examples are available.

Figure 6 portrays the smoothed output of the geophysical model in hindcast representing global temperature change since 1920. It also shows the simultaneous smoothed of the U.S. economy (inverted for easier correlation) from Batri (1987). As would be expected from figure 5 and the agricultural link, the U.S. economy has marched inversely in step with the global temperature hindcast by the model: the hotter the global temperature, the drier the mid-west agricultural (corn) belt and the worse the economic situation - as forecast by the model for the 1990's.

3.2. RELATION OF SEA LEVEL RISE TO GEOPHYSICAL MODEL

Sea level has changed over all timescales both globally (eustatic) and locally (isostatic/tectonic). The essential thesis is that higher global temperature, for whatever reason, induces a rising eustatic sea level as a result of density decrease (thermal expansion) and ice melting: the impact of isostasy is not part of this argument.

In the more recent past, the calculation of eustatic sea level change by Barnett (1984) as a steady rise, in keeping with expected greenhouse growth, over the last century has been challenged by Pirazzoli (pers. comm.) and Klige (pers. comm.) both of whom recognise

PRECIPITATION

Figure 5. Precipitation change with global warming

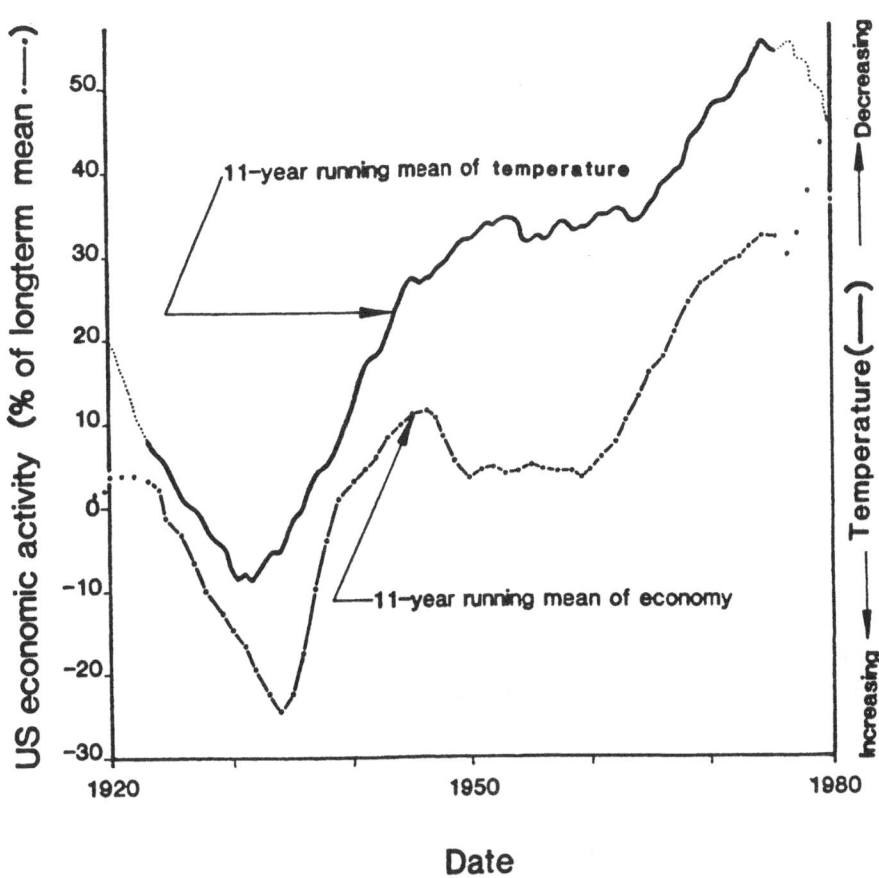

Figure 6. Economic response to global temperature change

a fairly steady rise to about 1930 followed by fluctuations about a stable mean thereafter. Pirazzoli's and Klige's reviews match parallel observations of temperature change (e.g., Jones and Wigley, 1980) to show that sea level change is, as expected, in step with temperature change. The separation of the greenhouse effect from natural climate change by Denness (1984 a) explains these observations and illustrates the reality of both the greenhouse effect and natural climate change with respect to sea level rise.

3.3. GREENHOUSE REVISITED

The previous cases involved sea level rise in response to a natural climatic signal. What is the scale of the recent manmade overlay? To gain a sense of anthropogenic proportion one might first refer to the natural changes in the atmospheric greenhouse gas CO_2 recorded in various marine and glacial successions over much of geological time. Denness (1984 a) provided a brief compendium showing that there has been a close association of global atmospheric CO_2 with the geophysical model over all timescales from a few thousand years or less to hundreds of millions of years.

Figure 7 describes the rapid increase of atmospheric CO_2 (much faster than any rise in the geological past) that followed the inception of the Industrial Revolution, which started in the early eighteenth century in Europe and encapsulated the whole world by the 1870's. The longer term trend from the mid 1800's, after Machta (1979) is seen to exhibit two accelerations in growth rate about 1870 and 1950; the shorter term covered by the more precise data of Callander (1958) and Keeling et al. (1982) can show only the 1950 breakpoint. Both of these trends establish the increase in atmospheric CO_2 to be real (even if only about half of that calculated to result from fossil fuel burning, etc.) and increasing as global population increases according to the models derived empirically by Meyer and Vallee (1975) and analytically in the Appendix here.

4. Reacting to Deterministic Forecasts

In order to react sensibly it is necessary to know with confidence what one is reacting to. Both physical and social change factors are considered here before addressing possible preventative and mitigating actions.

4.1. CLIMATE FORECASTS AND IMPLICATIONS FROM THE GEOPHYSICAL MODEL

It is stressed that the following represents the minimum changes that can be expected with no account taken of the aggravating impact of possible additional population growth and that of the greenhouse effect beyond that of natural climate change has not been considered with regard to the geography of drought of social implications.

4.1.1. *Physical Changes*. The main physical changes expected to result from global climate change, whatever the cause, concern sea level rise and drought. Other factors such as floods,

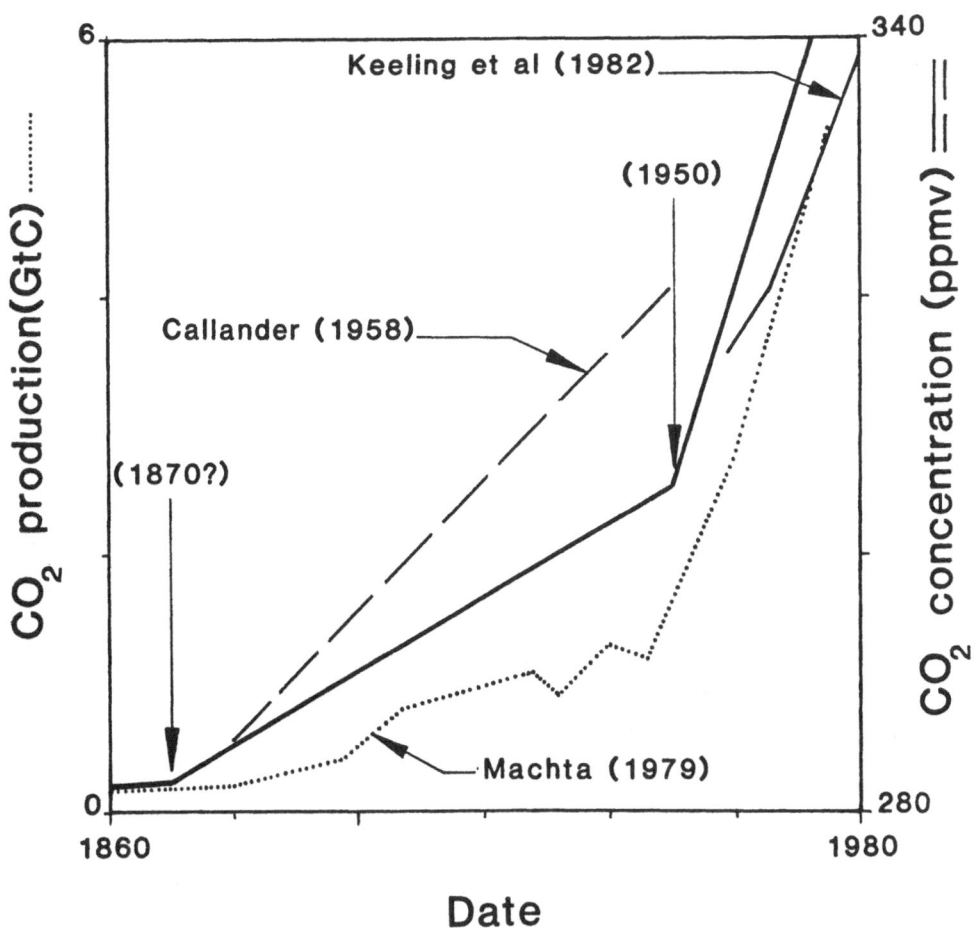

Figure 7. Acceleration of global CO_2.

hurricanes and other weather phenomena are thought to be less significant in the long term.

Albeit that sea level has fallen as well as risen in the past, we are here concerned with a future in which only a sea rise seems likely from both greenhouse effect and natural climate components of global temperature change forecast by the geophysical model, first published in the present context by Denness (1984 a), are set to increase from the late 1980's into the twenty first century. As the twentieth century has so far seen a close match of sea level change with global temperature (Both from observation and from the model) it is entirely logical to expect a resumption and acceleration of eustatic sea level rise in the very near future.

Historical aspects of the geography of drought in relation to global temperature change and its description by the natural component of the geophysical model have been discussed in Section 3.1. Figure 5 was invoked to describe precipitation changes that can be expected from the rising global temperature that is expected from both the greenhouse effect and natural change over the coming decades. Among the areas forecast to become drier are the mid-west of the U.S.A., Kazakhstan in the U.S.S.R., and parts of north-west Europe, between them producers of most of the world's grain. If the decreasing precipitation approaches drought proportions the outcome is clear; already drought has afflicted the U.S. as expected.

4.1.2. *Socio-economic and Political Implications*. The socio-economic and political consequences of rising sea level and drought are likely to be widespread and serious. Although they cannot all be itemised in detail some are already clear. An example is the need to divert resources into coastal protection or to abandon coastal areas in response to sea level rise: the choice will depend largely on the scale of the existing investment, the political influence of the afflicted nation and the degree of confidence in forecasts of future rises. The scale of diversion of resources from elsewhere in the global economy will be immense, probably exceeding the proportion currently spent on defence.

Another example, relating to drought, is the future of Western Europe and the so-called Superpowers, the U.S.A. and U.S.S.R. in relation to the developing world. As these parts of the developed world, accustomed as they are to policing the less developed, find themselves economically constrained and with contingent social problems it is interesting to muse upon their likely reaction to the emergence of major powers from the burgeoning developing world, much of which will be enjoying far mor clement climatic conditions.

4.2. PREVENTION/MITIGATION OF CLIMATE-DRIVEN PROBLEMS

There is a variety of ways in which the greenhouse effect (already underway) can be mitigated: its extension could be prevented by simple politico-economic decisions in the developed world followed by similar moves in the developing world.

The generation of energy from fossil fuel or wood is the most serious contributor to the greenhouse effect. Yet it is only the bad building and transport habits of the developed world in temperate climates, accustomed to cheap and harmless energy, that is responsible for much of this problem. Insulation is poor, energy is not conserved and waste ensues causing the unnecessary generation of more energy and releasing more CO_2. Switching to renewable energy resources from wind, waves, the sun and so on volunteers a long term communal solution. So does the introduction of endothermic space and water heating: this basically involves reversing the household refrigerator principle by taking heat from the outside to warm the inside of a dwelling - at about a quarter of the cost of a conventional heating system and little more installation cost, i.e. it does not require communal decisions but is available to the individual. Universally used this proven system could alone achieve all the greenhouse targets of the developed world for the next 20 years without curtailing industry at all.

It is certain that countermeasures to solve greenhouse problems will themselves, generate largely unanticipated spin-off problems. One thinks, for example, of the obvious need for energy insulation in buildings in temperate climates (a need incidentally that was sparked off by the energy crises of the early 1970's - politico-economic event - but has lived on into the "greenhouse concern age") which caused major condensation problems. These problems may in themselves be trivial but will deter democratic people from responding wholeheartedly and urgently to governmental anti-greenhouse schemes as necessary unless they too can be solved readily, as has the condensation problem by a simple cheap device that does not create a third generation problem, for example.

5. Conclusions

A deterministic geophysical model has been developed to describe the variation of natural global temperature over all timescales. The model can incorporate a greenhouse component over the last century or so and by so doing replicates measured global temperature for that period better than any other model known to the writer.

The geophysical model matches past sea level changes via its global temperature representation and forecasts significant rise in the near future. It also explains the variation of economic progress on national and global bases over historical and recent timescales via an agricultural link and forecasts imminent global depression.

Preventative and mitigation measures to combat the greenhouse effect are within the capacity of existing technology. Whether genuine encouragement of their introduction is within the capacity of existing political machinery remains to be seen.

6. Acknowledgments

The writer is pleased to acknowledge the moral support for this work in early days, when its study was less fashionable than now, provided by many of his friends, especially Ann Hayton and Ralph Whittington. He is also grateful for various suggestions from Rhodes Fairbridge.

7. References (Climate)

Barnett, T.P. (1984) "The estimation of 'global' sea level change: a problem of uniqueness", J. Geophys. Res. 89, pp. 7980-7988

Batri, R. (1987) "The great depression of 1990.", Simon and Schuster, New York

Burns, J.R., & Denness, B. (1985) "Climate and social dynamics: the Tripolitanian example, 300 BC - AD 300.", *In town and Country in Roman Tripolitania*, Buck, D.J. and Mattingley, D.J. (eds.), British Archaeological Reports, Int. Ser. 274, Oxford, pp. 201-225

Callander G.S. (1958) "On the amount of carbon dioxide in the atmosphere.", Tellus, 10, pp. 243-248

Croll, J. (1875) *Climate and time*, Appleton, New York

Denness, B. (1981) "How to build an ocean.", Proc. IEEE Conf. Oceans '81, Boston, pp. 341-344

Denness, B (1982) "Slope stability - Are rainfall-induced landslips predictable?", Proc. 7th Southeast Asian Geot. Conf., Hong Kong, 1, pp. 71-72

Denness, B. (1983 a) "From seabed to the skies.", Northern Executive, 1, pp. 77-78

Denness, B. (1983 b) "The economy/climate link in 2000 AD.", Northern Executive, 2, pp. 117-118

Denness, B. (1983 c) "Economic climate.", Financial Weekly, 252, p. 10

Denness, B. (1984 a) "The greenhouse affair.", Marine Pollution Bull., 15 (10), pp. 355-362

Denness, B. (1984 b) "An analytical climate model: application to the southern hemisphere Quaternary period.", Vogel, J. (ed.) Proc. SASQUA Int. Symp. on *Late Cainozoic Palaeoclimates of the Southern Hemisphere*, Swaziland, Balkema, Rotterdam, pp. 35-42

Denness, B. (1984 c) "The coincidence of a general climate model and historical climatic observations in east Asia.", Whyte, R.O. (ed.) Proc. Int. Conf. on *The Evolution of the East Asian Environment, 1: Geology and Palaeoclimatology*, Hong Kong, Centre of Asian Studies, University of Hong Kong Press, pp. 199-219

Denness, B. (1984 d) "The energy-economy-climate link.", Energy Exploration and Exploitation, 3 (1), pp. 61-69

Denness, B. (1985 a) "Water resource forecasting.", Proc 4th IAHR Conf. on *Water Resources Development and Management*, II, Chiang Mai, Thailand, pp. 1177-1194

Denness, B. (1985 b) "The greenhouse dilemma.", in *Nature*, 318, p. 596

Denness, B. (1986 a) "Water: supply and demand forecasting.", Proc. Conf. *World Water 86*, ICE, London

Denness, B. (1986 b) "Water Level.", New Scientist, 1459, p. 59

Denness, B. (1987) "Sea level modelling: the past and the future.", Prog. Oceanog., 18, pp. 41-59

Denness, B. (1989 a) "The variation of the Universal Gravitational Constant and the distribution of earthquakes in China through time and space.", Seismology and Geology, The Seismological Press, Beijing, II (2), pp. 53-65

Denness, B. (1989 b) "Evolution and environmental determinism.", Modern Geology, 13, pp. 283-286

Hansen, J., Johnson, D., Lacis, A., Lebedeff, A., Lees, P., Rind, D., & Russell, G. (1981)

"Climate impact of increasing atmospheric carbon dioxide.", in Science, 213, pp. 957-966

Imbrie, J., & Imbrie, K.P. (1979) *Ice ages: solving the mystery.*", MacMillan Press, New York, 224 p.

Jelgersma, S. (1966) "Sea level changes in the last 10,000 years.", in Sawyer, J.S. (ed.) Proc. Int. Symp. on *World Climate from 8000-0 BC*, Royal Met. Soc., London, pp. 54-69

Jones, P.D., & Wigley, T.M.L. (1980) "Northern hemisphere temperatures, 1881-1979.", Climate Monitor, CRU, University of East Anglia, Norwich, 9, pp. 43-47

Keeling, C.D., Bacastow, R.B., & Whorf, T.P. (1982) "Measurement of the concentration of carbon dioxide at Mauna Loa Observatory, Hawaii.", in Clark, W.C. (ed.) *Carbon Dioxide Review: 1982*, Oxford University Press, New York, pp. 377-385

Klein, W.H. (1982) "Detecting carbon dioxide effects on climate.", in Clark, W.C. (ed.) *Carbon Dioxide Review: 1982*, Oxford University Press, New York, pp 215-242

Klige, R.L.K. (pers. comm., 1989) Advice at NATO Research Workshop, Fuerte Ventura, Spain, 1-7 March, 1989.

Lamb, H.H. (1977) "Climate: present, past and future.", 2, *Climatic History and the Future*, Methuen, London, 835 p.

Liss, P.S., & Crane, A.J. (1983) *Manmade carbon dioxide and climatic change: a review of scientific problems*, Geo Books, Norwich

Machta, L (1979) "Atmospheric measurements of carbon dioxide.", in Elliott, W.P., and Machta, L (eds.) *Workshop on the Global Effects of Carbon Dioxide from Fossil Fuels*, U.S. Dept. of Energy, Washington, D.C., pp. 44-50

Meyer, F., & Vallee, J. (1975) *Technological forecasting and social change*, 7, pp. 285-300

Milankovitch, M. (1938) "Astronomische Mittel zur Erforschung der Erdgeschichtlichen Klimate.", Gutenberg, B (ed.) *Handbuch der Geophysik*, 9, Berlin, pp. 593-698

Moore, B., Boone, R.D., Hobbie, J.E., Houghton, R.A., Medillo, J.M., Petersen, B.J., Shavers, G.R., Vorosmarty, C.J., & Woodwell, G.M. (1981) "A simple model for analysis of the role of terrestrial ecosystems in the global carbon budget.", in Bolin, B. (ed.) *Scope 16 - Carbon Cycle Modelling*, John Whiley, Chichester, pp. 365-385

Mörner, N.A. (1980) "The Fennoscandian uplift: geological data and the geodynamical implication.", in Mörner, M.A. (ed.) *Earth Rheology, Isostasy and Eustasy*, John Whiley, Chichester, pp. 251-284

Pirazzoli, P. (pers. comm., 1989) Advice at NATO Research Workshop, Fuerte Ventura, Spain, 1-7 March, 1989

Shackleton, N.J., & Cita, M.B. (1979) "Oxygen and Carbon isotope stratigraphy of benthic foraminifera at Site 379: detailed history of climate change during the late Neogene.", in *Initial Reports of the Deep Sea Drilling Project*, U.S. Governmental Printing Office, Washington, D.C., 47, pp. 433-445

Wetherald, R.T., & Manabe, S. (1981) "Influence of seasonal variation upon the sensitivity of a model climate.", J. Geophys. Res., 86, pp. 1194-1204

Wigley, T.M.L., Jones, P.D., & Kelly, P.M. (1980) "Scenario for a warm, high CO_2 world", in *Nature,* 283, pp. 17-21

APPENDIX: POPULATION GROWTH ACCELERATION AND ITS RELEVANCE TO THE GREENHOUSE EFFECT

A.1 Synopsys

The youthful growth stages of individual animals and plants exhibit expansion according to the cube of time. Extending earlier work which has shown marked consistency between growth rates for individual plants and animals and whole communities, it can be shown that the human population as grown in a series of acceleration expansion phases separated by steps of cultural evolution with their attendant technological revolutions, each phase exhibiting a cubic-time growth pattern.

The relation between technological progress and population growth is analysed here for the population as a whole and the archaeological and historical record is examined to substantiate the association of the beginning of cubic-time growth phases with technological revolutions . It is shown that the historical growth record for which reliable population estimates are available, i.e., since the fifteenth century, corresponds closely with the theoretical pattern which, if uninterrupted, would culminate in an infinite population by 2030 AD. However, it is concluded that such an expansion could equally well be transformed into the steady an rapid cultural evolution of a stable population able to depend on several technological revolutions in the individual lifetime and contain the greenhouse effect by so doing.

A.2 Introduction

Homo sapiens derives his very name from his most distinguishing characteristic, wisdom. Wisdom is born of intelligence, the intellectual skill which enables the growth of a knowledge base from step-by-step learning, reasoning and remembering. One of the ways in which man has employed intelligence is reflected in the historical record of the growth rate of the overall population. Every time he has conceived a sufficiently bright idea to lead to a technological revolution he has invested the material rewards flowing from it upon procreation; population growth has accelerated every time.

This is the generalised embodiment of the notion behind Malthusianism, an essentially laissez-faire philosophy which sees the technology-population link as inevitable. However, the analytical framework for the entire growth pattern is more confusing.

First it is necessary to consider the growth of a population which is not beneficiary of intelligence, one which does not therefore engage in cultural evolution or generate technological revolutions. By invoking the Gaia concept, which sees the whole of the biosphere as one living entity so that a whole population is to the biosphere as an individual is to the population or a cell within the individual is to the host, we can investigate the possibility that each size unit, such as an individual or a population, shares the same growth pattern; Gaia was described at length by Lovelock (1979).

Individual animals, including humans, exhibit a growth pattern according to the cube of time during their youthful stage as illustrated by Tanner (1978). Therefore, we might expect that the total population of an animal species would also grow according to the cube of time during this expansion phase if the Gaia concept should apply here and if the intelligence element were excluded. The converse is not necessarily true, i.e., the demonstration of cubic-time population expansion does not prove the existence of Gaia as noted by Denness (1986 a).

After exploring the possibility of simple cubic-time population growth, without leaning too heavily on the Gaia hypothesis, the impact of intelligence can be added. This then leads to a theoretical model which mirrors almost exactly the actual growth record of mankind especially since accurate estimates could be made.

A.3 Simple Cubic-Time Population Growth

Let us consider a population in which the average mass of the individual is constant; this implies the reasonable assumption that over the long term the population contains the same proportion of individuals at all stages of development. In an expanding population the number of individuals grows. Therefore, for a given environment of fixed energetics constraints the volume of the cumulative (and interactive for mobile species) exclusion zones of all individuals grows in order to postpone competition as long as possible; the exclusion zone is that space surrounding the individual required to provide it with sufficient sustenance. This is the most energy-efficient system as the avoidance of conflict reduces its potential energy to a minimum. After competition ensues, following the exhaustion of available free space, subsequent population expansion requires either a change in energetics or is subject to the constraints of the -3/2 law of distribution and growth as previously applied to plants by Shinozaki and Kira (1956), Todaki and Shidei (1959), Yoda et al. (1963), White (1981) and Whittington (1984) and to animals by Hayton (1984) and Denness (1986, b).

Davidson (1985) showed that for three nations of humans, USA, Brasil and Australia, each until recently being able to expand into sparsely occupied territory within its own boundaries to avoid internal competition, the full available historic records exhibit a general population expansion according to the cube of time. Denness (1986 c) provided theoretical reasoning for this observation for the simple case in which the impact of cultural evolution is unnecessary to avoid competition or is otherwise absent. Such cubic-time growth runs counter to the common, but demonstrably erroneous, assumption of exponential expansion. However, the popular "exponential" assumption appears to owe less to mathematical exactitude than to literary licence, which allows its use to imply merely "at a great rate".

The population of mankind as a whole has not experienced the freedom from competition enjoyed by these relatively recent colonised nations whose initial growth rates ran ahead of the rest in unfettered expansion. The overall population has not expanded continuously according to the cube of time. For the growth of the whole population the impact of cultural evolution has been more evident.

A.4 The Influence of Cultural Evolution on Growth Rate

By applying intelligence the changing of energetics constraints becomes possible, i.e., the species that can learn from its experiences or from inspiration can improve its efficiency at that most critical of its basic functions, reproduction. It seems reasonable to assume that the chance of a community producing a worthwhile idea within a given time is proportional to the number of individuals within it who are available for thinking; in other words, the time taken to produce an idea is inversely proportional to the population. Therefore, as an intelligent population expands the time taken for it to produce knowledge reduces so that it is more frequently able to enhance its reproductive efficiency of that is the way it chooses to apply that growing knowledge reservoir.

It might be assumed that an intelligent species such as man would recognise what it is doing and thereby interrupt this sequence to divert its knowledge reservoir into activities other than procreation. However, Malthus the notable eighteenth century economic philosopher, well reported by Flew (1957), believed it inevitable that the rewards of each technological revolution, an expression of a step in cultural evolution, would be absorbed by a consequent increase of population thus preventing any rise in the general standard of living (or, in ecological terms, an improvement in energetics efficiency). Let us now examine the facts.

Population statistics tend generally to be somewhat scattered through partisan national literature and to depend heavily on subjective interpretation prior to the recent historical record. However, McEvedy and Jones (1978) provided a convenient and broadly founded compendium from which the world population figures used here are taken: these are summarised graphically in Figure 1. McEvedy and Jones derived figures for early man based on analogy with the distribution of equivalent modern primates followed by estimates from archaeological and historical reconstructions and finally modern census abstracts.

For cube-time growth we can anticipate an early linear relationship between population and time at a slope of 3 on a double logarithmic graph until man had his first constructive idea. Thereafter, for each step in knowledge acquisition leading to a technological revolution, the graph should reveal a step to a more efficient population growth rate, i.e., faster expansion denoted by a transfer of the relation of gradient 3 to the left. Each time this happens is equivalent to starting the population growth afresh at a faster pace and from a pre-existing base equal to that at the end of the previous phase.

On the purely graphical presentation in Figure 2 the essential feature is that all the data points from Figure 1 have been included and the steps in the plot are the result of ensuring that they all fall on lines at a slope of 3. The steps therefore differentiate successive growth stages from man's earliest intellectual progress to the most recent advances, moving from bottom right to top left in Figure 2. No attempt has been made to control the position of the steps so that they are the natural outcome of assuming the growth stages are constrained to cubic-time expansion. It thus seems appropriate to examine whether any anthropological or historical significance can be attributed to the points at which the growth rate changed gear; according to the above reasoning these should represent major technological advances.

63

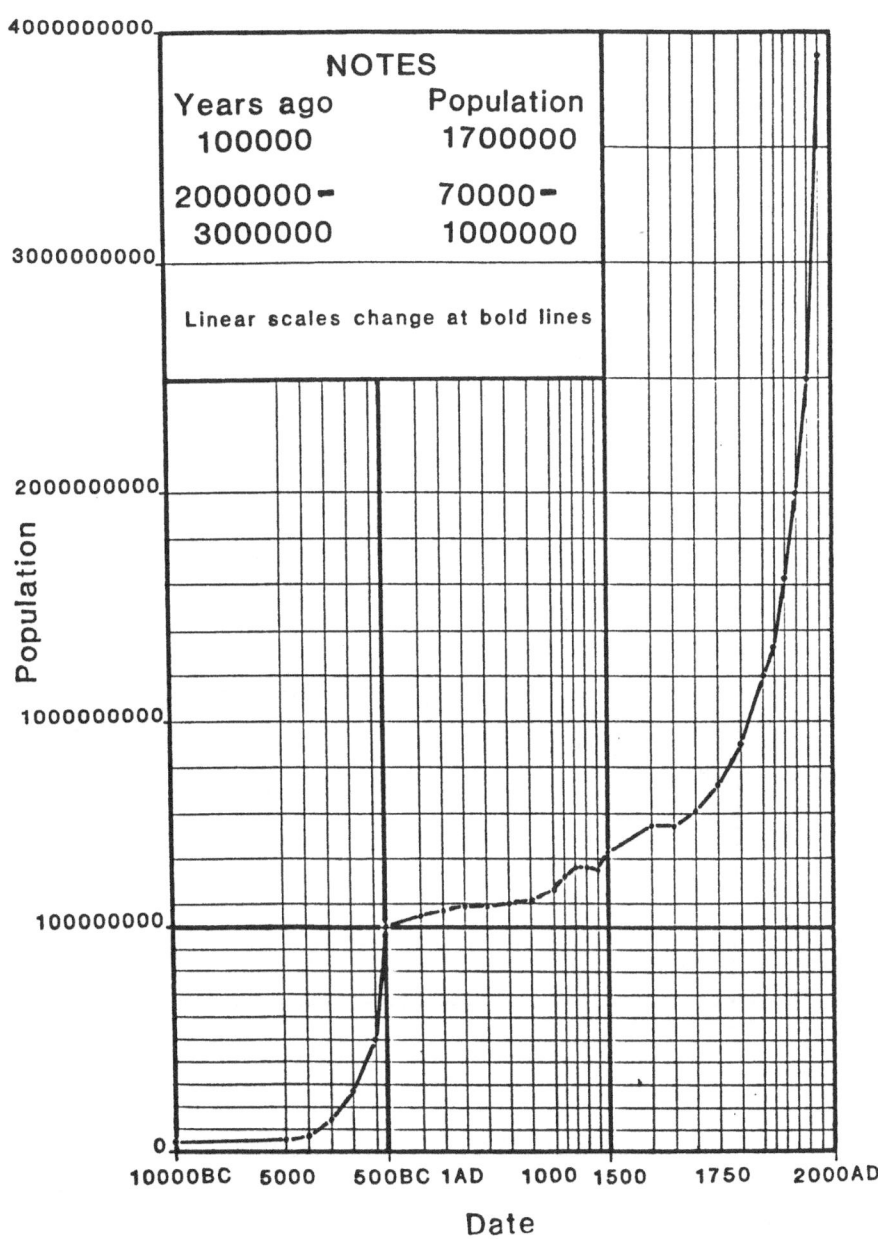

Figure A.1. World population growth

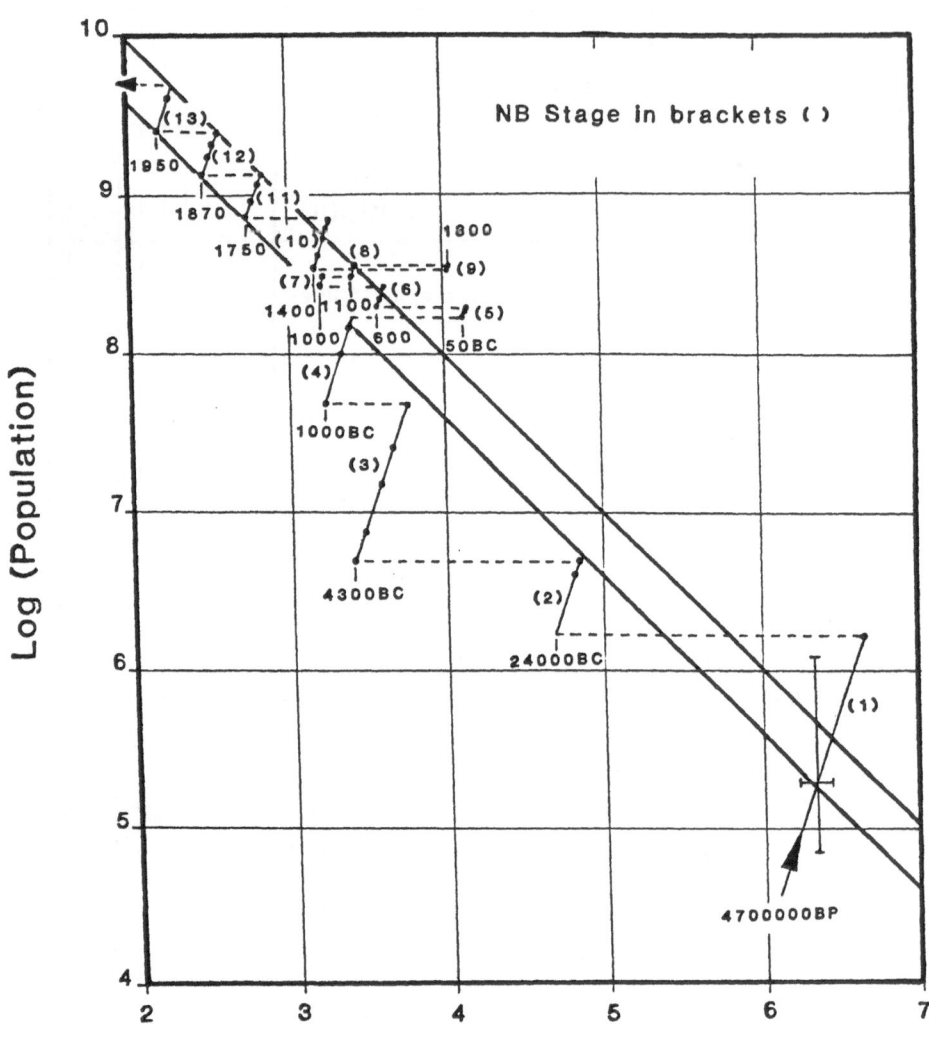

Figure A.2. World population growth steps (log-log)

A.5 The Archaeological and Historical Record

The earliest stage of development (Stage 1) started with the appearance of the first hominid, **Australopithecus** who is commonly known to have appeared in Africa about 5 million years ago with a cranial capacity of about 600 cc - little more than that of a gorilla. By analogy with modern chimpanzee and gorilla populations his successor, **Pithecanthropus**, numbered between 70,000 and well over 1,000,000 between two and three million years ago by which time his cranial capacity had expanded to 900 cc. From Figure 2 that nevertheless appears to have been insufficient tot result in the long overdue breakthrough that eventually came, according to the data from McEvedy and Jones, about 24,000 BC, somewhat after the final increase in cranial capacity to the 1450 cc average of **Homo sapiens** who appeared about 100,000 years ago and is considered to have numbered about 1.7 million. In passing it is interesting to note that the origin of the earliest stage of expansion - the non-intellectual stage - appears, by applying the cubic-time expansion model to the mean of McEvedy and Jones estimates, to have been about 4.7 million years ago; this is remarkably close to the commonly accepted date of 5 million years.

Stage 2, the second recognisable from the estimates condensed by McEvedy and Jones, took the sluggish population of 24,000 BC and expanded it to three times its precious size within about 20,000 years. Whatever the invention that allowed that progress its impact on man, beleaguered as he was by the depths of the last ice age, was immense and resulted in his population growth rate, seen in the light of ecological energetics, taking its greatest single step.

To continue the cubic expansion requires changing to Stage 3 in about 4300 BC, which is consistent with the general onset of the Neolithic agricultural revolution. That appears to have been the second most substantial advance in man's evolution and also that leading to the greatest proportional increase of population (tenfold) of his intellectual period. Within that steady cubic-time expansion the Bronze Age came and went without apparently registering its impact. Not so the Iron Age, however; about 1000 BC man embarked on Stage 4 equipped with durable tools. However, a millennium of progress in this new ecological niche was evidently too much for him. There followed a series of alternating recessions and recoveries, including one actually retrograde period, before the expansion entered the uninterrupted series of stages leading to the renowned "population explosion" of the second half of the present century.

About 50 BC, according to Figure 2, there began a 650 year during recession during which growth proceeded more slowly than during the Neolithic period several thousand years before. Here termed Stage 5, this corresponds to the so-called Dark Ages normally attributed to some measure to the fall of the Roman and Han Empires as political unifiers leaving the field open to relative anarchy as society lost its structure throughout about half the world's population. However, let us now explore a possible ecological explanation as an alternative to this social rationale. Figure 2 describes a general stepwise trend from bottom right to top left at an overall gradient of -1, a theoretical explanation for which is given later. Although Stage 1 had overshot this boundary while awaiting the development of sufficient cranial capacity to generate a technology the succeeding Stage 2, 3 and 4 had gone ahead of the general trend. Therefore during these three preceding the population had acquired a potential beyond the capacity of this development model. Seen in that light the Dark Ages are but a dissipation of that overdeveloped potential; as is common to most physical systems the subsequent relaxation resulted in rebound beyond the general trend before homing in on it again.

Stage 6 brought a measure of recovery starting about 600 AD, effectively resuming the Iron Age progress. Then about 1000 AD comes the next real step, perhaps rooted in the development of effective shipping in Europe and agricultural advances in China. This Stage 7, was relatively short-lived, however, as the subsequent Stage 8 followed on within a century illustrating the fragility of the Stage 7 accomplishments. Stage 8 reverted yet again to similar progress to that in the Iron Age.

Then the bad news. The Black Death swept through Europe shortly after China succumbed to waves of Mongols. Each lost about one third of its population to a force beyond its influence. In each case the source was the same - Mongolia. The Bubonic Plague was certainly beyond human control and whatever possessed the Mongolian Khans was equally remarkable as the Chinese outnumbered the Mongols by about 150:1. The resultant Stage 9 of the fourteenth century marks the only recorded decline in global population and the last time at which there was a significant departure from the bounds of the general stepwise trend.

About 1400 the world began the steady stepwise progress of the modern population era. Stage 10 was heralded by the invention of guns and marked improvement in ships to enable discovery and colonisation of pastures new. With the speeding up of technological advances contingent on population number the next phase, Stage 11, was not long in following. In that case it was the second agricultural revolution, seen from Figure two to have become effective about 1750. Hot on its heels came Stage 12, coincident with the industrial revolution, here seen to influence population growth (and of course the greenhouse effect) globally from about 1870.

Finally (to date) we have the current population explosion of Stage 13 which has been underway since 1950. Not surprisingly that date is representative of the twentieth century communications revolution.

At this state it is appropriate to emphasise that each of the steps in Figure 2 is simply the result of applying the ecological cubic-time expansion model to the world population data given by McEvedy and Jones. That those steps are then seen to be coincident with commonly agreed technological revolutions is to be expected from the innovation rate hypothesis introduced earlier and developed further as follows. Were that hypothesis not available, the coincidences would be totally inexplicable.

A.6 An Analytical Growth Model

Figure 3 describes a small generalised part of the stepwise trend developed in Figure 2. At the beginning of the nth growth stage the population P_{bn} and its effective age is t_{bn}; at the end of the stage the values are P_{en} and t_{en} respectively. Similarly for the $(n+1)^{th}$ stage, etc. Let the intercepts on the Log P axis at zero on the lof t axis be respectively

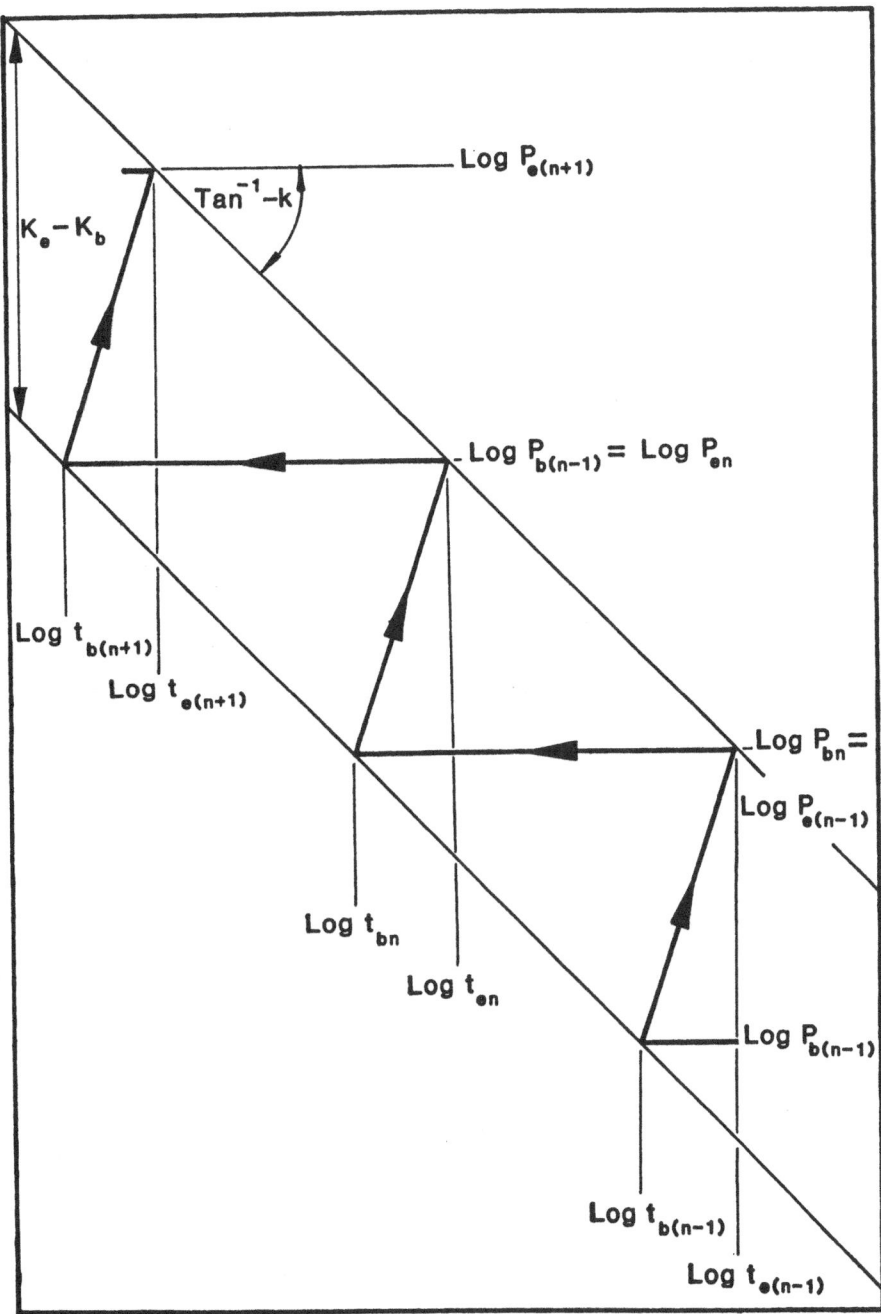

Figure A.3. Schematic representation of three typical growth steps.

K_b and K_e for the parallel lines bounding the beginning and end of the growth stages. Finally let the slope of those boundary lines be $\tan^{-1} -k$.

$$\therefore \log P_e - K_e - k\log t_e$$

$$i.e.\ t_e - \left(\frac{10^{K_e}}{P_e}\right)^{1/k} \tag{1}$$

Now let the innovation stage:

$$t_e - t_b - t_i \tag{2}$$

According to the cubic-time expansion model:

$$3(\log t_e - \log t_b) - \log P_e - \log P_b \tag{3}$$

But from Figure 3,

$$\log P_e - \log P_b - (K_e - K_b) - k(\log t_e - \log t_b) \tag{4}$$

Putting (4) in (3):

$$3(\log t_e - \log t_b) - (K_e - K_b) - k(\log t_e - \log t_b)$$
$$i.e.\ (3 + k)(\log t_e - \log t_b) - K_e - K_b$$
$$\therefore \log\left(\frac{t_b}{t_e}\right) - \frac{K_b - K_e}{3 + k} \tag{5}$$
$$\therefore t_b - t_e \cdot 10^{\frac{K_b - K_e}{3 + k}}$$

Introducing (5) into (2):

$$t_i - t_e\left(1 - 10^{\frac{K_b - K_e}{3 + k}}\right) \tag{6}$$

Substituting t_e from (1) in (1):

$$t_i - (\frac{10^{K_e}}{P_e}) \cdot (1 - 10^{\frac{K_b - K_e}{3+k}}) - (\frac{1}{P_e})^{1/k}) \cdot (10^{\frac{K_e}{K}} \cdot [1 - 10^{\frac{K_b - K_e}{3+k}}]) \tag{7}$$

$$i.e. \; t_1 - A(\frac{1}{P_e})^{1/k} \tag{8}$$

$$where \; A - 10^{\frac{K_e}{k}} (1 - 10^{\frac{K_b - K_e}{3+k}}) - constant$$

Returning to the concept of intelligence leading to technological revolution, we have already seen the reasonable argument that the chance of a population producing a worthwhile idea in a given time is proportional to the number of thinking individuals within it. If we assume that the proportion of thinking to non-thinking individuals remains constant this is equivalent to the time taken to produce a worthwhile idea being inversely proportional to the population.

The time taken to produce a technologically significant idea in terms of population growth stages is t_i (Figure 3), i.e. the duration of one steady state, cubic-time expansion step. The population at the time the idea is produced is P_e, i.e. that when the revolutionary step is made at the end of the stage. Therefore,

$$t_1 - \frac{B}{P_e} \qquad \text{(9), where B is constant.}$$

Dividing (9) by (8):

$$1 - \frac{B}{A} \cdot P_e^{-(1 - \frac{1}{k})} ; \; i.e. P_e^{-(1 - \frac{1}{k})} - \frac{B}{A} - constant \tag{10}$$

As P_e is a variable, (10) has a singular solution:

$$k - 1 \tag{11}$$

That is to say the slope of the boundary lines on Figure 3 is at a gradient of -1. Boundaries at this theoretical slope have been superimposed on Figure 2 and are seen to be almost exactly consistent with the observed stepwise-cubic progress of world population growth in the modern era (since about 1400 AD) and generally in keeping with it since **Homo sapiens** arrived on the scene.

With k = 1 from (11), then from Figure 3:

$$\log t_{bn} - \log t_{b(n+1)} - \frac{3}{4}(K_e - K_b)$$

$$\therefore t_{b(n+1)} - \frac{t_{bn}}{10^{\frac{3}{4}(K_e - K_b)}} \tag{12}$$

But from (5) and (11):

$$t_e - t_b \cdot 10^{\frac{K_e - K_b}{4}} \tag{13}$$

And (13) in (2):

$$t_i - t_b\left(10^{\frac{K_e - K_b}{4}} - 1\right) \tag{14}$$

$$\therefore t_{i(n+1)} - t_{b(n+1)}\left(10^{\frac{K_e - K_b}{4}} - 1\right) \tag{15}$$

Putting (12) in (15):

$$t_{i(n+1)} - \frac{t_{bn}}{10^{\frac{3}{4}(K_e - K_b)}} \cdot \left(10^{\frac{K_e - K_b}{4}} - 1\right) \tag{16}$$

But from (14):

$$t_{in} - t_{bn}\left(10^{\frac{K_e - K_b}{4}} - 1\right)$$

$$\therefore t_{bn} - \frac{t_{in}}{\left(10^{\frac{K_e - K_b}{4}} - 1\right)} \tag{17}$$

Putting (17) in (16):

$$t_{i(n+1)} = \frac{t_{in}}{(10^{\frac{K_e - K_b}{4}} - 1)} \cdot \frac{(10^{\frac{K_e - K_b}{4}} - 1)}{10^{\frac{3}{4}(K_e - K_b)}}$$

$$= \frac{t_{in}}{10^{\frac{3}{4}(K_e - K_b)}} = \lambda t_{in}, \text{ where } \lambda < 1 \tag{18}$$

So:

$$\sum_{r=1}^{\infty} t_{ir} = t_{in}(1 + \lambda + \lambda^2 + \ldots) = \frac{t_{in}}{1-\lambda} \tag{19}$$

where:

$$\lambda = \frac{1}{10^{\frac{3}{4}(K_e - K_b)}} \tag{20}$$

Of special interest here is the case in which $\lambda = 0.5$. Then from (19):

$$\sum_{r=n}^{\infty} t_{ir} = 2t_{in} \tag{21}$$

For this to be so:

$$10^{\frac{3}{4}(K_e - K_b)} = 2 \quad \textit{from (20)}$$

$$\textit{i.e. } \frac{3}{4}(K_e - K_b) = \log 2 = 0.30103 \tag{22}$$

$$\textit{i.e. } K_e - K_b = 0.40137$$

Let us now return to Figure 2 on which the step boundaries have been superimposed at a spacing considered to be a best fit "by eye" to the extremes of the most recent, and hence best documented and most reliable, steps. The vertical separation of these beginning and ending boundaries, i.e. $K_e - K_b$, is 0.40 within the accuracy of interpretation possible from this graphical representation. Therefore, it would appear that the world human population has, for at least the whole of modern times (since 1400 AD) and possibly before, been expanding in a series of accelerating steps, consistent with equation (21), which can be summed with interesting consequences.

From Figure 2 the last whole step, Stage 12, took 80 years for completion. Therefore, from its beginning in 1870 the sum of the duration of all subsequent steps, according to equation (21), is only 160 years leading to a theoretically infinite population in the year 2030 AD. Among the milestones on the way would be a relatively modest 5 billion in 1990 as observed at the end of current stage, 8 billion by the turn of the century, 20 billion (at present considered to be the maximum supportable with modern technology) by 2020 and instability thereafter.

A.7 Discussion

This is a sobering thought, the immediate reaction to which is the nervous assertion that mankind is not so crazy as to misdirect the rewards of its intellectual achievements solely into extending its procreation-oriented technology which can never quite sustain its numbers (according to Malthusianism). But that might well also have been the reaction several growth stages (and decades or centuries) ago.

The philosophy is not new - only its detailed analysis now receives a high profile. Meyer (1958) first focused upon this mode of global population growth. Meyer and Vallee (1975) drew particular attention to the mode of expansion as $P = 200 \times 10^9 (2026 - t)^{-1}$, where P is the global population and t is the date AD (the date BC being a negative AD date). Von Foester et al. (1960) almost repeated Meyer's empirical analysis but with less causal sensitivity and Taagepera (1976) developed the same theme further but began to loose touch even more with the real growth mechanism, explained by Denness (1986 d and 1987). It is particularly interesting to note that the expansion mechanism, first empirically observed by Meyer (1958) and theoretically justified by Denness (1986 d) and in more detail here, has described almost exactly the global population growth for about the last 30 years - conventional means of growth estimation have failed to do so (c.f. innumerable United Nation, World Bank and national sources) - and if uninterrupted will lead to an untenable population shortly after 2020 AD according to the expression $P = 2 \times 10^{11} (2030 - t)^{-1}$ using the above definitions (Denness, 1987).

Do we have to wait until it happens before we believe the evidence which **Homo sapiens** has drawn up over the last 100,000 years - indeed, **Homo** since the evolution of **H. habilis** 2.5 million years ago(Dobzhanksy, 1977)? Or can we really justify the term "sapiens" and apply the results of our intellectual agility to the betterment of the way of life of a steady-state population and amelioration of the greenhouse effect, rather than dissipating it on continuing the headlong dash to self-destruction - apparently the only alternative - within the next 44 years?

There is some evidence in recent years of a decline in the overall population growth rate. This is commonly held to imply that expansion is now under control and that the population will level off at perhaps about three times the present size over the next couple of centuries. However, reference to Stage 13 of Figure 2 shows that its cubic-time expansion progressed from an effective age of 150 to 185 from 1950 to 1985; this implies the slowing down of the growth rate from 2.0% in 1950 to 1.6% simply by ageing within one simple cubic-time growth phase consistent with observations. The same was true for all the previous phases and has no implication for the rate of growth after the next universal step of cultural evolution with its attendant technological revolution.

Nevertheless there is abroad today a mood of awareness of man's condition not previously evident in history. There is also on hand the fruits of the last technological revolution, the Communications Revolution. This has equipped modern man with the necessary device to bring the whole population into the global village to hear the words of wisdom from sages regarding the need for procreative restraint, a unifying capability never previously available. It is to be hoped that it is used.

If reason prevails, as surely it must, instead of squandering the rewards of technological revolution on procreation, the will arise the opportunity for a population of constant size to improve its standard of living and eventually quality of life at similar rates to those currently associated with population growth. Extending Figure 2 to Stage 14 between the bold sloping parallel lines would see a population of five billion in 1990 with an effective age of 75 years (for the population **not** the individuals) capable of expanding at a rate of 4.1% per year, slightly less than Kenya already enjoys! Spread across the whole world with an emphasis on the current underprivileged quarters this represents an exciting prospect if channelled into development rather than population growth: if not so channelled, the greenhouse effect is unstoppable.

A.8 Conclusions

It has been shown that the historical record of population growth is closely matched by an expansion model which encompasses cubic-time expansion phases separated by rejuvenation steps. The steps reflect cultural evolution with attendant technological revolutions.

The future appears to hold the options of continuing adherence to the hitherto uninterrupted expansion mechanism leading to an unsupportable number of people in the 2020's or of diverting procreational energies into development consistent with ameliorating the greenhouse effect. The latter would lead to the improvement of the global standard of living at about 4% per year for a stable population of five billion in perpetuity if other environmental issues were also resolved.

A.9 References (Population)

Davidson, S.M. (1985) "Population growth.", in *Nature*, 314, p. 398

Denness, B. (1986 a) "The Earth is alive and well - fact or fiction?", Proc. Audubon Living Earth Symp., Amherst, Mass., Aug. 1985, 23, pp. 1-9

Denness, B. (1986 b) "A general ecological distribution law.", Proc. Audubon Living Earth Symp., Amherst, Mass., Aug. 1985, 21, pp. 1-16

Denness, B. (1986 c) "Population growth: the human example.", Proc. Audubon Living Earth Symp., Amherst, Mass., Aug. 1985, 22, pp. 1-20

Denness, B. (1986 d) "A low priority for demographic study at home.", I.W. County Press, **5280**, p. 11

Denness, B. (1987) "Sea level modelling: the past and the future.", Prog. Oceanog., **18**, pp. 41-59

Flew, A. (1957) Australasian Journal of Philosophy, **35**, pp. 1-20

Hayton, A.F. (1984) "Of plants and men.", in *Nature*, 310, p. 178

Lovelock, J.E. (1979) *Gaia: a new look at life on Earth*, Oxford University Press, Oxford

McEvedy, C. & Jones, R. (1978) *Atlas of World Population History*, Allen Lane, London

Meyer, F. (1958) *L'Encyclopedie Française*, Paris, 20, p. 24

Meyer F., & Vallee, J. (1975) *Technological forecasting and social change*, 7, pp. 285-300

Shinozaki, K., & Kira, T. (1956) J. Inst. Polytech., Osaka City Univ., **D7**, p. 35

Taagepera, R. (1976) "Crisis around 2500 AD? A technology-population interaction model.", General Systems, **XXI**, pp. 137-138

Tanner, J.M. (1978) *Foetus into man: physical growth from conception to maturity*, Open Books, London.

Todaki, Y., & Shidei, T. (1959) J. Jap. For. Soc., **41**, p. 341

von Foester, H., Mora, P.M., & Amiot, L.W. (1960) Science, 132, pp. 1291-1295

White, J. (1981) J. Theor. Biol., **89**, pp. 475-500

Whittington, R. (1984) "Laying down the -3/2 law.", in *Nature*, 311, p. 217

Yoda, K., Kira, T., Ogawa, H., & Homuzi, K. (1963) J. Biol., Osaka City Univ., **14**, p. 107

ATMOSPHERIC, OCEANIC AND CLIMATIC RESPONSE TO GREENHOUSE AND FEEDBACK EFFECT

SASKIA JELGERSMA
Geological Survey of The Netherlands
P.O. Box 157
2000 AD Haarlem
The Netherlands

ABSTRACT. Global warming due to the greenhouse effect should result in a sea-level rise. Estimates for the next 100 years range from 20-140 cm. This range is due to the fact that the input to the models has many uncertainties.

An important record about climatic changes is available in the geological past, but the evidence should be refined by geologists and used as an input to the forecast of future climatic changes.

Deep sea cores provide proxy evidence of changes in the ocean surface temperatures. The ratio of the oxygen isotopes reflect accumulation and decay of the continental ice caps. These 100,000 years cycles correspond nicely to the calculated solar insolation curves of Milankovitch.

Superimposed on global cycles are those of shorter duration which are probably not global. Climatic changes in the Late Glacial are likely to be more sudden and time-restricted than can be derived from the deep sea record. Our present interglacial, the Holocene, is characterized by fluctuations that include a climatic optimum and periods called "little ice ages".

The paleoclimatic record indicates that changes in temperature do not respond in a gradual way but responds in sharp jumps. A second conclusion is that long-term climatic variations over the past 400,000 years and extrapolations for the coming 60,000 years should lead to a beginning of the next ice age. How the predicted greenhouse warming will interact with the natural trend is not known.

1. Introduction

An increasing body of evidence suggests that in the coming decades a global warming due to the greenhouse effect may lead to a substantial rise in sea level. The amount of future sea-level rise is still uncertain but it certainly should be higher than observed by mean tide gauge readings of the last 100 years.

About half of the world population is living in deltaic areas, coastal lowlands and river plains; these areas are zones of intense economic and agriculture activity.

Many coastlines are subject to erosion at the present time, not only due to a recently

75

R. Paepe et al. (eds.), Greenhouse Effect, Sea Level and Drought, 75–84.
© 1990 *Kluwer Academic Publishers.*

observed rise in sea level but mainly to human interference in the coastline and to the river valleys and basins. These human activities interfere with the natural system in a negative way. Accordingly the social and economic impacts of a future sea-level rise on the coastal lowlands of the world will be profound and widespread.

It must be stressed upon that beside CO_2, the trace gases nitrous oxide (N_2O), methane (CH_4), ozone (O_3) and chlorofluorocarbons (CFC's) are also increasing. The quantities of these gasses are much smaller compared to CO_2 but they contribute to the global warming just as much as CO_2-global warming. Especially, the emission of CH_4 from decay of carbon held in the organic matter of soils and in wetlands is likely to speed up by a global temperature rise. All scientists agree that a reduction of CO_2 and other greenhouse gases could reduce the future global warming.

The topic of this conference is to manipulate the hydrological cycle in such a way that large amounts of water are to be stored on the land in areas with a serious deficiency of water. The result should appear in two ways: to reduce the future sea-level rise in order to save coastal lowlands, and to provide water for the deficiency areas.

The present paper will discuss future climatic changes and the expected sea-level rise, data about geological and historical evidences of climatic changes and evidence of climatic changes during the last 100 years.

2. Future Climatic Changes and Sea-Level

The amounts of CO_2 and CH_4 in the atmosphere are closely linked to the surface temperature. This is proved by investigation of bubbles of air trapped in a 2000 m thick ice core recovered by a Russian drilling project in Antarctica (Lorius, 1988). Results for the last 160,000 years, the last interglacial and the last glacial, indicate that CO_2 was 25% higher and CH_4 100% higher during the last interglacial (see Fig. 1, evidence derived from gasses trapped in an Antarctic ice core (redrawn after Lorius et al. 1988)). Over the last 200 years the CO_2 content of bubbles of air trapped in glacier ice shows a steep rise of CO_2 concentration. This rise is thought to be caused by emission of CO_2 due to the industrial revolution (Fig. 2). Accordingly there is a general consensus that the measured increase in CO_2 and other greenhouse gases will effect the radiation balance of the earth and will cause climatic changes. The main result will be a rise in temperature due to heat trapping.

There is no debate among scientists about the reality of the greenhouse effect but there is controversy about the following problems:
1. What will be the rise in temperature with a doubling of CO_2?
2. Over how many decades will this occur?
3. How will the climate change in relation to variations in precipitation, cloudiness and storms? Changes in the ocean circulation system are not unlikely.

The expected rise in sea level is closely related to the above- mentioned uncertainties. During the Villach conference (1985) it was agreed that a doubling of CO_2 from pre-industrial levels could occur as early as 2030 AD. The associated increases of the global mean equilibrium surface temperature could be between 1.5 and 4.5°C. Schneider (1989) believed

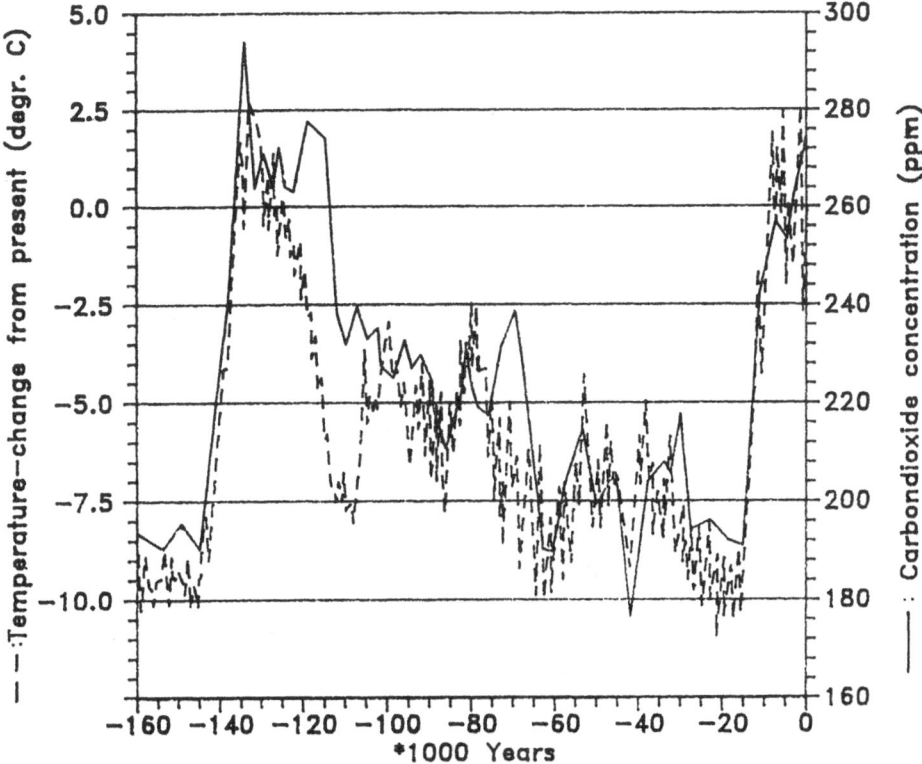

Figure 1. Temperature and carbon dioxide concentration over the past 160,000 years. Temperature indicated in hatched line, carbon dioxide concentration in straight line. Evidence derived from gasses trapped in an Antarctic ice core (redrawn after Lorius, 1988).

that the doubling of CO_2 should occur between 2030 and 2080 AD with a rise in temperature between 3 and 5.5°C. General Circulation Models (GCM's) are used to forecast changes in climate on global scale. The shortcomings of these computer models make forecasting of future climate not yet reliable.

The estimated sea-level rise in the coming 100 years is caused by two factors: the expansion of the ocean waters and the melting of mountain glaciers. A contribution of the Antarctic and the Greenland ice caps is thought to be of little or no importance in the given time span.

If the global warming continues for several centuries the melting of parts of the Greenland and the Antarctic ice sheets is likely to occur and should result in an important rise in sea level.

The estimates for the expected future rise in sea-level show important variations. The US EPA (1983) gives an estimation between 56 and 368 cm, Hoffman et al. (1983). During the Villach conference, Robin (1985) calculated the future rise in sea level to be between

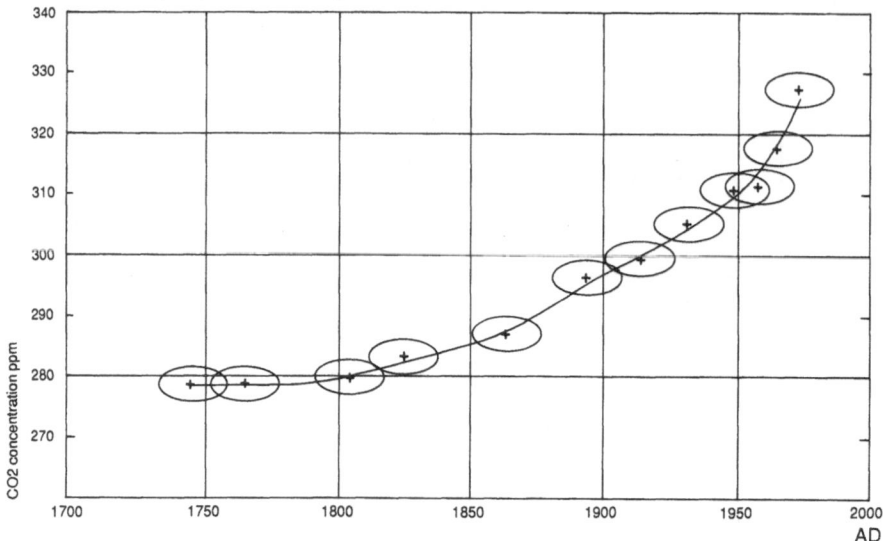

Figure 2. Atmospheric CO_2 concentration measured in glacier ice formed during the last 200 years (redrawn after Neftel et al., 1985).

20 and 140 cm. Wigley and Raper's best guess (1988) is between 10 and 35 cm. These high variations in the various estimates is due to the fact that the input to the models has a lot of uncertainties. Robin expressed this problem as follows: "With our lack of knowledge of the hydrological cycle and the dynamics of the oceans and the polar ice sheets, forecasting of global changes of sea level involves considerable extrapolation and speculation".

3. Evidence of Recent Climatic Change

The record of the last 100 years indicates an increase of CO_2 by 25% above the interglacial level and a rise in temperature of 0.4°C. This rise in temperature is hotly debated; how reliable is it? is it only a "noise" or does it indicate the beginning of the greenhouse warming? It must be admitted that we do not know.

The recent rise in sea level of 15-20 cm/century, observed by 100 years of tide gauge readings, and the erosion of shorelines has been attributed to this rise in temperature of 0.4°C.

It must be mentioned however, that the tide gauges readings are influenced by tectonic movements: low-lying shorelines and deltas are in most cases subject to both tectonic subsidence and compaction. The shoreline erosion, observed in many parts of the world, is not only caused by the recent sea- level rise but mainly by human interference in the shoreline and the river valleys (Bird, 1985).

Some mountain glaciers have retreated as much as 1200 m during the last 100 years. The observed phenomena are commonly attributed to the rise in temperature. In a recent article (Oerlemans 1988) published a study where an attempt was made to simulate historic variations of glacier fronts by mean of a simple climate glacier model. His conclusion is that the 0.4°C rise in temperature is only responsible for half of the observed retreat. If he includes the radiation balance that is influenced by the so-called volcanic dust veil then the full observed retreat comes out of the model (Fig. 3). It must be concluded that the

Effect of the greenhouse warming during the last 100 years

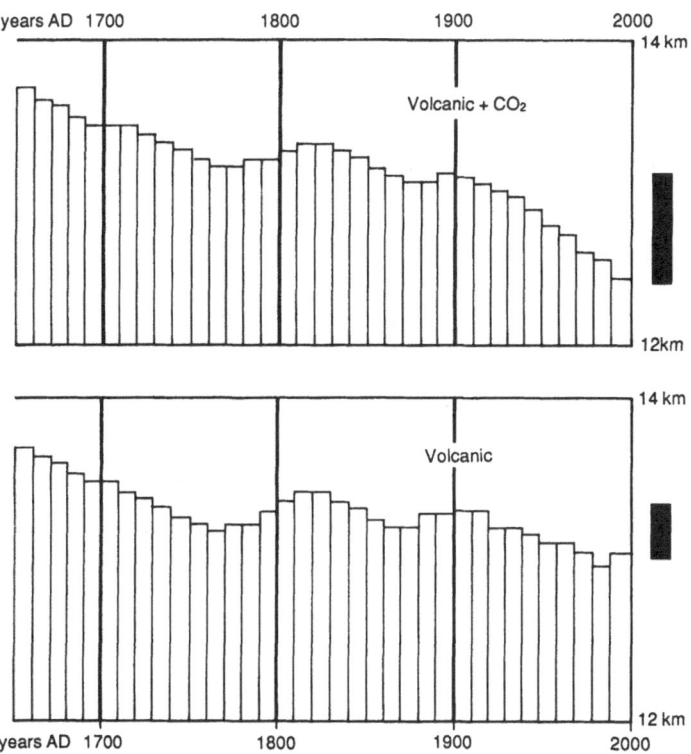

Figure 3. The effect of the greenhouse warming explains about half of the retreat of the mountain glaciers during the last 100 years (redrawn after Oerlemans 1988).

two lines of evidence, sea level rise and glacier retreat can only be partly explained by the recent global warming.

4. Geological Record

The geological record contains rich and abundant evidence of past climatic changes, especially the Quaternary with its cycles of accumulation and decay of ice caps. During this period the great changes in climate are dramatically reflected into the geological processes. The sedimentary record shows an alternation of soils, the occurrence of periglacial phenomena and alternating marine and continental deposits. The palaeobotanical record disclosed vegetation shifts from barren tundra to deciduous forest. The same conclusions about climatic shifts can be derived from other biological evidence like molluscs and bones of animals. The most complete record of climatic changes can be found in the deep sea cores (Emiliani, 1955, Shackleton, 1977, Shackleton et al 1973, Imbrie et al 1984).

Analyses of tests of fossil foraminifera, when subjected to isotopic determination can supply a proxy record of temperature and ice volume. The latter is derived from the ratio of O^{18} to O^{16}: the curve represents the waxing and the waning of continental ice sheets. The global ice volume measured by the oxygen isotopes shows a record of cycles having a period of about 100,000 years. These cycles show an excellent time correspondence with

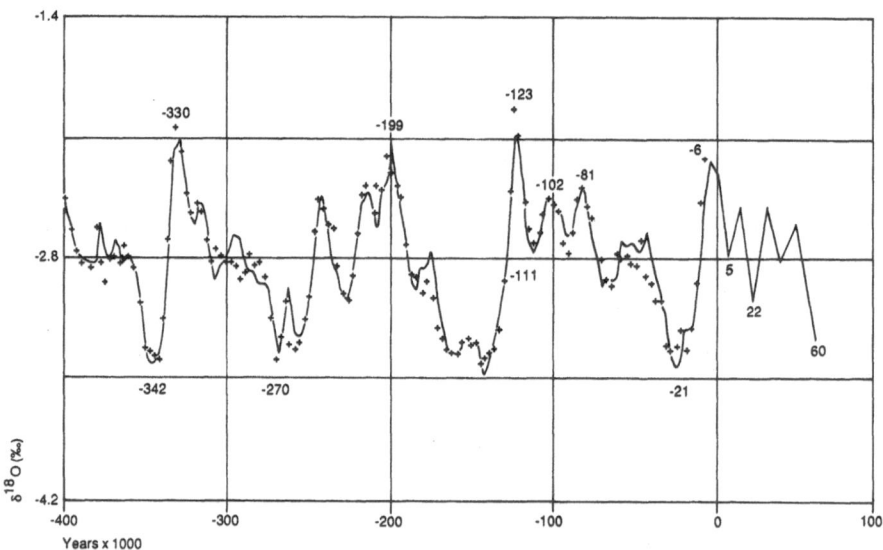

Figure 4. Oxygen isotope record representing temperature and ice volume of the last 400,000 years and calculated for the future (redrawn after Berger 1981).

the calculated solar radiation curve of Milankovitch. The curves (Fig. 4) are asymmetric and the warm periods (the interglacials) are of much shorter duration than the cold periods (glacials).

The curves of the oxygen isotope records also provide an indication of a gradual and smooth transition from full glacial to full interglacial conditions. This smooth and gradual transition is however caused by the time lag in the melting down of the ice caps. The record

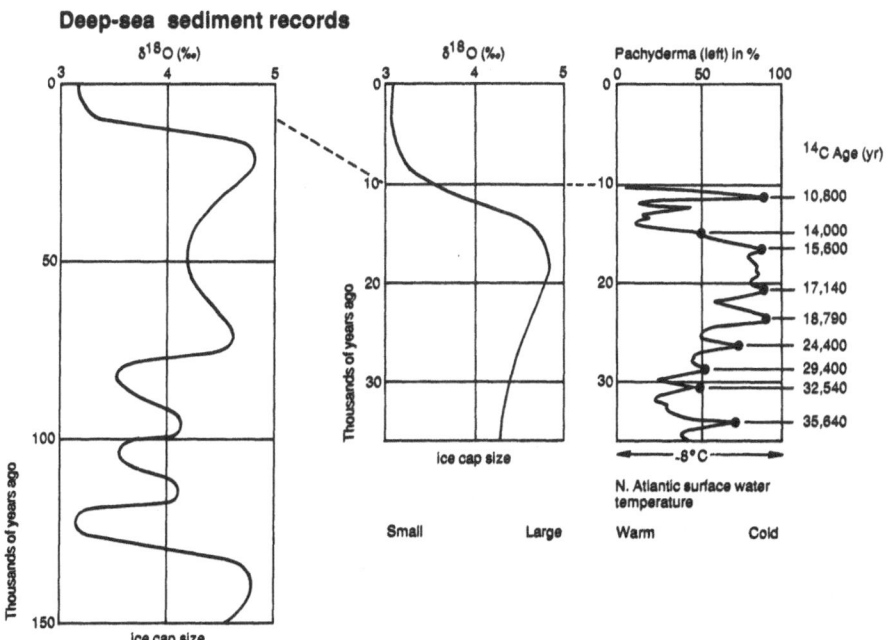

Figure 5. The two diagrams on the left indicate a gradual and smooth transition from full glacial to interglacial; based on the oxygen isotope record. The diagram on the righ is based on analyses of shells of *Pachyderma* in a North Atlantic deep sea core (redrawn after Ruddiman et al, 1981 and Broecker, 1987))

of a core from a deep-sea sediment obtained west of the British Isles indicates an abruptly warming at about 15,000 B.P. (Fig. 5).. The dominance of the planktonic foram *Pachyderma* indicates cold water which changed in a few hundred years to warmer water. An abrupt cooling, in less than a few hundred years, occurred during the Younger Dryas (Ruddiman et al., 1981 and Broecker, 1987). The same evidence of abrupt warming and cooling derived from the study of fossil beetles is observed in Great Britain (Atkinson et al., 1987). It may be concluded that climatic changes occur much more abruptly and quicker than thought before. More details can be found in "Abrupt Climatic Changes, evidences and implications" (edited by W.H. Berger and L.D. Labeyzi, 1987).

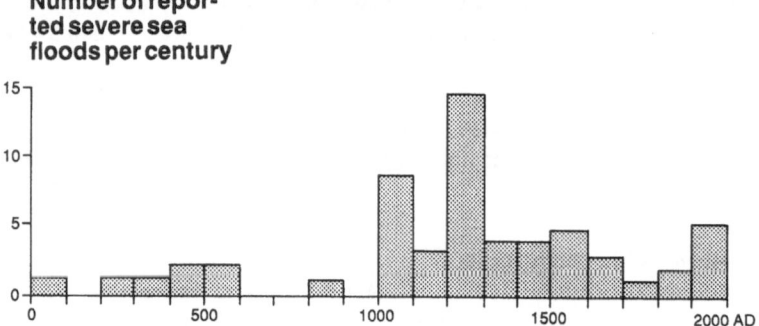

Figure 6. Reports of severe storm floods on the coasts of the North Sea and the English Channel (redrawn after Lamb 1982).

Another important conclusion derived from the fossil isotopes data from deep sea cores is that long-term climatic variations over the past 400,000 years and predictions for the coming 60,000 years should then lead to a beginning of the next ice age (Berger, 1981, Fig. 4).

5. Historical Record

During the recent period, the Holocene, the climatic optimum occurred more than 6000 years ago. The July temperature as derived from palaeobotanical evidence was a few degrees higher as today.

In historical time a warm period occurred between 900 and 1300 AD, this is called the medieval warm epoch. A marked cold epoch followed, the Little Ice Age, with greatest extremes between 1500 and 1800 AD.

According to Lamb (1984) between 1675 and 1704 AD the water temperature off the Faeroes was 5°C lower than today. This low temperature indicates changes in the penetration of polar water to the south. The boundary between the cold Arctic water and the warm Gulf Stream was pushed southwards during this period. This could also explain the occurrence of severe storms during the little Ice age as they are thought to be related to an enhanced thermal gradient between latitudes 50° and 65°N. The Little Ice Age is thought to be a regional and local climatic event that affected the countries especially around the North Sea and Scandinavia and parts of the rest of Europe. It is sometimes thought that this cold period is related to the historical high "Dust Veil Index" of Lamb. Volcanic eruptions inject in the stratosphere both dust and sulphur compounds. These dust veils have a screening effect on solar radiation. The Dust Veil Index of Lamb can be correlated with the Greenland ice core acidity. More details can be found in *Climatic Changes on a yearly to millennial basis* (edited by N.A. Mörner and W. Karlén, 1984).

6. Conclusions

1. The paleoclimatic record indicates that change in temperature do not occur in a gradual smooth way but responds in sharp jumps.
2. The paleoclimatic record indicates 100,000 years cycles, superimposed on which are cycles of shorter duration e.g. 40,000, 23,000 and 19,000 years.
3. Long-term climatic variations over the past 400,000 years and extrapolation for the coming 60,000 years suggest a beginning of the next ice age.
4. The doubling of pre-industrial CO_2 in the atmosphere is thought to take place between 2030 and 2080 AD. The associated rise in global temperature is thought to be between 1.5 to 4.5°C; other ideas point to a 3 to 5.5°C in temperature.
5. For the next 100 years the rise in mean sea level is calculated between 20 and 140 cm. This rise is composed out of 2 factors: the expansion of the ocean water and the melting of mountain glaciers.

7. References

Atkinson, T.C., Brifo, K.R., and Coope, G.R. (1987) "Seasonal temperatures in Britain during the past 22,000 years, contstructed using beetle remains", *Nature* Vol. 325, pp. 587-592.

Berger, A. (editor) (1981) *Climatic Variations and Variabilities: Facts and Theories*, Reidel Publ. Comp., Dordrecht.

Bird, E.C.F. (1985) "Coastline changes", Wiley Interscience, Chichester.

Broecker, W.S. (1987) "Unpleasant surprises in the greenhouse?", Nature Vol. 328, pp. 123-126.

CLIMAP Project Members, (1976) "The surface of the ice-age Earth", Science, 191, pp. 1131-1144.

CLIMAP Project Members, (1984) "The last interglacial ocean", Quaternary Research, 21, pp. 124-224.

Emiliani, C. (1955) "Pleistocene temperatures", Journal of Geology, Vol. 63, pp. 538-578.

Gornitz, V., Lebedeff, S. and Hansen, J. (1982) "Global sea-level trend in the Past Century", Science, 215, pp. 1611-1614.

Hays, J.D., Imbri, J., & Shackleton, N.J. (1976) "Variations in the earth orbit: face makers of the Ice age", Science, 194, pp. 1121-1132/

Hoffman, J.S., Keyes, D., and Titus, J.G. (1983) "Projecting Future Sea-level rise", U.S. GPO 055-000-0236-3. Washington, D.C.: Government Printing Office.

Imbrie, J., Hays, J.D., Martinson, D.G., McIntyre, A., Mix, A.L., Morley, J.J., Pizias, N.G., Prell, W.L., and Shackleton, N.J. (1984) "The orbital theory of Pleistocene climate: Support from a revised chronology of the marine 0-18 record", *Milankovitch and climate; Understanding the response to Astronomical Forcing* (1984) part I and II. Edited by A. Berger, J. Imbrie, J. Hays, G. Kukla and B. Saltzman. D. Reidel, Dordrecht, Boston,

Lancaster. pp. 269-305.

Lamb, H.H. (1984) "Climate and history in northern Europe and elsewhere", *Climatic changes on a yearly to millennial bases* (1984). Edited by N.A. Mörner and W. Karlén. D. Reidel, Dordrecht, Boston, Lancaster. p. 225-241

Lamb, H.H. (1984) "Some studies of the Little Ice Age of recent centuries and its great storms", *Climatic changes on a yearly to millennial bases* (1984). Edited by N.A. Mörner and W. Karlén. D. Reidel, Dordrecht, Boston, Lancaster. pp. 309-311

Lorius, C., Barnola, J.M., Legrand, M., Petit, J.R., Raynaud, D., Ritz, C., Barkov, N., Korotkevich, Y.S., Petrov, V.N., Genthon, C., Jouzel, J., Kotlyakov, V.M., Yiou, M., & Raisbeck, G. (1989) "Long-term climatic and environmental records from Antarctic ice" in A. Berger, R.E. Dickinson, & J.W. Kidson, (editors), *Understanding Climate Change*, American Geophysical Union, pp. 11-16.

Milankovitch, M.M. (1941) "Canon of Insolation and the Ice Age Problem", Königlich Serbische Akademie, Beograd. English translation by the Israel Program for Scientific Translations, published for the United States Department of Commerce and the National Science Foundation, Washington D.C.

Neftel, A., Moor, E., Oeschger, H., and Stauffer, B. (1985) "The increase in atmospheric CO_2 in the last two centuries", Evidences from polar ice cores, Nature Vol. 315, pp. 45-47.

Oerlemans, J. (1988) "Simulation of historic glacier variations with a simple climate-glacier model", Journal of Glaciology Vol. 34, pp. 333-341.

Raper, S.C.B., Warrick, R.A. and Wigley, T.M. (1988) "Global sea-level rise: past and future", SCOPE Workshop on sea-level rise. in press.

Ruddiman, W.F. and McIntyre, A. (1981) "The North Atlantic Ocean during the last deglaciation", Palaeogeogr., Palaeoclimatol. Palaeocol. 35, pp. 145-214.

Robin G. de Q. (1986) "Changing the sea-level", *Projecting the rise in sea-level caused by warming of the atmosphere. The greenhouse effect climatic change and ecosystems*, Edited by B. Bolin, Bo R. Doös, Jill Jäger and Richard A. Warrick. SCOPE 29, John Wiley, Chichester.

Schneider, Stephan H. (1989) "The changing climate", Scientific American, September 1989, pp. 38-47.

Shackleton, N.J., and Opdyke, N.D. (1973) "Oxygen isotope and palaeomagnetic stratigraphy of equatorial pacific core, V28-238: Oxygen Isotope Temperature and Ice Volumes on a 10^5 and 10^6 year scale".

Shackleton, N.J. (1977) "The oxygen isotope stratigraphic record of the late Pleistocene", Philosophical Transactions, Royal Society 280 pp. 169-179.

Wigley, T.M., Jones, P.D., and Kelly, P.M. (1986) "Empirical climate studies" *Warm world scenarios and the detection of climatic change induced by radiatively active gasses. The greenhouse effect climatic change and ecosystems* (1986). Edited by B. Bolin, Bo R. Doös, Jill Jäger and Richard A. Warrick. SCOPE 29, John Wiley Chichester.

ICE SHEETS AND SEA LEVEL CHANGE AS A RESPONSE TO CLIMATIC CHANGE AT THE ASTRONOMICAL TIME SCALE

A. BERGER, TH. FICHEFET, H. GALLEE, I. MARSIAT,
CH. TRICOT and J.P. van YPERSELE
Université Catholique de Louvain
Institut d'Astronomie et de Géophysique G. Lemaître
2 Chemin du Cyclotron
B-1348 Louvain-la-Neuve

ABSTRACT. Understanding how and why global climate is changing is investigated at the astronomical time scale related to the glacial-interglacial cycles of the Quaternary Ice Age.

A 2-dimensional physical model taking into account the coupling between the atmosphere, the upper ocean, the sea-ice, the ice-sheets and the continental surfaces has been forced by the long-term variations of the insolation induced by the astronomical changes in the elements of the Earth's orbit. The low frequency part of the ice volume and sea-level changes have been correctly reproduced in agreement with the deep sea and ice cores records and with the climatic reconstructions made from multiple geological observations. However, after 6 kyr BP, the remaining ice volume of the Greenland and northern American ice sheets is overestimated in the simulation, probably because of the absence of an interactive carbon cycle providing a time-dependent atmospheric CO_2 concentration.

Extrapolation has been made for the next 100,000 years assuming no human interference at this time scale : the next ice age is expected to occur before 60,000 years AP, the cooling rate between now and then being roughly 0.01°C per century. The maximum amount of ice to be expected in the northern hemisphere is $27*10^6$ km^3 representing a 70 m sea-level drop in 55,000 years, i.e. slightly more than 12 cm per century in average.

1. Astronomical Theory of Paleoclimates

The longest astronomical cycles which might influence climate and which can be tested by appropriate statistical investigation of the long time series available in the terrestrial records are those involving variations in the elements of the Earth's orbit (Berger et al., 1984). Associated with the caloric seasonal insolation, these elements are at the origin of the Milankovitch theory (1941). For a glacial age to occur, this theory requires that northern high-latitude summers must be cold to prevent the winter snow from melting, in such a

R. Paepe et al. (eds.), Greenhouse Effect, Sea Level and Drought, 85–107.
© 1990 *Kluwer Academic Publishers.*

way as to allow a positive value in the annual budget of snow and ice, and to initiate a positive feedback cooling over the Earth through a further extent of the snow cover and a subsequent increase of the surface albedo.

On the assumption of a perfectly transparent atmosphere, that hypothesis thus requires a minimum in the northern hemisphere summer insolation at high latitudes. As a consequence, the semi-major axis and the solar constant being considered constant over the whole Quaternary period, there remain three astronomical parameters which are of primarily importance in paleoclimatology. These are still at the basis of modern versions of the Milankovitch theory, but their numerical values have been greatly improved (Berger, 1984; Berger and Loutre, 1988) and details of the explanation of why these cycles should affect climate have been deeply modified over the past years (Berger, 1988).

First, there is the eccentricity, e, which affects the relative intensity and the duration of the seasons in the different hemispheres but also the difference between maximum and minimum insolation received in the course of one year, a difference which can amount as much as 30% for the most extreme elliptical orbit during Quaternary. Second, the variation of the longitude of the perihelion relatively to the moving equinox, $\tilde{\omega}$, shows that the season of closest approach to the Sun varies. In fact, $\tilde{\omega}$ is mainly used through the precession parameter $e \sin \tilde{\omega}$. This parameter plays an opposite role in both hemispheres and is namely a measure of the difference in length between half-year astronomical seasons and of the difference between the Earth-Sun distances at both solstices. Third, the effect of the varying obliquity, ε, must be considered. A greater obliquity reduces the geographical contrast of the annual irradiations and intensifies the difference between summer and winter insolation in both hemispheres. According to the Milankovitch theory, it is possible to demonstrate that the "insolation glacials" would be associated with high eccentricity, northern summer at aphelion, and low obliquity.

Long term changes in orbital parameters change the total energy received by the whole Earth over the entire year very little, through the varying mean distance from the Earth to the Sun, but it changes its redistribution over latitudes and months (see for example Figure 1) by as much as 13% (Berger, 1978 a, 1979, 1988). At 6 kyr BP, from July to October, practically all of the globe receives more insolation than today and less from December to April. The maximum increase amounts 35 Wm^{-2} in northern polar latitudes. Southern polar latitudes receive more insolation than today (15 Wm^{-2} at the maximum) only during October and November, this hemispheric asymmetry in the maxima reflecting the present-day astronomical situation of the Earth according to the different seasons. The maximum depletion is in the intertropical regions with a peak of -25 Wm^{-2} in February. Seasonal contrast measured by the difference between July and January has increased everywhere with a maximum at the equator of 45 Wm^{-2} decreasing poleward to reach 35 Wm^{-2} and 10 Wm^{-2} at the north and south poles, respectively. On an annual basis, latitudes between 40° N and 40° S were receiving 1 to 2 Wm^{-2} less than now and polewards in both hemispheres, latitudes received up to 5 Wm^{-2} more.

At 125 kyr BP, all latitudes are receiving more insolation than today from May to October with a maximum of 60 Wm^{-2} in high polar northern latitudes at the summer solstice. At

87

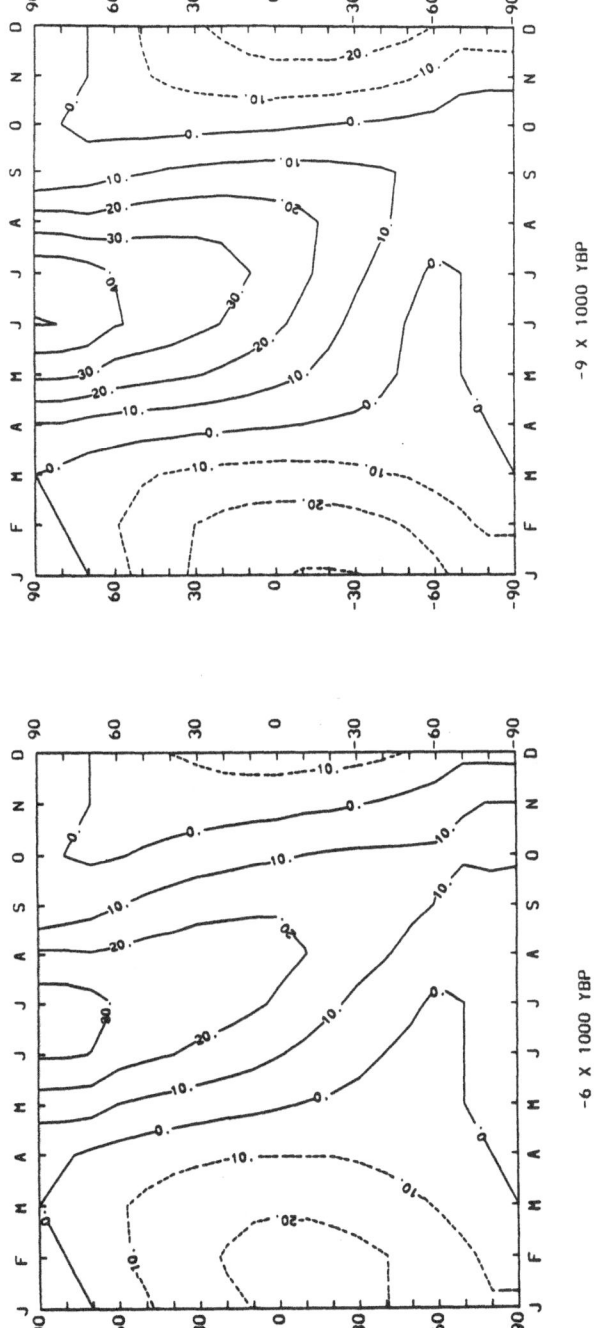

Figure 1. Seasonal and latitudinal pattern of insolation changes relative to present-day values for 6, 9, 122 and 125 kyr BP in Wm⁻² (Berger, 1978, 1979).

Dashed and full lines represent insolation respectively lower and higher than today.

88

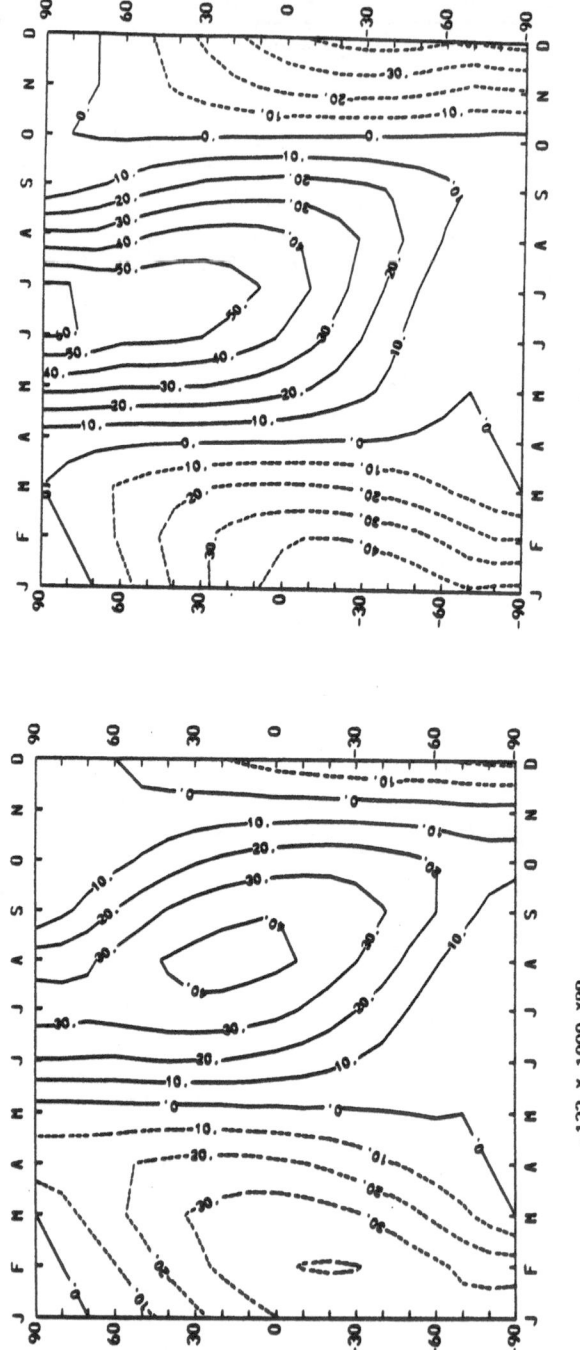

-125 x 1000 YBP

-122 x 1000 YBP

Figure 1. Continued.

the same time a negative deviation reaches its maximum in southern tropical latitudes in January (-50 Wm^{-2}). Seasonality has thus greatly increased in equatorial regions (+85 Wm^{-2}) and in the temperate northern hemisphere (+60 Wm^{-2}) mainly; in southern polar latitudes it increased by 45 Wm^{-2}. At the annual scale, the deviations from today values for each latitudinal belt are less than at 6 kyr BP.

At 6 kyr BP, the latitudinal gradient is reduced in northern hemisphere through out the year, except in early fall, with a maximum of 20 Wm^{-2} in July and February. In the southern hemisphere, it is increased in June to September and is reduced the rest of the year (the change is about 15 to 20 Wm^{-2}). A similar pattern characterizes 125 kyr BP but the changes are more intense reaching 45 Wm^{-2} in the winter hemisphere (decrease in the northern and increase in the southern hemisphere). The general latitudinal and seasonal insolation pattern of 125 kyr BP is much more similar to 9-10 kyr BP than to 6 kyr BP; in any case, the deviations of monthly insolation from today values are larger at 125 kyr BP than at 10 kyr BP by 20 Wm^{-2} during the Eemian compared to the peak of the Holocene.

Indeed, at 9 kyr BP, high polar latitudes receive more energy than today in June and July in northern hemisphere (40-50 Wm^{-2}) and in October in southern hemisphere (but much less). On the other hand, subtropical southern latitudes in January and high northern polar latitudes in December are receiving 30 Wm^{-2} less than today. Seasonality has therefore increased from temperate southern latitudes to the north pole. The southern polar latitudes were characterized by much less change in seasonality but by a large increase in southern spring insolation not compensated by any decrease in insolation in other seasons.

Since summers in the northern hemisphere get underinsolated when those in the southern become overinsolated, the evidence that ice ages occur essentially synchronously in both hemispheres was long taken as refuting the model. However, as the high latitude caloric insolation (Berger, 1978 b) is mainly dependent on obliquity, whereas that of low latitudes is essentially dependent on precession, and as the obliquity effect is the same in both hemispheres, whereas the precession effect is the opposite, the nature of the Milankovitch model itself implies a minimal (almost non-existent) asymmetry between hemispheres for all latitudes higher than 70° (the Milankovitch critical latitudes). Moreover, the model also implies compensation of negative summer deviations by positive winter deviations. Under such conditions not only would the summer temperatures in northern high latitudes be such as to prevent snow and ice from melting, but the implied milder winters would allow a substantial evaporation in the intertropical zone and more abundant snowfalls in temperate and polar latitudes, this humidity being transported there by an intensified general circulation due to the maximum in the latitudinal insolation gradient (Berger, 1976).

The Milankovitch model passes also severe statistical tests. Power spectra of climate variations (Hays et al., 1976) revealed by analysis of the isotopic composition of marine shells in deep-sea cores (Imbrie et al., 1984) and in ice cores (Lorius et al., 1985; Genthon et al., 1987), show clear peaks corresponding to the astronomical cycles (Berger, 1989). The 23 and 19 kyr periods are not significantly different from the periods associated with the largest amplitude terms in the series expansion of the precessional parameter (Berger, 1977); the periods about 41 kyr are essentially identical with the most important term in the expansion of the obliquity and the peaks in the range of 100 kyr, containing most of the variance, might be regarded as either a contribution from eccentricity or as a beat effect of precessional periods (Wigley, 1976).

Difficulties in explaining how the relatively small changes in Milankovitch caloric insolation (4% maximum) could be sufficient to initiate or end glacial ages, are now being resolved by analysing in details the latitudinal and seasonal patterns of insolation changes. In particular, Kukla (1978) demonstrated that the Earth's surface temperature is most affected by seasonal change of irradiation and Berger (1979) showed that insolation signatures of Quaternary climatic changes can be found in monthly values which can depart from their long-term mean values by as much as 13%. Moreover, several models exist, from simple to sophisticated ones - but considering only parts of the climate system - which allow the spread of ice till 50° N and which generate variations of zonal mean temperatures in reasonable agreement with the patterns of major climatic changes of the recent geological past, if summer, or better, monthly insolation falls by a percentage well within the range of astronomical variations (e.g., Suarez and Held, 1979, Prell and Kutzbach, 1987; Kutzbach and Gallimore, 1988).

But a symmetrical behaviour of the two hemispheres as far as climate changes are concerned does not necessarily mean an absolute synchroneity in the events. The synchronism of the climatic variations between the hemispheres is indeed a function of the time scale. At the 10 to 100 kyr time scale, glacials and interglacials appear to be roughly in phase in the two hemispheres, which does not seem to be the case at the 1000-yr time scale and is certainly not at higher frequencies. The following scenario would therefore not contradict the paleoclimatic reconstructions from proxy data. An insolation decrease, earlier in southern hemisphere than in the northern hemisphere, would create there an early cooling as a direct response of the atmosphere. This cooling will, however, be damped by the inertia of the oceans and the already existing huge Antarctic ice sheet provides a very small potential for self-amplification (except maybe through an increase in the sea ice cover around Antarctica). On the contrary, a similar insolation decrease, which would appear next in the northern hemisphere, would generate there a stronger response mainly because of a lesser oceanic influence and the possibility of initiating large ice sheets on the continents. As soon as these start to impact seriously the northern hemisphere climate, the northern hemisphere cooling will become larger than the southern one. As a possibility exist to transfer this cooling into the southern hemisphere, the southern hemisphere climate will be forced, from this moment on, to behave according to the northern hemisphere ice sheets forcing.

Indeed, Suarez and Held (1979) demonstrated with a simple energy balance model that, although variations in the Earth's orbital parameters may be responsible for the large fluctuations in the extent of northern hemisphere ice sheets during the Pleistocene, it is nevertheless necessary to look for mechanisms other than the interhemispheric exchange of heat in the atmosphere in order to explain the low temperatures of the southern hemisphere during glacials. It was proposed that, among the potential mechanisms for this cooling of the southern hemisphere, are the cross-equatorial heat transport by the ocean circulation (Manabe and Broccoli, 1985) and/or the fluctuations in the concentration of carbon dioxide (Broecker, 1982) as deduced from ice cores from the Antarctic (Barnola et al., 1987) and Greenland (Neftel et al., 1982) ice sheets, or in the loading of aerosols in the atmosphere.

2. Modelling Paleoclimates

Major advances in the theory of climate are under way. Key ingredients to this progress include the large variety of paleoclimatic indicators (geochemical, botanical, biological), the accuracy of dating, the expanding geographic coverage of measurements, and the application of statistical techniques to extract the climatic signal from other signals - and noise - contained in geologic records. Simultaneously, a hierarchy of climate models of increasing accuracy, dimensionality, and versability are being developed and applied. These models are invaluable tools for the quantitative testing of theories and for predicting or suggesting phenomena that can subsequently be studied with observations. Records of past climates can allow a broad testing of the capabilities of the climate models over a range of known changes in orbitally induced radiation. These tests may reveal weaknesses of the models; they will also suggest further process studies which will help improve the integrated understanding of the climate system (Hecht, 1985).

Climate models can be divided into two very broad categories: (1) general circulation models (GCMs), and (2) statistical-dynamical models, which include energy balance models (EBMs) and 2-D zonally averaged dynamical models (Kutzbach, 1985). Both GCMs and statistical- dynamical models have been used for the modelling of paleoclimates. GCMs have been used for simulating geographic features of paleoclimates and for including, in the most explicit form, processes such as precipitation that depend on details of the atmospheric flow. However, because of the large computational costs that are involved, GCMs have only been used to provide "snapshot" views of the climate in equilibrium with some boundary conditions. These conditions include the seasonal and latitudinal changes of solar radiation at the top of the atmosphere (resulting from modifications in the Earth's orbit about the Sun), as well as variations in sea-surface temperatures (CLIMAP, 1976, 1981), major ice sheets, atmospheric composition and turbidity, locations of continents (Barron and Washington, 1984) and topography (Ruddiman and Raymo, 1988) for simulations distant enough in the past.

For example, Prell and Kutzbach (1987), using a GCM, have investigated monsoon variability over the past 150,000 years. The agreement between simulated and observed paleoclimatic time series suggests that both orbitally produced solar radiation changes and glacial age boundary condition changes are necessary to explain the major regional features of monsoon climates at millennial or longer time scales. Royer et al. (1984) simulated the annual cycle of climate at 125 kyr and 115 kyr BP using a GCM with the appropriate seasonal insolation distribution and boundary conditions similar to the present ones. The change of insolation from 125 kyr BP to 115 kyr BP has produced in the model a cooling and an increase of soil water content over eastern Canada. These are favourable conditions for the accumulation of an ice cover, provoking an abortive glaciation. However, it must be mentioned that the GISS model failed to maintain snow cover through the summer at location of suspected initiation of the major ice sheets, despite the astronomical reduced summer and fall insolation (Rind et al., 1989).

On the other hand, the ice-age simulations of Gates (1976 a,b) and Manabe and Hahn (1977) indicate that global-averaged surface air temperature was 4-5 K below present for

August 18 kyr BP. Consistent with the generally lower temperature, an appropriate 15 % reduction in the reduction in the intensity of the hydrological cycle was simulated by the models, i.e., less precipitation and less evaporation. For both models, the simulated decrease in temperature over land exceeds the prescribed decrease of ocean temperature. Moreover, the larger cooling of the land relative to the ocean tends to weaken the northern hemisphere summer monsoon circulations and, more generally, to produce a tendency for anticyclonic outflow from the continents. On the contrary, experiments with both low- and high-resolution GCMs show that the increase of seasonality at 9 kyr BP resulted in continental interiors that were warmer in summer and cooler in winter than they are today. This intensified the summer monsoon circulation of Africa and southern Asia, because ocean temperatures during these seasons were much the same as today (Kutzbach, 1981; Kutzbach and Otto-Bliesner, 1982; Kutzbach and Guetter, 1984 b).

Combining the changes in solar radiation due to orbital forcing with the changes in prescribed boundary conditions (the ice sheets configuration, sea-ice extent, ocean temperature, atmospheric CO_2 concentration and aerosols), paleoclimate "snapshots" were obtained from 18,000 yBp to present at 3000-year intervals for both January and July conditions (Kutzbach and Street-Perrott, 1985; Kutzbach and Guetter, 1984 a, 1986). This modelling study is part of the Cooperative Holocene Mapping Project (COHMAP) which is producing global maps of paleoclimatic conditions for the past 18,000 years at 3000-year intervals and comparing observed paleoclimates and the simulated climates. Some comparisons between the model results and the data reveal areas of discordance. For example, the model showed moisture increases in the northern tropics at 15,000 and 12,000 yBP when the data indicated relatively dry conditions (Kutzbach and Street- Perrott, 1985). Work is now in progress to study these areas and times of discordance and to use them to suggest improvements of the data, of the model, of the boundary conditions used and of the methods for comparing the data to the model results (Webb et al., 1987).

On the other hand, statistical-dynamical models have a great potential for long-period climate simulations. As their dimensionality is reduced, they are less computer time-consuming, which allows to take into account all the different parts of the climate system in an interactive and time-dependent way. Therefore, these models can be used to simulate the transient response of the climate system to any "external" forcing or internal perturbation. For example, some of them have been coupled to slow responding portions of the climate system (e.g., ice sheets) in order to investigate the temporal evolution of climate on all time scales.

3.   Transient Response of the Climate System

In addition to the calculation of the Earth's climate which is in equilibrium with a particular insolation pattern and other boundary conditions (Kutzbach, 1985), the simulation of the transient response of the climate system to orbital variations must allow a better understanding of the physical mechanisms involved in the relationship between the astronomical forcing and climate. As suggested earlier (Berger, 1976), the time-evolution of the latitudinal distribution of the seasonal pattern of insolation is the key driving factor behind the behaviour of the climate system where the complex interactions between its different parts take turns with this orbital perturbation.

It is because of that dynamical behaviour of the seasonal cycle that time-dependent coupled climate models are required to test whether or not the astronomical forcing can, by itself, drive the long-term climatic variations. Such a time-dependent climate model including the atmosphere, the hydrosphere, the cryosphere and the lithosphere and their complex interactions has been built (Berger et al., 1989 a and b). It is forced only by the astronomical variations of monthly insolation for each latitude, the so-called boundary conditions used in equilibrium atmospheric general circulation model experiments (ice sheets size and area, sea surface temperature, albedo ...) being all generated by the climate model itself at each time step of the numerical integration.

In order to simulate the waxing and waning of the ice sheets over the last glacial-interglacial cycle, the model considers all parts of this climate system and the main physical processes and interactions concerned with the formation and melting of the ice sheets. Moreover, to investigate the importance of the role of some physical mechanisms in explaining these long-term climatic variations, the numerical values of the constants used in the model have been kept within their range of physical plausibility. At the same time, each of the subsystems was given a degree of complexity which could reflect its relative role in the real climate system within the frame of a Milankovitch experiment. It means, in particular, that this model must parameterize the evaporation and the transport of water to the high polar latitudes where part of it falls as snow building up the main ice sheets. The heat balance at the surface of the ice sheets, their dynamics and the bedrock adjustment to the ice loading will then govern their evolution. The land, ice sheets, ocean and sea-ice surfaces must interact with the atmosphere which requests the surface processes to be properly taken into account.

Therefore, a "zonally and sectorially averaged" seasonal climate model for the northern hemisphere, with its atmospheric dynamics and a simplified hydrological cycle, was developed to compute the annual snow accumulation rate over the modelled ice sheets for a particular seasonal insolation pattern. In each zonal band, an energy budget at the surface of the oceans, of the sea-ice, of the continent and of the ice sheets is explicitly computed. This atmosphere-land-ocean model is then run asynchronously with the ice sheets model.

As the model was conceived to study the transient response of the whole climate system to the astronomical forcing alone, the atmospheric greenhouse gas concentration was kept constant at a typical interglacial value. On the other hand, the experiments that we have done so far, consider the northern hemisphere only. Before a model for the entire globe be available, we were forced to assume that the variations in ice volume in the northern hemisphere dominate those of the southern hemisphere, that the variations of the energy at the top of the atmosphere in the southern hemisphere will be partly damped by the ocean and that the variations in the southern hemisphere ocean, sea ice and ice sheet will not affect significantly the long term climatic variations in the northern hemisphere. In the version of the model used for paleoclimate reconstruction, there is no deep ocean yet, although we recognize that deep water temperature has changed over the last glacial-interglacial cycle (Labeyrie et al., 1987) and that the change in heat stored by the ocean can be important.

Our paleoclimate model (PCM) is, therefore, latitude, altitude and time-dependent. It takes into account both the ice-sheet dynamics and the bedrock response (i.e. the long-memory part of the climate system) and the atmosphere, the upper mixed layer of the ocean, the sea-ice, and the land surface (covered or not by snow and ice). The latter components of the climate system all have short response times (up to some decades) and form what we call the 2.5-D climate model (BCM). It is also used to study the response of the climate system to man's induced changes in greenhouse gases concentration (Fichefet et al., 1989; Tricot, 1989).

A first version of the whole paleoclimate model is described in Berger et al. (1989 a), Tricot et al. (1989) and Fichefet et al. (1989). The robustness of such a model was tested by improving the physics of some of the parameterizations (infrared transfer, sea-ice formation, precipitation, continentality and slope effects over the ice sheets, Hadley heat transport).

The results obtained from such a modified version remain broadly the same which tends to support the importance of the processes that were selected to simulate the response of the climate system to the astronomical forcing (Berger et al., 1989 b).

The last paper (Berger et al., 1989 b) also contains a description of the ice sheet-lithosphere model that we use and the upper-ocean model is explained in Berger et al. (1989 c). An extended description of a new version of the whole model is available in Gallée et al. (1989 a and b).

4.   The Last Glacial-Interglacial Cycle

The Milankovitch experiment has been carried out using a simplified but qualitatively realistic geography with three ice sheets for the northern hemisphere representing respectively the Laurentide and the Cordilleran ice sheets, the Eurasian ice sheets and the Greenland ice sheet.

The BCM and ice-sheet-lithosphere model were coupled asynchronously and run from 122 kyr to present. After discussions with experts and after tests were performed to analyse the sensitivity of the model results to the initial size of the Greenland ice sheet, a size of 2/3 its present-day value was assumed at the beginning of the simulation (the present-day value is taken to be 2.5×10^6 km^3).

Figure 2 shows the deviation of the simulated total continental ice volume from the observed present-day value (assumed to be 30.4×10^6 km^3, i.e. 27.9×10^6 km^3 for Antarctica and 2.5×10^6 km^3 for Greenland - Hughes et al., 1981), as a function of time compared to the Chappell-Shackleton sea-level curve (1986). These deviations are computed by adding to the Northern hemisphere ice volume changes simulated by the model, the changes in Antarctica reconstructed by assuming (i) that the Antarctic ice volume at 122 kyr BP and from 6 kyr BP to the present was the same as now ; (ii) and that 18 kyr ago, it was $9.8 \times {}^{10}$ km^3 larger than today (Hughes et al., 1981).

The Chappell-Shackleton sea-level changes have been directly transformed into ice volume, taking into account variation in the area of the ocean surface at 18 kyr BP, according to

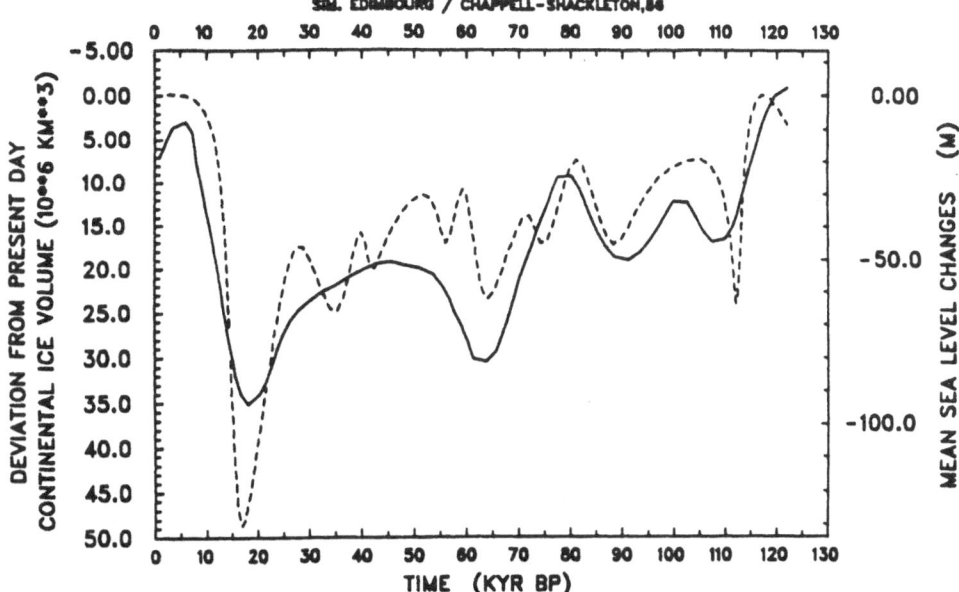

Figure 2. Long-term variations over the last glacial-interglacial cycle. (Berger et al., 1989 b (full line, left scale) and Chappell & Shackleton, 1986 (dashed line, right scale).

a calculation made by Marsiat and Berger (1990 and Annex 1). This relation implies that the Chappell-Shackleton minimum sea-level of -130 m at 18 kyr BP corresponds to a maximum ice volume increase of 48.6 x 10^6 km^3. Moreover, the melting of these ice sheets in roughly 12,000 years indicates an average rate of sea level rise of 1 cm per year (one meter per century !) between the glacial peak and the Altithermal.

The main characteristic of this figure 2 is the general agreement between the reconstructed and the simulated global ice volumes at the 10 kyr time scale. However, one must point out that our maximum ice volume in stage 2 is underestimated by 13.3 x 10^6 km^3. This feature may be partly explained by the fact that the model does not include changes in greenhouse gas concentration, nor sea level changes, thus not allowing further extent of the ice sheets over the continental shelves.

Moreover, the model underestimates the amount of ice sheet melting during the peak of the Holocene interglacial, leaving a surplus of roughly 3 x 10^6 km^3 of ice in the northern hemisphere. Consequently, the ice albedo-temperature feedback, together with the insolation decrease after 8 kyr BP lead to a present-day residual northern American ice sheet of 3.3 x ^{10} km^3 and to the Eurasian ice sheet of 2.4 x 10^6 km^3, and a Greenland ice volume in excess of 2.8 x 10^6 km^3 over the observed present-day value. This discrepancy prevents us to simulate more accurately the 100 kyr cycle. This is probably due to neglecting additional physical processes which can affect significantly the surface energy balance (for example, variation in atmospheric greenhouse gas concentrations as reconstructed from the Vostok core - Lorius et al., 1988; Saltzman, 1987) or which can change the ice dynamics.

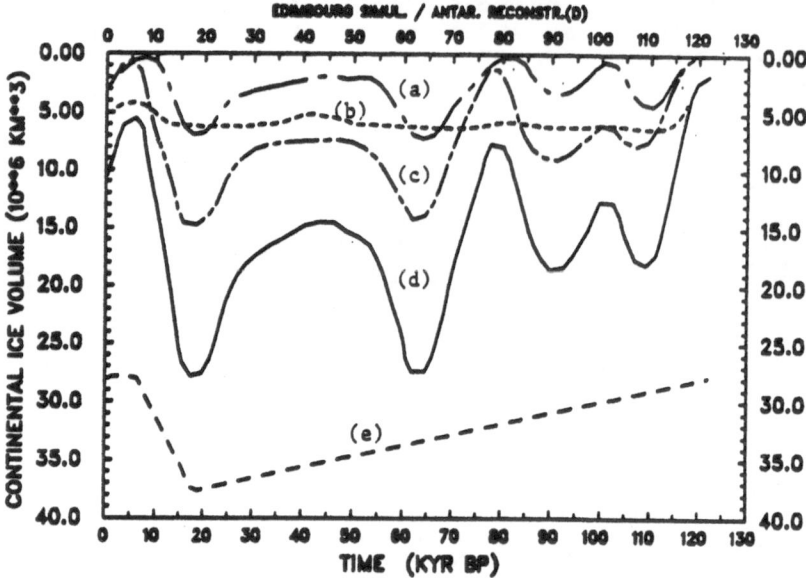

Figure 3. Long-term variations over the last glacial-interglacial cycle of the ice-volume of individual ice sheets simulated by the model described by Berger et al. (1989 b).
(a) is the Eurasian Ice Sheet; (b) the Greenland Ice Sheet; (c) the North American Ice Sheet; and (d) the northern hemisphere ice volume (a + b + c); (e) represents our assumption for the Antarctic Ice Sheet

Nevertheless, it seems very significant that the low frequency part of the global ice volume is well reproduced and more importantly that the simulated waxing and waning of the three individual ice sheets (Figure 3) are in phase with independently obtained geological reconstructions, as those made by Boulton et al. (1985) and Mangerud (1987). The first 2,000 years of the simulation show a very slow increase of the Greenland ice sheet. At 120 kyr BP the summer insolation conditions are such that the continent begins to accumulate snow at latitudes higher than 75° N on an annual basis. So the northern American ice sheet begins to form whereas, 2 kyr later, at 118 kyr BP, the Eurasian ice sheet is initiated. During stage 5, the model simulates small global ice volumes with maxima at 108 and 90 kyr BP and minima at 100 and 80 kyr BP. These last two dates agree well with the timing of the Barbados II and I high sea levels. The rapid and strong variations in the seasonal insolation pattern before 70 kyr BP prevent the Eurasian and northern American ice sheets from reaching a significant size, but the Greenland ice sheet reaches about its maximum volume as early as 112 kyr BP.

Although the chronology of the fluctuations of the Eurasian ice sheets is less well known, the minimum Eurasian ice volumes around 100 and 77 kyr BP agree pretty well with the reconstruction by Mangerud (1987) which shows that Norway was free of ice from 105 to 94 kyr BP (isotopic stage 5c) and from 85 to 70 kyr BP (isotopic stage 5a).

Between 80 (75 for the Eurasian ice sheet) and 64 kyr BP, significant growth occurs in all three ice sheets (modelled and observed). After the ice sheets first reach an appreciable size during a period which fits well stage 4 (Shackleton and Opdyke, 1973; Martinson et al., 1987), a slow decrease of the ice volume is simulated between 64 and 50 kyr BP. This is followed by a slow increase to 30 kyr BP related to the fact that the variations of insolation are small and that when the ice sheets get higher than 2 km, air starts the be cold, and its saturation water vapour pressure and precipitations decrease considerably. Finally, an insolation decrease from 30 kyr BP to 23 kyr BP leads to a subsequent increase in the ice volume up to the last glacial maximum peaking at 18 kyr BP.

The simulated ice volumes at 18 kyr BP are 6.2, 14.8 and 7.0 x 10^6 km^3 respectively for the Greenland, northern American and Eurasian ice sheets. These volumes can be compared with mid-range values obtained directly from reconstructions made by Hughes et al. (1981) and by Fisher et al. (1985). At about 18 kyr BP, all the ice sheets started to melt.

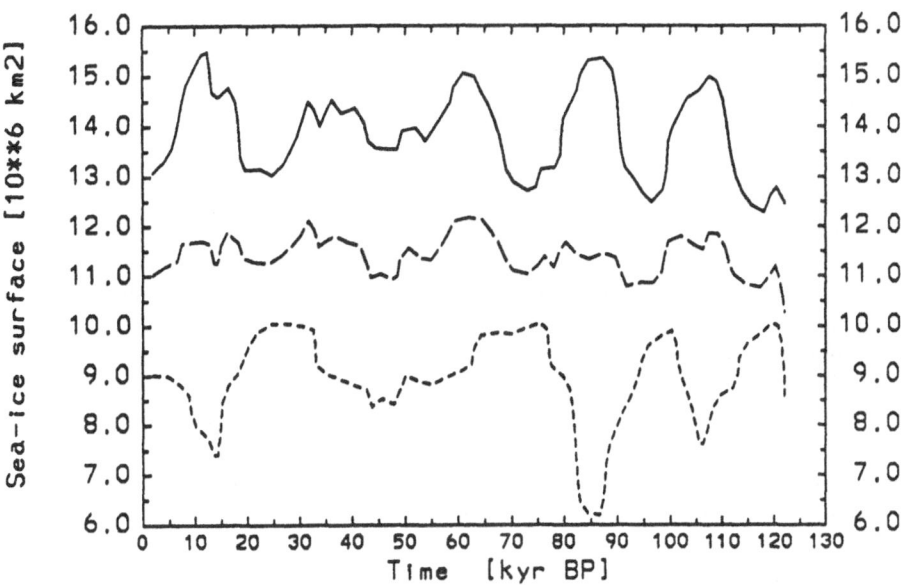

Figure 4. Long-term variations of the northern hemisphere sea-ice extent simulated by the Paleoclimate Model described by Berger et al. (1989 b).

In our model, rapid deglaciation up to 6 kyr BP is a consequence of the feedback mechanisms initiated by a slight increase in insolation at a time when the maximum extent of ice make the ice sheets most vulnerable. The Eurasian ice sheet disappears almost totally just after 8 kyr BP. A few thousand years later, the Greenland and northern American ice sheets reach their minimum extent. This timing corresponds well qualitatively with the geological observations according to which (i) the European (Andersen, 1981; Boulton et al., 1985) and Asian (Andersen, 1981) ice sheets disappeared totally around 8 kyr BP, and (ii) the northern American ice sheets approached their present configuration (i.e. the Barnes and Penny Ice Caps, relics of the Laurentide ice sheet - Dyke and Prest, 1987) after 6 kyr BP. However, the simulated remaining ice volumes of the Greenland and northern American ice sheets are overestimated. Consequently, after 6 kyr BP, when the summer insolation decreases again, the existence of higher surface albedos in the model allows the ice sheet to start growing, contrary to observations.

The time evolution of the February, August and annual mean total sea-ice area simulated by the paleoclimate model is displayed in Figure 4. In this Figure the upper, middle and lower curves give the values for respectively February, year (annual mean) and August. A newer version of the model (Gallée et al., 1989 a, b) displays less high frequencies but a similar general behaviour over the last 125,000 years. The figure indicates that the mean annual sea-ice area oscillates between 10.5 and 12 x 10^6 km^2 with a period which is very close to that of the precession cycle (in mean 22 kyr). This variation is rather weak and only represents 15 % of the present-day mean sea-ice coverage. It is very interesting to note that maxima in the annual mean sea-ice area do not coincide necessarily with maxima in the continental ice volume, although, the largest sea-ice extent ($\geq$ 12 x 10^6 km^2) takes place during the isotopic stage 4 at 62 kyr BP, when the predicted continental ice volume is at a maximum (see Figure 2). It should be observed that this large maximum in the mean annual sea-ice area is associated with large values both in the winter and summer sea-ice extent. Figure 4 also shows that the amplitude of the seasonal cycle of the sea-ice cover varies markedly over the last 122,000 years. The amplitude is largest slightly after the isotopic stage 5d at 106 kyr BP (7.5 x 10^6 km^2), during the transition between isotopic stages 5b and 5a at 86 kyr BP (9 x 10^6 km^2), and during the deglaciation at 13 kyr BP (8 x 10^6 km^2). All these periods are characterized by relatively high summer and low winter insolation which enhance the seasonal cycle of the ice pack.

5. Future Climate

An integration for the next 80,000 years, for the northern hemisphere ice sheets only, was performed. The results are shown on figure 5. A global cooling is expected for the next 55,000 years followed by a warming to 70 kyr AP. The northern hemisphere ice volume would increase from the present day 2.6 x 10^6 km^3 to 6 x 10^6 km^3 within 10,000 years and to 7 x 10^6 km^3 at the peak of the next glaciation. A cold peak is also visible at 25 kyr AP. These results are in agreement with previous calculations based on more simple models (Berger et al., 1990). This tendency for our natural climate to be steered

astronomically towards the next ice age leads to an average cooling of 0.01°C per century. Although this is negligible when compared to the 1 to 5°C warming expected from the greenhouse gas concentrations increase in the course of the 21th century, it is not the case for the related sea level change. The build up of 27 x 10^6 km^3 of ice in the northern hemisphere would indeed be responsible for a eustatic sea-level drop of about 70 m which represents a falling of the sea level at an average rate of 10 cm per century. This is the same order of magnitude that the 15 cm rise over the past century, but in the reverse direction. If we assume that this long term trend in equivalent sea level fall is already underway, since climate has reached its Holocene peak 6,000 years ago, the absolute rise over the past century - whatever its cause may be - must have been even larger. This may have implications on the prediction for the next century. It is indeed expected that thermal expansion and the partial melting of the temperate glaciers and possibly also of the Greenland ice cap lead to a sea level rise the "best estimate" of which may lie between 20 and 60 cm on top of this basal trend (Pirazzoli, 1989; Warrick et al., 1989, Jones and Henderson-Sellers, 1990), these values including the possible growth of the Antarctic ice sheet due to an increase of snow fall
under warmer conditions in this region (Oerlemans, 1989).

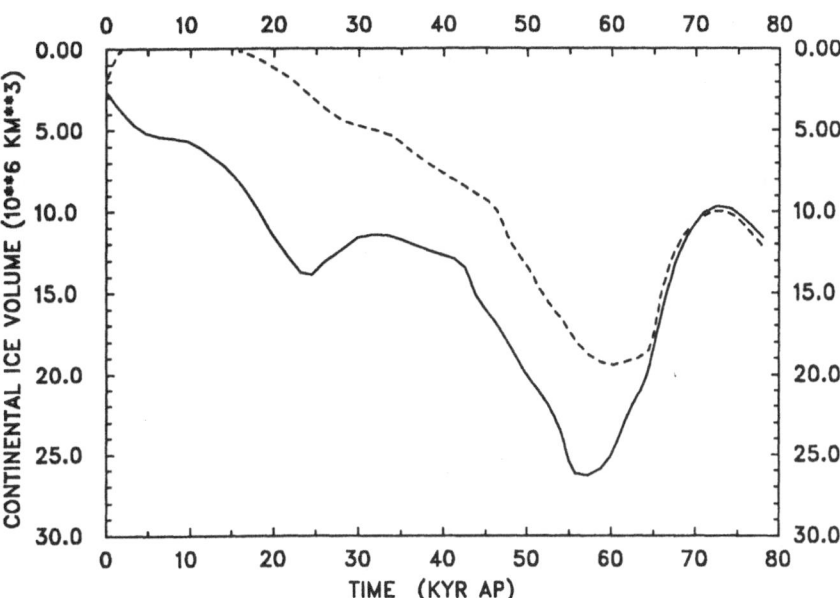

Figure 5. The future climate at the astronomical time scale without anthropogenic disturbances (full line) and assuming no Greenland ice sheet to day (dashed line) from Berger et al. (1989 a, b, 1990 and Gallée, 1989).

In order to test the sensitivity of the future astronomical climate to the possible impact of man's activities, it was assumed that the greenhouse warming of the 21th century would melt the Greenland ice sheet totally. An integration of the PCM was therefore carried out over the next 80 kyr with this initial condition (again we recall that the Antarctic and southern hemisphere ice sheets are not included and that the greenhouse gases concentrations were kept constant at their present-day value). The main results of such a simulation (Gallée, 1989) are that the ice sheets are not going to reappear in the northern hemisphere before 15 kyr AP and the next glaciation is delayed to 60 kyr AP and is less extensive (19 instead of 27 x 10^6 km^3. The main difference between the two simulations, as far as the behaviour of the individual ice sheets is concerned, arises mainly for the Fennoscandian ice sheet which appears only 50 kyr AP and reaches only half its "natural" size (4 instead of 7 x 10^6 km^3; the Greenland and Laurentide ice sheets reappear 15 kyr AP in the no-initial-Greenland ice sheet run, the Laurentide ice sheet reaches a maximum size of 11 instead of 14 x ^{10} km^3 and the Greenland ice sheet recovers its 5 x 10^6 km^3 quite rapidly.

6. Conclusions

Simulation of the waxing and waning of the ice sheets over the past 1 million years seems to indicate that the climate is sensitive to the orbital parameter variations, or at least that these are able to trigger feedbacks which then induce significant climatic variations at that time scale. Yet no general circulation model has been able to generate such long-term climatic variations. This is because the complexity of the different parts of the climate system themselves and of their mutual interactions does not only request considerable computer time and space if first principles are used, but also requests that all the most important processes be taken into account with their relative weight. The non-linearities in the climate system can indeed easily introduce spurious response (which can be masked sometimes by compensating errors) if one process is overweighted as compared to the others (including those missing).

Simplified 0-D and 1-D models have thus largely been used to simulate these long-term climatic variations. The problem with such a basic approach is that we could obtain almost any desired sensitivity is desired by tuning procedures involving parameterizations not tied firmly to first principles (although their general formulation derives sometimes directly from such first principles). This is why a zonally-sectorially averaged model of intermediate complexity was used, which considers the atmosphere, the ocean and the cryosphere in an interactive way. The first experiment considered the northern hemisphere only. Our 2.5-D climate model was therefore forced with the astronomical insolation only over the last glacial-interglacial cycle. A good agreement was found between the low frequency parts of the simulated long-term climatic variation and of the most recent reconstructions of sea-level and ice volume changes from geological observations. In addition to the physical processes included in the model which play an important role in amplifying the astronomical signal, it is likely that the model reproduces well the data because it accounts correctly for the seasonal and spatial character of the insolation changes.

However, discrepancies between the simulated and observed climates remain, in particular the underestimate of the ice volume at stage 2 and the excess of ice left over at the peak of the Holocene interglacial. These are due to neglecting important physical factors, as for example realistic topography (North et al., 1983), the variations of greenhouse gas concentration and the deep ocean circulation. It is also related to the lack of generating powerful feedback processes in temperate and low latitudes as, for example, the one related to evaporation and to the important role played by water vapour in the greenhouse effect (Raval and Ramanathan, 1989). Incorporation of these factors is expected to improve our simulation of the 100,000-year cycle.

7. Acknowledgments

Th. Fichefet is supported by the National Fund for Scientific Research (Belgium) and H. Gallée by the Antarctic Programme of the Prime Minister's Office for Science Policy Programming. This research was sponsored partly by the Climate Programme of the Commission of the European Communities under Grants No. CLI-076-B (RS) and No. EV4C-0052-B (GDF), and by the National Fund for Scientific Research (Belgium) who provided supercomputer time.

8. Annex 1

8.1. ICE VOLUME AND SEA LEVEL OVER THE LAST GLACIAL CYCLE

Geological evidence and oxygen-isotope variations in deep-sea cores provide valuable information about the sea-level variations of the past. Ice-volume equivalent is usually computed by using a constant oceanic area. In a recent paper (Marsiat and Berger, 1990), a relationship is developed between the continental ice-volume variation and the sea-level drop by taking into account the sea-floor topography and, therefore, the variation of the oceanic area.

Oceanic area is indeed not constant with time. As sea level goes down, for example, there is a regression releasing part of the continental shelf. Therefore, a relationship between ice volume and sea level has been obtained by computing the water volume equivalent for every slice of 1-m thickness, from the surface to the bottom of the ocean using the present-day bathymetry data of CLIMAP (1981). The equivalent ice volume was obtained by using a density ratio of 0.9 between glacial-ice and sea-water. The vertical adjustment of the continental shelves related to the changing water thickness above them and to the isostatic impact of the ice sheet on the continent itself was not taken into account. In fact, the shape of the decrease of the oceanic area leads us to assume that this correction would have little effect on the estimates of the last glacial maximum ice volume. Indeed, the ocean area decrease (Figure A1) is particularly important at the beginning of the glaciation, when

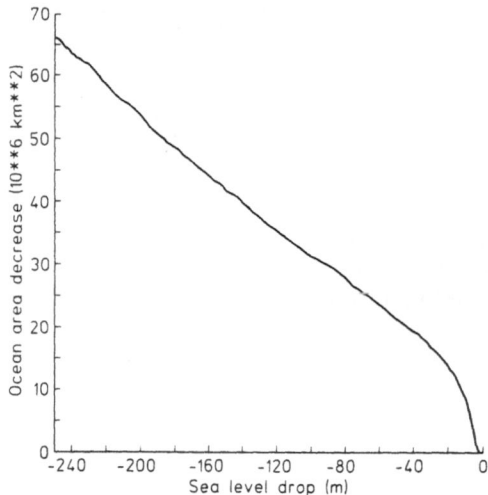

Figure A1. Oceanic-area decrease as a function of the sea-level drop (not taking into account the isostatic process)(Marsiat and Berger, 1990).

the isostatic response of the oceans and continents is not very important. Presently, a sea-level drop of 40 m (corresponding to 15.5×10^6 km^3 of ice) reduces the oceanic area by 20 $\times 10^6$ km^2, more than 5% of its present-day value. At the last glacial maximum, a sea-level drop of 130 m would reduce the oceanic area by 7.3×10^6 km^2, i.e., a little bit more than

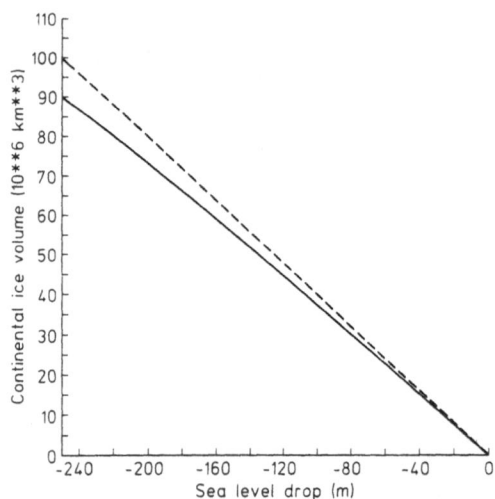

Figure A2. Relation between continental ice volume (in million of km^3 and sea-level drop (in m)(Marsiat and Berger, 1990).

10% of the present area: a subsequent drop of 90 m has thus the same effect as the initial drop of 40 m.

The results (Figure A2) show that the relationship between ice volume and sea level is a non-linear one reducing, for a given sea-level drop, the ice-volume equivalent computed with a constant oceanic area. For instance, the last glacial maximum sea-level drop of 130 m estimated by Chappell and Shackleton (1986) corresponds to an ice volume of 48.6 x 10^6 km^3, 7% less than the 52.1 x 10^6 km^3 computed with a constant ocean area. In Figure A2, the dashed line shows the continental ice volume and sea-level drop if the oceanic area is constant , taken as its actual value, whereas the full line represents the same if the variation of the oceanic area is taken into account.

9. References

Andersen, B.G. (1981) "Late Weichselian ice sheets in Eurasia and Greenland.", in Denton, G.H. and Hughes, T.J. (Eds.) *The Last Great Ice Sheets*, Wiley-interscience Publ., USA, pp. 1-65

Barnola, J.M., Raynaud, D., Korotkevich, Y.S. & Lorius, C. (1987) "Vostok ice core provides 160,000-year record of atmospheric CO_2", *Nature* 239(6138), pp. 408-414

Barron, E.J., & Washington, W.M. (1984) "The role of geographic variables in explaining paleoclimates: Results from Cretaceous climate model sensitivity studies.", J. Geophys. Res., 89, pp. 1267-1279

Berger, A. (1976) "Long-term variations of daily and monthly insolation during the last Ice Age.", EOS, 57(4), p. 254

Berger, A. (1977) "Support for the astronomical theory of climatic change.", *Nature* 268, pp. 44-45

Berger, A. (1978 a) "Long term variations of daily insolation and Quaternary climatic changes.", J. Atmos. Sci., 35(12), pp. 2362-2367

Berger, A. (1978 b) "Long-term variations of caloric insolation resulting from the Earth's orbital elements.", Quat. Res., 9, pp. 139-167

Berger, A. (1979) "Insolation signatures of Quaternary climatic changes.", Il Nuovo Cimento, 2C(1), pp. 63-87

Berger, A. (1984) "Accuracy and frequencies stability of the Earth's orbital elements during the Quaternary.", in Berger, A.L., Imbrie, J., Hays, J., Kukla, G and Saltzman, B. (Eds.) *Milankovitch and Climate*, Part I, Reidel Publ. Company, Dordrecht, Holland, pp. 3-40

Berger, A. (1988) "Milankovitch theory and climate.", Review of Geophysics, 26(4), pp. 624-657

Berger, A. (1989) "Pleistocene climatic variability at astronomical frequencies.", Quaternary International, 2, pp. 1-14

Berger, A., Imbrie, J., Hays, J., Kukla, G. and Saltzman, B. (eds.)(1984) *Milankovitch and Climate.*, Reidel Publ. Company, Dordrecht, Holland, 895 p.

Berger, A., & Loutre, M.F. (1988) *New insolation values for the climate of the last 10*

104

million years., Sc. Report 1988/13, Institut d'Astronomie et de Géophysique G. Lemaître, Université Catholique de Louvain, Louvain-la-Neuve

Berger, A., Gallée, H., Fichefet, Th, Marsiat, I., & Tricot, Ch. (1989 a) *Testing the astronomical theory with a coupled climate-ice sheet model. Global and Planetary Change* (in press)

Berger, A., Fichefet, Th., Gallée, H., Marsiat, I., Tricot C., & van Ypersele J.P. (1989 b) "Physical interactions within a coupled climate model over the last glacial-interglacial cycle.", Philosophical Transactions of the Royal Society of Edinburgh, Earth Sciences

Berger, A., Fichefet, Th., Gallée, H., Tricot, Ch., Marsiat, I., & van Ypersele, J.P. (1989 c) "Astronomical forcing of the last glacial-interglacial cycle.", in Crutzen, P.,Gérard, J.Cl., and Zander, R. (eds.) *Our Changing Atmosphere*, Université de Liège, Institut d'Astrophysique, Cointe-Ougrée, pp. 353-382

Berger, A., Gallée, H., & Mélice, J.L. (1990) "The Earth's future climate at the astronomical time scale.", in Goodess, Cl. and Palutikof, J. (eds.) *Future Climate Change and Radioactive Waste Disposal*", University of East Anglia, Norwich (in press)

Boulton, G.S., Smith, G.D., Jones, A.S & Newsome, J. (1985) "Glacial geology and glaciology of the last mid-latitude ice sheets.", J. Geol. Soc. London, 142, pp. 447--474

Broecker, W.S. (1982) "Glacial to interglacial changes in ocean chemistry.", Prog. Oceanog., 11, pp. 151-197

Chappell, J. & Shackleton, N.J. (1986) "Oxygen isotopes and sea level.", *Nature* 324, pp. 137-140

CLIMAP Project Members (1976) "The surface of the Ice-Age Earth.", Science, 191, pp. 1131-1136

CLIMAP Project Members (1981) "Seasonal reconstructions of the earth's surface at the last glacial maximum.", Geol. Soc. Am. Map Chart Ser. MC-36

Dyke, A.S. & Prest, V.K. (1987) "Late Wisconsinan and Holocene history of the Laurentide ice sheet.", Géographie Physique et Quaternaire, 41(2), pp. 237-263

Fichefet Th., & Gaspar Ph. (1988) "A model study of upper ocean-sea ice interactions.", J. Phys. Oceanogr., 18, pp. 181-195

Fichefet, Th., Tricot, Ch., Berger, A., Gallée, H., & Marsiat, I. (1989) "Climate studies with a coupled atmosphere-upper ocean-ice sheets model.", Philosophical Transactions of the Royal Society of London, A239, pp. 249-261

Fischer, D.A., Reeh, N., & Langley, K. (1985) "Objective reconstructions of the late Wisconsinan Laurentide Ice Sheet and the significance of deformable beds.", Géographie Physique et Quaternaire, 39(3), pp. 229-238

Gallée, H. (1989) *Conséquences pour la prochaine glaciation de la disparition éventuelle de la calotte glaciaire recouvrant le Groenland.*, Recherche faite sous contrat CEA/BC-4561, France, Sc. Report 1989/7, Institut d'Astronomie et de Géophysique G. Lemaître, Université Catholique de Louvain, Louvain-la-Neuve

Gallée, H., van Ypersele, J.P., Fichefet, Th., Marsiat, I., Tricot, Ch., & Berger, A. (1989 a) *Simulation of the Last Glacial Cycle by a Coupled 2-D Climate-Ice Sheet Model.*, Part (1): The climate model, Scientific Report 1989/1, Institut d'Astronomie et de Géophysique G. Lemaître, Université Catholique de Louvain, Louvain-la-Neuve

Gallée, H., van Ypersele, J.P., Fichefet, Th., Marsiat, I., Tricot, Ch., & Berger, A. (1989 b) *Simulation of the Last Glacial Cycle by a Coupled 2-D Climate-Ice sheet model*, Part 2: "Response to insolation and CO_2 variations", Scientific Report 1989/3, Institut d'Astronomie et de Géophysique G. Lemaître, Université Catholique de Louvain-la-Neuve

Gates, W.L. (1976 a) "Modelling the Ice Age climate", Science, 191, pp. 1138-1144

Gates, W.L. (1976 b) "The numerical simulation of Ice-Age climate with a global general circulation model.", J. Atmos. Sci., 33, pp. 1844-1873

Genthon, C., Barnola, J.M., Raynaud, D., Lorius, D., Jouzel, J., Barkov, N.I. & Korotkevitch, Y.S. (1987) "Vostok ice core: climatic response to CO_2 and orbital forcing changes over the last climatic cycle.", *Nature* 239(6138), pp. 414-418

Hays, J.D., Imbrie, J., & Shackleton, N.J. (1976) "Variations in the Earth's orbit: Pacemaker of the Ice Ages.", Science, 194, pp. 1121-1132

Hecht, A. (1985) "Paleoclimatology. A retrospective of the past 20 years.", in Hecht, A. (ed.) *Paleoclimate Analysis and Modelling*, Wiley, New York, pp. 1-26

Hughes, T.J., Denton, G.H., Anderson, B.G, Schiling, D.H., Fasthook, J.L. & Lingle, C.S. (1981) "The last great ice sheets: a global view.", in Denton, G.H. and Hughes, T.J. (eds.) *The last great ice sheets.*, Wiley Interscience Publ., USA, pp. 275-317

Imbrie, J., Hays, J., Martinson, D.G., McIntyre, A., Mix, A.C., Morley, J.J., Pisias, N.G., Prell, W.L., & Shackleton, N.J. (1984) "The orbital theory of Pleistocene climate : support from a revised Chronology of the marine $\delta^{18}O$ record.", in Berger, A.L., Imbrie, J.,Hays, J., Kukla, G., and Saltzman, B. (eds.) *Milankovitch and Climate.*, Reidel Publ. Company, Dordrecht, Holland, pp. 269-305

Jones, M.D.H., & Henderson-Sellers, A. (1990) "History of the greenhouse effect.", Progress in Physical Geography, 14(1), pp. 1-18

Kukla, G. (1978) "Recent changes in snow and ice.", in Gribbin, J. (ed.) *Climatic Change*, Cambridge University Press, pp. 114-129

Kutzbach, J.E. (1981) "Monsoon climate of the early Holocene: Climatic experiment using the earth's orbital parameters for 9,000 years ago.", Science, 214, pp. 59-61

Kutzbach, J.E. (1985) "Modelling of paleoclimates.", Advances in Geophysics, 28A, pp. 159-196

Kutzbach, J.E., & Guetter, P.J. (1984 a) "Sensitivity of late-glacial and Holocene climates to the combined effects of orbital parameter changes and lower boundary condition changes: 'Snapshot' simulations with a general circulation model for 18, 9, and 6 kyr BP.", Ann. Glaciol., 5, pp. 85-87

Kutzbach, J.E., & Guetter, P.J. (1984 b) "The sensitivity of monsoon climates to orbital parameter changes for 9000 years BP: Experiments with the NCAR general circulation model.", in Berger, A., Imbrie, J., Hays, J.,Kukla, G. and Saltzman, B. (eds.) *Milankovitch and Climate*, Part 2, Reidel Publ., Dordrecht, Netherlands, pp. 801-820

Kutzbach, J.E., & Guetter, P.J. (1986) "The influence of changing orbital parameters and surface boundary conditions on climate simulations for the past 18 000 years.", J. Atmos. Sci., 43, pp. 1726-1759

Kutzbach, J.E., &Otto-Bliesner, B.L. (1982) "The sensitivity of the African-Asian monsoonal climate to orbital parameter changes for 9000 years B.P. in a low-resolution general circulation model.", J. Atmos. Sci., 39, pp. 1177-1188

Kutzbach, J.E., & Street-Perrott, F.A. (1985) "Milankovitch forcing of fluctuations in the level of tropical lakes from 18 to 0 kyr BP.", *Nature* 317, pp. 130-134

Kutzbach, J.E., & Gallimore, R.G. (1988) "Sensitivity of a coupled atmosphere-/mixed layer ocean model to changes in orbital forcing at 9000 years BP.", J. Geophys. Res., D1, pp. 803-821

Labeyrie, L.D., Duplessy, J.Cl., & Blanc, P.L. (1987) "Variations in mode of formation and temperature of oceanic deep waters over the past 125,000 years.", *Nature* 327, pp. 477-482

Lorius, Cl., Jouzel, J., Ritz, C., Merlivat, L., Barkov, N.I., Kofogkevich, Y.S., & Kotlyakov, V.M. (1985) "A 150,000-year climatic record from Antarctic ice.", *Nature* 316, pp. 591-596

Lorius, Cl., Barkov, N.I., Jouzel, J., Korotkevitch, Y.S., Kotlyakov, V.M., & Raynaud, D. (1988) "Antarctic ice core: CO_2 and climatic change over the last climatic cycle.", EOS, 69 No. 26, pp. 681, 683-684

Manabe, S., & Hahn, D.G. (1977) "Simulation of the tropical climate of an Ice Age.", J. Geophys. Res., 82, pp. 3889-3911

Manabe, S., & Broccoli, A.J. (1985) "A comparison of climate model sensitivity with data from the last glacial maximum.", J. Atmos. Sci., 42 No. 23, pp. 2643-2651

Mangerud, J. (1987) "The last glacial history of Scandinavia between the last interglacial and the last glacial maximum.", in Frenzel, B. (ed.) *The beginning of an inland glaciation - facts and problems of climate dynamics.*, Proceedings Mainz Symposium

Marsiat, I., & Berger, A. (1989) "On the relationship between ice volume and sea level over the last glacial cycle.", Climate Dynamics, 4(2), pp. 81-84

Martinson, D.G., Pisias, N.G., Hays, J.D., Imbrie, J., Moore, T.C. & Shackleton, N.J. (1987) "Age dating and the orbital theory of the ice ages: development of a high-resolution 0 to 300,000-year chronostratigraphy.", Quat. Res., 27(1), pp. 1-29

Milankovitch, M. (1941) *Kanon der Erdbestrahlung und seine Anwendung auf das Eiszeitenproblem.* Royal Serbian Sciences, Spec. pub. 132, section of Mathematical and Natural Sciences, Vol.33, Belgrade, 633 pp. ('Canon of Insolation and the Ice Age Problem .',English Translation by Israel Program for Scientific Translation and published for the U.S. Department of Commerce and the National Science Foundation, Washington D.C., 1969)

Neftel, A., Oeschger, H., Schwander, J., Stauffer, B., & Zumbrunn, R. (1982) "Ice core sample measurements give atmospheric CO_2 content during the past 40,000 yr.", *Nature* 295, pp. 220-223

North, G.R., Mengel, J.G. & Short, D.A. (1983) "Simple energy balance model resolving the seasons and the continents, application to the astronomical theory of the Ice Ages.", J. Geophys. Res., 88(C11), pp. 6576-6586

Oerlemans, J. (1989) "A projection of future sea level.", Climatic Change, 15(1/2), pp. 151-174

Pirazzoli, P.A. (1989) "Present and near future global sea level changes.", Global and Planetary Change, 1(4), pp. 241-258

Prell, W.L., & Kutzbach, J.E. (1987) "Monsoon variability over the past 150,000 years.", J. Geophys. Res., 92, pp. 8411-8425

Raval, A., & Ramanathan, V. (1989) "Observational determination of the greenhouse effect.", Nature 343, pp. 758-761

Rind, D., Peteet, D., & Kukla, G. (1989) "Can Milankovitch orbital variations initiate the growth of ice sheets in a general circulation model?", Journal of Geophysical Research, 94, D10, pp. 12.851-12.871

Royer, J.F., Deque, M., & Pestiaux, P. (1983) "Orbital forcing of the inception of the Laurentide ice sheet.", Nature 304, pp. 43-46

Ruddiman, W.F., & Raymo, M.E. (1988) "Northern Hemisphere climates regimes during the last 3 Myr: possible tectonic forcing.", Phil. Trans. Roy. Soc. Lond., B318, pp. 411-430

Saltzman, B. (1987) "Carbon dioxide and the $\delta^{18}O$ record of late-quaternary climatic change : a global model.", Climate Dynamics, 1, pp. 77-85

Shackleton, N.J. & Opdyke, N.D. (1973) "Oxygen isotope and paleomagnetic stratigraphy of Equatorial Pacific core V28-238: oxygen isotope temperatures and ice volumes on a 10^5 years and 10^6 years scale.", Quat. Res., 3(1), pp. 39-55

Suarez, M.J., & Held, I.M. (1979) "The sensitivity of an energy balance climate model to variations in the orbital parameters.", J. Geophys. Res., 84(C8), pp. 4825-4836

Tricot, Ch. (1989) "The transient response of climate to greenhouse gas concentration changes : a preliminary study with a two-dimensional coupled atmosphere-ocean model.", in Crutzen, P, Gérard, J. Cl., and Zander, R. (eds.) Our Changing Atmosphere, Université de Liège, Institut d'Astrophysique, Cointe-Ougrée, pp. 333-338

Tricot, Ch., Gallée, H., Fichefet, Th., Marsiat, I., & Berger, A. (1989) "A simulation of the long-term variations of the global ice volume over the past 122,000 years: a test of the astronomical theory.", in Lenoble, J. and Geleyn, J.F. (eds.) IRS'88 : Current Problems in Atmospheric Radiation, A. Deepak Publishing, pp. 338-341

van der Veen, C.J. (1988) "Projecting future sea level.", Surveys in Geophysics, 9, pp. 389-418

Warrick, R.A., Barrow, E.M., & Wigley, T.M.L. (1989) The greenhouse effect and its implications for the European Communities., Climatic Research Unit, School of Environmental Sciences, University of East Anglia, Norwich, UK

Webb, T., Street-Perrott, F.A. & Kutzbach, J.E. (1987) "Late-Quaternary paleoclimatic data and climate models.", Episodes, 10, pp. 4-6

Wigley, T.M.L. (1976) "Spectral analysis and astronomical theory of climatic change.", Nature 264, pp. 629-631

PRESENT, PAST AND FUTURE PRECIPITATION: CAN WE TRUST THE MODELS?

G. KUKLA

Lamont-Doherty Geological Observatory of Columbia University
Palisades, New York
10964 USA

ABSTRACT: Results of several current atmospheric general circulation models were compared. The models show poor skill in reproducing observed precipitation, and differ considerably in their precipitation forecasts for the doubled CO_2 world. Driven by the approximately modified insolation impact, they are unable to simulate an onset of a glaciation. Improved representation of oceanic heat and mass transports is necessary for reliable prediction of future precipitation shifts.

1. Introduction

Much attention is currently being paid to the man-made increase of the carbon dioxide concentrations in the atmosphere. Mathematical models of atmospheric circulation project a substantial global warming in the near future resulting from the increase. This warming is expected to be accompanied by rising sea levels and by intensified droughts in parts of the continents. In this paper we examine to what degree is the current generation of climate models capable of projecting the precipitation of the CO_2 rich world.

Current state-of-the-art climate models are the general circulation models (GCMs) with mixed layered ocean. These models have a realistic seasonal insolation cycle, partly prescribed seasonally varying sea surface temperatures, large scale condensation and cumulus convections, predicted although not fully interactive clouds, and predicted sea ice and snow extents. They can simulate the mass and energy transfers in the atmosphere and the vertical heat exchange in the uppermost 50-65 m thick layer of the ocean.

The models, however, do not allow any changes from the present conditions in the horizontal and deep vertical heat and mass transfer in the oceans. In other words, while they do allow the winds to change, they do not allow the ocean circulation to adjust. This limitation was imposed on climate models by a scarcity of observations on the deep ocean circulation, by the complicated feedbacks operating in the sea-air interactions and by a relatively small scale of the near surface ocean eddies, which are by and order of magnitude

109

R. Paepe et al. (eds.), Greenhouse Effect, Sea Level and Drought, 109–114.
© 1990 *Kluwer Academic Publishers.*

smaller than those operating in the free atmosphere.

Description and the results of five general circulation models were recently published in sufficient detail to allow inspection of the most important aspects of the hydrologic cycle. They are the results of the National Center of Atmospheric Research (NCAR) described by Washington and Meehl (1984), those of Geophysical Fluids Dynamics Laboratory (GFDL) described by Wetherald and Manabe (1986) and Manabe and Wetherald (1986), those of the Goddard Institute of Space Studies (GISS), described by Hansen et al. (1984), those of the United Kingdom Meteorological Office (UKMO), described by Wilson and Mitchell (1987) and finally those of the Oregon State University (OSU), described by Schlesinger and Mitchell (1985) and Schlesinger and Zhao (1989 in press).

Principal representation of the physics of the climate system is common to all five models. Although some models are handing some aspects of the climate system in more detail than the others, the advantage is usually compensated by other processes represented less accurately than the other models.

So by and large, non of the current GCM predictions can be singled out as substantially more reliable than the other.

There are three tests which can be used to judge the utility of the model results. They are:
1) degree of success in simulating the current observed climate
2) degree of mutual agreement among the models in the prediction of the increased CO_2 impact
3) degree of success in the simulation of past climates and in particular the initiation of a glacial.

2. Comparison with Observed Precipitation

The first question we raise is to what degree do the numerical models succeed in describing the current climatic means of precipitation. Several earlier studies addressed this problem. They are among others those of Schlesinger (1989), Schlesinger et al. (1985), Cess et al. (1989), Kellogg and Zhao (1988), and Zhao and Kellogg (1988).

All of the above listed models were reasonably successful in describing in qualitative terms the major geographic features of the precipitation fields. They correctly identify the location and the season of major precipitation maxima and minima. The models are less successful in quantitative aspects of the simulations. The global precipitation totals are by 10 to 25% higher than those observed (Jaeger, 1976). In general, the models overestimate precipitation in the low latitudes especially over the tropical oceans and in the Asian monsoon belt. Considerable discrepancies also occur in the high latitudes. It can be argued that over the oceans the observations are scarce and that the long term climatological averages may be biased. However, departures of considerable magnitude are also observed in the continental interiors, where the observations are sufficiently abundant.

3. Comparison of the Precipitation in the Doubled CO_2 model world

The second test involves the mutual comparison of the differences which the models predict in the precipitation rates and soil moisture between the present climate and that of the world with doubled CO_2 concentrations.

All models predict increased winter precipitation over the tropical oceans and in the high latitudes of the Northern Hemisphere and decreased precipitation over northern Africa. All expect decreased winter precipitation in the southern USA and increased in the central and northern parts of Asia and North America.

Less agreement exists for the continental interiors especially in summer. For example the NCAR model expects a substantial reduction of summer precipitation and soil moisture over central Australia where the GISS, OSU and UKMO models foresee a substantial increase.

While all models expect higher precipitation rates over parts of the tropical oceans, and lower rates over the mid-latitude oceans, the NCAR and the GISS models predict an large precipitation drop in Southern Asia where the GFDL and OSU models expect a considerable increase. In the grain belt of central North America, the GFDL model expects a decrease, while GISS and NCAR models expect an increase in precipitation.

In zonal mean precipitation the GFDL model shows relatively modest differences from the present throughout the year whereas the NCAR model foresees a large drop of precipitation between about 20° N and 30° S in all seasons accompanied by an increase in the high latitudes.

The GISS model predicts that the global annual precipitation rate will increase by 11% accompanied by a temperature increase of 4.2°C and the UKMO model foresees an even larger 15% increase and 5.2°C warming. The relatively conservative NCAR model predicts only 7.1% increase of annual precipitation and a temperature rise of 3.5°C. The intercomparison of individual results of the CO_2 rich world thus shows agreement in some general aspects of the expected precipitation trends, but also some serious qualitative and huge quantitative differences in the predictions made on regional scales.

4. Modelling the Onset of the Last Glaciation

Last of the tests looks at the success of the general circulation models in simulating an onset of a glaciation.

It is abundantly documented that during at least the last 2.5 million years the earth surface experienced enormous environmental shifts between relatively warm interglacial modes and relatively cold glacial ones. The earth is currently in the latest phase of an interglacial mode (Kukla and Matthews, 1972). The shifts are periodical and are either caused or at least modulated by the perturbations of the earth orbit around the sun (Milankovitch, 1941; Hays et al., 1976)

During the last 350 millennia which are best dated by radioisotopes, each culmination or near culmination of the obliquity (tilt of the earth axis toward orbital plane) accompanied by a June perihelion (the closest approach of earth to the sun on its annual orbit) was followed by warmings, whereas about opposite configurations marked the onset of coolings (Kukla et al., 1981). The last glacial started about 115 thousand years ago (Lorius et al., 1989; Kukla, 1980). Rind et al. (1989) used the GISS general circulation model to test whether it would increase the snow accumulation rate or a duration of snow and sea ice covers when forced by insolation shifts known to have taken place during the transition into a cold glacial mode. This was done by increasing 5 times the difference of the monthly mean insolation values between 114 and 116 thousand years ago. The modified insolation field shows increased irradiation to the low and middle latitudes in April. It shows a greatly decreased irradiation in July and August in the high latitudes of the Northern Hemisphere and in October through December in the high latitudes of the Southern Hemisphere. It has to be mentioned that the current insolation trends, although less intense, have a similar geographic and seasonal structure (Kukla, 1982).

Several variants of the experiment were run. None was able to increase the annual snowfall or to maintain snow at the locations where the last few glaciations started. Even when a 10 metres thick ice sheet was inserted over those parts of the continents which were glaciated 18,000 years ago and the surface albedo and temperature were correspondingly adjusted, the model was unable to maintain the ice. Only when the sea surface temperature was cooled down to the reconstructed 18 ka B.P. values did some of the ice in Baffin Island survive the summer. The most puzzling feature of the experiment is that a decrease rather than an increase of precipitation was observed over the areas where the ice in the last glaciation started to accumulate. Most paleoclimatologists assume that an increase rather than a decrease of precipitation accompanied the initial build-up of the ice sheets. It is also sufficiently well documented that the oceans surface in the middle latitudes remained relatively warm during the early stages of the ice build-up (Ruddiman et al., 1980)

It is unlikely that any other equilibrium general circulation model or the use of any other past insolation distribution would succeed in simulating the climate forcing of a glacial onset since the key parameterizations and mechanisms of the GISS model are in principle the same as in other general circulation models.

5. Discussion

It is obvious that the result of the three tests are far from satisfactory. Although the general mean global increase of precipitation is predicted by all the models, the differences among the individual predictions of regional precipitation rates in the CO_2 enriched atmosphere are such that none can be trusted.

The more serious question is raised by the fact that the models are consistently overestimation current precipitation in the low latitudes. Could this mean that some of the processes operating in the real world climate system such as for instance convection are incorrectly represented in the models, forcing them to simulate unrealistically high precipitation? If so, isn't it possible that the predicted precipitation increase in the CO_2 rich world is exaggerated as well?

It is also disquieting that the past insolation input was unable to increase precipitation in the areas where the ice started to build up at the onset of the last glaciation.

Since precipitation is intimately related to the cycle of atmospheric water vapour, which is by far the most important greenhouse gas and since the sensitivity of surface temperature to the atmospheric CO_2 depends closely to the water vapour pressure (Raval and Ramanathan, 1989) it is only natural to ask whether the GCMs temperature forecasts for the CO_2 rich world could not be flawed as well.

It is necessary to keep in mind that only a small fraction of the solar energy received by the earth system is absorbed in the atmosphere. The bulk of it is received and redistributed in the oceans. The oceanic circulation did change substantially in the past. This is extensively documented by abundant paleoclimatic evidence. In our current climate models we presume that the pattern of oceanic circulation did not change in response to a major insolation shift or that it will not react to a major radiative perturbation such as the CO_2 doubling. This is an unrealistic expectation.

The reliable prediction of the future CO_2 rich climates, both in terms of precipitation as well as surface temperature requires significant upgrading of the modelling efforts. Only a new generation of models linking atmospheric and oceanic circulation can lead to reliable climate forecasts.

6. References

Cess, R.D., Potter, G.L., Blanchet, J.P., Boer, G.J., Ghan, S.J., Kiehl, J.T., Treut, H.L., Li, Z.-X., Liang, X.-Z., Mitchell, J.F.B., Morcrette, J.-J., Randall, D.A., Riches, M.R., Reockner, E., Schlese, U., Slingo, A., Taylor, K.E., Washington, W.M., Wetherald, R.T., & Yagai, I (1989) "Interpretation of cloud-climate feedback as produced by 14 atmospheric general circulation models.", Science 245, pp. 513-516

Hansen, J., Lacis, A., Rind, D., Russell, G, Stone, P., Fund, I., Ruedy, R., & Lerner, J. (1984) "Climate sensitivity: analysis of feedback mechanisms", in Hansen, J.E., & Takahashi, T. (eds.) *Climate Processes and Climate Sensitivity*, Washington, D.C., Geophys. Monogr. Ser., pp. 130-163

Hays.J.D., Imbrie, J. & Shackleton, N.J. (1976) "Variations in the earth's orbit: pace maker of the Ice Ages.", Science 194, pp. 1121-1132

Jaeger, L. (1976) "Monatskarten des Niederschlags für die ganze Erde.", *Berichte des Deutschen Wetterdienstes* 139, pp. 1-38

Kellogg, W.W., & Zhao, Z.C. (1988) "Sensitivity of soil moisture to doubling of carbon dioxide in climate model experiments. Part I: North America.", Journ. of Climate 1, pp. 348-366

Kukla, G. (1980) "End of the last interglacial: a predictive model of the future?", in van Zinderen Bakker, S.E.M. and Coetzee, J.A. (eds.) *Paleoecology of Africa and the Surrounding Islands*, A.A. Balkema, Rotterdam, pp. 395-408

Kukla, G. (1982) "Carbon dioxide and polar climates.", Harpers Ferry, W. Va.: U.S. Dept of Energy DOE/CONF 8106214, pp. 237-288

Kukla, G., Berger, A. Lotti, R., & Brown, J. (1981) "Orbital signature of interglacials.", in *Nature* 290, pp. 295-300

Kukla, C., & Matthews, R.K. (1972) "When will the present interglacial end?", Science 178, pp. 190-191

Lorius, C., Barnola, J.M., Legrand, M., Petit, J.R., Raynaud, D., Ritz, C. Barkov, N., Korotkevich, Y.S., Petrov, V.N., Genthon, C., Jouzel, J., Kotlyakov, V.M., Yiou, F., & Raisbeck, G (1989) "Long-term climatic an environmental records from Antarctic ice", in Berger, A., Dickinson R.E., & Kidson, J.W. (eds.) *Understanding Climate Change*, American Geophysical Union, pp. 11-16

Manabe, S., & Wetherald, R.T. (1986) "Reduction in summer soil wetness induced by an increase in atmospheric carbon dioxide.", Science 232, pp. 626-628

Milankovitch, M.M. (1941) "Canon of insolation and the ice-age problem.", Königliche Serbische Akademie, Beograd

Raval, A., & Ramanathan, V. (1989) "Observational determination of the greenhouse effect.", in *Nature* 342, pp. 758-761

Rind, D., Peteet, D., & Kukla, G. (1989) "Can Milankovitch orbital variations initiate the growth of the ice sheets in a general circulation model?", Journal of Geophysical Research 94(D10), pp. 12851-12871

Ruddiman, W.F., McIntyre, A., Niebler-Hunt, V., & Durazzi, J.T. (1980) "Oceanic evidence for the mechanism of rapid Northern Hemisphere glaciation.", Quaternary Research 13, pp. 33-64

Schlesinger, M.E. (1989) "Model projections of the climatic changes induced by increased atmospheric CO_2", in Berger, A., Schneider, S., and Duplessy, J.C. (eds.) *Climate and Geo-Sciences*, Boston, Kluwer Academic Publishers, pp. 375-415

Schlesinger, M.E., & Mitchell, J.F.B. (1985) "Model projections of the equilibrium climatic response to increased carbon dioxide.", in MacCracken, M.C., and Luther, F.M. (eds.) *Projecting the Climatic Effects of Increasing Carbon Dioxide* (DOE/ER-0237) Washington, D.C., U.S. Dept. of Energy.

Schlesinger, M.E., & Zhao, Z.-C. (1989 in press) "Seasonal climatic changes induced by doubled CO_2 as simulated by the OSU atmospheric GCM/mixed layer ocean model.", Journal of Climate 2

Washington, W.M., & Meehl, G.A. (1984) "Seasonal cycle experiment on the climate sensitivity due to a doubling of CO_2 with and atmospheric general circulation model coupled to a simple mixed layer ocean model.", Journal of Geophysical Research 89, pp. 9475-9503

Wetherald, R.T., & Manabe, S. (1986) "An investigation of cloud cover change in response to thermal forcing.", Climatic Change 8, pp. 5-23

Wilson, C.A., & Mitchell, J.F.B. (1987) "Simulated climate and CO_2 induced climate change over western Europe.", Climatic Change 10, pp. 11-42

Zhao, Z.-C., & Kellogg, W.W. (1988) "Sensitivity of soil moisture to doubling of carbon dioxide in climate model experiments. Part II: The Asian monsoon region.", Journal of Climate 1, 367-378

PART II: SEA LEVEL

SEA LEVEL

RHODES W. FAIRBRIDGE
Dept. of Geological Sciences
Columbia University and NASA-GISS
2880 Broadway, New York 10025

SASKIA JELGERSMA
Geological Survey of the Netherlands
P.O.box 157, 2000 AD Haarlem, The Netherlands
and President INQUA Shorelines Commission

ABSTRACT: Global variations of sea level are reviewed in the light of their behaviour in the recent past and possible consequences of a future rise that may be accelerated by the greenhouse effect. The world's coastlines are partly in uplift regions, (glacio-isostatic or tectonic), and here sea level is falling. Most regions, however, are subsiding, so that sea level is always rising; these subsiding regions are also the most heavily populated, so that the threat of a CO_2-induced acceleration of sea-level rise is real. At the present time extreme human hazards due to specific storm and tide surge events are believed to be more serious than secular rise.

Possible steps to mitigate coastal hazards are considered. Geohydrological control of world sea level is now seen to offer an inadequate potential, especially inasmuch as it can do very little to help the heavily-populated subsiding areas. Mitigation of the hazards fall into three categories: (a) long-term planning of refuge and evacuation systems, (b) short-term warning systems, and (c) scientific prediction systems.

Scientific knowledge of sea level/climate relationships, and of geological factors concerned with crustal uplift or subsidence, is still considered to provide a quite inadequate basis for reaching many national and international policy decisions during either the decade of the 1990's or indeed for the 21st century. Recommendations are made here for (a) global expansion of tide-gauge monitoring; (b) increased effort to analyze sea level behaviour in its oceanographic sense; (c) energetic attention to the long-term geological behaviour of the world's coastlines (over millennial periods or more). A series of scientific experiments are needer to check out untested assumptions and a few suggestions are proffered.

1.   Introduction

At a given point around the coastlines of the world the state of sea level may be mechanically

117

R. Paepe et al. (eds.), Greenhouse Effect, Sea Level and Drought, 117–143.
© 1990 *Kluwer Academic Publishers.*

measured by a continuously recording tide-gauge or mareograph. This instrument damps out the short-term oscillations of wave action and its daily variations are usually presented as monthly of annual means. For many years the International Association of the Physical Sciences of the Ocean (IAPSO, a member of IUGG) has maintained an Advisory Committee for Tides and Mean Sea Level to supervise the work of collecting world data, and its archives are maintained at the Bidston Observatory (of IOS), near Birkenhead (U.K.). The relative stability of individual tide stations can be established form time to time by levelling from a national geodetic grid; unfortunately such reference surveys exist for only a few areas in the world and elsewhere the relative stability of gauging stations must be judged on geological reasoning which is bound to be subjective. Satellite monitoring of the sea surface has been initiated, but is not yet at a level of precision that is needed for systematic studies. Accurate geographic (geodetic) positioning of selected tide stations has already begun (Carter et al., 1986).

Scientific analyses of the trends and patterns of sea-level change in different parts of the world were begun more than half a century ago (e.g. Thorarinsson, 1940; Gutenberg, 1941) and, after careful filtering, a mean global figure of 1.2 mm/yr sea-level rise in the first half of the 20th Century was obtained by Fairbridge and Krebs (1962). With much more data and refined techniques the work was expanded two decades later by Gornitz et al. (1982, 1987), also obtaining the mean value of 1.2 mm/yr. Regionalisation of the data analyses was attempted globally by Pirazzoli (1986, 1988), leader of IGCP-200 (an international project for sea-level specialists). He reached the conclusion that each region was so different from the next that no global averaging was possible. Attempts by others to solve this problem (e.g., Barnett, 1984; Aubrey & Emery, 1983) have generated a disparate array of figures which seem, by inference, to support the deduction of Pirazzoli.

Theoretically, if the Earth's surface at the site of a given tide station were perfectly stable and there was no net shift in the global hydrologic balance, there would be no change in annual mean sea level (MSL). Seasonally, due to atmospheric pressure, wind directions, temperature, salinity, tides, currents and other variables, there are appreciable changes of MSL. But if there were no net shift over several decades, we would conclude that there could be no significant climate change. Each tide station record, however, rends to differ, one from another. Studies of closely-spaced stations show considerable coherence, fortunately, so that regional trends can be discerned. Formerly glaciated regions like Scandinavia are systematically rising and deltaic regions such as the Netherlands or Mississippi delta are systematically sinking. Furthermore, within a given region, the minor fluctuations tend to be in phase; thus in the southeastern Pacific, every few years, MSL rises and fall corresponding to the succession of El Niño/La Niña climatic/oceanic phenomena.

These large interannual variances, while often detected over broad areas, are not by any means synchronous over the whole globe like the water level in one's domestic bath. Not only are there regional oceanographic factors, as noted above, but also there are hydrologic factors (on land) which affect the annual exchange cycle. The world's rivers (major ones) now have gauging stations, so that one can judge how the annual discharge varies, not only from one region to another, but also for the whole world. Although the number of stations is still not adequate, it seems likely that the annual variance in river discharge is of the order of 5% of the mean (about 40-50 x 10^6 km^3). The progressive damming of many rivers and withholding of normal discharge probably results in a progressively decreasing supply of water from some rivers, whereas increasing melt-rates of mountain glaciers from some others may augment the hydrologic supply. Calving rates form the great Antarctic and Arctic ice caps do show some interannual variance but in the present century suggest no secular trend.

A familiar question is, then: what is the meaning of MSL as a geohydrologic monitor? The simplistic answer has been to average all the world's tide-gauge results, correct for regional concentrations, filter out apparently unreliable sites, adjust for tectonic factors, and so on. But to provide a realistic answer we need far more data, in four categories: (a) sea level history, (b) its contemporary behaviour, (c) fluvial input, and (d) glacial-melt input. Related information concerns some attention to groundwater or artesian seepage, primary production of "juvenile water" from mid-ocean ridges and volcanic sources, rates of overturn as controlled by geostrophic wind systems and their astronomic forcing, and finally by tidal and current systems. It seems like an endless list of desiderata, but all are somehow involved.

2. Sea Level and the Greenhouse Theory

In his "State of the World 1988", Lester Brown wrote (p. 17):
"Somewhat more predictable results of a hotter Earth is a rise in sea level. As the water in the ocean warms it will expand accordingly. In addition, the warming will reduce the amount of water trapped in glaciers and ice caps. Projections by the Environmental Protection Agency (EPA) show a rise in sea level by 2100 of between 1.4 and 2.2 m (4.7 to more than 7 feet). This would hurt most in Asia where rice is produced in low-lying river deltas and flood plains. Without heavy investments in dikes and sea walls to protect the rice fields from salt-water intrusions, even a relatively modest 1 m rise (about 1 feet) would markedly reduce harvests ..."

This sort of analysis is commonly taken as the "conventional wisdom" of the contemporary scientific establishment. The original theoretical argument presented at about the start of the 20th century by Svante Arrhenius claimed that a rise or fall of atmospheric CO_2 will affect radiation and the thermal characteristics of the Earth's troposphere.

This theory is good science, and the measured rise of CO_2 during the 20th century has been persuasively paralleled by a rise in mean global temperature. A rigorous interlocking of cause and effect is still not proven, however, and it has been reasoned that there is (a) a natural rise of global temperature following the recovery from the Little Ice Age (roughly 1300-1850); and (b) the expected temperature rise should be much greater, unless explained as a buffering effect by some negative feedback processes. Thus, there are still important questions, as yet unanswered.

Several things, nevertheless, seem to be perfectly clear. There is already an unusually large CO_2 build-up in the atmosphere, exceeding the maximum of the last interglacial (as shown by air bubble analysis of Antarctic ice), and a doubled level is expected by the middle of the 21st century. Secondly, the rate of fossil fuel consumption, coal, oil, natural gas) cannot be effectively reduced, at the present state of technology, without impoverishing the entire globe. As third world countries gradually industrialize, the trend is likely to be up, *not* down. Thus we must conclude that a high CO_2 atmosphere is here to stay, at least for several centuries.

This fact, of a semi-permanent high CO_2-level, may after all be a blessing in disguise, because astronomic trends suggest that the Earth should be headed into a serious cooling phase (Fairbridge and Shirley, 1987). A second and even more catastrophic Little Ice Age may thus be serendipitously avoided. What is still impossible to predict is how much the warming tendencies will be able to offset the cooling trend.

Some degree of warming, such as experienced during the Norman-Viking period (AD 900-1300) could be beneficial to many high latitude regions such as Canada, Iceland, Scandinavia and northern U.S.S.R. The mean temperatures in northern lands were up to 2.5°C above those of today less than one millennium ago. Mean sea level was approximately 0.5 m higher than today and this is a good, reliable analog that can be used in our planning. It is not without interest that there was a *natural rise* in the atmospheric CO_2 in the 11th century that was at a rate comparable with that of the 20th century. A similar rise since the Little Ice Age would be expected even without an anthropogenic contribution. We might suspect that there may be some sort of unanticipated CO_2 sink that is in operation.

3. Multiple Forcing of Sea-Level Changes

Any world-wide change of sea level is defined as eustatic. (The Random House Dictionary, 1987, defines EUSTASY as "any uniformly global change of sea level that may reflect a change in the quantity of water in the ocean, or a change in the shape and capacity of the ocean basins." (Spelling of the Greek root ($\sigma\tau\alpha\sigma\iota\varsigma$) calls for spelling of eustasy, as in isostasy, ecstasy ... and *not* "eustacy" favoured by some writers).

As clearly established by Kuenen (1950) there are three principal mechanisms for changing sea level eustatically:

(a) By changing the shape, or capacity, of the oceanic basins ("tectono-eustasy"); besides the tectonic deformation of the "container" (the ocean basins), there is a second aspect, the Archimedean displacement of water by partial filling of the container with sediment and volcanic material, as along the mid-ocean ridges, a process called "sedimento-eustasy" (Fairbridge, 1961); a third aspect is "hydro-isostasy" the sluggish vertical adjustment of the Earth's crust to the load of water added or removed; these are all very slow processes and irrelevant for studies of the 20th century changes.

(b) By changing the geohydrologic balance, this is to say, the ratio of total volume of water in the ocean versus that on the land (mainly as ice, "glacio-eustasy", but partly as groundwater, in vegetation, etc.).

(c) By altering the total mass of water, in any state, on the Earth's surface, such as by addition of "juvenile water" from volcanoes and mid-ocean seeps, or by mineralogic conversion; these are also extremely slow processes, so that only category (b) is relevant for short-term studies. In principle, these are well-established processes, but quantitatively they are very difficult to monitor at a level of high precision, so that monitoring should be made a high priority for concerned international organisations.

Some of the qualitative relationships between these varied eustatic potentials are illustrated in Figure 1. Besides eustatic disturbances of the sea surface , form place to place around the world there are dynamic oscillations and deformations of the ideal surface of the geoid that tend to be short-term and more or less seasonal or cyclic.

On Figure 1 two heavily-outlined interacting conditions are indicated by (a) a rectangle (Global Mean Sea Level) and (b) a circle (Local MSL). Below, left and right, are shown processes that affect the land surface. At the bottom left are indicated Tectono- and Sedimento-Eustasy and Hydro-isostasy (discussed above), all having a long-term effect on global MSL. To the right is indicated the most significant long-term effects that influence local MSL, the subsidence or rise of the land surface; these processes include neotectonics, i.e., subsidence crustal motions, up and down, and tow processes that are both only one-way, downward, i.e., subsidence due to sediment compaction; this can be due to natural loading and man-made loading, such as buildings, highways, dikes, and fluid withdrawal of ground water, oil and natural gas.

It is strictly the Local MSL that interacts, both ways, with Coastal Landforms and Mad-Made Structures (see blocks on right). From varied points of view, from the individual property owner to the regional or national authorities who are responsible for harbours, dikes, power stations and urban services, this is a very important fact that is sometimes forgotten in the face of intense discussions on the questions of global MSL, most of which are totally secondary to the fast-acting processes that affect local MSL.

The upper parts of Figure 1 are entirely devoted to questions of *causality*, which is divided in two. On the left is exogenetic energy, i.e., for the most part external to the various systems and processes of the atmosphere, hydrosphere, biosphere and pedosphere. On the right is endogenetic energy which concerns the internal or inherent systems and dynamics applicable to the spheres mentioned above.

Placed at the center, top, is CLIMATE CHANGE (circle), which includes both oscillatory change and longer-term, secular change. This change can be scientifically recorded on any scale from monthly to the million-year level.

To the left is indicated SOLAR ACTIVITY, the principal variable. Satellite monitoring has now shown that solar radiation is not a physical constant as formerly believed, and the extraterrestrial cosmic-ray-engendered C-14 isotopic flux that is modulated by solar energy has now been measured (in tree rings back for more than 8000 years. The Sun's activity is modulated by its own angular momentum which is constantly changing due to the changing gravitational attractions and torques applied by the planets. Of direct and immediate interest to the inhabitants of planet Earth is that the next low solar activity cycle is due to begin about AD 2990 (Fairbridge & Shirley, 1987). To what extent this cooling trend will be offset or reversed by the greenhouse effect of rising CO_2, CH_4, etc., remains to be seen.

Two principle forms of solar radiation, electro-magnetic and particulate, (solar winds),

122

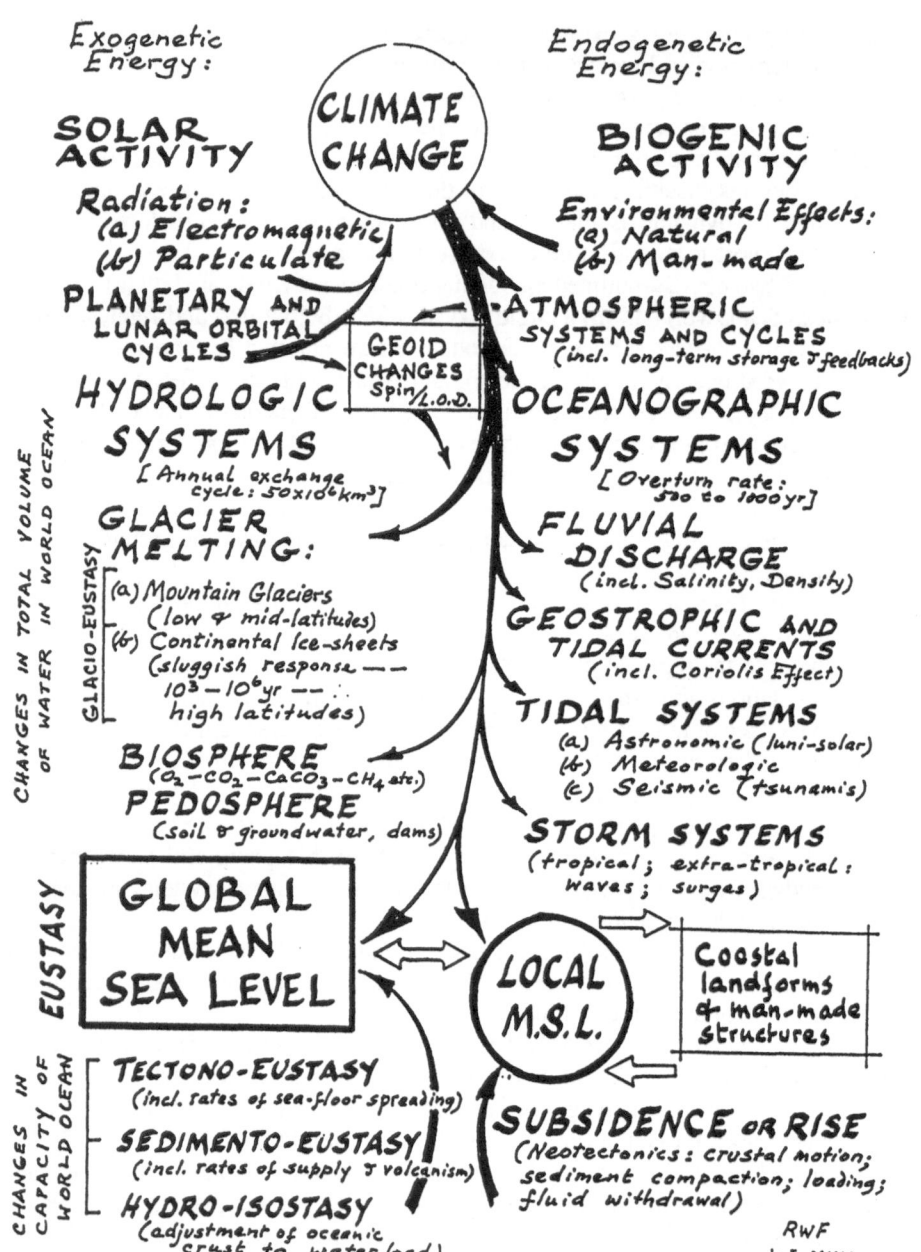

Figure 1. A qualitative flow chard, illustrating the interrelationships between CLIMATE CHANGE (top, center) and GLOBAL MSL (bottom right), with their respective causative and feedback systems.

are linked to the Earth's atmosphere, both through chemical changes (e.g., ozone) and circulation effects. Furthermore, there are constant changes in all of the planetary orbits, which not only affect the Sun, but also the Earth-Moon system. The latter is influenced through the Milankovitch periodicities (on time scales of 100,000-200,000 yr), but also by very short-term effects, influencing the planet's spin rate and L.O.D. (length of day). Both change the shape of the geoid (see box, center): e.g., ice loading near the poles shifts the moment of inertia and raises spin rate. A rise of spin rate causes MSL to rise near the equator and fall in polar regions (and vice versa). It may be commented that geoidal (or geodetic) shifts of sea level cause regional changes in opposite senses but do not alter either the total volume of water of the capacity of the ocean basins. The expression "geoidal eustasy" coined by Mörner (1980, 1981) is a confusing contradiction in terms.

Climate change affects primarily the *hydrologic systems* (evaporations, precipitation, runoff, river discharge, subsurface storage, and vegetative cover). In a most dramatic and important way, these involve first of all changes to the mass-balance dynamics of glaciers; the low and mid-latitudes are most sensitive to changes over a few years. Long-term warming leads to melting, but some negative feedback may occur, because warming of offshore waters often leads to increased precipitation of snow, so that some glaciers may advance while others retreat. High-latitude ice sheets are extremely sluggish, for the most part; over the past 2-3 million years, during interglacial warming cycles of 10,000 years or more, there has *never* been a general deglaciation of Antarctica of Greenland. The albedo of ice and snow at high latitudes is extremely high and there is an appreciable thermodynamic problem involved in raising the internal temperature of about 27×10^6 km^3 of ice from a mean of about -17°C to the melting point during brief summer seasons; during the 6-months of winter darkness, the process is of course reversed.

The collective term for all glacier melt/growth effects on global MSL is glacio-eustasy, which with *steric changes* (thermal expansion or contraction of the water body) are the major controls of the total volume of water in the world ocean. A minor contributor is the net variance in fluvial discharge, but its significance has not been exhaustively researched. Consideration of the long-term history of eustatic rise and fall due to the 100,000 yr alternation of glacial and interglacial intervals (controlled by the Milankovitch orbital factors), will show that the glaciers involved are very largely the mid-latitude ice sheets of North America (which reached 45° N), Scandinavia, Alps, etc. Thus it should be realized that although those mid-latitude ice sheets wasted back very rapidly during each rapid global warming that marked the interglacial intervals, the high-latitude ice sheets were very little involved except near the margins. It is therefore highly unrealistic to postulate that a global warming due to rising CO_2 would cause wholesale melting or "runaway surges" of the great polar icecaps.

The ice volume required to raise global sea level by 2.5 mm is approximately 1000 km^3. The world's smaller ice caps and mountain glaciers (including all of Arctic Canada, Iceland and Spitsbergen) amount to no more than 200,000 km^3, sufficient to raise the ocean level by 50 cm. Many of these lie within the Arctic Circle, where the same conserving mechanisms apply as for Antarctica and Greenland. With contributions from the marginal areas of those continental ice sheets, the time taken to raise world sea level by 50 cm is of the order of

300 years. This estimate is based on the last known warming cycle (the "Little Climatic Optimum" of Lamb, 1977) that lasted from about AD 800-1300. Mean summer temperature reached about 2°C above today's in latitudes 50-60° N, but higher in the far north so that the Vikings were able to colonize southern Greenland.

Referring once more to Figure 1, the right-hand side summarizes the endogenetic energy processes, among which (certainly at the present time) *biogenetic activity* is of paramount significance. Its effects are divided in two categories:

(a) Natural Biogenic Activity: CO_2 is primarily controlled by the photosynthetic activity of plants. In the ocean, the principle CO_2-control is by the phytoplankton. These flourish particularly in glacial periods (and cool cycles in general), because the equator-pole thermal gradient is high, wind systems are strengthened and accelerated upwelling brings enhanced nutrients to the phytoplankton. Atmospheric CO_2 is therefore low in glacial times; this is proven by analyses of trapped gas bubbles in ice cores. On land, where the plants are annuals, as in grasslands for example, no net CO_2 shift is involved from year to year, but when grasslands shift to forest (as during a glacial/interglacial transition) there is a net change to carbon storage, in the wood, which provides a negative feedback at a time of general CO_2 rise.

A second but equally important greenhouse gas is CH_4 (methane). It is liberated by termites, ruminant cattle, and other organisms, but by far its most important source is in the world's coastal wetlands, the salt marshes and mangroves. Studies of Antarctic ice cores by the French Grenoble Laboratory (Jouzel et al., 1989) show that the atmospheric loading of CH_4 rises dramatically during eustatic-high intervals and drops abruptly when sea level falls. This fluctuation reflects the expansion of coastal wetlands during high sea level and stabilization periods. The rise in CH_4 is a positive feedback mechanism, expanding the natural greenhouse effect established by the interglacial climatic mode, when wind systems are weaker, upwelling is reduced and there is a net shift to CO_2 into the atmosphere.

(b) Man-made (Anthropogenic) Environmental Effects: mankind, with its advanced technology and often-blind interference in the Earth's natural systems, has undoubtedly achieved some remarkable environmental changes on our planet. Much of this activity is directed towards the improving of mankind's sustenance and creature comforts, so that extremely careful examination of all of its aspects is urged before embarking on any major corrective strategies. Several aspects of anthropogenic pollution are highly undesirable (e.g., urban smog, water and foodstuff contamination, acid rain); such problems should unquestionably receive high priority for national and international action. Some other aspects of anthropogenic interference with the environment need more cautious consideration and cost-benefit ratios need to be carefully examined.

Following the Biogenetic Activity column (on Figure 1) follow two closely related systems, each characterized by cyclical changes, some forced by the exogenetic agencies, some dynamic and intrinsic to the system itself. Of the two, the atmosphere is clearly mercurial and rapidly changing but, even so, it possesses the capacity to store energy (such as via the ocean or via glacier ice or via artesian water withdrawal). The ocean has been described as "the great flywheel of the global heat engine", inasmuch as it constitutes a sink or a source for heat and materials that may have an overturn rate or 500 to 1000 yr. Climate change also leads to variations in evaporation/precipitation rates and thus to Fluvial Discharge (and thence to shifts in the ocean salinity and density relations). Both of these affect sea level in general, while the location of a river mouth often plays a remarkable role in the pattern of local MSL rise and fall.

Less well known are the influences of climatic changes on geotropic and tidal currents. These are not easy to measure and for past changes a large amount of interference or proxy evidence must be used. Nevertheless, it is apparent, as mentioned above, that changes in the global temperature gradients, as well as displacements of the principal atmospheric pressure cells by the growth of ice caps (with their high albedo), must result in shifts of some geostrophic circulations and changes in upwelling patterns. Palaeontologists working on the microfauna from deep-sea cores have been able to contribute greatly to this proxy evidence of such changes.

Tidal systems fall into three categories: (a) astronomic (luni-solar); (b) meteorologic (the "inverted barometer effect"); and (c) seismic (tsunamis). The first category is a particularly strong influence on the global and local MSL effects. Not only on the diurnal, fortnightly and annual levels, considerable evidence has accumulated to support the theory of Pettersson (1912, 1930) that the 18.6 yr period (nodal and nutation), with its long-frequency harmonics, such as 93, 186, 558 yr, etc., play an important role in sea-level variance; theoretical and observational support have been provided by Currie (1976) and Wood (1985), and have been recently introduced into the general astronomic framework by Fairbridge (1989). Although the 18.6 period is well-established in tide spectra, it is not strictly a tide-raising force, but influences the latitude of tidal potential and thus the coastal current systems, even the Gulf Stream. The principal long-term tide cycles are 1.1318 yr and 18.03 yr (The Saros" period); the latter and its various beat frequencies reaches its peak at certain perigee-syzygy perihelion epochs. Wood (1985) has shown that during the last 1000 yr every instance of major dike breaks and coastal flooding along the southern North Sea coasts were in this category (see Table 1). In other words, coastal hazards with major loss of life (other than man-made breaks such as in time of war) could have been astronomically anticipated, although not specifically predictable because they must also be related to a coincident crescendo of several "unfortunate" processes. Secondly, and importantly from the global warming viewpoint, these were *all* "event" disasters. No one is drowned by a 1 mm rise of sea level.

(b) Meteorologic tides: these can be the predictable result of the passage of a low pressure system across a coastline, one of the essential "unfortunate" coincidences mentioned above. Inasmuch as the peak tide cycles will recur fortnightly for 2-3 days over several winter months and the low pressure systems pass over about once every 5 days, the chances of a disastrous coincidence rises appreciably. Augmenting the simple "inverted barometer effect" are on-shore winds and surge phenomena, which tends to localize the most hazardous spot.

(c) Tsunamis or "tidal waves": these are not luni-solar but are triggered by seismic action, submarine landslides, and volcanic explosions. While generally insignificant in the Atlantic, they are constant in the North Pacific. Over the last 1000 years, it has been estimated that

Table 1.

MILLENNIUM PROJECT: COASTAL FLOODING (EXTRA-TROPICAL)

Date (Greg)	Old Style	
1099.846(Nov 6)	Nov 12	Southern North Sea "St. Martins". 100,000 drowned.
1134.731(Sept 25)	Oct 1	Start of late medieval sea-level rise.
1164.112(Feb 10)	Feb 17	"First Julian Flood"; 20,000 drowned.
1219.025(Feb 10)	Jan 16	"First Marcellus Flood"; 36,000 drowned.
1287.950(Dec 9)	Dec 14	"Lucia Flood"; 50,000 drowned; opening of Zuiderzee; also UK floods.
1334.874(Nov 15)	Nov 23	"Clemens Flood"; many victims.
1362.022(Jan 8)	Jan 16	"Second Marcellus Flood". "Great drowning".
1362.663(Aug 31)	Sept 8	Southern North Sea.
1373.748(Oct 1)	Oct 9	"Dionysius Flood"; town of Westeel lost.
1375.750(Oct 2)	Oct 10	German coast.
1377.767(Nov 7)	Nov 15	Zeeland flood.
1393.009(Apr 23)	May 1	Schleswig-Holstein flood.
1421.854(Nov 9)	Nov 18	"Elizabeth Flood"; Maas River changed course; Javerland flood.
1464.877(Nov 16)	Nov 25	"Catherine Flood" (Friesland); "Winkel Flood"; Javerland flood (II).
1509.712(Sept 17)	Sept 26	Opening of Jadebusen, N. Germany.
1511.016(Jan 7)	Jan 16	"Antonius Flood" or "Ice Flood" (sea-ice invasion)
1530.843(Nov 5)	Nov 15	Zeeland flood "St. Felix"; 400,000 drowned.
1555.717(Sept 20)	Sept 30	Thames flood (U.K. Whitehall inundated).
1570.805(Oct 22)	Nov 1	"All Saints' Day Flood", Friesland; 50,000 drowned.
1606.025(Jan 10)	Jan 20	Severn floods, U.K., and Holland.
1634.775(Oct 11)		"Second Great Drowning"; 90,000 drowned in Germany; 11,000 in Denmark.
1663.985(Dec 17)		Thames flood (U.K. Whitehall inundated)
1717.984(Dec 25)		"Christmas Flood"; 18,140 drowned.
1723.181(Mar 6/7)		Netherlands flood; also Mass. to Maine
1825.090(Feb 3/4)		"Greatest North Sea Flood" of 19th century.
1839.956(Dec 15)		Cape Cod - Boston, Mass.
1851.287(Apr 14/16)		Massachussetts and N.H.
1869.762(Oct 5)		Maine and Nova Scotia
1875.870(Nov 15)		Thames flood (U.K.)
1894.060(Jan 22)		Cape Hatteras, N.C.
1906.197(Mar 13)		Southern North Sea.
1914.887(Nov 20)		Quebec, Canada.
1928.901(Nov 25)		Thames flood (U.K.), Belgium, Holland
1931.173(Mar 4/5)		New Jersey to Halifax, N.S.
1953.085(Jan 31/Feb 1)		"Greatest Flood for Several Centuries" (U.K. + Netherlands)
1954.975(Dec 22)		Maadesiel Dike-break, North Sea
1962.129(Feb 16/17)		"Second Julian Flood", southern North Sea.
1962.178(Mar 6/7)		Mid-Atlantic (US) "Worst in History".
1978.029(Jan 9/12)		California, NE-US, Belgium & U.K.
1980.814(Oct 25)		Japan and US mid-Atlantic.
1981.945(Dec 11)		Venice (Italy), California.
1983.079(Jan 27/31)		California-Oregon (+ El Niño); mid-Atlantic.
1987.001(Jan 1/2)		California; US east coast.

one tsunami every 100 yr generates a shore wave of the order of 30 m (one hundred feet); although a localized event, in a heavily populated area it could be totally catastrophic.

Finally, on Figure 1, comes the question of storm systems. A rising CO_2 level is predictably accompanied by a mid-latitude warming of the sea-surface temperature (SST), although it has not yet been detected. This should increase the frequency and intensity of tropical cyclonic storms (hurricanes, typhoons). These are liable to result in the same sort of tidal surges and "event" disasters mentioned above. They are a particular threat to the low-relief, subsiding coasts characteristic of the great deltas like the Ganges-Brahmaputra (in Bangladesh) and the Mississippi, as well as almost the entire Atlantic and Gulf coasts of the United States. Due to surges and funnel-shaped estuarine enhancements, local sea level has been known to reach more than 10 m above high spring tide.

Concerning the extra-tropical (polar frontal) storms, most of the same surge and tidal enhancement apply to create another mid-to-high latitude hazard. The coastal disasters listed on Table 1 all fall into this category. It may be readily perceived that the great majority of them occurred in mid-winter, in the months just before and after the perihelion. If their incidence is reviewed in the light of eustatic-steric rise and fall, it can be readily appreciated that the 12th and 13th century breaks, including the opening of the Zuiderzee, came near the culmination of the "Little Climatic Optimum" high sea level (about 50 cm above today). However, with the climatic deterioration of the 14th century (beginning of "Little Ice Age"), there was a systematic southward shift of the jet stream and the frontal storm tracks. Swiss glaciers began to readvance in the late 13th and early 14th centuries. These were accompanied by a catastrophic fall in temperature, so that in the decade 1305-1315 approximately 50% of the population of Bohemia died of starvation.

A dramatic rise in frequency of storms occurred at the same time as the Little Ice Age eustatic-steric *fall* of MSL (Bakker, 1957). This apparently paradoxical contrast underlines the principal message in this review: that it is not the slow, secular eustatic change of sea level that is hazardous, but the frequency and amplitude of discrete "event" storminess. During a cooling cycle there is a *decrease in frequency* of extratropical storms in N.W. Europe with an increasing influence of the Scandinavian or Central European high pressure systems. This fluctuating pattern of North Sea storminess and sea level oscillation appears to persist through the whole Holocene Epoch (Duphorn, 1976; Linke,), being accompanied by variable peat accumulation and river discharge (Jelgersma, 1966, 1979; Van de Plassche, 1982). In Arctic Canada, it is found that the formation of sea ice during increasing HP (cold) cycles prevents the building of beachridges, which are thus favoured by the warm, open water, stormy cycles (Fairbridge and Hillaire-Marcel, 1977).

It may be appropriate at this point to mention once more the dramatic regularity of storm-generated beach ridges in both high latitudes (Alaska, Canadian Arctic, northern Norway, the Baltic) and in low latitudes (e.g., Gulf of California, western Florida, West Africa, Western Australia). While those in the high latitudes tend to conform to long-period lunar/solar periodicities (just as do the spectra in the ice cores), the ridges in lower latitudes display higher frequencies, often approximating the 11-yr solar cycle. A tentative chronology has been presented (Table 1, in Fairbridge, 1989, p. 87).

4. Neotectonics, Subsidence, and Compaction

As indicated in Figure 1, the major factor in characterizing local MSL over long periods is likely to be the deformation of the Earth's crust, a process known as "neotectonics" (a term coined by Obruchev in 1948) definition and history: Fairbridge, 1981). Attention to this subject has been stimulated by the seismologists and the increasing interest in earthquake prediction; indeed, the study of neotectonics is almost on the way to becoming an experimental science (Vita-Finzi, 1986), and the IUGG maintains an active Commission on Recent Crustal Movements. Such movements are of two sorts: first are those associated with abrupt shifts of the crust in vertical or horizontal directions, and accompanied by powerful earthquakes. During the 1960's, earthquakes in Chile resulted in 1 to 1.5 m subsidence of a coastal belt 600 x 30 km in area (Weischat, 1963). Local MSL rose by 57 cm during the 1908 earthquake in Messina, Italy.

The second sort are associated with secular and essentially imperceptible changes, like the gradual subsidence of the western Netherlands at about 1.5 mm/yr, or the possible isostatic uplift of Scandinavia, where the Gulf of Bothnia is rising at up to 9 mm/yr; the eastern Hudson Bay is rising at the same rate. These contemporary uplift and downwarping data are determined from tide-gauge records

Most serious from the point of view of large coastal populations are the slow but inexorable subsidences associated with deltas and actively accumulating geosynclinal belts like the Gulf of Mexico or eastern United States. In the Mississippi delta, sustained rates of subsidence exceeding 10 mm/yr are measured and many points along the east coast are subsiding at more than 3 mm/yr.

In all regions of high sedimentation, the eroded material carried down by rivers is on the one hand unloading the Earth's crust in the watershed region, and on the other hand transferring that load to the deltas and offshore region. The two regions are very unequal in area, the stream catchment areas being much larger than the deltas. A map (Figure 2) of the principal deltas of the world shows that they are widespread, but for the most part absent from the young orogenic belts (a useful reference is SEPM sp. publ. 15 "Deltaic sedimentation, modern and ancient", ed. Morgan, 1970).

An important feature of freshly deposited muds (clay content > 40%) is that they carry over 80% of interstitial water, i.e. corresponding to their porosity. De-watering begins as soon as fresh sediment accumulates on top and compaction continues, at am exponentially decreasing rate over several million years. Thus it is almost true to say that all sedimentary coasts are subsiding, if nothing else, due to compaction. Besides simple de-watering, there are also mineralogical changes, the organic decay and oxidation of peats, and consolidation due to vibration (seismic or wave pounding in a shallow gulf).

The average "Wilson-cycle" sequence of more or less continuous sedimentation lasts about 100-200 million years and, as a result of the crustal heating associated at the beginning of each cycle with crustal rifting, there will be secular cooling and tectonic subsidence throughout the cycle. Thus, a typical "trailing edge" or Atlantic" type of geosyncline will be marked throughout its history by tectonic subsidence and sediment compaction. Disregarding essentially short-term eustatic oscillations of sea level during this cycle, the net effect

Location map of coastal instability

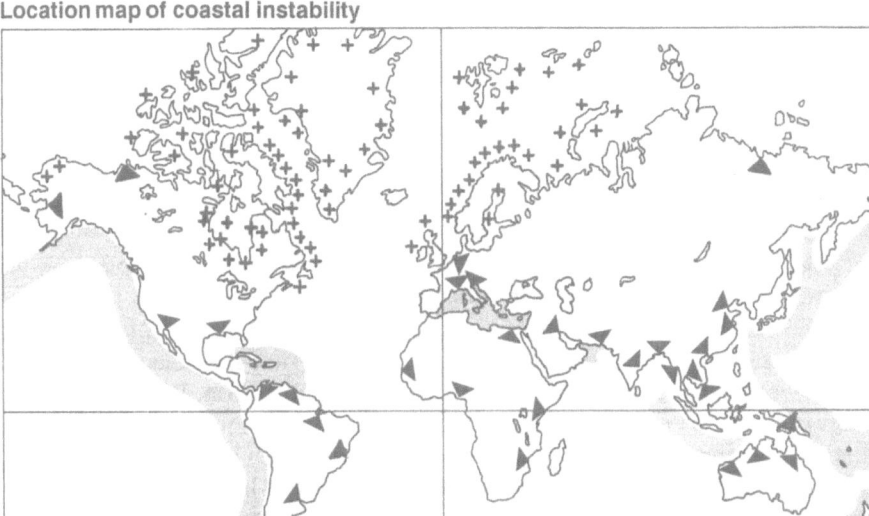

▶ Large delta
+ Isostatic uplift
   Areas of plate tectonics

Figure 2. Sketch map showing the location of the principal deltas of the world. Note that they are principally distributed around the "Atlantic-type" coasts and very little along the currently active orogenic coasts.

is of a constantly rising MSL. Rates of subsidence from this source are normally of the order of 1 mm to 1 cm per century.

The point justifying this brief excursion into historical geology is precisely this: the normal behaviour of local sea level is to rise. We should keep this in mind when considering the evidence of world wide tide gauges. If we exclude glacio-isostatic rising areas, and youthful orogenic coasts, we must expect most tide gauges to present a signal of rising MSL, for most of the time. In geological terms that means something like four billion years.

In the 20th century it is necessary to consider in addition the anthropogenic or man-induced subsidence in deltaic areas and coastal lowlands. Loading by man-made structures, such as building highways and dikes, greatly accelerates normal de-watering and compaction. At La Guardia Airport in Flushing Bay, New York, the natural intertidal organic silt was built up by fill material to a thickness of 7.5 m. During 25 yr the surface subsided by 2.5 m in response to this loading.

Vibration, such as occurs naturally under seismic action, greatly accelerates subsidence along highways and railways. This is exacerbated where the clays are thixotropic, a condition often associated with former marine clays, with a high salt content, in postglacial uplift areas, e.g., in Norway, Sweden, and Quebec. The vibration leads to a sudden liquification of the clays causing spectacular landslides.

Lowering the groundwater by drainage or pumping during dry spells, or during an interval of lowered sea level, can also lead to important subsidence. In the polder-lands of the Netherlands, close attention is paid to this groundwater withdrawal rate in order to avoid potentially catastrophic subsidence.

Following the construction of dikes in the Netherlands, formerly water-saturated sediments have to be leached of salt and reclaimed. Considerable research was done by Bennema et al. (1954) on the resultant compaction. Clays with 35% of their particles less than 2 microns have a porosity of more than 80%. After reclamation they will be compressed to about 50% of their original thickness over a 100 yr period. Peats will be compressed more: to about 1/9 of their original thickness. On the other hand the Holocene sands (either in former river beds or dunes) are subject to very little compaction. Where highways and buildings can be sited over these sandy foundations a great economy can be achieved. Some striking examples of differential compaction can be seen in the Netherlands in peat lands that were reclaimed many centuries ago. Where the peat had been previously eroded in places by creeks and the creek beds were later infilled by sand and silt, due to the differential compaction, these former gullies eventually became ridges that stand up to 2 m above the surrounding landscape underlain by peats and clay. On the other hand, some creeks became filled in with peaty mud; in this case in only 75 years a local compaction of 2.5 m was measured. In the adjacent polder-land the compaction amounted to only 1-1.5 m during the same period.

A serious man-induced geological hazard is land subsidence due to the excessive withdrawal of groundwater in areas underlain by youthful sedimentary basins. Detailed information about this problem can be derived from the *Proceedings of the International Symposia on Land Subsidence* and the *UNESCO Guidebook to Studies of Land Subsidence Due to Groundwater Withdrawal.*

Surface and subsurface layers of subsiding basins in coastal areas consist of young, unconsolidated material deposited in alluvial, lacustrine and shallow marine environments. Depending on their lithology, there permeabilities may vary from very low to high. Accordingly, subsurface strata generally consist of confined and semi-confined aquifers consisting of gravel and sand of high permeability and low compressibility, alternating with aquitards consisting of clays of low vertical permeability and high compressibility. Groundwater withdrawal will cause the artesian head to decline resulting in compaction due to increased effective stress. The compaction of these aquifers is immediate and is chiefly recoverable if fluid pressure is restored.

This compaction is, however,, small compared to the compaction that occurs in the aquitards. The decrease in head in the aquifer creates a hydraulic gradient from the clays of the aquitards to the aquifers. The resulting compaction is then large and irreversible. Summarizing, it may be concluded that the compaction in the aquifers occurs directly if pore pressures decay but the amount is small. Compaction in the aquitards is much more time-dependent due to the slow vertical escape of water but the ultimate compaction is large and mostly permanent.

The problems associated with subsidence due to groundwater withdrawal are various. The most important hazard in coastal areas is that of salt-water intrusion in there aquifer and the flooding of human settlements due to disturbed drainage.

In the Houston-Galveston region of Texas the subsidence of the land surface amounts to as much as 2.8 m since 1915 in the area of highest groundwater withdrawal. There is a close relation between the observed decline of the water table in the wells and the observed subsidence of the benchmarks. The rise of local MSL has been quite exceptional. Because of increased subsidence, some areas have become subject to inundation by normal tides and even larger areas may be subject to catastrophic flooding by hurricane tides.

In Japan some 59 areas, most of them situated in the coastal zone, are subject to more or less serious land-surface subsidence due to groundwater withdrawal. One of these areas is the Nobi fluvial and deltaic plain where groundwater withdrawal has caused a surface lowering of the land between 20 and 140 cm in 15 years. Due to this subsidence, 248 km^2 of the area is now below mean sea level. In the northeastern part of Tokyo the over-pumping of the aquifers has resulted in a maximum subsidence of 4.57 m between 1920 and 1975 and a groundwater lowering of 60 m. Due to this subsidence the area has become a high risk for flooding during the tidal flood of a typhoon. Consequently, dikes had to be built to protect the area.

In Taiwan, land subsidence in the Taipeh river basin amounted to 2 m in 22 years of heavy groundwater withdrawal. This situation has hampered the normal drainage by heavy rainfall and causes flooding, from time to time. Finally, the example of increasing land subsidence in Bangkok is a classical one. The excessive withdrawal of groundwater has resulted in the last 20 years in a surface lowering of 12 cm/year.

Subsidence due to gas and oil production is also well known. A good example is the Wilmington oil field in California where a rapid increase in subsidence occurred in the Wilmington/Long Beach harbour area. Between 1928 and 1968 a subsidence of 9 metres took place along an axis corresponding to the crest of the oil-bearing structure. This subsidence has been controlled and partially reversed by fluid reinjection of the wells.

It may be concluded that the observed amounts of subsidence due to groundwater withdrawal are much higher than those predicted for future sea-level rise.

5. Planning and Testing of Data

Scientific progress always requires mote data acquisition, the questioning of assumptions and the analysis and testing of the accumulated information. Because of its global and multi-faceted nature, the phenomenon of sea level change demands multi-disciplinary studies. Any person, institution, or commission that innocently imagines that it can reach a well-balanced judgement on such a topic unfortunately suffers either from incredible naivety or from an arrogance tinged with hubris. In approaching these challenges we all need to be as open minded as possible and to take scientific modesty as a *sine qua non*.

The above homily is not intended in any way to suggest pontification, but rather as a guideline towards the problems we face. No "crazy scheme" should be rejected out of hand

and no amateur should be excluded because he or she does not possess gilt-edged credentials. Often it is the old salt-water fisherman who knows more practical things about the tide than does the professor of particle physics.

The data-gathering, testing and experimentation ideas need to be organized in several areas, and in different areas. Each plan need to be examined on a basis of team selection and technical qualifications required. Economy of talent and funds must also be kept constantly in mind. The following organization into categories of data acquisition and of experimental studies may be worth consideration:

Category 1. *Geographical and Geological Classification of World Coastlines.*

Every coastal authority, down to the level of the individual riparian householders, needs to be informed about the "type of coast" that is their concern. Information about projected rise of MSL may be totally irrelevant, for example, to the harbour-master at Post-de-la-Baleine (Great Whale River, Hudson Bay) where the mean glacio-isostatic rise of the Earth's crust is 9 mm/yr. What if global MSL rises at 5 mm/yr? He should not be greatly concerned. In contrast, the situation in a suburb of New York, where the land is sinking at more than 10 mm/yr, is totally different.

Accordingly, a global cartographic analysis of coastal characteristics is essential (Jelgersma, 1987). This classification need not to be excessively complex. Historical attempts at a classification of the world's coastlines have been summarized by Fairbridge (1968) and by Davies (1972); a recent review of coastal change with reference to categories has been completed by Bird (1985).

The five criteria that we recommend are as follows:
(a) material (hard and soft rock, resistance to wave erosion; abundance of loose material, prograding or retreating);
(b) local subsidence and compaction (natural and man-made);
(c) oceanographic setting (fetch, wave regime, currents);
(d) geodetic and tectonic setting;
(e) younger geologic history (back to about 100 kyr)

Several attempts in the past have been make to construct a global map, or map series (e.g., about 30 years ago by the Office of Naval Research, Washington; but not formally published; by Valentin (1952) at the University of Berlin, Geographical Institute; by the INQUA Shorelines Commission; and others). None of these maps are in print today and a new one (or series) is urgently needed.

Serious inherent difficulties existed that prevented earlier maps from reaching a useful level of precision. Several areas require internationally-coordinated efforts by national organizations. Many questions involve long-term trends that cannot be securely detected by satellite monitoring of tide-gauge analysis. Although "secular" and "long-term", the rates of change may exceed 1 mm/yr, so that their cumulative effect should quickly become noticeable (Fig. 3. Geodetic relevelling has shown that the Appallachian mountain belt is rising today at about the same rate, so that the apparent sea-level rise can be accounted for by tectonic subsidence. The tide-gauge itself is situated on crystalline rock which is not susceptible to de-watering or compactional subsidence. Note that the MSL rise displays no acceleration that might indicate recent rise of global temperature or accelerated glacial

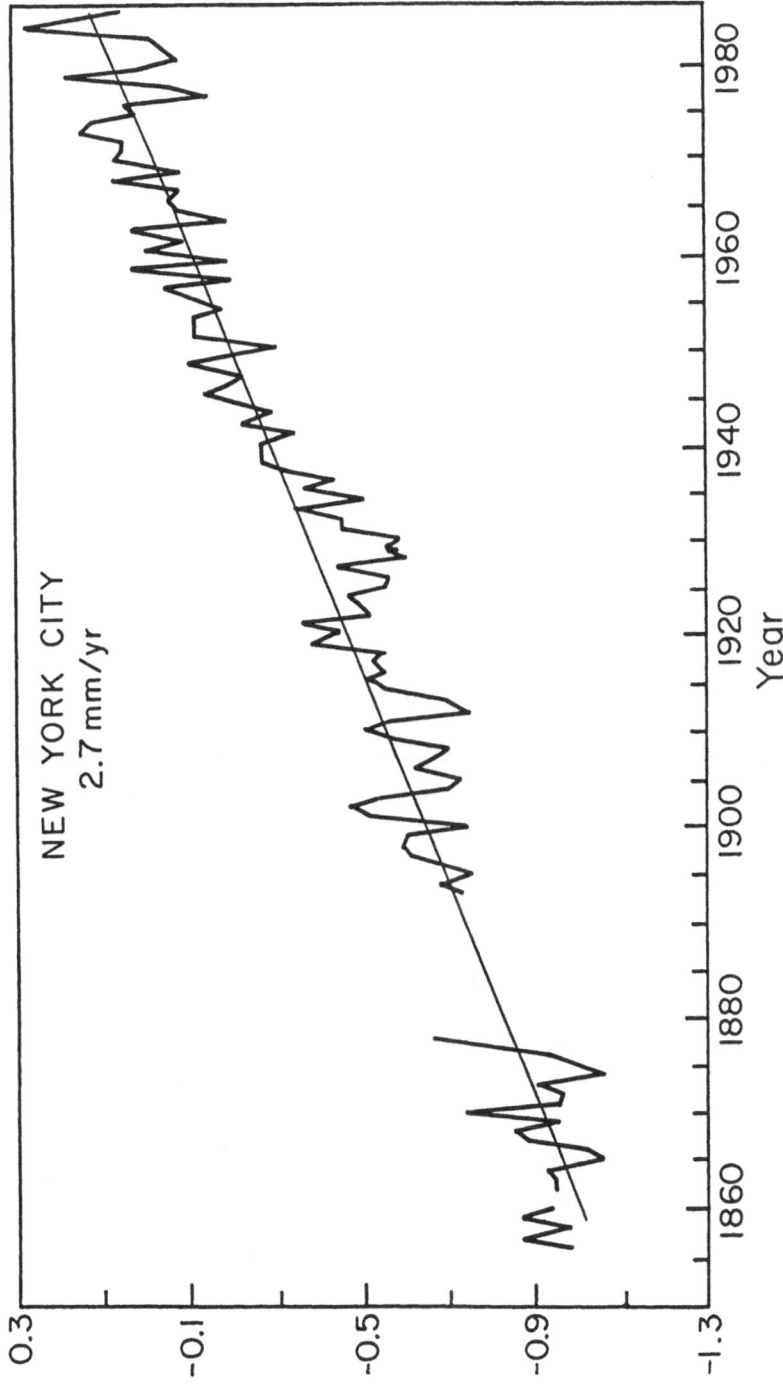

Figure 3. A long-term annual MSL record for New York City, showing a mean relative rise of 2.7 mm/yr.

melting).

One aspect of coastal classification that is often liable to be forgotten of ignored is the crustal stability problem. If we could find a perfectly stable platform, a tide gauge fixed to it could be used as a reliable indicator of genuine changes in MSL, be they eustatic, steric, oceanographic or whatever. While public attention (or funds) often focus on the *least stable* situations, e.g., the Mississippi delta, it is also important to establish where are the *most stable* settings, so that the rapidly changing dynamic variables may be tested in a most favourable and unequivocable setting.

Experiment Series A: Geographically and geologically, those more stable situations are generally to be found at low latitudes (far removed from recent glacier loading); preferably on coral or eolianite reefs (erosion and biological indicators tend to zero in on mean low tide level); where the climate rends to be dominated by subtropical high-pressure systems (hurricane or tsunami deposits are easily distinguished from the usual accumulations); where rainfall tends to be sparse (thus reducing the volume of terrigenous sediment, and its role in crustal loading - which of course leads to a secular rise of local sea level).

Experiment Series A should explore the globe for those most stable situations where two questions can be treated: (a) how do the tide-gauge records reflect the last 50 years? (b) how does the geological stratigraphy and geomorphology reflect the last 5000 years? If the answer to question (b) is displayed in a field simulation where a series of erosional or progradational steps are found to be repeated at the same elevations over a wide geographic area, then that region may be judged relatively stable, and the tide-gauge record (a) may be taken to be relatively free from long-term secular components. Series A experiments need to be repeated around the globe in geologically and oceanographically analogous situations.

Experiment Series B: the tectonic stability required in series A is not achievable world-wide, but in the subarctic regions of Canada, Scandinavia, and the northern USSR there are ancient Precambrian terrains that show an extremely steady secular uplift (glacio-isostatic uplift). Where geodetic surveys confirm that rate, the present-century MSL can be distinguished by simply subtracting the mean emergence rate from geodetic uplift rate. A particularly favourable region for scientific study which is now also favourable for international cooperation is around the Baltic Sea. It is excellent for sea-level studies because it has a very low astronomic tide and the former beach lines of the last 5000 years have been extensively mapped and dated. Several recent studies (e.g., tide-gauge records in Stockholm and Hanko, Finland) shoe that the region is responding in a highly stable manner and therefore the interannual variations can be interpreted as a time signal of regional climatic character; in this region the role of freshwater runoff from the river network is relatively important. This region is also provided with a comprehensive geodetic network and has not only precise temperature records back to 1750 (thermometers designed and installed at the instigation of Prof. Celsius himself), but also a series of tree-ring-based climatic records that go back more than one millennium. In northern latitudes ting-widths reflect temperature (not precipitation as they do in the American southwest). It is thus an ideal region for carrying out experiments in long-term MSL-climatic relationships.

It must be evident that a cartographic basis for appraisal of world-wide coastal criteria

Table 2. List of perigee-syzygy-perihelion peak tidal dates for the 1990's (Wood, 1985). It shoud be understood that high tides by themselves are not likely to be destructive but only during coincidence with extreme storm events.

Year	Date	Hour (e.s.t)
1988	2/17	0730
1988	8/27	0900
1988	9/25	0630
1989	3/7	2000
1989	4/5	1830
1989	10/14	1830
1989	11/12	1700
1990	4/25	0530
1990	5/24	0230
1990	12/2	0430
1990	12/31	0430
1991	6/12	1300
1991	7/11	0930
1991	12/21	1700
1992	1/19	1630
1992	2/17	1600
1992	7/29	2100
1992	8/27	1730
1993	2/7	0500
1993	3/8	0400
1993	4/6	0200
1993	9/16	0400
1993	10/15	0200
1994	3/27	1530
1994	4/25	1330
1994	11/3	1400
1994	12/2	1300
1995	5/15	0100
1995	6/12	2130
1995	12/22	0100
1996	1/20	0100
1996	7/1	0800
1996	7/30	0430
1997	2/7	1300
1997	3/8	1200
1997	8/18	1500
1997	9/16	1230
1998	3/28	0000
1998	4/25	2200
1998	10/5	2330
1998	11/3	2200
1999	5/15	0830
1999	6/13	0500
1999	11/23	0930
1999	12/22	0930

136

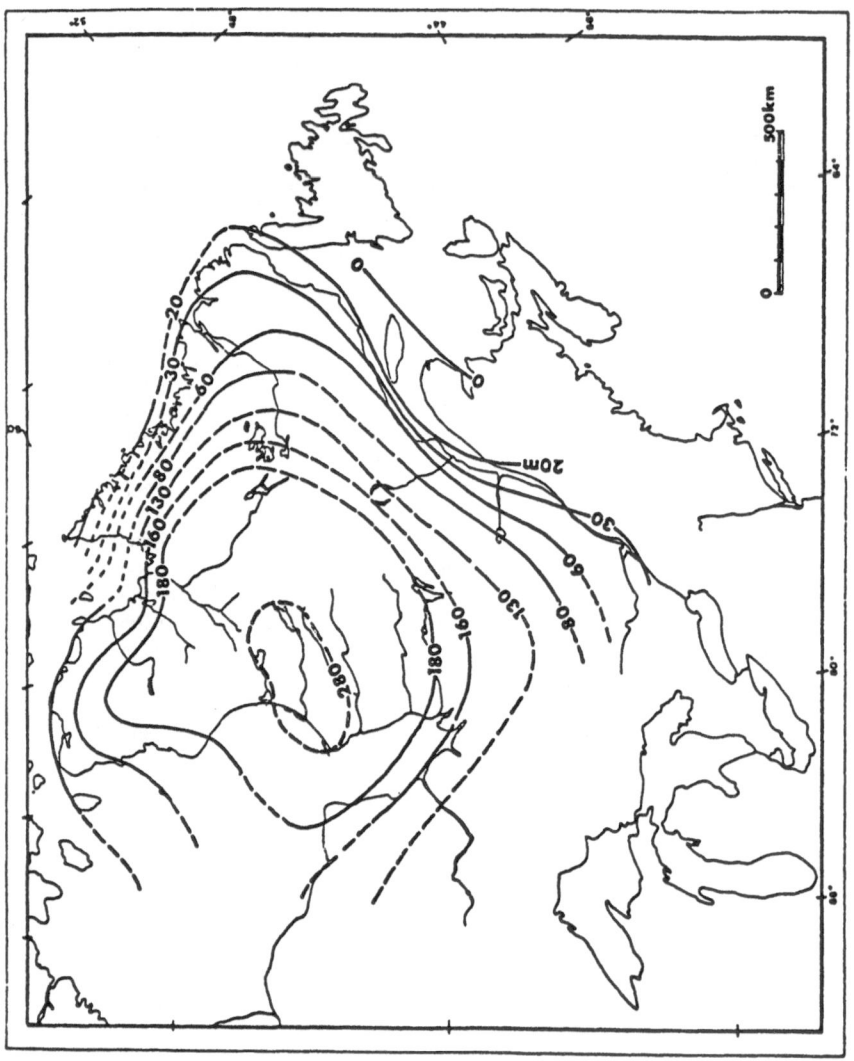

Figure 4. Map of part of the Canadian shield (Quebec; courtesy of C. Hillaire-Marcel), showing uplift since about 7000 yr BP.

could also be a basis for planning *coastal hazard* preparedness, which need not be emphasized further at this point.

Category 2: *Contemporary & Near-Future Global MSL*, to be subjected to interdisciplinary study including continuous monitoring by satellite (remote sensing); to be coordinated with "ground-truth" data acquisition, by means of tide-gauges, SST observations, meteorologic data and astronomic resources.

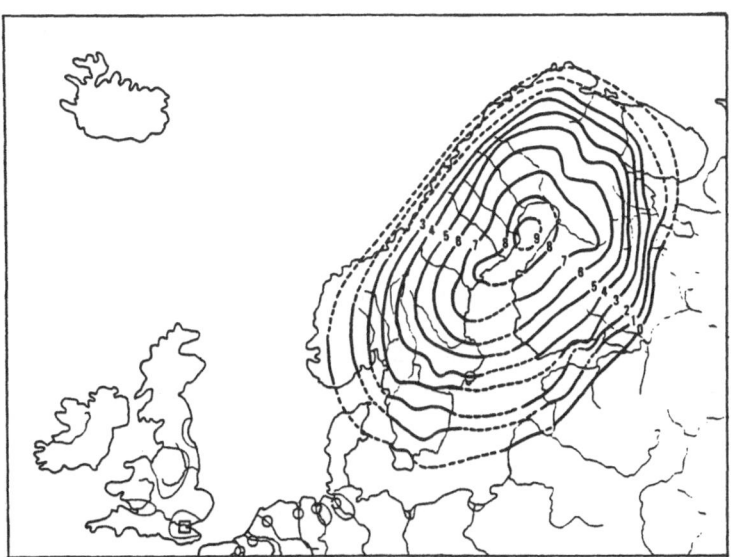

Figure 5. Observed land uplift in the Fennoscandian area in mm/yr. After J. Kakkuri (Geological Survey of Finland, special paper No. 2).

(a) Intra-annual variance experiments are needed to test amplitudes and response rates. We need this information to discriminate between "noise" and secular trends in MSL.

(i) Experiments with respect to seasonality (with changing rates of geostrophic circulation, notably Gulf Stream, Benguela Current and Kuro Shio), and with respect to specific storm patterns (tropical, polar frontal).

(ii) A particularly complex, interdisciplinary experiment could test the immediate effect on MSL of a major solar flare outburst (following the Bucha model); stratospheric chemistry, pressure and temperature would require rocket or balloon monitoring; tropospheric reactions (including notable jet-stream perturbations) could be coordinated through a chain of Arctic and mid-latitude weather stations; simultaneous MSL departures, both positive and negative, are to be expected (some new or temporary gauging stations would be needed, especially in regions subject to sea-ice formation).

(iii) Several El Niño/La Niña type experiments are currently in progress. Some of these could be expanded to take in the interrelationships (a) with the Indian and S.E. Asian monsoons, as well as the Australian, Ethiopian, and West African monsoons; (b) teleconnection with the winter snowfall (albedo and global cooling) created by the area and duration of such precipitation in North America and Siberia/Tibet (e.g., monitored by Kukla: pers. communication); (c) relationships between all of the above and major drought-prone areas in the world (e.g., American mid-west, African Sahel, Brazil's Nordeste, eastern Australia, grain-belt of the USSR, etc.).

(b) *The Pole Tide, the Chandler Wobble and Quasi-Biennial Oscillation.* The QBO (approximately 2.2 yr) is, according to Lamb (1977), the one climatic cyclicity of more than 12 months that is universally recognized; it is prominently displayed in most annual tide-gauge analyses. Question: how does it interrelate to the 1.1318 yr luni-solar tide, to the Chandler Wobble (already known to be present in the GCM dynamics from work by Currie and Hameed), and to the Chandler/Spin Beat Frequency (about 6.2 yr)?

(c) *Luni-Solar Tides.* According to Wood (1985), long-term luni-solar tides of more than 12 months, all resolve into some harmonic or combination of the two fundamental periodicities: the 1.1318 yr tide and the nutation cycle of 18.604 yr. The latter is locked-in to the Chandler periods (18.6 = 3 x 6.2).

Wood (1985, p. 480) has calculated the precise days and hours of peak perigee-syzygy-perihelion tides for the next decade (see Table 2). A specific international team should monitor these events globally (and issue *hazard warnings* when socio-political advisable). The next ten years could thus constitute the "grand experiment" for MSL.

6. Conclusions and Recommendations

1. Tide-gauge records reflect patterns of local and regional behaviour, but global distribution of recording stations are inhomogeneous and inadequate. By far the majority are located in mid-latitudes of the northern hemisphere, so that truly global representation is still not adequate. We recommend the establishment of many more gauging stations, with specific attention to the following areas:
   (a) high Arctic and sub-Antarctic;
   (b) mid-Indian Ocean and mid-Pacific atolls;
   (c) coasts of West and East Africa;
   (d) islands of West and East Indies.

2. Global tide-gauge records individually display either a rising or falling trend during the course of the last 100 yr. Fall of MSL (land emergence) is associated principally with glacio-isostatic crustal uplift. Such regions are marked by rocky coasts, but often with a veneer of postglacial debris, as in much of Canada, Greenland, Spitsbergen, northern Britain, much of Scandinavia, northern USSR, and many parts of the subantarctic coasts.

   Land uplift is also observed extensively along the younger organic belts, (parts of

Japan, New Guinea, New Zealand, southern Alaska, Crete, Calabria, etc.) but the rising areas are not rising uniformly and are frequently side-by-side with sinking areas. We recommend the preparation of a global inventory of coastal regions, recognizing their long and short-term trends. Rise of MSL (land submergence) is almost universally associated with deltas, estuaries and characterized by low muddy or sandy coastlines.

3. It is recommended that the global map project begun by the late Professor Helmut Valentin (University of Berlin) should be restarted with input of data already assembled by the INQUA Shorelines Commission. Funding should be sought so that the map series could be printed and distributed at modest prices to all third-world coastal countries, where the engineering information of the maps could be invaluable for long-term economic planning.

4. The principal message of this review is that the secular rise and fall of global MSL has little to do with the contemporary human/economic problems of sea level. Major loss of life and property result from episodic "events", i.e., crescendo storm surges. We recommend closer studies of their causes and development.

5. In the light of increasing density of coastal populations, and in view of the probability of a CO_2-induced rise in frequency of tropical cyclonic storms, we recommend that a Commission for Coastal Hazard Preparedness be established by a responsible international authority. This commission could prepare guidance plans for appropriate national bodies to put in place. Such plans could suggest the construction of diked refuges, safe command posts, early warning systems, monitoring procedures, legal constraints on unrestricted coastal developments, and other precautionary steps within a framework of recognized engineering, scientific, economic and legal practice.

7. Acknowledgments

This chapter could not have been prepared without the generous support of NATO for the Advanced Research Workshop and the tireless work of Prof. Roland Paepe, E. van Overloop, Isabelle Demolin and Dolores Fairbridge, who collectively helped to make it a resounding success. The manuscript has been kindly reviewed and improved by Vivian Gornitz (NASA; GISS) and Charles Finkl (Coastal Education and Research Foundation).

Additional research funding for both R.W.F. and the Workshop in general was furnished by the U.S. National Science Foundation (Critical Engineering Systems Program), and logistic support has been furnished by the generosity of NASA-Goddard Institute for Space Studies (Dr. James Hansen).

Scientific support for S.Jelgersma was provided by courtesy of the Geological Survey of the Netherlands.

8. References

Aubrey, D.G., & Emery, K.O. (1983) "Eigenanalysis of recent United States sea levels", Continental Shelf Research 2, pp. 21-33

Bardach, J.E. (1989) "Global warming and the coastal zone", Climatic Change 15, pp. 117-150

Barnett, T.P. (1984) "The estimation of global sea level change: a problem of uniqueness", J. Geophys. Res. 89(C5), pp. 7980-7988

Baulig, H. (1935) "The changing sea level", Institute of British Geographers 3, pp. 1-46

Bennema, J., Geuse, E.C.W.A., Smits, H., & Wiggers, A.J. (1954) "Soil compaction in relation to Quaternary movements of sea level and subsidence of the lands especially in the Netherlands", Geol. en Mijnbouw 16, pp. 173-178

Bremer, H., & Clayton, K.M. (1989) "Coasts: erosion and sedimentation", Zeitschr. Geomorph. N.F., Suppl. - Bd. 73, pp. 1-183

Bird, E.C.F. (1985) *Coastline Changes: A Global Review*, John Wiley and Sons, Chichester

Carter, R.W.G., et al. (1987) "Sea-level, sediment supply and coastal changes: examples from the coast of Ireland", Progress in Oceanography 18, pp. 79-101

Cheney, R.E., & Miller, L. (1990) "Recovery of the sea level signal in the western tropical Pacific from Geosat Altimetry", J. Geophys. Res. 95(C3), pp. 2977-2984

Currie, R.G. (1976) "The spectrum of sea level from 4 to 40 years", Geophys. Jour. Roy. Astron. Soc. 46, pp. 513-520

Davies, J.L. (1972) *Geographical Variation in Coastal Development (Geomorph. Text 4)*, Oliver & Boyd, Edinburgh

Eddy, J. (1982) "The solar constant and surface temperature", in Reck, R.A. and Hummel, J.R. (eds.) *Interpretation of Climate and Photochemical Models, Ozone and Temperature Measurement*, Am. Inst. Physics (AIP Conf. Proc. 82), New York, pp. 247-262

Ekman, M. (1988) "The world's longest continued series of sea level observations", Pure Appl. Geophys. 127, pp. 73-77

Fairbridge, R.W. (1961) "Eustatic changes in sea level", *Physics and Chemistry of the Earth*, Pergamon Press, London, 4, pp. 99-185

Fairbridge, R.W. (ed.) (1968) *The Encyclopedia of Geomorphology*, Van Nostrand Reinhold, New York

Fairbridge, R.W. (1981) "The concept of neotectonics, and introduction", Zeitschrift für Geomorphologie, Supplement-Band 40, vii-xii

Fairbridge, R.W. (1989) "Quasi-equilibrium in Holocene climatic change", Quaternary International 2, pp. 83-89

Fairbridge, R.W., & Krebs, O.A. (1962) "Sea level and the southern oscillation", Geophys. J., Roy. Astron. Soc. 6(4), pp. 532-545

Fairbridge, R.W., & Shirley, J.H. (1987) "Prolonged minima and the 179-yr cycle of the solar inertial motion", Solar Physics 110, pp. 191-220

Faure, H. (1987) "Mécanisme d'amplification du cycle climatique global: L'effet de couvercle de la glace de mer controle le CO_2 atmospherique", C.R. Acad. Sci. 2, 305, pp. 523-527

Gabrysch, R.K. (1984) "Groundwater Withdrawals and land-surface subsidence in the

Houston-Galveston region, Texas, 1906-1980", Texas Dept. Water Res. Rpt., 287, pp. 1-64

Godschalk, D.R., et al. (1989) "Catastrophic coastal storms: hazard mitigation and development management", Duke University Policy Studies, pp. 1-275

Gornitz, V., & Kanciruk, P. (1989) "Assessment of global coastal hazards from sea level rise", Proc. Sixth Symposium on Coastal and Ocean Management/ASCE, Charleston, S.C., pp. 1345-1359

Gornitz, V., Lebedeff, S., & Hansen, J. (1982) "Global sea level trend in the last century", Science 215, pp. 1611-1614

Gornitz, V., & Lebedeff, S. (1987) "Global sea-level changes during the past century", in Nummedal, D., et al. (eds.) Sea Level Fluctuation and Coastal Evolution, Soc. Econ. Paleont. Miner., Spec. Publ. 41, pp. 3-16

Gottschalk, M.K.E. (1971, 1975, 1977) Stormvloeden en Rivieroverstromingen in Nederland: I - De periode voor 1400; II - De periode 1400-1600; III - De periode 1600-1700, van Gorcum, Assen

Gutenberg, B. (1941) "Changes in sea level, postglacial uplift, and mobility of the Earth's interior", Bull. Geol. Soc. Am. 52, pp. 771-772

Hameed, S., & Currie, R.G. (1989) "Simulation of the 14-month Chandler Wobble in a global climate model", Geophys. Res. Letters 16(3), pp. 247-250

IAHS (1969, 1976, 1984) Land Subsidence, IAHS publ. 88 & 89, 121 and 151, Tokio, Anaheim (Calif.), Venice, resp.

Jelgersma, S. (1966) "Sea level changes during the last 10,000 years", in Sawyer, J.S., et al. (eds.) World Climate from 8000 to 0 BC, Roy. Meteorol. Soc., London, pp. 54-71

Jensen, H.A.P. (1953) "Tidal inundations past and present", Weather 8, Part I, pp. 85-89; Part II, pp. 108-113

Jouzel, J. et al. (1989) "Global change over the last climatic cycle from the Vostok Ice core record (Antcarctica)", Quaternary International 2, pp. 15-24

Klige, R.K., & Dobrovolsky, S.G. (1988) "The ocean level and models simulating its oscillation", J. Coast. Res. 4, pp 273-278

Kuenen, P.H. (1950) Marine Geology, J. Wiley and Sons, New York

Lamb, H.H. (1972/1977) Climate: Present, Past and Future, Methuen, London

Lamb, H.H. (1980) "Climatic fluctuation in historical times and their connection with transgressions of the sea, storm floods and other coastal changes", in Berhulst, A., and Gottschalk, M.K.E. (eds.) Transgressies en occupatiegeschiedenis in de kustgebieden van Nederland en Belgie, Belgisch Centrum voor Landelijke Geschiedenis, Pub. No. 1 66, pp. 251-284

Lorius, C., Barkov, N.I., Jouzel, J., Korotkevich, Y.S., Kotlyakov, V.M., & Raynaud, D. (1988) "Antarctic ice core: CO_2 and climate change over the last climatic cycle", Eos, June 28, pp. 681-684

Meier, M.F. (1984) "Contribution of small glaciers to global sea level", Science 226(468), pp. 1418-1421

Morgan, J.P. (ed.) (1970) "Deltaic sedimentation, modern and ancient", Sea. Econ. Paleont. Min., Tulsa, Oklahoma, sp. publ. 15

142

Mörner, N.A. (1969) *The Late Quaternary History of the Kattegat Sea and the Swedish West Coast*, Sver. Geol. Unders. Afh. Serie C 640, 478 p.

Mörner, N.A. (1976) "Eustatic changes during the least 8,000 years in view of radiocarbon calibration and new information from the Kattegat region and other northwestern European areas", Pal. Pal. Pal. 19, pp. 63-85

Mörner, N.A. (1980) "Eustasy and geoid changes as a function of core/mantle changes", in Mörner, N.A. (ed.) *Earth Rheology, Isostasy and Eustasy*, John Wiley and Sons, Chichester, pp. 533-553

Mörner, N.A. (1980) "A 10,700 year's paleotemperature record from Gotland and Pleistocene/Holocene boundary events in Sweden", Boreas 9, pp. 283-287

Mörner, N.A. (1981) "Space geodesy, palaeogeodesy, and palaeogeophysics", Annales de Géophys., 37, pp. 69-76

Nakada, H. (1986) "Holocene sea level in oceanic islands: implications for the rheological structure of the earth's mantle", Tectonophysics 121, pp. 263-276

Newman, W.S., & Fairbridge, R.W. (1986) "The management of sea-level rise", *Nature* 320, pp. 319-321

Newman, W.S., Pardi, A.R., & Fairbridge, R.W. (1989) "Some considerations of the compilation of late Quaternary sea-level curves: a North American perspective", in Scott, D.B., Pirazzoli, P.A., and Honig, C.A. (eds.) *Late Quaternary Sea-Level Correlation and Applications*, (NATO ASI Ser. C, 256), Kluwer, Dordrecht, pp. 207-228

Peltier, W.R. (1987) "Mechanisms of relative sea-level change and the geophysical responses to ice-water loading", in Devoy, R.J.N. (ed.) *Sea Surface Studies*, Croom Helm, London, pp. 57-94

Peltier, W.R., & Tushingham, A.M. (1989) "Global sea-level rise and greenhouse effect: might they be connected?", Science 244, pp. 806-810

Petterson, O. (1912) "The connection between hydrographical and meteorological phenomena", Quart. J. Roy. Meteorol. Soc. 38, pp. 173-191

Petterson, O. (1930) "The tidal force", Geogr. Annaler 12, pp. 261-322

Pirazzoli, P.A. (1986) "Secular trends of relative sea-level (RSL) changes indicated by tide-gauge records", J. Coastal Res. S1(1), pp. 1-26

Pirazzoli, P.A. (1989) "Present and near-future global sea-level changes", Pal. Pal. Pal. (Global Planet. Change) 75, pp. 241-258

Plassche, O. van de (1986) *Sea-Level Research: A Manual for the Collection and Evaluation of Data*, Geo Books, Norwich

Poland, J. (ed.) (1984) *Guidebook for the Studies of Land Subsidence Due to Groundwater Withdrawal*, Prepared for the I.H.P. Working Group 8.4, UNESCO

Pugh, D.T., Spencer, N.E., & Woodworth, P.L. (1987) *Data Holdings of the Permanent Service for Mean Sea Level*, Permanent Service for Mean Sea Level, Bidston Observatory, Birkenhead

Ramanathan, V. (1988) "The greenhouse theory of climate change: a test by an inadvertent global experiment", Science 240, pp. 293-299

Rossiter, J.R. (1962) "An analysis of annual sea-level variations in European waters", Geophys. J. Roy. Astron. Soc. 12, pp. 259-299

Roy, P., & Connell, J. (1989) *Greenhouse: the Impact of Sea-Level Rise on Low Coral Islands in the South Pacific*, University of Sydney, RIAP Occasional Paper No. 6

Scott, D.B., & Greenberg, D.A. (1983) "Relative sea-level rise and tidal development in the Fundy tidal system", Can. J. Earth Sci. 20, pp. 1554-1564

Shennan, I. (1987) "Holocene sea-level changes in the North Sea", in Tooley, M.J., and Shennan, I. (eds.) *Sea-Level Changes*, B. Blackwell, Oxford, pp. 109-151

Smith, D.E., & Dawson, A.G. (1983) *Shorelines and Isostasy*, Institute of British Geographers Special Publication No. 16, Academic Press, London

Taira, K. (1980) "Holocene events in Japan: paleo-oceanography, volcanism and relative sea-level oscillation", Pal. Pal. Pal. 32, pp. 69-77

Tanner, V. (1933) "L'étude des terrasses littorales en Fennoscandie et l'homotaxie intercontinental, in Comptes Rendus Cong. Int. Géog., Paris 1931, Vol. 2, pp. 61-76

Thomson, R.E., & Tabata, R. (1987) "Steric height trends at Ocean Station PAPA in the Northeast Pacific ocean", Mar. Geod. 11, pp. 103-111

Thorarinsson, S. (1940) "Present glacier shrinkage and eustatic changes of sea level", Geogr. Annaler, Stockholm, 22, pp. 131-159

Tooley, M.J., & Shennan, I. (eds.) (1987) *Sea-Level Changes*, B. Blackwell, Oxford

Valentin, H. (1954) *Die Küsten der Erde*, Veb. Geogr.-Kartogr. Anst. Gotha

Verhagen, H.J. (1990) "Coastal protection and dune management in the Netherlands", J. Coastal Res. 6(1), pp. 169-180

Vita-Finzi, C. (1986) *Recent Earth Movements: An Introduction to Neotectonics*, Academic, London and Orlando, Fl

Wegmann, E. (1969) "Changing ideas about moving shorelines", in Scheer, C.J. (ed.) *Toward a History of Geology*, MIT Press, Cambridge, Massechusetts, pp. 386-414

Weischet, W. (1963) "Further observation of geologic and geomorphic changes resulting from the catastrophic earthquakes of May 1960, in Chile", Seismol. Soc. Am. Bull. 53(6), pp. 1237-1257

Wigley, T.M.L., & Raper, S.C.B. (1987) "Thermal expansion of sea water associated with global warming", *Nature* 330, pp. 127-131

Wood, F.J. (1985) *Tidal Dynamics*, Reidel, Dordrecht

Wyrtki, K. (1975) "El Niño - the dynamic response of the equatorial Pacific Ocean to atmospheric forcing", J. Phys. Oceanogr. 5(4), pp. 572-584

Wyrtki, K., & Mitchum, G. (1990) "Interannual differences of Geosat altimeter heights and sea-level: the importance of a datum", J. of Geo. Res. 95, pp. 2969-2975

NON-EUSTATIC SEA LEVEL CHANGES FROM MEDITERRANEAN TIDE GAGES

V. GOLDSMITH
Department of Geology and Geography
Hunter College (City University of New York)
695 Park Avenue, New York, N.Y. 10021, USA

ABSTRACT: As scientists have collected data on sea level changes, more and more attention has been focused on non-eustatic effects. The Mediterranean is an excellent example of this (e.g., Pirazzoli, 1987). Two examples of non-eustatic effects revealed in recent tide gage analyses include meteorological/oceanic effects (Goldsmith and Gilboa, 1987) and local neotectonics (Emery et al., 1988). In both examples, these effects largely outweigh, and mask, eustatic effects as revealed in monthly mean sea levels determined from tide gages in the Mediterranean. Moreover, The importance, and even dominance, of the non-eustatic effects involving sea level seasonality, local tectonics, and local sea level cycles in the order of tens of years, is consistent with studies conducted in other areas.

1. Introduction

1.1. MEAN SEA LEVEL SEASONALITY

With increasing concern about the "Greenhouse Effect", attention has focused on measurements of sea level rise (SLR) from tide gages as a way of detecting and quantifying this effect. Recently, however, it has become clear that sea level measurements from tide gages represent a lot more than eustatic sea level changes.

Goldsmith and Gilboa (1985 and 1987) called attention to tide gage seasonality due to seasonal water temperature effects, and seasonal winds in the eastern and western Mediterranean respectively. Here, because the seasonal sea level changes were in the order of tens of cm, these authors suggested that sea level changes predicted to be in the order of mm/year from the "Greenhouse Effect" would be difficult to detect. Pirazzoli (this volume) presents a discussion of these problems, and suggests a solution for differentiating eustatic SLR from other factors.

Thompson (1980) conducted an analysis of British monthly mean sea levels from tide

145

R. Paepe et al. (eds.), Greenhouse Effect, Sea Level and Drought, 145–152.
© 1990 *Kluwer Academic Publishers.*

gages. He calculated that over 80% of the monthly mean sea level variance could "...be related to seasonal changes, the static pressure effect, and the influence of winds," and another 10% to pressure effects. Thompson (1980) further suggested that the seasonal variation could be approximated by an annual tide with an amplitude of 7 cm, which he attributed to "...the steric oscillation of the adjacent North Atlantic."

Woodworth (1984) discussed the worldwide distribution of sea level seasonality, pointing out that in some areas (e.g., parts of Asia) the seasonality may reach a range of 1 m, and thus be larger than the daily variation. With this range, the seasonality affects the annual beach erosion cycle, and certainly must be taken into account when detecting adverse coastal effects from the expected "Greenhouse Effect". In summarizing, Woodworth (1984) attributed the mean sea level (MSL) seasonality in the mid-latitudes to temperature changes in the upper layers of the oceans, peaking in late summer, and in the tropics to seasonal variation in heat content throughout the oceanic water column. He divides the Mediterranean into two regions, with the eastern portion having a MSL peak between months 4.5 and 1.5, and a range of 5 to 7.5 cm, and the western portion peaking between months 1.5 and 4.5 with a range of 2.5 to 5 cm. The seasonal ranges in the Mediterranean appear to be consistent with ranges in other parts of the world, even though the Mediterranean daily tide ranges are low relative to other areas.

1.2. LOCAL TECTONIC EFFECTS

Woodworth (1987) inferred a north-south tilting of the UK mainland from differences in sea level trends of tide gages spanning a common time (1916-1982). This is a different interpretation of the latitudinal slope of UK tide gages than Thompson (1980), who attributed this slope to a systematic error in geodetic levelling. However, Woodworth's interpretation is consistent with data from other areas, indicating tilting of large land blocks to the extent that this movement is easily detected in contemporary tide gage records (e.g., Japan and Australia in Aubrey and Emery, 1986a and 1986b). Woodworth (1987) suggests that the inability to distinguish between long-term sea level changes and absolute land movements in tide gage records presents a major analysis difficulty. This is further complicated by the large drop in sea levels in the mid 1970's observed in tide gages along the European Atlantic coastline (Woodworth, 1987; Pirazzoli, this volume).

Emery et al. (1988) detected large variations in 31 Mediterranean tide gages which they attributed to such coastal neo-tectonics as "...emergence along the coast of Israel and at Alexandria, ...probable plate underthrusting in Greece and Turkey, to volcanism near Mount Etna, to deltaic compaction at the mouth of the Nile and at Izmir, and to deltaic compaction coupled with water pumping at the Po Delta." (see Figure 1). The importance of the local tectonic signal in dominating Greek tide gages is supported by the studies of Flemming and Woodworth (1988). They showed a consistency in scale and direction of land movement between secular trends interpreted from 16 Greek tide gages and tectonic movement detected from archaeological and geological data.

Pirazzoli (1987), in a study of sea level changes over a geologic time scale, emphasizes the importance of mean sea surface topographic variations in the Mediterranean. He refers

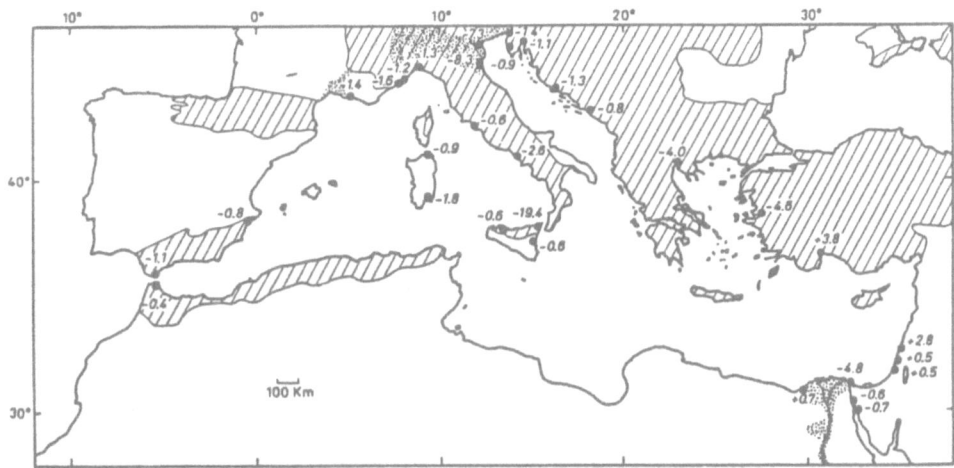

Figure 1. Relative movement of land in mm/yr at selected tide gage stations in the Mediterranean (from Emery et al., 1988). Diagonal shading - areas of Alpine folding; stippled areas - deltas.

to differences in sea surface elevation detected by SEASAT of as much as 50 m, between a low in the Eastern Mediterranean adjacent to the Hellenic Arc and a high situated in the Gulf of Lyons. Although this variation is attributed to variations in density of the underlying earth structure, the role of local tectonics in affecting shoreline sea levels is not clear, nor its temporal variation.

1.3. OTHER SEA LEVEL EFFECTS

Other processes influencing sea levels as measured by tide gages include currents (Pirazzoli, 1987; Thompson and Pugh, 1986) barometric pressure (Pugh and Thompson, 1986), and shelf waves (Huthnance, 1987). Although these effects are relatively small, they probably need to be taken into account, and deducted, in order to better discern the secular sea level trend.

2. Seasonality in Mediterranean Tide Gages

2.1. INTRODUCTION

Figure 1 (from Emery et al., 1988) shows the large variations in secular sea level trends as detected in 31 Mediterranean tide gages. In Emery et al. (1988) it was shown that there were large variations in sea level changes in any given year across the Mediterranean. Some of these differences can be attributed to regional variations in MSL seasonality (Fig.

148

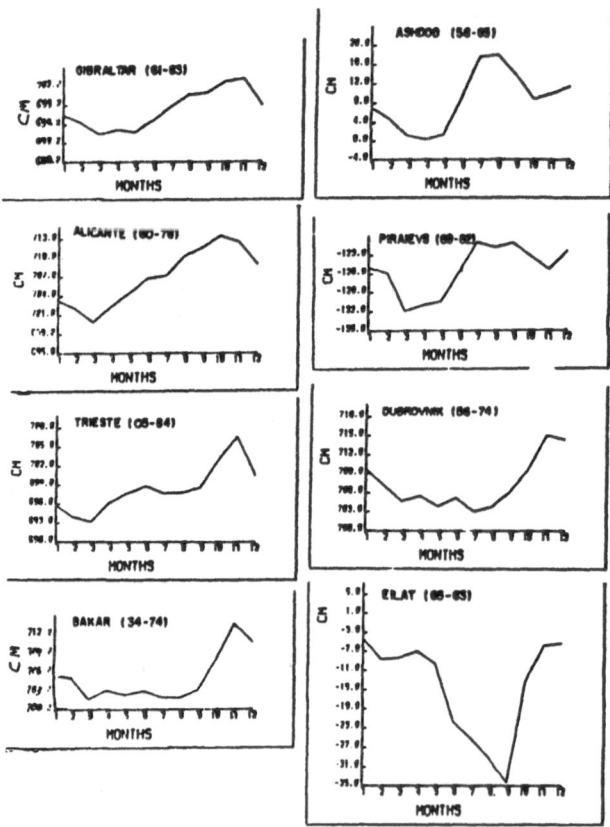

Figure 2. Monthly averages of mean sea level at seven tide gage stations in the Mediterranean and at Eilat, northern Red Sea (from Goldsmith and Gilboa, 1987)

2). For example, Piraeus, Greece and Ashdod, Israel display similar seasonal patterns, so do Gibraltar and Alicante, Spain, as doe the three Adriatic Sea locations (Trieste, Bakar and Dubrovnik). Thus, regional differences in MSL seasonality are quite apparent. Now, let's look at some of these regions.

2.2. SEASONALITY ALONG THE ISREALI COAST

Along the Israeli coast there is a strong tidal seasonality with an average range of 20 cm (Fig. 3). This is approximately equal to the neap tide range, and 1/3 of the spring tide range. Highest sea levels occur at the end of summer (July-September) while lowest sea levels occur at the end of winter (March-April). Heat flow data from the water column

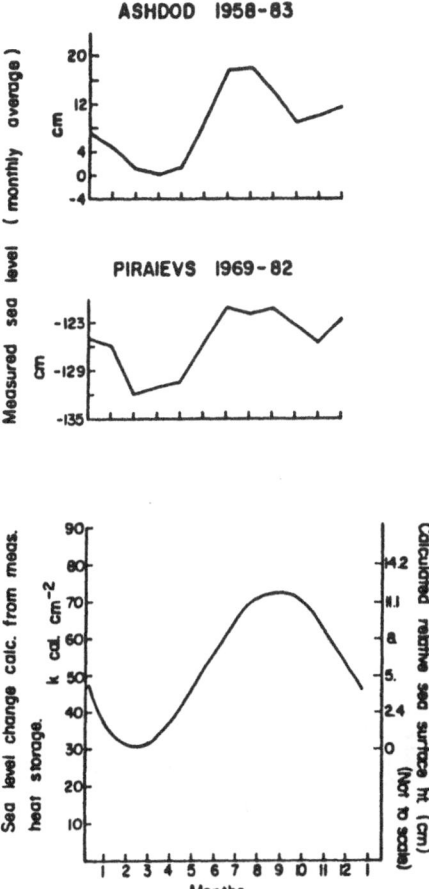

Figure 3. Comparison of monthly averages of mean sea level at Ashdod, Israel and Piraeus, Greece with monthly mean sea level computed from heat storage measurements made during the same time period in the southeastern Mediterranean.

obtained over many years along this continental shelf display a similar pattern. Moreover, computations of theoretical sea level changes based on *seasonal water temperature changes alone* accounts for 2/3 of the seasonal sea level changes measured by the tide gages (Fig. 3). Analyses of more than 20 years of records from the 3 Israeli tide gages shows that the amplitude of this seasonal pattern varies from year to year (Goldsmith and Gilboa, 1985 and 1987).

A somewhat similar pattern is found in the Western Mediterranean; e.g., Alicante, Spain and Gibraltar have a seasonality with an average range of 15 cm with the highest sea levels in October-November, and the lowest sea levels in March (Fig. 4).

150

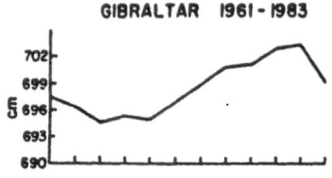

MONTHLY TIDE HT. VS. WIND DIR.
GIBRALTAR & ALICANTE, SPAIN

GIBRALTAR 1961 - 1983

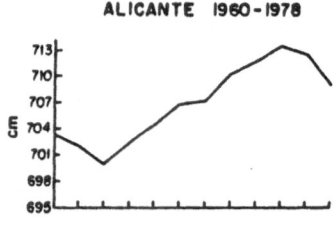

ALICANTE 1960 - 1978

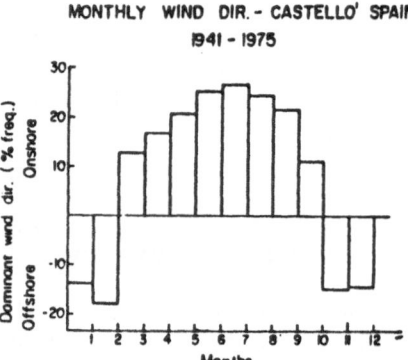

MONTHLY WIND DIR. - CASTELLO' SPAIN
1941 - 1975

Figure 4. Comparison of monthly averages of tide height at Gibraltar and Alicante, Spain with onshore/offshore winds.

(Note that when the onshore winds become dominant in March, the monthly mean tides at Alicante start to rise, and continue to rise until the offshore winds again dominate in November).

Here, the seasonality is not so regular, nor is it as sharp as in the southeastern Mediterranean. Also, the peak level in the western Mediterranean occurs about two months later than the eastern Mediterranean. This is due to wind effects along the Spanish coast. The role of seasonal onshore/offshore winds in increasing/decreasing tide levels, respectively is clearly documented in Figure 4. Here, in the western Mediterranean, unusually high monthly mean sea levels occur in the summer months, which relates to onshore winds that occur during this time of the year (Goldsmith and Gilboa, 1985 and 1987).

Another factor in the seasonal sea level changes involves local wind effects, especially on/offshore winds. In the southeastern Mediterranean, very low monthly mean sea levels occur occasionally in the winter months (i.e. 20 cm below MSL). This is attributed to strong offshore easterly winds which blow continuously for many days, but which do not occur every year. The importance of the wind in lowering sea levels is especially significant at Eilat, in the Gulf of Eilat, northern Red Sea (Fig. 2). Here, strong offshore winds (i.e. to the south) blow 85% of the time. As a result, monthly mean sea levels are always below mean low water (MLW), with the lowest monthly level at -35 cm in September.

3. Neotectonics

Local tectonics, including active faulting, delta subsidence, and plate tectonics, clearly dominate sea level trends from tide gage records in the Mediterranean. The data show Israel and Alexandria, Egypt to be emergent, whereas the Nile Delta and Gulf of Suez rift are submergent (Fig. 1). Other examples of the dominance of local effects are southern Turkey, Messina, and the effects from deltas (Venice, Porto Corsini and Izmir). However, there does appear to be an area of relative stability in the northwestern Mediterranean as indicated by consistent tide gage submergence rates of approximately 1.2 mm/yr. However, this could be misleading because of the limited time scale of tide gage records, and the size of the crustal blocks involved.

4. Climatic Cycles

Climatic cycles in the order of tens of years (discussed in several papers in this volume) can influence coastal erosion and development by affecting sand supply. This is well documented by Jelgersma et al. (1970) and is also discussed in Goldsmith (1985). It is complicated by anthropomorphic influences (Jelgersma et al., 1970; Denness, 1987).

5. Importance for Geohydrologic Management

Gornitz and Kanciruk (1989) describe a scheme for a global coastal hazards data base that contains information for delineating high risk areas threatened by the hypothesized sea level rise from global warming. Criteria for high risk areas include "high wave/tide energies". However, as demonstrated here, any schemes proposed for hydrogeologic management should take into account annual variations in local sea level seasonality. For example, Woodworth (1984) points out the importance of tidal seasonality in beach erosion.

Although seasonality in monthly mean tide gage levels occur throughout the Mediterranean, the range, pattern, and regularity vary widely (Fig. 2). This seasonality is more complex than the simple eastern/western Mediterranean classification of Woodworth (1984, Fig. 18). This is because the seasonal sea level changes appear to be related to a combination of local temperature and wind effects in the Mediterranean, which vary widely from year to year. Thus, eustatic sea level determinations become quite difficult to interpret.

On a longer time scale, tide gage analyses reveal the possibility of local sea level cycles in the order of tens of years, that appear to be regionally related (Emery et al., 1988). The possibility that these cycles are related to local sea temperature and/or wind effects, and their role in determining regional sea level changes, requires further investigation.

6. References

Aubrey, D., & Emery, K.O. (1986a) "Relative sea levels of Japan from tide gage records", Geol. Soc. Am. Bull. 97, pp. 194-205

Aubrey, D., & Emery, K.O. (1986b) "Australia - an unstable platform for tide gage measurements of changing sea levels", J. Geol. 94, pp. 699-712

Denness, B. (1987) "Sea level modelling: the past and the future", Prog. Ocean. 18, pp. 41-59

Emery, K.O., Aubrey, D.G., & Goldsmith, V. (1988) "Coastal neo-tectonics of the Mediterranean from tide gage records", Marine Geology 81, pp. 41-52

Goldsmith, V. (1985) "Coastal dunes", in Davis, R., *Coastal Sedimentary Environments*, Springer-Verlag, pp. 303-378

Goldsmith, V., & Gilboa, M. (1987) "Mediterranean sea level changes from tide gage records.", Proc. 20th Inter. Coastal Eng. Conf., A.S.C.E., N.Y., pp. 223-231

Gornitz, V., & Kanciruk, P. (1989) "Assessments of global coastal hazards from sea level rise", Proc. of the 6th Coastal and Ocean Management/A.S.C.E., pp. 1345-1359

Huthnance, J. (1986) "The subtidal behaviour of the Celtic Sea - III. A model of shelf waves and surges on a wide shelf", Cont. Shelf Research 5, pp. 347-377

Jelgersma, S., de Jong, J., Zagwijn, W.H., & Van Regteren Altena, J.F. (1970) "The coastal dunes of the western Netherlands: geology, vegetational history and archaeology", Mededelingen Rijks Geologische Dienst, Nieuwe Serie No. 21, pp. 93-167

Pirazzoli, P.A. (1987) "Sea level changes in the Mediterranean", in Tooley, M.J., and Sherman, I. (eds.) *Sea Level Changes*, Basil Blackwell, N.Y., pp. 152-181

Pugh, D., & Thompson, K. (1986) "The subtidal behaviour of the Celtic Sea - I. Sea level and bottom pressures", Cont. Shelf Res. 5, pp. 293-319

Thompson, K. (1980) "An analysis of British monthly mean sea level", Geophys. J. R. Astr. Soc. 63, pp. 57-73

Thompson, K. & Pugh, D. (1986) "The subtidal behaviour of the Celtic Sea - II. Currents", Cont. Shelf Res. 5, pp. 321-346

Woodworth, P. (1984) "The worldwide distribution of the seasonal cycle of mean sea level", Inst. of Ocean. Science Rpt. No. 190, 91 p.

Woodworth, P. (1987) "Trends in U.K. mean sea level", Marine Geodesy 11, pp. 57-85

PRESENT AND NEAR-FUTURE SEA-LEVEL CHANGES: AN ASSESSMENT

P.A. PIRAZZOLI
URA-141 du CNRS, Laboratoire de Géographie Physique
1, Place Aridstide Briand
92190 Meudon-Bellevue
France

ABSTRACT: The present paper attempts to answer two questions:
1. Is the global sea level rising, and how far? 2. Should we expect a global sea-level rise during the next century?

The answer to the first question is yes; the sea level has risen since the end of the last century, but this rise has been two to three times less than what is generally reported in the literature. In addition, during the last 40 years there has probably been no global sea-level rise at all.

The answer to the second question is much less definite. A significant sea-level rise during the next century is a dangerous possibility that must not be ignored. However too many points are still unclear and it is difficult to say when and to what extent this rise may occur.

1. Introduction

The possible influence on sea level of the increasing "greenhouse" CO_2 effect has gained widespread attention with recent warming scenarios predicted by computer models. In this paper, after a critical review of previous methods of estimating the sea-level changes which have occurred during the last century, a new method, aimed at removing the influence of local effects, is proposed.

It is also shown that estimations of the possible sea-level rise during the next century change with different authors, reflecting considerable uncertainties.

2. Previous Estimates of Recent "Global" Sea-Level Changes

Since the pioneer work by Gutenberg (1941), values between 1.0 and 1.5 mm/year are often quoted in the literature for the present-day rate of eustatic sea-level rise (Table 1).

R. Paepe et al. (eds.), Greenhouse Effect, Sea Level and Drought, 153–163.
© 1990 *Kluwer Academic Publishers.*

The most frequent approach has consisted in averaging records of tide-gauge stations, after having omitted areas of glacio-isostatic of tectonic uplift (Fennoscandia, Northern Canada, Alaska, etc.), but including wide adjacent areas of less spectacular subsidence (North Sea, Western Europe, Atlantic coast of North America, etc., see Table 2).

Table 1. Estimates of global (average) sea-level rise from tide-gauge records

Authors	Number of stations	Period of time considered	Average rate of sea-level rise (mm/yr)
Gutenberg, 1941	69	1807-1937	1.1
Polli, 1952	110	1871-1940	1.1
Cailleux, 1952	76	1885-1951	1.3
Valentin, 1954	253	1807-1947	1.1
Lisitzin, 1958	6	1807-1943	1.1
Fairbridge & Krebs, 1962	unspecified	1860-1960	1.2
Kalinin & Klige, 1978	126	1900-1964	1.5
Emery, 1980	247	1850-1978	3.0
Gornitz et al., 1982	193	1880-1980	1.2
Barnett, 1983	9	1903-1969	1.5
Barnett, 1984	152	1881-1980	1.4
		1930-1980	2.3
Pirazzoli, 1986	229	1807-1984	indeterminable
Gornitz & Lebedeff, 1987	130	1880-1982	0.9-1.2
Pirazzoli, 1989	58	1920-1950	0.75
	(Europe)	1950-1980	0.2
		1880-1980	0.52

Other authors have attempted to palliate the very uneven distribution of tide-gauge stations with sufficiently long records (which are situated mainly around the North Atlantic) by assuming that isolated stations could represent the crustal movements occurring in wide areas or even in continents. This second approach is unacceptable, since global isostatic models and neotectonic field studies have shown that recent vertical movements of the earth surface are likely to differ from one area to the other, especially near continental shelves.

Consequently, the average of all secular trends is biased towards an apparent sea-level rise rather than the eustatic factor, the latter remaining undetermined in the absence of an absolute reference datum. This bias can only be enhanced by estimations in which areas of glacio-isostatic uplift are neglected, whereas areas of less spectacular subsidence remain included in the average.

Only 13% of the stations with sufficiently long records indicate a sea-level rise between

Table 2. Methods for calculating the global eustatic factor.

METHOD	REMARKS
1 - Averaging tide-gauge data (most authors)	■ Uneven geographical distribution of data. ■ The assumption that single stations can represent average neotectonic movements in wide regions is unacceptable. ■ The assumption that crustal movements can be compensated by using a sufficiently large number of stations biases the averages towards an apparent sea-level rise. ■ Ocean dynamics effects disregarded
2 - Subtracting geological trends from tide-gauge data (Gornitz et al., 1982; Gornitz & Lebedeff, 1987)	■ Can provide good results locally ■ Few geological sea-level data are available in the same area as a tide-gauge station ■ Inappropriate in active tectonic areas ■ Ocean dynamics effects disregarded
3 - Applying a correction factor deduced from global isostatic models (Peltier, 1985, 1987, 1988)	■ Applicable to all sites. ■ Simplifying assumptions in the model may cause systematic deviations from field data. ■ Tectonic trends other than glacio-isostatic ones disregarded. ■ Ocean dynamics effects disregarded.
4 - Computing steric sea-level trends with oceanographic data (Thompson & Tabata, 1987)	■ Short records, strong noise. ■ Glacio-eustatic and ocean dynamics factors disregarded.
5 - Subtracting the long-term trends from tide-gauge data and adding an eustatic factor estimated from comparisons between temperature changes and very long tide-gauge records (Pirazzoli, 1989)	■ Land movement and long-term ocean dynamics are removed. ■ The eustatic factor can be estimated only at very few sites.
6 - Using satellite altimetry	■ This method is the only likely to provide an absolute datum. ■ The accuracy of altimetric measurements needs to be improved.

1.0 and 1.5 mm/year. On the other hand, 38% of the stations indicate a rise higher than 1.5 mm/year, 20.5% a rise lower than 1.0 mm/year, 1% a relative stability, and 27.5% a sea-level drop (Pirazzoli, 1986). The overall average represents a rise between 1.0 and 1.5 mm/year, but this should be interpreted as an average of crustal movements (including eventually a eustatic component of unknown amount) rather than a purely eustatic rise. For geological reasons, subsidence is expected to occur more frequently than uplift in oceanic and coastal areas (Table 3).

Table 3. Coastal subsidence is more frequent than coastal uplift.

CAUSES	EFFECTS
River sedimentation ($= 20 \times 10^9$ t/yr)	- Global sea-level rise: 0.02 mm/yr - Local isostatic adjustments: up to several mm/yr
Cooling of oceanic lithosphere	- Hawaii: 2 to 4 mm/yr - Tahiti: 0.4 mm/yr - Atolls: 0.05 to 0.1 mm/yr - Continental margins: 0.03 mm/yr
Hydroisostasy	- Outer Great Barrier Reef: 0.3 mm/yr
Human action	- Locally important, especially in coastal planes

Gornitz et al., (1982) have used radiocarbon-dated sea-level indicators in an attempt to remove local geological trends from tide-gauge data. This method, the most logical one, comes up however against the difficulty that sufficient geological data are seldom available in the same area as a tide gauge station. As many stations have been located in delta or estuary areas, where local subsidence effects due to sediment load or compaction may differ from those of nearby areas, geological data reported from the neighbourhood should be used with caution. At any rate, this method is not appropriate in actively seismic areas, where the long-term trend of vertical movements can be contrary to movements occurring at the time of earthquakes of during interseismic periods (several examples of this have been reported from Japan, Greece and other seismic areas). In addition, this method is unable to remove sea-level changes produced by oceanic dynamics effects such as those mentioned by Fairbridge (1989).

Peltier (1985, 1987, 1988) has proposed to apply a correction factor deduced from global isostatic models based on viscoelastic earth assumptions, which are aimed at describing the interactions between ice sheets, continental masses and oceans. This method, which is not independent from the preceding one since the isostatic models have to calibrated with geological sea-level data, has been used, for instance, on the U.S. coast. It has the great advantage of being applicable to all sites, providing expected isostatic trends even where no geological sea-level data are available. The model depends however on a number of simplified assumptions that may cause systematic deviations from the data observed in the field. Furthermore, it cannot take into account ocean dynamics effects, steric changes, or neotectonic phenomena other than glacio-isostatic ones.

A fourth method, aimed at computing steric sea-level trends during the period 1956-1983, has been used by Thompson and Tabata (1987). With oceanographic data from station PAPA (Northeast Pacific region). The trend calculated has been a sea-level rise of about 1 mm/year. However, a critical examination of the results made by the authors concluded that sea-level changes of such small amplitude would be masked by the great (1-10 cm) interannual variability of open ocean steric heights.

3. A New Approach

In the absence of absolutely stable areas in the world a new, more realistic approach has been chosen (Pirazzoli, 1989), consisting of three steps.

A. The local long-term trends have been filtered from the data, in order to remove crustal movements and obtain short-term sea-level fluctuations.

B. The inspection of very long records (e.g., Amsterdam from 1682 to 1930, and Brest since 1807) has shown that an acceleration of 0.9 mm/year occurred between the end of the 19th century and the middle of the 20th century. As this acceleration can be correlated with a recent change in the global temperature (a 0.5°C rise between 1880 and 1950, with less significant changes before and after this period[1], Jones et al., 1986; Hansen & Lebedeff, 1987) and with a marked melting of mid-latitude glaciers during the same period, it has been interpreted as corresponding to a eustatic rise.

C. The value +0.9 mm/year between 1880 and 1950 has been added to the average local sea-level fluctuations resulting from A. in order to obtain eustatic changes.

This method does not aim at establishing a single value for the global ocean volume change, which in the absence of an absolute datum remains a chimera, but at assessing on a sound basis regional eustatic changes in sea-level, free from neotectonic movements and from long term ocean dynamics effects.

The critical limitation of this method is indeed step B, since almost uninterrupted records starting earlier than 1850 are extremely rare. The results from the North Atlantic area are shown in Fig. 1.

On the European side, when land movements are removed, the sea-level rise of the last century has been only 4 to 6 cm. This value, which is two to three times smaller than the estimation of 10 to 15 cm claimed by most authors, is more consistent with current computations of the effects on sea level of the thermal expansion of the ocean water (2-5 cm, according to Wigley & Raper, 1987) and of the melting of small glaciers (1.4-5 cm

---

[1]. In fact, there was an increase in global temperature of 1 + °C during about 1820-1860 (Lamb, 1977, pers. comment Denness).

according to Meier, 1984, if the period 1880-1950 is considered). On the American side, the sea-level variability is higher, probably in relation with changes in the Gulf Stream and in the Atlantic circulation. In both areas the average sea-level has remained almost stable or has dropped slightly since 1950.

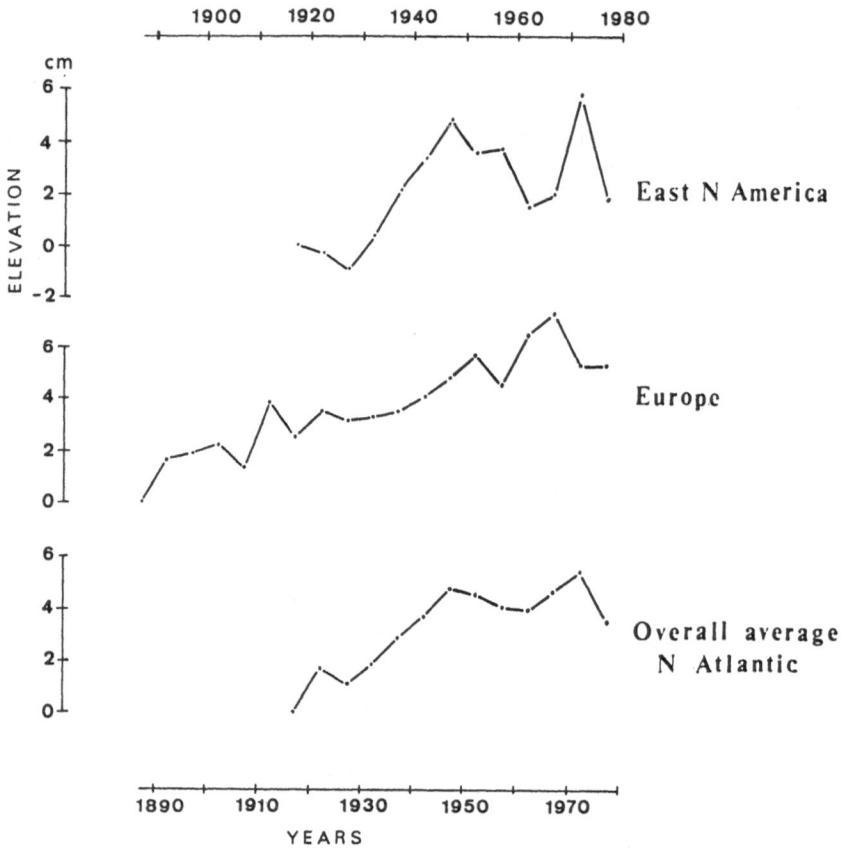

Figure 1. Regional average 5-yr sea-level variation in the North Atlantic. The local secular trends have been subtracted from the data, but the eustatic component in them has been estimated independently and added (from Pirazzoli, 1989).

The only method which could establish and refer to an absolute datum (the centre of the earth's mass), thus providing an unprecedented improvement in sea-level studies, is satellite altimetry. Unfortunately, the accuracy of satellite altimetric measurements is still inadequate to measure the present-day small eustatic changes.

4. Near-Future Sea-Level Trends

Estimations of the possible sea-level change during the next century, in relation to the growing content of CO_2 and other "greenhouse" gases in the atmosphere, vary with different authors from a 7 cm sea-level drop to a rise of over 3.5 m (Table 4).

Furthermore, the varying estimations are hard to assess, due to a confusing mixture of assumptions and methods and to the fact that several kinds of feedback phenomena have not yet been taken into account. It is well known, for example, that about half the CO_2, injected into the atmosphere since the middle of the last century has disappeared, probably absorbed in the ocean. The limits of the capacity of absorbtion by the ocean, and possibly by the biosphere, are largely unknown, however, and makes linear or exponential extrapolations into the future of the recent atmospheric CO_2 trends unsound.

Table 4. Estimates of total future global sea-level rise (cm).

Authors	Year			
	2000	2025-2030	2050	2085-2100
Gornitz et al., 1982			40/60	
Revelle, 1983				70
Hoffman, 1984	4.8/17.1	13/55	23/117	56/345
Polar Res. Board, 1985				10/160
US Dept. Energy, 1985				10/90
Hoffman et al., 1986	3.5/5.5	10/21	20/55	44/368
Thomas, 1986				64/230
Jeager, 1988	-5/19	-4/52	-7/138	
Clim. Res. Unit., 1988		6/34		
Robin, 1986		25/165	(year unspecified)	

Analysis of the air bubbles entrapped in polar ice cores has shown that during the late Quaternary the atmospheric content of CO_2 increased about 40% between the last glacial maximum and the pre-industrial Holocene period (Lorius et al., 1985). In pre-industrial times (between about 1650 and 1850 AD) atmospheric CO_2 was about 260-270 ppm (Raynaud & Barnola, 1985). As much CO_2 has been added between 1850 and 1985 as had been by the normal increase over thousands of years between a glacial stage and an interglacial stage. In contrast, while the eustatic change has been of the order of 10^2 m since glacial times, there has been negligible (10^{-1} m or less) change in the last century (Fig. 2).

Consequently, changes of atmospheric CO_2 doe not seem to be a main direct cause of climatic change during the late Quaternary. If not merely a response to these changes,

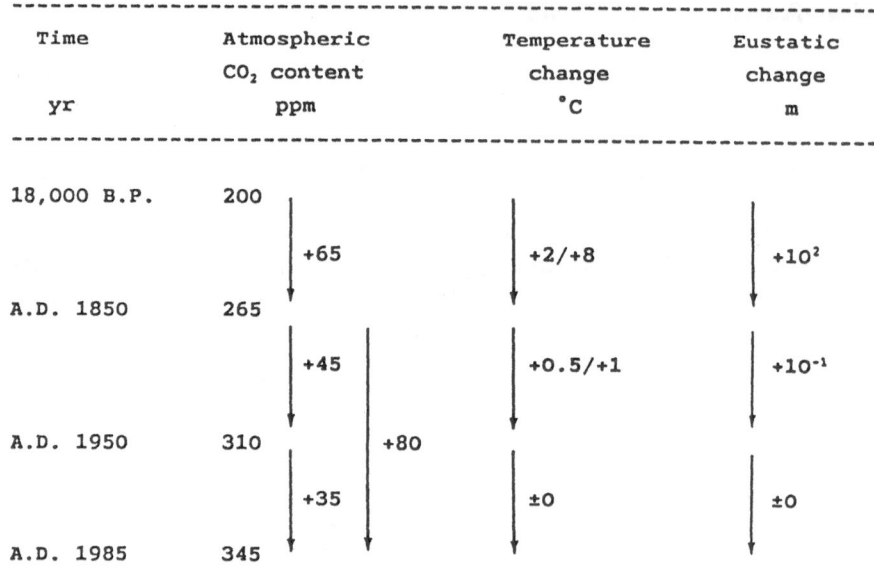

Time	Atmospheric CO₂ content	Temperature change	Eustatic change

In LaTeX table form:

Time yr	Atmospheric CO_2 content ppm	Temperature change $°C$	Eustatic change m
18,000 B.P.	200		
	+65	+2/+8	$+10^2$
A.D. 1850	265		
	+45	+0.5/+1	$+10^{-1}$
A.D. 1950	310 (+80)		
	+35	±0	±0
A.D. 1985	345		

Figure 2. Variations in atmospheric CO_2 content, average temperature and sea level since the last glacial maximum, and during the last century.

they may be at most a part of the system by which orbital variations induce changes in climate.

Absorption of atmospheric CO_2 by the oceans depends on temperature and salinity of the surface ocean and on the presence of plankton (IGBP, 1988) and is greater at high latitudes (Faure, 1987). Here an increase in temperature would result in longer seasonal periods of ice-free polar seas, with increased absorption of CO_2 by the ocean, and therefore in a decrease of the "greenhouse" effects.

At low latitudes, where absorption by the atmosphere of oceanic CO_2 may increase, a rise in temperature would result in more evaporation and cloud cover, and therefore in less absorption of solar heat by the earth and the ocean (Ramanathan et al., 1989). This would cause the temperature to decrease.

Uncertainties on future "greenhouse" effects are increase by the fact that it is still obscure whether the slight rise in the average air surface temperature observed since 1880 should be ascribed to "greenhouse" effects, or merely to minor and random fluctuations to be expected in our complex climatic system. In particular, the slightly falling or almost constant temperature observed between about 1940 and 1980, still within a period of rapid increase in atmospheric CO_2, is especially puzzling to those who ascribe the slight rise in temperature during the first part of this century to "greenhouse" effects. Far from fitting their models, this near stability suggests rather that no distinct relation between CO_2 and temperature can yet be detected.

Although some indications of warming in the 1980's are unequivocal (Hansen & Lebedeff, 1988), they may have been enhanced in part by the exceptional 1982-1983 El Niño and a clear greenhouse-induced pattern is still not evident, since the greatest warmings in the 1980's appear at mid-latitudes, not at high latitudes as predicted by computer models (Kerr, 1988).

5. Conclusions

As concerns sea level, one may anticipate that after a relative stability since about 1950, the sea level will start rising again during the next years, if not already underway. This will be a consequence of the slight increase in temperature observed since the middle of the 1960's and of the time lag of 18 to 19 years existing between changes in temperature and the advance or retreat of small glaciers, and also of the thermal relaxation time for the upper layers of the ocean.

Whether this expected small rise (1 to 2 cm) will come to an end two decades later, or continue or even accelerate, will depend on the future evolution of the global air-surface temperature. Sea level will be expected to follow the same evolution, with a time lag of 18 to 19 years.

In conclusion, the relationship between the atmospheric CO_2 content, temperature and sea level is far from being demonstrated regarding the recent past. However, in spite of many uncertainties and of the weakness of many present-day approaches, the possibility that an atmospheric increase in CO_2 and other gases could finally lead to a "greenhouse" increase in temperature and a rapid sea-level rise is ominous enough not to be set aside or ignored easily. The consequences of a possible sea-level rise would be dramatic for mankind, since two-thirds of the world's population are living within the coastal zone, as defined by the United Nations experts. As suggested by Pirazzoli et al. (1989), further studies and new interdisciplinary approaches are necessary, in order to test assumptions and verify models.

6. References

Barnett, T.P. (1983) "Recent changes in sea level and their possible causes", Clim. Change 5, pp. 15-38

Barnett, T.P. (1984) "The estimation of 'global' sea level change: a problem of uniqueness", J. Geoph. Res. 89, C5, pp. 7980-7988

Cailleux, A. (1952) "Récentes variations du niveau des mers et des terres", Rev. Générale d'Hydraulique 48, pp. 310-315

Emery, K.O. (1980) "Relative sea levels from tide-gauge records", Proc. Nat. Acad. Sci. USA 77, 12, pp. 6968-6972

Fairbridge, R.W. (1989) "Letter to Nature", (submitted)

Fairbridge, R.W., & Krebs, O.A.Jr. (1962) "Sea level and the Southern Oscillation", Geoph.

162

J., 6, pp. 532-545

Faure, H. (1987) ""Mécanisme d'amplification du cycle climatique global: l'effet de couvercle de la glace de mer contrôle le CO_2 atmosphérique", C. R. Acad. Sci. II, 305, pp. 523-527

Gornitz, V. (1989) "Mean sea-level changes in the recent past", Intern. Workshop on *Climatic Change, Sea level, Severe Tropical Storms and Associated Impacts*, Norwich, Sept. 1987 (in press)

Gornitz, V., & Lebedeff, S. (1987) "Global sea-level changes during the past century.", in Nummedal, D et al. (eds.) *Sea-Level Fluctuation and Coastal Evolution*, Soc. Econ. Paleont. Miner., Special Publ. n° 41, pp. 3-16

Gornitz, V., Lebedeff, S., & Hansen, J. (1982) "Global sea level trend in the past century.", Science, 215, pp. 1611-1614

Gutenberg, B. (1941) "Changes in sea level, postglacial uplift, and mobility of the Earth's interior.", Bull. Geol. Soc. Am. 52, pp. 721-772

Hansen, J., & Lebedeff, S. (1987) "Global trends of measured surface air temperatures.", J. Geophys. Res. 92, D11, pp. 13,345-13,372

Hansen, J., & Lebedeff, S. (1988) "Global surface air temperatures: update through 1987.", Geophys. Res. Lett. 15, 4, pp. 323-326

Hoffman, J.S., Keyes, D., & Titus, J.G. (1983) "Projecting future sea level rise: methodology, estimates to the year 2100, and research needs.", U.S. Environmental Protection Agency, Washington D.C., 121 p.

Hoffman, J.S., Wells, J.B., & Titus, J.G. (1986) "Future global warming and sea level rise.", in Sigbjarnason, G. (ed.) *Iceland Coastal and River Symposium*, Reykjavik, pp. 245-266

IGBP (1988) *A plan for Action*, IGBP Secr., R. Swedish Acad. Sci., Stockholm, 200 p.

Jeager, J. (1988) "Developing policies for responding to climatic change.", World Climate Programme, Impacts Studies, WCIP-1, WMO/TD-No. 225, 53 p.

Jones, P.D., Wigley, T.M.L., & Wright, P.B. (1986) "Global temperature variations between 1861 and 1984.", in *Nature* 322, 6078, pp. 430-434

Kalinin, G.P., & Klige, R.K. (1978) "Variation in the world sea level", in *World Water Balance and Water Resources of the Earth, Studies and Reports in Hydrology*, No. 25, Unesco, pp. 581-585

Kerr, R.A. (1988) "The weather in the wake of El Niño.", Science, 240, p. 883

Lisitzin, E. (1958) "Le niveau moyen de la mer.", Bull. Inf. Comité Central d'Océanogr. et d'Etudes des Côtes (COEC) 10, pp. 254-262

Lorius, C., Jouzel, J., Ritz, C., Merlivat, L., Barkov, N.I., Korotkevich, Y.S., & Kotlyakov, V.M. (1985) "A 150,000 year climatic record from Antarctic ice.", in *Nature* 316, pp. 591-596

Meier, M.F. (1984) "Contribution of small glaciers to global sea level.", Science, 226, 4681, pp. 1418-1421

Peltier, W.R. (1985) "Climatic implications of isostatic adjustment constraints on current variations of eustatic sea level", in *Glaciers, Ice Sheets, and Sea Level: Effects of a CO_2-Induced Climatic Change*, U.S. Dept. Energy, DOE/ER/60235-1, pp. 92-103

Peltier, W.R. (1987) "Mechanisms of relative sea-level change and the geophysical responses to ice-water loading.", in Devoy, R.J.N. (ed.) *Sea Surface Studies*, Croom Helm, London, pp. 57-94

Peltier, W.R. (1988) "Global sea level and earth rotation.", Science, 240, pp. 857-956

Pirazzoli, P.A. (1986) "Secular trends of relative sea-level (RSL) changes indicated by tide-gauge records.", J. Coast. Res., Sp. Issue 1, pp. 1-26

Pirazzoli, P.A. (1989) "Recent sea-level changes in the North-Atlantic.", in Scott, D.B., Pirazzoli, P.A., and Honig, C.A. (eds.) *Late Quaternary Sea-Level Correlation and Applications*, Kluwer, Dordrecht, NATO ASI Series C, 256, pp. 153-167

Pirazzoli, P.A., Grant, D.R., & Woodworth, P. (1989) "Trends of relative sea-level change: past, present and future.", Quaternary International (in press)

Polli, S. (1952) "Gli attuali movimenti delle coste continentali", Ann. Geof. 5, 4, pp. 597-602

Ramanathan, V., Cess, R.D., Harrison, E.F., Minnis, P., Barkstrom, B.R., Ahmad, E., & Hartmann, D. (1989) "Cloud-radiative forcing and climate: results from the Earth Radiation Budget Experiment.", ...

Raynaud, T., & Barnola, J.M. (1985) "An Antarctic ice core reveals atmospheric CO_2 variations over the past few centuries.", in *Nature* 315, pp. 309-311

Revelle, R. (1983) "Probable future changes in sea level resulting from increased atmospheric carbon dioxide.", in *Changing Climate*, Nat. Acad. Sci., Washington, D.C., pp. 433-447

Robin, G. de Q. (1986) "Changing the sea level.", in Bolin, B., et al. (eds.) *The Greenhouse effect. Climatic Change and Ecosystems*, Scope 29, Wiley, pp. 323-359

Thomas, R.H. (1986) "Future sea level rise and its early detection by satellite remote sensing.", in Titus, J.G. (ed.) *Effects of Changes in Stratospheric Ozone and Global Climate*, vol. 4, Sea Level Rise (quoted by Titus, J.G. (ed.) *Greenhouse Effect, Sea Level Rise and Coastal Wetlands*, U.S. EPA, Washington, 1988)

Thompson, R.E., & Tabata, S. (1987) "Steric height trends at Ocean Station PAPA in the Northeast Pacific ocean.", Mar. Geod. 11, pp. 103-111

United States Department of Energy (1985) *Glaciers, Ice Sheets, and Sea Level: Effect of a CO_2-Induced Climatic Change*, DOE/ER/60235-1, 348 p.

Valentin, H. (1954) "Die Küsten der Erde.", Veb. Geogr. -Kartogr. Anstalt Gotha, 118 p.

Wigley T.M.L., & Raper, S.C.B. (1987) "Thermal expansion of sea water associated with global warming.", in *Nature* 330, pp. 127-131

INFLUENCE OF GLOBAL CLIMATIC PROCESSES ON THE HYDROSPHERE REGIME

R.K. KLIGE
Water Problems Institute of the
USSR Academy of Sciences
13/3 Sadovo-Chernogriazskaya
103064 Moscow, USSR

ABSTRACT: During the last century geophysical changes, presumably natural energetic processes, resulted in gradual climatic warming. It has caused intensification of natural water exchange, continental drying, glacier retreat and ocean replenishment. Increasing anthropogenic influence makes the water exchange period longer. All these processes have resulted in a global sea level rise at a rate of 1.5 mm per year. Future warming could amplify this process significantly.

Studies of the most recent decade indicate that quite significant climatic changes are connected with geophysical processes, particularly fluctuations of natural energetic factors.

Gradual increase in the mean global surface air temperature occurred during the last century in the Northern and Southern hemispheres, both on continents and in oceans (Fig. 1). Global climatic changes during the period of observations were determined by a complex of factors, the input of solar radiation to the Earth's surface being the primary one.

Recent climatic processes are closely connected with man's influence: greenhouse gases ejection, oil film formation in the ocean, vegetational successions, water consumption, agricultural activities, etc., which in turn, influence not only human activity but the conditions of human existence.

As a result of climatic warming intensity of natural water exchange increases, continents become drier, glaciers retreat and oceans extend. Growth of anthropogenic influence intensifies water vapour exchange, increases its duration and modifies natural water quality.

The considerable influence of the Earth's planetary evolution, i.e. the geological processes changing the Earth surface, on recent natural process variations should also be taken into account. Geological studies demonstrate topographical contrasts and a hydrosphere volume increase. This conclusion can be well shown by the analysis of ocean level fluctuations during the last 300,000 - 500,000 years. Its average fall rate is 0.3 mm per year until the end of the last ice age.

Ocean level observations show a general tendency towards sea level rise during the last 250 - 300 years at a rate of 1 mm per year with relative falls and rises occurring with a

R. Paepe et al. (eds.), Greenhouse Effect, Sea Level and Drought, 165–181.
© 1990 *Kluwer Academic Publishers.*

period of some 33 years. During the recent century the rate of sea-level rise has reached 1.5 mm per year. The most intensive rise at a rate of more than 3 mm per year occurred from 1924 to 1948 (Fig. 2). Now the sea level is falling at a rate of more than 2 mm per year on 4.7 % of the world shoreline and up to 2 mm per year on 7.9 %. It is relatively stable at 24 % of stations and is rising on 48.2 % at a rate of less than 2 mm per year. On 14.3 % of the world shoreline it is rising at a rate of more than 2 mm per year.

Interannual sea level rise is determined by changes in the global water exchange system, which, in turn, is associated with thermal fluctuations (Fig. 3). Sea level has risen simultaneously with a global temperature rise of 1°C but with a 19-year time lag.

Global temperature rise and consequent activation of evaporation are accompanied by an increase in precipitation over both oceans and continents (Fig. 4). It results in humidification, which is more pronounced in coastal and nearby continental slopes. On the whole, the last century is characterized by a precipitation increase over continents (excluding Antarctica) at a rate of 0.25 mm a year (31 km^3 or 0.03 % per year).

Thermal fluctuations have caused a significant change in the global atmospheric circulation pattern. Total evaporation from the continents changed by some 0.3 mm (or 0.07%) per year; maximum changes were 6.5 % per year. Maximum values were typical for the 20's and 50's of the twentieth century; minimum values for the 80's and 90's of the nineteenth century (Fig. 5).

In the period under consideration the inland regions of continents generally tended to become drier due to the fact that the climate tended to become warmer. Thus, the surface runoff into undrained regions of Asia reduced by some 150 km^3, or 34 % of its average value. Runoff into undrained regions of Europe fell by some 50 km^3 (or 16 %) (Fig. 6).

Runoff decrease and evaporation increase have provoked a decrease in humidity, which is most noticeable in inland regions (Fig. 7) as a level and volume decrease of inland lakes. The present period is characterized by a significant fall in groundwater resources.

Temporal variations of groundwater level and volume can be observed in many areas (33 million km^2 or 25 % of the area of the Northern Hemisphere). According to this, groundwater level has fallen since the beginning of the twentieth century, the most reliable data relating to its fall during the period 1919-1950, particularly with the acceleration of this process in 1930, 1936-1939, 1947, 1950 (Fig. 8). A slight indication of a trend towards sea level rise began in 1951. In general, on continents it was falling by 8 mm per year so that its global volume fell by 108 km^3 per year.

Calculations show that during 1880-1975 a significant change occurred in the volume of mountain glaciers (Fig. 9). During the 1880's the main trend was towards the reduction in glacier mass due to reduced accumulation from precipitation. In 1900 for a short period of time glacier balance was stable (1902, 1910, 1924) with even positive accumulation in certain years (1913, 1917). This was caused by a gradual increase in precipitation and an inadequately developed warming trend. Critical disturbance of glacier balance was recorded in the 1930's and 1940's, when maximum increase in summer temperatures coincided with a sharp (8 %) fall of precipitation. By the 60's and 70's this process slackened as a result both of summer temperature fall and an increase in precipitation. On the whole, during the period of 1882-1975 the resulting water balance of continental mountain glaciers was negative (minus 23-25 km³/year, or 0.06 % of the total glacier mass).

Continental mountain glacier water volume diminished during the period of 1894-1975 by 1,886 km³ or 5 %. During 1928-1958 the water content of mountain glaciers decreased on a yearly basis by 58 km³, of Arctic island glaciers by 37 km³, of Greenland by 275 km³ (Fig. 10), and of Antarctic glaciation by 770 km³. At present the rate of glacier melting is 170 km³ per year.

Human activity exerts an increasing influence on the hydrological regime of the continents. This results in a perceivable increase in the water balance and water transfer through evaporation during irrigation development and expansion of reservoir surface areas. At the same time a steadily growing water volume (69 km³ per year) is impounded in artificial reservoirs (Fig. 11).

In general, as a result of both natural and man-induced processes, the total volume of surface waters and groundwater of the continents from 1894 to 1975 diminished by 320 km³/year. This process accelerated to 1361 km³/year in some years during 1918-1952 (Fig. 12).

Changes in the continental water balance are closely related to general global water balance fluctuations, including the balance of the World Ocean. From the end of the nineteenth century, when anomalies of potential evaporation reached -40 cm, till the 40's, the total evaporation from the ocean surface increased by 74 mm or almost 27,000 km³ (5 %) from its mean value for the period of 1894-1975 to a value of 507,153 km³. Thereafter up to the 60's evaporation diminished by 2 %.

The increase in the ocean volume changed during the period under consideration from -10,194 km³/year in 1916 to +9,244 km³/year in 1914 and 1941, with a mean value of 316 km³/year. For the period of 1908-1958 thermally induced increase in the ocean volume was 1,103 km³/year. Bearing in mind the processes of artificial reservoir construction, the natural increase was about 1,118 km³/year.

The volume of the oceans increased by about 56,253 km³ during 1894-1975. As a result of anthropogenic influence, the global temperature will continue to rise, thus resulting in glaciers' melting and partial decay. As a result the ocean level will probably rise by 3 - 5 m and some 1 million km² of land will be inundated in coastal regions.

Table 1. Changes in the Earth Water Components for the Period of 1900-1985.

Water Exchange Components	Total volume, km³	Per Cent of the whole hydrosphere	Changes in volume km³/year	Total changes km³	Per Cent of the change for whole period	Per Cent of the total water balance change
Lakes	176.4×10^3	0.01	-63	-4.79×10^3	-2.72	-10.31
Groundwater	34.2×10^6	2.47	-136	-1034×10^3	-0.03	-22.26
Antarctic Glaciation	21.6×10^6	1.55	-315	-23.94×10^3	-0.11	-51.56
Greenland Glaciation	2.3×10^6	0.17	-82	-6.23×10^3	-0.27	-13.42
Arctic Island Glaciers	83.5×10^6	6×10^{-3}	-12	-0.91×10^3	-1.09	-1.96
Mountain Glaciers	40.6×10^3	3×10^{-3}	-3	-0.23×10^3	-0.57	-0.49
Reservoirs	5×10^3	0.36×10^{-3}	+69	$+5.24 \times 10^3$	+100	+11.29
World Ocean	1.34×10^9	96.5	+542	$+41.19 \times 10^3$	$+3.1 \times 10^{-3}$	+88.71

Table 2. Influence of Water Exchange Variations on the Global Sea Level (1900-1985).

Water Balance Components	Volume Changes km'/year	Total Volume Changes km'	Global Sea Level Changes km'/year	Per Cent of Total Global Sea Level Changes
Lakes	-63	-4788	-0.17	+10.3
Groundwater	-136	-10336	-0.38	+22.3
Antarctic Glaciation	-315	-23940	-0.87	+51.6
Greenland Glaciation	-882	-6232	-0.32	+13.4
Arctic Island Glaciation	-12	-912	-0.03	+13.4
Mountain Glaciers	-3	-228	-0.01	+0.5
Water Reservoirs	+69	+5244	+0.19	-11.5
World Ocean	+542	+41192	+1.50	+88.7

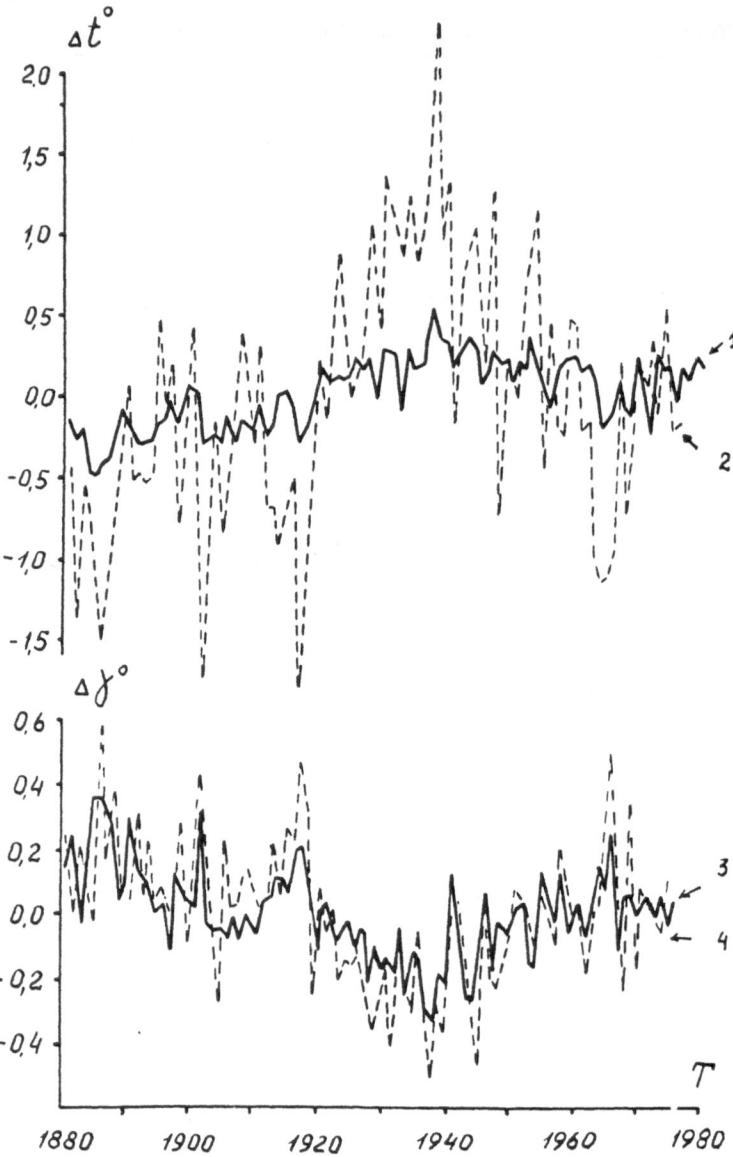

Figure 1. Air temperature changes in anomalies for latitudes: 17°5' - 87°5' N (2); variations of annual (3) and winter (4) latitudinal temperature gradient for 25° - 75° N.

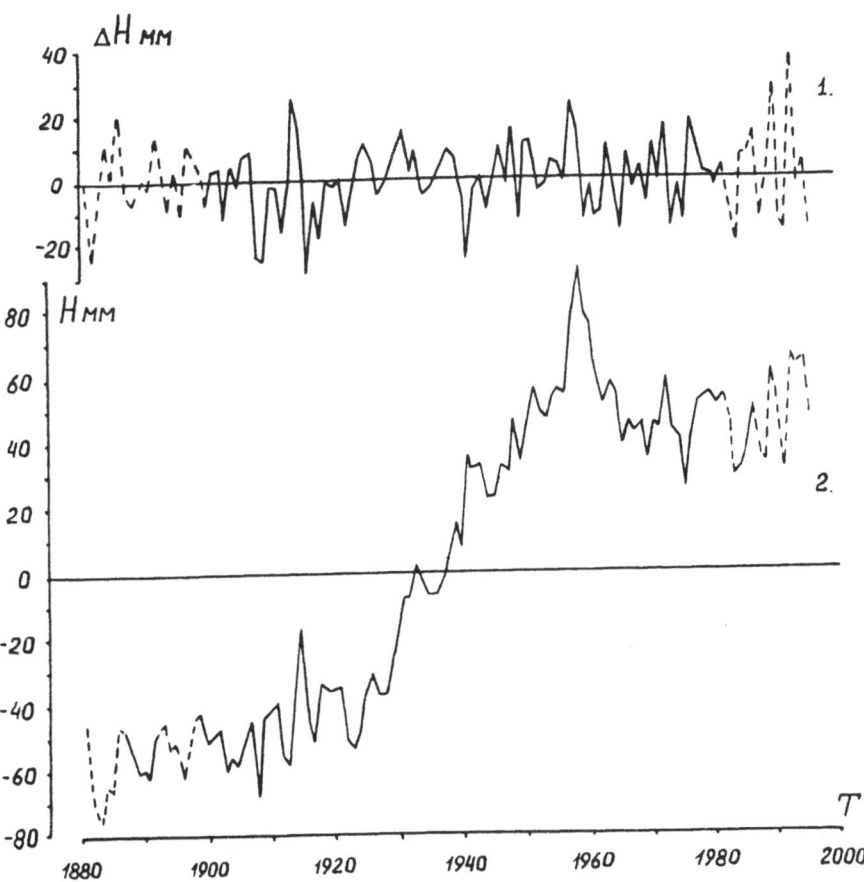

Figure 2. Annual augmentations and mean changes of the World Ocean Level.

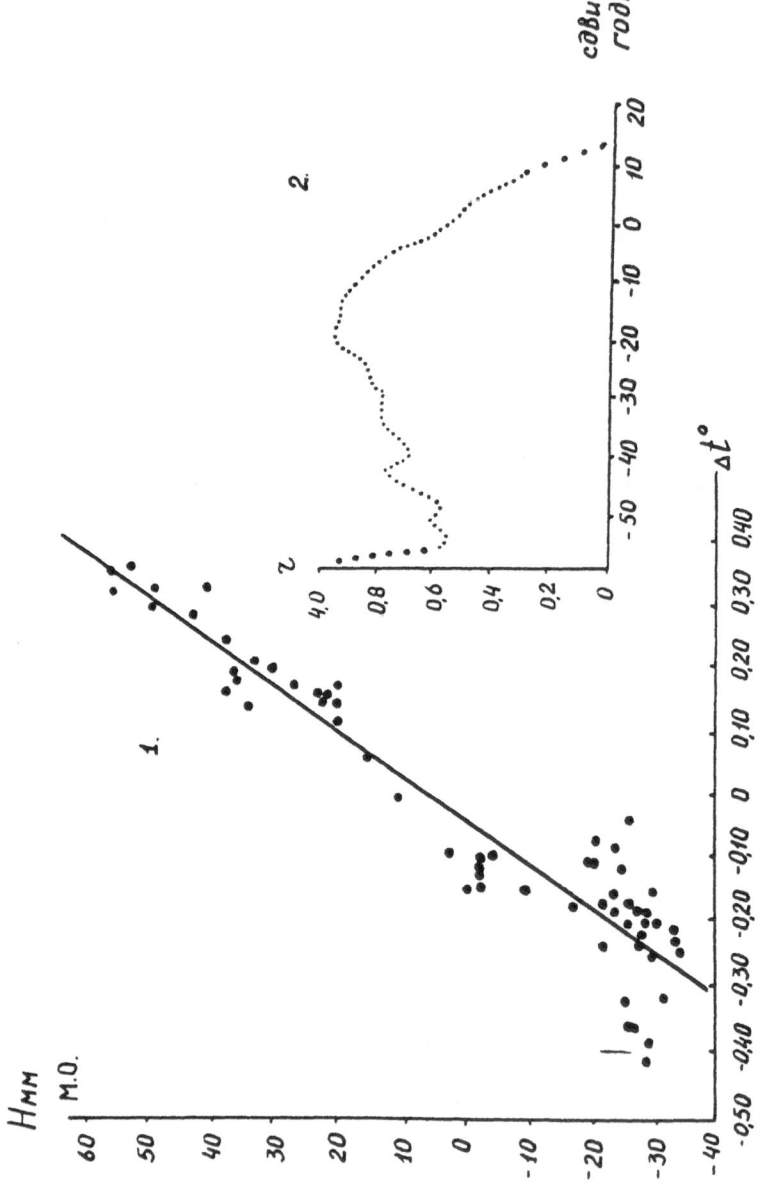

Figure 3. Ocean level vs. air temperature anomalities with 19-year las (1) and autocorrelation function (2).

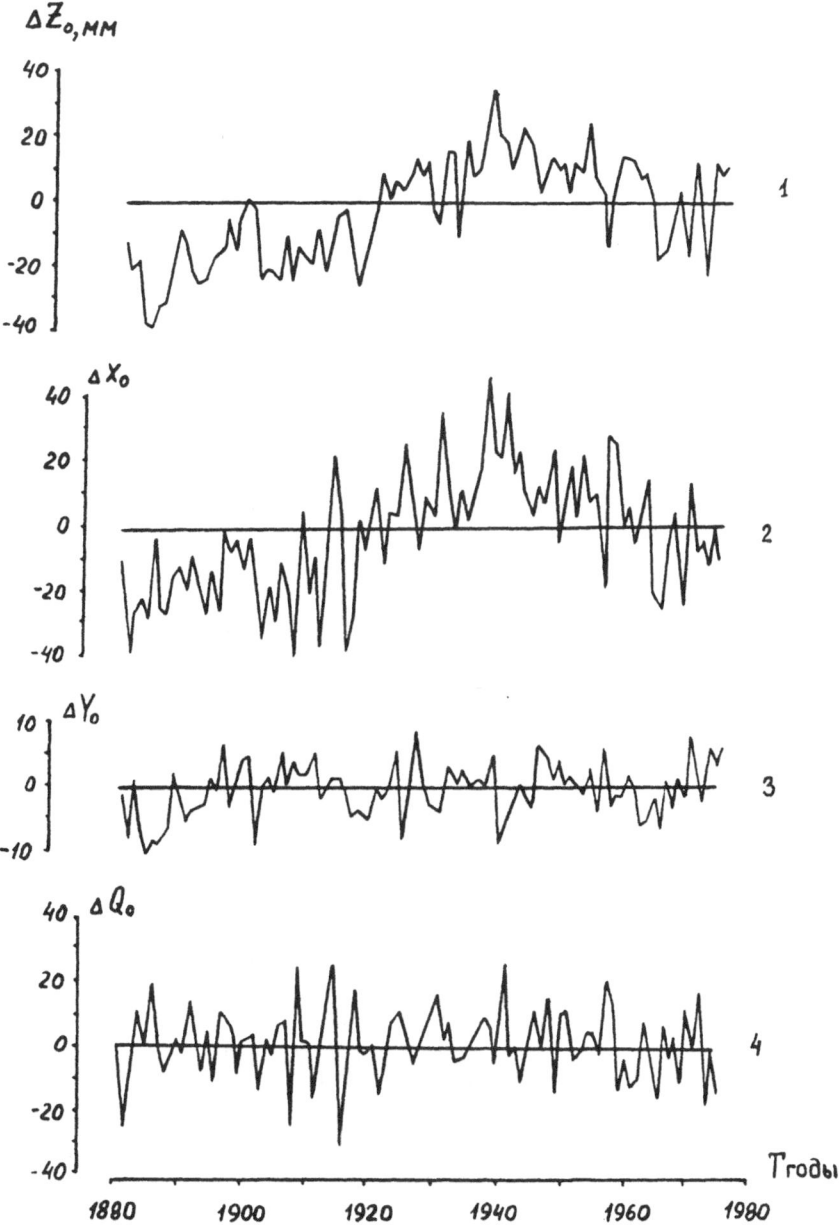

Figure 4. Changes of the World Ocean water balance components in anomalies; evaporation (1), precipitation (2), river and glacier runoff (3), annual ocean volume augmentation (4).

174

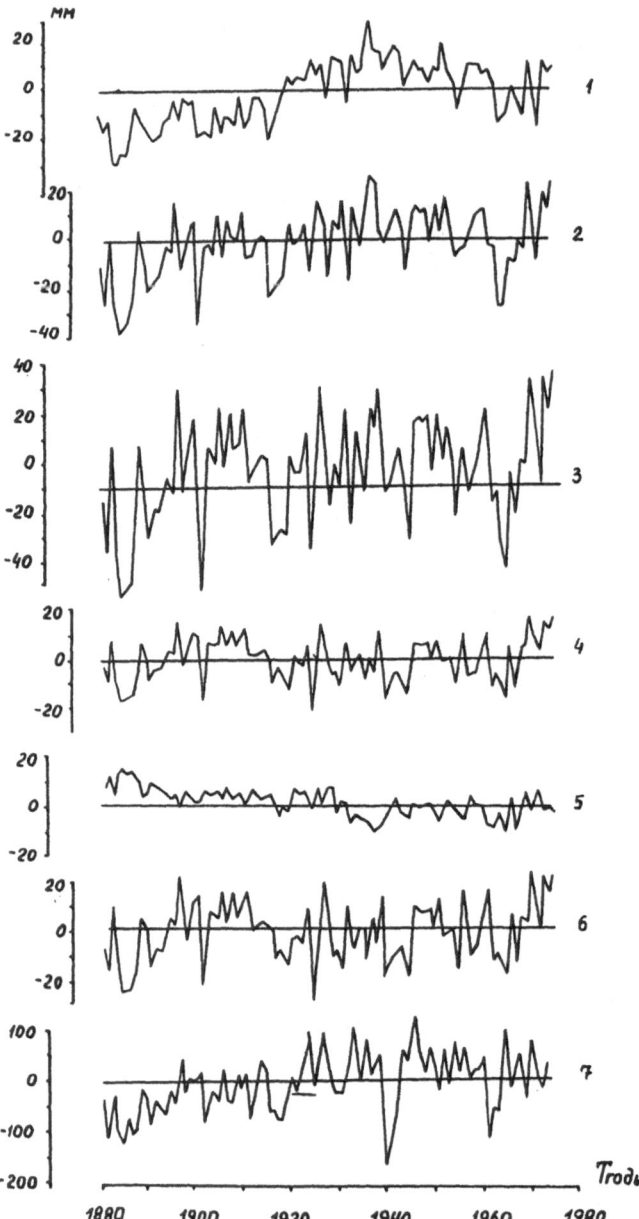

Figure 5. Changes in inland water balance components: maximum possible evaporation (1), actual evaporation (2), precipitation (3), total continental river runoff (4), enclosed areas runoff (5), surface and groundwater runoff from continents (6) and islands (7).

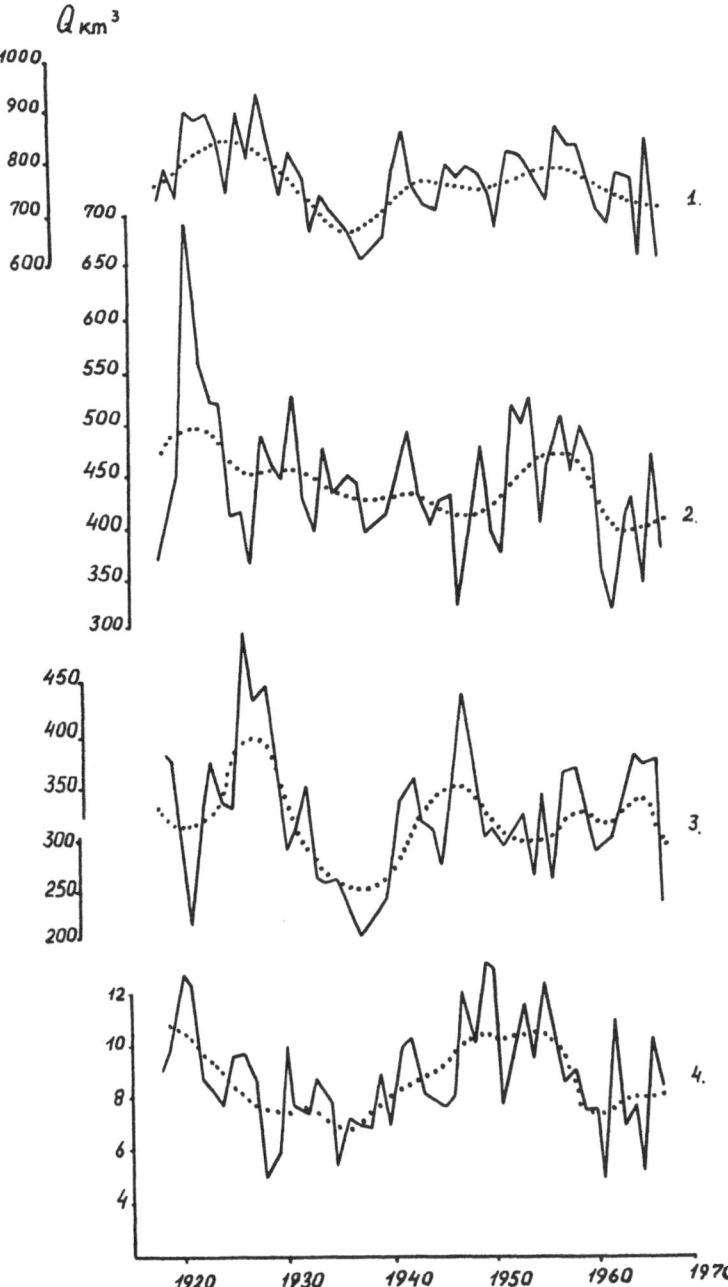

Figure 6. River runoff of enclosed areas.

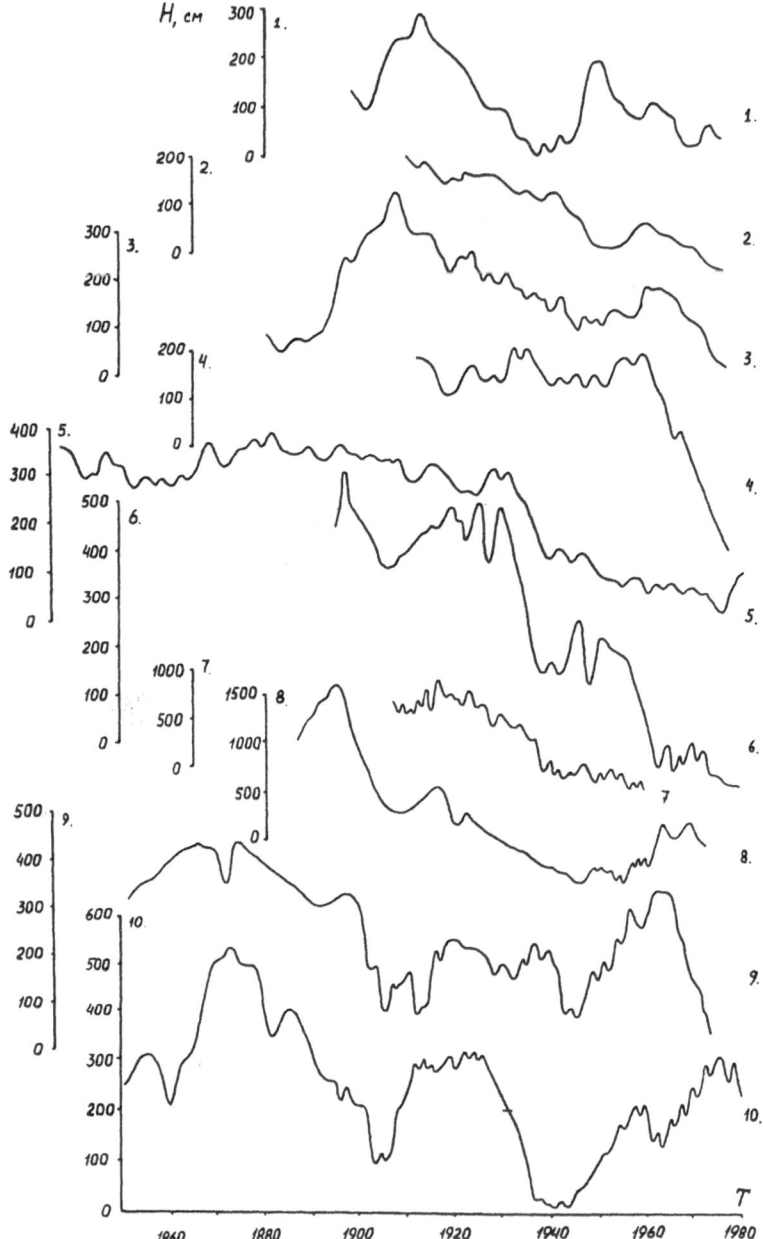

Figure 7. Changes in lake levels: Chang Lake (1), Issyk-Kul Lake (2), Baklash Lake (3), Aral Sea (4), Caspian Sea (5), Dead Sea (6), Neivasha Lake (7), Rudolf Lake (8), Lake Chad (9), and Great Salt Lake (10).

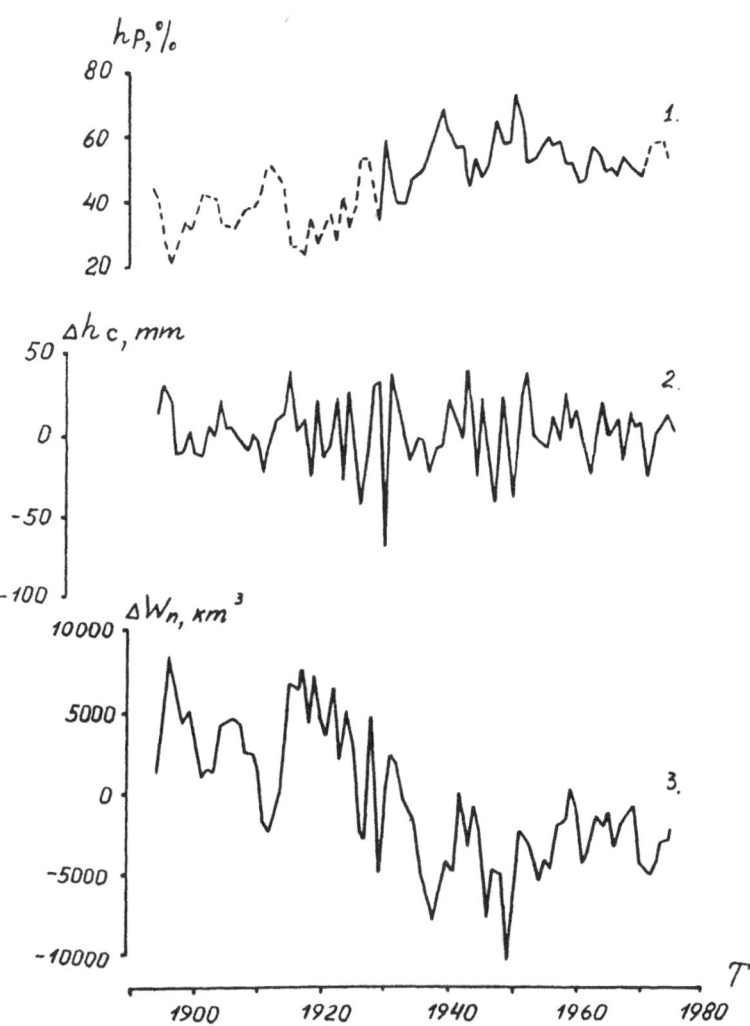

Figure 8. Changes of groundwater level in per cent of predictability (1), its annual augmentation (2), total volume augmentation for the continents (3).

178

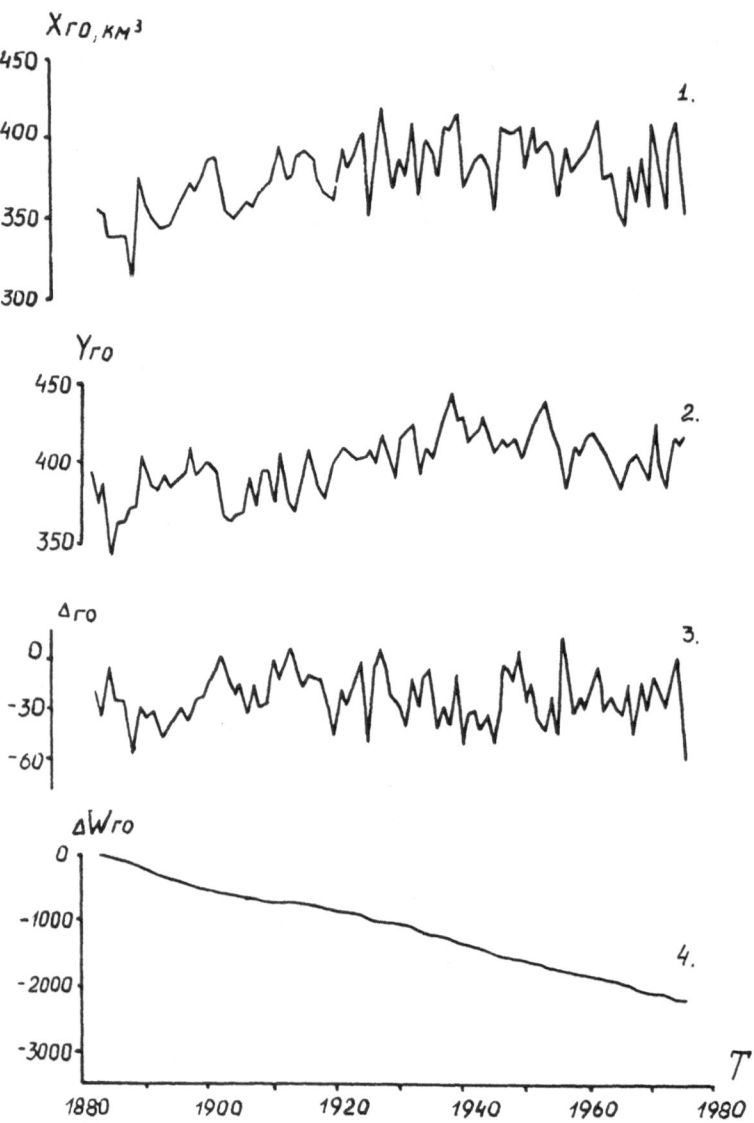

Figure 9. Changes in precipitation: (1), meltwater runoff (2), balance, volume augmentation (4) in mountain glacier regions.

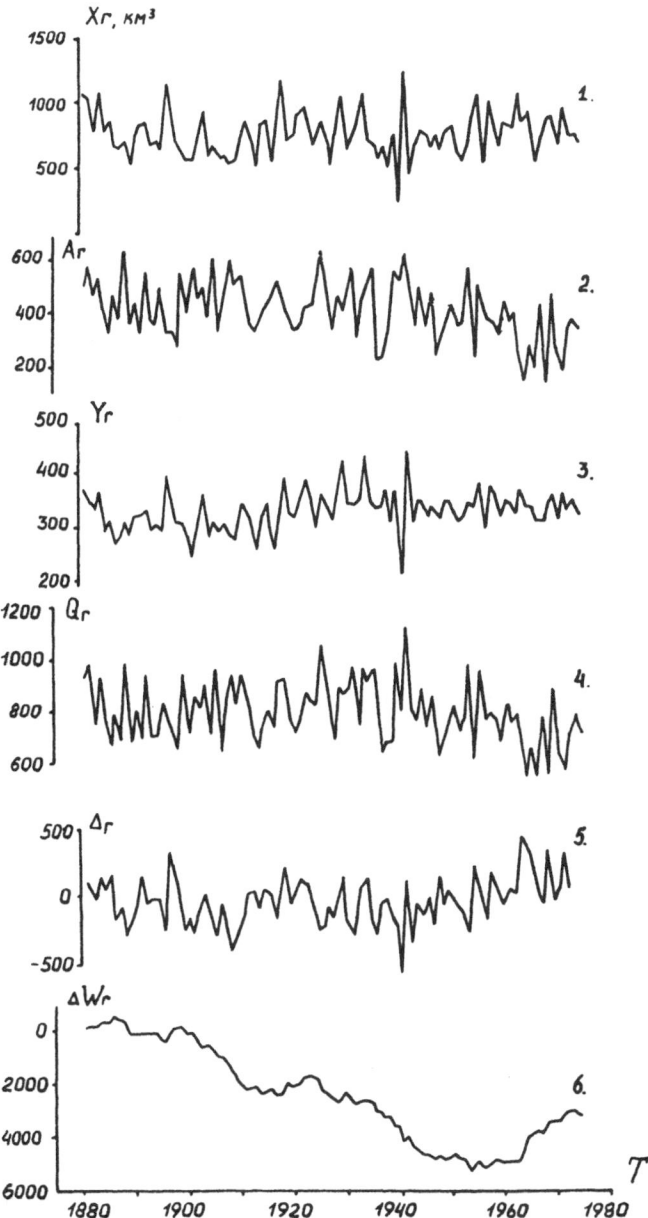

Figure 10. Greenland water balance changes: precipitation (1), iceberg flow (2), total water discharge (3), total ice/water discharge (4), resulting water balance (5), ice/water supply augmentation (6).

180

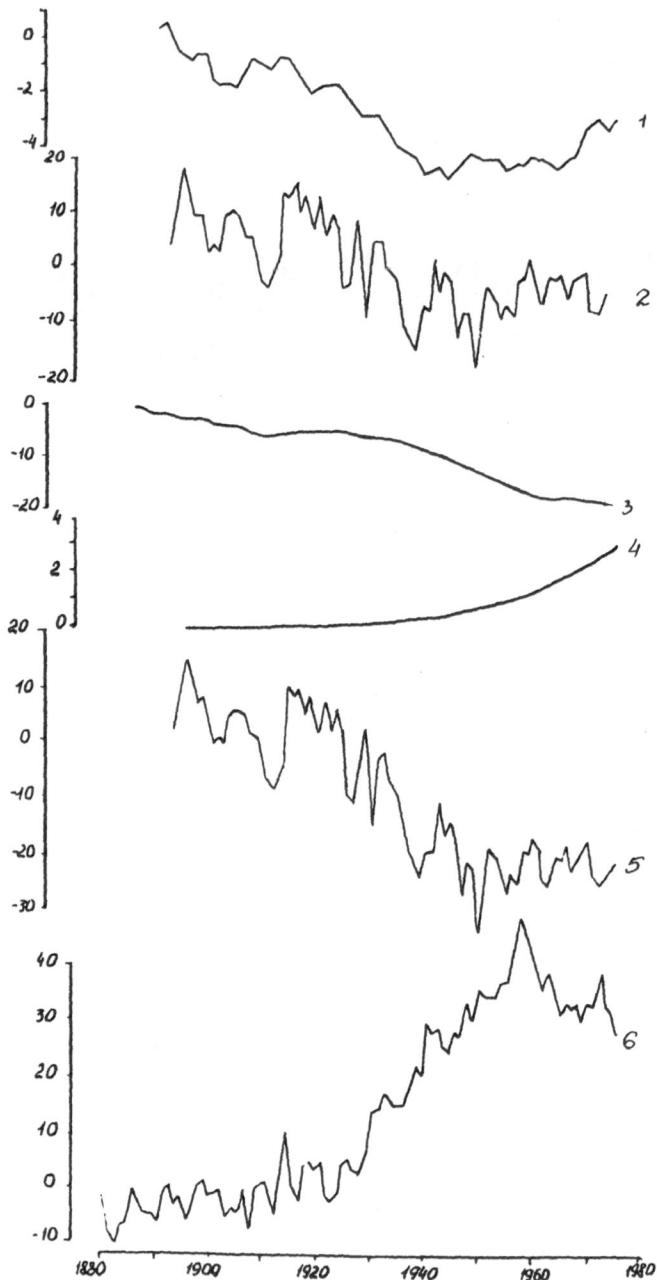

Figure 11. Changes in the world water consumption (Q), consumptive water losses (q), additional precipitation (Δx), additional surface runoff (Δy).

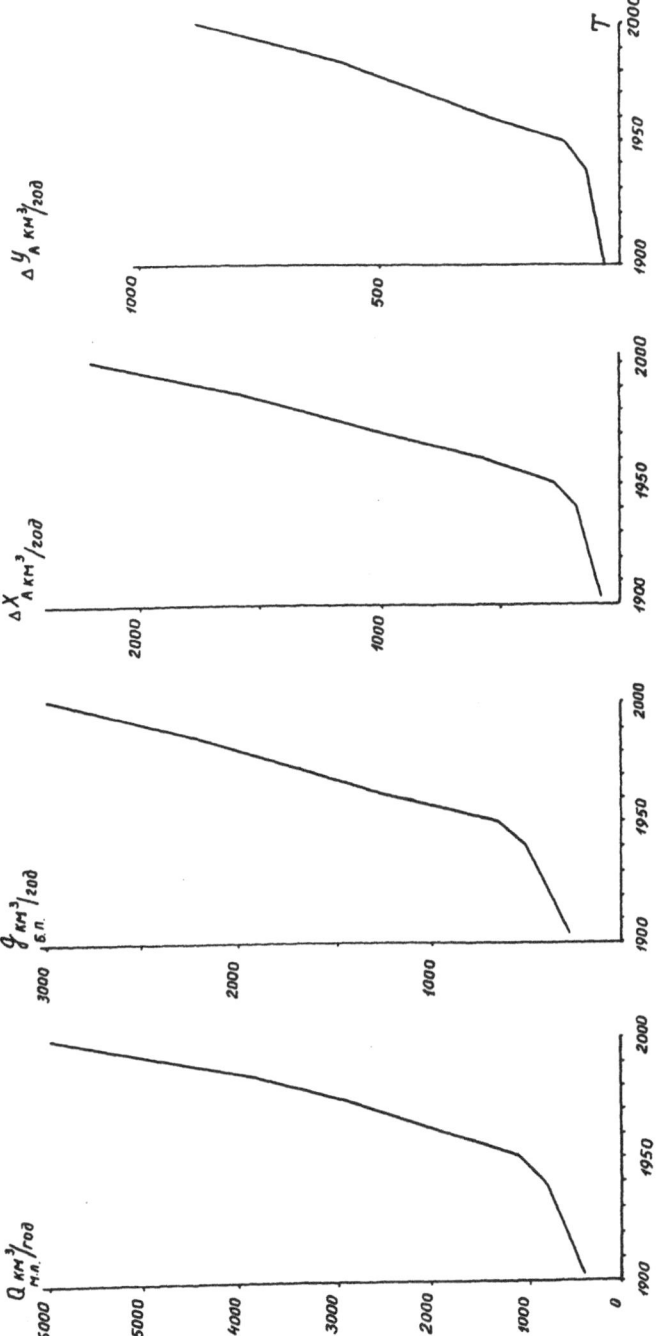

Figure 12. Changes in the water supply: lakes (1), groundwater (2), glaciers (3), water reservoirs (4), land waters on the whole (5), ocean waters (6).

PART III: DROUGHT AND WATER DEFICIENCY

WATER DEFICIENCY VERSUS WATER EXCESS: GLOBAL MANAGEMENT POTENTIAL

RHODES W. FAIRBRIDGE
Columbia University and NASA-GISS
2880 Broadway, New York, NY 10025

1. Discussion Statement

The availability of water, in liquid form, is one of the essential metabolic requirements for almost all forms of life, from the simplest unicelled organism up to the human race. Unlike other creatures, however, human being have the intellectual capacity to utilize far more water than is required for mere survival, and dedicates this excess to various purposes relating to such things as personal comforts (e.g. air conditioning), hygiene (e.g. washing, bathing), industrial processes (e.g. paper manufacture, food preparation, mining operations, petrochemicals), agriculture (e.g. irrigation and livestock watering), and hydropower (e.g. electricity production).

This very incomplete list is simply to indicate the remarkably diverse ways that fresh water is employed by **Homo sapiens** in a manner that is not matched by any other member of the animal or plant kingdoms. The quantity of water employed in this way is out of all proportion to the population level when compared with the uses made by other creatures.

Not only are the proportions grossly lopsided, but clean and pristine water supplies are constantly contaminated or polluted by many of the various activities of the so-called civilized world. In the more primitive societies, the hunter-gatherers or even the subsistence farmers, in region still isolated from the mixed blessings of western culture, the local water supplies are rarely contaminated by the inhabitants in their activities related to the basic needs of survival.

2. Facts of Life

Two socio-biological anomalies are thus identified: (a) Mankind's unique role in the world of living creatures as a user of water; and (b) Mankind's prodigal misuse and pollution of water. These are what may be simply regarded as "facts of life"; they are facts that must nevertheless be faced by the human race. At the present moment we propose only

185

R. Paepe et al. (eds.), Greenhouse Effect, Sea Level and Drought, 185–197.
© 1990 *Kluwer Academic Publishers.*

to consider some pragmatic aspects of the problem: how to live with those "facts of life"?

This bring up two geographical anomalies: (a) Large numbers of human beings live in regions where there is either a rainfall deficiency where they suffer periodic droughts and starvation, or water demand is so high that even moderate precipitation is inadequate for the needs of sophisticated populations. (b) Other groups of people live in regions with rainfall excess, but it is often so seasonal in its occurrence that excessive flooding may account for serious loss of life, destruction of livestock, shelter and property. In the year 1988, the catastrophic floods in Bangladesh, China and Sudan contrast with equally catastrophic drought in the United States and elsewhere. These events are *not* unique, but are matched many times over the last millennium.

3.   Unsolved Problems: The Climate Question

Emerging from these basic facts about which there can be little argument are two further problems, about which a great deal of controversy prevails: (a) climate change; (b) sea-level change. Clearly, a large expenditure of funds and research energy will be needed before definitive statements will be feasible. At the present phase, although much good work has been, and is continuing to be done, the two fields remain highly speculative.

As to climate change, there are basically two schools of thought. One school submits that inasmuch as the level of CO_2 in the atmosphere is being carefully monitored and shows a systematic rising trend, and as the quantities involved are compatible with the volume of CO_2 generated or liberated by human activity (burning of fossil fuels, deforestation, etc.), it is reasonable to assume that a "greenhouse effect" should be building up in the Earth's atmosphere, and that eventually, if not now, this trend will lead to a mean global rise in temperature.

A second school of thought considers rather the long-term historico-geological record. Even on a decadal of century basis, the Earth's climate is known to fluctuate considerably, controlled by feedback mechanisms, acting within a 4-million-year steady-state framework (mean temperature: $18 \pm 5°C$). In the current millennium the droughts of the Sahel are nothing new for African history. Those of the American Southwest have been documented archaeologically for centuries, long before the introduction of the internal combustion engine. Recent analyses of air bubbles in glacier ice show a sudden rise in CO_2 level quite comparable to that of today, but in the 12th Century AD. It is true that over the last 100 years there has been a fluctuating rise in temperature in many parts of the world, but the historical record would indicate that this no more than a normal rise, following a long-term cooling cycle characterized as the "Little Ice Age" (Lamb, 1977; Groves, 1988). It was preceded by several centuries of mild to warm climates, often characterized as the "Viking Time", when Norsemen had settlements in Southern Greenland and where the mean temperatures were 4-5°C above those of the present day.

None of the above observations should be construed to suggest that the climatic problem is anything but a wide-open, interdisciplinary research field replete with enigmas and unsolved problems, which should be challenging for the great minds of science.

4. The Sea-Level Question

The second area of unsolved problems is sea level. Like climate, it also calls for interdisciplinary research, and indeed this is recognized by the great international unions (IUGG, geophysics and oceanography, IGS, geography and coastal environment, and INQUA, geological history and dynamics of sea-level change). The International Geographical Correlation Programme has recently concluded an invaluable global project on sea-level correlation, IGCP-200 (Pirazzoli, 1988). A successor project has been initiated, dedicated to coastal evolution in the Quaternary, IGCP-274 (led by Van de Plassche, who edited an excellent manual on sea-level research, 1986; its first meeting was in Amsterdam, September, 1988). SCOR, the oceanographic research group also has an on-going study of contemporary sea-level behaviour. IHB, the international hydrographic organization, actively pursues questions of global water problems.

It is gradually becoming recognized that both climate and sea level are merely important aspects of the overriding question of global change in general, which has been made the first priority of the International Geosphere-Biosphere Programme (of ICSU, the International Council of Scientific Unions, which represents the world's topmost scientific consultative body). To focus on this global change the main efforts are being directed towards climate and atmosphere. However, the interactions with many geological and oceanographical processes are not entirely forgotten, and the IUGS has set up a Task Force on Global Change (under K.J. Ksü, sea news report in *Episodes*, 11(2), 1988). Furthermore, INQUA has its own Committee on Global Change (headed by H. Faure); as its first approach, Faure has outlined a draft on "Continental Biogeoflux Changes". Jointly with IGCP, the UNESCO has recently launched a new "subprogramme" entitled "Quaternary Geosciences and Human Survival; its first meeting was in Bangkok in November, 1988.

As with climate, the sea-level question divides broadly into two schools of thought. A statistically weighted and average analysis of worldwide tide gauge data suggests that MSL in the 20th Century is rising at about 1.2 mm/year (Fairbridge & Krebs, 1962; Gornitz et al., 1982, 1987). Several other specialists have approached the problem, with varying degrees of agreement, but reaching a general consensus, viz., that MSL is globally rising. It seemed, to many scientists that it would be a reasonable deduction to assume that the world's glaciers were melting and that sea level was rising eustatically by the net input of meltwater. Meier (1984) reported the recent retreat of many small mid-latitude mountain glaciers. From the great polar ice-masses, Antarctica, Greenland, etc., the mass balance is more difficult to determine, but the consensus at the present time is essentially "no change". An important, but strictly limited variable is the steric factor: the volumetric expansion of sea water due to warming of the surface waters (Gornitz et al., 1982; Wigley & Raper, 1987); over decade- to century-length periods this can only explain a few cm of the rise.

The second approach to the sea-level question has been to examine the geological structure of the tide-gauge sites, i.e. the neotectonic setting and the global geodetic dynamics (Mörner, 1987). This was the procedure adopted by Fairbridge and Krebs (1962) in removing data from tectonically active areas, but a more comprehensive survey by Pirazzoli (1986) strongly suggests that, even in the relatively stable regions, there is often a slow, secular tendency for crustal subsidence. This is particularly apparent when one considers that tide gauges are normally installed in or near harbours for the benefit of shipping, and that those ports are frequently on river estuaries where sediment-loading and long-term crustal subsidence combine to generate geodetically negative land surfaces. Not surprisingly, the measured sea level displays a progressive rise (Fairbridge, 1987).

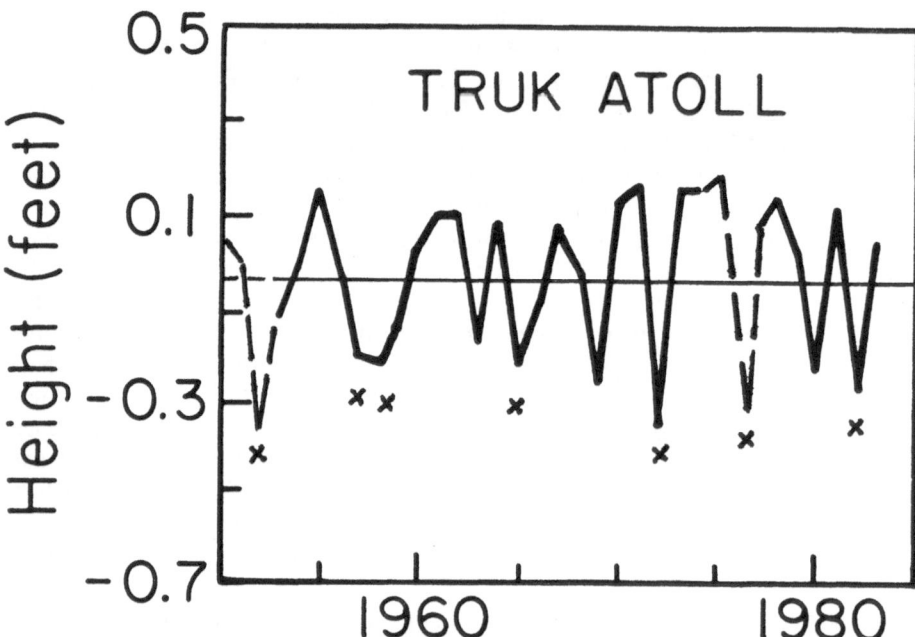

Figure 1. Annual mean sea level for Truk Atoll (E. Caroline Is.), Central Pacific, at lat. 07°26.8' N., long. 151°50.7' E. Broken line = incomplete record, x x x = El Niño years.

Comment on Figure 1: *No statistically significant rise or fall of MSL is evident. The location is far removed from plate-tectonic boundaries, glacio-isostatic influences, the major geostrophic currents and sedimentation sites. Equatorially, there is no secular steric effect. Geologically, the islands disclose no Holocene uplift or subsidence.*

The writer would invite inspection of the tide-gauge records for the few stations that are totally removed from areas of sediment loading, volcanism, plate boundaries and so on. This only leaves a few "clean" records, but a wide area of the central Pacific is studded by atolls that seem to be largely free from disturbance. The MSL trace for Truk Atoll, for example, discloses only minor fluctuations for mor than 40 years. The fluctuations are due to the well-known ENSO or El Niño effects, and disclose no trend whatever (Fig. 1).

5. Mankind's Options

The question that needs to be posed now is this: regardless of what is the cause of relative (i.e. local) sea level rise, is there anything that can be done about it? That is to say, in a management or engineering sense.

In a report of the U.S. National Academy of Science, entitled "Responding to changes in sea level: engineering implications" (Chm, Dean, 1987), to which the present author also contributed, the conclusion was reached that there were three potential strategies: (a) retreat, legislative; (b) defensive, dyke-building; and, only briefly mentioned (c) geohydrologic management (i.e. to stop the sea level from rising). Clearly, the simplest procedure would be to legislate against people building homes and commercial structures on the fringe territories of subsiding events. For cultural and human investments of great and irreplaceable value (e.g. like the cities of Amsterdam and London), defensive structures are certainly justified.

In the NAS report there was firm consensus that for subsiding coasts, an added rise from CO_2 induced melting would only exacerbate an existing problem. Nevertheless, on a 10 to 50-year basis, coastal engineering skills were adequate to meet most threats. Certainly, nothing could be done against geological (crustal) subsidence, although in the case of man-made subsidence due to water, oil or gas withdrawals, there are feasible re-injection techniques and other engineering solutions. Much depends on the rates of relative MSL-rise. A somewhat alarmist series of scenarios presented in earlier NAS reports are addressed in an interesting book by Barth and Titus (1984), which at least served to help draw world attention to the problem. Even if the CO_2 scenarios were found to be grossly exaggerated, the one fact that is inescapable is that perhaps 50% of the world's coastlines are subsiding, some faster, some slower. Thus, anything feasible that can be devised which can alleviate or mitigate the coastal hazard of rising sea level deserves most serious consideration.

Newman and Fairbridge (1986) pointed out that during the 20th century the construction of major dams and diversions were already playing an increasing role in slowing the relative sea level (RSL) rise. It was shown that in 50 years RSL would have risen approximately 0.75 mm/year more than it has, were it not for the reservoir construction. This evaluation provides the quantitative basis for the thesis presented in that paper that geohydrologic management was not only feasible, but should be actively explored.

6. Water Deficiency Areas

There are two categories of water deficiency: (a) developed areas that want to develop further, but are constrained by inadequate water supplies (e.g., parts of southern California, Arizona and New Mexico); and (b) under-developed areas that are inhibited if not periodically devastated by water deficiency.

In this study, the first category can for the most part be conveniently set aside for attention by the regional hydrologists, engineers and legislators. But there is one aspect that should be given very serious attention and this could contribute in a valuable way to the global management program. This is the matter of storm runoff and treated-sewage wastewater, which at present is largely conducted to open-flow rivers or directly to the ocean. It is known from experience that reinjection into the groundwater is feasible for such waters, provided they are not contaminated by non-biodegradable substances such as PCBs and heavy metals. Thus it becomes a twofold problem, both essentially political or legislative in nature, that involve (a) sophisticated pollution control, and (b) a willingness to spend public funds to pump the surplus waters to the sites of injection wells. Relatively short transit distances in porous aquifers are normally required to purify biological wastewaters. Thus, for example, the drinking water well can be safely sited at only 33 m from a septic tank employed by the average house-hold, provided that a reasonably porous alluvial or coastal sandy aquifer is available. In short, in the first category (developed regions), every means possible should be investigated that could usefully recycle existing water supplies.

There is a second category of advanced or highly organized communities that for special reasons rely entirely on desalination. Salt water is pumped from the sea, is subjected to extremely costly purification and then the wastewaters are commonly returned to the ocean. Communities of this sort have economic justifications such as for tourism (e.g., Canary Islands, Bermuda, Bahamas), for the oil industry or for national reasons (e.g., Saudi Arabia, South Yemen, Persian Gulf emirates). With increasing technological skills, a proliferation of desalinized water supplies can be forecast for littoral communities. Reinjection of their wastewaters into the coastal sand formations helps sustain date-palm groves and oasis-style agriculture. The total volume of hydrologic transfer is not large but it is in the right direction: removal from the ocean and storage or recycling on land.

Most attention is now being directed to the needs of the underdeveloped world: this is for several reasons. First and perhaps most urgent are the humanitarian reasons. The civilized world does not relish the repeated news of droughts, hardships and starvation that constantly emerge from the water-deficient areas. Much effort has been concentrated on the multiple problems. Regrettably, a great deal of this activity has been characterized by emotional hand-ringing, without the guidance and determination needed to do something really constructive about it. To murmur "kismet" or "how sad" is all very well, but does little to fill empty stomachs. In part the problem is tied up in complex knots involving religion, politics, administrative responsibility and a widespread ignorance of what is involved. In part the do-nothing approach is locked in the "Act-of-God" philosophy, a reasoning of despair that is based on having no obvious alternative solution. To argue that there is indeed a solution, viz. geohydrologic management, would only be a cruel hoax however, if a large number related problems were not simultaneously addressed: birth control, education, rationalized regional development, and so on. At this point, we can only address the geohydrologic management aspect.

For purposes of this investigation some rather arbitrary cutoffs will be made in order to limit the field to the most needed regions. Geographically, the world's largest areas of water-deficient underdeveloped environments are located in Africa, in central Asia, and in parts of South America. Water-deficient areas exist in North-America, Australia, and elsewhere, but these call for rather different approaches.

Three sources of fresh water may be considered: (a) Seasonal supplies, e.g. monsoons, that in part of are under-utilized; (b) Through-running (allogenic) river supplies; (c) As yet unexploited sources in adjacent watersheds (rainfall-excess regions).

6.1. THE CLIMATIC VARIABLE

As concerns category (a), the seasonal supply, there is a basic fact of climatology that is inescapable: the interannual variability of precipitation in the high-pressure subtropics and monsoon regions is 10 to 100 times greater than in the temperate or equatorial belts. Vigorous studies are in progress in climatological circles to learn more about this tremendous interannual variability.

In part, the problem is philosophical. Many climatologists firmly believe in the Lorentz school of thought that climate is not predictable on more than about a 5-day lead time; in short, it is stochastic process, a product of complex random interactions. If this is true, then the management authorities must engineer their plan accordingly. A great deal of water storage would be indicated, and many years would go by when it was under-utilized because it would not be possible to predict in any way a region's needs from year to year.

Another view, however, is beginning to gain ground. According to Lamb (1977) the clearest and most universally observed climatic cycle is the QBO, the quasibiennial oscillation, expressed in atmospheric pressure, upper-air winds, and precipitation. It has recently been detected (i.e. simulated) in Global Circulation Models (Sperber et al., 1988). Basically, it means that one may expect a high-low-high-low alternation of various annual climate parameters. It is especially well developed during strong sun spot cycles. In weak ones it tends to break down or resolve into a quasitriennial cycle (Schove, 1987).

In the past there has been a good deal of rather simplistic correlation between the sunspot cycles and various climatic indicators. Attempts to forecast on this basis have in general been woeful. To explain this disastrous record it is only necessary to consider that the length of the "cycle", from maximum to maximum, is around 11 ± 6 years, thus ranging from 7 to 17 years. Furthermore, its amplitude at maxima ranges from about 10 to 200 spots (the "Wolf Number"). Various climatic trend patterns have a disarming habit of reversing during cycles that exceed 80 spots. It is not surprising that the weather forecasters shake their heads at the solar cycle.

Light, however, is beginning to emerge on the long-term prediction scene. Recently Labitzke (1987) by simply taking into account the QBO and reanalysing the high-latitude upper air temperature data, found a powerful solar periodicity. Evidently, there is a vast amount still to be investigated but predictable events can now be calculated with a measure of precision, based on astronomical data. It should be emphasized that these potentials are only just beginning to be considered, and have not yet been incorporated into any formulations of "establishment" meteorology.

A general review of the astronomical approach was presented by Fairbridge and Sanders (1987) and the long-term relations between sunspots and weather was summarized by Schove (1987). The key to understanding the Sun's variable behaviour is a long series of plots made (by the writer) of the solar orbit around the barycentre of the solar system that have been carried back several thousand years. The Sun described an inertial orbit around the barycentre in the shape of an epitrochoid, a big loop around a small one. As reported at the COSPAR meeting in Helsinki, Charvátová (1988) observed that when the Sun passes across the intersections it builds to a peak of sunspot (and luminosity) activity. The Sun's velocity and angular momentum vary quite remarkably, from a low at peribac (closest approach to the barycentre) to a high at apobac (Fairbridge & Shirley, 1987; Shirley, 1988). Near the transition points there is a rapid exchange of torque applied by the "tide-raising planets" (Jupiter, Earth and Venus) according to Landscheidt (1988) which influences the convective turbulence and luminosity of the Sun, as well as the streams of high-energy subatomic particles that radiate out as the "solar wind".

The geometry of the epitrochoid would be quite symmetrical if there were only two planets, the two giants Jupiter and Saturn, that together carry 86% of the system's angular momentum and then the sunspot cycle would be easy to predict. The writer has established eight different orbital patterns associated with the configurations of other planets but these recur in rather nicely predictable sequences. The mean periodicity is 19.86 ($\pm$ 0.88) years, and in longer groupings at 59.6 years, 178.8 years, and so on (Fairbridge & Sanders, 1987; Jakubcová & Pick, 1987; Bucha et al., 1985). The closest approaches to the barycentre and slowest inertial motions are reached by the Sun according to a 178.8 year average. These epochs are associated with very low activity cycles and often with "Little Ice Age" climates on Earth (Fairbridge & Shirley, 1987).

In semi-arid Africa those low activity periods are often marked by protracted cycles of drought. Thus a considerable effort of further study should be devoted to this astronomical connection.

One should hasten to add that the Sun and the planetary motions are not the only climatic forcing agencies. Sophisticated spectral analyses disclose that the Moon's periodicities also have their place in long climatic series, especially the 18.6 year lunar declination (and nodal) cycle (Currie, 1987 a). It is believed to operate through its influence on the atmospheric standing tidal wave affects diurnal air pressure and therefore the rainfall. However, it also has an influence through ocean currents, especially on the Continental Shelf (Kaye & Stuckey, 1973; Loder & Garrett, 1978).

On long time series, such as the flood records of the Nile it is found that in certain periods the sunspot cycle is quite important, but then it fades and is overtaken by the 18.6 year lunar cycle (Fairbridge, 1984; Hameed, 1984; Currie, 1987 b). This curious interaction between these two extraterrestrial forcing agencies is found in other time series, even at high latitudes (Guiot, 1987). In the case of the Senegal River discharge and the Sahel droughts a 31 year beat frequency of luni-solar origin has been detected (Faure & Gac, 1981).

In this "excursion" into astronomical forcing, one should not lose sight of the multiple shorter cycles that are beginning to emerge. Already the QBO is well established. Related cycles like the ENSO (El Niño) periodicities are gradually becoming known (Fairbridge, 1990), and more will follow.

6.2. REGIONS OF WATER EXCESS

Regions of abundant rains are located generally within three fundamental climatic categories: the temperate belts of the prevailing westerlies, the equatorial belt and the various subtropical monsoon regions. To be useful as potential suppliers for the water-deficient regions, *juxtaposition* is the key word. Most heavy rainfall areas of the prevailing westerlies have no application in subsaharan Africa. Exceptions to this statement are found in the potential westerly rain resources of Morocco, Algeria, eastern Anatolia and the Zagros. For the most part, therefore, interest is concentrated in the equatorial tropics and monsoonal subtropics, especially those with orographic enhancement. In the case of Africa, these prolific areas are concentrated in the following regions:

(1) The mountains bordering the Western Rift zone of central Africa that feed the Nile, flowing east and north; and the Zaire system flowing westward. (2) Mountains of the Cameroons and neighbouring parts of Nigeria, Gabon, Congo and Central African Republic; all with some streams draining to the Atlantic and some north to the endorheic Chad Basin. (3) Fouta Djalon of West Africa, with streams draining south and west to the coast, and partly northeast to the Niger system. (4) Uplands of southern Zaire, eastern Angola and the southern Rift zone, draining into the Zambesi or the Okavango system. (5) Highlands of Ethiopia, draining radially, into the Nile, into the Rift (Omo system) and other endorheic areas in Eritrea and Ogaden.

The global statistics of precipitation are impressive, and the fact that about 120,000 km^3 falls on land makes it a large potential storage resource. After deduction for evaporation and evapotranspiration, there remain a total runoff (from rivers, seepage, and glacial melting) of nearly 50,000 km^3 (Starkel, 1989). If human intervention made it possible to intercept only 1 to 2 percent of this volume, two highly desirable objectives could be approached: (a) a 1.2 to 2.4 mm reduction in annual absolute sea-level rise; and (b) some 500-1000 km^3 of fresh water could be made available for transfer to the water-deficient areas (Newman & Fairbridge, 1986).

As remarked above, the key word in the transfer operation is "juxtaposition". All of the African water-excess areas mentioned above are very under-utilized, at least seasonally, and accordingly could benefit from engineering management programs. Series of small, interconnected dams can be fed into an arterial system of pipelines, leading eventually to the semi-arid regions where injection wells would supply the extensive underground aquifers.

Before moving on from a consideration of "regions of water excess" caution prompts a reminder of long-term climatic trends, about which very little consideration has been given. While the next "ice age" is several thousand years ahead, century-long fluctuations should not be disregarded. During the last glacial period (maximum: around 20,000 years ago), the savanna lands and even large parts of the rain forest were completely arid. Drifting sand dunes from the Sudan reached down from the north, and others from the south, at times reached the Zaire River, near the equator (Fairbridge, 1964). The African "ice-age aridity" is now widely recognized and comparable low-latitude aridification is encountered in South-America, India and Australia. The practical aspect is that during excessive cool intervals of low solar activity, in the present non-glacial phase of Earth history, even if persisting only for a few decades, could have very serious consequences in what are the high rainfall areas of today. In short, the very long-term precipitation variances need to be studied, even for places that appear to possess and over-abundance at the present time. A 1360 year series of Nile flood data, for example, studied by Hassan and Stucki (1987) shows cycles of very large variance. Indeed, over the last 60,000 yr, Paulissen and Vermeersch (1989) show that the Nile has undergone four complete switches from humid to hyper-arid and back again.

7. Environmental Concerns

Any human interference into natural systems is bound to have environmental impact. The ambient organizations must be treated with both scientific understanding and respect. Cost-benefit ratios and compensation would call for mutual understanding and education on a giant scale.

Three distinctive site categories require in-depth study: 1. Watershed areas, especially the high mountain valleys; 2. Transit areas that will be crossed by pipelines; 3. Recharge and artesian access supply areas.

1. The collecting areas, we believe, should be planned with a "small is beautiful" philosophy. Large dams have proved highly disruptive to regional ecologies. Small dams have several advantages, notably that they can employ small, local workforces who can benefit from the fishing and freedom provided by the reservoir from seasonal vagaries of climate; in the event of mishaps, the damage caused by a small structural failure or ecological miscalculation is less serious than for a giant installation.

2. The transit areas should preferably be across regions of relatively low relief. For many of the possible selections in central Africa this condition is fortunately rather widespread, but inevitably there are some deep valleys where siphon and pumping methods will need to be included in the plans. Pipeline construction can be considered in a cost-relationship to open-channel methods. The relatively low cost of the latter is offset by the much longer meandering course which follows the contour lines and calls for extensive aqueduct bridging; a further disadvantage is the barrier effect, which obstructs movement of stock, migrational animals and so on. Open channels are also subject to evaporation and to growth of water plants which may in fact entirely block the channel unless constantly cleaned out.

3. In the semi-arid recipient areas there are two distinct operation envisaged: (a) the drilling of large numbers of injection walls, connected by feeder pipelines. Depending upon the types of aquifer (high and low porosity sands and sandstones for the most part), there will be a wide variance in spacing and intake rates. Subsurface migration of the new groundwater is also a variable. (b) Drilling of user-convenient wells (villages, agricultural land). Two types of aquifer can be considered - deep artesian supplies located in structural basins and provided by sandstone aquifers mainly of Tertiary and Mesozoic age; and near-surface alluvial (or dry wadi) tracts which are widespread in Subsaharan Africa, but are restrictive (narrow) being limited to former water-courses (McCauley et al., 1982)

8. Conclusions

1. Mankind is unique in the biosphere as both a user and contaminator of water. There is no substitute for water in biological metabolism. How can our species learn to live with these "facts of life"?
2. Two population anomalies present global problem areas: one group living in rainfall-deficient regions where they face hazards of catastrophic droughts; and the second living in rainfall-excess regions, where they face hazards of catastrophic flooding.
3. Two related but unsolved problems are: (a) climate change, and (b) sea-level change. The writer, controversially, claims that climate will be ultimately predictable thanks to increasing knowledge of the Sun, planets and the Moon and, also controversially, that sea-level is rising in many places by mainly because of crustal subsidence.
4. Regardless of unsolved problems (item 3), a large-scale geohydrologic management program could go far to alleviate or at least mitigate human sufferings listed under item 2, by damming headwater streams and transferring waters to subsurface (aquifer) storage and re-use in the semi-arid lands.

9. References

Barth, M.C., & Titus, J.G. (eds.)(1984) *Greenhouse Effect and Sea Level Rise*, New York, Van Nostrand Reinhold, 325 p.

Bucha, V., Jakubcová, I., & Pick, M. (1985) "Resonance frequencies in the Sun's motions.", Studia Geophys. et Geodet., 29, pp. 107-

Charvátová, Ivanka (1988) "The solar motion and the variability of solar activity.", COSPAR Conference, Helsinki, in press

Currie, R.G. (1987 a) "Examples and implications of 18.6- and 11-year terms in world weather records.", in Rampino, M.R. et al., (eds.) *Climate: History, Periodicity, and Predictability*, New York, Van Nostrand Reinhold, pp. 378-403

Currie, R.G. (1989 b) "On bistable phasing of 18.6-year induced drought and flood in Africa since AD 650.", Jour. Climatology, 7, pp. 373-389

Dean, R.G. et al. (eds.)(1987) *Responding to Changes in Sea Level: Engineering Implications*, Washington, D.C., National Academy Press, 148 p.

Fairbridge, R.W. (1964) "African ice age aridity.", in Nairn, A.E.M. (ed.) *Problems in Palaeoclimatology*, London, Wiley Interscience, pp. 356-363

Fairbridge, R.W. (1984) "The Nile floods as a global climate/solar proxy.", in Mörner, N.A., and Karlén, W. (eds.) *Climatic Changes on a Yearly to Millennial Basis*, Dordrecht, D. Reidel, pp. 181-190

Fairbridge, R.W. (1987) "The spectra of sea level in a Holocene time frame.", in Rampino, M.R. et al. (eds.) *Climate: History, Periodicity, and Predictability*, New York, Van Nostrand Reinfold, pp. 127-142

Fairbridge, R.W., & Krebs, O.A. Jr. (1962) "Sea level and southern oscillation.", Geophys. Jour., v. 6, pp. 532-545

Fairbridge, R.W., & Sanders, J.E. (1987) "The Sun's orbit, AD 750-2050: basis for new perspectives on planetary dynamics and Earth-Moon linkage.", in Rampino, M.R. et al. (eds.) *Climate: History, Periodicity, and Predictability*, New York, Van Nostrand Reinhold, pp. 446-471 (bibliography, pp. 475-541)

Fairbridge, R.W., & Shirley, J.R. (1987) "Prolonged minima and the 179-year cycle of the solar inertial motion.", Solar Physics, 110, pp. 191-220

Faure, H., & Gac, J.Y. (1981) "Sahelian drought to end in 1985?", in *Nature* 291, pp. 475-478

Gornitz, V., Lebedeff, S., & Hansen, J. (1982) "Global sea level trend in the past century.", Science, 215, pp. 1611-1614

Gornitz, V., & Lebedeff, S. (1987) "Global sea-level changes during the past century.", Soc. Econ. Pal. Min., sp. publ. 41 (*Sea-Level Fluctuation and Coastal Evolution*), pp. 3-16

Grove, J.M. (1988) "The Little Ice Age", London, Methuen, 498 p.

Hameed, S. (1984) "Fourier analysis on Nile flood levels.", Geophys. Res. Lett., 11, pp. 843-845

Hassan, F.A., & Stucki, B.R. (1984) "Nile floods and climatic change.", in Rampino, M.R., et al. (eds.) *Climate: History, Periodicity, and Predictability*, New York, Van Nostrand Reinhold, pp. 37-46

Jakubcová, I., & Pick, M. (1987) "Correlation between solar motion, earthquakes and other geophysical phenomena.", Annales Geophysicae, 5, pp. 135-

Kaye, C.A., & Stuckey, G.W. (1973) "Nodal tidal cycle of 18.6 years: its importance in sea-level curves of the east coast of the United States and its value in explaining long-term sea-level changes.", Geology, 1, pp. 141-144

Labitzke, K. (1987) "Sunspots, the QBO, and the stratospheric temperature in the north polar region.", Geophys. Res. Lett., 14, pp. 535-537

Lamb, H.H. (1977) "Climate history and the future.", London, Methuen (and Princeton U.P.), 835 p.

Landscheidt, J. (1988) "Solar rotation, impulses of the torque in the Sun's motion, and climatic variation.", Solar Physics, 12, pp. 265-295

Loder, J.W., & Garrett, C. (1978) "The 18.6-year cycle of sea surface temperature in shallow seas due to variations in tidal mixing", Jour. Geophys. Res., 83 (C-4), pp. 1967-1970

McCauley, J.R., et al. (1982) "Subsurface valleys and geoarchaeology of eastern Sahara revealed by Shuttle radar.", Science, 218, pp. 1004-1020

Meier, M.F. (1984) "Contribution of small glaciers to global sea level.", Science, 226, pp. 1418-1421

Mörner, N.-A. (1987) "Eustasy, geoid changes and dynamic sea surface changes due to the interchange of momentum.", in Qin, Y., and Shao, S. (eds.) *Late Quaternary Sea-Level Changes*, Beijing, China Ocean Press, pp. 26-39

Newman, W.S., & Fairbridge, R.W. (1986) "The management of sea-level rise.", in *Nature* 320, pp. 319-332

Pirazzoli, P. (1986) "Secular trends of relative sea-level (RSL) changes indicated by tide-gauge records.", Journal Coastal Research, Sp. Issue 1, pp. 1-26

Schove, D.J. (1987) "Sunspot cycles and weather history. ", in Rampino, M.R., et al. (eds.) *Climate: History, Periodicity, and Predictability*, New York, Van Nostrand Reinhold, pp. 355-377

Shirley, J.H. (1988) When the Sun goes backward: solar motion, volcanic activity and climate.", in *Cycle Linkage: Planetary-Solar-Terrestrial*, Irvine, California, Foundation for Study of Cycles, Proceedings, pp. 85-92

Sperber, K.R., Hameed, S., Gates, W.L., & Potter, G.L. (1987) "Southern oscillation simulated in a global climate model.", in *Nature* 329, pp. 140-142

Van de Plassche, O. (ed.)(1986) *Sea Level Research: A Manual for the Collection and Evaluation of Data*, Norwich, Geobooks, 618 p.

Wigley, T.M.L., & Raper, S.C.B. (1987) Thermal expansion of sea water associated with global warming.", in *Nature* 330, pp. 127-131

EXTRATERRESTRIAL IMPACTS, VOLCANOES, CLIMATE AND SEA LEVEL

F.J. AYALA CARCEDO
Instituto Tecnológico GeoMinero de España
Ríos Rosas 23
28003 Madrid
Spain

ABSTRACT: The subjects of climate and sea level change are highly complex systems influenced by many factors grouped in several types. One of these types is major hazards: extraterrestrial impacts and internal geodynamic hazards (volcanoes, earthquakes and tsunamis). In this paper, extraterrestrial impacts and volcanic eruptions are analyzed. Astronomic, geological and historical evidence of great extraterrestrial impacts is presented. The paper analyses also the probability and characteristics of bolides and their kinematics as well as the effects of impacts on land or open sea, dust veil, ozone changes and tsunamis. The volcanic eruptions are analyzed with regard to the mechanics, nature and influence of volcanic aerosols and their level of influence on the climate system.

1. Introduction

Climatic and sea level changes are topics of major concern in the geosciences today. Sea level change during the last two million years has been strongly influenced by cyclic climatic changes. These climatic changes are primarily driven by insolation changes forced by modifications in the Earth's orbit due to astronomical laws (Berger, 1988), according to the theory of Milankovitch (1920). Therefore, sea level change is dependent on climatic change.

There are several reasons for the present concern with this topic. In the field of climatic changes these relate to:

a) Evidence of decadally important climatic changes, i.e. temperature decrease during the sixties and catastrophic droughts in the Sahel in the early seventies.

b) The continuous CO_2 increase measured at Mauna Loa from 1959 and the awareness of an anthropogenic increase of natural greenhouse effect.

c) The ozone hole over Antarctica detected in recent years, linked to knowledge of the role of CFC aerosols in ozone depletion.

R. Paepe et al. (eds.), Greenhouse Effect, Sea Level and Drought, 199–216.

d)   Research in the role of atmospheric nuclear explosions and nuclear winter.

In the field of sea level rise, the main cause of concern is the increasing trend of sea level during the last century linked to the anthropogenic increase of natural greenhouse effect (Wind Ed., 1987). Sea level rise would affect lowlands where there are important populations as in Bangladesh, The Netherlands, etc. Major changes of sea level in geologic times have probably been caused by changes in the volume of the oceanic ridge system resulting from increases in its length and spreading rates (Turcotte et al., 1978), especially during the Quaternary from changes in the ice-caps caused by insolation changes. Actually, most of sea level rise is caused by thermal expansion of the upper oceanic layer due to sea surface temperature (SST) increases.

What might be the influence of major hazards like great extraterrestrial impacts and volcanic eruptions on climate and sea level? This will be explored in this paper.

2.   Great Extraterrestrial Impacts, Climate and Sea Level

2.1. POSSIBLE IMPACTING OBJECTS: ASTEROIDS AND COMETS

There are two kinds of great impacting objects in the Solar System: asteroids and comets.

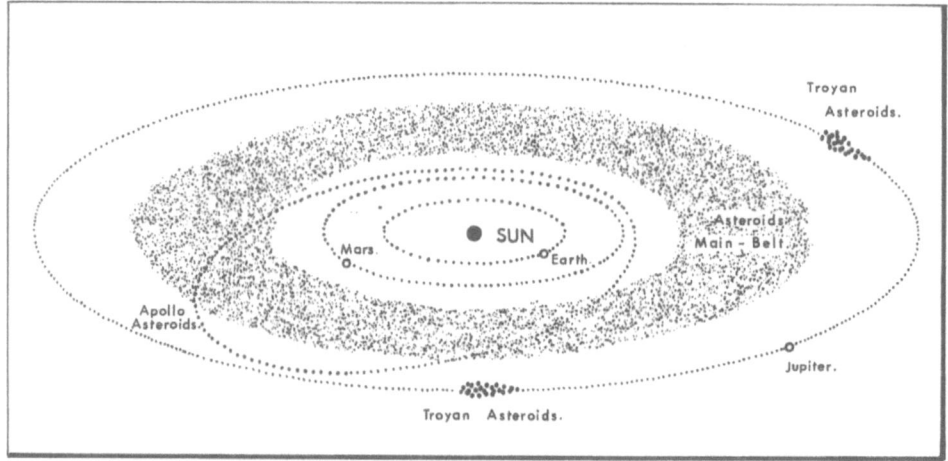

Figure 1. Asteroids in the Solar System.

Asteroids or planetoids (Fig. 1) are solid objects moving in general elliptical orbits. There are three main groups: Apollo-Amor objects, Main-Belt and Troyans. All have periods of revolution from 2 to 12 years and rotate in "direct orbits" in the same way as do the planets. Distinction between Earth-crossing bodies (Apollo) and non-Earth-crossing bodies (Amor) is not fundamental from the point of view of possible impacts because many of the Amor objects are repeatedly perturbed into Earth-crossing orbits on a ~ 10^4 year time scale whose impact probabilities are comparable to those of Apollo objects (Wetherill et al., 1982). Troyans are in Jupiter's orbit in the Lagrange points. According to analyses performed by Shoemaker (1977) and Grieve and Dence (1979), impacts of Earth-crossing objects greater than 0.5 km that may produce craters of 10 km diameter occur with a frequency of one in 105 years. The largest Apollo object known is probably 2212 Hephaistos with a diameter of 9 km. The largest Amor object is 1036 Ganymede of about 40 km.

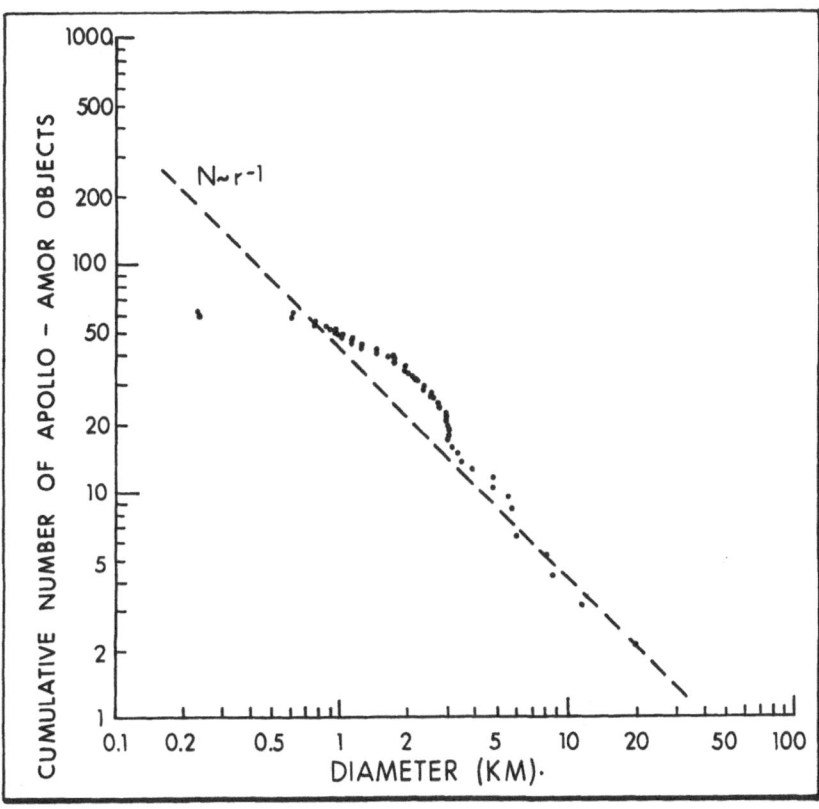

Figure 2. Statistical distribution of Apollo-Amor objects (Mod. from Wetherill and Shoemaker, 1982, The Geological Society of America).

The cumulative diameter distribution of the observed ApolloAmor objects, is given in fig.2 (Wetherill and Shoemaker, 1982).

In 1981, there were 31 Apollo and 29 Amor objects. There are 100 or more Apollo-Amor Marscrossing objects with perihelion beyond 1.3 A.U. (1 A.U. = 1.496 x 10^8 km). Most of Apollo-Amor objects have, like Main-Belt Asteroids, two types of composition: a) Lighter coloured S-asteroids (silicates) have mainly pyroxene and probably olivine, like terrestrial and lunar mantles. b) Darker C-asteroids (carbonaceous) are probably similar to carbonaceous meteorites.

Not only the Apollo-Amor objects can produce impacts on Earth. Besides cometary objects, a small fraction of the million or so bodies of the Main-Belt of Asteroids may be converted into Earth-crossing objects as a consequence of resonance between their motion and that of Jupiter and Saturn.

Velocities of possible impacting objects from the Solar System may reach the Earth's escape velocity, about 70 km/s. The minimum velocity of entrance into Earth's gravity field is the escape velocity, 11.2 km/s. Velocity is a function of mass (Kondratyev, 1988). Velocities along entrance trajectories for several bolides are given in Fig. 3.

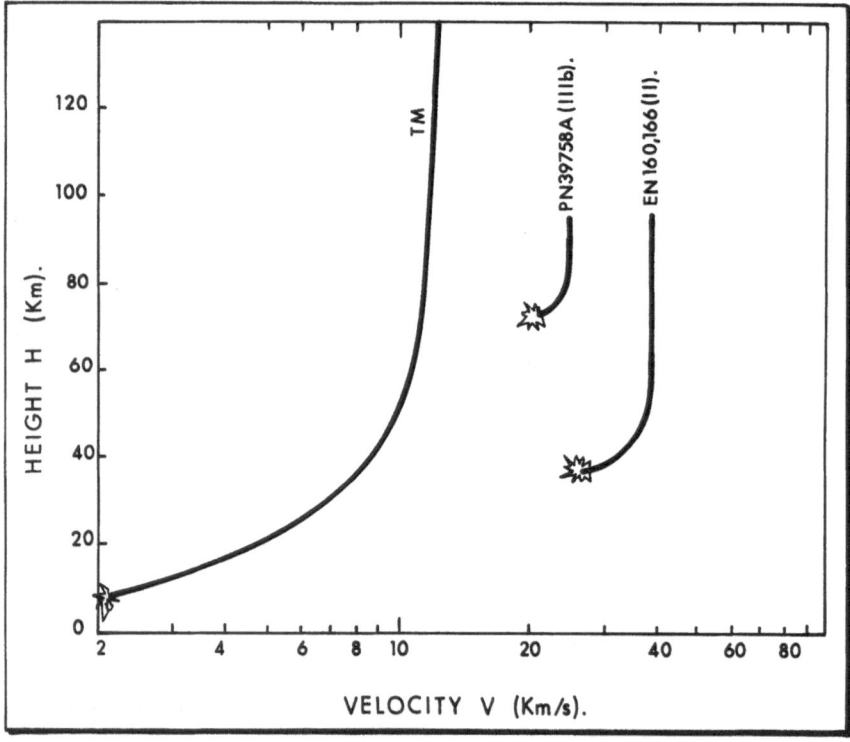

Figure 3. Entrance velocity of extraterrestrial objects in the Earth's atmosphere. TM: Tunguska Meteor (Kondratyev).

Comets are the other possible source of impacts on Earth and planets. They have a nucleus one to ten km in diameter. Nuclei are icy conglomerates of dust (chondritic and graphite) and volatiles (primarily water ice with CO, CO_2 and HCN) with a density from 1 to 1.3 g/cm^3. When they are activated by the solar wind they develop from the nucleus a core (104-105 km) and, away from the Sun, a plasma tail (up to more than 108 km). There are two groups of comets: a) a Short Period (SP) of less than 200 year return period (i.e. Encke's and Halley's Comets), and b), Long Period (LP), greater than 200 years. LP comets come, with near parabolic orbits randomly oriented, from the Oort Cloud (Fig. 4), a big cometary reservoir with about 1012 comets (Weissman, 1982). Motion of these LP comets is activated by random passing stars. End-states for LP comets is the ejection from the solar system or physical splitting (Weissman, 1979). SP comets commonly have low-inclination orbits and their lifetimes, due to loss of volatiles, is 10^3-10^4 years (Weissman, 1982). Marsden (1979), lists orbits for 545 LP and 113 SP comets. According to Weissman (1982), most probable impact velocities are 56.6 km/s for LP and 29.8 km/s for SP comets. About 15% of total cratering observed in terrestrial impacts (astroblemes), with substantial uncertainty, may be due to LP and SP comets (Weissman, 1982). Probably, an important fraction of small meteorites arise from the passage of Earth throughout the dust-tails of comets.

2.2. ASTRONOMICAL, GEOLOGICAL AND HISTORICAL EVIDENCE OF IMPACTS

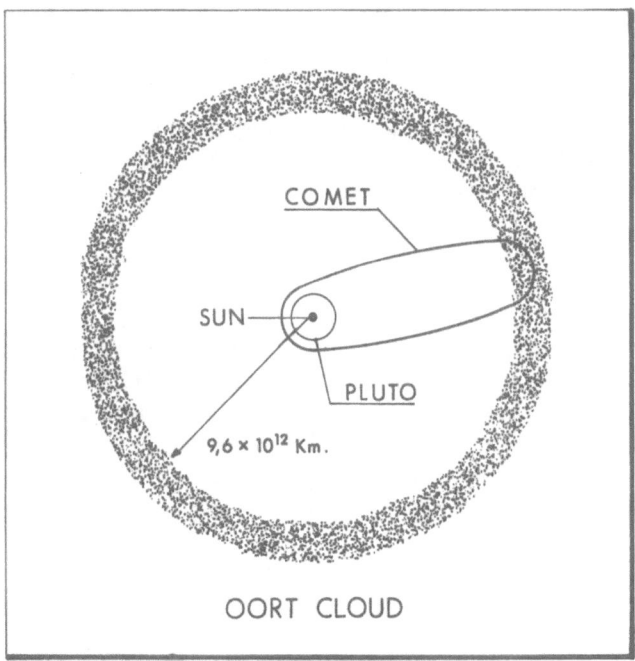

Figure 4. The Oort Cloud, a spheric cloud origin of comets with about 10^{12} cometary nuclei.

The only large solid body without impact structures observed in the Solar System, is Io, as satellite of Jupiter which exhibits important volcanic activity, probably responsible for the disappearance of craters. Big craters are very common on Mercury, Mars and the Moon. Mercury has a crater, Caloris, 1,600 km in diameter and the Moon has about 10,000 visible craters (e.g. Tycho, 85 km in diameter and about 100 M.y. old). Big craters in other rocky planets and satellites are well conserved due to the absence of "geodynamic" external processes in the last billion years, but despite this appearance, the probable rate of great impacts per unit mass was similar to that on Earth during geological times. The difference is that on Earth erosion has destroyed most of craters and 2/3 of its surface is ocean so that the discovery of craters is difficult.

Since the availability of satellite images, much evidence of astroblemes have been discovered on Earth besides well known cases like Barringer or Meteor crater in Arizona. About 100 impact structures have been identified raging from a few tens of meters to about 140 km in diameter. During only the last million of years there have been five probable impacts, one almost certainly in Chutotsk, USSR (1M.y. old, 3 km), Chubb Lake, Ontario (0.02 M.y. old) the afore mentioned Barringer Crater (25.000 years old, 1.2 km) and the explosion over Tunguska (Siberia) in 1908. About 35 astroblemes are less than 100 M.y. old. This bias toward younger ages is the result of erosion processes. A well studied case is the Ries crater, 26 km in diameter, in southern Germany (Silver, 1982). Analysis performed point towards the impact of an asteroid 1-2 km in diameter at 25 km/s and a mass ratio of material ejected from the crater to asteroid greater than 100. The impact took place 14.7 ± 0.6 M.y. ago. The hypothesis of Alvarez et al. (1979) about a big impact in Cretaceous-Tertiary Boundary is well known.

The only entrance of a big bolide into the atmosphere known in historical times was over the Tunguska area (Siberia) on 30th of June 1908. The rise of the resulting fire column was about 20 km, explosions were heard up to 1,000 km, the Transsiberian train was affected by surface earthquake waves 600 km away and there was tsunami waves in rivers. According to Fesenkov (1961), the impacting object was a microcomet. Kondratyev (1988) has estimated the initial size between 0.6 and 0.92 km for an assumed effective density of 0.6 g/cm3 and a mass entering the atmosphere about 105 tons (~ 1.5 g/cm3), with an initial diameter of about 23 m. At the end point (~ 9 km, fig. 3) it had a velocity less than 14 km/s. At near 20 km (Voznesensky, 1925) it was broken into two parts. The end was a giant explosion of a detonating gas-air mixture with abundant hydrocarbons, triggered by powerful electric discharges. the exploding body was about 5 km in diameter and length 25 to 30 km, according eyewitness accounts, with a final angle of impact about 50°. The energy released by the explosion was about 5 x 1023 ergs (Kondratyev, 1988), a half that of St. Francisco's Earthquake in 1906.

Probably on 25th June, 1178, five British monks witnessed a big impact on the Moon (Gervasins of Canterbury Chronicle, in Sagan, 1980). The crater has been studied by Hartung and is called Giordano Bruno.

2.3. EFFECTS ON CLIMATE

The effect on climate of asteroids and comets is different, according to researches on the Tunguska microcomet performed by Kondratyev (1988) and Turco et al. (1981).

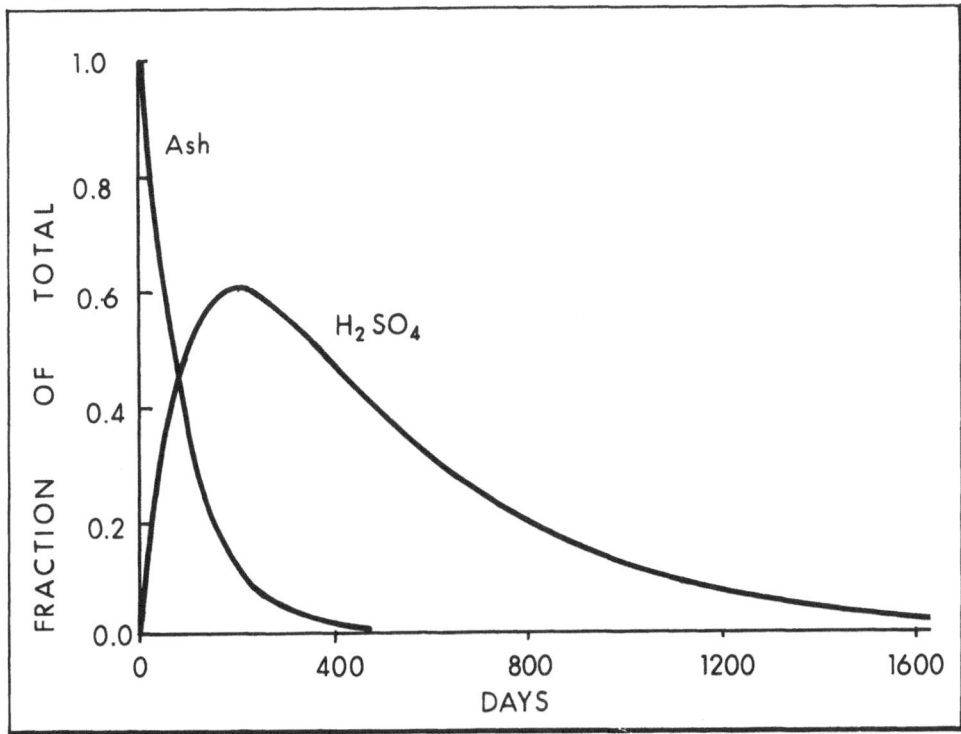

Figure 5. Evolution of volcanic aerosols in the stratosphere (Bryson, 1982).

Comets, due to their hydrocarbon and water content, change under "wet" conditions, the photochemical processes in the stratosphere. In these conditions, high-speed reactions of NO with HO_x, prevent O_3 destruction. In fact (according to Kondratyev [1988]), the Tunguska microcomet, compensated for the ozone depletion caused by a high-speed pre-Tunguska meteor. The effect on ozone is like that of nuclear explosions in the atmosphere. Besides this, the main impact was a warming, due to an increased greenhouse effect. This increase was produced by the introduction into the stratosphere and upper troposphere of large amounts of greenhouse gases derived from cometary matter and air: H_2O (lifetime: 1.5-2 years), CO_2, CH_4 (7-11 years) and NO_x (3-4 years). According to back-analysis performed by Kondratyev (1988), dust from the comet did not compensate for the greenhouse effect. In summary, the effect of impacting comets on climate is a global warming for a few years and an ozone increase. The main source of chemical and physical components introduced in the stratosphere is from the cometary matter and air.

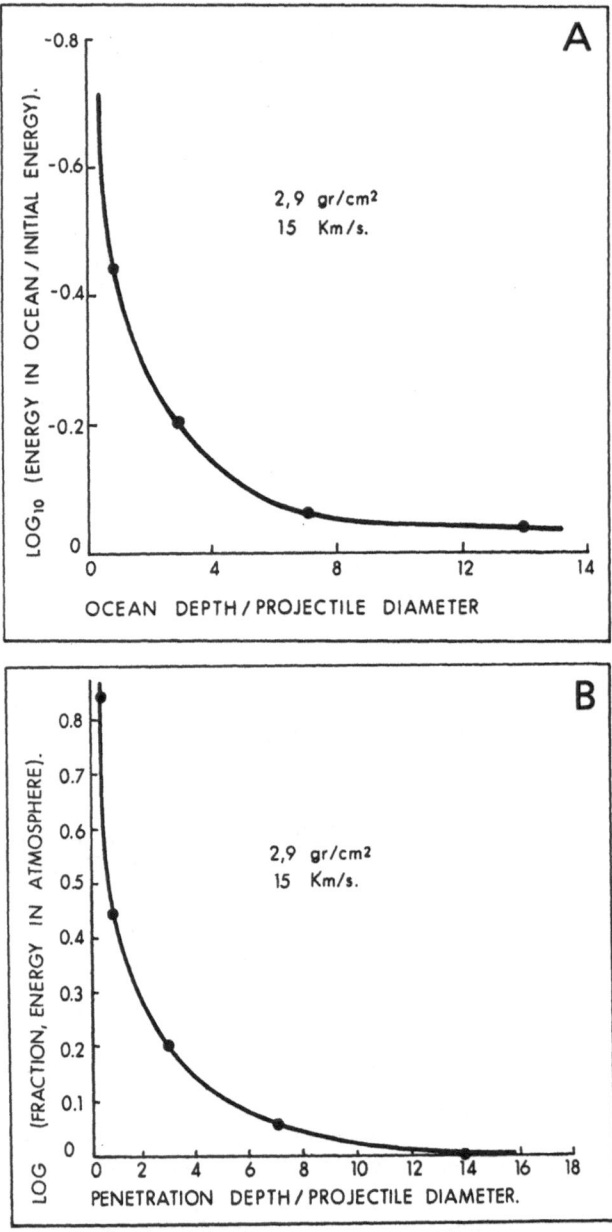

Figure 6. Transfer of energy from bolid to ocean (A) and atmosphere (B) (Mod. from O'Keefe and Ahrens, 1982, The Geological Society of America).

In the case of rocky bolides, the effects are different. Asteroids produce temporary cylindrical atmospheric holes in their wake, much more important than comets. Ejected matter may rise to the stratosphere and is composed of land matter and vaporized projectile when it has impacted on land. Production of large quantities of NO_x may produce important ozone depletion in the stratosphere and extensive acid rains leading to an impact on land and marine biota by defoliation, solution of $CaCO_3$ and changing the Ph. In addition, the finest dust (< 0.5 μ) may diminish solar radiation entrance for two to six months (see Fig. 5). For an asteroid of 10 km in diameter, with a volume of 523,6 km^3, if we assume an ejected volume up to the stratosphere of only ten times greater, there would be a net ejected introduction greater than 5.000 km^3 (total volume erupted in the greatest eruption of last century, the Katmai in 1912, was 15 km^3). This volume is much higher than a typical great ignimbrite eruption with high plume. The effect might be a global cooling of several degrees. this effect might be reinforced by large quantities of soot produced in large fires, because "soot absorbs sunlight more effectively and settles more slowly than does rock dust" (Wolbach et al., 1988). Calculations performed by Toon et al. (1982), show that ocean cooling by dust is only half of land cooling due to the thermal inertia of seas. CO2 released from marine and land carbonates and fires at medium-term, when dust was settled, would perhaps compensate progressively for related cooling. Direct transfer of heat from the bolide to air (see Fig.6) and great fires may warm the air for up to several weeks.

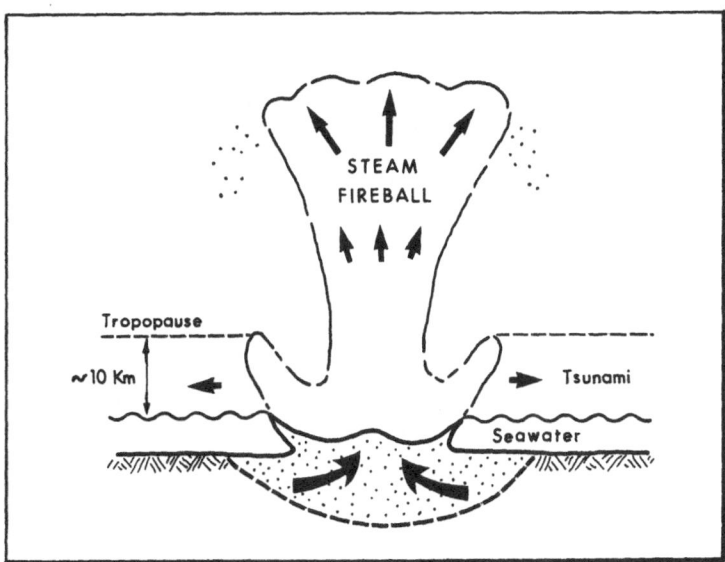

Figure 7. Final state of a big impact on sea-floor. See the big tsunami (Mod. from Melosch, 1982, Geological Society of America).

Large asteroids impacts in the ocean produce a great steam explosion (Fig.7). This steamdust bubble produces a vapour-dust plume that rises many kilometres into the atmosphere. Such a plume is a highly efficient system for driving a global cloud of fine dust and ice particles which can darken the Earth for months (Toon et al., 1982). Then, global cooling would probably be quicker than for land impact. Otherwise, Emiliani et al. (1981) have suggested global warming from water vapour and its induced greenhouse. Perhaps, the first effect would be a short warming followed by almost total darkness for up to 50 days (Toon et al., 1981) followed by 2-4 months of partial darkness and cooling.

In Fig.8 are shown possible climatic impacts from the hypothetical bolide of Alvarez et al. (1979) (O'Keefe and Ahrens, 1982).

2.4. EFFECT ON SEA LEVEL

Many scientists think thermal expansion of the upper layers of oceans is probably responsible for most of the increase in sea level observed during the last century (e.g.Barnett, 1984; Roemmich and Wunsch, 1984), but "no study has demonstrated that global warming might be responsible for an accelerated rate of sea level rise" (Titus, 1987). Notwithstanding the global change of sea level due to ice ages of about one hundred meters is well known (Don et al., 1962).

Global glaciation is improbable after a great impact (Toon et al., 1982) but it is highly

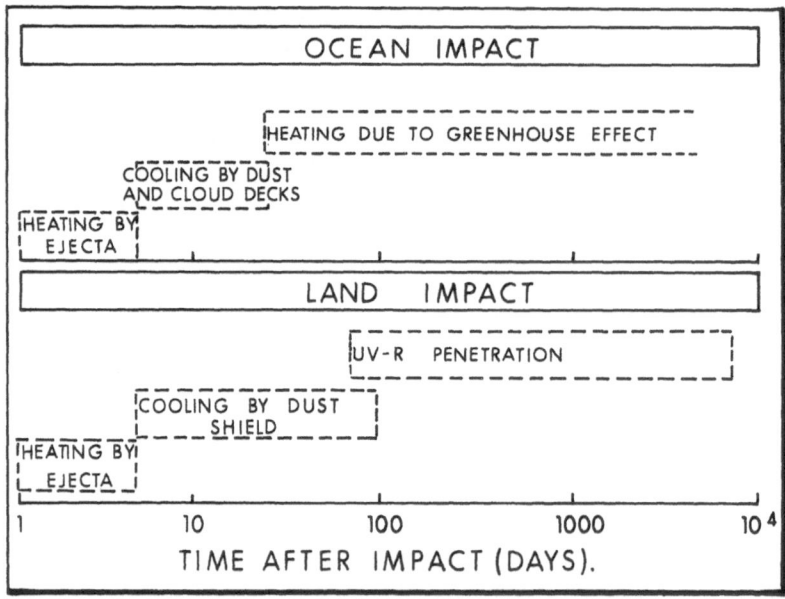

Figure 8. Possible climatic effects of the hypothetical Cretaceous-Tertiary boloid (Mod. from O'Keefe and Ahrens, 1982, The Geological Society of America).

probable that an impact could cause an extension of ice and snowfields. Due to the thermal inertia of oceans the possible effect of an initial warming on sea level would be negligible and possible temperature decrease due to the following cooling would be delayed. But a decrease due to snowfield growth should be observable (?) although it is very difficult of estimate. Probably a decrease of sea level would be, combining cooling of ocean and snowfield growth a minimum of 0.5-1m.

Nevertheless, the main impact on sea level would arise from big tsunamis produced by a big impact in the oceans. From the ratio of energy delivered to the oceans (O'Keefe and Ahrens, 1982, Fig. 6) and initial kinetic energy of an asteroid of $g = 2.9$ g/cm3, 1 km in diameter with $v = 15$ km/s about 1.6×1027 ergs would be transferred to the oceans (average depth: 5 km). From Iida (1963), for a tsunami of magnitude 5, with an energy of 25.6×10^{23} ergs, the probable wave height is 32 m at the coast. The process of generation of a tectonic tsunami is different because energy is communicated upwards. Tsunami energy is about one tenth of forcing earthquake energy. Tsunamis from an impact are more similar to the volcanic ones like Krakatoa in 1883, or others (for comet explosions) produced by a volcanic explosion in the atmosphere with a velocity of shock waves similar to the velocity of a tsunami, i.e. about 220 m/s for an ocean deep of 5 km (Press and Earkrider, 1966) that produces a compound sea-air wave. Comparison between energies activating tsunamis suggests that probably the main tsunamis are those generated by impact of large asteroids in the oceans. Such tsunamis would traverse the world several times probably destroying almost all coastal biota on Earth. Probably, such phenomena in the past have left geological signatures.

Apart from sea level changes triggered by a large impact on the oceans from thermal changes and tsunamis, earthquakes triggered by these impacts may induce relatively regional sea level changes. Energy released in the San Francisco Earthquake in 1906, with a magnitude $MS = 8,25$, was about 10^{24} ergs in 6 seconds. Energy transferred to land from a continental impact from an asteroid 1 km in diameter at 15 km/s is about 10^{27} ergs. Energy estimated, in the maximum earthquake observed of Ms = 8.9 was about 10^{25} ergs, i.e. 100 times less. Its power was much smaller. Such an impact in a tectonically unstable region, especially near coastal zones, may produce a regional subsidence of several metres due to activated tectonics or compaction of soft zones in deltas or coastal lowlands, with generalized liquefaction processes. Both kinds of process cause sudden catastrophic relative sea level changes.

3. Volcanoes, Climate and Sea Level

3.1. VOLCANOES AND CLIMATE

There is historical evidence for the impact of volcanic eruptions on climate. The Roman writer Plutarco and the Chronicles of the Han Dynasty in China recorded the veil that covered the Sun about 42 B.C. as well as the bad crops several years after. The climatic impact (decrease of average temperature) has been recorded in the rings of Californian pine-trees.

The cause was the Etna eruption. The same phenomena were observed after the volcanic eruption of Krakatoa in 1883.

Quantitative records of temperature have been performed after 1760 and these are very useful for back-analysis. Recently systematic measurements with Laser (LIDAR) have improved our knowledge about volcanic aerosols in the stratosphere (Agung 1963-72, Fuego 1974-77 etc), as well as analysis on ice-cores, specially of H_2SO_4 content, a product of photochemical effects on SO_2 in the stratosphere.

Analysis of a quantitative data from the last two centuries have shown (Rampino and Self, 1984):

a) A surface air temperature decrease, ΔT between the year before the eruption and the lowest temperature 1-3 years after the eruption for the Northern Hemisphere (NH) ranging from -1°C (laki, 1783, 64°N) to 0 - 0.1°C (Mt. St. Helens, 1980, 46°N).

b) ΔT is roughly proportional to $M_D^{1/3}$ (M_D = Global stratospheric aerosol mass loadings M_D = 1.5 x 10^{14} τ_D (grams), τ_D = Global - average peak optical depth).

In Fig.9 (Shönwiese, 1988), the relationship between Mean NH Air Temperature (TNH) anomalies (annual and 10 yr Gaussian lowpass filtered), Dust Veil Index DVI and Greenland Acidity Index AI, may be observed. The effect on 1-3 years after temperature are well defined. Correlation coefficients DVI-TNH from 1760-1940 are between -0.3 and -0.5, and less (-0.1 up to -0.4) for AI. Rampino and Self (1984 b) have pointed out that high latitude eruptions probably affect the AI more, and also caused the high AI during the Little Ice Age from 1350 to 1700 A.D. Regional effects of volcanism are quite different, according to Schönwiese (1988).

The increase in stratospheric albedo produced by aerosols injected into the stratosphere (albedo of H_2SO_4 is 0.99 at 75%), increases the stratospheric temperature for 1-3 years. Rampino and Self (1984b) have pointed out the possible influence of volcanic eruptions on the phenomena ENSO (El Niño-Southern Oscillation), an abrupt change in oceanic and atmospheric circulation in the Pacific Ocean near the end of a number of years. The two anomalous episodes from 1950 (1963 and 1982-83) coincided with two explosive eruptions with related climatic effects: Agung (1963) and El Chichón (1982). The warming of the stratosphere and reducing temperature gradient between the stratosphere and the surface in tropical latitudes would have diminished oceanic and atmospheric circulation. After El Chichón, the equatorial stratospheric temperature increased by up to 4°C.

Another effect of explosive eruptions is the depletion of O_3 in the stratosphere due to HCL and HF injected up to the stratosphere. Less than 10% of annual global volcanic fluxes of HCL and HF (0,4 - 11 x 10^{12}g and 0.06 - 6 x 10^{12}g respectively in total), are injected into the stratosphere (Symonds et al., 1988). Mankin and Coffey (1984) have measured in the column of EL Chichón (1982) above 12 km an increase of about 40% in HCl.

CO_2 and H_2O from volcanoes do not play a significant role in global cycles of C and H_2O respectively. SO_2 from volcanoes is about 9% of the total worldwide annual SO_2 flux from man-made emissions and 4% of the total sulphur emissions to the atmosphere (Stoiber et al. 1987).

Dust from volcanic eruptions favours rain and snow during the first one to six months

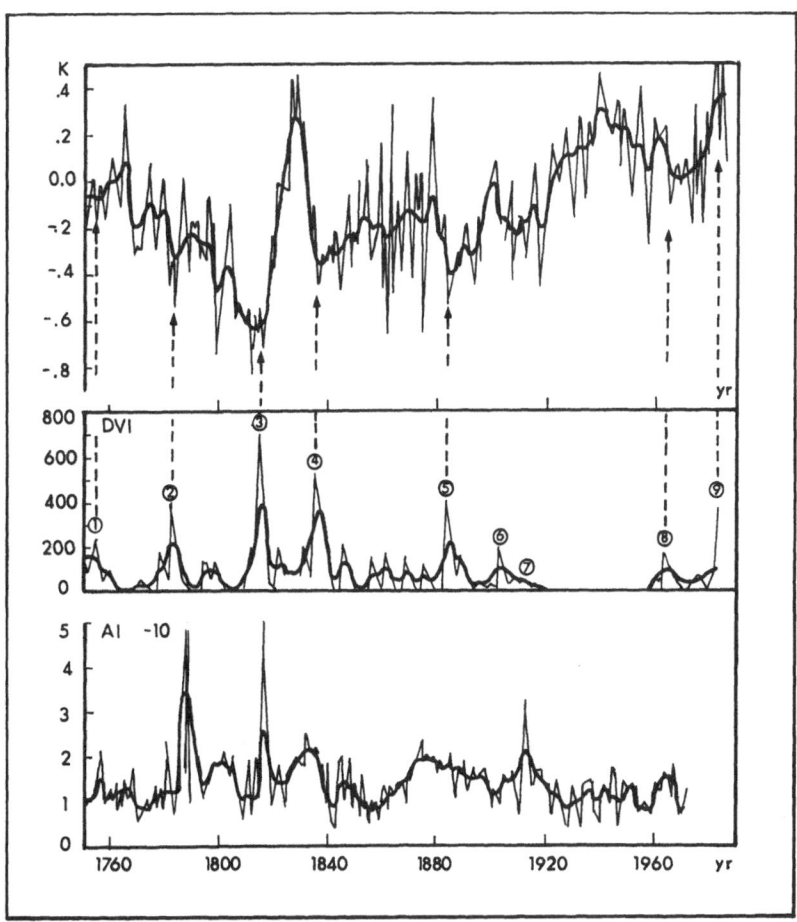

Figure 9. Mean Northern Hemisphere Air Temperature Anomalies (annual and 10 yr Gaussian low-pass filtered), Dust Veil Index and Acidity Index in ice-core from Greenland (1760-1980).

Numbers represent main explosive eruptions (Mod. from Schönwiese, 1968)

after an eruption due to its role in the condensation of water vapour enhancing the effect of cooling.

3.2. VOLCANIC PROCESSES AFFECTING CLIMATE

Nowadays there are 530 subaerial active volcanoes known and several thousands of kilometres of submarine volcanic ridges. The number of known eruptions has increased dramatically during the last two centuries. Before 1800 A.D. the number of known eruptions was 246; in 1981 it was 627 (Simkin et al 1981).

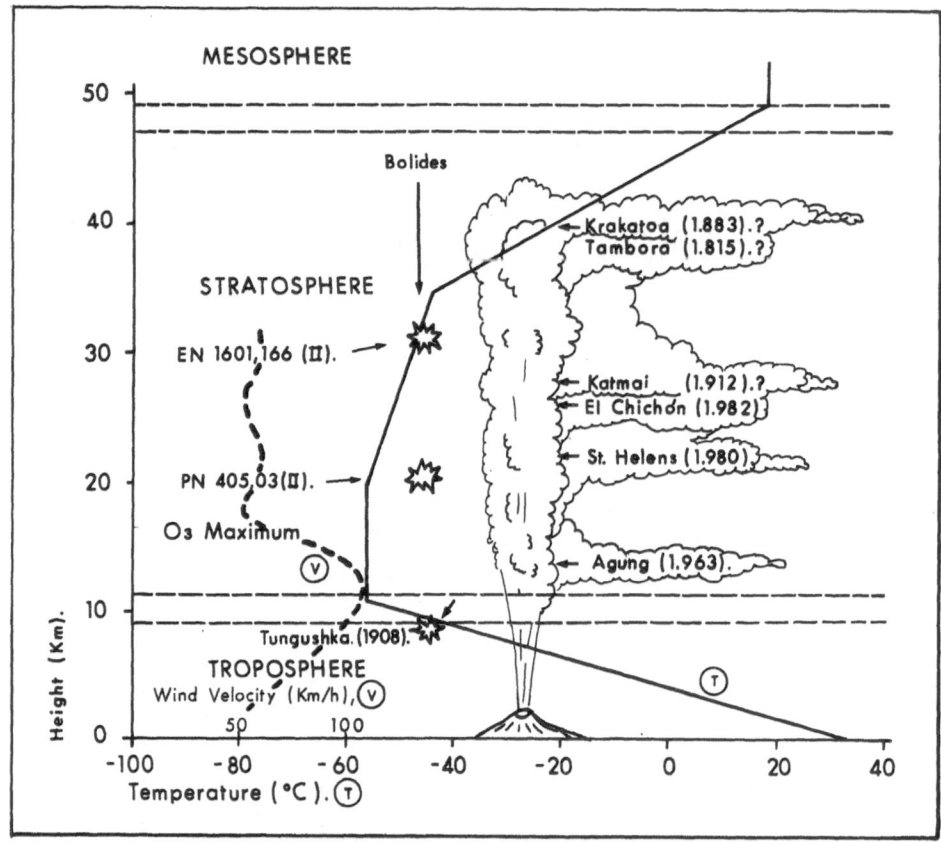

Figure 10. Height of main eruptive columns in the last two centuries, and average velocity of winds with height (Shaw et al., 1974: velocities)

The two conditions for volcanic processes to affect climate are: emission up to the lower stratosphere and high sulphur content of emission. These two conditions have proven to be the main ones after detailed studies performed on recent volcanic eruptions, e.g. El Chichón in 1982 (Rampino and Self, 1984a). In Fig. 10 is shown the height of eruptive column in the most significant eruptions of the last two centuries.

In the troposphere, there are active processes of vertical mixing due to decrease of temperature with altitude and active air-cleaning due to rains; in the stratosphere, the gradient of temperature is inverted and the vertical stability is high. In such conditions, stratospheric winds (Fig. 10), with high velocities (East-West frequently), spread aerosols in latitudinal bands. In Fig. 11 is shown the spreading of Krakatoa aerosols in 1883. Measurements with LIDAR performed by Langley Center (NASA) after the El Chichón eruption in 1982, show a sharp peak in the spread of aerosols at a height of 25 km (in Rampino and Self, 1984b).

To be spread, dust and gases must be erupted up to the stratosphere. This occurs in explosive eruptions with a Volcanic Explosivity Index (VEI) greater than 3 (scale 0-7). These explosive eruptions are, ranged from less to more: Vulcanian (with obstruction in conduit which is destroyed), Plinian (open conduit) and Ultraplinian.

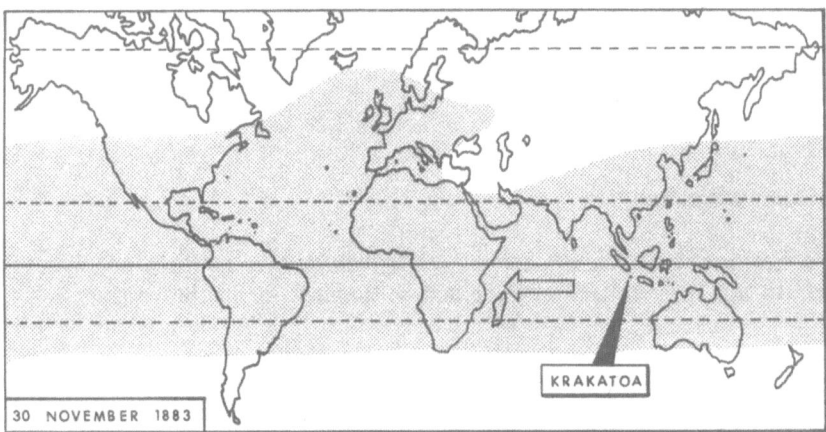

Figure 11. Spread of Krakatoa cloud according to the Royal Society.

Sulphur content is a critical factor for impact on climate. Small sulphur-rich eruptions (basaltic-andesitic) may produce the same climatic effect as much larger ones (dacite to rhyolite). Sulphur-poor (Rampino and Self, 1984a) types of sulphur-rich eruptions are vulcanian to subplinian. Rampino and Self (1984) suggest, with data from several authors that large-volume basaltic fissure-type eruptions such as Laki (1783) could be the main volcanic events with climatic impact. In this hypothesis, extensive basaltic plateau deposits such as Karro and Dekan, probably have had a significant effect on climate in geological times.

The effect of volcanic ashes is maximum in the first months after eruption; them after the fall of the ash, the H_2SO_4 droplets are the main factor affecting the radiation equilibrium (see Fig. 11). Sizes of dust for the El Chichón cloud (eruption at the end of March), ranged from 3-6μ in May (average) to 1-2μ in July. The smallest particles have an important role as condensation nuclei for H_2SO_4 droplets formation. The droplet size grows with time.

3.3. VOLCANOES AND SEA LEVEL

The contribution of volcanic driven decrease of temperatures to sea level changes must be small, delayed and short. However, volcanic tsunamis have been highly disastrous. The Krakatoa eruption triggered a tsunami with 30.000 victims in the Sunda Straits. Explosivity decreases rapidly with sea depth and only shallow submarine eruptions may trigger tsunamis. The main processes in these shallow eruptions generating tsunamis are ring explosions, collapse of pyroclastic columns and formation of calderas (Araña and Ortiz, 1984). Coupling of air and sea waves from aerial explosions is another cause previously noted.

In summary, for asteroidal impacts or cometary explosions as well as for volcanoes, the main impacts on sea level are tsunamis.

In geological times, however, submarine eruptions in middle-oceanic ridges, have probably had a strong influence on sea level corresponding to the rate of ocean-expansion and volume erupted (Turcotte and Burke, 1978). Uncertainties about our knowledge on true sea level during geological times (Mörner, 1987) poses important questions about this hypothesis.

4. Acknowledgments

To my colleagues M. Hernández and L.F. Carballal for their contribution to bibliographic research; to Ma R. de Lara for typing and F. Ramirez for the drawings.

5. References

Alvarez, V., Alvarez, L.W., Asaro, F., & Michel, H.V. (1979) "Experimental evidence in support of an extraterrestrial trigger for the Cretaceous-Tertiary extinctions.", in EOS (American Geophysical Union Transactions), v. 60, p. 734

Arana, V., & Ortiz, R. (1984) *Volcanology*, Rueda-CSIC, Madrid, 510 p. (in Spanish)

Barnett, T.P. (1984) "The estimation of 'global' sea level change: a problem of uniqueness.", Journal of Geophysical Research 89 (c5), pp. 7980-7988

Berger A. (1988) "Milankovitch Theory and Climate.", Rev. of Geophys., 26-4, pp. 624-657

Bryson, R. (1982) "Volcanoes and climate.", Mundo Científico, 18 (Spanish edition of La Recherche), pp. 948-958

Don, W.L., Farrand, W.R. & Ewing, M. (1962) "Pleistocene ice volumes and sea level lowering.", Journal of Geology 70, pp. 206-214

Emiliani, C., Kraus, E.B., & Shoemaker, E.M. (1981) "Sudden death at the end of the Mesozoic.", Earth and Planetary Science Letters, v. 55, pp. 317-334

Fesenkov, V.G. (1961) "On the cometary nature of the Tunguska meteorite", Astron, J. 28(4), pp. 577-585

Grieve, R.A.F., & Dence, M.R. (1979) "The terrestrial cratering record II. The crater production rate.", Icarus 38, pp. 230-242

Iida, K. (1963) "Magnitude, energy and generation mechanisms of tsunamis of earthquake", in Tsunamis Meetings Associated with the 10th Pacific Science Congress, Monograph. 24, pp. 7-18, I.U.G.G.

Kondratyev, K. Ya. (1988) *Climate Shocks, Natural and Anthropogenic*, Wiley. USA, 296 p.

Mankin, W.G., & Coffey, M.T. (1984) "Increased stratospheric HCl in the El Chichón Cloud.", Science, 226, pp. 170-172

Marsden, B.G. (1974) "Comets.", Annual Review of Astronomy and Astrophysics, 12, pp. 1-21

Melosch, H.J. (1982) "The mechanics of large meteoroid impacts in the Earth's oceans.", in Silver, L.T. and Schultz, P.H. (ed.) *Geological Implications of Impacts of Large Asteroids and Comets on the Earth*, The Geological Society of America, pp. 121-129

Milankovitch, M.M. (1920) " Théorie mathématique des phénomènes thermiques produits par la radiation solaire.", Acad. Yugos. des Sc. et des Arts de Zagreb, Gauthier-Villars, Paris

Mörner, N. (1987) "Pre-Quaternary Long-Term Changes in Sea Level.", in Devoy, R.N.J. (ed.) *Sea Surface Studies, A Global View.*, Croom Helm and Methuen Inc. NY, pp. 233-242

O'Keefe, J.D. and Ahrens, T.J. (1982) "The interaction of the Cretaceous Tertiary Extinction Bolide with the atmosphere, ocean and solid Earth.", in Silver, L.T. and Schultz, P.H. (ed.) *Geological Implications of Impacts of Large Asteroids and Comets on the Earth*, The Geological Society of America, pp. 103-121

Press, F., & Harkrider, D. (1966) "Air-sea waves from the explosion of Krakatoa.", Science, 154, pp. 1325-1327

Rampino, M.R., & Self, S. (1984a) "Sulphur-rich volcanic eruptions and stratospheric aerosols.", *Nature*, 310, pp. 677-679

Rampino, M.R., & Self, S. (1984b) "Atmospheric effects of El Chichón.", Investigación y Ciencia, 90 (Spanish edition of Scientific American), pp. 22-34

Roemmich, D., & Wunsch, C. (1984) "Apparent changes in the climatic state of the deep North Atlantic Ocean.", *Nature* 307, pp. 47-450

Sagan, C. (1980) - Cosmos, USA

Shaw, D., Watkins, N., & Huang, T. (1974) "Atmospherically transported volcanic glass in deep-sea sediments: theoretical considerations.", J. Geophys. Res. 79, pp. 3087-3094

Schönwiese, C.D. (1988) "Volcanism and air temperature variations in recent centuries.", in Gregory, S. (ed.) *Recent Climatic Change*, Belhaven Press, London, pp. 20-30

Shoemaker, E.M. (1977) "Astronomically observable crater-forming projectiles.", in Roddy, D.J., Pepin, R.O. and Merrill, R.B. (eds.) *Impact and explosion cratering: planetary and terrestrial implications*, New York, Pergamon Press, pp. 617-628

Silver, L.T. (1982) "Introduction", in Silver, L.T. and Schultz, P.H. (eds.) *Geological Implications of Impacts of Large Asteroids and Comets on the Earth*, The Geological Society of America, XIII-XIX

Simkin, T., Siebert, L., McClelland, L., Bridge, D., Newhall, C., & Latter, J. (1981) *Volcanoes of the world, a regional directory, gazetteer and chronology of volcanism during the last 10000 years*, Hutchinson Ross, Strondsburg, 232 p.

Symonds, R.B., Rose, W.I., & Reed, M.H. (1988) "Contribution of Cl and F-bearing gases to the atmosphere by volcanoes.", *Nature*, 334, pp. 415-418

Stoiber, R.E., Williams, S.N., & Huebert, B. (1987) "Annual Contribution of SO_2 to the atmosphere by volcanoes.", J. of Volcan. and Geoth. Res. 33, pp. 1-8

Titus, J.G. (1987) "The Causes and effects of sea level rise.", in Wind, H.G. (ed.) *Impact of Sea level Rise on Society*, Balkema, pp. 104-125

Toon, O.B. et al. (1982) "Evolution of an impact-generated dust cloud and its effects on the atmosphere.", in Silver, L.T., and Schultz, P.H. (ed.) *Geological Implications of*

Impacts of Large Asteroids and Comets on the Earth, The Geological Society of America, pp. 187-201

Turcotte, D.L., & Burke, K. (1978) "Global Sea Level Changes and the Thermal Structure of the Earth.", Earth and Plan. Sc. Lett., 41, pp. 341-346

Turco, R.P., Toon, O.B., Park, C., Whitten, R.C., & Noerdlinger, P. (1981) "Tunguska meteor fall of 1908: effects on stratospheric ozone", Science 214 (45126), pp. 19-24

Voznesensky, A.V. (1925) "The 30 June 1908 meteor fall in the Upper Khatanga river.", Mirovedenie 14 (1), pp. 25-36

Weissman, P.R. (1982) "Terrestrial impacts rates for long and short-period comets.", in Silver, L.T., and Schultz, P.H. (ed.) *Geological Implications of Impacts of Large Asteroids and Comets on the Earth*, The Geological Society of America, pp. 15-25

Weissman, P.R. (1979) "Physical and dynamical evolution of longperiod comets.", in Duncombe, R.L. (ed.) *Dynamics of the Solar System*, Hingham, Massachussets, D. Reidel, pp. 277-282

Wetherill, G.W., & Shoemaker, E.M. (1982) "Collision of Astronomically observable bodies with the Earth.", in Silver, L.T., and Schultz, P.H. (ed.) *Geological Implications of Impacts of Large Asteroids and Comets on the Earth*, The Geological Society of America, pp. 1-15

Wind, H.G. (ed.) (1987) *Impact of Sea Level Rise on Society*, Balkema, Rott, 191 p.

Wollbach, W.S. et al. (1988) "Global Fire at the Cretaceous-Tertiary boundary.", *Nature* 334, pp. 665-669

RECENT DEVELOPMENTS IN RESEARCH ON THE LOESS IN CHINA

LIU Tungsheng
Institute of Geology, Chinese Academy of Sciences
P.O.Box 634, Beijing 100011, China
&
Xian Laboratory of Loess and Quaternary Geology
Chinese Academy of Sciences
and
HAN Jiamao
IFAQ, Vrije Universiteit Brussel,
Pleinlaan 2, 1050 Brussels, Belgium

ABSTRACT: Loess is well developed at the middle latitude of the Eurasian continent. I occupies nearly 400,000 km² in North China, mainly in the middle reaches of the Yellow River forming the famous loess plateau. The loess plateau is situated also longitudinally at the zone where the monsoon from the east meets the drier west wind. The distributional characteristics of the loess with other sediments are generally developed in parallel belts from the NW to the SE, as the Gobi desert, the desert, and the loess. Loess itself shows also a gradational distribution.

Recent studies have shown that loess has a higher depositional rate in the west of the loess plateau than in the east. A 240 m thick loess sequence in Lanzhou has been deposited since the Olduvai subchron (1.8 m.y.) while to the east in Luochuan, the similar sequence has a thickness less than 135 m.

Loess deposition in China has a long time span as well as a large spatial distribution. Recent studies in Baoji about 200 km west of Xian in the Weihe River Valley in the southern part of the loess plateau have indicated that it has a time span no less than 2.5 million years.

The profile in Baoji shows that there are better developed loess-palaeosol sequences than other loess profiles. It has altogether 37 cyclic climatic pairs. The cyclic nature of these deposits in such a long time span is useful for the understanding of global climatic changes.

1. Introduction

In his book "Children of the Yellow Earth", J.G. Andersson (1934), one of the pioneers of Quaternary research in China paid much attention to the younger Upper Pleistocene

R. Paepe et al. (eds.), Greenhouse Effect, Sea Level and Drought, 217–224.
© 1990 *Kluwer Academic Publishers.*

218

Malan Loess which he considered to be a deposit of dry climate.

Teilhard de Chardin and Young (1930) first established the stratigraphy of the older loess and called it "Reddish Clay". They noticed the climatic significance of the "reddish clay" which are palaeosols intercalated within the loess. James Thorp described these palaeosols and discussed the climatic condition under which they formed in his book "Geography of the Soil of China" (Thorp, 1936).

Fundamental research on the loess in China since then has been concentrated on its spatial and temporal characteristics. The time-space relationship of the loess system in China gradually emerges as a multi-disciplinary study on the interaction of the physical, chemical and biological processes together with the impact of mankind which plays an important role in the development of the loess plateau.

2. Distributional Characteristics of the Loess in China

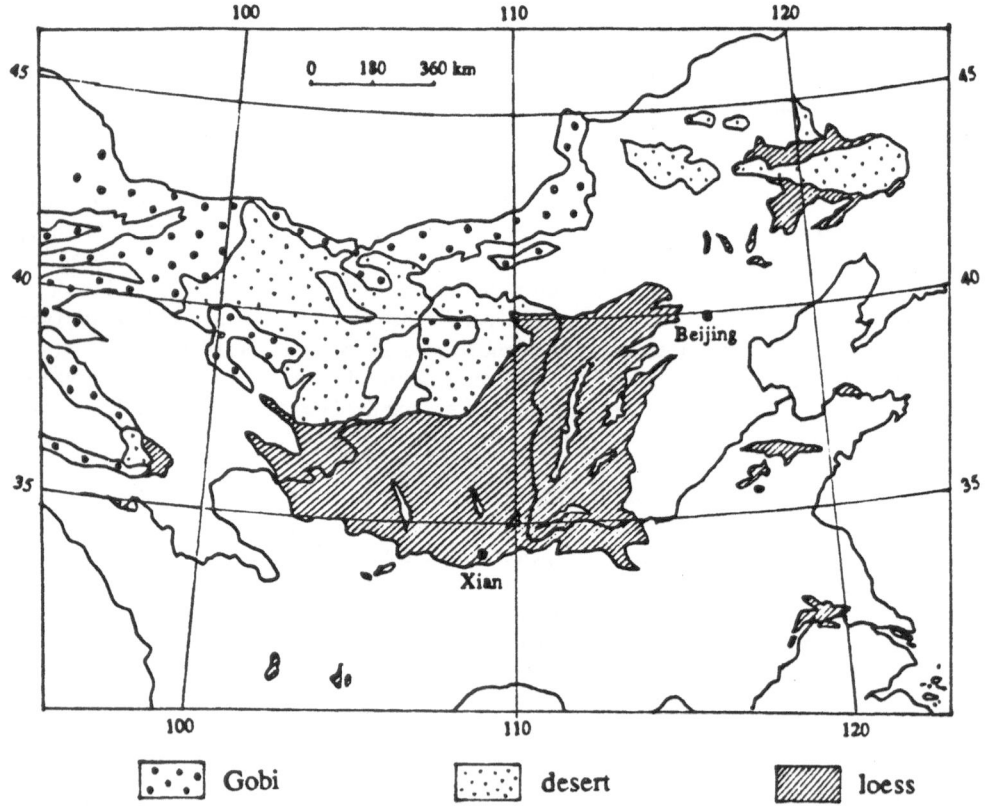

Figure 1. The relationship between Gobi, desert and loess (modified from Liu et al., 1985).

Loess is well developed in the arid an semi-arid regions in the middle latitude of the Eurasian continent. It occupies nearly 440,000 km² in North China (Liu et al., 1964). In the middle reaches of the Yellow River loess appears as a continuous thick mantle with an area of 350.000 km² forming the loess plateau. The loess plateau is situated also longitudinally at the zone where the monsoon from the east meets the drier west wind. The wide areal distribution of the loess makes it favourable for the study of issues which can only be understood in a large spatial range. The distributional characteristics of the loess with other sediments are generally developed in parallel belts from the NW to the SE, as the Gobi desert, the desert, and the loess (Fig. 1). Loess itself shows also a gradational distribution. It is coarser in grain size in the NW and finer in the SE. Three zones can be distinguished i.e. the sandy loess, the loess, and the clayey loess (Fig. 2).

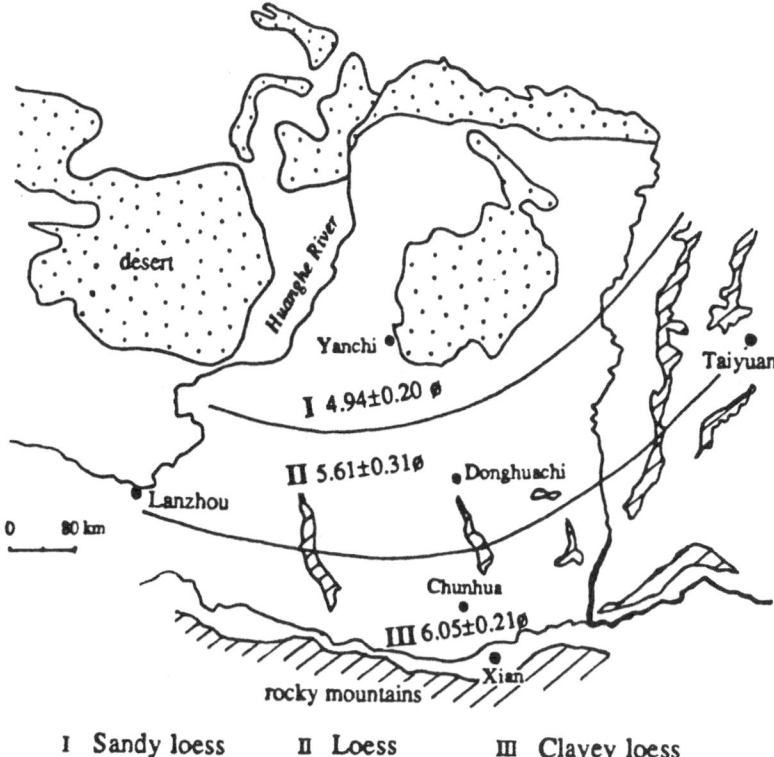

I   Sandy loess   II Loess   III Clayey loess

Figure 2. Zonation of grain size of loess (after Liu et al., 1965, 1985).

These background geological traverses across the east-west and north-south of the loess plateau have brought fruitful results. Studies have shown that in the western part of the loess plateau the depositional rate of wind blown silt (loess) was higher than that in the east. In Lanzhou, a profile of 240 metres thick was deposited over a period between 1.8-2.4 m.y. (Burbank and Li, 1985; Derbyshire et al., 1987). The mean depositional rate at Lanzhou calculated by Ding (1988) is 0.23 m/ky. But some 600 km away to the east in Luochuan and Baoji, the loess with a thickness about 135 metres has a time span of more than 2.45 m.y. The mean depositional rate is 0.05-0.06 m/ky (Ding, 1988).

Geochemical studies from the west to the east at the loess plateau have revealed that chloride particles are common in the loess in Qinghai, while to the east in Gansu concretions of gypsum nodules could be observed at the older loess and palaeosols, further east in Shaanxi and Shanxi, nodules of calcium carbonate are common.

The geochemical behaviour of the soluble salts deserves a careful study to elucidate the mechanism of the physical, chemical and biological processes which have occurred along with climatic variations in the loess plateau.

Isotope geochemistry research has also yielded promising results. Oeschger and Shen's study on [10]Be concentration in the Malan loess enabled them to compare the fluctuation of [10]Be with the oxygen isotope curves of the deep sea cores (Shen et al., 1987). Zheng et al. (1987), Gu (1987) and Xia have obtained a useful interpretation of isotopes of the loess and palaeosols. Chen and Li worked on oxygen isotopes of land snails both recent and fossil, with an aim to study their palaeoclimatic significance. Jia et al. (1987) studied organic components in the loess and palaeosols. Jia and Liu have shown that there might exist the possibility that pedogenic organic components contribute to the enhancement of the magnetic susceptibility of the palaeosols. Hus and Han of IFAQ, VUB are also interested in this issue. Heller and Liu; An and Kukla are also working on this problem.

Spectral analyses of data from the loess profile of Luochuan by Lu and An revealed that quasi-0.08 m.y and quasi-0.04 m.y. are dominant oscillations (Liu et al., 1985). Ding (1988) found that the cycle of 0.10 m.y. is also prominent in the loess and palaeosol sequence. All these works concentrated on the issue of the application of magnetic susceptibility variation of the loess-palaeosol profile (Heller & Liu, 1982) to compare and correlate with deep sea core oxygen isotope studies.

Intercalation of loess and palaeo-sanddune in the transitional zone of the loess and the desert in the northern part of the loess plateau is also of interest in recent studies of climatic changes during the last 15,000 years.

Geochemical studies on the relationship between prevalence of endemic diseases and the abundance of trace elements os of particular interest for the health of the people living in the loess plateau. Fluorosis disease has a higher prevalence in the transitional zone of desert and loess (Liu et al., 1985). Southward at the interior of the loess plateau Keshan disease - heart muscle necrosis due to deficiency of selenium - appears among the people relying on a local diet. At the southern part of the loess plateau there is Kachin Beck disease - bone disease due to deficiency of selenium and molybdenum in drinking water and the soil. Goitres are also common in the loess covered mountainous region.

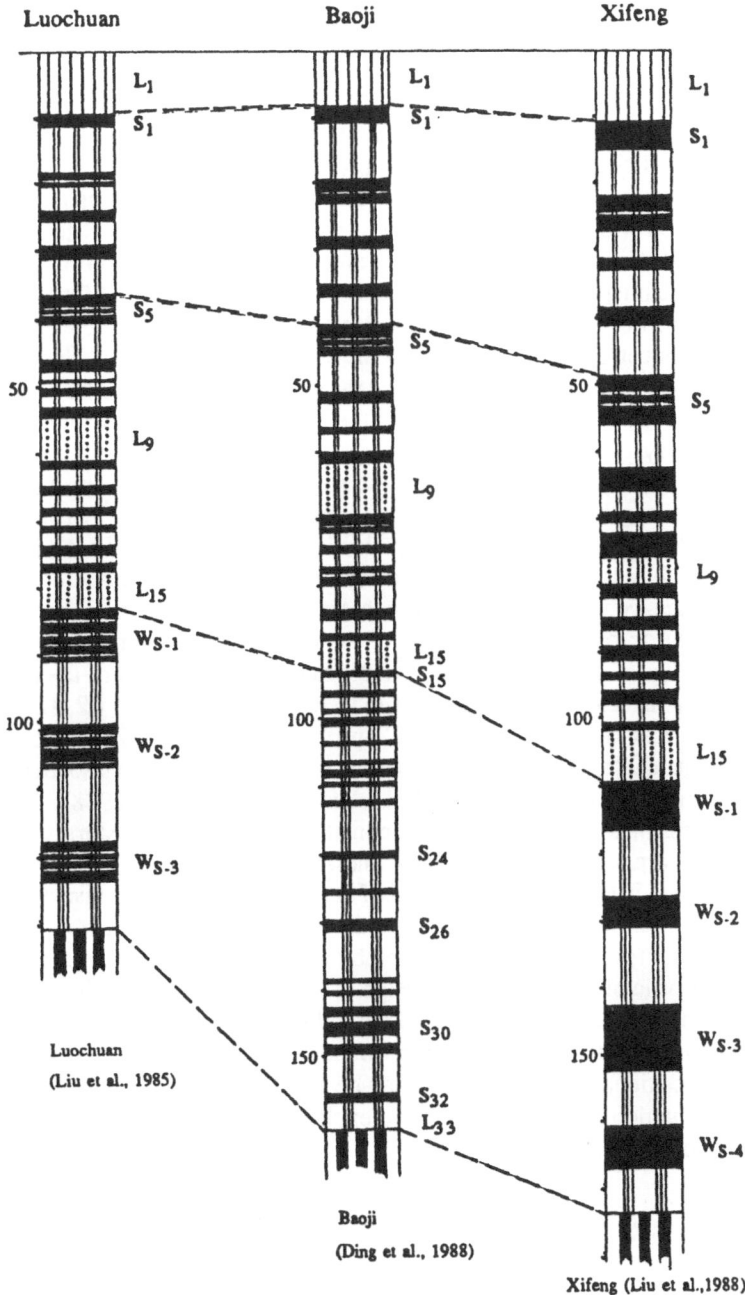

Figure 3. Comparison of stratigraphical structure.

3. Temporal Characteristics of the Loess in China

Loess in China has been depointed over a long time span as well as its large spatial distribution. Recent studies in Baoji about 200 km west of Xian in the Weihe River Valley in the southern part of the loess plateau exhibit a better stratigraphical sequence than other profiles described above. In the Luochuan profile loess and palaeosol sequences are well developed at the upper part of the profile. The profile in Baoji shows that its upper part can be correlated with other profiles such as that in Luochuan and Xifeng (Fig. 3), and also with a well exposed lower part which extended to the Gauss chron at its base. This enables us to establish a better cyclic sequence of loess and palaeosol for the last 2.5 m.y. It has altogether 37 cyclic pairs (Fig. 4). We have tentatively correlated the curve constructed according to palaeomagnetism with that of the deep-sea core oxygen isotope curves. Since there are no good records published yet for the deep-sea cores older than the Olduvai the study of loess in Baoji might be useful for the reconstruction of the climatic curve if the entire Quaternary.

This may refine earlier results. From the study at the Luochuan profile we can only divide the Wucheng Loess into broad groups such as WS-1 (soil of Wucheng Loess) and Wl-1 (loess of Wucheng loess). This designation is good in field studies but it is nor consistent with the soil and loess (S + L) system which attempts to reflect climatic conditions. From recent knowledge we would like to propose that a new scheme of loess-palaeosol sequencing could be helpful to obtain a better view of the depositional history of loess in China.

Thick loess layers denote higher wind force and longer duration of the eolian process, which could be detected in the loess profile in Baoji. They are L1, L2, L3, L4, L5, L6, L9, L13, L15, L24, L25, L26, L27, L32, and L33 in Figure 4. These thick loess layers, indicating 15 major stronger wind-blown events, occurred at the cold episodes of the 100,000 years of warm and cold cycles through the last 2.5 m.y. (Ding, 1988).

Original loess materials or the silt component deposited by dust storms in recent years show that they are similar to Pleistocene and Holocene loess.

A dust storm (Liu et al., 1982) is a kind of natural hazard; loess is also a kind of natural hazard deposit. During the Quaternary the climate changed from time to time. Sometimes it is stronger, sometimes it is weaker, thereby producing loess and palaeosol sequences.

From the profile of loess palaeosols (Figs. 3 & 4) there exist two kinds of sedimentary boundaries. The boundary at the bottom of the soil with the loess underneath are gradational, usually due to the pedogenic origin. The boundary beneath the loess is a kind of abrupt change of depositional situation which may be truncated and then covered by loess. The sudden accumulation of the loess could be explained as a geological hazard event which is of catastrophic nature.

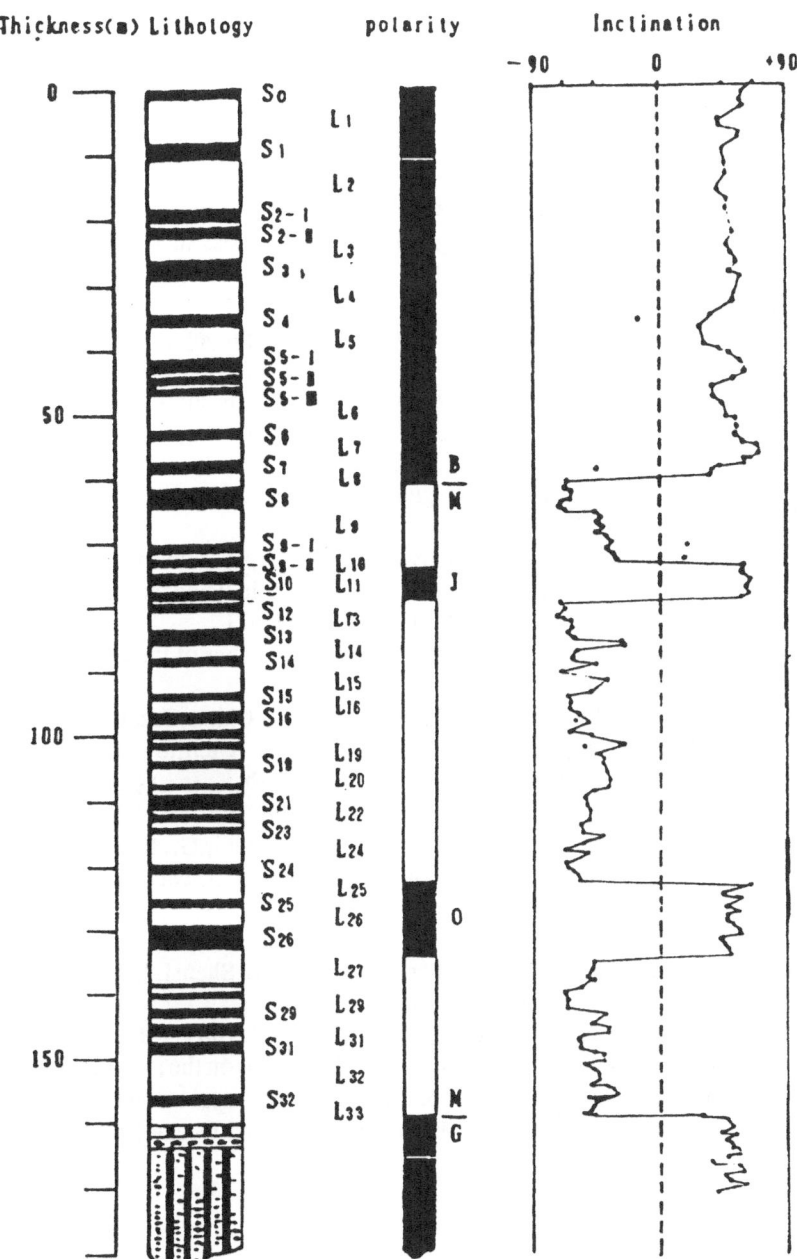

Figure 4. Magnetostratigraphy of Baoji loess section (after Ding et al., 1988).

4. References

Andersson, J.G. (1934) "Children of the Yellow Earth: studies in prehistoric China.", Kegan Paul, Trench, Trubner & Co. Ltd., London

Burbank, D.W., & Li Jijun (1985) "Age and palaeoclimatic significance of the loess of Lanzhou, North China.", in *Nature* 316, pp. 429-431

Derbyshire, E., Wang Jingtai, Shaw, J., & Rolph, T. (1987) "Interim results of the studies of the sedimentology and remanent magnetisation of the loess succession in Jiuzhoutai, Lanzhou, China.", in Liu Tungsheng (ed.) *Aspects of Loess Research*, China Ocean Press, Beijing, pp. 175-192

Ding Zhongli (1988) "Investigation and division of pedostratigraphy and climatostratigraphy of the China loess.",Thesis for doctor's degree, Institute of Geology, Academia Sinica.

Gu Zhaoyan, Liu Tungsheng, & Zheng Shuhui (1987) "A preliminary study on quartz oxygen isotope in Chinese loess and soils.", in Liu Tungsheng (ed.) *Aspects of Loess Research*, China Ocean Press, Beijing, pp. 291-302

Heller, F., & Liu Tungsheng (1982) "Magnetostratigraphical dating of loess deposits in China.", in *Nature* 300, pp. 431-433

Jia Rongfen, Liu Tungsheng, & Yuan Baoyin (1987) "A preliminary study on lipids in loess and palaeosol of Luochuan section, China.", in Liu Tungsheng (ed.) *Aspects of Loess Research*, China Ocean Press, Beijing, pp. 311-321

Liu Tungsheng et al. (1964) "The loess along the middle reaches of the Huanghe (Yellow) River.", Since Press, Beijing (in Chinese)

Liu Tungsheng et al. (1985) *Loess and the Environment*, China Ocean Press, Beijing

Liu Tungsheng, Chen Mingyang, & Li Xiufang (1982) "A satellite images study on the dust storm at Beijing on April 17-21, 1980.", in Liu Tungsheng (ed.) *Quaternary Geology and Environment in China*, China Ocean Press, Beijing, pp. 49-52

Shen Chengde, Liu Tungsheng, Beer, J., Oeschger, H., Bononi, G., Suter, M., & Wolfli, W. (1987) "[10]Be in Chinese Loess.", in Liu Tungsheng (ed.) *Aspects of Loess Research*, China Ocean Press, Beijing, pp. 277-282

Teilhard de Chardin, P., & Young, C.C. (1930) "Preliminary observations on the preloessic and post-potian formations in Western Shansi and Northern Shensi.", Mem. Geol. Surv. China, Ser. A, 8

Thorp, J. (1936) "Geography of the soil of China.", Geol. Surv. China

Zheng Shuhui, Wang Yang, & Chen Chengye (1987) "Studies on the isotopes in carbonates in Luochuan loess section: applicability of the Ca daduies as palaeoclimate indicators.", in Liu Tungsheng (ed.) *Aspects of Loess Research*, China Ocean Press, Beijing, pp. 283-290

DESERTIFICATION OR DESERT RECLAMATION? CHANGE IN CLIMATE OR IN HUMAN BEHAVIOUR?

M. DE BOODT
Laboratory of Soil Physics
Centre for Desert Science (Eremology)
Rijksuniversiteit Gent
Belgium

ABSTRACT: The recent dry spells in North Africa during the period 1968-1986 were characterised by their sudden appearance for which no reliable forecast system exists up to now. Contrary to the prevailing opinions a few years ago it is now believed that the desertification observed in the boundary regions of the desert is for a major part due to climatological changes rather than to human activities.

It is observed that the today's generation handled the dry spells less well than their ancestors did. The reasons for this are discussed. The catastrophe could have been less harsh if proper insight and technological know-how had been used to mitigate the drought. Its most appropriate aspects are discussed both in general terms and on the specific plant level. During the dry spells the best care and the most efficient use of water for crop production are needed to maximize the chances for the survival of plants, animals and man.

1. Introduction

Water is the key factor for initiating or sustaining life in desertic areas. In such regions any fluctuation in water availability has an immediate impact on life.

Due to prolonged geo-climatical conditions some deserts have existed more or less continuously for millions of years, e.g. the Peruvian desert along the West coast of Latin America. Others are only a few thousand years old or even of almost historical age like the ones in North Africa where after the great pluvials of five to six millennia ago, the savanna turned into the Sahara desert. The engravings in the caves of Tassili dating back to the Neolithic are the mute witnesses of this change.

Today the desert encroachment which is going on in many parts of the world but especially in Africa it is of much concern, not only to the scientists but also to the people. The last decades it has been extending at an even greater speed than ever before. Is it necessary to recall that at the beginning of this century the capital of Sudan, Khartoum, was still

225

R. Paepe et al. (eds.), Greenhouse Effect, Sea Level and Drought, 225–240.
© 1990 *Kluwer Academic Publishers.*

surrounded by savanna vegetation? Today one needs to travel more than 100 miles before the first meagre groups of trees can be found. In North Africa the desertification progresses in many places at an average of 5 km per year. Since the first botanical mapping in those areas was carried out around the turn of this century, it can be verified that the desert extended by more than 1,000,000 km^2 or almost 35 times the area of Belgium. This example is not unique. In all the boundaries of the great deserts, encroachment is a common phenomenon both in the Asian as well as in the American deserts. The average yearly progress of the aridity in some places can be measured by km-sticks.

The boundaries of the deserts are involved most because the population pressure there is strongest. The poorer the people, the more they are pushed to the least fertile lands, but the harder it is to survive. Many children often mean an insurance for survival as they can work the land and take care of the older people. In order to survive in arid conditions they often have to take more from nature than nature can restore.

World opinion was shaken and alerted when the balance of the dry years in the seventies and eighties was made up. In Africa six million people have died from starvation, fifty million are threatened by it, and a hundred and fifty million suffer from malnutricion. The children will bear the consequences, which are both physical and intellectual, for life, even if the food supply would change drastically tomorrow. In Niger and Mauritania the herds are halved, while in Mali the decrease is 30 %. The list can go on, but what does it matter if no adequate remedy can be found. Food supply is one thing, basic understanding of the phenomena followed by education and proper management of natural resources is better. Will mankind be able to bring about more lasting solutions, or is there a real change in climatological conditions?

2. Desertification Defined

Are climatological changes to blame for the widespread desertification? To answer this question, a clear definition must be given first of what is meant by desertification. Let us look at the edaphological meaning of the word. As the metabolism of the plant in those parts of the globe is mainly determined by temperature and available moisture, both parameters also define the growing season. When the growing season over a number of years is reduced from about 150 days to less than that so that crops, even meagre ones, cannot be produced desertification takes place. Deserts are those hot areas where most of the years plant growth is impossible through lack of available water. In such a case one observes that the growing season is too short. Hence it is said that in a desert the growing season is almost next to zero. The fringe of the deserts is characterised by a growing season of 150 days maximum and 0 days minimum (see Figure 1). This FAO map was published in the middle of the last dry period (1968-1986). The 150 days line to limit the border of the Sahel has on average moved already more than 100 km south. On this map the so-called half deserts of East Africa, the Namibian and Kalahari deserts are shown. They are characterised by an average of about 150 growing days but with frequent crop losses due to sudden lack of water. The desertification has progressed there too although less dramatic dry spells have occurred

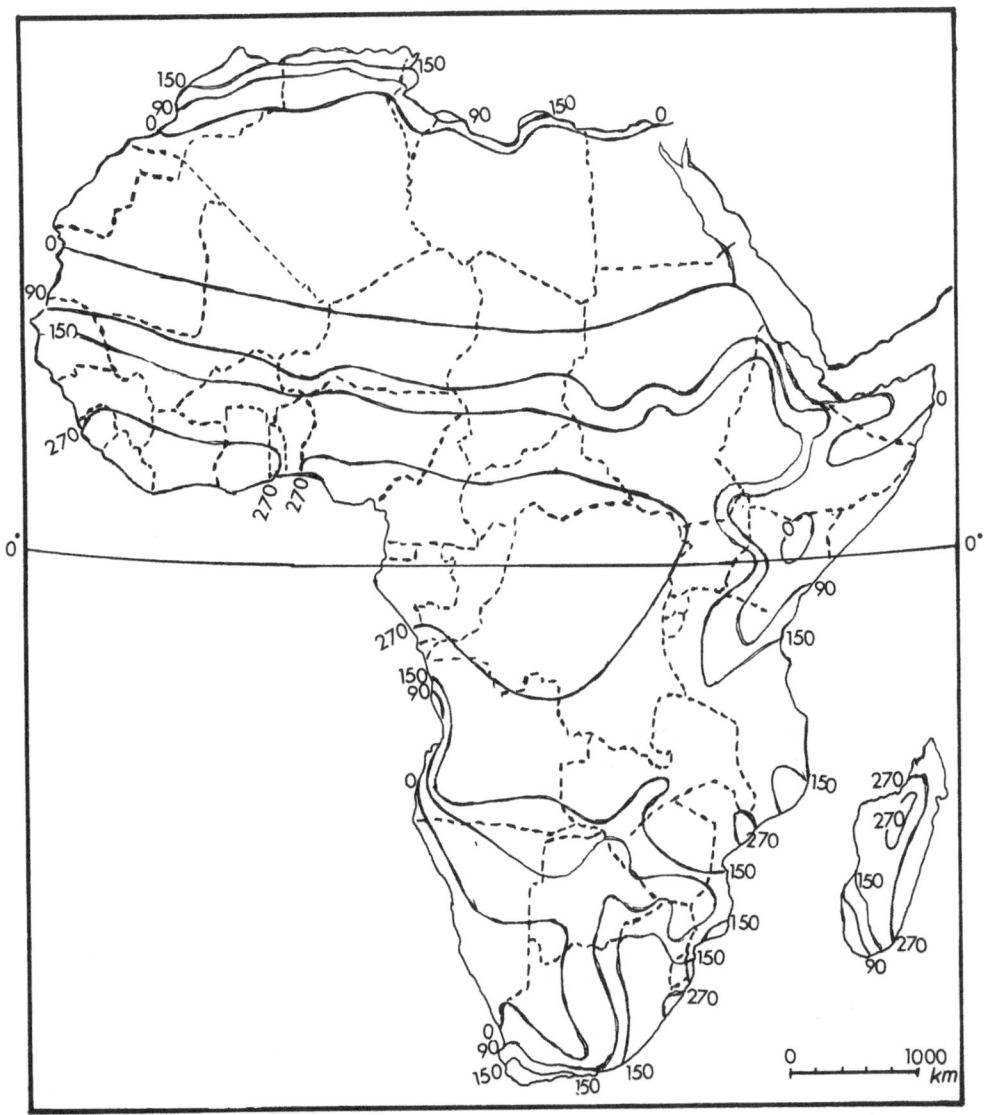

Figure 1. Map of Africa published by FAO (1977), indicating the length of the growth period for the different zones. If it is less then 150 days the area is considered to be desertic. In the Sahel in 1986 the limit of the 150 days growth period has moved on average ± 100 km more South.

there during the last decades while the survival of the population has been satisfactory. It shows the interdependence of human behaviour and climatological conditions with respect to the preservation of nature and good environmental conditions.

3. Climate or Man?

According to the available data it is not yet possible to tell if there has been a real and dramatic change in the climatological conditions to explain the desertification over the past three decades or if there is more to it. The idea is often heard that a change in ecological factors due to human behaviour i.e. cutting trees, overgrazing by herds, not replacing the trees and the grasses etc... are the major causes of the quasi permanent progress of desertification. Can the recently observed changes around the globe such as the so called greenhouse effects induced by man (the increase of CO_2 in the atmosphere which is supposed to cause the destruction of the ozone layer) trigger a definite trend changing the spatial distribution of air-temperature and rainfall? The last point is important as in many well defined parts of the globe rainfall is now higher than the recorded average while in other parts the water deficit has never been so large.

The climatological records of the Sahelian countries show that dry spells are returning much more often in recent times than ever before. According to Rognon (1989) catastrophic dry spells in the last centuries have occurred between 1681-1687, 1738-1756, 1828-1839, 1910-1915, 1939-1944 and 1968-1986. However, a real periodicity cannot be detected. That is why neither the duration nor the short intervals of the climatological change which is now going on could be predicted. This last aspect is more exceptional than the dry spells themselves. It is found recently that in the semi deserts of Southern Africa the rainfall periodicity was correlated with the surface temperature of the oceans with a lag time of 2 or 3 months. Can this explanation help to inprove the prediction of the growing season for these areas? The shorter the periods concerned, the more geographical factors seem to be of importance. They might overshadow the astronomical ones. Hence it becomes more and more difficult to find the real causes of the recent climatological changes even for relatively large areas like the Sahara desert. As long as the primary causes are not fully detected, any forecast will be fortuitous. In general the astronomical factors (see Berger in his contribution to this book) and, for the recent changes, also the solar spot activity cycles must be taken into account. They are characterised by quasi regular periods oscillating between 9.9 and 11.2 years, which in their turn are associated with a cycle of 22 years, correlated with the reversion in the magnetic fields of the sun spots. According to some other scientists the periodicity of 30 years, overshadowed by accidental but local events, should be closer to the real phenomena which the world witnesses for the last three centuries.

As described by Berger et al (1984) Milankovitch proved the link between the variation in the solar insolation and the changes of the terrestrial orbit. A solid theory is still to be developed to explain climatological changes over smaller periods, say over a few decennia which we witness today and which are so important to human life.

4.   Human Behaviour During the Recent Dry Spells

The most striking fact of the ongoing desert encroachment in the last twenty years is the great number of victims. Never before dry spells have been so harsh to the human race.

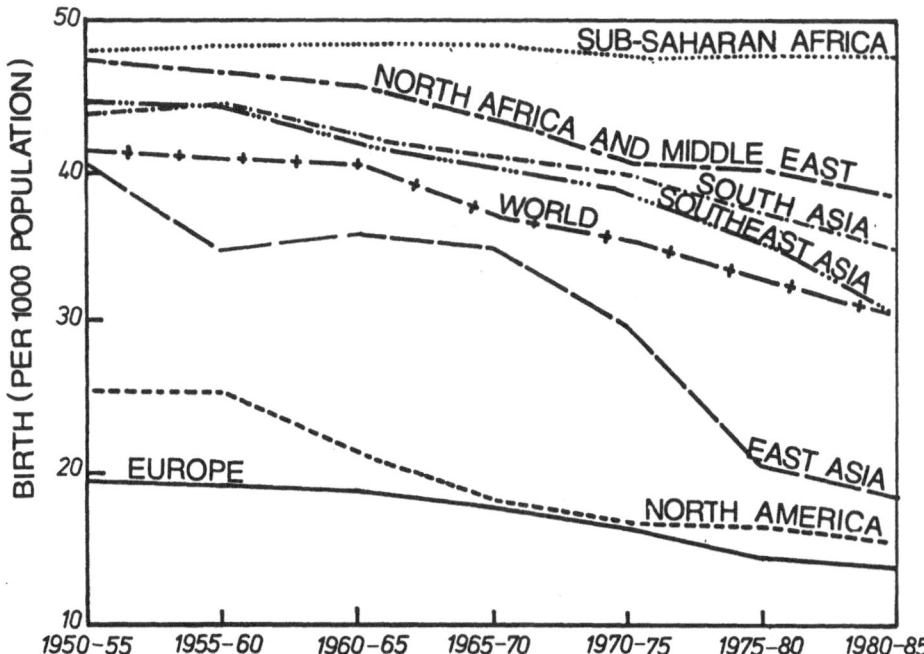

Figure 2. Since 1950 birth rates in the world have declined with the exception to this trend in sub-Saharan Africa, even during the dry spells 1968-1986 (after J.C. Caldwell & P. Caldwell, 1990).

The question arises if the local people are now less well-armed to survive than their ancestors. First there is the much larger number of people to be supported by a shrinking amount of water and ditto available surface of cultivated land. In the boundary areas of the desert the population is now between 6-12 times larger than at the beginning of the century. Sudan, with a population of about 2 million in 1900 jumped to almost 27 million today. As can be seen from Figure 2, birth rates since 1950 in Egypt and in sub-Saharan Africa are above 40/1000 inhabitants. This means that in a life span of about 35 years the population can increase 4 times.

As was pointed out by J.C. Caldwell and P. Caldwell (1990) there exists in these regions a socio-religious system that is not more traditional, primitive or backward than those prevailing in Europe, Asia and the Americas, but it accounts for the persistent high birth rate. In the sub-Sahara society the emphasis lies on ancestry and descendance. It reflects the strength of the ties based on family and lineage. In such a system there is an overpowering need for descendants to ensure the survival of that lineage. Those who live now are the caretakers. They will be blessed or cursed through the intervention of the ancestors, dead or alive, according to their contribution to the survival of that lineage which is after all the respect due to the ancestors from whom they descend.

In such a fast growing society agriculture and livestock production could hardly follow. Through the development effort of the richer nations new methods were introduced and old ones abandoned. The custom of building up reserves for unexpected catastrophes was either not possible because of lack of goods or was not done as money was spent on expendable goods to follow the European way of live. Good husbandry and cunning of the land were just in a transition period when the dry spells came. The nomadic way of life changed to a sedentary one perhaps too rapidly for too many people. They had just started to cultivate the land using fertilisers as shifting cultivation was no longer possible. Human labour was being replaced by animal traction to cultivate the land. Adequate chemical fertilisation was still unknown in many cases and the land was exhausted. Fertilisers, pesticides etc. were not yet common goods. The old system which could bear prolonged dry spells was overthrown and modern technology was not yet fully adapted. This in-between stage, together with the population explosion might be the main reason why the drought was so harsh on the generation of today and why they could not resist the adverse situation as well as their ancestors. On top of that, civil wars were going on in some of the most affected countries. It meant an unprecedented disaster and a blame for mankind as a whole.

5. Desert Reclamation

Is there a way out? A return to the old system is not possible. Only improved knowledge, education and a better understanding of new methods are meaningful. How to save water, how to detect rechargeable water at the fringe of the desert at reasonable depths, how to bring surface water from places of abundance to the dry areas by proper canalization and tunnelling are sensible approaches to improve the situation.

The logical consequence to deal with is to increase the efficiency of the water use. New irrigation systems with minimal losses like drip and low sprinkling irrigation, new drought resistent varieties, the prevention of salinisation, new treatments and feeding systems of the herds etc. all are aspects which have to be properly implemented. Above all, regarding the problem of the population explosion, social and economical issues have to be tackled as well. These items should be studied. This is the major problem.

6. Mitigation of Drought

From the above-mentioned, it is obvious that mitigation of drought covers a vast complex of actions to be carried out in many fields. For the discussion, the essential points should be highlighted. They are: the amelioration and the proper management of the water provision on the macro-, the medium as well as on the micro-scale.

Macro-scale improvements mean gigantic undertakings such as building trans-Sahelian canals, huge dams etc. They are very expensive, and often require international assistance. Among those undertakings, long distance transfer of water is the most important. Successful examples exist in California, Egypt, Israel, and Turkmenistan. A number of huge works are stopped underway or still in the planning stage as their realisation is more than once postponed not only because of the tremendous amounts of money involved but because of the unknown impact on the environment.

In the very recent history, such gigantic works nevertheless are proven to be very adequate. If the Assuan Dam in Egypt had not been constructed about 20 years ago, transporting water from its reservoir over more than 1000 km to the Delta in the north, be it in the bed of the natural river Nile, would have been unreliable and the dry spells in Africa would have had even more dramatic consequences. An additional 40 million people would have been in the starvation zone, which is double the number of those threatened during the last dry spell.

The technology of water transfer has made enormous progress in recent times. It has become a technology on its own. This subject has been discussed in a separate contribution during this symposium.

Much cheaper is the drilling for water at intermediate depth i.e. between $\pm$ 60 and $\pm$ 120 m. Very often it is renewable water. It is much more abundant in the fringe areas of the deserts than ever thought before. On both sides of major rivers like the Nile, the Senegal and Niger etc. often the water carrying layers extend over many tens or hundreds of kilometres. So called "dry valleys" at medium depth are carrying seizable amounts of water. This can be caught in a relatively easy way. Water provided on a medium scale often carries a few grams of salt per litre which is not considered to be optimal, but modern technology knows how to handle such problems.

In the period immediately after the last world war, it still took months to dig a well of 50-60 m deep, mostly of the cone-shape type. With the modern equipment it is only a matter of a few days. On top of that, water from the wells is proven to be the cheapest way to provide the needed moisture to the soil to grow more and better crops. To lift the water, it costs less than a few dollar cents per m^3 per 10 meter. It is most recommended as around each well easily 50 to a few hundreds hectares of agricultural lands can be developed. In general, the owners of the well are also the users of the water. They manage it carefully and feel a responsibility. It makes a complete difference with the huge water supply organisations where the care for the water is often an anonymous issue. At depths beyond 3-400 meter fossil water can be detected. Special studies in North Africa have proved that the reserves are often impressive. In general, the salt content of the water can be relatively high. For the time being these resources are practically not exploited. The level of knowledge on how to use water very efficiently must be well established before it will be justified to make use of this unique human resource.

How efficient the supply of water as a whole might be, it only can bear fruit when the water is used efficiently on the farmer's level. This brings us to the discussion of the drought resistant plants and modern technology for water use on the micro-scale or field level.

7. Efficient Irrigation Systems

From experience, also shared by FAO (1980), is learned that the three major types of irrigation even when carried out in an efficient way still differ from each other in output because of the shortcomings inherent to each system.

The overall irrigation efficiency Ep for border, sprinkler and drip irrigation is respectively 0.32, 0.45 and 0.61 % from the ideal, which is close to the theoretical approach. In many desert areas, however, the efficiency drops to 0.10 %, especially in the Near East and North Africa where irrigation water is free of charge or only a lump sum has to be paid independently of the amount of water received. Traditionally border irrigation can even lead to complete floodings when close control is lacking.

8. Drought Resistant Plants

Efficiency of water use can be increased by choosing the right crop. Some varieties are much more efficient than others. Plant breeding in the common sense of the word has provided crops having a water efficiency which is 20 % better than the former breeds. The recent acquisitions through genetic engineering are not mentioned here as the new plants are too expensive for the farmers of poor countries.

When the amount of water for plant growth is the limiting factor and not the area which potentially can be irrigated, one often gets a greater benefit by irrigating more land with less water. Under such circumstances not all crops react the same. Therefore the yield response coefficient ky of the plant under water stress must be known. This parameter has been developed during recent years in a well documented literature and published by FAO (1977-1980).

The lower the ky factor the less the plant suffers when the water availability goes down. A low ky factor indicates that the crop is resistant to drought. Low ky plants are groundnuts, cotton, grapes, alfalfa, olives, safflower, sorghum, soybeans, sunflower etc. Amongst the medium ky ones, following plants can be mentioned: barley, wheat, green bean, cabbages, citrus fruits, melon, peppers, potatoes, tomatoes etc.

9. Salinization as a Major Problem

While irrigating the water penetrates the soil. Its first action is to dissolve the salts. Indeed, one of the characteristics of any desertic soils is its high salinity. In moist soil the pores are filled with the water enriched with the solutes. Due to the capillary rise and the high temperatures the salts migrates with the water to the soil surface where it evaporates and leaves the salt behind.

Salt precipitation near the plants kills the vegetation due to the high osmotic pressure, which means that the majority of the water, both from the plants and the soil, is sucked into the salt. First the plant growth is slowing down, ultimately the plant is killed (see

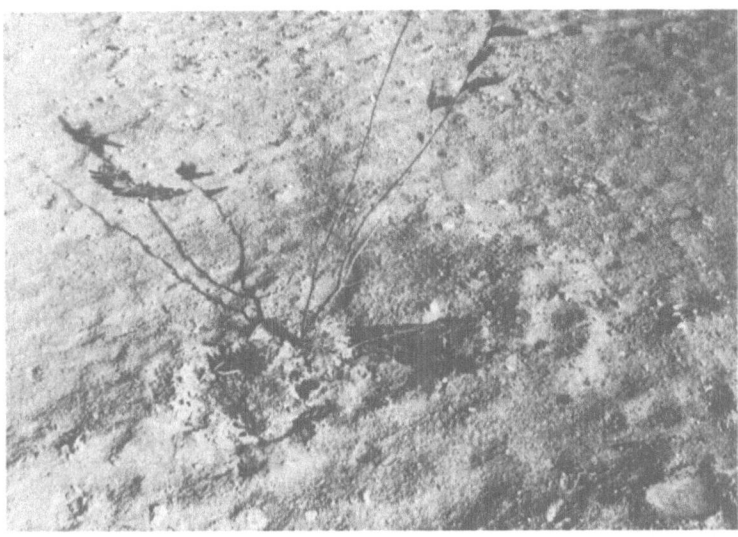

Photo 1. Drip irrigation applied on a desertic sandy soil near Cairo when growing citrus trees. Without soil conditioning, resulting in salt crust formation killing the trees.

Photo 2. With soil conditioning: luxurious growing trees are observed. Evaporation from around the tree is cut avoiding salinisation while adequate water distribution in the root zone is assured as explained in Figure 3.

photos 1 and 2). A new technology recently developed at the State University of Gent has made it possible to overcome this failure.

As mentioned before, drip irrigation is beyond any doubt in many cases one of the most efficient and economic ways of supply water to the plants. Its application is limited because of the risk of salinisation, especially around the plants and the drippers. Therefore special precautions, known as soil conditioning, have to be taken.

In this respect, sandy soils differ in many respects from loamy soils (Figure 3). The pores in sandy soils are larger due to the presence of coarse sandy particles. The water retention capacity is small. Due to gravity forces most of the irrigation water proceeds vertically downwards. There is no lateral suction of water because in large pores the capillary forces are nearly non-existing.

The behaviour of the different soil types is illustrated in Figure 4. The natural infiltrating water profiles in a loamy and a sandy soil are showing a normal and an elongated onion shape respectively. In the case of a sandy soil the water drops move in a narrow beam downwards. In such a case the root system of the plant will only develop in the moist part of the soil. It will be long and elongated with little ramifications. The yield of a plant with such a root is mostly low.

In general it can be said that the more voluminous the root develops the higher the yield will be. Soil conditioning promotes a wide soil moisture front and volume, as shown in Figure 3. That way yields can be doubled.

In Figure 4 on the left the case of a loamy soil is given. Loamy soils are composed of small soil particles. This means a high capillary suction power due to the fine pores present. The pores have a diameter of between 50 and 100 μm. When introduced in such a soil, the water is subjected not only to gravity, promoting a vertical downwards movement, but also to lateral suction. As at the top the soil is drier, the water pull in the pores will be highest. Thus the water moves at the beginning only in the top layer with a priority for lateral movement close to the surface. This water evaporates easily leaving a salt crust. Specific soil treatments are necessary to prevent this.

Soil conditioners have to fulfil a triple purpose:
a. to keep the water in the upper layers where the root system normally develops;
b. to prevent the evaporation of the water close to the surface;
c. to avoid the salt crust formation.

It can be done by spreading cheap synthetic polymers, the so-called soil conditioners, on the surface (Figure 5). Their action is based on the change of the capillary properties starting from the soil surface. To prevent deep infiltration and to fix the water in the root zone another effective water adsorbent must be used. This is done by introducing a powder, being hydrolysed polyacrylamide, known as a hydrogel. In such circumstances it adsorbs $\pm$ 300 times its own weight. It can be mixed at the appropriate depth (root zone) in the soil to retain the water (see Figure 6a and 6b) exactly where it is needed by the plant roots.

Those products were quite expensive some 10-15 years ago, but mass production, even in developing countries made the prices fall. Another important development has been that some waste plastic foils could be recuperated and put into an emulsion. When derived from polyethylene, polypropylene, P.E.G. etc. of which the chemical formulae are given in

A. Sandy and coarse sandy soils without capillary pull.

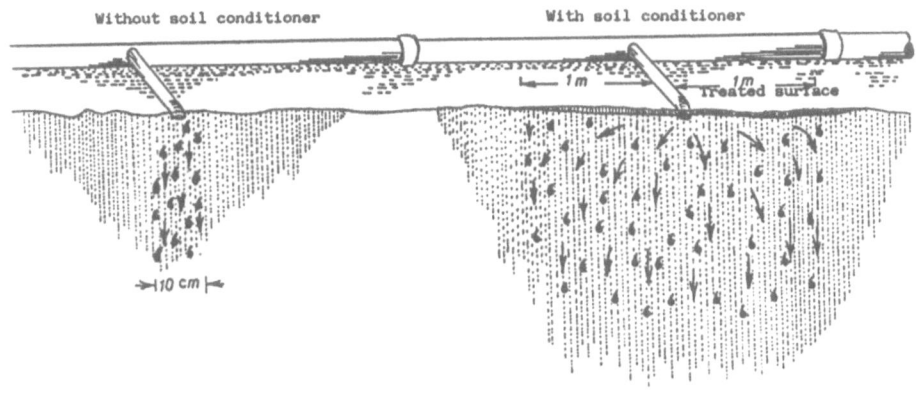

B. Loamy soils with capilary pull.

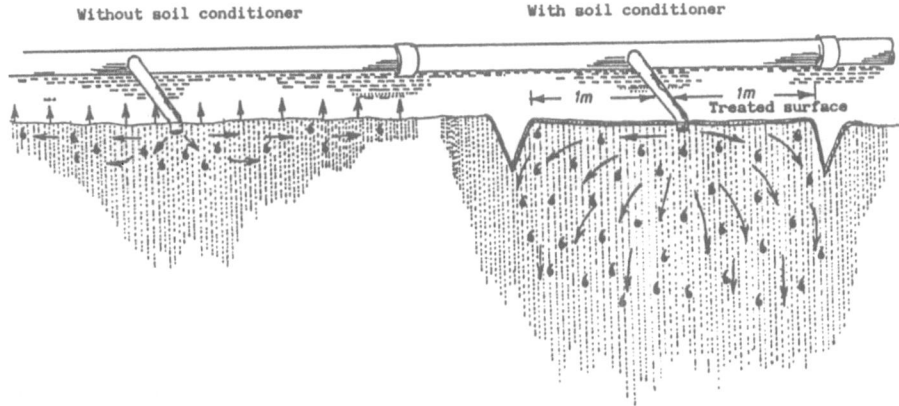

Figure 3. Water infiltration in desertic soils can be manipulated by applying soil conditioners. On sandy soils most effective results are obtained using emulsions of which the hydrophobic and hydrophylic parts get oriented respectively to the atmosphere and to the wet soil. So evaporation from the soil surface is cut and the lateral extension of the water to moisten the root zone is promoted. On heavy soils the emulsions will prevent evaporation as well as salt crust formation.

236

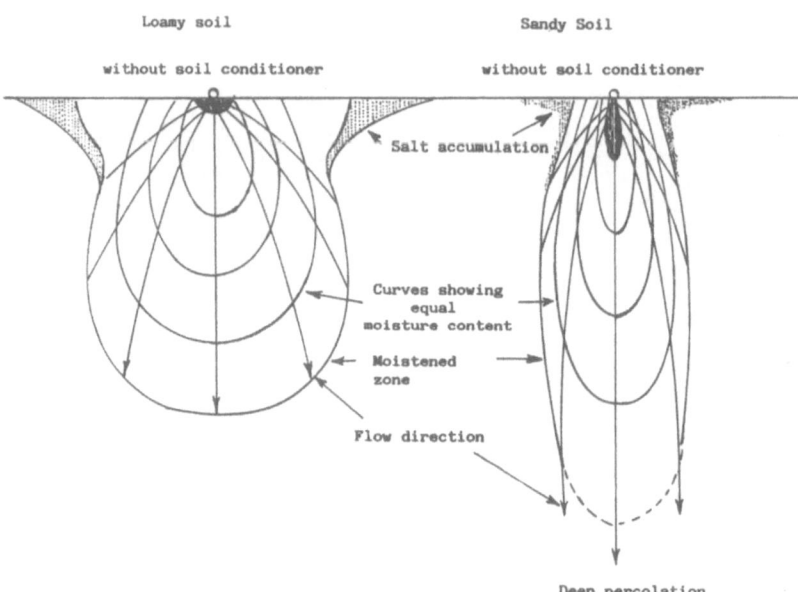

Figure 4. Complements Figure 3. It compares the normal water distribution and salt accumulation in loamy and sandy soils when drip irrigation is applied without soil conditioning.

Figure 5, cheap and effective products can be developed. They can be applied around the plants and the drippers to cut the evaporation. They are hydrophobic preventing capillary rise and hence salt crust formation.

Other conditioners mentioned at the top of Figure 5 which can be put into solution are hydrophilic. They can fix relatively high amounts of water. These soil conditioners in combination with chemical fertilizers and pesticides can be mixed together in the plough layer in order to promote good plant growth.

Soil conditioners can be spread over the seeding lines as a mulch using a normal pesticide spraying machine. The polymers, being adhesives, when brought into contact with the soil particles will aggregate them. In surface treatment, the dose is normally 20-40 g active material per m². When planting trees these chemicals are mixed in the plant pit to promote water storage in the root zone. A dose of not more than 0.1 % active material by weight calculated on the amount of soil involved is needed to be effective.

The size and depth of the waterfront in the soil profile and the amount of available

A. POLYMERS SOLUBLE IN WATER : HYDROPHYLIC SOIL CONDITIONERS

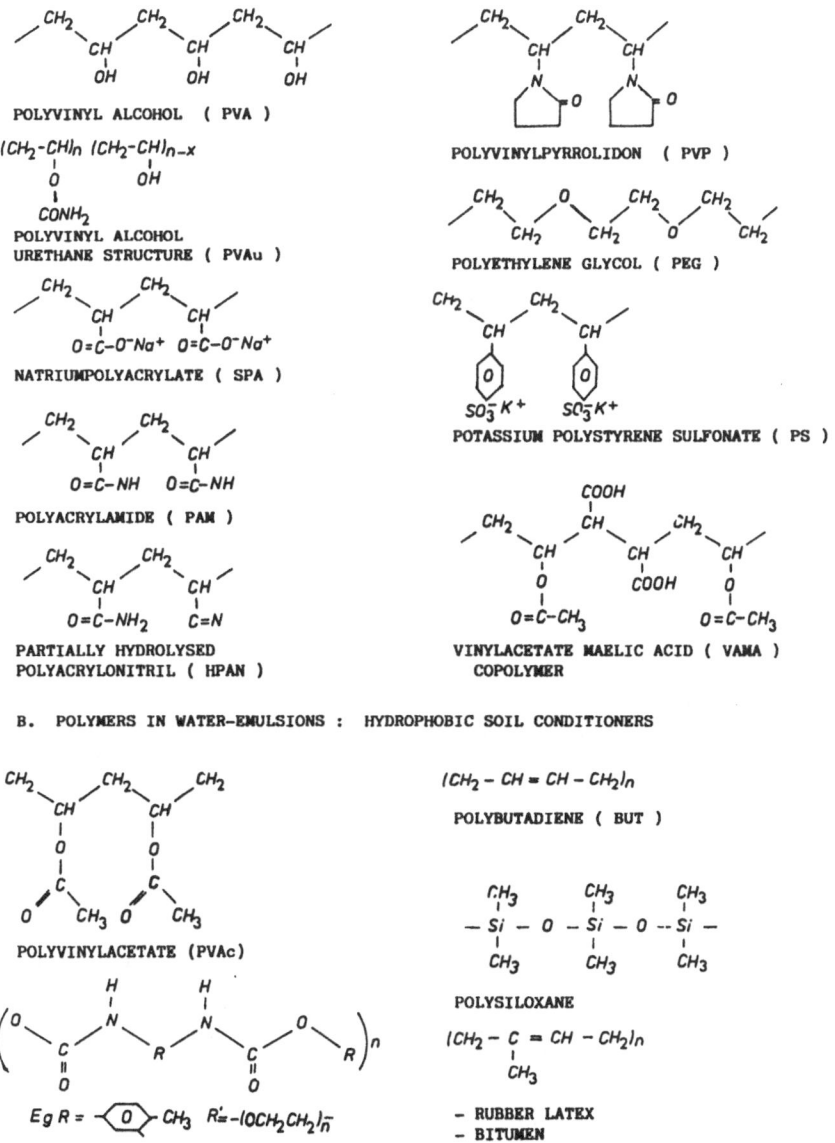

POLYVINYL ALCOHOL (PVA)

POLYVINYL ALCOHOL
URETHANE STRUCTURE (PVAu)

NATRIUMPOLYACRYLATE (SPA)

POLYACRYLAMIDE (PAM)

PARTIALLY HYDROLYSED
POLYACRYLONITRIL (HPAN)

POLYVINYLPYRROLIDON (PVP)

POLYETHYLENE GLYCOL (PEG)

POTASSIUM POLYSTYRENE SULFONATE (PS)

VINYLACETATE MAELIC ACID (VAMA)
COPOLYMER

B. POLYMERS IN WATER-EMULSIONS : HYDROPHOBIC SOIL CONDITIONERS

POLYVINYLACETATE (PVAc)

POLYURETHANE

POLYBUTADIENE (BUT)

POLYSILOXANE

- RUBBER LATEX
- BITUMEN

Figure 5. Formulae of soil conditioners which are the most frequently used with either predominantly hydrophobic or hydrophilic properties.

OPTIMIZING SOIL CONDITIONS USING HYDROGELS

A) EXHAUSTED SANDY OR LIGHT SOILS

a) NOT TREATED   b) TREATED

B) ARID CLAYEY OR HEAVY SOILS

a) NOT TREATED

b) TREATED

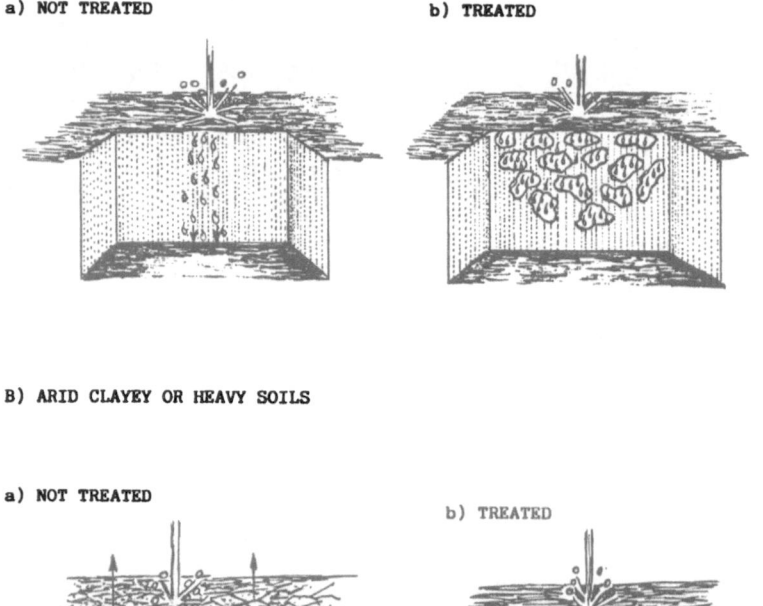

Figure 6. Hydrophilic soil conditioners, called hydrogels, can also be produced and applied in the dry form. The increase in water holding capacity in the root zone is obtained by mixing the product in the soil as illustrated both in a sandy or light soil and in a heavy soil. It shows how deep-percolation can be prevented.

water can be regulated when drip irrigation is applied and the output of the drippers is adjusted. The way it works is shown in Figures 3a and 3b, and in Figures 6a and 6b.

10. Efficient Water Management

From the foregoing it becomes obvious that there exist two major types of polymeric organic chemicals which can interact in two ways on the water distribution in the soil and on the water availability for plant growth. When using hydrogels the powdery chemical is thoroughly mixed with the soil (Figure 6). When water is applied, even in huge quantities, it is adsorbed in the soil profile at the height of the root development. The other type of chemical is hydrophobic. This is broadcasted as a diluted synthetic polymer on the soil surface. It can be either a mulch or a solution mixed superficiously with the soil particles to facilitate their aggregation. In this way slaking, evaporation and salt crust formation are avoided.

When a hydrophilic polymer is applied on the wet surface of a sandy soil, conglomeration and aggregation of the soil particles could also occur. The pore size of the loose soil will diminish, the average diameter will drop from $\pm$ 50 μm to $\pm$ 5 μm. It means that most of the water transmission pores will be replaced by smaller water storage pores. Thus the irrigation water can be retained in the aggregated upper soil layer. As more water is applied by drip irrigation the soil pores will be filled up and a downwards water movement will start. A hemispherical waterfront will develop in the root zone as shown in Figure 3. The depth will depend on the amount of water applied. Deep seepage as normally occurs in untreated soils does not happen because the water is held up by the porous suction in the top layer. As mentioned above the other, more effective but also more expensive, way of avoiding deep penetration is by mixing the hydrogels with the soil.

Measurements on experimental fields in Egypt and Saudi Arabia have shown that approximately 50 % of the irrigation water could be kept in the ground which otherwise would have been lost through evaporation or deep infiltration.

There is still another phenomenon to be mentioned. The water vapour immediately underneath the treated soil layer is subjected to a temperature gradient. The water vapour near the surface has a higher potential than at the cooler but deeper layers. This causes a reflux of the water also known as a distillation-condensation effect in the root zone. The condensed water contains only minute amounts of salt. In this moistened layer situated mostly at a depth of about 15-20 cm a vigourous root growth can often be observed.

In summary, the deep seepage and salt accumulation pattern all too often occurring in the classical irrigation system in semi-arid and arid regions can be mastered and replaced by an economic and efficient water management system with little or no salinization hazards.

240

11. References

Berger, A., Imbric, J., Hays, J., Kukla, G. & Saltzman, B. (1984) *Milankovitch and Climate*, NATO-ASI Series, Reidel Publish., 510 p.

Caldwell, J.C. & Caldwell, P. (1990) "High fertility in sub-Saharan Africa.", Ed. Scientific American Inc, New York, 265, pp. 82-88

De Boodt, M. (1987) "Long distance water transfer and its possibilities for fighting desertification." *Proceedings International Water Supply Association*, Symposium Drought and Famine, London 1986, Water Supply, London, 5, pp. 179-188

De Boodt, M. (1989) "Appropriate technologies to reclaim desert soils.", in De Boodt, M. and Hartmann, R (eds.) *Proceedings of Symposium on Desert Science (Eremology)*, Gent, Belgium, August 31-September 25, 1987, Faculty of Agric. Sciences, State University Gent, pp. 11-33

De Boodt, M. (1990) "Application of polymeric substances as physical soil conditioners." in De Boodt, M. et al. (eds.) *Soil Colloids and their Association in Soil Aggregates*, Plenum Publishing Corporation, London, New York, 19, pp. 580-592

Doorenbos, I. and Pruitt, W.O. (eds.) (1977) *Crop Water Requirements*, Irrigation and Drainage Paper No. 24, FAO, Rome

Doorenbos, I. and Kassam, A.H. (eds.) (1980) *Yield response to water*, Irrigation and Drainage Paper No. 33, FAO, Rome

Rognon, P. (1989) *Biographie d'un désert*, Collection scientifique synthèse, Plon, Paris, 347 p.

Taylor, A.S. & Ashcroft, G.C. (1972) "Physical Edaphology.", in W.H. Freeman and Co (ed.) *The physics of irrigated and non-irrigated soils*

NATURAL PROTECTION AND VOLUNTARY EXTENSION OF THE TROPICAL AFRICAN FOREST COVER

M. LEROUX
Université Jean Moulin
CNRS UA 260 - PICG 252
B.P. 638, 69239 Lyon - France

ABSTRACT: The tropical African forest cover prevents an excessive warming above it, and pushes to its margins the heat lows, lows occupied by low-level wind discontinuities (Meteorological Equator and Inter-Oceanic Confluence); thus the forest keeps out the dry airstreams, such as the continental trades, which are restricted to its margins. Because it extends as far as the ocean, it maintains the maritime advection permanently over itself; so that the forest ensures its own protection. The aerological influence of a compact forest cover concerns only low-levels; as soon as the surface influence disappears, the Meteorological Equator recovers its usual structure and the distribution of heavy rains, connected with its mid-level I.C.Z. structure is then no longer modified. Present and past conditions (especially during the Last Glacial Maximum and the Holocene Climatic Optimum) show that its abilities to protect itself are limited, its survival depending on an always precarious equilibrium between opposing streams, on the one hand the maritime airstream and on the other hand, in the opposite direction, the dry continental trades blowing from north and south towards the forest margins.

The recent evolution of meteorological conditions with heavy rainfall distribution in a narrow belt closer to the Equator, connected with the strengthening of the continental trades, makes the equilibrium of most forest margins more fragile and susceptible to becoming abruptly destabilised. Consequently it appears that the protection of the present forest cover is really the first priority.

The tropical forest is at present a major concern and we have to protect the "lung of the planet", which is often evoked in relation to rainfall (the contribution of transpiration allowing several reutilisations of precipitable water potential (Leroux, 1983; Monteny, 1986; Pouyaud, 1986)), and more recently in relation to the absorption of carbon dioxide; however, the tropical forest is rarely considered as a meteorological factor able to determine the surface pressure field and the wind field at low-levels.

When the destruction of forest is evoked, responsibility normally rests with either a decrease of rainfall or exploitation by man; but the cause is not sought in the increasing power of continental trade winds, precisely characterized by a strong saturation deficit.

Our purpose is, to observe the interactions between a compact forest over and its aerological environment, in order to understand how the forest ensures itself - at the present - its own protection, and also why the forest was - in the past - unable to stop the dry air at its boundary, and consequently

R. Paepe et al. (eds.), Greenhouse Effect, Sea Level and Drought, 241–252.
© 1990 *Kluwer Academic Publishers.*

Figure 1. The Guineo-Congolese forest.

to maintain itself. Then we observe the present state of the Guineo-Congolese evergreen forest (Fig. 1) and chiefly the conditions required by nature for the establishment and survival of this tropical forest cover.

1. The Self-Protection of the Forest

The self-protection of the forest is primarily explained by the surface thermal field, the surface pressure field and the low-level wind field above and around the forest area.

A compact tropical forest (in the Congo basin, as in the Amazon basin) modifies the earth surface response to solar radiation, by way of its thermal behaviour, comparable to that of an ocean; as a consequence the *real* shore of Africa is quite different from the *effective* shore, which coincides with the forest margins.

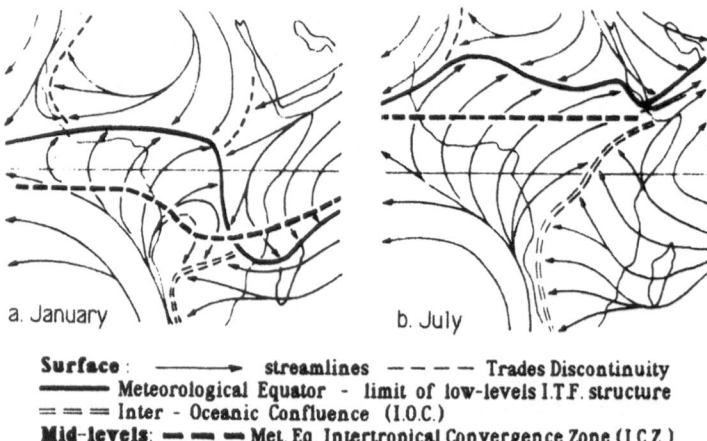

Figure 2. Discontinuities and surface airstreams - present situation.

1.1. SURFACE PRESSURE FIELD AND WIND FIELD

The forest cover (like the ocean) makes an excessive warming above it impossible; thus the forest pushes to its margins the warm thermal lows (or heat lows) at the surface and low-levels, heat lows occupied by low-level discontinuities of the wind field, such as, in tropical Africa, the Meteorological Equator (M.E.), or in the southern part the Inter-Oceanic Confluence (I.O.C.), a type of Trades Discontinuity (Leroux, 1983).

As a consequence, in boreal winter (Fig. 2-a), the surface Meteorological Equator is stopped on the northern limit of the forest area, because of low pressure maintained on the continent where the highest temperatures are observed. In austral winter (Fig. 2-b), the Inter-Oceanic Confluence is stopped on the southern limit of the forest area, even when the lowest pressure is observed north of the Equator, in the Saharan belt; however, in Eat Africa the Meteorological Equator can easily cross the Equator, northwards.

As a result, in tropical Africa, according to the longitude, three distinctive situations exist (Fig. 3):

a. in East Africa the Meteorological Equator moves with relative freedom northwards and southwards, nothing being able to stop at low-level the low pressure field associated with solar zenithal motion,

b. in West Africa the Meteorological Equator moves over the continent associated with zenithal motion during the summer period, but during the winter period (from December to February) its migration is stopped from Liberia to Cameroon, precisely on the northern limit of the evergreen forest,

c. in Central Africa (between 10 and 30 degrees East)the evergreen Congolese forest does not allow the free motion of the zenithal thermal maximum; as a result two surface heat lows exist at the same time, either by direct or indirect thermal effect, according to whether the zenith occurs in one hemisphere or in the other.

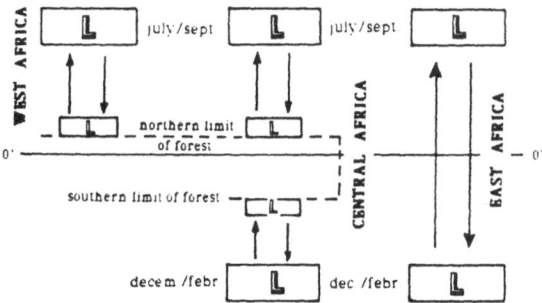

Figure 3. Annual migration of surface heat lows.

Consequently, because the forest does not allow the establishment above it of deep heat lows and, because it is therefore surrounded by a succession of direct and indirect lows, the forest keeps out the dry continental trades. A dense forest needs a reduced evaporation; precisely because it extends as far as the ocean, the forest maintains permanently over itself a westerly maritime advection. Trees are thus always covered by an airstream with a low saturation deficit which, in connection with the return of hydric storage by vegetation (transpiration), maintains the humidity of the air. This moisture supply is particularly important during the two to three months of the winter season without significant rainfall, precisely when the aerological conditions external to the forest area are severe.

Therefore, the dense forest ensures its own protection.

1.2. AEROLOGICAL STRUCTURE AND RAIN-MAKING CONDITIONS

Aerological action of a compact forest concerns only low levels; as soon as the thermal influence of the ground weakens and disappears (on average 1,500/2,000 metres above the surface level), the structural conditions of rain-making recover their usual pattern as is the case for the Meteorological Equator: the spatial distribution of heavy rains connected with its I.C.Z. structure of mid-levels is then not modified.

It is necessary to emphasize the differences between levels (Fig. 4):

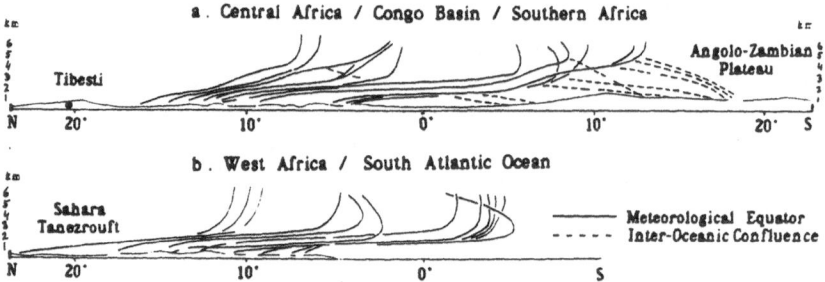

Figure 4. Annual migration of the Meteorological Equator: monthly meridian cross-sections in low and mid-levels.

- at low levels the forest commands the pressure and wind fields in Western and Central Africa,
- at mid-levels the dynamic conditions are quasi-independent from the substratum; for this reason the distribution of the more important rains is commanded by conditions external to the forest. These conditions, the same for the whole of Africa (at these latitudes), are connected with the general circulation of the troposphere and are therefore quasi-independent from the surface conditions; the speed and amplitude of annual migration are very different from the low-layer: at mid-levels the annual migration of the Meteorological Equator I.C.Z. structure is slow, with small amplitude, roughly

between the latitudes 10-12 degrees North and 10-12 degrees South of the Equator. Consequently the meteorological influence of forest is restricted to the lowest layer. In winter it maintains a low-level quasi-horizontal structure: the Meteorological Equator I.T.F. structure in the north and the very similar vertical structure of the Inter-Oceanic Confluence in the south (Leroux, 1983). These structures are crossed by squall-lines from East to West, disturbances which only bring a little amount of short duration and stormy rains (by comparison with the heavy rains of mid-levels I.C.Z. structure).

The forest influence, its limits, and the conditions of its survival, are also observed in African palaeoenvironments.

2. African Palaeoenvironments

Forest survival is strongly dependent on the equilibrium between, on the one hand the power of heat lows to attract the maritime airstream and on the other hand the power in the opposite direction of the dry continental trades, blowing from north to south. It is precisely during the period when the thermal lows have an indirect character (and a weak attracting power), at the moment when the low-level discontinuity (M.E. or I.O.C.) is "stopped" along the forest margin, that this equilibrium is the most precarious.

Two extreme past meteorological conditions (Leroux, 1986) are then associated with two very different extents of the forest cover.

2.1. LAST GLACIAL MAXIMUM (18-25 KYRS B.P.)

During the Last Glacial Maximum period, forest was only present in three refuges (Fig. 5-a & Fig. 6): Liberia, Cameroon-Gabon and Zaire; the former forest area being then covered by savanna (Hamilton, 1976; Maley, 1987). During the cooling maximum temperature had slightly decreased on the near Atlantic Ocean (from 1 to 3°C, but from 5 to 8°C on the continent), rainfall had also decreased below normal by at least 25%, but precipitation remained relatively abundant in equatorial regions. The most important event was the strengthening of tropical airstreams and in particular - to explain the forest partial disappearance - the strengthening of continental trades (Hooghiemstra, 1986; Leroux, 1987). This power was attested by sand-dune building (Sarntheim, 1978; Van Zinderen Bakker, 1980, 1982), shown by the southward Ogolian-Kanemian dune extension (Talbot, 1984) and the northward Namibian-Kalaharian dunes invading the western side of the Congo basin (Heine, 1982).

At this time the attraction by heat lows (shallow) was reduced so that the dry continental airstreams then easily drove back the atlantic monsoon airstream and passed over the forest area. Trees were unable to support for long the growing length of the dry season. The forest refuges were precisely located on sheltered areas of relief (western side of Guinean range, or Cameroon-Gabon range), where the humid atlantic airstream at low-levels was a protection against the dry continental harmattan (these are the present natural conditions of survival of the Guineo-Liberian dense forest, during the winter period; Fig. 7).

2.2. HOLOCENE CLIMATIC OPTIMUM (9-6 KYRS B.P.)

The forest reconquest started from these coastal refuges, spreading progressively like a "contagion", as heat lows became deeper (following the improvement of insolation), increasing the advection of atlantic moisture. But the first reason for the reconquest was still external to the forest, related to the dry opposing continental trades becoming gradually slower, by causing the general circulation of the troposphere to move into a long period of slowing down (Leroux, 1986). During the Holocene Climatic Optimum, between about 9,000 and 6,000 B.P., the spread of tropical forest reached its maximum (Fig. 5-b).

After about 4,500 B.P. the extensive forest retreated from the conquered area, because rains decreased again and also because the troposphere was entering again into a new mode of rapid general circulation.

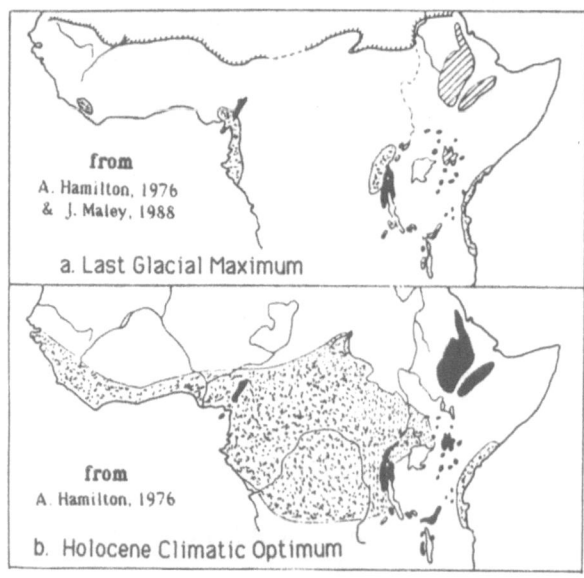

Figure 5. Extensions of the African evergreen forest.

The present Guineo-Congolese forest is therefore already a "retreating" forest, and there is nothing to suggest that such a natural and long established process - for more than 4,000 years - is able to stop by itself.

Moreover, in the near future, the present anthropic conditions are not really favourable to the forest (Bertrand, 1983; Henderson-Sellers et al., 1984; Gornitz et al., 1985). Consequently, with the evidence of the past and the observation of the present, we have to consider what is now necessary to preserve the dense African forest, according to two possible future scenarios.

3. Present Situation and Future Evolution

3.1. GLOBAL WARMING SCENARIO

The first scenario is the most favourable. It is a question of waiting for the global warming

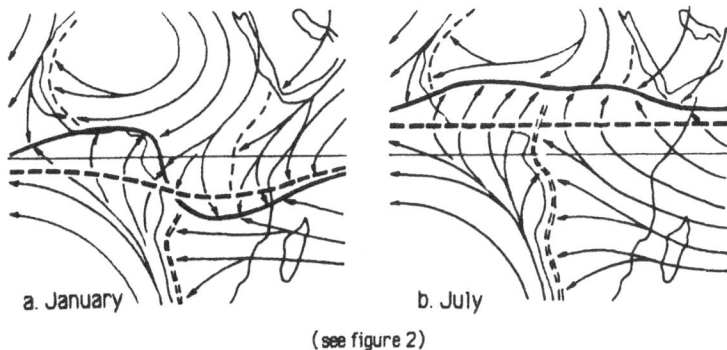

a. January

b. July

(see figure 2)

Figure 6. Discontinuities and surface airstreams during the Last Glacial Maximum (18,000-15,000 yrs B.P.).

trend, as predicted by Jones et al. (1988), to become reality. If this warming occurs it will greatly benefit tropical forest areas, such as the Sahelo-Sudanian belt. In such an event, with the general circulation (theoretically) entering into a slow mode (Leroux, 1988 b, 1989), the tropical zone will dilate itself northward and southward, the annual displacement of tropical discontinuities and rain-making structures will become more flexible, and rainfall will generally increase, especially in the tropical margins, such as the Sahel. However, some reservations are necessary: a global temperature trend seems unable to explain the intensity of the general circulation, the polar latitudes being the most important to world climatic changes (Leroux et al., 1989); but in high latitudes, warming does not appear to be the present trend, with summer temperatures between 65° and 85° N decreasing slightly since the 1930's (Barry et al., 1985). We can also observe, for example, that the insolation values decreased by 6% between 1950 and 1980 in the temperate boreal zone (Loginov et al., 1983) and that in the United States the trends "indicate a rise of temperature until the 1930's" but "a decrease since that time" (Karl et al., 1989).

Unfortunately the recent situation in Africa as in the whole tropical zone, is not symptomatic of a warm scenario and a slow mode; on the contrary the general circulation seems gradually to be entering into a rapid mode, in accordance with a progressive cooling of high latitudes (Leroux, 1989).

3.2. HIGH LATITUDES COOLING SCENARIO

Recent climatic evolution since the 1950's is characterized by a trade-winds acceleration; this strengthening is for example observed in the Atlantic Ocean (Servain et al., 1987), in the Pacific Ocean (Whysall et al., 1987; Rasmusson, 1987), or in the southern hemisphere Tropics (0°-25° S) where Diaz et al. (1989) note a steady increase in scalar wind speed during the 1950-1986 period. In Sahelo-Saharan Africa the morphogenic action of the wind is already responsible for new and strong aeolian building (Courel et al., 1987).

We can also observe, during recent years, a vigorous increase of dust haze (Bellec et al., 1985; Gac, 1984, 1985; Gordon et al., 1988), not only due to drought but also to a strengthening of the wind. At the same time the northern Tropical High Pressure belt is stronger and has a more southern location and the surface Meteorological Equator has pushed southwards as in January 1983 to 5° South of Equator (Buisson, 1986); harmattan and dust haze reached Sao Tome (3° N) in January 1983 and 1984 (Piton, 1987), Brazzaville (Congo) in January 1983, Libreville (Gabon) during January 1987 (Guillot, 1987) and in January 1989 ... these recent events do not have the same value as means established over a long period but they confirm what appears to be the present regime.

In the same way, drought in Africa is becoming worse, both in the north (Sircoulon, 1986; W.M.O., 1987) and in the south where, for example during 1982-1983, according to Rasmusson et al. (1984), "the drought in Southeast Africa was among the worst of the century". Since the sixties, rains connected with the mid-levels I.C.Z. structure, that is to say the heavy rains, have become concentrated in a narrow belt, closer to the Equator, both from north and south (Leroux, 1988 b). As a consequence, a decrease of rainfall in the Sahelo-Sudanian belt is associated with the increase in the equatorial zone; Burke et al. (1989) estimate for the 1980's a Sahelian rainfall decrease of almost 30%, "while equatorial precipitation has been 10% greater". This modification is attested by the rise of lake Victoria after the extraordinarily rainy season of 1962 (Flohn, 1987) and similar rises of the lakes Turkana, Tanganyika, Rukwa, and Malawi; the hydrological regimes of equatorial rivers, such as the Ogooué River (Mahé, 1988), have also shown a progressive modification of the main rainy season since the sixties; the studies, of Janicot (1985) and Nicholson et al. (1986) revealed the two modes of distribution of rains associated with I.C.Z. structure: its widening during the fifties and especially its latitudinal narrowing since the sixties.

Such a recent evolution of rainfall distribution, connected with the strengthening of the continental trades, makes the equilibrium of the forest margins more fragile, and susceptible to an abrupt destabilisation.

4. Conclusion

Long before an eventual reafforestation, the first priority is consequently the protection of the present forest cover.

Preservation begins with the respect of the forest self-protection process, in order to avoid the destabilisation of a precarious equilibrium.

The *meteorological tool* (surface and low-layer observations) makes it already feasible to observe a contingent change of temperature and pressure fields and of the wind velocity around the forest area.

It is naturally unrealistic to propose a complete transformation into evergreen forest, particularly for economic reasons. But it appears necessary to determine the vegetation density most appropriate relations "tree/plantation/cultivated land/air moisture" able to sustain - during crucial months - a uniform response of the effective surface of the ground to solar radiation.

The *satellite tool* allows the stipulation of a minimum cover density, and the radiation response over and around the forest area to maintain the attraction of peripheral heat lows and the advection of atlantic moisture.

The eventual voluntary extent of the forest must follow the process of natural reconquest by means of a progressive "contagion".

On the margins of the present occupied area it is necessary to supply artificially the transpiration of the trees but only during the rainless period; the main fraction of air moisture (or weakness of saturation deficit) is naturally provided by the resulting inland extension if the atlantic airstream. It is thus not unrealistic to expect a gradual increase of precipitable water potential (recirculation being naturally and artificially ensured) and probably a related (but slight) complement of precipitation, with the maintenance of a low-level aerological structure, the I.T.F. structure being favourable for the formation and water-supply of squall-lines.

Outside the evergreen forest area, the rare opportunities to force the inland penetration of maritime moisture have first to be considered. For example, any reafforestation project concerning the Senegalese territory (victim of drought, but perhaps more so of an intensive exploitation of the tree capital) must begin with safeguarding the Casamance forest, particularly in the western Casamance which is covered during the whole year by an atlantic advection, maritime trade or monsoon. Past locations of the Liberian forest refuge during the glacial period (Fig. 4a), the azonal Guinean species found even now in the coastal Senegalese area of "Niayes" and the present aerological conditions of Guineo-Liberian forest survival (Fig. 7) show clearly that the first reimplantations have to be made along the Senegalese

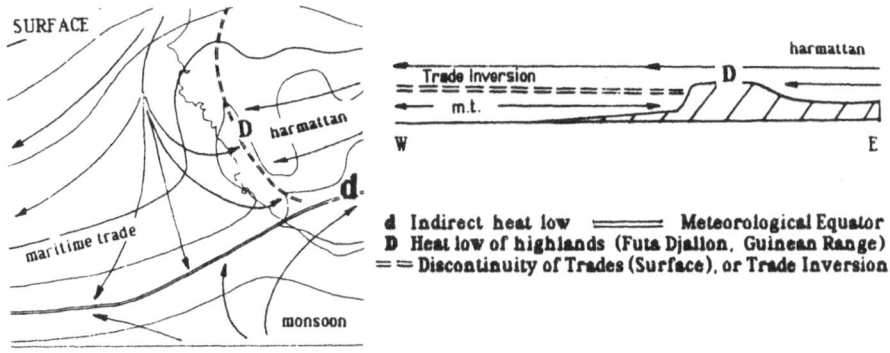

Figure 7. The meteorological shelter position of the Guineo-Liberian dense forest.

coast where the easterly continental stream, the harmattan (warmer and lighter), rarely reaches the shore because it must rise above the cooler maritime trade-winds. A gradual intensification of vegetation would push the discontinuity of trades inland and consequently extend the area covered by the atlantic air; coastal regions would expect a weaker saturation deficit, no rainfall improvement due to the drastic aerological stratification (Leroux, 1983) but, for compensation, nocturnal condensations, often abundant especially during the winter period.

In summary, even the tropical forest is not actually as important as sometimes suspected, especially concerning the control of CO_2 (by comparison with the importance of the ocean), it appears urgent to protect its present cover.

An integrated survey programme must then examine the different influence of opposing factors, chiefly the vegetation density, which may be "meteorologically dangerous" for the forest cover survival.

5. References

Barry, R.G., Henderson-Sellers, A., & Shine, K.P. (1984) "Climate sensitivity and the marginal cryosphere", *Climate Process & Climate Sensitivity*, Geophysical Monograph n° 29, pp. 221-237

Bellec, B., & Guillot, B. (1985) "Brumes sèches et nuages de sable sur l'Océan Atlantique et en Afrique de l'Ouest.", Veille Climatique Satellitaire n° 5, ORSTOM/CMS, Lannion, pp. 17-20

Bertrand, A. (1983) "La déforestation en zone de forêt en Côte d'Ivoire", Bois et Forêts des Tropiques, n° 202, pp. 3-17

Buisson, A. (1986) "Tendances climatiques pour l'évaluation de ce que pourrait être la saison des pluies en Afrique occidentale pour l'année 1986.", Veille Climatique Satellitaire n° 15, ORSTOM/CMS, Lannion, pp. 38-45

Burke, K., & Wells, G. (1990) "Latitudinal fluctuations of the African hydrologic budget during geologic and recent history", (this volume)

Courel, M.F., & Chamard, Ph. (1987) "Apparition de nouvelles formes dunaires dans l'Azawad et le Gourma septentrionel (Mali)", Photo-Interprétation (3), fasc. 3, Paris, pp. 25-33

Diaz, H.F., Bradley, R.S., & Eisched, J.K. (1989) "Precipitation fluctuations over global land areas since the late 1800's", Journal of Geophysical Research, vol. 94, no. D1, pp. 1195-1210

Flohn, H. (1987) "East African rains of 1961/62 and the abrupt change of the White Nile discharge", in Balkema, A.A. (ed.) *Palaeoecology of Africa*, vol. 18, pp. 3-18

Gac, J.Y. (1985) "Le phénomène des brumes sèches au Sénégal en 1984-1985", Veille Climatique Satellitaire n° 7, ORSTOM/CMS, Lannion, pp. 31-35

Gordon, L.W., & Middleton, N.J. (1988) "The alteration of land surface cover across the western Sahel recorded by orbital photography 1965-1986", Intern. Satell. Land Surface Clim. Proj., 2nd Results Meeting, Niamey, Niger, April 1988

Gornitz, V, NASA (1985) "A survey of anthropogenic vegetation changes in west Africa during the last century - climatic implications.", Climatic Change no. 7, pp. 285-325

Guillot, B. (1987) "Champs thermiques de surface en Afrique de l'Ouest, de novembre 1986 à février 1987. Remarques sur la signification de certaines situations", Veille Climatique Satellitaire n° 16, ORSTOM/CMS, Lannion, pp. 14-17

Hamilton, A. (1976) "The significance of patterns of distribution shown by forest plants and animals in tropical Africa for the reconstruction of upper Pleistocene Palaeoenvironments: a review.", in Balkema, A.A. (ed.) *Palaeoecology of Africa*, vol. 9, pp. 63-97

Heine, K. (1982) "The main stages of the late Quaternary evolution of the Kalahari region, southern Africa", in Balkema, A.A. (ed.) *Palaeoecology of Africa*, vol. 15, pp. 53-76

Henderson-Sellers, A., & Gornitz, V. (1984) "Possible climatic impacts of land cover transformations with particular emphasis on tropical deforestation.", Climatic Change no. 6, pp. 231-257

Hooghiemstra H. (1986) "Changes of major wind belts and vegetation zones in N.W. Africa 20,000-5,000 yr B.P. as deduced from a marine pollen records near Cap Blanc.", in Ward, J. (ed.) *Review of Palaeobotany and Palynology*, sp. issue

Janicot, S. (1985) "Analyse spatio-temporelle du champ des précipitations annuelles sur l'Afrique de l'Ouest et l'Afrique Centrale", Veille Climatique Satellitaire n° 10, ORSTOM/CMS, Lannion, pp. 9-19

Jones, P.D., Wigley, T.M.L., Folland, C.K., Parker, D.E., Angell, J.K., Lebedeff, S., & Hansen, J.E. (1988) "Evidence for global warming in the past decade.", in *Nature* 332, p. 790

Karl, T.R., & Jones, P.D. (1989) "Urban bias in area-averaged surface air temperature trends", Bull. Am. Met. Soc., Vol. 70 no. 3, pp. 265-270

Leroux, M. (1983) "Le climat de l'Afrique tropicale.", Ed. Champion/Slatkine, Paris/Genève, t.1: 636 p., 349 fig., t.2: notice, 250 cartes

Leroux, M. (1986) "Les mécanismes des changements climatiques en Afrique", Symp. INQUA Dakar, Coll. Trav. & Doc. n° 197, ORSTOM, Paris, pp. 255-260

Leroux, M. (1987) "L'Anticyclone mobile Polaire, relais des échanges méridiens: son importance climatique", Géodynamique 2 (2), ORSTOM, pp. 162-167

Leroux, M. (1988 a) "La variabilité des précipitations en Afrique Occidentale. Les composantes aérologiques de problème", Veille Climatique Satellitaire n° 22, ORSTOM/CMS, Lannion, pp. 26-45

Leroux, M. (1988 b) "Les conditions structurales de la variabilité pluviométrique de l'Afrique tropicale", Colloque d'Aix en Provence, Publication de l'Association Internationale de Climatologie vol. 1, pp. 179-180

Leroux, M. (1989) "Circulation générale de la troposphère et variations climatiques.", Annual Workshop IGCP 252, Cassis, January 1989

Leroux, M., Petit-Maire, N, & Davies, O.K. (1989) "Differential insolation at North and South latitudes explains palaeoclimatic changes is tropical Africa for the last 30,000 years", (to be published)

Loginov, V.F., & Pivovarova, Z.I. (1983) *Meteorology and Hydrology*, no. 8, Moscow, pp. 55-60

Mahé, G., Olivry, J.C., & Lerique, J. (1988) "La variabilité du régime des tributaires de Golfe de Guinée: Indices de crises ou de changements climatiques?", in Olivry, J.C. (ed.) *Géodynamique de l'hydrosphère continentale*, Rapport scientifique 1983-1987, ORSTOM, pp. 115-121

Maley, J. (1987) "Fragmentation de la forêt dense humide de africaine et extension des biotopes montagnards au Quaternaire récent: nouvelles données polliniques et chronologiques. Implications palaeoclimatiques et biogéographiques.", in Balkema, A.A. (ed.) *Palaeoecology of Africa*, vol. 18, pp. 307-329

Monteny, B.A. (1986) "Forêt équatoriale, relais de l'océan comme source de vapeur d'eau pour l'atmosphère", Veille Climatique Satellitaire n° 12, ORSTOM/CMS, Lannion, pp. 39-51

Nicholson, S., & Entkhabi, D. (1986) "The quasi-periodic behaviour of rainfall variability in Africa and its relationship to the Southern Oscillation", Archiv für Meteor., Geophys. und Bioklim., Ser. A, 34, pp. 311-348

Piton, B. (1987) "Les anomalies océanographiques et climatiques de 1983 et 1984 dans le Golfe de Guinée.", Veille Climatique Satellitaire n° 16, ORSTOM/CMS, Lannion, pp. 18-31

Pouyaud, B. (1986) "L'évaporation, composante majeure du cycle hydrologique", *Climat et Développement*, Coll. et Sémin., Ed. ORSTOM, Paris, pp. 130-139

Rasmusson, E.M., & Arkin, Ph.A (1984) "El-Niño/Southern Oscillation and large-scale tropical drought.", *Meteorological Aspects of Tropical Droughts*, TMP rep. ser. no. 15, WMO, Geneva, pp. 71-76

Rasmusson, E.M. (1987) "Tropical Pacific variations", in *Nature* 327, p. 192

Sarntheim, M. (1978) " Sand deserts during glacial maximum and climatic optimum", in *Nature* 272, no. 5648, pp. 43-46

Servain, J, & Seva, M. (1987) "On relationships between tropical Atlantic sea-surface temperature , wind stress and regional precipitation indices: 1964-1984", *Ocean-Air Interactions*, vol. 1, pp. 183-190

Sircoulon, J. (1986) "La sécheresse en Afrique de l'Ouest. Comparaison des années 1982-1984 avec les années 1972-1973", Cahiers ORSTOM, série Hydrologie, vol. XXI n° 4, pp. 75-86

Talbot, M.R. (1984) "Late Pleistocene rainfall and dune building in the Sahel", in Balkema, A.A. (ed.) *Palaeoecology of Africa*, vol. 16, pp. 203-214

Van Zinderen Bakker E.M.Sr. (1980) "Comparison of late Quaternary climatic evolutions in the Sahara and the Namib-Kalahari region", in Balkema, A.A. (ed.) *Palaeoecology of Africa*, Vol. 12, pp. 381-394

Van Zinderen Bakker, E.M.Sr. (1982) "African palaeoenvironments 18,000 yrs BP.", in Balkema, A.A. (ed.) *Palaeoecology of Africa*, vol. 15, pp. 77-99

Whysall, K.D.B., Cooper, N.S., & Bigg, G.R. (1987) "Long-term changes in the tropical Pacific surface wind field", in *Nature* 327, pp. 216-219

W.M.O. (1987) "Rainfall in the western Sahel for the period 1896 to 1987. Modified and updated from Farmer and Wigley, 1985.", in *Water Resources and Climatic Change: Sensitivity of Water-Resource Systems to Climate Change and Variability*, Norwich, U.K., WMO/TD-no. 247, p. 10

RIVER AND SOILS CYCLICITIES INTERFERING WITH SEA LEVEL CHANGES

ROLAND PAEPE & ELFI VAN OVERLOOP
Vrije Universiteit Brussel (VUB)
Earth Technology Institute (ETI)
& Belgian Geological Survey

ABSTRACT: Palaeosoils and river deposits alternate at specific levels within continental lithostratigraphic sequences of various parts of the globe. They attest to cyclicities through time of global importance. This study is divided into a number of chapters especially regarding the last 2,4 Ma years, the last 130K-years and the last 10K-years, which may all be read almost independently from the others. Without some lengthy chapters about the long term, middle term, and short term geosoil cycles it was quite impossible to come to any reasonable conclusion about the number of soils occurring in the Quaternary and the time stability of the geosoils in such sequences, especially of the 100K interglacial soil. Four main groups of cycles of periodicities have been detected. Two long range cycles, of 100K and 400K respectively, dominate the interglacial/glacial soil sequences of the Quaternary. These reveal the continental evidence of Imbrie's 100K cycle and Hays' 400K cycle. It is quite remarkable that the recorded soils in these continental systems all are relict soils of interglacial age. In the coastal fringe area these soil sequences may interfere periodically with interglacial marine deposits as well. From the North Sea and Mediterranean regions all palaeosoil development can be proved to have occurred after, or towards the end of each of the marine transgressions.

Another question was whether these marine transgressions result from climatic forcing. The decreasing amplitude of the interglacial marine high level stands and glacial low stand fluctuations towards the Holocene MSL (Mean Sea Level) infer that major regression/transgression cycles are tectonically biased movements rather than climatically controlled ones.

The next two periodicity cycles are the middle range last 100K cycle (the Upper Pleistocene or Last Glacial) and the short range 10K cycle (the Holocene). Within the last 100K a complete warm/cold cycle was achieved. During this period one interglacial (warm climatic) soil has developed at the beginning of the cycle, although after the maximum of the marine transgression. During the following cold part of the cycle at least ten other interstadial (colder climatic) soils may have developed. Fluctuations of sea level within this cycle seem mainly related to climatic changes. Finally the 10K cycle of the Holocene shows rapid fluctuations of the 40 m sea level rise during the interglacial after the maximum soil development. Before this maximum soil development some 10,000 years ago, sea level had already been rising for more than 60 m since the maximum cold of 18,000 BP.

253

R. Paepe et al. (eds.), Greenhouse Effect, Sea Level and Drought, 253–280.
© 1990 *Kluwer Academic Publishers.*

The present mean sea level (MSL) should naturally have attained its equilibrium as a result, first, of the long term tectonic rheology cycles of lithosphere, second, of the shorter climatic cycles. Geotraverses from the North Sea Basin towards the Mediterranean into the equatorial Belt proved that the tectonic cycles were of global intensity and frequency whereas the climatic cycles are increasing in both frequency and intensity towards the Equator.

1. Introduction: Deep Sea Versus Land Data

The validity of standard sequences obtained from terrestrial deposits is still questionable, especially for those who prefer to work with the deep sea core results as standard scale. Terrestrial sequences are believed to be:
- composed of non-continuous sections
- incomplete because of sedimento-stratigraphic hiatuses
- of high variability as to sedimentation rates
- without stable marker horizons
- of no global response (i.e. with no climatic signal in the equatorial belt)
Deep sea records are considered to represent:
- continuous stratigraphic sections
- continuity in sedimentation
- generally uniform sedimentation rates
- stable marker horizons with microfossils
- a global response from the Poles to the Equator
Moreover, land data represent restricted dating possibilities:
- long-term palaeomagnetic dating as B/M, M/G boundaries
- occasional K/Ar dating on basalt and tuff
- radiocarbon dating for middle and short terms
- unstable TL datings
- prehistoric and archaeological dating
In evaluating these statements, especially with regard to sedimentological-stratigraphical continuity (which is an absolute necessity for establishing curves about climatic changes on a global scale) the following remarks may be put forward.
For the deep sea record:
- the B/M (Brunhes/Matuyama) Boundary occurs sometimes in a cold stage (Shackleton & Opdyke, 1973), sometimes in a warm phase (Imbrie, 1979).
- the rate of bioturbation is high.
- only global changes in ice-volume are recorded without specifying which particular glacier ice is dealt with and without any evidence of changes in other landmark features as e.g. forest cover density.
For the terrestrial record :
- continuity of palaeosoil sequences in windborne deposits as eolian loess is generally accepted; this statement, however, is applicable to other depositional series in between soils as e.g. loess.
- landmarks as palaeosoils are within limits of variance and of time transgressivity of

geographical spreading of palaeosoils, synchronous from one geographical area to another as well as with the deep sea oxygen-isotope record (see section 2)
- there is a firm possibility of a step by step correlation between sequences of widely varying geomorphological entities such as basin sequences, terrace sequences and plateau sequences.

2. Palaeosoils as Stable Stratigraphical Key Horizons

As in all other geological sections, stable time bound key horizons are needed to establish the timescale of a given geological sequence. Indeed, if absolute dating materials are not available, one is to produce relative dating methods such as the relative age determination of palaeosoils. Dating of soils is useful as long as the soil signals used for this purpose show a certain degree of stability of occurrence in the stratigraphic record i.e. they should occur at definite time levels within reasonable small boundary intervals of time.

Palaeosoils in the geologic sequence offer the possibility of such stratigraphic stability. However, the geological status of palaeosoils of the Pleistocene and of the Holocene Series remains controversial, especially amongst geologists and pedologists. Indeed, unlike guide fossils showing a definite taxonomy, palaeosoils most often do not. Actually, a large range of fossil soils appear as monolayered, truncated soil horizons and hamper soil solum studies. Others show multicyclic (polycyclic) soil horizons and also appear as single sediment layers instead of a well developed soil solum or even a soil catena. Palaeosoils may not even find their equivalent in the global soil zonation and may reflect environmental climatic conditions of the Past which do not longer exist today.

These peculiarities all render the study of soil dynamics and soil genetics precarious or even impossible when dealing with palaeosoils. But then the question arises whether a complete soil solum is mandatory for the purpose of dating a soil as a stratigraphic key horizon. Since palaeosoils show continuous sequences from one area to another, this simple fact may already indicate that palaeosoils are time bound and stable time stratigraphical units.

Nevertheless, despite their time stability, fossil soils remain difficult to date; they are seldom built up by autochthonous parent-material unless they are of an organic (like peat and humic horizons) or chemical (like calcrete horizons) origin. More often, as fossil soils (like modern soils) are developed in foreign parent material different in age and origin from the soil, the time relationship between the palaeosoil and the sediment remains dubious. If humic or calcrete horizons are suitable for immediate dating or environmental assessment (e.g. by pollen analysis or stable isotope geochemistry), most other fossil soil horizons are not since they mostly reflect only mineralogical weathering in nature.

Yet, palaeosoils exist and occupy specific time stable stratigraphic positions in sediment sequences; moreover, they may be followed from one sedimentary series into another, thus serving as marker beds or key horizons in distant geographical belts. Hence, palaeosoils are formal Members of a "normal" stratigraphic sequence and as such, are very useful geologic time indicators as well as good indicators for assessment of the environment.

Although the term "palaeosoils" is commonly used, palaeosoils as members of a lithostratigraphic sequence should be simply considered and named "geosoils" in the sense

256

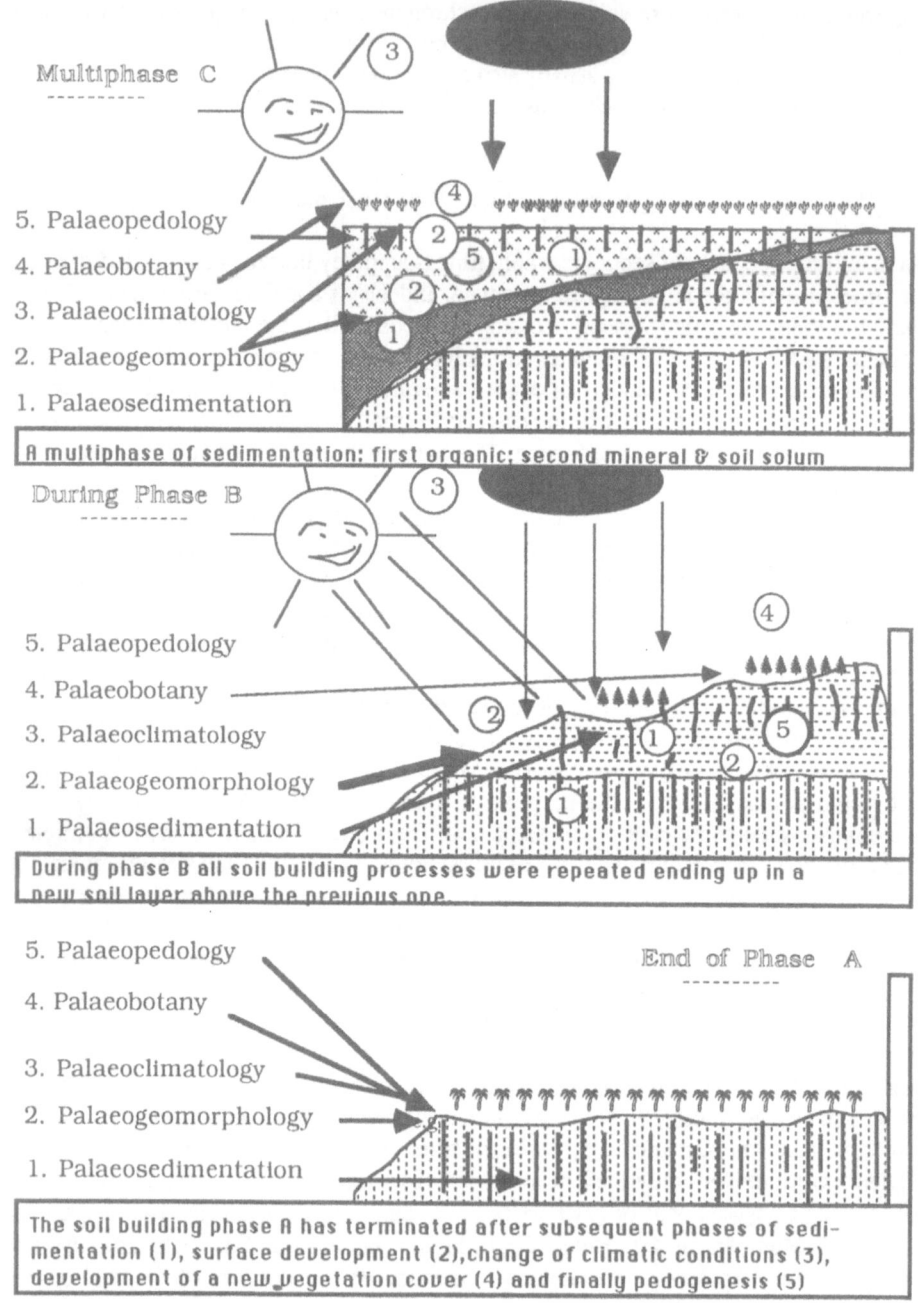

Multiphase C

5. Palaeopedology
4. Palaeobotany
3. Palaeoclimatology
2. Palaeogeomorphology
1. Palaeosedimentation

A multiphase of sedimentation: first organic; second mineral & soil solum

During Phase B

5. Palaeopedology
4. Palaeobotany
3. Palaeoclimatology
2. Palaeogeomorphology
1. Palaeosedimentation

During phase B all soil building processes were repeated ending up in a new soil layer above the previous one.

5. Palaeopedology
4. Palaeobotany
3. Palaeoclimatology
2. Palaeogeomorphology
1. Palaeosedimentation

End of Phase A

The soil building phase A has terminated after subsequent phases of sedimentation (1), surface development (2), change of climatic conditions (3), development of a new vegetation cover (4) and finally pedogenesis (5)

Figure 1. The five moments in the landscape building process.

given by Morrison (1965). This restricts the occurrence of a palaeosoil level to its strictly geological and geomorphological expression without any other further bias to pedology.

From this point of view five important steps or moments may be ascribed to fossil geosoils occurring in a lithostratigraphic sequence (Fig. 1):

1 **the palaeosedimentation moment** which is the phase of building up the parent material which may vary from sedimentary and metamorphic to volcanic and plutonic deposits.
2 **the palaeogeomorphological moment** of shaping land surfaces corresponding to specific processes of landscape development of the past prior to the pedological moment or even synchronous with it.
3 **the palaeoclimatic moment**: aside from the foregoing climatic phases including sedimentation and landscape development, one or more climatic phases may enhance weathering (eventually involving all or some of the aforementioned moments of the Past) which previously was impossible.
4 **the palaeobotanical moment**: one or more sequences of vegetational development of the past following the previous moment of initial weathering.
5 the palaeopedological moment: a fossil soil (buried or relict; with modern equivalent or not) inferring a moment of standstill in the sedimentological aggradational process of the Past.

It is quite clear that geosoils, combining all these natural moments, should no longer be considered as occasionally occurring features but instead as stable horizons in both time of occurrence and periodicity. Furthermore, all five moments point to climatic changes as the common generator of the global forcing involved; henceforth, geosoils are true stable climatic indicators as well, operating at given periods of time. Global climatic changes in terrestrial sequences may consequently be detected from the soil stratigraphic sequences with the same degree of accuracy and resolution as from deep sea sequences.

3. The 100K, 200K & 400K Periodicities Along the North-South Geotraverse

3.1. THE SOIL SEQUENCES

The time stability and periodicities of geosoil levels become clear when comparing soil stratigraphic sequences from various global belts. At first sight such comparison may look too simple from the methodological point of view. Biohorizons as *Stylotractus universus* occurring at isotope stage boundary 11/12 in the deep sea records are compared over long distances as well, because of their time stability. For the same reasons, geosoils giving proof of such time stability may then be compared over long distances as well in an attempt to establish global continental lithostratigraphic correlation.

The North-South Geotraverse from the North Sea Basin (Belgium, 50°N.) through the Eastern Mediterranean (Greece, 36°N.) towards the Equator (Burundi, 4°S.) is a first attempt to put such long distance correlation into focus. It reveals remarkable evidence of geosoil stability (Fig.2). For the last 800,000 years (Middle and Upper Pleistocene encompassing the Brunhes polarity zone) eight geosoils of the interglacial type regularly appear at 100Ky

258

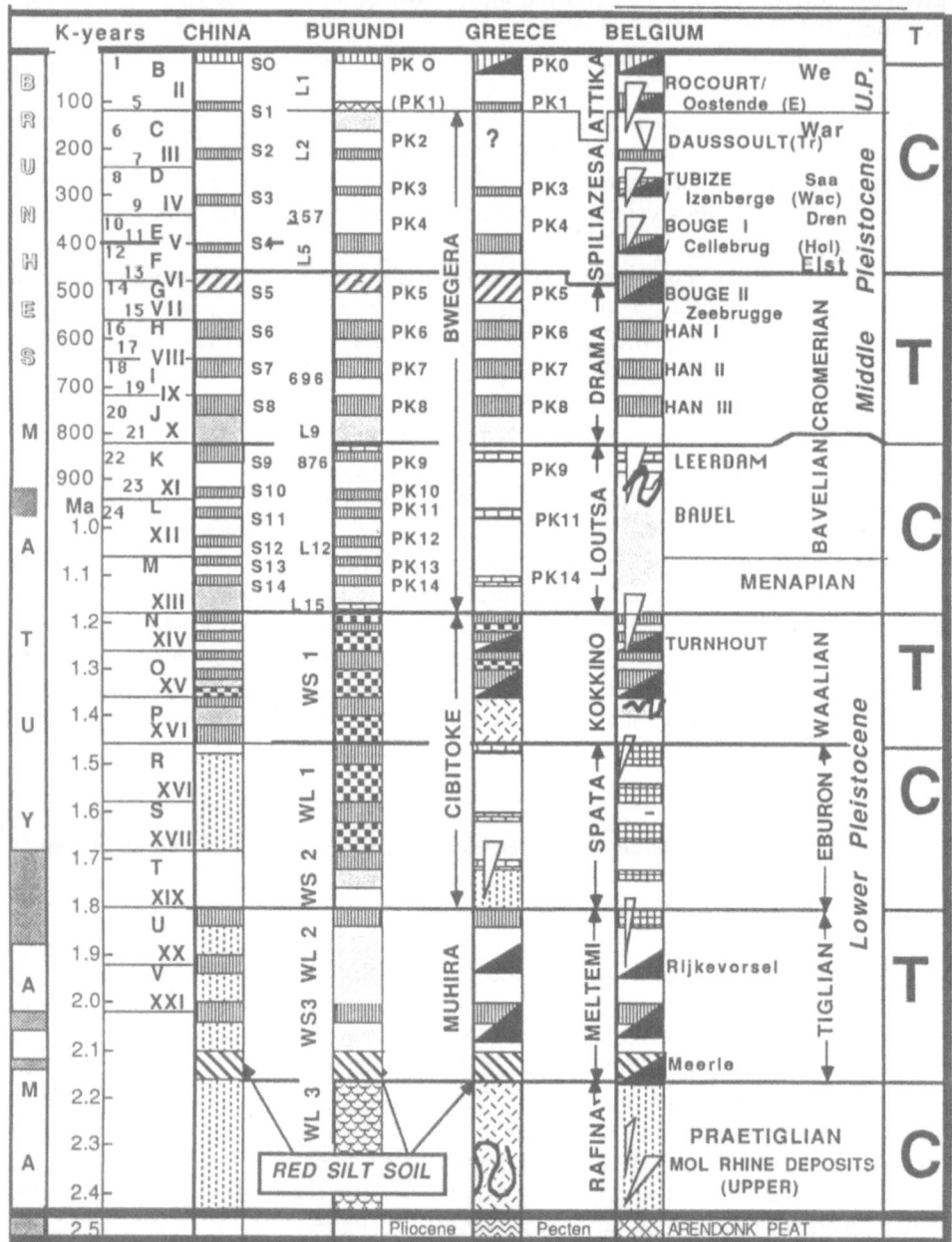

Figure 2. North-South Geotraverse: North Sea belt - Equatorial belt.

time intervals in the stratigraphic column of the North Sea Belt. The soil levels interfere with colder and dryer periods of loess or similar windborne deposits together with periglacial features such as frostwedges (indicated by a V-shaped symbol). Quite often the interglacial soil may be underlaid with a marine deposit of the same interglacial age. From before 800Ky to about 2.2 Ma years ago, another 13 soils occur at regular 100K distances.

As most of the sections are in loess or loesslike deposits correlation of the Westernmost geotraverse with the Eastern loess section at Luochuan in China (36°S.) was inevitable and remains by and large perfectly feasible. The stability of the loess-geosoil sequences of China reinforce the principle of both geosoil stability and the possibility of long distance lithostratigraphic correlation.

3.1.1. *Geosoils of the Last 800K-Years.*

In all of the sequences considered, geosoils are best developed within the time interval between 800,000 and 450,000 years with a maximum development at the Eastern level of geosoil S5 at Luochuan and at the corresponding pedocomplex PK5 of the Western levels. This time interval corresponds to the classical Cromerian Complex of Zagwijn (1975) with four distinctive interglacials and corresponding interglacial geosoils (Paepe, 1975). They are corresponding with the Drama Complex of Greece (Paepe et al., 1986) and with the middle part of the Bwegera Formation in Burundi/Zaire (Ilunga Lutumba, 1984).

The interglacial geosoils of the next 400K-years are less well developed in all of the geologic sequences considered and show a number of hiatuses in the soil sequence. This time interval corresponds with the complicated glacio-lithostratigraphic subdivisions of Northern Europe, i.e., the Elst, the **Holsteinian**, the Drenthe, the **Wacken**, the Saale, the **Treene**, the Warthe, the **Eemian** and the Weichselian Stages (the bold ones being the interglacials). The sequence is indisputably the most complete again in Luochuan and in Burundi/Zaire i.e. in the Tropics and Subtropics.

The less well developed nature of the last 400K series of geosoils led to the assumption that the climate was generally cooler and dryer than during the previous 400K period. It also leads to the recognition of long-term warm periods (Thermomers, T) with generally warmer interglacials and long-term cooler/dryer periods (Cryomers, C) with generally cooler interglacials. This 400K-years cyclicity (Paepe et al.,1986) is similar to what was shown by HAYS et al.in 1981 for the 400 K periodicity from the deep sea core studies.

Thus may be concluded that for the last 800,000 years geosoils on a global scale from North to South and from East to West, reveal two imbricated long terrestrial cyclicities: a 100K-years periodicity composed of eight warm (interglacial)/cold (glacial) cycles which can be split up into two 400K-years periodicities: one initial warmer cycle starting at about 800K and a generally cooler one starting at about 400K and including the very last warm/cold cycle of the Upper Pleistocene (starting some 127K-years ago). A new Thermomer has just started some 10K-years ago with the beginning of the Modern Stage or Holocene of which the Neolithic Marathon soil is the first expression (see hereafter). Today's worldwide recognised so-called greenhouse effect is trapped within this new long-term generally warmer natural phase.

3.1.2. *Geosoils of the 800K/2.4 Ma Time Span.* Beyond the 800,000 years boundary in the Lower Pleistocene Subseries correlation becomes less obvious. In Europe, cold cryomer conditions prevailed hampering the regular development of geosoils over a time of another 400K-years which started some 1.2 Ma years ago. It is the original cold Menapian Stage which recently has been subdivided into a Menapian Substage and a Bavelian Stage (Zagwijn, 1987) in which two warmer Substages occur namely the Bavel and the Leerdam Substages. Only a few calcrete soil horizons are developed in Belgium and in Greece at this level whereas soils of the same stratigraphic position double or even triple in number in the Loess Belt of China and in the Equatorial Belt of Africa (Burundi/Zaire). Indeed, over less than 400,000 years at least six, sometimes seven geosoils occur. The immediate effect is that correlation in the Lower Pleistocene Subseries of tropical and monsoonal regimes is not hampered by a lack of evidence but by an abundance of evidence throughout the same time span.

This geosoil doubling effect in tropical and subtropical belts reaches its maximum with again at least six new geosoils in the next time interval located roughly between 1.2 and 1.5 Million years. It corresponds to the well known thermomer encompassing the Waalian Stage in the North Sea Belt stratigraphic classification. Usually four well developed organic geosoils characterised by the arrival of the fern *Azolla filiculoides* in the pollen spectrum occur. The same occurs in Greece in the so called Kokkino (Limanaki) Formation and in Burundi/Zaire in the so-called Cibitoke Formation (upper part).

Similarly one finds during the next Cryomer (referred to as the Eburonian Stage in the North sea Belt) organic peat bogs of the tundra which developed during the interglacial phases of that Super Stage. In the Eastern Mediterranean only a few calcrete horizons occur within the equivalent Spata Stage, and an even much weaker development is observed in the Equatorial Belt (Burundi/Zaire) and not at all in the monsoonal region of the Loess Plateau (Xian Province) in China.

The next warm thermomer tallies with the famous Tiglian Stage of the North Sea Belt with Azolla tigliensis. It shows still more geosoil development in the monsoonal (China loess) belt with at least four well developed ones within the Wusheng Loess 3 (WS 3) covering the time interval of 1.8 to 2.2 million years. Less developed are the equivalent soils in the Meltemi Stage of Greece and in the Muhira Formation of Burundi/Zaire.

3.1.3. *The Red Silt Soil.* Except for the North Sea Belt a well developed Red Silt Soil (of the Latosoil type) forms the lowermost soil in China, Greece and Burundi. There is no equivalent of such soil within the above lying Pleistocene geosoils series. That is why investigators in the various regions considered this Red Silt Soil originally to represent the Neogene/Pleistocene Boundary. However, in Greece Paepe, Hus and Lin (in press) found in 1981 at the type locality of Kokkino Limanaki that the Red Silt Soil was older than the Olduvai Event (palaeomagnetic evidence) and younger than the first cold oscillation of the Quaternary (foraminiferal evidence). Below that soil, lagoonal marls with periglacial features occur in turn overlying a Pecten Crag which has been definitely proved to be of Upper Pliocene Age (Christodoulou, 1969). The real End-Tertiary soil occurs immediately underneath.

At Luochuan, a similar situation occurs where the Matuyama/Gauss Boundary of 2.43 Ma was found to exist immediately under the Red Silt Soil whereas below the palaeomagnetic boundary other well developed latosols occur. These may be correlated with the Pliocene Arendonk Peat Member of Belgium occurring in the Late Pliocene North Sea Belt (Paepe & Van Hoorne, 1976).

Three basic questions arise from the evidence as given. These are:

1 Why is the beginning of the Quaternary characterised by a soil reflecting vegetational, climatical and geomorphological conditions of the Tertiary even after the first severe cold phase took place already during the first Cryomer dating from the beginning of the Pleistocene and lasting from roughly 2.4 to 2.0 million years?

2 Why did none of these traditionally Tertiary-bound conditions of soil development re-appear after that period so that subsequent typical Quaternary soil development becomes totally different and generally of lesser development?

3 Why is this the first soil of the first Quaternary Thermomer encompassing the Tiglian/Meltemi Stage and not a separate one?

The conclusion of this chapter deals essentially with the stratigraphical significance of the 100K and 400K cyclicities.

(a) It is remarkable that geosoils (i.e. land surfaces and relevant vegetations) of the interglacial stages can be found to be piled up one upon another without interruption by sedimentary sequences revealing the presence of interstadial soils. These sequences show the prerequisite position for the interglacial soil landscape development. It is a specific geomorphological position which most probably coincides with a plateau type feature. No evidence of any interstadial soil development nor of related glacial lithostratigraphic sequences is shown between such positions. If such intermediate sediments have ever existed, they have either been eroded away or have been preserved only in specific relict type positions (see section 4).

This geomorphological parameter of the interglacial soil landscape adds to the time stability of the interglacial geosoil. It may be concluded that the 100K and related 400K long term Geosoil Lithostratigraphic Cycles show over the last 2.5 Million years an uninterrupted interglacial landscape sequence of extensive, widely developed landmarks of similar nature. In these features the geomorphological link with the past interglacial landscape shows maximum preservation. It therefore becomes possible to use these geosoils for global correlation and for the land locking of the Brunhes/Matuyama and Matyama/Gauss polarity reversals.

(b) According to the higher resolution of geosoil frequency in the tropics than in higher latitudinal regions it may be questioned whether Glacial Periods resulted from disturbances in the Global Circulation System (GCS) in the Pole areas or in Equatorial zones which seem to have responded more rapidly and more frequently to GCS changes.

3.2. THE RELATIONSHIP BETWEEN 100K, 200K & 400K GEOSOILS AND SEA LEVEL CHANGES

One of the first relations established by early geomorphologists was the principle that all geomorphological evolution was intimately related to a base level of erosion. As oceans occupy 4/5 of the global surface, the Mean Sea Level (MSL) represents the most important base level of erosion and hence of landscape modelling.

If the previously discussed interglacial soil-landscape sequences have been developed on privileged geomorphological positions as plateau-surfaces, the areas around these "interfluvial" regions were subject to frequent cyclic remodelling as a result of the sea level changes. The land-/seascape dynamic ratio continuously changed between two stand-extremes of low and high sea level, i.e. in the most geo-dynamic part of the global surface: between the isolated plateau's on the one side and the deep sea beyond the continental shelf, on the other.

Moreover, within the narrow marginal zone of changing coastlines attesting to the highest and the lowest stands of the MSL, marine and continental deposits rapidly alternate. Such marginal sedimentation areas were studied intensively in the periphery of the Southern Bight of the North Sea Basin, on the East coast of Attica (Greece) for the Aegean Sea (Eastern Mediterranean) and for Lake Tanganyika in the Ruzizi Plain (Zaire-Burundi) as described above.

The land-sea relationship for the Southern Bight of the North Sea was studied from the interglacial geosoil position in between the marine transgressive deposits. The Lower Pleistocene sequence was studied along the Dutch-Belgian border (Paepe & Vanhoorne, 1978), the Middle Pleistocene sequence along the French-Belgian border (Sommé & Paepe, 1981) and the Upper Pleistocene along the Flemish Valley (Tavernier, 1943; Paepe & Van Hoorne, 1968), the southernmost fossil estuary on Belgian territory of the Delta system in the Southern Netherlands.

3.2.1. Sea Level Changes during the Last 800K-Years. The span-interval of the last 800K-years (Upper- and Middle Pleistocene combined) shows the best correlation (Fig. 3). Along the Flemish Valley from Melle (Ghent) towards the North Sea (Zeebrugge) MSL fluctuations within the time span of 800 Ky to 400 Ky were recorded. At Melle two marine highstands respectively at +5 m and +12 m above MSL were recorded with overlying continental peat horizons as belonging to the beginning of the Cromerian Complex (Paepe, 1974). These initial high MSL stands (occurring after undetermined lower stands during the preceding cold dry Bavelian Stage) coincide with isotope stage peaks 19 and 17 which are separated by a double peaked isotopic trough corresponding to a low MSL stand at -15 m approximately. Another much deeper isotopic trough occurs just before isotope stage 15 most probably coinciding with a yet undetermined major low MSL stand of presumably -50 m right in the Middle of the Cromerian Stage. The last mentioned stage 15 again is composed of two high isotope peaks corresponding with two high stands of MSL in the Zeebrugge Member (Paepe et al., 1981). However, at a considerably lower level than those occurring at the beginning of the Cromerian Complex, respectively at -10 m and -25 m MSL.

From this first analysis of the first 400K-years of the Middle Pleistocene (totally

SEA LEVEL CHANGES
DURING THE LAST 800.000 YEARS
IN THE SOUTHERN BIGHT OF THE NORTH SEA

[France. Belgium and the Southern Netherlands]

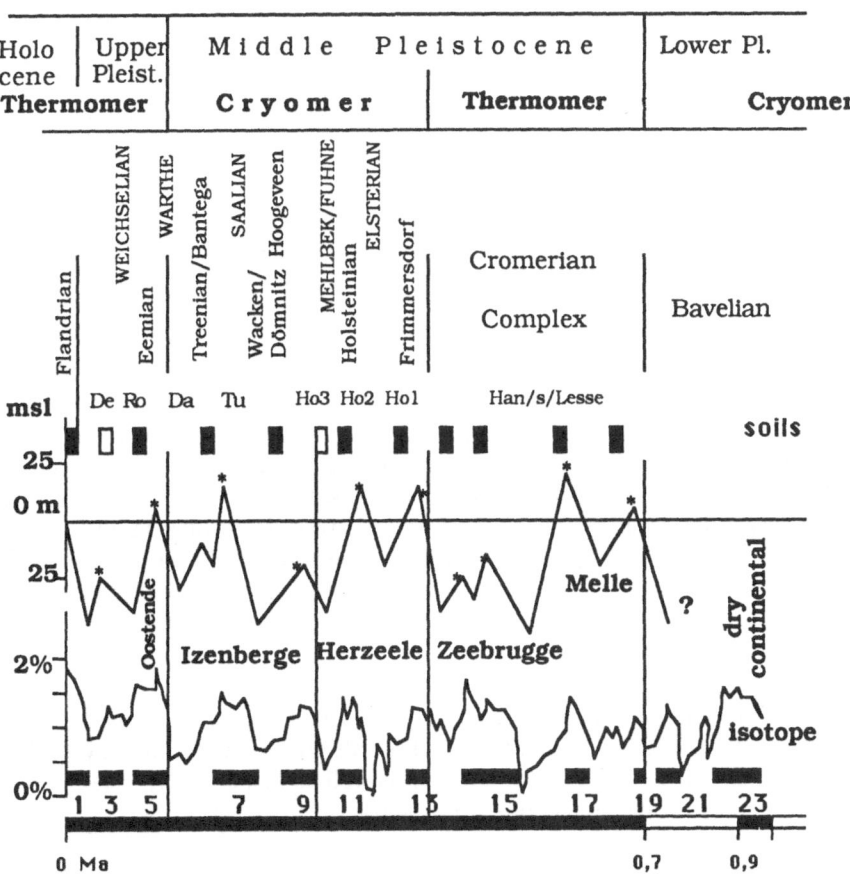

Fig. 3 : The Flemish Valley in the north of Belgium is the southern-most outlet of the Zealand Schelde, Maas and Rhine estuary which filled up completely during the last 800.000 years. Peaks and lows of the oxygen isotope curve correlate with the high and low inter-glacial MSL stands. As soils occur on top of the marine deposits they are indicating the end phase of the interglacial after the maximum transgression of the sea.

corresponding to the Cromerian Complex Stage) four high stands of MSL are recorded of which the two earlier ones (Melle) are generally 20-50 m higher than the two later ones (Zeebrugge). It also means that the earlier ones are above today's MSL and the later ones below. Between the earlier and later Cromerian transgressions the deep stand (at least -50 m) of which no evidence is found except for the deep valleys incised in some parts of the Flemish Valley (De Moor;Paepe, 1963).

Each of the high stands is succeeded by an organic soil layer totalling four geosoil levels dating of the Cromerian Complex Stage. Equally, along the terraces of the Lesse (more specifically at Han-sur-Lesse) four truncated textural B-soil horizons also occur in the same time span of the Cromerian Complex Stage and were indicated as Han Soils (Paepe,). These soils by correlation with the Eemian and Holocene situation (see p.) are considered as post-marine. This conclusion corresponds with the opinions of Iversen (1958) and Grycuk (1965) proclaiming a generally drier climate towards the end of an interglacial, with little if any sedimentation and soil development (Ložek, 1975) during this depositional standstill.

At Herzeele (French-Belgian border) isotope stages 13 and 11 correspond with two peaks of marine highstands respectively occurring before Holsteinian Stage peat layers Ho1 and Ho2 (Sommé;Paepe et al., 1978; Van Hoorne & Geysels, 1987). These Herzeele highstands, with an average of +15 m MSL correspond with the Melle highstands.

During isotope stage 9 the next highstand reaches only -25 m MSL whereas during the next isotope stage 7 it occurs around +15 m MSL. These highstands have been observed at Izenberge (Belgium) not far from Herzeele (France) along both sides of the famous river of the Great War, the Yzer. Once again, peat bogs occur above the marine layers indicating that the end of each transgression was followed by a slightly dryer period inducing land aggradation with soil development (corresponding with the Tubize and Daussoulx Soils on the continent). The Izenberge high type transgression continues within the Ostend highstand at +5 m MSL (isotope stage 5) seconded by a low type transgression at -25 m MSL (isotope stage 3) during the Upper Pleistocene namely the Denekamp Interstadial (see section 4).

Each of these transgressions is followed again by soil forming processes as the Rocourt Soil (Gullentops, 1954; Paepe, 1963) and the Zelzate Soil (Paepe, 1964) at 110 Ky and 30 Ky respectively. The MSL of the Holocene Flandrian Transgression occupies the lowest position of all former highstands. Actually a general lowering of the highstand MSL from an average of +20 m MSL some 800 Ky ago towards today's zero MSL may be observed. Absolute low stands seem to have diminished as well so that one may speak of a general amplitude decrease in the course of the last 800 Ky.

3.2.2. The Last 2.4 Ma Time Span. If one extends the above mentioned observations to the end of the Pre-Tiglian Stage i.e. at about 2 Ma (Fig. 4), the amplitude of high and low MSL stands is increasing from about 50 m for the end of the Middle Pleistocene to about 100 m for the beginning of the Lower Pleistocene.

The frequency of occurrence of high and low MSL stands varies as well throughout the Quaternary. At first inspection peaks of sea level highstands are less compatible with isotope stage peaks and one single peak seems to cover two or even a group of corresponding isotope peaks. For example the MSL highstand at +35 m. of the Tiglian Stage located at about 1.7 Ma occurred in the middle of the Olduvai polarity event. The latter range is composed of at least two, perhaps three isotope peaks each of which may correspond with the sole MSL highstand. The deeper isotope troughs occurring before the beginning (1.77 Ma) and at the end (1.57 Ma) of the Olduvai event coincide with two low MSL stands at -75 m and -50 m respectively. The whole covers a time of almost 200K-years.

The next 200K-years climb up in time to about 1.37 Ma encompassing on the isotope curve a series of five weakly developed peaks of a weaker intensity than all previous ones. Only the troughs at the beginning and at the end of this 200 K-years time span are naturally reflected into two low MSL stands at -50 m. and -65 m respectively. In between, two high level stands occurring at about -15 m each may possibly correspond with the group of intermediate isotope peaks. Even when based on this minimum assumption these highstands occur almost at a position 50 m lower than the Olduvai - Tiglian MSL highstand at +35 m. According to the classical Southern North Sea Chronostratigraphic Classification this period encompasses a greater part of the cold Eburonian Stage within the second Cryomer of the Lower Pleistocene.

The following 200 K-years lasting to about 1.1 Ma show a succession of four MSL highstands at only +10 m and shallow MSL lowstands at -10 m between two major troughs: the aforementioned at -60 m and a new deep low stand at -30 m. Except for the first one, it is difficult to make any correlation with the peaks and troughs of the isotope curve in this reach. Nevertheless, these higher highstands in comparison to the previous series point to a possible correlation with the Waalian Stage marine transgressions.

From the -30 m. level on, a steady rise of the sea level can be followed so that this latter lowstand may well encompass the Menapian Cold Stage as recently redefined by Zagwijn & De Jong (1988). In the same line of thought the next two MSL highstands at respectively +5 m and +25 m, just before the deep trough on the oxygen-isotope curve, are correlated with the Bavel and Leerdam interglacials of the Dutch Bavelian Stage. The deep trough represents the extreme cold phase separating the Lower from the Middle Pleistocene viz. the Bavelian Stage from the general warmer Cromerian Complex Stage. As stated before, the landscape evidence of the deep isotopic trough may be represented by the deep gullies eroded at the base of the Flemish Valley (Paepe, 1981) and the steady retreat to the North of Holland of the estuaries of the Rhine and Meuse rivers (Zagwijn, 1975; Paepe, 1981).

Despite these assumptions of a correlation between the lithostratigraphic and chronostratigraphic subdivisions of the North Sea which formerly was assumed to be one of the most complete of quaternary classifications, it becomes clear that such a subdivision today is far from complete when compared to the standard deep sea record and the geosoil record of the monsoonal subtropics and the tropics. The main reason for this incompatibility is due to the fact that the (chronostratigraphic) North Sea Belt Stages absolutely did not encompass oxygen-isotopic stages during the Lower Pleistocene. Indeed, the "continental/coastal fringe" classification of the higher Middle Latitudes does not show a sufficiently continuous record to follow and detect all climatic variations and it is not at all clear what the impact of the neotectonics at each of these Stage levels is in that region. Therefore chronostratigraphic "Stages" have erroneously grouped one or more 100K-years intervals thus rendering correlation extremely difficult if not impossible.

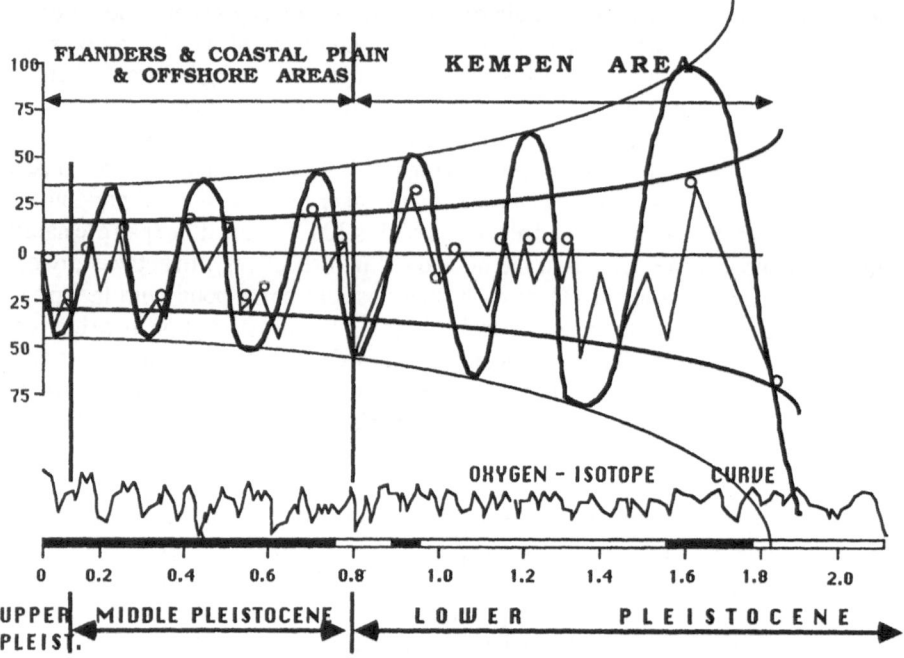

This figure shows the possible correlation between sea level high stands and low stands during the last 2.4 Ma years of the Quaternary..

o   : Fixed measured ordnance datum level of sea level highs and lows

: Asymptotic curve of high and low MSL (mean sea level).

: Periodicity graph of high and low MSL in comparison with the oxygen-isotope curve of the last 2.4 Ma years.

Figure 4. Sea level changes of the last 2 million years (southern bight of the North Sea).

Even if one should be careful in making such correlations, it cannot be denied that a number of high and low MSL stands exist and are seemingly following patterns of climatic changes within the limits of neotectonic movements.

When the maximum and minimum stands of Pleistocene sea level changes of the two above discussed graphs (Fig.4) are connected something quite striking is happening. For the last 800 K-years, starting at present zero MSL, one not only finds the 200 K periodicity but an asymptotically increasing amplitude between the maximum and minimum levels of the coastline stands. In the present coastal plain area maximum amplitude differences for the Last Glacial/Holocene figure about 25-30 m and at 800 K-years about 75 m. This trend continues beyond the 800 K-year (i.e. the Brunhes/Matuyama boundary) with an ever widening trend to about 130 m (almost doubling the 75 m amplitude) over slightly more than another 800 K-years, at the beginning of the Pleistocene. From this asymptotically double dimming trend of both maxima and minima around the present MSL the question arises whether this amplitude will reach zero in the near future. This very question withholds further consideration about changes in MSL namely whether these are climatically biased or solely due to tectonic movements, including seafloor movements as well. Paepe (1963, 1981) has suggested the possibility of terrace building as periodically induced tectonic terrace steps. Indeed, many of the terrace systems developed around the Southern North Sea are, climatically speaking, polycyclic since they are composed of sediment series formed under widely differentiated climates (Paepe, 1981; Paulissen, 1977). This brings one to a further consideration that if terraces, due to cyclic induced tectonic movements (enhancing high and low MSL stands i.e. higher and lower base levels of erosion) correlate with climatic stages of the deep sea record and (as was stated above) with climatic soil phases, does that not point to climatic changes rather being a result of sea floor movements than vice versa? At least it teaches us at this stage of the investigation that sea level changes and related landscape morphological evolution are not necessarily solely the result of climatic Global Change alone.

4. The 40K and 20K Periodicities Along the North-South Geotraverse

In the previous chapter only 100K-years interglacial soil levels have been considered in lithological sequences where little if any other deposits than such soils are present. In particular situations some others, especially periglacial and windborne (e.g. loess) deposits, have been preserved as relatively thin layers between the soils. In some particular conditions, however, complete warm/cold sediment sequences may be found trapped in fossil valleys or on broad plateaus. Most of these complete sequences are dating mainly back to the Last Interglacial/Glacial cycle (Upper/Late Pleistocene) although others may be found as well.

In such sequences interglacial palaeosoils occupy the lowest position whereas in the layers above them other fossil soils of different nature occur. The latter usually are called "interstadial" palaeosoils in contrast to the "stadial" sediments which reflect cold phases. "Interstadials" are of a milder climate than the vigorously cold glacial "stadials" but never as mild as an interglacial. Interstadial geosoils are therefore of weaker development than interglacial geosoils.

Many such Late Pleistocene sections have been described all over Europe over the last thirty years; yet most of them were located north of the Pyrenees and of the Alps. Only in the last five years sections of Late Pleistocene Age have been described from regions south of the European mountains in Italy, Spain and Greece. A jump-over to the Equatorial Belt was made possible thanks to mapping of the Ruzizi plain in Burundi/Zaire by Ilunga Lutumba. In Fig.5 sections from Zelzate (Belgium) in the North Sea Belt (Paepe 1966), from Koroni (Greece) in the Southern Peloponnesus (Paepe, Mariolakos & Van Overloop, 1981) and from Gihungwe (Burundi) in the West African Rift (Ilunga Lutumba, 1984) have been selected to establish a Late Pleistocene North-South geotraverse. This middle range geotraverse parallels the long range one with only interglacial soils represented. Again the sections are projected against the Late Pleistocene oxygen isotope stages of Shackleton & Opdyke (1973).

The Zelzate section has developed entirely above marine layers of the Last Interglacial (Eemian Stage). Immediately above it is the so-called Rocourt Soil of Eemian Age. Since its introduction by Gullentops (1954), this soil has been generally considered to form the base of the Late Pleistocene in the typical Loess Belt of Western and Central Europe. In the Zelzate section it was made clear by Paepe (1964) that this interglacial soil was developed after the maximum of the marine transgression of the same Last Interglacial which is generally believed to correspond to isotope stage 5e. The Rocourt Soil which in most of the sections north of the Alps occurs as a single red soil textural horizon is proved to be polycyclic thus forming a complex which developed at a different stage following isotope stage 5e i.e. after 115K-years. Since the studies carried out at La Grande Pile (Woillard & Mook, 1982) confusion has risen as to the boundary between the Last Interglacial and Last Glacial viz. whether the interglacial should be extended until the upper limit either of isotope stage 5a or 5e. This will be discussed at full length later in this paper.

Above the Rocourt Soil another series of soil levels are found to which the names Amersfoort (68K-years, Zagwijn, 1967), Brørup (65K-years, Andersen, 1965) and Odderade (55K-years, Averdieck, 1968) were given. As in many places the three soil layers are telescoped together into one complex soil horizon the name Warneton Soil Complex for such a grouping was introduced by Paepe (1963). The whole series of soils correspond with weakly developed steppelike soils. Therefore they are considered interstadial. Henceforth they encompass isotope stage 4 (dated 73K through 61K-years) according to the former datings and in our opinion these soil horizons should not be confused with any possible soil development of an age older than 75K-years. In fact they are clearly separated from the Rocourt Soil by important periglacial features (such as cryoturbations and frost wedges) whereas immediately above them, large frost wedges usually occur along a line of strong erosional unconformity indicating a severe cold phase of some 50K-years ago.

In comparing the above discussed part of the Zelzate section with the sections in Koroni and in Gihungwe the statements about the Rocourt and Warneton Soil Complexes gain stronger evidence. Actually, in Koroni the equivalent of the Rocourt Soil split up in three red textural B horizons labelled as Koroni Soils of which Koroni Soil 2 can still be subdivided into two horizons. Between Koroni Soil 1 and the underlying Last Interglacial marine deposits, frostlike wedges occur. Above Koroni Soil 3 follows a series of three weaker developed brown soils in a sediment series which, as is the case in the Zelzate series, is again strongly eroded at the top. Similarly in Gihungwe a threefold subdivision for the interglacial soils and a threefold subdivision for the overlying interstadial soils is maintained.

Between 50K-years (first extreme cold) and 26K-years (second extreme cold) another threefold system of well dated interstadial soils develop at respectively the Poperinge/Moershoofd level (45K), the Hoboken/Hengelo level (34K) and finally at the Zelzate/Denekamp level (29K) (Zagwijn & Paepe, 1968). They represent mild phases of the full glacial period (isotope stage 3) so that their degree of development is weaker still than the previous interstadial soils of the Warneton Soil Complex. The 45K and the 34K soil levels are quite often lacking due to erosional activities going on at this stage of the Last Glacial (as is

LATE PLEISTOCENE SOIL-STRATIGRAPHICAL SECTIONS

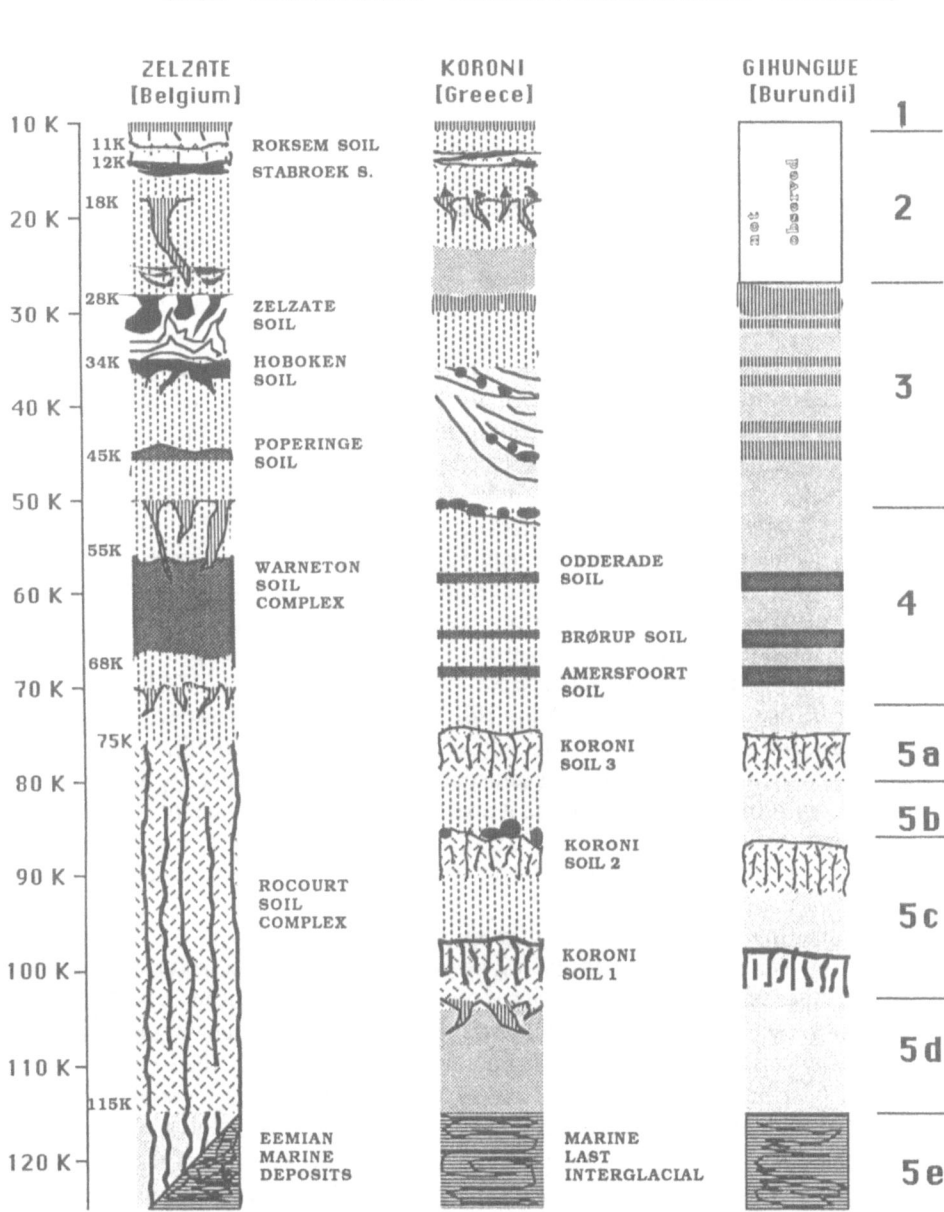

Figure 5. Late Pleistocene soil-stratigraphical sections.

observed in Koroni). However, at Gihungwe in the Equatorial Belt one observes againa doubling of these interstadial soils.

After the maximum cold at 18K-years, again three interstadial humic soil levels were reported in the Zelzate section at about 17K (Zulte/Lascaux level), 12K (Stabroek/Bølling level) and at about 11K (Roksem/Allerød level) of which the equivalents of the uppermost two are found in Koroni (this part was not observed in Gihungwe). The whole is covered with the next interglacial soil of the Holocene.

A few preliminary conclusions may be drawn at this stage (Fig. 6):

(a) The so called "Interglacial Soil" represents a fivefold complex soil development of which part was formed during warm phases corresponding with the end of isotope stage 5e, beginning, middle, and end of isotope stage 5c, and finally of isotope stage 5a.

(b) All soil developments are posterior to the maximum of the interglacial marine transgression (Zelzate & Koroni) viz. the maximum lake level rise (Gihungwe).

(c) Only the 5e soil development is considered to be fully representing the Last Interglacial and indicated as an interglacial pedocomplex (PK) marking the end of the Last Interglacial after either a marine or lake transgression; certainly before the start of the first cooling or drying of the climate.

(d) Above the interglacial soil type formations occur. From 75K-years ago three groups of three interstadial soils each, between 68K and 55K-years, 45K and 28K-years, and 17K and 11K-years respectively. These nine interstadial soils together with the four weaker interglacial soil types represent 13 Last Glacial Soils (GS).

(e) By controlling the time interval between a number of date limits as 73K against 95K and 115K one may readily detect the 20K and 40K-years cyclicity which is proved to be a result of obliquity and equinox respectively (Berger, 1978).

(f) The position of a interglacial soil is indisputably located at the very beginning of each warm/cold 100K cycle. Moreover, as it occurs between the maximum of the interglacial transgressions (high MSL) and the first next cold, it can only be a fraction of 10,000 years considering the time span intervals of both continental Holocene and 5e isotope stage.

(g) Two other questions are arising from the above considerations:

- Would an interglacial include the whole period of sea level rise starting immediately after the maximum cold? If answered positively then the Holocene would be lasting almost twice as long i.e. 20,000 years from 17K BP to the beginning of the next glacial period in the near future about 5000 A.D. Likewise, the space of time of the Last Interglacial should be enlarged as has been done by some American Quaternarists by shifting the beginning of the Last Interglacial to about 145K BP.

- Should not, in the light of the above mentioned, the position of the interglacial palaeosoil be determined with even greater precision? This question will be answered in the following chapter on the study of the palaeosoil cycle during the Holocene.

5. Palaeosoil Cycles of the Holocene

The Holocene sequence of geosoils within the duration of the current interglacial, i.e. the last 10,000 years, has been unravelled thanks to the geological-archaeological studies carried out in Greece since 1975. In numerous excavation sections of the Attica peninsula namely in the Marathon Plain, down along the East Attica coastal plain to Cape Sounion and in the Kifissos Valley in the Western part between Athens and Pireefs, at least twenty Holocene Soils (HS) of different soil types have been revealed. It points to a soil cyclicity of an average of one in every 500 years. As is shown in Fig.6 the frequency of this continuous series of twenty soils seems to amplify as they become more recent and weaker in development. This will be discussed in due detail hereafter.

Athens and its surroundings is indeed famous for its abundance in continuity of vestiges of antiquity all over the past 7000 years viz. since the Neolithic Period. The relationship between the occupation of land by man and the natural landscape evolution may therefore be studied in that area at the highest possible resolution. It was possible to establish litho-stratotypes (Fig.6) i.e. complete lithological sequences of the rock layers and of the soils in the famous battlefield of Marathon (Holostratotype) and in the Kifissos Valley linking Athens and Pireefs (Hypostratotype).

Deposits from the ancient river Haradros have build up a fossil alluvial fan in the middle of the Marathon Plain. In Fig. 6 the Marathon Holostratotype shows above the substratum of Mesozoic shales a series of sands, gravels, loams and clays separated by geosoils. The great number of cherts in these deposits proved that one is dealing here with a complete series of Holocene deposits.

Indeed, above the shales one finds the first (HS1) and most developed strong red geosoil which therefore has been specially labelled with the name Marathon Soil. This soil paves the fossil valley which deepens towards the sea so that the assumption is then that the it shapes the erosion valley which was formed during the sea level low stand of the previous Last Glacial Stage. It points at an Early Holocene Age for its development.

The sand covering the Marathon Soil is sealed off by another weakly developed soil (HS2) upon which follows a thick series of coarse well rounded gravels. The fluviatile sands between HS1 and HS2 testify to the fluviatile filling up of the valley due to the post glacial sea level rise which hampered the evacuation of the valley waters. As this very first phase of sea level rise came to an end the valley bottom was again overgrown by vegetation so that HS2 developed and sealed off the very first part of the valley filling.

Early Neolithic remains and cherts found above these soil series confirm their Early Holocene Age. Both are believed to be older than 8000 BP (Before Present), the start of the Neolithic in Greece being located at about 6000 BC (Before Christ). This period of two thousand years corresponds in archaeological terms with the Mesolithic while in geological terms it encompasses the relatively dry phases of the Pre Boreal and the Boreal. The high degree of development of the Marathon Soil (HS 1) inplicates deciduous forest conditions which mean high temperature and a certain amount of moisture higher than during the dry Pre Boreal. The most probable position for such climatic conditions is the transition of the Pre Boreal towards the Boreal about 9000 BP. The weaker HS2 brown soil developed

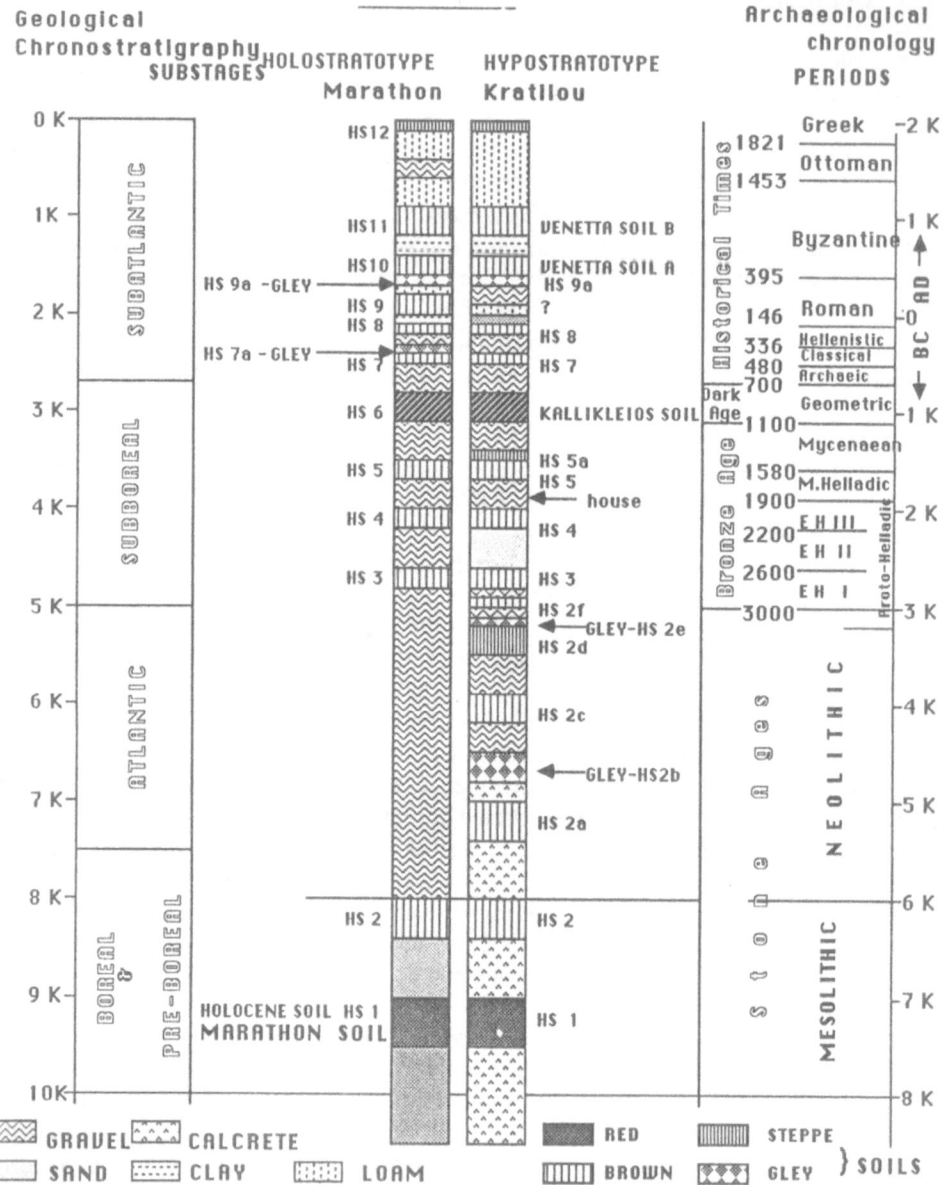

Figure 6. Litho-Stratotypes of the Holocene in Attica (Greece).

in slightly similar conditions 1000 years later at the end of the dry Boreal Substage when climatic conditions were gradually transgressing into the moister Atlantic Substage. The sharp contrast with the overlying thick coarse gravel body indeed points to a definite change towards a more energetic fluviatile activity inferring moister climatic conditions; the thickness of the gravel body points to a strong rise in sea level which also occurred as a result of the Atlantic Climatic Optimum.

Above the gravel body, four geosoils HS3, HS4, HS5 and HS6 occur at regular time intervals of roughly 500 years, covering the whole space of time of the Bronze Age. This archaeological period encompasses roughly the time interval of 5000 BP to 3000 BP which corresponds to 90% of the new dry Sub-Boreal Substage. Fluviatile activity seriously decreased and was periodically interrupted by even dryer phases thus offering the possibility for vegetation to overgrow valley bottoms at each of the dry/wet transitions and for soils to form.

Towards the end of the Bronze Age and at the beginning of the Geometric Period about 1000 BC (some 3000 years ago) the postglacial valleys, basins and coastal plains were completely filled in until today's topographical level and sealed off by a new strongly developed geosoil HS6. As this soil level locates a very precise landmark in the evolution of the land machinery it is connoted again with a specific name: the Kallikleios Soil. For the second time in the Holocene, soil development at least points to an overwhelming vegetation which in turn points to higher moisture conditions announcing clearly the dawning of the moister Sub-Atlantic Sub Stage. Conditions of life must have changed dramatically by both the complete filling up and the bevelling of the landscape as well as by a the tremendous growth of the forest.

After 700 BC (some 2700 BP) these newly established lowlands offered less flooded landscape conditions together with abundant woodland which allowed the highly developed civilisations of Historical Times develop. Nevertheless, periods of reduced river activity and partial reforestation occurred, the latter being the origin of eight geosoils of most different soil types varying from brown soils (HS7,HS8,HS9,HS10,HS11) to steppe (HS12) and gley soils (HS7a,HS9a) developed above the Kallikleios Soil (HS6). Several cyclicities seem to be mixing in this very last substage of the Holocene: a 1000 years cycle for the brown soils interfering with a 500 years, or even a 250 years cycle if the weaker soils are taken into account. It means that during Historical Times, woodlands, steppe and wet lands alternated more frequently than was hitherto accepted, interfering with low energetic river activity phases resulting in the deposition of mainly sandy loams and loamy sands.

Further to the soil cyclicities in the Marathon plain the grouping of HS1, HS2, and HS3 to HS6, and HS7 to HS12 subdivides the Holocene sequence into the four well known and above mentioned geoclimatic substages of 2500 years each in duration i.e. the Suess carbondioxide cycle. It should be stressed that these four major geologic subdivisions correlate extremely well with the Mesolithic, the Neolithic, the Bronze Age and the Historical Times periods respectively.

Similar conclusions are drawn from the hypostratotype of Kratilou in the Kifissos Valley, west of Athens. However, here the number of soils and sediment facies changes in the tranquil depositional environment of the Kifissos river reveals a much greater variety than

274

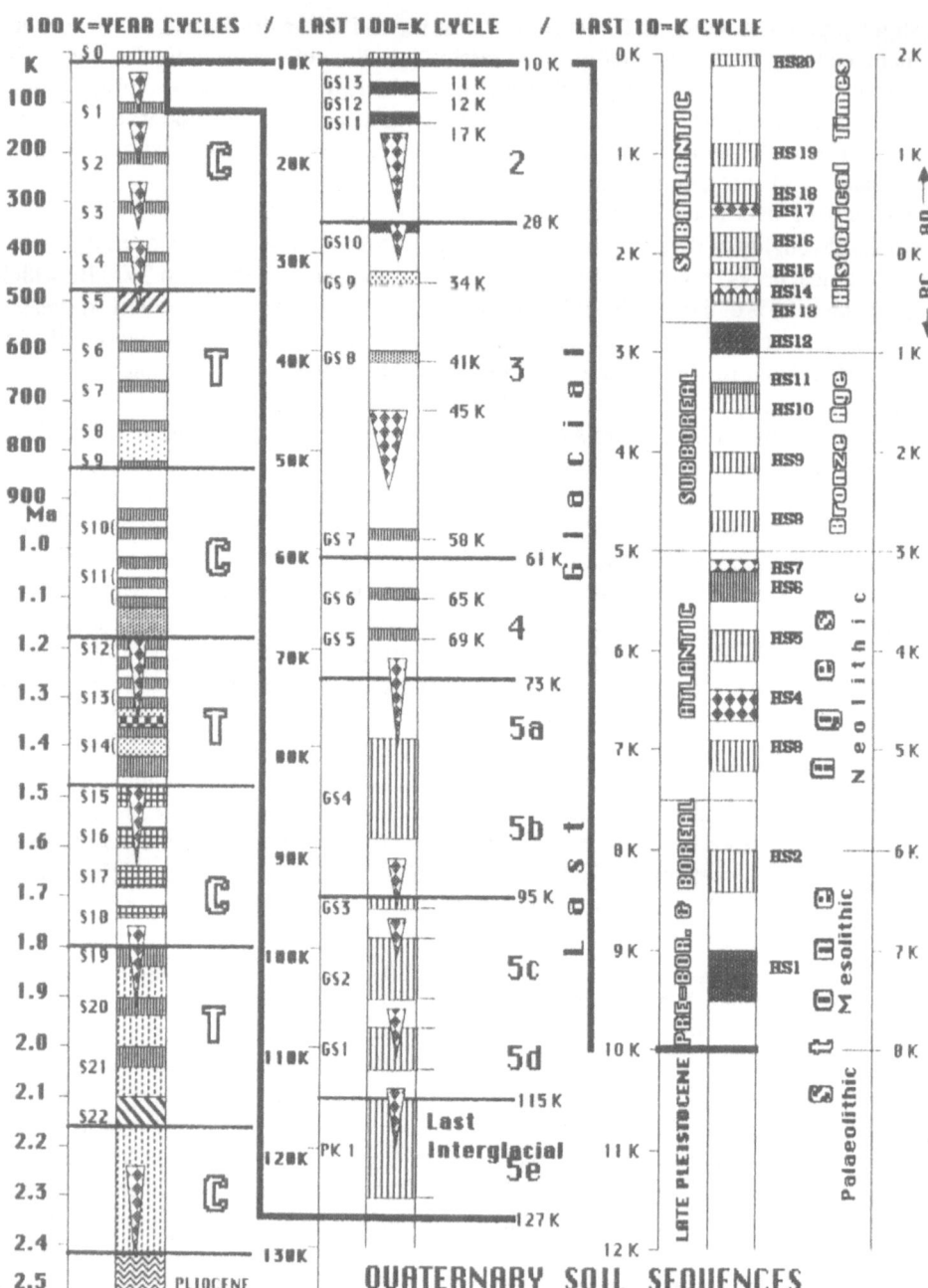

Figure 7. Quaternary soil sequences.

in the restless alluvial fan sequence of the Marathon Plain. Especially within the Atlantic/Neolithic Substage, in which no such subdivision could possibly be made in the Marathon Plain all types of soils, like deciduous brown soils, steppe soils and gley soils, equal to those occurring in the Sub Boreal and Sub Atlantic Substages of the Marathon Holostratotype, are now represented. These additional soil levels reconfirm the existence of cyclicities of 1000, 500 and 250 years. A total of twenty soils, as stated above, have finally been recorded (Fig. 7).

The revealed 1000 years cycle of the strong brown/red soils of deciduous forest development and the 500 and 250 years cyclicity of the intercalated steppe and gley soils point at quick changes in the moisture/temperature balance within the major 2500 years cycle.

Surprisingly again these smaller soil cyclicities are repeated in the sequence of civilisations of the Bronze Age (the 500 years cycle) and of the Historical Times (the 250 years cycle) so that one could almost speak of an environmentally determined historical evolution.

Moreover, the 1000 years cycle reveals a recurrent cyclicity of extreme drought (Paepe, 1986) which probably became effective since the beginning of the Holocene, but certainly since the 8th century BC (the Agora drought). Recurrences are found at the 2nd century A.D. (the Roman drought), the 12th century A.D. (the Akominatos drought) and finally today in the 20 century A.D. (the Sahel drought). These interfere in time with rises in sea level as was found from many places along the North Sea and the Mediterranean shores.

6.   Conclusive Remarks on the Soil Cyclicities

The great number of soil cyclicities which exist throughout the Quaternary at the long term, middle term, and short term level prove at the same time the complexity of the drought and sea level changes. All three levels of cyclicities are operating in a kind of sequential system, the first step being an interglacial of the long term cycle followed by the short term cycle whereafter the middle term cycle is enhanced until finally the next interglacial shows up and the system is repeated again.

Without precise knowledge of the time boundaries within which the cycles operate it will remain difficult to forecast which type of cyclicity is generating a drought or sea level change or any other natural hazard. The prediction of the re-enforcing effect of natural climatic changes on the greenhouse effect or vice versa henceforth remains difficult.

The final discussion will focus on the position and the time stability of the Interglacial Soil in the long term, middle term, and short term cycles, which are essential parameters to be determined. The interglacial soil represents indeed the land surface which in the past was covered to its maximum extension by the most dense forest vegetation including the tropical forests, several times. Its rhythm of disappearance and reappearance within certain boundaries of time, as stated at the beginning of this paper, covers so many moments of the landscape evolution that it should obviously be regarded as one of the most important counterparts to the deep sea records to be used in the detection of climatic changes.

6.1. POSSIBLE STRATIGRAPHIC POSITIONS OF AN INTERGLACIAL SOIL

Unlike the Last Interglacial Pedocomplex (PK) at Koroni (Greece) or Zelzate (Belgium) the Marathon Soil is not underlain by marine deposits in the Marathon plain. Instead, the thick gravel bed which is definitely a consequence of the maximum sea level rise of the Holocene Atlantic Climatic Optimum (the Flandrian transgression) covers both the Marathon Soil (HS1) and HS2. Indeed, there is a possible confusion between the Marathon Soil which was formed some 3000 years before the Climatic Optimum below the gravel sedimentation, and the Kallikleios Soil (HS12 of Fig.6) of comparable intensity of development as the Marathon Soil, which was formed 3000 years after and above the gravel body. Both red clay soils should be considered in the sequence of the twenty Holocene Soils, as typical interglacial soils whereas the other ones, as the brown soils, steppe soils, and gley soils of much weaker development, should be interpreted as interstadial soils of the same interglacial stage.

The very question to be answered is with which of the stratigraphical positions the interglacial soils of the long term and middle term cycle should be stratigraphically compared: with the Marathon Soil (HS1) or with the Kallikleios Soil (HS12) at the beginning or in the middle of the Holocene interglacial respectively? In other words: with the one before, or after the maximum of the interglacial transgression?

It should also be pointed out that in the Marathon Plain both the Marathon and Kallikleios Soils are converging together into a complex soil entity with increasing distance from the sea (which in the Marathon Plain is less than 2 km.). More inland, on the surrounding heights outside the plain, where this complex soil entity is resting on pre-holocene or even pre-quaternary deposits, it is once more converging with the Holocene most recent topsoil (HS20), thus representing only one single Holocene interglacial soil composed of the Marathon Soil, the Kallikleios Soil, and the topsoil (in Fig.6 labelled as HS1, HS12 and HS20 respectively) together.

With regard to the possible stratigraphical position of the Last Interglacial soil in the Holocene sequence, its equivalent of the middle term record of 127K-years and the interglacial soils of the 2.4 Ma years long term Quaternary record may be of a fourfold origin:
- the single interglacial soil of the beginning of each interglacial stage;
- the single interglacial soil of the middle interglacial stage;
- the soils of the beginning and the middle interglacial stage combined into an interglacial complex;
- the interglacial complex combined with the present topsoil.

6.2. THE POSITION OF THE LAST INTERGLACIAL SOIL

- The Last Interglacial Soil is occupying a similar position as the single middle holocene soil when occurring above marine deposits of the Last Interglacial transgression maximum; in this position its age is covering only the upper 1/3 of the isotope stage 5e.
- The Last Interglacial also compares to a single holocene interglacial soil when it occurs at the basis of the marine interglacial deposits although its age is covering then more than 2/3 of the 5e isotope stage.
- Usually both Last Interglacial Soils at the base and at the top of the marine deposits

appear together in the same sequence.
- The Last Interglacial soil occupies a position of the complex soil (beginning and middle interglacial interpenetrating into one polycyclic soil) when occurring on a plateau in continental sequences as e.g. loess deposits; it is then also representing almost the entire interglacial time span if considered to be 10,000 years.
- Finally it seems not really to matter whether the Holocene topsoil is present or not as it lacks the status of a fully developed interglacial soil.

6.3. THE TIME STABILITY OF THE INTERGLACIAL SOIL

6.3.1. Taking the Holocene as an Interglacial Soil Stratigraphic Model of 10,000 Years Long, the Last Interglacial Soil Does Not Cover the Entire Isotope Stage 5e (Between 127K and 115K BP).
- Actually only the last 1/3 of the time when resting on marine deposits; geosoils of an age younger than 115K-years should then be totally integrated into the Last Glacial (GS) soil record.
- On land Last Interglacial soils may occasionally be interpenetrating with early Glacial Stage soils (GS soils) as e.g. in the case of loess soils like the Rocourt Soil. Even then only 2/3 of 5e is covered.

6.3.2. The Single Interglacial Soils of the Long-Term 100K Sequence May Very Well Form In Less than 500 Years (Probably Even In Less than 250 Years) Considering the Revealed Cyclicities in the Holocene Soil Sequence.
- The time stability of a single (monocyclic) interglacial soil may thus be estimated at the order of 0.5 % and 0.25 %.
- A complex interglacial soil being a combination of the interglacial soils from the beginning and from the middle of the Interglacial, thus covering a timespan of roughly 5000 years is still indicating a time stability of 2%.
- If the interglacial soil is composed of the base, the middle, and the topsoil, representing a 9000 years interval there is still a time stability of less than 10%. The latter should not be taken into account if the topsoil should not really be considered as an element of the interglacial soil process, as was stated above.
- Most of the interglacial soils being either of the double complex type or of the single type as they seem to occur above deposits of the maximum interglacial marine transgression (Fig.3 & 4) leads to the assumption that the time stability of occurrence of the interglacial geosoil is most probably of the order of 2%.
- Palaeosoils, or better geosoils, are stable fossil indicators of palaeoclimatic evidence. Even when only a few geochronological datings are available, geosoils may be fitted in at regular time intervals because of their high degree of time stability.

6.4. GEOSOILS AND GLOBAL CHANGE

- Unlike most other fossils, geosoils are very little time transgressive. The North-South

Geotraverses of the long term- and middle geosoil cycles prove that. Recent studies reveal a similar status for the Holocene soil cycles but more study is still needed.

- Geosoils, as stated before, are doubling or even tripling showing the sensitivity of the Equatorial Belt to the Global Circulation System. A time stability of 2% for these geosoils proves that the doubling or tripling are isolated, independent climatic, features and certainly no casual appearances of a local situation. The presence of two soils or more instead of one, indicates that the tropics were able to register the global climatic changes better than anywhere else on this planet.

- The myth of a high latitudinal interglacial being the equivalent of an equatorial interpluvial and the glacial being the equivalent of a pluvial is hereby defenitely banned. When it was dry at the higher latitudes it was also dry at the lower ones.

6.5. DEFORESTATION AND SOIL DEGRADATION

- This study may finally help to clarify the problem of "natural" deforestation and natural forest regeneration, both obviously bound to the same multiple series of cyclicities as the geosoils and the geomorphic surfaces on which the soils developed. With respect to the four recognised moments of geosoil development from sediment accumulation, to surface development, to vegetational development, and finally to soil development, these are substantial elements of all these cyclicities and become so to say automatically operational at the appropriate moment in the geosequence cycle leading to the geosoil development.

- As with geosoils, forests do disappear and rejuvenate naturally as a response to global climatic change in the course of geologic time. Deforestation and soil degradation should then be seen essentially as a man induced feature. The present demands of Mankind and its Society on its natural environment, the Biosphere and the Geosphere, are in a sharp disharmony with the possibilities of current natural regeneration possibilities of forest and soils. The long term geoclimate at the end of the current interglacial indeed sets back the well established landscape equilibrium of the last three thousands years (Kallikleios Soil). This process of disintegration is re-enforced by the natural desertification due to the 1000 years historical cycle. Mankind's demands for a natural equilibrium of its present environment stands in a sharp contrast with the point of cyclic evolution which has been reached in the dynamic equilibrium of the Biosphere and the Geosphere.
It is not at all clear in what stage the Greenhouse effect is influencing the dynamic equilibrium, at a moment in the geoclimate's evolution where biodiversity should normally be decreasing and landscape changes are dramatically different to those that were prevailing mainly during the last three millennia in which Mankind witnessed the dawn of its Historical Civilisations.

7. References

Andersen, B.G. (1965) "The Quaternary of Norway", John Wiley and Sons, London, pp. 91-138

Averdieck, F.R. (1967) "Die Vegetationsentwicklung des Eem-Interglazials und der Früh-Würm-Interstadiale von Odderade, Schleswig-Holstein", Fundamenta 2, pp. 101-125

Berger, A.L. (1978) "Long-term variations of coloric insolation resulting from the earth"s orbital elements", Quaternary Research 9, pp. 139-167

Christodoulou, G (1961) "Die Foraminiferen des Marinen Neogens (Astien) von Attika", Inst. Geol. Subs. Res. 8/1

De Moor, G (1963) "Bijdrage tot de kennis van de fysische landschapsvorming in Binnen-Vlaanderen", Belgische Vereniging voor Aardrijkskundige Studies XXXII, 2, pp. 329-433

Gricuk, B.P. (1965) *Paleogeografiya Severnoj Evropy v pozdnam pleistocene (Late Pleistocene paleogeography of Northern Europe)*, Poslednij europejskij pokrov. Moscow

Gullentops, F. (1954) "Contributions à la chronologie du Pléistocène et des formes du relief en Belgique", Mémoire de l"Institut géologique de l"Université de Louvain, tome XVIII

Hays, J.P., Imbrie, J. & Shackleton, N.J. (1976) "Variations in the earth"s orbit: pacemaker of the Ice Ages", Science 194, pp. 1121-1131

Ilunga L. (1984) *Le Quaternaire de la plaine de la Ruzizi. Etude morphologique et lithostratigraphique*, Thèse de doctorat, VUB

Imbrie J. and Imbrie, K.P. (1979) *Ice Ages: Solving the mystery*, Mac Millan, London

Iversen, J. (1958), "The bearing of glacial and interglacial epochs on the formation and extinction of plant taxa", Uppsala University Fennicae 29

Ložek, V. (1965) "The relationship between the development of soils and faunes in the warm Quaternary phases", Anthropozoïkum/Prague 3, pp. 7-51

Morrison, R.B. (1965) "The Quaternary of the U.S. Quaternary geology of the Great Basin", Princeton, pp. 265-285

Morrison, R.B. (1967) "Principles of Quaternary soil stratigraphy in means of correlation of Quaternary successions", in R.B. Morrison and M.E. Wright (eds.), Int. Assoc. Quat. Res. (INQUA), VII Cong., 1965, Proc. 9, pp. 1-69

Paepe, R. (1963) *Bouw and oorsprong van de vlakte van de Leie*, PhD thesis, Gent

Paepe, R. (1964) "Les dépôts quaternaires de la plaine de la Lys", Bull. Soc. Belg. de Géol. LXIII, 3, pp. 1-39

Paepe, R. (1966) "Stratigraphy of the River Scheldt and stratigraphy of the Flemish Valley", R. Van Hoorne (ed.) *IInd International Conference on Palynology*, Guidebook, Utrecht, pp. 1-17

Paepe, R. (1969) "Quelques aspects des dépôts quaternaires de la Famenne", Bull. Soc. belge Géol., Paléont., Hydrol., T. 78, fasc. 1, pp. 69-75

Paepe, R. (1974) "Correlation of Middle Pleistocene deposits with the aid of palaeosoilss in Belgium", *Quaternary Glaciations in the Northern Hemisphere*, Report nr 1, IGCP session, Cologne 1973, pp. 69-77

280

Paepe, R. and Van Hoorne, R. (1967) *The stratigraphy and palaeobotany of the Late Pleistocene in Belgium*, Toelicht. Verhand. Geol. Kaart en Mijnkaart van België 8

Paepe, R. and Van Hoorne, R. (1976) *The Quaternary of Belgium in its relationship to the stratigraphical legend of the geological map*, Toelicht. Verhand. Geol. kaart en Mijnkaart van België 18

Paepe, R. and Van Hoorne, R. (1978) "Décernation du "Prix Baron van Ertborn" pour leur mémoire: The Quaternary of Belgium", Bull. Acad. Sc. Belgique, Gl. Sc., 64/12

Paepe, R., Baeteman, C., Mortier, R. and Van Hoorne, R. (1981a) "The marine Pleistocene sediments in the Flandrian area", Geologie en Mijnbouw 60, pp. 321-330

Paepe, R., Van Molle, M. and Mortier, R. (1981b) "Quaternary stratigraphy of terrace systems of the Maas river basin", Sonderveröff. Geol. Inst. Univ. Köln 41, pp. 131-153

Paepe, R. and Mariolakos, I. (1984) "Paleoclimatic reconstruction in Belgium and in Greece based on Quaternary lithostratigraphic sequences", Proceedings of the E.C. Climatology Programme Symposium, Sophia Antipolis, France, 2-5 October 1984

Paulissen, E. (1973) "De morfologie en de Kwartairstratigrafie van de Maasvallei in Belgisch Limburg", Mémoires A.R.Sc.L.B.-A. de B. 35, 127p.

Shackleton, N.J. and Opdyke, N.D. (1973) "Oxygen Isotope and palaeomagnetic stratigraphy of Equatorial Pacific Core V28-238: Oxygen Isotope Temperatures and Ice Volumes on a 105 year and 106 year scale", Quaternary Research 3, pp. 39-55

Sommé, J. and Paepe, R. (1978) La formation d""Herzeele: un nouveau stratotype du Pléistocène Moyen marin de la mer du Nord", Bull. de l""Assoc. Franç. pour l""étude du Quat., pp. 81-149

Tavernier, R.(1943) "De kwartaire afzettingen in België", Natuurwetenschappelijk Tijdschrift 25, pp. 121-137

Van Hoorne, R, and Denys, L. (1987) "Further paleobotanical data on the Herzeele Formation (Northern France)", Bull. de l""Assoc. franç. pour l""étude du Quat., pp. 7-18

Woillard, G. and Mook, W.G. (1982) "Carbon-14 dates at grande Pile: Correlation of Land and Sea Chronologies, Science, 215: pp. 159-161

Zagwijn, W.H. (1961) "Vegetation, Climate and radiocarbon datings in the Late-Pleistocene of the Netherlands. Part I: Eemian and Early Weichselian", Meded. Geol. Stichting 14, pp. 15-45

Zagwijn, W.H. (1975) *Indeling van het Kwartair op grond van veranderingen in vegetatie en klimaat*, in Zagwijn, W.H. and Van Staalduinen, C.J., *Toelichtingen bij geologische overzichtskaarten van Nederland*, Rijks Geologische Dienst, pp. 109-114

Zagwijn, W.H. (1985) "An outline of the Quaternary stratigraphy of the Netherlands", Geologie en Mijnbouw 64, pp. 17-24

Zagwijn, W.H. and Paepe, R. (1968) "Die Stratigraphie der Weichselzeitlichen Ablagerungen der Niederlande und Belgiens", Eiszeitalter und Gegenwart 19, pp. 126-146

Zagwijn, W.H. and De Jong, J. (1984) "Die Interglaciale von Bavel und Leerdam und ihre stratigraphische Stellung in Niederländischen Früh-Pleistozän. Mededelingen Rijks Geologische Dienst, 37-3, pp. 155-169

NATURAL ARIDIFICATION OR MAN-MADE DESERTIFICATION? A QUESTION FOR THE FUTURE

N. PETIT-MAIRE
Laboratoire de Géologie du Quaternaire - CNRS -
Case 907 - Luminy - 13288 Marseille Cedex - France
IGCP # 252 (Past and Future Evolution of Deserts)

ABSTRACT: 1) In the last 150 ka, the tropical arid belt in North Africa (Sahara) and its southern margins (Sahel) successively underwent major changes in their latitudinal range:
  c 130 ka: extensive lakes throughout present Sahara
  c 18 ka: Saharan environments down to 14°N
  c 8-7 ka: Sahelian environments up to 22-23°N.
These continental processes coincide with sea level transgressive and regressive phases and the Milankovitch forcing.
2)   Since c 4.500 B.P., another expansion of the Sahara has taken place. The Sahel's northern limit, now at 17° N is still regressing southwards. These data demonstrate a trend towards an arid scenario and match the insolation curve since 10 ka ago and the future cold trend calculated by A. Berger (1981).
The natural climatic forcing therefore runs counter to an eventual greenhouse effect, as it would be once again associated with temperature and sea-level drops.
The evidence for climatic changes in the north african arid belt thus confirms the predictive model of Kukla (1980) and does not imply any responsibility of man, although deforestation and erosion locally emphasize the trend.
Changes in rainfall, increase in deforestation and erosion recorded both in the tropical belts and in the Mediterranean borders, the possible rise of oceanic vapour and precipitation due to a CO_2 greenhouse effect and the recent increase in strength and frequency of the dominant winds in the northern hemisphere are different factors which should be considered carefully and separately before one can imply any valuable theory on future global evolution. In the tropical arid belt enough field data are now available to allow a useful interpretation of environmental changes during the last global climate cycle and to define the part played by man in long trend variations and compare them with shorter term observations.
Successive extensions and shrinkings of the arid zone of North Africa (< 100-150 mm annual precipitation and 4-8 m evapotranspiration) are in evidence for the last 150 ka and are obviously related to marine isotopic trends (Hays and Imbrie, 1986) sea levels (Shackleton et al., 1981) and the astronomical Milankovitch insolation curve at 65°N (Berger, 1981, 1984).

R. Paepe et al. (eds.), Greenhouse Effect, Sea Level and Drought, 281–285.
© 1990 *Kluwer Academic Publishers.*

1.

During isotopic stage 5e, consequent upon an insolation maximum and related to the eemian temperature and sea-level rises (maximum at c 125 ka), large extensions of surface water existed in now hyperarid Sahara: in Libya, at 27°30'N, a 2 000 km^2 lake was Th/U dated at 173-90 ka, (Petit-Maire et al., 1980a and b; Gaven et al., 1981; Petit-Maire (ed.), 1982). The lacustrine conditions were shown to be due to aquifer rise and heavy run off from the nearby massifs. The summer and winter rainfall implied by the permanent presence of 50 billion m^3 of surface water (even if its salinity was between 3 and 10%) is totally inconsistent with the present annual mean of 30 mm. More recent observations have confirmed the existence of this lacustrine phase in the Sahara during the eemian interglacial: in Mali (Fabre, 1983; Petit-Maire, 1986, 1989), in Algeria (Fontes et al., 1987), in Tunisia (Ballais and ben Ouezdou, 1987) and in southern Egypt (Szabo et al., 1989).

By 90 ka ago, the Shafi libyan lake evaporated and only sporadically did small pools later exist on its north-eastern margin, fed by run off from the Harruj area. This pattern is consistent with astronomical hindcasts and sea level regression after the eemian optimum. No human action was involved in these hydrological changes.

To this day no reliably dated evidence has been shown for the Sahelo-Sahelian area for other than a progressive but very irregular climatic deterioration up to the last glacial maximum extension.

2.

During isotopic stage 2, the total disappearance of all surface water in the present Saharan belt coincides with the 130 m lowering of sea-level. Well orientated dune fields have been established at least down to 14°N, implying both strong, rapid, long acting winds and annual rainfall less than 100-150 mm (Talbot, 1984; Ousseini and Morel, 1989) in this presently semi-arid to semi-humid area.

Again, no human action is implicated in this vast extension of the desert in North Africa during the global cold phase which culminated at c 18 ka ago. This can be attributed only to profound changes in atmospheric circulation related to astronomical change.

3.

Very shortly after 17 ka ago rainfall patterns again varied and the Sahel developed northwards. Crater lakes in the Tibesti rose at c 15 ka (Faure, 1969). At 10,5 ka, most aquifers rose in all areas (Petit-Maire and Riser, 1980, 1983; Callot, 1984; Fontes et al., 1985) except for the very core of isolated Sahara basins where, in the lowlands along the Tropic of Cancer, Taoudenni depression to the West and Gilf Kebir to the East (Petit-Maire, 1986; Petit-Maire et al., 1987; Fabre and Petit-Maire, 1988; Kröpelin, 1987; Pachur and Kröpelin, 1987), the climatic optimum only began 9 ka ago. It deteriorated quickly. By 6 ka ago, strong droughts were recurrent in northern Mali, although the Sahelian biotope persisted for nearly a millennium. At 5-4.5 ka ago, the environmental saharisation was over at those latitudes and progressively stretched southwards to its present limit, 17°N (Petit-Maire, 1989).

Although neolithic groups lived in the whole Sahara, around the swamps and the lakes, between 7 and 4.5 ka ago, they were few in number, did not raise cattle and could by no means be blamed for the deterioration of the biotopes. In contrast, this fits the decrease in the insolation curve and could be related to a renewed natural profound change in atmospheric circulation. The intensification of winds, both in strength and frequency, and the subsequent reduction of rainfall over tropical Africa correspond to the onset of a rapid circulation scenario (Leroux, 1987), similar to that prevalent during the last glaciation with related extension of the margins of the Sahara and environmental degradation typifying the end of an interglacial (Kukla, 1980).

4.

Astronomical insolation trends are thus closely related to atmospheric patterns over tropical Africa. The present trend is towards a continuation of the insolation decrease probably heading towards a new glaciation (Berger, 1981, 1984) as witnessed by the previous cycle with an expected impact on extension of the cryosphere, regression of sea level and extension of the Sahara over the present Sahel.

This natural tendency runs counter to an eventual greenhouse effect. One cannot predict at the moment being whether this will reverse the trend.

5. References

Ballais, J.L. & Ben Ouezdou, H. (1987) "Formes et dépots du Quaternaire continental de la bordure présaharienne du Maghreb oriental. Essai de synthèse provisoire", Conférence P.I.C.G. 210, pp. 18-19

Berger, A. (1981) "The astronomical theory of palaeoclimates", in A. Berger (ed.) *Climatic variations and variability: facts and theories*, Dordrecht, Reidel, pp. 501-525

Berger, A. (1984) "Accuracy and stability of the Quaternary terrestrial insolation", in *Milankovitch and Climate*, 1, pp. 83-111

Callot, Y. (1985) "Dépôts lacustres et palustres quaternaires de la bordure nord du grand Erg Occidental (Algérie).", Comptes-Rendus de l'Académie des Sciences, Paris (D), 299, pp. 1347-1350

Fabre, J. (1983) "Esquisse sratigraphique préliminaire des dépôts lacustres quaternaires", in N. Petit-Maire and J. Riser, (eds.) *Sahara ou Sahel? Quaternaire récent du Bassin de Taoudenni*, pp. 421-441

Fabre, J. & Petit-Maire, N. (1988) "Holocene climatic evolution at 22°-23°N from two palaeolakes in the Taoudenni area (northern Mali)", Paleaogeography, Paleaoclimatology, Palaeoecology, 65, pp. 133-148

Faure, H. (1969) "Lacs quaternaires du Sahara", Mitt. intern. Verein. Limnol., 17, pp. 131-146

Fontes, J.C., Gasse, F., Callot, Y., Plaziat, J.C., Carbonel, P., Dupeuble, P. & Kaczmarska, I. (1985) "Freshwater to marine-like environments from Holocene lakes", in northern Sahara, Nature, 317, pp. 608-610

Gasse, F., Fontes, J.C., Plaziat, J.C., Carbonel, P., Kaczmarska, I., De Deckker, P., Soulié-Marsche, I., Callot, Y. & Dupeuble, P.A. (1987) "Biological remains, geochemistry and stable isotopes for the reconstruction of environmental and hydrological changes in the Holocene lakes form North Sahara", Palaeogeography, Palaeoclimatology, Palaeoecology, 60, pp. 1-46

Gaven, C., Hillaire-Marcel, C. & Petit-Maire, N. (1981) "A Pleistocene lacustrine episode in southeastern Libya", Nature, 290, pp. 131-133

Hays, J.D., Imbrie, J. & Shackleton, N.J. (1976) "Variations in the Earth's orbit: pacemaker of the Ice Ages", Science, 194, pp. 1121-1132

Kröpelin, S. (1987) "Palaeoclimatic evidence from early to mid-Holocene playas in the Gilf Kebor (southwest Egypt)", in Coetzee, J.A. (ed.) Palaeoecology of Africa, vol. 18, Balkema, Rotterdam, pp. 189-208

Kukla, G. (1980) "End of the last Interglacial: a predictive model for the future?", Palaeoecology of Africa, 12, pp. 395-408

Leroux, M. (1987) "L'Anticyclone mobile polaire, relais des échanges méridiens: son importance climatique", Géodynamique, vol. 2, ORSTOM, pp. 161-167

Ousseini, I. & Morel, A. (1989) "Utilisation de formations alluviales azoïques pour l'étude des paléoenvironnemnets du Pléistocène supérieur et de l'Holocène au Sud du Sahara: l'exemple de la vallée du fleuve Niger dans le Liptako nigérien", Bull. Soc. géol. France, 8, n° 1, pp. 85-90

Pachur, H.J. & Kröpelin, S. (1987) "Wadi Howar: palaeoclimatic evidence from an extinct river system in the south eastern Sahara", Science, 237, pp. 298-300

Petit-Maire, N. (ed.) (1982) Le Shati, lac pléistocène du Fezzan (Lybie), Paris/Marseille, CNRS, 118 p.

Petit-Maire, N. (1986) "Palaeoclimates in the Sahara of Mali: a multidisciplinary study", Episodes, 9/1, pp. 7-16

Petit-Maire, N., Casa, L., Delibrias, G. & Gaven, C. with an appendix by Testud, A.M. (1980a) "Preliminary data on quaternary palaeolacustrine deposits in the Wadi ash Shati, Libya", in Salem, M.J., and Busrewil, M.T. (eds.) The geology of Libya, 3, London, Academic Press, pp. 797-807

Petit-Maire, N., Fabre, J., Carbonel, P., Schulz, E. & Aucour, A.M. (1987) "La dépression de Taoudenni (Sahara malien) à l'Holocène", Géodynamique, 2/2, pp. 61-67

Petit-Maire, N. (1989) "Interglacial environments in presently hyperarid Sahara: Palaeoclimatic implications", in Proceedings, NATO Advanced Research Workshop on Palaeoclimatology and Palaeometeorology, Kluwer Academic Publishers (in press)

Petit-Maire, N., Delibrias, G. & Gaven, C. (1980) "Pleistocene lakes in the Shati area, Fezzan (27°30'N)", Palaeoecology of Africa, 12, pp. 289-295

Petit-Maire, N. & Riser, J. (1981) "Holocene lake deposits and palaeoenvironment in central

Sahara, northeastern Mali", Palaeogeography, Palaeoclimatology, Palaeoecology, 35, pp. 45-61

Petit-Maire, N. & Riser, J. (ed.) (1983) *Sahara ou Sahel? Quaternaire récent du bassin de Taoudenni (Mali)*, Marseille, 473 p.

Shackleton, N.J. et al. (1982) "The Deep-Sea Sediment Record of Climate Variability", Prog. Oceanogr., 11, pp. 199-218

Szabo, B.J., McHugh, W.P., Schaber, G.G., Haynes Jr, C.V. & Breed, C.S. (1989) "Uranium-Series Dated Authigenic Carbonates and Acheulian Sites in Southern Egypt", Science, 243, pp. 1053-1056

Talbot, M. (1984) "Late pleistocene rainfall and dune building in the Sahel", in Coetzee, J.A., and Van Zinderen Bakker, A.M. (eds.) *Palaeoecology of Africa*, 16, Balkema, Rotterdam, pp. 203-214

CLIMATIC OSCILLATIONS AS REGISTERED THROUGH THE RUZIZI PLAIN DEPOSITS (NORTH LAKE TANGANYIKA) ZAIRE - BURUNDI - RWANDA

LUTUMBA ILUNGA
ISP/Bukavu
BP 854 Bukavu
Zaire

ROLAND PAEPE
Vrije Universiteit Brussel (VUB)
Earth Technology Institute (ETI)
& Belgian Geological Survey
Pleinlaan 2
1050 Brussel

ABSTRACT: The Ruzizi Plain deposits show evidence of alternating savannah and semi-arid climates deduced from geomorphology (calcretes - ferricretes - stone-line), sedimentology (meandering river - braided river -sheetflood deposits), and pedology (Gley - Pseudogley - Alfisol - Calcrete palaeosoils).

1. Introduction

Climatic oscillations are important markers of the Quaternary. They are commonly studied thanks to the sensitivity of the micro- and macrofossils in them. The aim of the present work is to show how these oscillations have been registered through the generally non fossiliferous Quaternary deposits of the Ruzizi Plain which borders Lake Tanganyika in the North. Geomorphologic (calcretes - ferricretes -stone-line), sedimentologic (meandering river - braided river - sheetflood deposits) , and pedologic (Gley - Pseudogley - Alfisol - Calcrete palaeosoils) features are used.

2. Lithostratigraphy of the Ruzizi Plain Deposits

The first lithostratigraphic study of the Quaternary deposits of the Ruzizi Plain (Fig. 1 & 2) is that by Direction de Géologie (1974) done with the collaboration of J. Lepersonne and some geologists of the "Bureau de Recherche Géologiques et Minières" (BRGM) of France. This study gives a general lithostratigraphic (informal) and relative chronostratigraphic

R. Paepe et al. (eds.), Greenhouse Effect, Sea Level and Drought, 287–299.

288

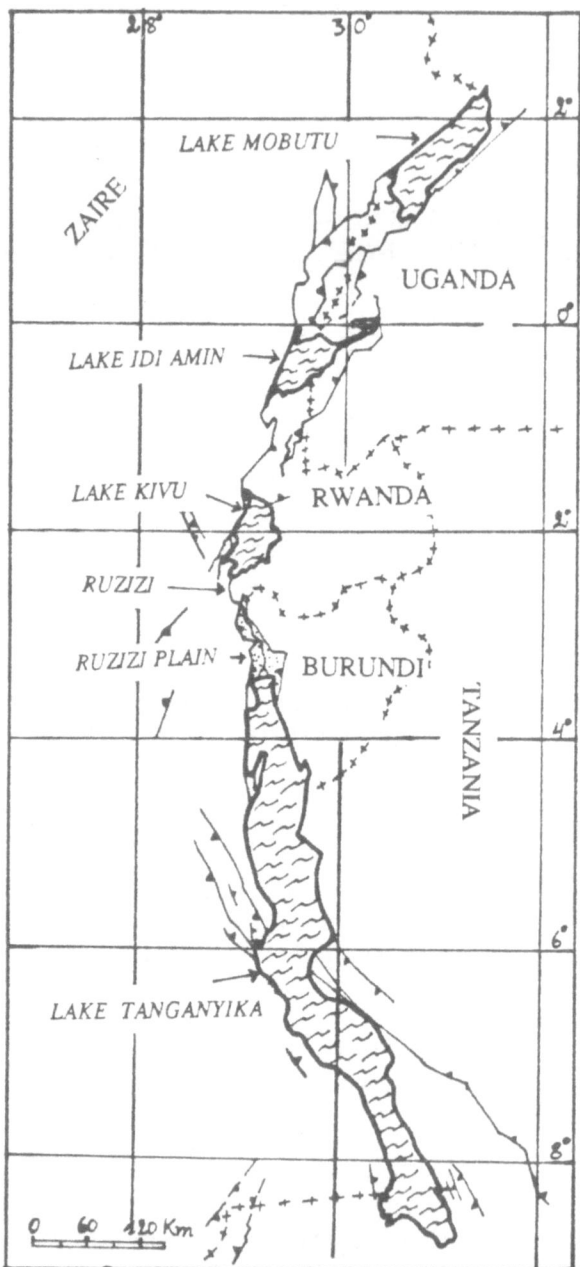

Figure 1. Ruzizi Plain in the Western Rift Valley.

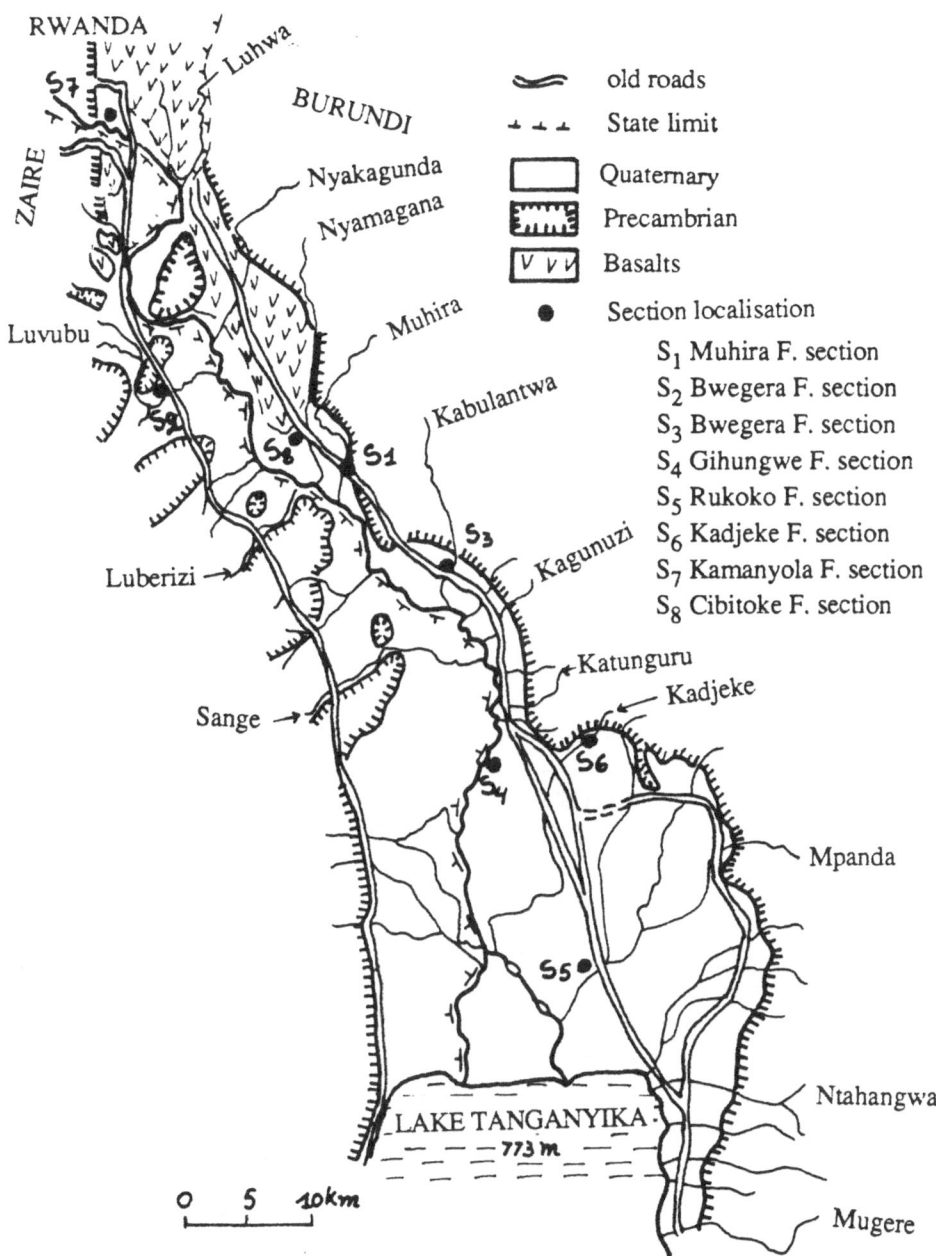

Figure 2. Ruzizi Plain Sections Map.

framework.

The second and more detailed study is that by ILUNGA and PAEPE (Geobound, in press) who subdivided these deposits into ten formal lithostratigraphic units (Table 1, Fig. 3 & 4) which are hereafter described and interpreted from a palaeoclimatic point of view.

Table 1. Lithostratigraphy of the Ruzizi plain deposits.

PROBABLE AGE	LITHOSTRATIGRAPHY		
HOLOCENE	KADJEKE F.		
	KAMANYOLA F.		RUKOKO F.
UPPER	NAOMBE F.		
PLEISTOCENE	GIHUNGWE F.	TSHAMATE GROUP	
LOWER TO MIDDLE PLEISTOCENE	LUVUNGI F.		
	BWEGERA F.		
LOWER	CIBITOKE F.		
PLEISTOCENE	MUHIRA F.		

2.1. THE MUHIRA FORMATION (LOWER PLEISTOCENE)

The Muhira Formation (Lower Pleistocene) consists of four faulted units which are from bottom to top:
- The lower unit shows an upward coarsening sequence of fine silt to sand suggesting a deltaic environment. It is a yellowish, thinly laminated deposit of about 10 m. thick displaying to the top a large scale tabular cross-lamination with contorted laminae, slump structure and reactivation surfaces. The latter feature represents minor erosion surfaces

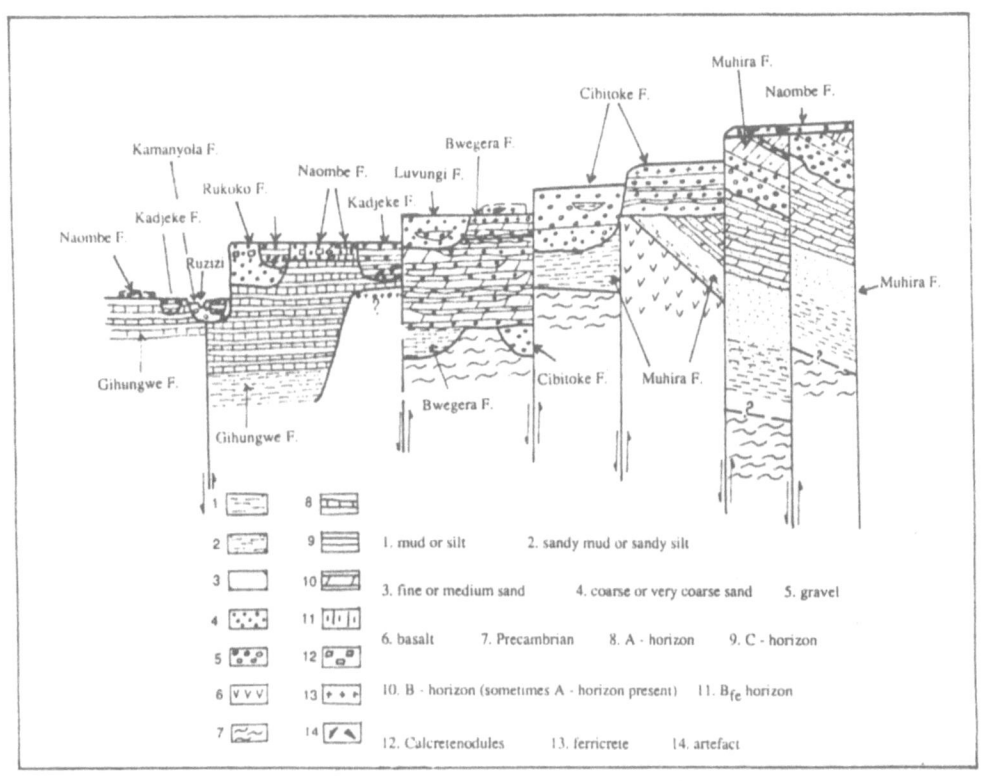

Figure 3. Ruzizi Plain Synthetic Lithostratigraphic Section.

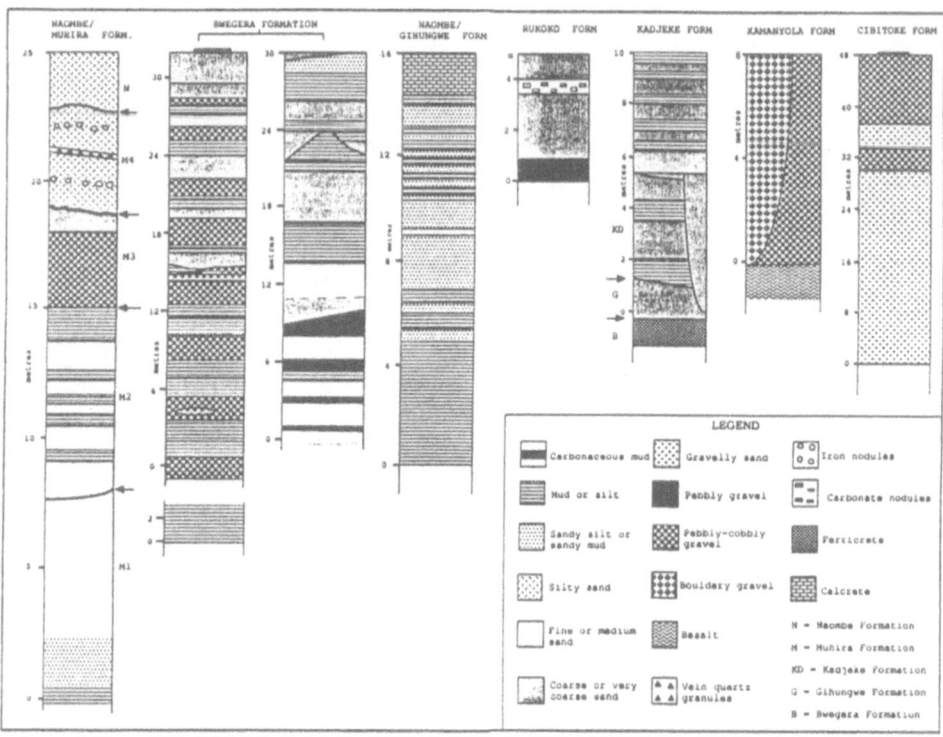

Figure 4. Ruzizi Plain Lithologic Sections.

on bedforms that were abandoned by a decrease in flow strength and then reactivated some time later (Collinson in Miall, 1985)
- The second unit shows cycles of yellow-brown fine sands alternating with grey-greenish muds in upward fining sequences suggesting a deltaic plain environment characterized by periodic flooding. These beds, especially the muds, are completely destratified, gleyified, and present burrow channels filled with red sandy material suggesting a humid climate (Duchaufour, 1977), probably strongly seasonal (Freyet, 1971).
- The third unit is made of gravel-sand facies probably of Allen's low sinuosity river type (Allen, 1965).
- The fourth unit is fluvio-colluvial made of two intensively bioturbated silty sand palaeosoils with iron nodules showing a predominance of illite 30% and smectite 25% over kaolinite 25% (Alfisol?) and separated by a stone-line of vein quartz granules mixed with lateritic concretions. The absence of the quartzite granules which constitute more than 90% of the Ruzizi plain gravel lithology and the presence only of vein quartz granules (residual) and iron nodules point to a colluvial process under semi-arid conditions on the Precambrean scarps. This semi-arid climate was preceded and followed by humid climates of the savannah type as suggested by the presence of iron nodules (Johnson, 1977).

2.2. THE CIBITOKE FORMATION (LOWER PLEISTOCENE)

The Cibitoke Formation consists of thick red or yellowish gravel deposits, with some intercalations of sand or mud layers of generally only a few dm. thick, capped by a ferricrete. Iron cementation of these gravel and finer beds can be seen from place to place where these gravels and mud facies become orthoconglomerates and mudstone lenses respectively.

These deposits present generally a G_m facies (massive or crudely stratified facies of MIALL, 1978), but sometimes a G_p (planar cross-stratified gravel) and a G_h (horizontal stratified gravel) facies. In some places a multistorey channel pattern is exposed displaying rapid alternation of lenticular S_p (planar cross stratified sands) and G_m facies capped by a S_h (horizontal stratified sands) facies extending out of the channel.

The river behaviour seems to have been braided. Indeed, the alternation of very coarse and fine deposits suggest fluctuating currents passing from the upper flow regime to the lower. The predominance of coarse clasts, the lenticular character of fine sediments as well as the presence of multistorey channels suggest a braided river environment (Costello & Walker, 1972). Such a behaviour contrasts with overbank flooding characterised by the predominance of fine materials and points to a dry phase probably of the semi-arid climate. This was probably followed by a savanna climate as suggested by the ferricrete cap well known to form under a savanna climate (Tricart, 1974).

2.3. THE TSHAMATE GROUP (LOWER TO UPPER PLEISTOCENE)

The name "Group" is used here for reasons of simplicity to put together all the sediments which represent some common features such as:
- younger than the Cibitoke Formation;

- a predominance of finer deposits (muds & sands);
- the greenish grey colour though locally some yellowish, dark and light colours are observed;
- the less exposed shale facies at the base of profiles;
- and the generally distributed carbonate concretions.

This group is composed, from bottom to top, of the Bwegera, Luvungi, Gihungwe and Naombe Formations.

2.4. THE BWEGERA FORMATION (LOWER TO MIDDLE PLEISTOCENE)

The Bwegera Formation consists of upward fining cyclothems capped by a ferricrete. These cyclothems are made of thickly to very thickly bedded white to grey quartzite pebbly gravel (sometimes absent) - coarse to fine sand - greenish grey muds intensively bioturbated and displaying root channels, prismatic structures, carbonate root casts, ferruginous and carbonate material aggregations and sometimes slickensides. The most striking features of these palaeosoils are however the gley and pseudogley mottling appearing separately or sometimes together, the pseudogley being overprinted by the gley and the coexistence of lepidocrocite with the carbonate nodules.

The sedimentary environment is characterized by the high proportion of fines, suggesting periodic fluviatile overbank flooding. The overprinting of the pseudogley by the gley mottling suggests first of all a fluctuating water-table at the sedimentation time, followed by a permanent water-table subsequent to the continuous soil burial (Wright, 1989). The presence of burrows, root channels and well developed aggregations of ferruginous materials and scattered carbonate nodules reflect a tropical, strongly seasonal climate (FREYTET, 1971).

2.5. THE LUVUNGI FORMATION (MIDDLE PLEISTOCENE)

The Luvungi Formation consists of a terrace deposit of fine to medium sand facies of white-grey, sometimes yellowish colour, displaying lenses of pebbly gravel facies and suggesting Allen's low sinuosity river type.

2.6. THE GIHUNGWE FORMATION (UPPER PLEISTOCENE)

The Gihungwe Formation is constituted of laminated silts (shale) at the base, followed by alternating dark and light medium to thickly bedded sandy silts and silts containing scattered carbonate nodules and locally displaying fine sand thin beds. The light beds are grey-green sands, sandy silts or silts with some iron oxidation spots while the dark beds are black and carbonaceous silts with root traces, and gley mottling.

The Gihungwe Formation probably took place in a shallow lake (euplanctonic diatoms: *Melosira ambigua*) which evolved into an interdistributary bay which fluctuated a lot as evidenced by the mud zonation of grey and grey-black. The presence of root traces and gley mottling means that mud beds have been colonized by vegetation under a humid, probably savanna climate.

2.7. THE NAOMBE FORMATION (LATE UPPER PLEISTOCENE)

The Naombe Formation consists of red sand deposits displaying a pebble-cobble gravel facies in the piedmont zone. Its thickness shows a good amount of variation. From a little more than 6 m to less than 1 m, the thickness diminishes from scarps to the center of the plain where it subsists now as patches because of subsequent erosion. The depositional process seems to have been streamflood suggesting a semi-arid climate which is here confirmed by the presence of calcrete nodules. Indeed, carbonate concretions, some several d_m thick can be observed in several profiles.

2.8. THE RUKOKO FORMATION (LATE UPPER PLEISTOCENE TO HOLOCENE)

The Rukoko Formation consists of white-grey medium to very course sand facies with interstratified brownish undulating podzolic very thin beds (clay & humus). These very thin beds are decimetrically spaced, subparalel but locally cross-stratified. Towards the base, the facies become even courser of granule. At about 0.5 m from the top a pedological carbonate concretion bed (B_{ca} horizon) can be seen which suggests a semi-arid climate. The Rukoko Formation probably took place in a high energy beach environment.

2.9. THE KAMANYOLA FORMATION (HOLOCENE)

The Kamanyola Formation is the deposit of the outflow of Lake Kivu to Lake Tanganyika through the Ruzizi River. It is composed of two members. The lower is made of pebbly to cobbly gravel interstratified with lenticular small bouldery gravel. The upper is a very thick massive bouldery gravel which distally diminishes to become sandy towards Lake Tanganyika. It seems to be the consequence of the Lake Kivu overflow under a humid climate at the beginning of the Holocene (Hecky, 1978).

2.10. THE KADJEKE FORMATION (HOLOCENE)

The Kadjeke Formation consists of a series of thinly to medium bedded interstratified whitish-grey to reddish-brown sands or pebbly sands and dark grey sandy silts. These look to have been formed in a humid alluvial fan by streamflood processes as suggested by the sheet-like form of individual units. However, the presence of a few sharp cutbanks are evidence that channelization took place but was not predominant. The occurrence of rootlets, which later on have been oxidised, suggests that the stream was sluggish or the channel dry for long time, allowing extensive plant colonization and thus, the just beginning of the soil pedogenesis as confirmed by the light pedoturbation of mud units. From a climatic point of view, we notice that the association of braided sheet sands with abundant plant debris, especially in channels, the generalised fine texture of the deposits, their black colour, and the absence of caliche nodules are hard to explain without invoking a wetter climate. The presence of pottery clasts in channels dates this climatic wet phase of Holocene, probably that of about 8,000-6,000 y BP (Roche et al, 1988).

3. Climatic Evidence

As described, the Ruzizi plain deposits show many features of climatic meanings. These have been subdivided into geomorphologic, sedimentologic and pedologic features.

With its annual means of about 975 mm/year of rainfall, a temperature of 24°C and a dry season of six to seven months, the present Ruzizi plain climate is of a dry savanna type. As the most humid palaeoclimate ever encountered in the deposits is of the savanna type, we can assume that we are dealing now with a relatively humid phase. It is probably less humid than the one which allowed the formation of the allochtonous ferricretes such as the ones covering the Cibitoke and Bwegera Formations. So the present climate can be considered as an intermediate one, however situated on the humid side. The dry side should be a semi-arid climate characterized by the excess of evapotranspiration over precipitation (Tucker, 1982) as suggested by calcrete nodules (Naombe Formation).

From a sedimentological point of view, observations of the present day behaviour of the Ruzizi plain rivers lead to the conclusion that they behave differently as a function of their distance from scarps. Generally of the straight type Ruzizi plain rivers, however, show a braided tendency near the scarps and a meandering one distally. Their general sedimentation is coarse with little overbank flooding. Such deposits correspond well with Allen's low sinuosity river model (Allen, 1965). The colour of the sediments is grey suggesting that the present climate produces enough humic acids to remove all the iron oxides. The analysis of the present behaviour leads to the conclusion that the development of the braided type, especially with red or yellow iron oxides colours may be interpreted as due to a drier climatic phase. This is the case for the Cibitoke Formation and for the proximal facies of the Naombe Formation. On the other side, the development of the meandering type favouring extensive overbank deposits with lenticular channel gravel and/or sand suggests a more humid climatic phase than the present one. This situation is encountered in the Muhira and Bwegera Formations. The sheet-like facies of the distal Naombe facies and of the Kadjeke Formation suggests a streamflood deposition which may occur under humid or dry phases. Here the colours of the sediments and the presence of plant debris can help to differentiate them.

Palaeosoils are also important tools for palaeoclimate recognition. In the Ruzizi plain, palaeosoils are more developed in floodplain muds which envelop subordinate to less ordinate gravel and/or sand. Five formations especially concerned are: the Muhira, Bwegera, Gihungwe, Naombe and Kadjeke Formations. In general, these deposits all show evidence of pedogenesis. In fact, pedoturbation is their common feature and is more pronounced in the overbank, than in the channel deposits. Many of the pedogenic features can also be observed. These are: the presence of organically enriched horizons (A-horizons) showing darkening by organic staining, mottling (gley, pseudogley and sometimes gley overprinted on pseudogley), and some biogenic features such as root traces, carbonate root casts and concretions, and iron rhizocretions. Prismatic structures with allochtonous materials from upward suggest expansion and contraction associated with wetting and drying. Aggregations of ferruginous materials as well as iron nodules can also be observed (Muhira and Bwegera Formations) A B_{ca}-horizon has been noted in the Naombe and Rukoko Formations (Late

Upper Pleistocene).

These features all have some climatic meaning:

- the presence of burrows, root channels and well developed aggregations of ferruginous materials and scattered carbonate concretions reflect a tropical, strongly seasonal climate (Freyet, 1971);
- the soils showing evidence for the illuviation of iron as well as carbonate compounds and perhaps clays suggest according to Johnson (1977) a savanna-woodland under the influence of a tropical seasonal climate;
- the caliche nodules (B_{ca}), where they are alone, are typical of semi-arid climatic areas where evapotranspiration exceeds precipitation (Tucker, 1982);
- a fluctuating water-table will result in pseudogley features developing in a buried soil, later overprinted by gley features as with continued burial (Wright, 1989). This hydromorphic situation suggests according to Duchaufour (1977) a humid climate.
- the presence of iron rhizocretions implies the presence of soluble iron in high concentration in the soil and thus of an acidic medium. The climate should be strongly seasonal and probably the savanna type. The carbonate root casts and segregations can be explained by the drying of the soil (evapotranspiration) leading to higher ion concentration and precipitation.

4. Cyclicities

Four types of cyclicities have been observed in the Ruzizi plain deposits. These are: the fluviatile, the fluvio-colluvial, the deltaic plain interdistributary bay and the alluvial fan or small river cyclicities.

The fluviatile cyclicity is encountered in the Muhira and Bwegera Formations. It is made of upward fining sequences of sand - mud for the Muhira Formation and of gravel (sometimes absent) - sand - mud for the Bwegera Formation. Gravel and sand beds are lenticular and represent channel deposits while mud beds are overbank deposits which have been totally transformed into soils so that C-horizons (undisturbed sediments) are no more observed. Only B-horizons, sometimes with A-horizons where they are not eroded can be observed.

Referring to the cyclicity cause, subsidence could have played a certain role as well as the to and fro lateral migration of river channels across their floodplains. The tectonic role should probably be reduced as the effects of its pulses are not evident. Indeed, Miall (1982) discusses the tectonic effects in generating large scale coarsening and fining upward cycles which have not been observed, the cyclothem textures being more or less the same throughout the sections.

However, in some of these palaeosoils we have noticed the coexistence of lepidocrocite and caliche nodules which normally should not form together. Indeed, Schwertmann and Taylor (1977) noted the absence of lepidocrocite in calcareous soils, although hydromorphic, which indicates that the introduction of CO_2 into the system favours formation of goethite rather than lepidocrocite. Such a situation probably implies that the Savanna climate responsible for the iron precipitation prevailed before the semi-arid one. If this is correct such palaeosoils can be considered to bear the marks of climatic oscillations and thus suggest a climatic cycle (savanna/semi-arid) respectively for each of the palaeosoils found at the Muhira and Bwegera type-sections.

The fluvio-colluvial cyclicity is encountered in the upper part of the Muhira Formation. It is made of two silty sand palaeosoils exhibiting iron nodules and separated by a stone-line of vein quartz and iron nodules (lateritic granules). The presence of iron nodules (Bfe-horizons) suggests a savanna climate while the stone-line pavement implies a semi-arid one (sheetflood deposit). So the whole sequence shows a climatic oscillation of the savanna/semi-arid type.

The deltaic plain interdistributary bay cyclicity is that of the upper part of the Gihungwe Formation which took place in an interdistributary bay which fluctuated a lot as evidenced by the mud zonation of grey and grey-black with the latter colour being due to organic matter content.

The presence of root traces and of gleyification means that these muds have been colonized by vegetation and evoluated at least as hydromorphic soils with A-horizons well marked covering C-horizons (undisturbed sediments). Thus, the deltaic plain interdistributary bay cyclicity is made of cyclothems of the A-C horizons soil type.

The alluvial fan or small river cyclicity is encountered along small rivers such as Katunguru and in the alluvial fan environment. Its deposits correspond to the Kadjeke Formation which is characterized by the sheet like form of individual units of sand and/or granule texture alternating with mud beds containing plant rootlets which have been later on oxidised in a red colour. However, because of the very light soil development on these muds, we consider them as C-soils which are eight time repeated at the type section.

5. Conclusion

The Ruzizi plain deposits show several indications of wet and dry climatic phases respectively corresponding to savanna and semi-arid climates. These have been deduced from geomorphologic (calcretes-ferricretes-stone-lines), sedimentologic (meandering river - braided river - sheetflood deposits) and pedologic (Gley - Pseudogley - Alfisol - Calcrete palaeosoils) features.

Four main types of cyclicities have been studied: the fluviatile for the second unit of the Muhira Formation and for the Bwegera Formation, the fluvio-colluvial for the fourth unit of the Muhira Formation, the deltaic plain interdistributary bay type for the Gihungwe Formation and the alluvial fan or small river type for the Kadjeke Formation.

The soil development in muds seems to be very different throughout the lithostratigraphy. From well developed soils with A and B-horizons or only B-horizons where erosion is important for the Lower and Middle Pleistocene deposits, we pass to the A-C horizon soil type for the Upper Pleistocene deposits (Gihungwe Formation) and we end with the C-horizon soil type for the Holocene (Kadjeke Formation). Such a different soil development can be related to pedogenic time duration and can lead one to think of a certain periodicity becoming shorter with time during the Quaternary.

6. References

Allen, J.R.L. (1965) "A review of the origin and characteristics of recent alluvial sediments.", Sedimentology 5, pp. 89-191.

Costello, W.R. & Walker, R.C. (1972) "Pleistocene sedimentology; Credit River, Southern Ontario: A new component of the braided river model.", J. sediment. Petrol., 42, pp. 389-400

Direction de Geologie (1974) *Notice explicative de la carte géologique du Zaïre au 1:2.000.000ème*

Duchaufour, P. (1977) *Pédologie, pédogenèse et classification*, Masson, Paris

Freytet, P. (1971) "Paléosols résiduels et paléosols alluviaux hydromorphes associés aux dépôts fluviatiles dans le Crétacé supérieur et l'Eocène basal de Languedoc.", Rév. Géogr. phys. Géol. dyn. 13, pp. 245-268

Hecky, R.E. (1978) "The Kivu-Tanganyika Basin: The last 14,000 years.", Pol. Arch. Hydrobiol., 25, pp. 159-165

Ilunga, L., & Paepe, R. (1990) *Lithostratotypes of the Ruzizi Plain deposits, North Lake Tanganyika, Zaïre-Burundi-Rwanda.*, Geobound (in press)

Johnson, G.D. (1977) "Palaeopedology of Ramapithecus-bearing sediments.", Geol. Rundsch. 66, pp. 192-216

Miall, A.D. (1978) "Lythofacies types and vertical profile models in braided river deposits: a summary.", in Miall, A.D. (ed.) *Fluvial sedimentology*, Can. Soc. Petrol. Geol. Mem., 5, pp. 597-604

Miall, A.D. (1982) "Analysis of fluvial depositional systems.", Ann. Assoc. Petrol. Geol., Tulsa

Miall, A.D. (1985) *Principles of sedimentary basin analysis.*, Springer Verlag, New York Inc.

Roche, E., Bikwemu, G., & Ntaganda, Ch. (1988) "Evolution du paléoenvironnement et quaternaire au Rwanda et au Burundi: Analyse des phénomènes morphotectoniques et des données sédimentologiques et palynologiques.", Inst. fr. Pondichéry, trav. Sec. Sci. tech., t. XXV, pp. 105-123

Schwertmann, V. & Taylor, R.M. (1977) "Iron oxides", in Dinauer, R.C. (ed.) *Minerals in soil environments*, Soil Sci. Soc. Am. Madison, Wisconsin USA, pp. 145-180

Street, F.A., & Grove, A.T. (1976) "Environmental and climatic implications of Late Quaternary lake-level fluctuations in Africa.", *Nature* 261, pp. 385-390

Tricart, J. (1974) *Le modelé des régions chaudes*, 2è. éd. Sedes, Paris

Tucker, M.E. (1982) *The field description of sedimentary rocks*, Open University Press, England

Wright, V.P. (1989) "Geomorphic and stratigraphic relationship of alluvial soils.", in Postgraduate Research Institute for Sedimentology (ed.) *Paleosols in silisiclastic sequences*, PRIS short course notes n° 001, University of Reading.

GLOBAL CHANGE: POPULATION, LAND-USE AND VEGETATION IN THE MEDITERRANEAN BASIN BY THE MID-21ST CENTURY

H.N. LE HOUEROU
Ph.D., D.Sc
CEFE & CNEARC
Montpellier

ABSTRACT: The likelihood of a temperature increase induced by the release of CO_2 and other warming gases in the atmosphere is discussed. A 3°C increase in the Mediterranean Basin by the mid-21st century is retained as a plausible assumption arising from a number of Global Circulation Models. The possible consequences of such a temperature rise on vegetation, crops and land-use are analysed. On the whole, in spit of some negative aspects, the consequences would seem rather beneficial to the North of the basin. However, they would be seriously detrimental to the South due to increased aridity. In the southern basin the consequences of a possible climatic change would however be trivial by comparison with the impact of the ongoing population growth. The present 285 million people in the 14 Afro-Asian Mediterranean countries considered in the study could reach, depending on different growth scenarios and projections, a low of 0.8 billion and a high of 1.95 billion by the year 2050. The impact of such a population explosion would be similar to that arising from geological considerations, further worsened by increasing aridity from rising evapotranspiration. The northern basin would thus undergo a phase of biostasis while the southern basin would be submitted to an acute phase of rhexistasis.

1. The Study Area and its Ecoclimatic Conditions

1.1. THE AREA

The Mediterranean Basin is taken to be as shown in Figure 1 and Tables 1 & 2; that is, countries or parts of countries having a Mediterranean type of climate neighbouring the Mediterranean Sea but not necessarily adjacent to it.

This area as a whole occupies a land surface of some 13 million km². We have arbitrarily discarded the Mediterranean parts of the USSR, Afghanistan, W. Pakistan and the true Mediterranean deserts with mean rainfalls below 50 mm per annum. The area under consideration is thus restricted to 7.4 million km².

The Mediterranean climate is characterized by rainfall occurring in the winter, with an

301

R. Paepe et al. (eds.), Greenhouse Effect, Sea Level and Drought, 301–367.
© 1990 *Kluwer Academic Publishers.*

intense and prolonged summer drought with high temperatures and potential evapotranspiration. The rainy season my last 9 to 10 months per annum in the moister parts but it may be less than one month in the Mediterranean deserts and sub-deserts of northern Africa and the Near East. Conversely the dry season may last an annual average of 2 to 12 months. As opposed to tropical or temperate-climate deserts, the scanty rains that fall always occur in the sinter in Mediterranean deserts.

The length and intensity of the dry season is a good indicator of moisture stress , which in turn is reversely related to the length of the growing season.

1.2. RAINY AND DRY SEASONS

The length and intensity of drought may be evaluated using two criteria, on a monthly basis for the sake of simplicity and clarity.

The rainy season is the period in which rainfall is above 1/3 of potential evapotranspiration (P > 0.35 PET). This discriminant threshold of 0.35 PET is not arbitrary since it has been experimentally shown to correspond, in most cases, to the direct evaporation from a bare, unsaturated ground surface (Doorenbos & Pruitt, 1975; Le Houérou & Popov, 1981). In other words, whatever amount of water is present in the soil above 0.35 PET is available for plant growth; This threshold is therefore a perfectly rational criterion for differentiation between rainy and dry seasons. The latter is thus the sum of periods where P < 0.35 PET. The intensity of the dry season is assessed by a very simple criterion: the ration between mean annual rainfall and mean annual evapotranspiration (P/PET) as in UNESCO's world map of arid zones (1979).

There are large areas with insufficient climatic data to make it possible to compute PET in a precise way, such as via Penman's equation (1948). In such cases, the empirical formula: P = 2t [1] of Bagnouls & Gaussen (1953), popularized by Walter & Lieth (1960), remains a good substitute since, as we have shown, the two threshold values P = 2t and P = 0.35 PET are closely correlated and exhibit quite similar absolute values in terms of mm of water in most climatic conditions (Le Houérou & Popov, 1981; Le Houérou, 1989)(Fig. 5). The rainy season is then appreciated as that period when P > 2t, and the dry season when P < 2t. The intensity of drought may also be assessed through the ration P/2t, where both factors are expressed in mm, either on a monthly or annual time scale.

1.3. THE COLD STRESS PERIOD

Cold stress may be indirectly recognized in various ways: mean annual temperature, mean temperature of the coldest month, mean daily minimum temperature of the coldest month (m), length of the frost occurring period, annual sums of temperature, periods with mean temperature below a given threshold (5, 10, 15°C), etc.

One of the most used, easy to handle and fairly precise criteria is the mean daily minimum temperature of the coldest month (January in the Mediterranean Basin). It has been shown

---

[1]. P = mean monthly rainfall, t = mean monthly temperature.

that this criterion is directly correlated with the length and intensity of winter frost and negatively correlated with the length of the potential growing season, as long as adequate moisture is available (Le Houérou, 1972, 1984; Le Houérou et al., 1975, 1979).

1.4. ECOCLIMATIC SYNTHESIS

Rainfall in the Mediterranean Basin may vary from 25 mm in the deserts to 2500 mm in moister conditions. Killing frost may occur from 0 to 200 days per annum.

The P/PET ratio may vary from 0.05 to 2.0 and "m" from -15 to +15°C. These new figures give an idea of the diversity of climatic conditions in the basin. Combining these criteria we reach the classification shown in Table 3.

Table 2 shows the various countries where these ecoclimatic zones occur.

The threshold values used in the classification were selected from a large number of research projects and studies in the basin over the past 50 years. They have been discussed in various scientific publications (Emberger, 1955; Le Houérou, 1959, 1969, 1074, 1982, 1986).

Using either the P/PET ratio or Emberger's pluviothermic index, since both are closely correlated, the mean daily minimum temperature for January, we have an ecoclimatic classification as shown in Table 2 and Figures 2, 3, 4, & 5. Figure 6 shows the ecoclimatic location of various cities in the basin in the present-day situation, whereas Figure 7 shows the shift in ecoclimatic zoning for a temperature rise of 3°C. In other words, sites move 3°C to the right in Figure 7 as compared to Figure 6.

In the 7.4 million km^2 of the study area:

$1.6 \times 10^6 = 22\%$ are semi-arid to hyper-humid (P > 400 mm),

$2.8 \times 10^6 = 37\%$ are arid steppe rangeland and cropland,

$3.0 \times 10^6 = 41\%$ are desert wasteland (50 > P > 25 mm).

Of the 2.8 million km^2 of arid steppe, around 50% is periodically cropped for subsistence cereal farming whenever autumn and early winter rains permit; however, those 1.4 million km^2 of cereals do not show up in official statistics, such as published by FAO, since only harvested crops are taken into consideration as crop failure is commonplace under these risky farming circumstances, where crop expectancy does not exceed 20 to 25%.

2. Vegetation: the Present Situation and Evolutionary Trends

2.1. GENERAL

Vegetation belts are laid along the latitudinal and altitudinal gradients, as both latitude and elevation are overriding discriminant factors influencing precipitation and temperature distribution patterns which in turn determine, to a large extent, vegetation and land-use.

Precipitation increases with latitude and altitude, whereas temperature follows the opposite pattern. Figure 8 shows the variation in elevation of some vegetation belts as a function of latitude and elevation. The increase of precipitation with altitude and latitude is a global trend. Many local factors interplay with this global trend and, modifying it, such as: exposure to rain bearing winds or rain shadows, the pattern of interface between sea and land masses in connection with dominant winds and frontal depression pathways.

However, roughly speaking, lowlands under the tropic of Cancer have a mean annual rainfall tending towards zero, whereas towards the 40-45° latitude N mean annual precipitation tends towards 800-1200 mm. That is a northward trend of some 40-60 mm for each degree of latitude or 0.35-0.55 mm per km (1° latitude = 110 km).

The relation between mean annual precipitation increase and elevation is more reliable: about 10% increase for each 100 m rise (Le Houérou, 1959; Le Houérou et al., 1975); in other words, precipitation tends to double for each km rise. Naturally, there are substantial departures from this overall average depending on aspect, topography, slope and other local features.

The temperature gradient is similar to that in other parts of the world, that is a decrease of 0.55° per 100 m rise. This lapse rate is also subject to local and seasonal variations, but much less so than the precipitation gradients: temperatures may be reliably predicted from elevation, provided some caution is exercised regarding site-specific conditions (temperature inversion, etc.).

The latitudinal gradient of temperature decreases northwards by about 0.60°C per degree of latitude, or 0.0055°C per km, at any given altitude.

2.2. NATURAL VEGETATION AND ITS DISTRIBUTION PATTERN

The present-day vegetation is a result of the complex interaction of a large number of different factors.

2.2.1. *Floristic factors*. The flora has been influenced by geological, palaeoclimatic and palaeogeographic history and biological evolution. The presently estimated number of flowering plants in our study area is about 15000 species. This flora is a mixture of recent palaeoartic and inherited palaeotropical elements (Quézel et al., 1980).

2.2.2. *Climatic factors*. The governing climate factors are:
- rainfall amount and distribution, drought stress,
- temperature and cold/frost stress,
- potential evapotranspiration and its relationship to precipitation (P/PET ratio).

2.2.3. *Edaphic factors*. Geology, petrography, physiography, topography, hydrology, erosion, soils, water and nutrient budgets.

2.2.4. *Anthropozoic activities*. The pressure of man and animals, particularly livestock, on ecosystems.

2.2.5. *Vegetation belts*. Broad climatic vegetation types may be correlated with climatic matrices using rainfall, temperature and potential evapotranspiration. Three such matrices are shown in Figures 2, 5, 6, 8-12: Bagnouls' & Gaussen's; Emberger's & Holdridge's.

These matrices have given birth to various classification systems: those of Bagnouls & Gaussen and of Holdridge are general whereas Emberger's is specifically designed for Mediterranean climates.

Other systems are based on elevation (Ozenda, 1976; Quézel, 1976; Gaussen, 1954; Tomaselli, 1976; Rivas-Martinez, 1982); but are only valid within narrow latitudinal limits as a function of latitude. For the Mediterranean Basin as a whole, one may consider the following very simplified outline in an order of decreasing aridity and temperature and of increasing elevation. However, this does not take into consideration the arid steppes or the Mediterranean deserts.

I - Xero-thermo-Mediterranean zone. This lowland, low latitude zone is predominantly semi-arid to sub-humid with mean annual precipitation ranging from 300 to 800 mm. The rainfall regime is Mediterranean but the temperature regime is tropical. Forest virtually never occurs and a number of tropical species are present: *Acacia Sp.Pl.*, *Prosopis*, *Hyphaene*, *Argania*, *Cactoid Euphorbiae Sp.Pl.*, *Phoenix*, *Maytenus*, *Nannorhops*, *Lycium Sp.Pl.*, *Periploca Sp.Pl.*, *Rhus Sp.Pl.*, *Juniperus phoenica* and others.

II - Thermo-Mediterranean zone. This is a mild winter zone with occasional light frosts in December/January. Rainfall may vary from less than 400 to over 1500 mm. Vegetation may be classified as from semi-arid to hyper-humid with dominant sclerophyllous shrubs and trees:

Ceratonia siliqua	*Olea europa var. oleaster*
Pistacia lentiscus	*Tetraclinis articulata*
Chamaerops humilis	*Periploca laevigata*
Rhus pentaphyllum	*Myrtus communis*
Euphorbia dendroides	*Quercus coccifera*
Juniperus oxycedrus	*Quercus calliprinos*
Juniperus phoenica	*Pinus pinaster mesogeensis*
etc.	*Pinus pinea.*

III - Meso-Mediterranean zone. This is perhaps the most typical Mediterranean zone, occurring over huge areas around the whole basin (Figs. 3 & 4). Winter temperatures are cool to cold ($-2 < m < 3°C$). Rainfall may vary from 350-400 to 1000-1500 mm: vegetation is semi-arid to hyper-humid. Dominant species are sclerophyllous or acicular, particularly four species of oaks and two pines and a large number of shrubs.

Quercus ilex s.l.	*(incl. Q. rotundifolia)*
Q. coccifera	*Q. suber*
Q. calliprinos	*Pinus halepensis*
	Pinus brutia

Among the numerous sclerophyllous shrubs:
Arbustus Sp.Pl., *Phyllirea Sp.Pl.*, *Erica Sp.Pl.*, *Rhamnus Sp.Pl.*, *Ulex Sp.Pl.*, *Pistacia Sp.Pl.*, *Genista Sp.Pl.*, *Rosmarinus Sp.Pl.*, *Styrax*, *Zelkhova*, etc.

IV - Supra-Mediterranean zone. Rainfall is rather high: 600-2500 mm and winter temperatures cool to cold. Natural vegetation is a deciduous and/or acicular forest.

Quercus faginea	*Quercus pubescens*
Q. afares	*Q. aegylops*

Q. trojana	*Ostrya carpinifolia*
Q. cerris	*Carpinus orientalis*
Q. toza	*Pinus brutia*
Q. frainetto	*Pinus pinaster mesogeensis.*

V - Alti-Mediterranean zone. Rainfall varies from 800-1500 mm, winter temperature is very cold with 3-4 months of hard frost and snow cover. Vegetation is predominantly coniferous and occasionally deciduous (chestnut, beech).

Pinus nigra s.l.	*Cedrus libani*
Pinus silvestris	*Cedrus brevifolia*
Fagus sylvatica	*Cedrus atlantica*
Fagus orientalis	*Abies Sp. Pl. (A.pinsapo, A.numidia,*
Castanea sativa	*A.salicica, A.maroccana, A.nebrodensis, A.cephalonica).*

VI - Oro-Mediterranean zone. Rainfall decreases in respect of the previous zone and winters are very cold. Vegetation is made up of scattered trees of Juniper.

Juniperus thurifera	in the western basin
Juniperus excelsa	in the eastern basin

Occasionally other Junipers may occur: *J. drupacea, J. sabina, J. communis.* Among scattered trees is a more or less continuous layer of custion-like spiny xerophytes:

Bupleurum spinosum	*Erinacea anthyllis*
Vella mairei	*Cytisus purgans*
Alyssum spinosum	*Astragalus Sp. Pl.*
etc.	

VII - Mediterranean-Alpine. Elevation varies from 1800 m to the North and over 3000 m to the South of the basin. There are over 6 months of hard frost. Vegetation is made of spiny pulvinate xerophytes with meadows and moors in depressions.

VIII - Cryo-Mediterranean. This is a high mountain zone with virtually no flowering plants above 2500 m to the North and 3500 m to the south.

IX - The arid zone vegetation is made up of various kinds of steppes:

a. **Grass steppes** dominated by perennial bunch grasses such as *Stipa tenacissima* in northern Africa and Spain.

b. **Dwarf shrub steppes** dominated by shrublets such as *Artemisia Sp. Pl., Hammada Sp. Pl.*

c. **Crassulescent steppes** of halophytic *Chenopodiaceae: Suaeda, Salsola, Atripex, Salicornia, Arthrocnemum, Halocnemon, etc.*

d. **Tall shrub steppes** located in favourable edaphic conditions: *Ziziphus, Rhus, Rotama, Calligonum, Nitraria, Acacia, Pistacia, Amygdalus, etc.*

3. Consequences of a Hypothetical Warming of the Atmosphere by 3.0 ± 1.5°C

3.1. DISCUSSION OF THE LIKELIHOOD OF A POSSIBLE TEMPERATURE RISE

The CO_2 content of the atmosphere increased from an estimated 280 ppmv (parts per million in volume) in the pre-industrial times of the mid 19th century. It rose to 314 ppmv when the Mauna Loa monitoring began in 1958 (Keeling et al., 1976). It has now reached 350 ppmv; it is expected to reach the 600 ppmv mark between the years 2030 and 2050 depending on whether the present exponential annual increase of 4% will continue or well be reduced to 2% (CNRC, 1982). Between 1958 and 1988, the increase has thus been 36 ppmv, which is equivalent to 68×10^9 tons of carbon. During the same period, about 126×10^9 tons of carbon have been emitted to the atmosphere by fossil-fuel combustion. It has been estimated that almost 200×10^9 of carbon (733×10^9 of CO_2) have been released to the atmosphere since the middle of the 19th century (adapted from NRC, 1979). Additional CO_2 has been released via deforestation; but this amount is not known with any accuracy to date.

Total and exchangeable carbon reservoirs are approximately as follows (WMO, 1977; US Dept. of Energy, 1979):

- Atmosphere	745×10^9 metric tons of C
- Total fossil-fuels & shales	12.0×10^{12} " " " "
- Recoverable fossil fuels & shales	7.5×10^{12} " " " "
- Terrestrial biosphere	2.8×10^{12} " " " "
of which:	
. Soil organic matter (including peat)	2.0×10^{12} " " " "
. Pants	0.8×10^{12} " " " "
- Oceans	40.0×10^{12} " " " "
of which:	
. Surface	0.6×10^{12} " " " "
. Intermediate & deep	39.4×10^{12} " " " "
- Carbonated sediments	2.0×10^{12} " " " "

N.B.: 1 t C = 3.66 t CO_2 = 2.5 t of Dry Organic Matter.

Fluxes are approximately (Slade, 1979):

- Fossil-fuel to atmosphere	5×10^9 m tons of C/year
- Biosphere-atmosphere exchange	56×10^9 " " "
- Ocean-atmosphere exchange	90×10^9 " " "
- Atmosphere-ocean sink	2.5×10^9 " " "

However, it is estimated that some 55% of the CO_2 emitted is sunk in the oceans and fixed in plants by photosynthesis (NRC, 1982).

Other radiatively active trace gases are being released to the atmosphere: chlorofluorocarbons ($CCl_2F_2 + CCl_3F$), nitrous oxide (N_2O), methane (CH_4), ozone (O_3). CO_2 has a thermal radiation similar to water vapour; it transmits a major fraction of incident sunlight but strongly absorbs a thermal radiation from the Earth. This radiative transmittance has been known for some 125 years as the "greenhouse effect" (Tyndall, 1863; Arrhenius, 1896; Chamberlain, 1899). A doubling of CO_2 in the atmosphere would amount to a global average increase of about $4.0 \pm$ W m^{-2}, all other atmospheric conditions remaining identical

to the present day situation (NRC, 1982). These 4 W m^{-2} could correspond to a global surface air temperature increase of about 1°C. But there are in addition:
- the combined effects of greenhouse trace gases whose doubling would add another 1°C.
- numerous feedback effects: increase of humidity of the atmosphere, change in cloud cover and optical properties, release of methane locked in the permafrost and decrease of global albedo from melting ice-caps.

A number of Global Circulation Model (GCM) scenarios (over 20) have been worked out over the past 15 years in order to try and determine temperature increase from the greenhouse effect (Manabe & Wetherald, 1975, 1980; Manabe & Stouffer, 1979; Hansen et al., 1978, 1980, 1981; Bach et al., 1985; Bach, 1986, 1987, 1988; Wigley, 1987; Petrivanov, 1987; Ramanathan et al., 1978; Gates et al., 1984). There are naturally some discrepancies between the above, but a tendency emerges towards an average increase of 3.0 ± 1.5°C for a doubling of CO_2 in Mediterranean latitudes (30-45° N) and 6-9°C in higher latitudes (Kellogg & Schware, 1981; Charney (CRB/NAS), 1979; WMO, 1981; NRC, 1979, 1982). We shall therefore work on the assumption of a 3.0°C increase by the year 2050.

This, however, calls for some caution.

* First, model scenarios are not predictions because, although they are based on the best available knowledge, there are major uncertainties regarding a number of parameters such as:
- the physics and chemistry of the atmosphere,
- the controversial role of aerosols,
- oceanic circulation modelling is still very defective,
- the magnitude of the role of the ocean as a CO_2 sink,
- the ocean-atmosphere interactions with thermal regulation,
- the role of change in cloudiness and its depth and optical properties,
- the evolution of CO_2 emission,
- there are only tests of model validation on specific aspects, not on the overall results,
- the magnitude of the role of forest clearance and of vegetation in general on the global CO_2 budget,
- the evolution of the release of greenhouse trace gases.

* Second: The temperature increase that should have already happened sine the beginning of the industrial times has not actually occurred so far. Global air surface temperature in the northern hemisphere increased by 0.5°C between the turn of the century and 1938; then it decreased by about the same till the 1970's. The opposite trends prevailed in the southern hemisphere (Van Loon & Williams, 1976; Kukla et al., 1977; Jones et al., 1982; Hare, 1985; Jäger, 1988).

* Third: Glaciers are expanding in the Alps and other mid-latitude high mountains.

* Fourth: The increase of nuclear fusion, expected between 2000 and 2020, might considerably reduce the burning of fossil fuel in the second quarter of the 21st century.

* Fifth: A number of leading scientists (climatologists and atmospheric physicists) hold the mainstream belief of a warming Earth; let us quote, for instance, a recent paper:

"In the global and annual mean, the net cloud effect is a cooling of the planet because of the albedo effect (54 W m^{-2}) dominates over the greenhouse effect (31 W m^{-2}) by 23 W m^{-2}" (Rockner et al., *Nature* 329, pp. 138-140, 1987).

* Sixth: An important but totally unpredictable cooling factor is the emission to the stratosphere of aerosols from volcanic explosions.

* Seventh: The effect of man-induced aerosols and dust from deserts and desertizing areas is a controversial issue: according to some physicists, they would increase rainfall and decrease temperature; according to others they would contribute to temperature rise.

* eighth: The increasing albedo in desertizing zones might compensate for its decrease in polar and sub-polar regions.

3.2. IMPLICATIONS FOR CLIMATIC VARIABLES

Presently available models do not provide reliable information on possible seasonal changes of temperature: for instance, whether the increase would be larger in winter or in summer, etc., which would have quite different effects on vegetation and crops. We shall therefor assume that temperature increase would be about uniformly spread over the whole year. It is suspected, but not definitely in evidence, that annual and seasonal temperature variability would somewhat increase. We shall therefore not assume any substantial change in variability.

GCMs do not provide any significant indication on change of rainfall in Mediterranean latitudes. Some indicate a slight upward trend, others a downward trend. Wigley (1987) and Petrivanov (1987) indicate a slight increase to the North of the basin and a weak decrease to the south. We shall therefore assume no significant change either in the overall amount of rainfall, seasonal pattern, or variability.

Currently available models do not provide any indication on Potential Evapotranspiration (PET). But, form the known relationship between temperature and PET (Griffith, 1972), one may fairly safely assume an increase of $3.0°C \times 58.93\ t_s{}^2 = 175$ mm/yr (Holdridge, 1947) to $3°C \times 68.57_s = 206$ mm/yr (Le Houérou & Popov, 1981). We shall therefore assume an annual increase of 200 mm in PET. But, when rainfed crops are concerned, PET increase through the growing season, assuming a potential 6 month growing period, would be 180 days which implies 98.6 mm in PET as the seasonal proportion. If the growing season is only 120 to 150 days, as for barley and wheat respectively, the increased water demand would be 65.7 and 82.5 mm respectively. Substantial yield decrease occurs when actual ET (ET_0) drops below 0.5 PET. The minimum additional water demand for barley and wheat would thus be 33 and 42 mm.

These amounts are quite significant, but of a magnitude that can be overcome by improved agricultural techniques (surficial tillage, tight weed control, crop rotation patterns, cultivar selection, more timely operations, better use of fertilizers and the like). In marginal water supply areas, however, i.e., between mean annual isohyets of 350 to 450 mm, crop yield expectancy could be significantly reduced and, naturally, still more so in subsistence farming zones with less than an average rainfall of 350 mm in northern Africa and the Near East.

---

[2]. t_s = Biotemperature = mean of total annual temperature above 0°C.

Increased water demand in permanently irrigated farming would be of the order or 160-180 mm (0.8 PET); that is, for the whole Mediterranean Basin, 23.5 million ha x 7000 m^2/ha/yr = 165 x 10^9 m^3/yr, which corresponds to a fictitious continuing discharge of 5 x 10^6 m^3 per second.

3.3. IMPLICATIONS FOR PLANT PHYSIOLOGY AND PRODUCTIVITY: "CARBON FERTILIZATION"

In areas where water is not a major limiting factor on plant productivity, i.e. in the sub-humid to hyper-humid ecoclimatic zones, "carbon fertilization" due to a higher concentration of CO_2 in the atmosphere would increase overall plant productivity as a result of enhanced net photosynthesis. However, it is difficult to foresee the overall increase in primary production, even if other factors such as water and nutrients are not limiting. About 10% would seem to be a conservative estimate, in the present state of the art. "Carbon fertilization" is known to enhance net photosynthesis, particularly in species having a C_3 carboxylation pathway (most Mediterranean species belong to the C_3 type). Both growth chamber and greenhouse experiments show a 30-65%, and sometimes much more, biomass production increase for a CO_2 increase on the atmosphere up to 600-700 ppmv or above (Lemon, 1983; Mortensen, 1983, Bolin, 1983; Houghton, 1986; Trabalka & Reichle, 1986; Crane, 1985; Shugart et al., 1985; Morison, 1985).

However, the rate of increase in the main useful (final) product, such as cereal grain, may be much less than in the overall biomass. The additional photosynthates are not evenly allocated to the various plant tissues. The factors governing this allocation are poorly understood. Moreover "carbon fertilization" is also known to increase Water Use Efficiency perhaps up to 30-50% (Morison, 1985), a fact that may compensate for the higher PET resulting from the temperature rise. The latter would furthermore accelerate soil organic matter oxidization, hence nutrient turn-over and therefore soil fertility.

3.4. IMPLICATIONS FOR NATIVE VEGETATION

A temperature increase of 3.0°C would, in principle, provoke an average upward shift of vegetation belts of 3°C x (100 m/0.55°C) = 545 m. In other words, each vegetation zone shown in Figure 3 would move about one step upwards. The Xerothermo Mediterranean zone would thus reach the 46° latitude N. Lowlands; the Thermo Mediterranean Eco-zone would reach 700 m of elevation at 40° N. and 1500 m at 30° N.; tree line would reach 2500 m at 46° N. and 3500 m at 30° N., etc.

Change in the botanical composition of plant communities and in their distribution pattern would obviously not occur suddenly, but rather progressively, for a particular climatic event of low probability. Some vegetation types will resist change more than others depending, among other factors, on their status in the local dynamic sequence and also on their stability and resilience and the impact of human activities.

How fast or how slow the changes take place is a matter of conjecture; we do not have any factual basis for any sort of generalization.

Annual biomass increase and productivity will be enhanced by carbon fertilization in those areas where water stress is not the major limiting factor to plant growth. That is in the sub-humid to hyper-humid ecoclimatic zones. Moreover CO_2-enhanced water-use efficiency could probably compensate for increased evapotranspiration; the winter-cold stress would be milder and the growing season would be longer. One may thus safely expect an increase of over 10% in the productivity of native vegetation in the moister Mediterranean zones as a result of a doubling CO_2 content in the atmosphere.

In the semi-arid, arid and hyper-arid ecoclimatic zones where water stress in a major limiting factor to plant growth, increasing PET would enhance aridity and therefore reduce plant productivity. Carbon fertilization would necessarily have a very limited effect due to water shortage in the growing season. Production might be reduced by 10% or more, with an average rain-use efficiency of 3 kg DM $ha^{-1}yr^{-1}mm^{-1}$ (Le Houérou, 1984). Furthermore worsening feedback effects would probably occur between permanent plant cover, biomass, and productivity as a result of the worsening of an already critical P/PET ratio. Under the combined effect of a worsened soil/water budget, erosion and a sharply increasing pressure by man and livestock, most of steppe rangeland vegetation would give way to man-made deserts. The upper limit of deserts may correspond by 2050 (or earlier) to the present lower limit of the semi-arid zone, i.e. the present 350-400 mm isohyet of mean annual precipitation; that is the foothills of the high, Mid and Tell Atlas and Tunisian Dorsal in northern Africa and of the main mountain ranges of the Near-Middle East (Taurus, Lebanon, Alaoui, Kurdistan, Zagros, Alborz).

The consequences of the desertization of the North African and Near-Eastern steppes would be very significant on the livestock (sheep) industry in these regions. Range production, which increased by an estimated 60 to 80% over the past 40 years (Le Houérou, 1985, 1986, 1989; Le Houérou & Aly, 1982), may be reduced to 10% of pristine condition with a very short growing season as vegetation would be reduced to tiny ephemerals (Tachitherophytes) and Ephemeroids of very low productivity (*Hordeum leporimum, Stipa capensis, Aegylops sp.pl., Trachynia distachya, Cynosurus coloratus, Elymus orientalis, E. delileanus, Schismus arabicus, S. perennis, Poa bulbosa, Poa sinaica, Carex pachystylis, C. physodes, etc.*). As probably over 50% of the sheep industry in the southern Mediterranean is traditionally kept in the Arid Steppe zone, sheep production systems would inevitably shift to a cereal grain concentrate based feeding system. This move has already begun to take place in the mid 1970's; at present, probably over 50% of sheep diet in this region consists of concentrate feed (Le Houérou & Aly, 1982; Le Houérou, 1985, 1989). This trend should increase in the future as range deterioration progresses. This implies that the sheep industry will increasingly rely on imported grain and concentrate feed and therefore on the international cereal market.

The cereal trade may thus become a strategic issue in the hands of cereal exporting countries, as the overall cereal production per inhabitant of the region has declined or, at best, stagnated over the past 40 years (Tables 15-19).

In the northern Mediterranean countries natural vegetation will expand as a result of marginal farmland abandonment. Over the past 30 years, forest in cultivated land area in Euro-Mediterranean countries is statistically consistent with the evolutionary trend in range and farmland hectareages. This issue will be further addressed later. In the southern

Mediterranean, forest and shrubland may disappear altogether by the mid-21st century under the fast increasing anthropozoic pressure.

Although the fact does not appear in official statistics, a number of field surveys combining remote sensing and ground control show that, in many areas, forest and shrub land receed by 1-2% per annum in spite of reafforestation programmes (Le Houérou, 1973, 1981; Bourbouze, 1982, 1986; Donadieu, 1985; Tomaselli, 1976). Again, this issue will be further discussed later under the heading of "Land-Use".

3.5. IMPLICATION FOR CROP DISTRIBUTION PATTERNS AND YIELDS

As for natural vegetation, cold sensitive crops will expand upwards and northwards. Olive cultivation zones will considerably shrink in the northern Mediterranean in spit of more favourable climatic conditions, because of the increasing cost of labour and the difficulty in developing mechanized harvesting. This trend has already been increasing over the past 20 years. Olive cultivation will thus be restricted to intensive-production, irrigated systems in lowlands, using dwarf clones and mechanical harvesting. Most of the present olive groves will return to the wild: pasture [3], rangeland, shrubland and forest.

In the present situation, commercial crops of citruses are restricted to Sicily, southern Italy, S. and E. coastal Greece, eastern, S.-E. and southern Spain and S. Portugal, South of 40° N. A 3°C increase of winter temperature would render the crop commercially feasible over large areas of lowlands in Greece, Italy, France, Spain and Portugal between 40 and 46° N., from which it is presently excluded by winter frost.

The area presently cultivated to citruses in Europe is some 250,000 hectares producing 8.6 metric tons per annum. This hectareage could be multiplied by a facto of 3 if the temperature increased by 3°C; the additional production could cover far beyond the european consumption thereby closing the european market to the 5 million tons of citruses presently produced over some 250,000 hectares of irrigated citrus groves in the South (2% of the presently irrigated 15 million hectares in the southern basin). But new markets may develop in eastern Europe.

A large proportion of the citrus groves of the southern Mediterranean might then have to be converted to other crops that are too cold-sensitive to be commercially grown in southern Europe, even with a 3°C increase of temperature, such as: avocado, mango, banana, pawpaw, sugar cane and other tropical species.

Vegetable crops would be little affected as large proportions are already grown under controlled conditions (greenhouses, etc.). This proportion will increase in the future to include virtually all the vegetable production both to the South and the North of the basin.

Rainfed agriculture will be affected in the semi-arid and arid zones, but virtually not at all under sub-humid to hyper-humid climates (with a mean annual rainfall above 600 mm) where water supply is ample during the growing season and will remain so with the amount of increased PET envisaged. The situation may be quite different in the semi-arid

---

[3] Pasture = Sown permanent grazing land.
   Rangeland = Extensively grazed natural vegetation.

and arid zones. Increasing PET will reduce water availability during the growing season where it is already a critical factor of production. Most of the shallow red-Mediterranean Oxysols overlying a shallow Pleistocene lime-crust will become inappropriate to cereal cropping between the 350-500 mm isohyets of annual rainfall. These soil types cover huge areas in Spain, southern Italy and Greece, probably over 8 million hectares. Cereal cropping on these soils is already, at present, quite marginal [4] and will become untenable with a worsened water balance particularly if, as expected, the price of cereals in the EEC countries progressively declines to the World Market's (it is about 50% above the World Market's at the present time). These estimated 8 million hectares would then have to be used differently; most likely in part for afforestation for timber production and in part for the extensive grazing of sheep and game (*Cervidae*) under low input conditions, perhaps in combination with shrubland and with cereals on the deeper soils with a favourable water budget.

To the South of the basin, the gamble cropping of cereals *in the arid zone* with harvest expectancies of 1/5 or less will continue and expand to the limit of the desert on all soils amenable to be tilled (i.e., all except rock outcrops and salt marshes). Rangeland will be restricted to stony hills and saline swamps. This will result from the demographic explosion (see Table 5, Fig. 15). However, the worsening of the P/PET ratio will make the present arid gamble-cropping still more risky. The P/PET ratio in the Mediterranean arid zone varies presently from 0.25 to 0.8 depending on local conditions; and increase of PET by 200 mm will drop this ratio to 0.20-0.06.

In the semi-arid North African and Near-Eastern zones crop expectancy will likewise decrease as a result of the worsening of the P/PET ratio. This ratio is at present 0.45 to 0.25 between 350-600 mm isohyets of annual rainfall; it will drop to 0.40-0.20. Crop expectancy which is now about 70 to 80% may drop to 50-60%. In other words, the lower part of the semi-arid zone will become similar to the present upper part of the arid zone (Figs. 6 & 7). Similar to the arid zone, most of natural vegetation will be cleared for cropping wherever the soil is soft and deep enough to be tilled, irrespective the slope or the erosion hazards. Given the tremendous demographic pressure, one cannot see how forestry regulations could be enforced any more. Most of the natural vegetation would disappear, either cleared for farming, eaten out by livestock or grubbed for fuel. These processes have been going on for a few decades already; they are likely to increase at the same rate as human and livestock populations.

To summarize: in the northern basin, forest and shrubland will expand steadily due to the abandonment of marginal farmland and to the moving of people from the countryside to cities. However, forest and shrubland wild-fires would grow exponentially (4.7% annually) putting a heavy burden on Mediterranean communities (650,000 ha burnt as an annual average from 1980 to 1986 around the basin with a global cost of 1 billion ECUs (1.2 billion US $); see Fig. 13).

Cropland will shrink, particularly cereal cultivation; most marginal farmland would have to be converted to extensive grazing, game and timber production (using highly productive

---

[4] Average yield of barley and wheat: 800-1000 kg/ha/yr for a cost of cultivation equal to the market value of 600-800 kg.

new clones of timber species).

In the southern basin, the gamble-cropping at present practised in the arid zone, would extend to the semi-arid zone. Most of the forest and shrubland would be cleared for cropping, or destroyed by heavy over-stocking and fuel collection.

3.6. IMPLICATIONS FOR EROSION AND SEDIMENTATION

As a result of the above-mentioned effects of increased temperature on land-use, crop distribution and natural vegetative cover, heavy wind and water erosion will take place in the South and East of the basin. Water erosion, which at present may average 5-10 t/ha/yr on medium to large sized watersheds (200-20,000 km^2)(Le Houérou, 1969), might multiply by a factor of five or more to reach average values of 25-50 t/ha/yr. Some large watersheds on shales and marls where most vegetation has been destroyed, have erosion rates of 30-60 t/ha/yr (Le Houérou, 1969) and up to 200-300 t/ha/yr on particularly sensitive substrates, such as the gypsiferous Miocene marls of S.E. Spain (Lopez-Bermudez et al., 1984).

It also has been shown that the clearing of a forest and its replacement by annual crops may increase runoff by a factor of 5 and erosion by a factor of 50 (Cormary & Masson, 1964).

The amount of sediments reaching the southern Mediterranean at present may be roughly estimated around 6.0×10^8 metric tons/year (reckoning an average erosion rate of 5 t/ha/yr over 1,200,000 km^2 of catchment); this may take the figure to 3.0×10^9 metric tons/year by the mid-21st century. The thickness of the soil layer eroded would then shift from a present 0.4 mm to 2.0 mm over these 1,200,000 km^2.

The floor of the Mediterranean having an approximate surface of 130 million hectares, the annual added sediment would make a layer some 1.8 mm thick against 0.36 mm at present; that is a sedimentation of 18 cm in 100 years or 1,800 m in one million years. The state of *Rhexistasis* (Erhard, 1956) caused by the destruction of virtually all natural vegetation would thus become a phenomenon of major geological significance, even if the erosion rate were only half of the above figures.

To the north of the basin very little sediment will reach the sea due to the expansion of forest and shrubland over the watersheds. Even of 2 to 5% of the forest and shrubland are burnt each year, the sediment load would be little. In addition, major hydraulic systems have been dammed by chains of reservoirs that trap the silt load. For instance, the Rhone River which at the 19th century carried some 20×10^6 m^3/yr of silt i.e. 50×10^6 t/yr, dropped to 5.5×10^6 t/yr by 1957, and to 2.2×10^6 t/yr by 1977 (Corre et al., 1988). Similar facts are reported from the Ebro, the Po and other major hydrological systems of southern Europe.

The situation prevailing to the North of the basin by the mid-21st century would thus be exactly the opposite to that occurring to the South; the North would be in a state of *Biostasis* due to its good vegetative cover; erosive phenomena would thus be mainly restricted to chemical processes (dissolution of carbonates, etc.) according to Erhard's theory of the *Biorhexistasis*. In other words, the North will be the subject of chemical sedimentation in relatively clear waters (if anthropic pollution is controlled as one may expect), whereas the South will be undergoing heavy detritic sedimentation in highly turbid and polluted waters due to the intense state of overland *Rhexistasis*, resulting from extreme, man-made pressures.

The impact of CO_2 increase on crop yields would be negative for arid and semi-arid zones (100-600 mm) rainfed agriculture because of the lessening of the P/PET ratio as suggested above.

In the sub-humid zones (600-800 mm of annual rainfall) the impact would probably be negligible or slightly positive, because water availability would no longer be a major limiting facto on crop production and productivity may be enhanced by "carbon-fertilization".

In the humid (800-1200 mm) to hyper-humid (P > 1200 mm) zones, the combined effects of lowered P/PET ratio, a prolonged growing season and the increased CO_2 content in the atmosphere, would boost photosynthesis and hence primary production significantly. Yet it is difficult to forecast a figure; the increase might perhaps be 10% for all other conditions than temperature, CO_2, PET and agricultural techniques remaining as at present. However, the combined effects of improved climatic production parameters of higher yielding varieties and advanced farming techniques might greatly increase productivity.

The same scenario would apply to irrigated agriculture which amounts to a present 23.8 million hectares in the basin (see Table 14). Over the past 20 years the hectareage of irrigated land increase by an average of 1.27% per annum (20 million ha in 1970 to 23.8 million ha in 1986). This growth rate cannot continue for very long. Probably within 25 years from now, by the year 2015, most water resources (surface and deep aquifers) will be fully tapped. In many areas, deep aquifers are already being overexploited (i.e. discharge > recharge). The total irrigated area in the basin would then reach 30 to 35 million hectares, i.e., an agricultural water use of some 2.6×10^{11} m^3 per annum, equating to a fictitious permanent discharge of 8×10^6 m^3 per second (for an average discharge of 7500 m^3/ha/yr) by comparison with 1.4×10^{11} and 4.3×10^6 m^3 per annum and per second respectively at the present time.

However, the expansion of drip irrigation and of greenhouse farming will considerably increase both Water Use Efficiency and yields. Irrigated agricultural production may thus increase by a factor of 5 to 10 with the doubling of the present water discharge before the mid-21st century even without the genetic improvement of cultivars.

4. Population: Demographic and Socio-Economic Factors and their Implications for Natural Vegetation and Land-Use.

4.1. AFRO-ASIAN MEDITERRANEAN COUNTRIES

To the South of the basin, population grows by an average of 3.2% per annum; that is an exponential rate of 2.8% and a doubling period of 25 years, which is 5.6 times faster than in Euro-Mediterranean countries.

As mentioned above all attempts to curb population growth in the region (Tunisia, Egypt) have failed so far. There is actually no example of population growth control in any muslim country of the world to date, including N.-W. China (where the growth rate is still 2.8% per annum in the islamic minorities, Le Houérou, 1987).

The total population of the 14 countries under study was about 44 million at the beginning of the present century but it will reach 372 million by the year 2000 whatever corrective action might be taken during the last 10 years of the century.

Projections (see Table 5 and Fig. 15) show a high 1950 million (almost 2 billion) and a low 800 million by 2050. Both are unlikely to as the most likely figure is the intermediate one shown in Table 5, i.e. 1.45 billion. For the latter projection we have used the present mean growth rate of the countries having the slowest growth (Turkey, Tunisia and Egypt). Several countries of the region have present growth rates between 3.2 an 7.0% per annum: Algeria (3.7), Morocco (3.2), Libya (7.0), Saudi Arabia (5.8), Syria (4.9), Iran (3.7), Iraq (3.5).

The agricultural population remained approximately stable to the South between 1970 and 1986 in absolute terms: 87.7 million in 1970 and 88.3 million in 1986 (see Table 6). However, in relative terms it decreased from 54% of the total population in 1970 to 35.5% in 1986. That is a decrease of 1.5% per annum in the Near and Middle East and 2% in northern Africa.

4.2. EURO-MEDITERRANEAN COUNTRIES

To the North of the basin the human population growth is at present about 0.5% per annum (see Table 4, Fig. 14). It grew from 150 million people in 1950 to 197 million in 1985. At the present growth rate of 0.5% per annum, the doubling period is 140 years. Growth is projected to decline to 0.3% by 2050, or before, with a doubling period of 233 years. The proportion of rural population is declining steadily (see Table 6). The agricultural population decreased from 35 million in 1970 to 24 million in 1986, that is a 31% reduction in 16 years and almost 2% per annum. The actual percentage of agricultural population in the total was 20% in 1970 and 12% in 1986, ranging from a high of 52% in Albania and a low of 6% in France. By comparison, the proportion of agricultural population in 1986 was as follows: U.K.: 2%; U.S.A.: 3%; Switzerland and W. Germany: 4%; Sweden: 5%; E. Germany: 9%; Bulgaria: 13%; Hungary and Ireland: 14% and Poland and Rumania: 21%.

4.3. GROSS DOMESTIC PRODUCT PER CAPITA (GDPPC)

The Gross Domestic Product Per Capita (GDPPC) as shown in constant 1980 US $ [5], changed as follows in the 21 countries under study between 1980 and 1987 (see Table 22):

a.   It decreased by a regional average of -22% in northern Africa (= -2.75% per annum).
b.   It decreased by a regional average of -11% in the Near East (= -1.38% per annum).
c.   It increased by a regional average of 48% (= +6% per annum) in the Euro-Mediterrane-

---

[5] 1.0 US $ 1980 = 1.40 US $ 1987.

an countries.

The decrease in GDPPC real terms in the South of the basin was largely due to the fact that population grew faster than economy. Naturally the important drop in oil revenues played a role but not necessary the major one, sine countries like Morocco, Lebanon, Syria, Jordan, Turkey and Tunisia, showed either a stagnation or a decline in GDPPC. While some oil/gas producing countries like Algeria, exhibited increasing gas production, Iran and Iraq showed a slight increase in spite of their 8-year war. Libya underwent a considerable drop of 2600 $ due to both the fall in oil price and the Chad war. Saudi Arabia and Lebanon exhibited a slight decrease for different reasons (oil prices and civil warfare respectively).

Socio-economic conditions have considerably worsened over the past few decades in the South of the basin as a result of the extremely high demographic growth. In spite of huge oil and gas production, GDPPC stagnated or declined. Unemployment is high (often 30-50%) and the proportion of the population engaged in agriculture is still over 35% in Afro-Asian Mediterranean countries by comparison to 12% in Euro-Mediterranean countries.

To the North of the basin, GDPPC is growing steadily in most countries (Albania and Yugoslavia are exceptions to the rule). At the same time, the agricultural population decreased from 40% in 1950 to 12% today (6.8 and 13% in France and Spain respectively). This situation, in turn, created new problems tied to marginal land abandonment: the desertion of the countryside. The desertion of the countryside, in combination with a 7% annual increase in Tourism and a sharp decrease in extensive grazing, provoked a fast expansion of the areas of forest and shrubland burnt annually by wild-fire (see Fig. 13). These expanded from an average annual mean of 200,000 hectares in the 1960's to an average of 650,000 hectares in the 1980's. The cost in 1987 was estimated at about 1.2 billion US $ (1.0 billion ECUs, Le Houérou, 1987).

New methods of fire prevention will have to be introduced, in particular prescribed burning combined with the reintroduction of extensive grazing by browsing animals (goats, *cervidae, camelidae*) in order to avoid fuel build-up and thus reduce fire hazards.

5. Conclusions

5.1. EURO-MEDITERRANEAN COUNTRIES

The impact of a possible doubling of the CO_2 content in the atmosphere, to 700 ppmv in the middle of the 21st century, would be moderate in the northern basin; and most likely on the whole rather beneficial, as a result of enhanced net photosynthesis and hence primary production.

Cold-sensitive tropical crops such as citruses, avocados, bananas, papaws, sugar cane, etc., would expand.

Cereal cultivation would be eliminated from 8 to 10 million hectares presently farmed in Mediterranean Europe. Those lands would have to be converted to low input rangelands for livestock breeding, game management, hunting, forestry, tourism and amenities.

The forest and shrubland would grow from the present 53 million to about 76 million hectares in Euro-Mediterranean countries, if the present rate of expansion (0.7% per annum) were sustained over the next 60 years (see Tables 8 & 9). The afforestation rate would thus grow by some 68% from a present 28% to 47% in 2050. If the present rate of shrinking of farmland goes on unabated (0.4% per annum), cropland would decrease to some 51 million hectares against the present 67.5 million; that is a loss of 16.5 million hectares of cropland.

The coastal areas devoted to tourism would experience a severe impact as the industry is currently growing by an average 5.6% per annum (Baric & Gasparovic, 1988). It reached some 106 million visitors in 1986: 30 million in Spain, 36 in France, 25 in Italy, 8 in Yugoslavia and 7 in Greece. National tourism represents about 45% of the visitors. The number of visitors to the northern shores of the Mediterranean basin would during the figure to 220 million by the year 2000 if the current growth rate remains as it has been over the past 20 years.

Tourism along the southern shores represents a little under 10% of the number of visitors per annum to the North, in 1986: Morocco 2.2 million, Turkey 2.0, Egypt 1.7, Tunisia 1.7, Israel 1.2, Syria 1.1, Cyprus 0.9.

The impact of tourism is not only very strong on a narrow belt of less than 5 km from the shores (touristic urbanization), but it also contributed quite significantly to the exponentially increasing areas of forest and shrubland burnt each summer (Le Houérou, 1973, 1987).

Vegetation would evolve towards more thermophylous and more sclerophyllous types. The various vegetation belts shown in Figure 8 would move about one step upward and northward; that is 550 ± 200 m upward in elevation and 5 to 6 degrees (550-660 km) northward in latitude at any elevation [6]. In other words, the temperature presently prevailing on the southern shores of the Mediterranean Sea would occur on its northern shores, will considerably expand up to 45-46° N. to the mountain-foothills of the Apennines, southern Alps, Cevennes and Pyrenees. The limits of the Mediterranean region, however, are not expected to change as long as the seasonal distribution of rainfall remains unchanged.

5.2. AFRO-ASIAN MEDITERRANEAN COUNTRIES

The situation strongly contrasts with the Euro-Mediterranean countries in terms of land-use trends and demographic and socio-economic factors.

The impact of CO_2 and temperature increase will be extremely important, not because of their direct consequences but because these consequences will worsen to a major extent provoked by the human population explosion. The population will have multiplied 8.5 fold during the 20th century; it will multiply another 3.9 times by the year 2050; that is a 33 fold increase in 150 years. The whole Afro-Asian Mediterranean region represents a non-desert area of some 1.51 million km^2 of a total land area of 11.26 million km^2 i.e. 13%. The population density on non-desert land thus grew as follows - ( ) = forecast:
- 29 inh/km^2 in 1900   108 inh/km^2 in 1970   182 inh/km^2 in 1988

---

[6] The average altitudinal and latitudinal lapse rates being about 0.55°C per 100 m upward and 100 km northward respectively.

| - 62 | " | 1950 | 142 | " | 1980 | (189 | " | 1990) |
| - 82 | " | 1960 | 162 | " | 1985 | (246 | " | 2000) |

It would reach the 530 inh/km^2 by the year 2050 for the lowest possible scenario, and 960 inh/km^2 if the present growth rate remains unabated, as it has for the past 40 years. This is an urban density incompatible with preserving any natural vegetation. For the past three decades, forest and shrubland vegetation has been receding by 1 to 2% per annum, natural steppe vegetation in the arid rangelands has been destroyed to a very large extent over the past 30 years, giving way to desert encroachment for more than 2% per annum.

By the year 2050, it would seem that only Turkey will have some forest remnants left in its Euxinian (non-Mediterranean) and perhaps Morocco on the High and Middle Atlas mountains exposed to Atlantic moisture.

Virtually all land except true desert, rock outcrops and salt marshes will be subjected to gamble-cropping of staple cereals.

Erosion, sedimentation and flooding will expand freely and catastrophic events will become more and more frequent as the environment deteriorates. Any climatic change resulting from rising temperature will then be a trivial part of the man-made continuing catastrophe. Unless the demographic explosion is mastered shortly, the disaster is bound to happen. As a matter of fact, it should be kept in mind that any attempt to curb demographic growth takes at leas 25 years to produce a significant impact on population numbers (see Fig. 15).

This suggests that, unless strong and determined action is taken between now and the year 2025, the catastrophe is arithmetically unavoidable. Social unrest and all sorts of extremes would flourish: they have already begun. European countries would not remain unscathed; they would undergo extreme pressure from the hungry multitudes on their southern flank. Such pressure could in turn lead to social unrest, racism and probably totalitarian political regimes. The symptoms already are present in some countries.

6. References

Abi-Saleh, B. (1982) "Altitudinal zonation of vegetation in Lebanon", Ecol. Medit., VII, pp. 355-364

Arrhenius, S. (1896) "On the influence of carbonic acid in the air upon the temperature of the ground", Philos. Mag., 41, p. 237

Assadollah, F., Barbero, M., & Quézel, P. (1982) "Les écosystèmes préforestiers et forestiers de l'Iran", Ecol. Medit., VII, pp. 365-380

Bach, W. (1987) "Scenario analysis", Proc. European Workshop on *Interrelated Bioclimatic and Land-Use Changes*, Noordwijkerhout, the Netherlands, 9 p. mimeo

Bach, W. (1988) " Development of climatic scenarios from General Circulation Models.", in Parry, M.L., et al. (eds.) *The Impact of Climatic Variation on Agriculture*, Sect. 4, pp. 125-158

Bagnouls, F., & Gaussen, H. (1953 a) "Saison sèche et indice xérothermique", Labor. Forestier, Fac. Sces., Univ. de Toulouse, 47 p.

Bagnouls, F., & Gaussen, H. (1953 b) "Périodes de sécheresse et végétation", Optes.

320

Rend. Acad. Sces., 236, Paris, pp. 1076-1077

Bagnouls, F., & Gaussen, H. (1957) "Climats biologiques et leur classification", Ann. de Géogr., 355, LXVI, pp. 193-220

Baric, A., & Gasparovic, F. (1988) "Implication of climatic changes on the socio-economic activities in the Mediterranean coastal zone.", UNEP, Split/Athens, 88 p. mimeo

Bolin, B., Degens, E.T., and Kletner, P. (eds.)(1979) *The Global Carbon Cycle*, SCOPE, Study no. 13, ICSU, Paris, 491 p.

Bourbouze, A. (1982) "L'élevage dans la montagne marocaine: Organisation de l'espace et utilisation des parcours par les éleveurs du Haut Atlas", INA, Paris-Grignon, 345 p.

Bourbouze, A. (1986) "Adaptation à différents millieux des systèmes de production des paysans du Haut Atlas", Techniques et Cultures, 7, pp. 59-94

Chamberlain, T.C. (1899) "An attempt to frame a working hypothesis on the cause of glacial periods on an atmospheric basis", J. Geol., 7, p. 545

Charney, J.G. (1979) "Carbon dioxide and climate: a scientific assessment", Nat. Acad. of Sci., Washington, D.C., 22 p.

C.E.E. (1985) *Perspectives d'une Politique Agricole Commune: Le Livre Vert de la Commission*, CEE, Bruxelles

Cormary, Y., & Masson, J. (1964) "Etude de conservation des eaux et du sol au Centre de Recherches du Génie Rural de Tunis. Application à un projet-type de la formule universelle de perte en sol de Wischmeier.", Cahier ORSTOM, Sér. Pédologie, II(3), pp. 1-26

Corre, J.J. & 22 others (1988) "Implication des changements climatiques dans le Golfe du Lion", PNUE, Athènes/Split, 137 p. multigr.

Cranc, A.J. (1985) "Possible effects of rising CO_2 on climate", Plant, Cell and Environment, 8, pp. 371-379

Donadieu, P. (1985) "Géographie et écologie des végétations pastorales méditerranéennes", Ecole Nat. Sup. du Paysage, Versailles, 324 p. multigr.

Emanuel, W.R., Shugart, H.H., & Stevenson, M.P. (1985) "Climatic change an the broad scale distribution of terrestrial ecosystem complexes.", in Parry, M.L. (ed.) *The Sensitivity of Natural Ecosystems and Agriculture to Climatic Change*, Kluwer Scientific Publ., Dordrecht, The Netherlands, pp. 29-43

Emberger, L. (1930) "La végétation de la région méditerranéenne, essai d'une classification des groupements végétaux", Rev. Gén. de Bot., 42, pp. 641-662, 705-721

Erhard, H. (1956) "La genèse des sols en tant que phénomène géologique", Masson, Paris, 90 p.

Gates, W.L., Han, Y.J., & Schlesinger, M.E. (1984) "The global climate simulated by a coupled Atmosphere-Ocean general circulation model: preliminary results", Climatic Research Institute, Report no. 57, Oregon State University, Cornwallis, Oregon

Gentile, S. (1982) "Zonation altitudinale de la végétation en Italie Méridionale et en Sicile (Etna exclu)", Ecol. Medit., VII, pp. 323-338

Griffith, J.G. (1972) *Report on the agrolimatic conditions in S.W. Spain*, FAO, Rome, 20 p.

Hansen, J., Johnson, D., Lacis, A., Lebedeff, S., Lee, P., Rind, D., & Russell, G. (1981) "Climate impact on increasing atmospheric carbon dioxide", Science, 213, p. 957

Hare, F.K. (1985) "Climatic variability and change", in Kates, R.W., Ausubel, J.H., and Berberian, M. (eds.) *Climate Impact Assessment*, SCOPE no. 27, John Wiley & Sons, New York, ch. 2, pp. 37-68

Holdridge, L.R. (1947) "Determination of world plant formations from simple climatic data", Science, 105, pp. 367-368

Holdridge, L.R., & Tosi, J.A. (1967) "Life zone ecology", revised edition, Tropical Science Center, San Jose, Costa Rica, 206 p.

Houghton R.A. (1986) "Estimating changes in carbon content of terrestrial ecosystems from historical data", in Trabalka, J.R., and Reichle, D.E. (eds.) *The Changing Carbon Cycle, a Global Analysis*, pp. 175-193

Idso, S.B. (1980 a) "The climatological significance of a doubling of Earth's atmospheric carbon dioxide concentration", Science, 207, p. 1462

Idso, S.B. (1980 b) "Carbon dioxide and climate", (reply to Schneider et al., 1980 and Leovy, 1980), Science, 210, p. 7

Imai, K., & Murata, Y. (1978) "Effect of carbon dioxide concentration on growth and dry matter production of crop plants", Jap. J. Crop Sci., 47(2), pp. 330-335; 47(4), pp. 587-595

Jäger, J. (1988) "Development of climatic scenarios: background to the instrumental record", in Parry, M.L., et al. (eds.) *The Impact of Climatic Variation on Agriculture*, Sect. 4, pp. 159-181

Jones, P.D., Wigley, T.M.L., & Kelley, P.M. (1982) "Variations in surface air temperatures. Par I, northern hemisphere, 1881-1980", Monthly Weather Review, 110, pp. 59-70

Keeling, C.D., Becastow, R.B., Bainbridge, A.E., Ekdahl, C.A., Guenther, P.R., & Waterman, L.S. (1976) "Carbon dioxide variation at Mauno Loa Observatory, Hawaii", Tellus, Vol. 28, 538 p.

Kellogg, W.W. (1977) "Effects of human activities on global climate", Tech. Note 156, WMO publ. no. 486, WMO, Geneva, 47 p.

Kellogg, W.W. (1978) "Is mankind warming the Earth?", Bull. of the Atomic Scientists, 34, p. 17

Kellogg, W.W., & Schware, R. (1981) *Climate Change and Society. Consequences of Increasing Atmospheric Carbon Dioxide*, Westview Press, Boulder, Colorado, 178 p.

Kukla, G.J., & 8 others (1977) "New data on climatic trends", in *Nature* 270, pp. 573-580

Lamb, H.H. (1977) *Climates: Present, Past and Future*, 2 Vols., Methuen, London

Legg, B.J. (1985) "Exchange of carbon dioxide between vegetation and the atmosphere", Plant, Cell and Environment, 8, pp. 409-416

Le Houérou, H.N. (1959) *Recherches écologiques et floristiques sur la végétation de la Tunisie méridionale*, 2 Vol., 54 Tab., 4 cartes H.T., Bibl. 530. Mém. H.S. Inst. Rech. Sahar., Univ. d'Alger, 510 p.

Le Houérou, H.N. (1962) "Les pâturages naturels de la Tunisie aride et désertique", Inst. Sces. Econ. Appl., Paris, 106 p.

Le Houérou, H.N. (1968) "La désertation du Sahara septentrional et des steppes limitropi-

322

phes", Ann. Algér. de Géogr., 3(6), pp. 1-27

Le Houérou, H.N. (1969) "La végétation de la Tunisie steppique (avec références au: végétations analogues d'Algérie, de Libye et du Maroc)", Ann. Inst. Nat. Rech. Agron. de Tunisie, 42, 5, 645 p.

Le Houérou, H.N. (1970) "North Africa: past, present and future", in Dregne, H. (ed.) *Arid Lands in Transition*, Amer. Assoc. for the Advanc. of Science, Washington, D.C. pp. 227-278

Le Houérou, H.N. (1973 a) "Ecologie, démographie et production agricole dans le pays méditerranéens du Tiers-Monde", Options Méditerranéennes, 17, pp. 53-61

Le Houérou, H.N. (1973 b) "Fire and vegetation in the Mediterranean Basin", Proc. Ann. Tall Timbers Fire Ecology Conf., 13, Tall Timbers Research Station, Talahassee, Florida, pp. 237-277

Le Houérou, H.N. (1975 a) "Problèmes et potentialités des terres arides de l'Afrique du Nord", Options Méditerranéennes, 26, pp. 17-36

Le Houérou, H.N. (1975 b) "Le cadre bioclimatique des recherches sur les herbages méditerranéens", I Georgofili, XXI, 7, pp. 57-67

Le Houérou, H.N. (1977 a) "Fire and vegetation in North Africa", in Mooney, H.A. and Conrad, C.E. (eds.) *Proc. Intern. Symp. on the Environmental Consequences of Fire and Fuel Management in Mediterranean Ecosystems*, Report WOZ, USDA Forest Service, Washington, D.C., pp. 334-341

Le Houérou, H.N. (1977 b) "Plant sociology and ecology applied to grazing lands research, survey and management in the Mediterranean Basin", in Krause, W. (ed.) *Application of Vegetation Science to Grassland Husbandry*, Handbook of Vegetation Science, XIII, Junk publ., The Hague, pp. 213-274

Le Houérou, H.N. (1981) "Impact of man and his animals on Mediterranean vegetation", in Di Castri, F., Goodall, D.W., and Specht, R.L. (eds.) *Mediterranean-Type Shrublands*, Ecosystem of the World, Vol. 11, Elsevier publ., Amsterdam, ch. 25, pp. 479-521

Le Houérou, H.N. (1982) "The arid bioclimates in the Mediterranean isoclimatic zone", Ecologica Mediterranea, VII, pp. 115-134

Le Houérou, H.N. (1984) "Rain-Use Efficiency: a unifying concept in arid-land ecology", J. of Arid Envir., 7, pp. 1-35

Le Houérou, H.N. (1985 a) "The impact of climate on pastoralism", in Kates, R.W., Ausubel, J.H., and Berberian, M (eds.) *Climate Impact Assessment*, SCOPE no. 27, J. Wiley and Sons, New York, ch. 7, pp. 155-186

Le Houérou, H.N. (1985 b) "La génération des steppes algériennes", Inst. Nat. Rech. Agron. (Relat. Extér.) et Minist. Relat. Extér., Paris, 45 p. multigr.

Le Houérou, H.N. (1987 a) "Vegetation wildfires in the Mediterranean Basin: evolution and trends", Proc. Workshop on *the Ecological Aspects of Forest Fires in the Mediterranean Basin*, Europ. Sces Foundation & Forest Ecosystems Res. Network, Ecologia Mediterranea, XIII, pp. 13-24

Le Houérou, H.N. (1987 b) "Ecological guidelines to control land degradation in European Mediterranean countries, Proc. Int. Conf. on *Policies to Combat Desertification in Europe*, Valencia, Spain, July 1987, EEC, Brussels, in press, 20 p.

Le Houérou, H.N. (1987 c) "Agro-forestry and Sylvo-pastoralism to combat land degradation in the Mediterranean Basin: old approaches to new problems", Proc. Madrid Symp. on Desertification, EEC, Brussels, in press, 15 p.

Le Houérou, H.N. (1988 a) "Considérations biogéographiques sur les steppes Nord-Africaines", Cpte-Rend. Vème Coll. Intern. de l'Assoc. Franç de Géogr. Phys.: *Biogéographie, Environment, Aménagement*, CNRS, Univ. Paris VII, sous presse, 23 p.

Le Houérou, H.N. (1988 b) "The grazing lands of the Mediterranean Basin", in Coupland, R.T. (ed.) *Natural Grassland Ecosystems*, Ecosystems of the World, Vol. 8, Elsevier Publ., Amsterdam, in press

Le Houérou, H.N., & Aly, I.M. (1982) *Perspective and Evaluation Study on Agricultural Development: The Rangeland Sector*, FAO, Tripoli, Libya, 77 p.

Le Houérou, H.N., Haywood, M., & Claudin, J. (1975) *Etude Phytoécologique du Hodna (Algérie)*, carte coul. 1/200 000, 3 feuilles, FAO, Rome, 154 p.

Le Houérou, H.N., & Popov, G.F. (1981) "An ecolimatical classification of inter-tropical Africa", Plant Production and Protection, paper no. 31, FAO, Rome, 40 p. 3 maps

Le Houérou, H.N., Pouget, M, Claudin, J. (1977/79) "Etude bioclimatique des steppes algériennes", Bull. Soc. Hist. Nat. de l'Afrique du Nord, 68(31-4), pp. 33-74, 3 cartes 1/1 000 000, 9 Figs.

Lemon, E.R. (1983) *CO_2 and plants: the response of plants to rising levels of atmospheric carbon dioxide*, Westview Press Publ., Boulder, Colorado, 280 p.

Lopez-Bermudez, F., Romero-Diaz, A., Fisher, G., Francis, C., & Thornes, J.B. (1984) "Erosión y ecología en la España semi-árida (cuenca de Mula, Múrcia)", Cuadernos Investig. Geográfica, X, 1-2, pp. 113-126

Manabe, S., & Stouffer, R.J. (1979) "A CO_2-climate sensitivity study with a mathematical model of the global climate", *Nature* 282, p. 491

Manabe, S., & Stouffer, R.J. (1980) "Sensitivity of a global climate model to an increase of CO_2 concentration in the Atmosphere", J. Geophys. Res., 85, p. 5529

Manabe, S., & Wetherald, R.T. (1975) "The effects of doubling the CO_2 concentration on the climate of a general circulation model", J. Atmos. Sci., 37, p. 99

Mass, C., & Schneider, S.H. (1977) "Statistical evidence on the influence of sunspots and volcanic dust on long-term temperature records", J. Atmos. Sci., 34, p 1995

Mitrakos, K. (1982) "Winter low temperature in Mediterranean-type Ecosystems", Ecologica Mediterranea, VII, pp. 95-102

Morison, J.I.L. (1985) "Sensitivity of stomata and Water-Use Efficiency to high CO_2", Plant, Cell & Environment, 8, pp. 409-416

Mortensen, L.M. (1983) "Growth response of some greenhouse plants to environment, x-long-term effect of CO_2 enrichment on photosynthesis, photorespiration, carbo-hydrate content and growth of *Chrysanthenum morifolium* Rawat.", Medlinger fra Norges Land-bruskshøgskole, 62, 12, pp. 1-11

National Research Council (1979) *Carbon Dioxide and Climate: a Scientific Assessment*, Nat. Acad. of Sces, Washington, D.C., 22 p.

National Research Council (1982) *Carbon Dioxide and Climate: a Second Assessment*,

Nat. Acad. of Sces., Nat. Acad. Press, Washington, D.C., 72 p.

Ozenda, P. (1970) "Sur une extension de la notion de zone et de l'étage sub-méditerranéen", Cpte-Rend. Somm. Séances Sco. Biogéogr., 47, pp. 92-103

Parry, M.L., Carter, T.R., & Konijn, N.T. (eds.)(1988) *The Impact of Climatic Variation on Agriculture*, Kluwer Acad. Publ., Dordrecht, Vol. 1, 876 p.; Vol. 2, 764 p.

Penman, H.L. (1948) "Natural evaporation from open water, bare soil and grass", Porc. Roy. Soc. London, 193, pp. 120-145

Petrovanov, S.E. (1987) "A climatic scenario based on the Vinnikov/Groisman's approach", Proc. European Workshop on *Interrelated Bioclimatic and Land-Use Changes*, Noordwijkerhout, The Netherlands, 28 p. mimeo

Poli-Marchese (1982) "Zonation altitudinale de la région fr l'Etna comparée avec celle des autres volcans", Ecol. Medit., VII, pp. 339-354

Quézel, P. (1976) "Les forêts du pourtour méditerranéen", Notes Techniques du MAB, n° 2, UNESCO, Paris, pp. 9-33

Quézel, P, & Barbero, M. (1982) "Definition and characterization of Mediterranean-type ecosystems", Ecologica Mediterranea, VII, pp. 15-27

Quézel, P., Gamisans, J. & Gruber, M. (1980) "Biogéographie et mise an place des flores méditerranéennes", Naturalia Monspeliensia, Actes du Colloque sur la mise en place et la caractérisation de la flora et de la végétation circum-méditerranéennes, organisé par La Fondation L. Emberger à l'Institut de Botanique de Montpellier, les 9-10 avril 1980, pp. 41-51

Ramanathan, V. (1975) "Greenhouse effect due to chlorofluorocarbons: climatic implications", Science, 190, p. 50

Ramanathan, V. (1980) "Climatic effects of anthropogenic trace gases", in Bach, W., Pankrath, J., and Williams, J. (eds.) *Interactions of Energy and Climate*, D. Reidel, Boston, Mass. pp. 269-280

Ramanathan, V., & Coakley, J.A.Jr. (1978) "Climate modelling through radiative-connective models", Rev. Geophys. Space Phys., 16, p. 465

Rivas-Martinez, S. (1982) "Etages bioclimatiques, secteurs chlorologiques et séries de végétation de l'Espagne Méditerranéenne", Ecologica Mediterranea, VII, pp. 275-288

Rockner, E., Schless, U., Biercamp, J., & Loewe, P. (1987) "Cloud optical depth feedback and climate modelling", *Nature* 329, pp. 138-140

Rotty, R.M. (1979) "Uncertainties associated with global effects of atmospheric carbon dioxide", Orau/IEA Report no. 79-6, Oak Ridge Associated Universities.

Schneider, S.H. (1975) "On the carbon dioxide climate confusion", J. Amer. Sci., 32, p. 2060

Schneider, S.H., Washington, W.M., & Chervin, R.M. (1978) "Cloudiness as a climatic feedback mechanism: effects on cloud amounts of prescribed global and regional surface changes in the NCAR.GOM", J. Atmos. Sci., 35, p. 2207

Shugart, H.H., & Emanuel, W.R. (1985) "Carbon dioxide increase: the implications at the ecosystem level", Plant, Cell & Environment, 8, pp. 381-386

Slade, D.H. (1979) "Summary of the carbon dioxide effects", Research and Assessment Program, US Dept. of Energy, Washington, D.C., 36 p.

Theys, M. (1986) *L'agriculture française et l'environnement dans les 20 prochaines années: vers un nouvel equilibre?*, Ministère de l'Equipement et de l'Environnement, Paris, 59 p.

Tomaselli, R. (1976) "La dégradation du maquis méditerranéen", Notes Techniques du MAB, n° 2, UNESCO, Paris, pp. 35-76

Trabalka, J.R., & Reichle, D.E. (eds.)(1986) *The Changing Carbon Cycle, a Global Analysis*, Springer-Verlag, Heidelberg, 592 p.

Tyndall, J. (1863) "On radiation through the Earth's atmosphere", Philos. Mag., 4, p. 200

Walter, H., & Lieth, H. (1960) *Klimadiagram Weltatlas*, G. Fischer Verlag, Iena, 200 p.

Wigley, T.M.L. (1987) "Climate scenarios", Proc. European Workshop on *Interrelated Bioclimatic and Land-Use Changes*, Noordwijkerhout, The Netherlands

Wigley, T.M.L., Jones, P.D., & Keely, P.M. (1980) "Scenario for a warm high-CO_2 world", in *Nature* 283, pp. 17-21

WMO (1986) *Report of the International Conference on the Assessment of the Role of Carbon Dioxide and of Other Greenhouse Gases in Climatic Variation and Associated Impacts*, Villach, Austria, 9-15 October 1985, WMO n° 661, WMO, Geneva, Switzerland, 78 p.

7. List of Figures

8. List of Tables

328

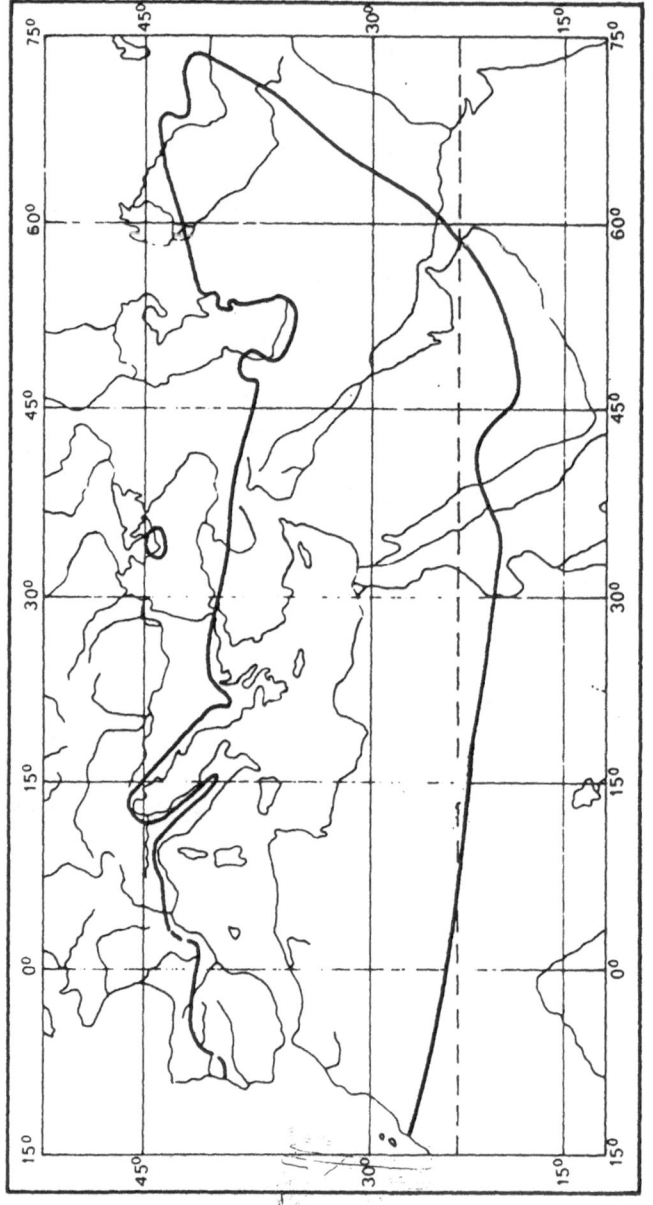

Figure 1. Geographical distribution of Mediterranean climates in the basin.

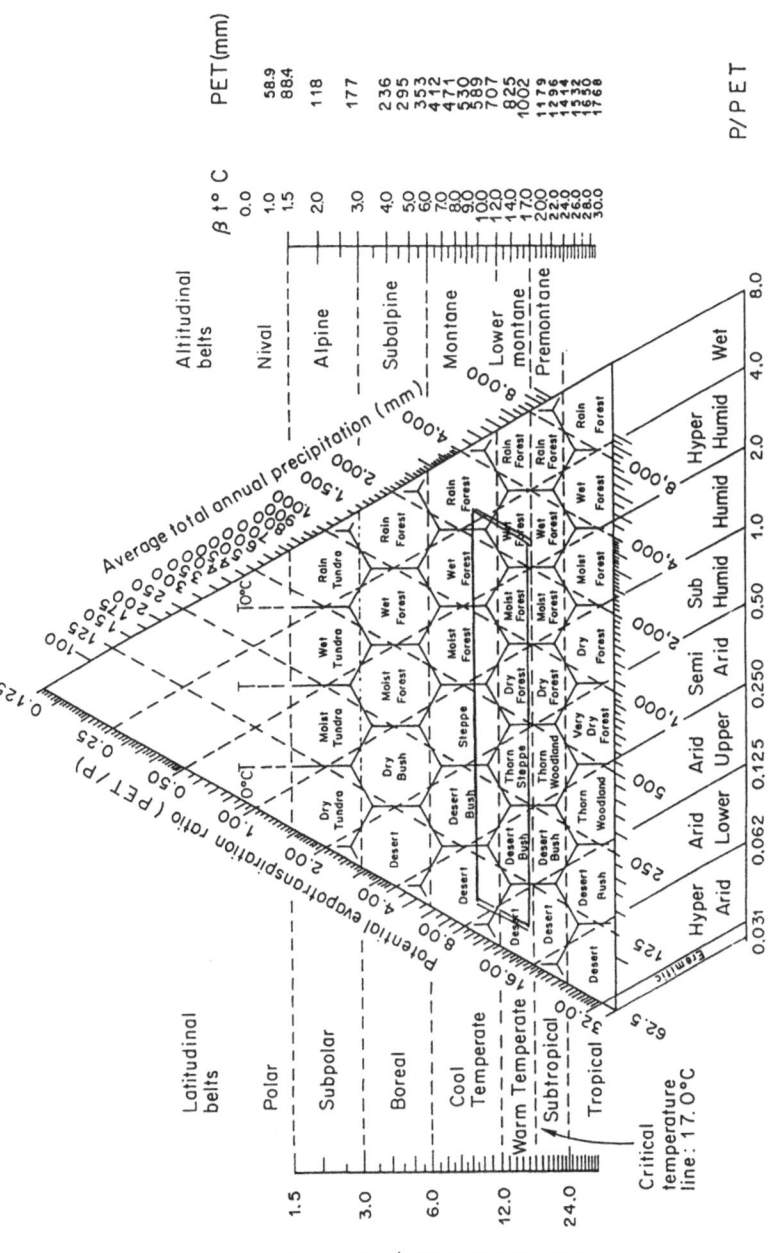

Figure 2. The Mediterranean area in Holdridge's life zone diagram.

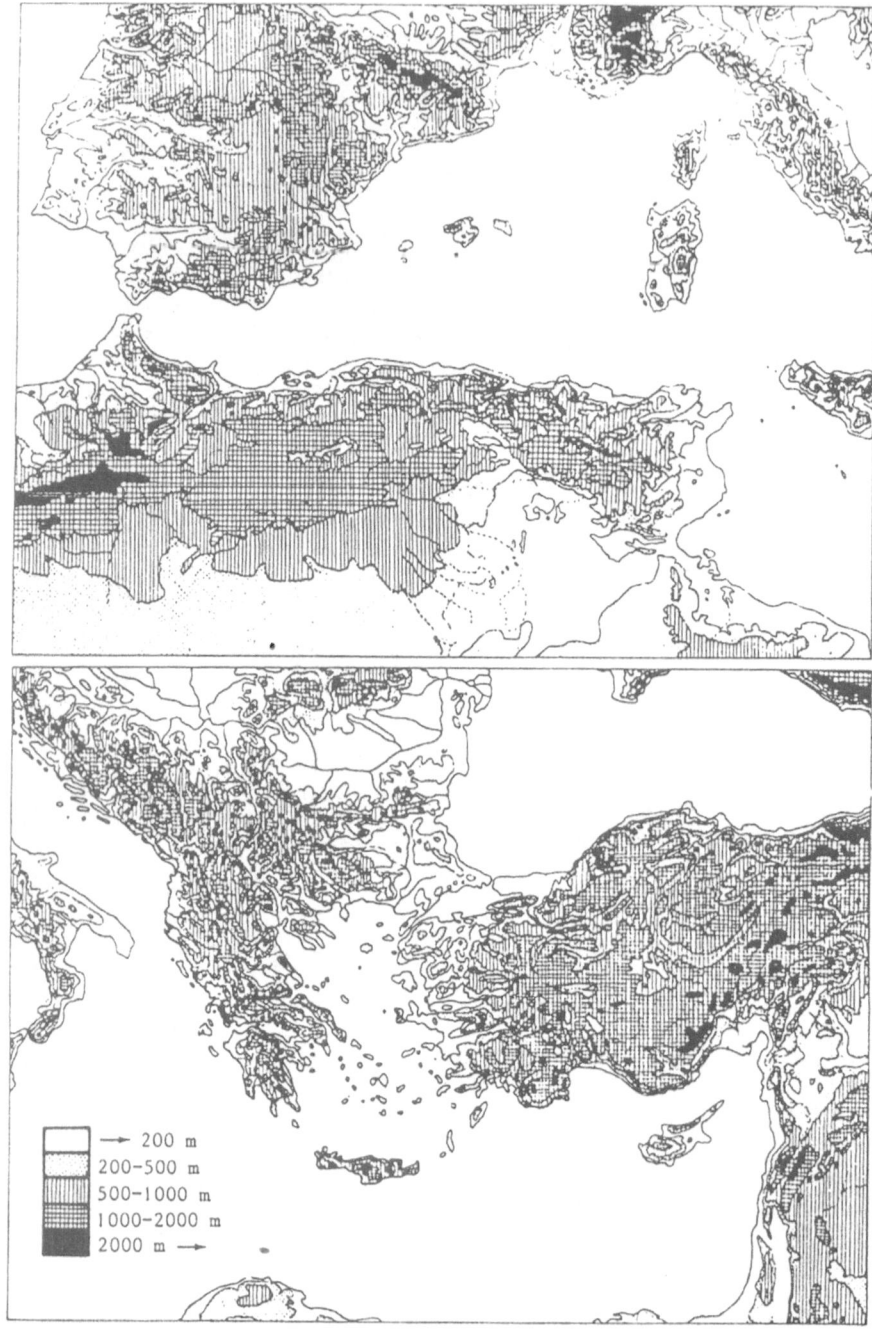

Figure 3. Elevation of circum-Mediterranean countries (Tomaselli, UNESCO).

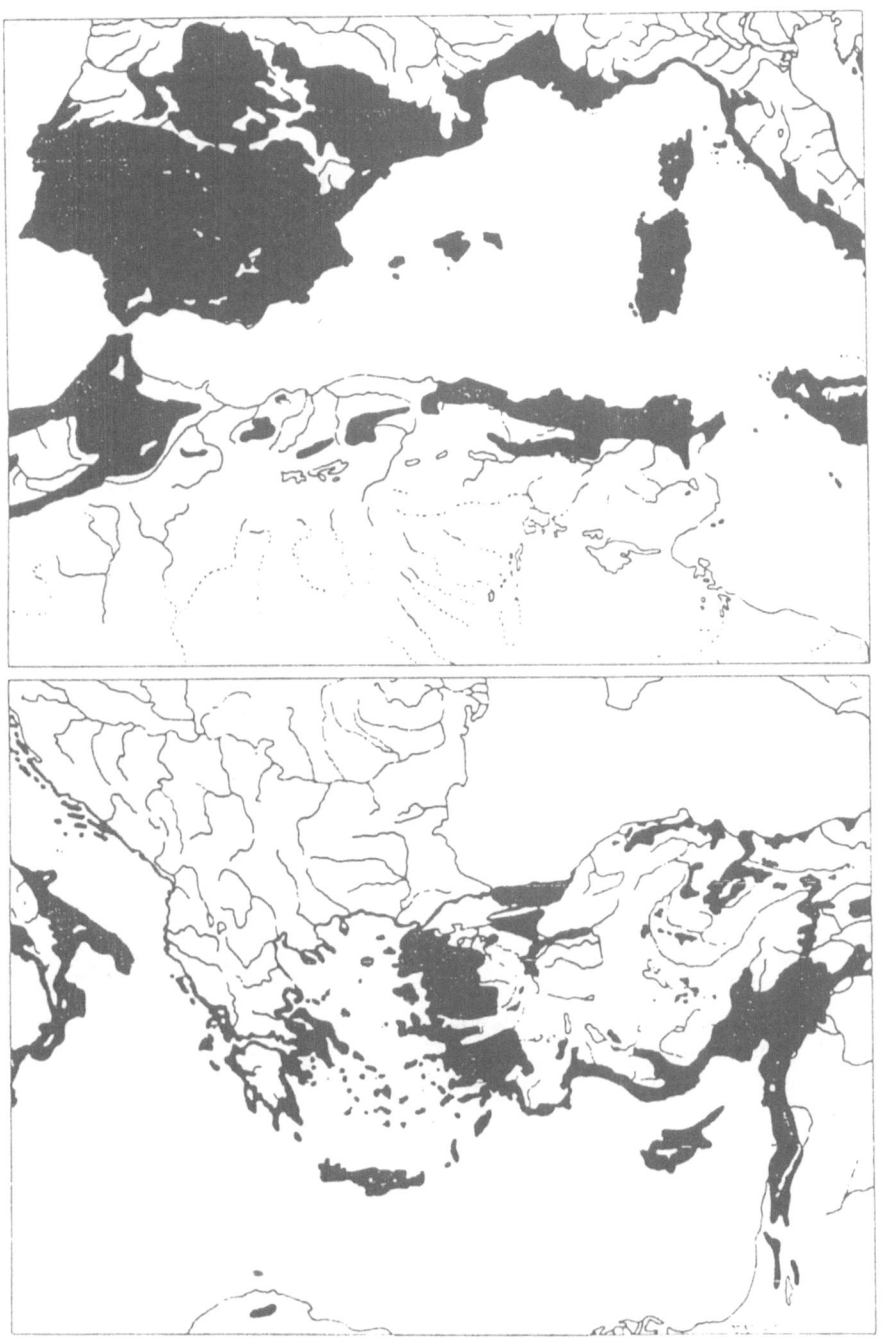

Figure 4. Geographical distribution of Mediterranean-type shrublands around the basin.

332

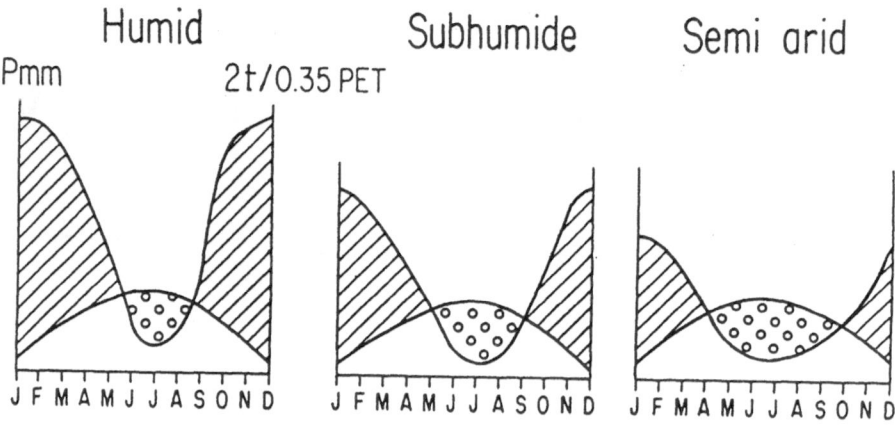

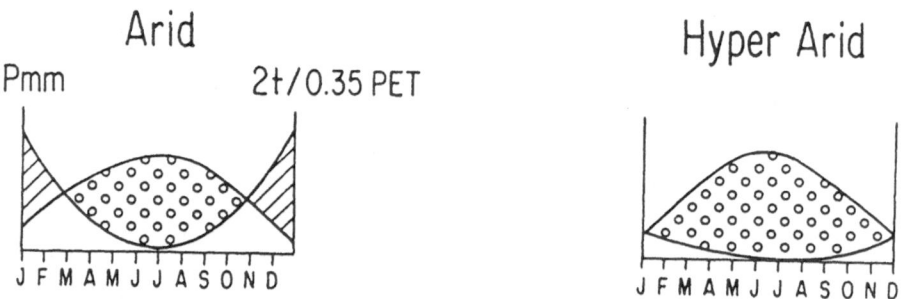

Figure 5. Ombrothermal diagrams: length of dry and rainy seasons under various types of Mediterranean climates and ecological zones.

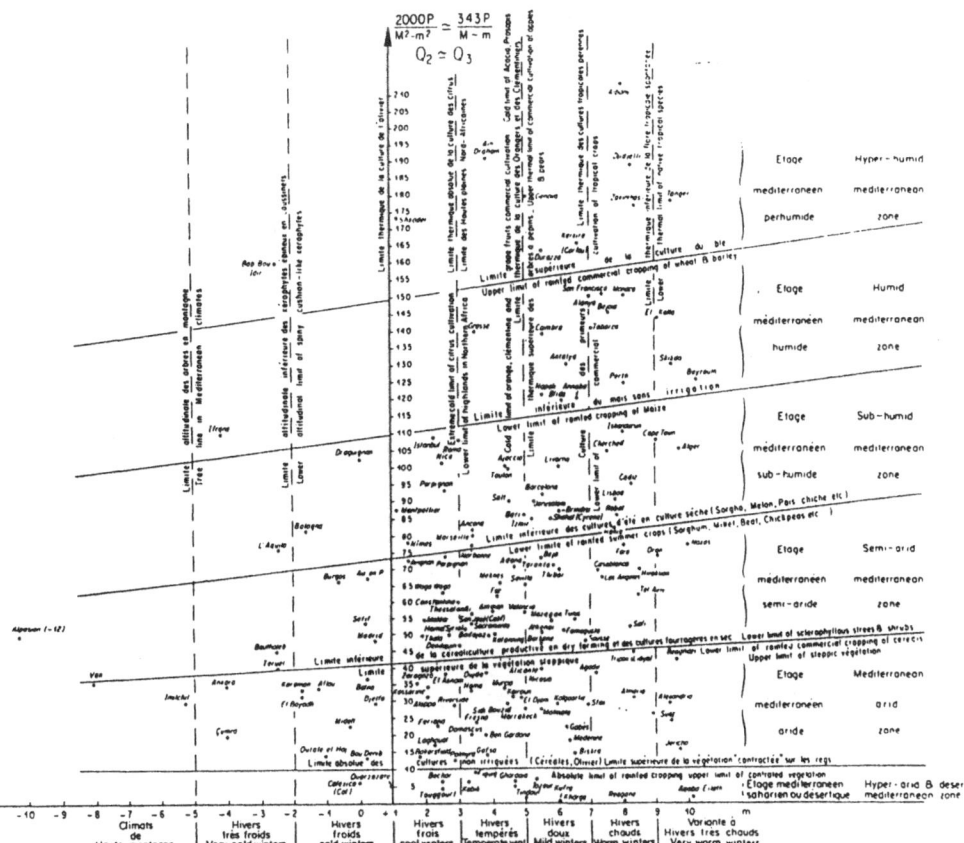

$$\frac{2000P}{M^2-m^2} \simeq \frac{343P}{M-m}$$
$$Q_2 \simeq Q_3$$

ZONE ISOCLIMATIQUE MÉDITERRANÉENNE CLIMAGRAMME D'EMBERGER ET ZONES HOMOGÈNES (HN LE HOUEROU 1973)
MEDITERRANEAN ISOCLIMATIC ZONING ECOCLIMATIC CLASSIFICATION (EMBERGER'S METHOD)

Figure 6. Geographical distribution of Mediterranean ecological zones and of some critical bioclimatic and agroclimatic thresholds under present-day conditions (Le Houérou, 1973).

334

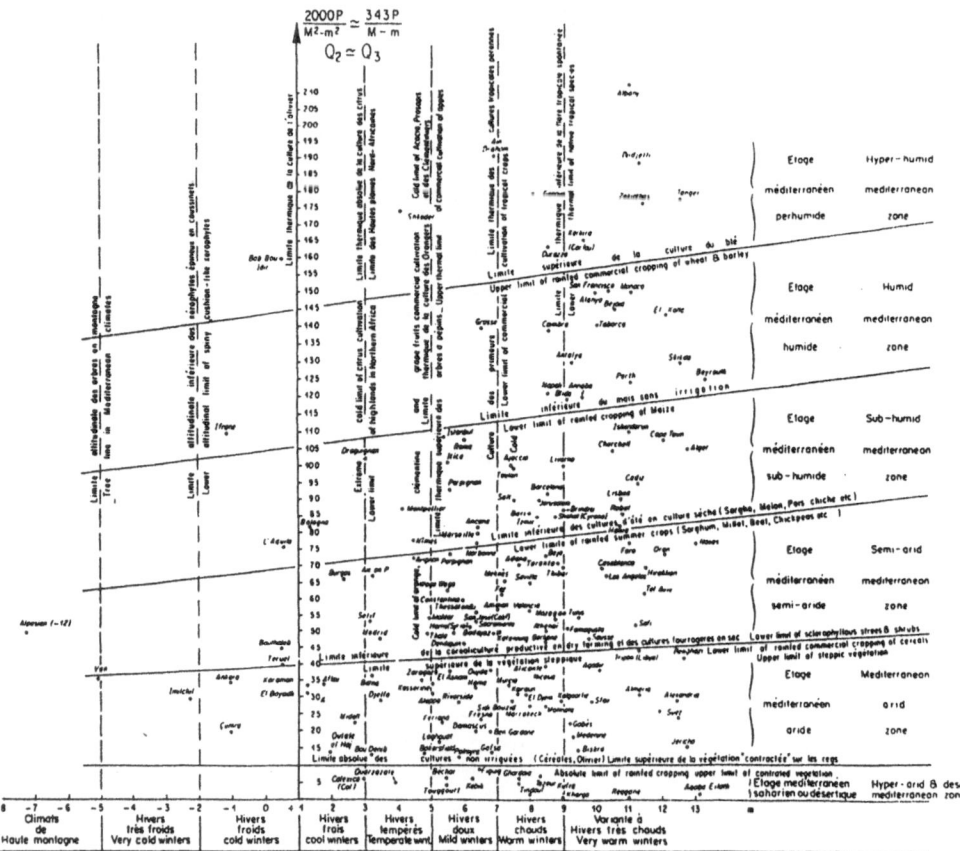

ZONES ISOCLIMATIQUES MÉDITERRANÉENNES DANS L'HYPOTHÈSE D'UN ACCROISSEMENT DES TEMPÉRATURES DE 3.0°C $\pm$ 1.5 INDUIT PAR CO^2 VERS L'ANNÉE 2050.

MEDITERRANEAN ISOCLIMATIC ZONING IN THE ASSUMPTION OF CO^2 INDUCED RAISE OF 3.0°C $\pm$ 1.5 IN TEMPERATURE TOWARDS THE YEAR 2050. (LE HOUEROU 1988)

Figure 7. Zonal shift of the same belts and thresholds under a 3°C rise of temperature.

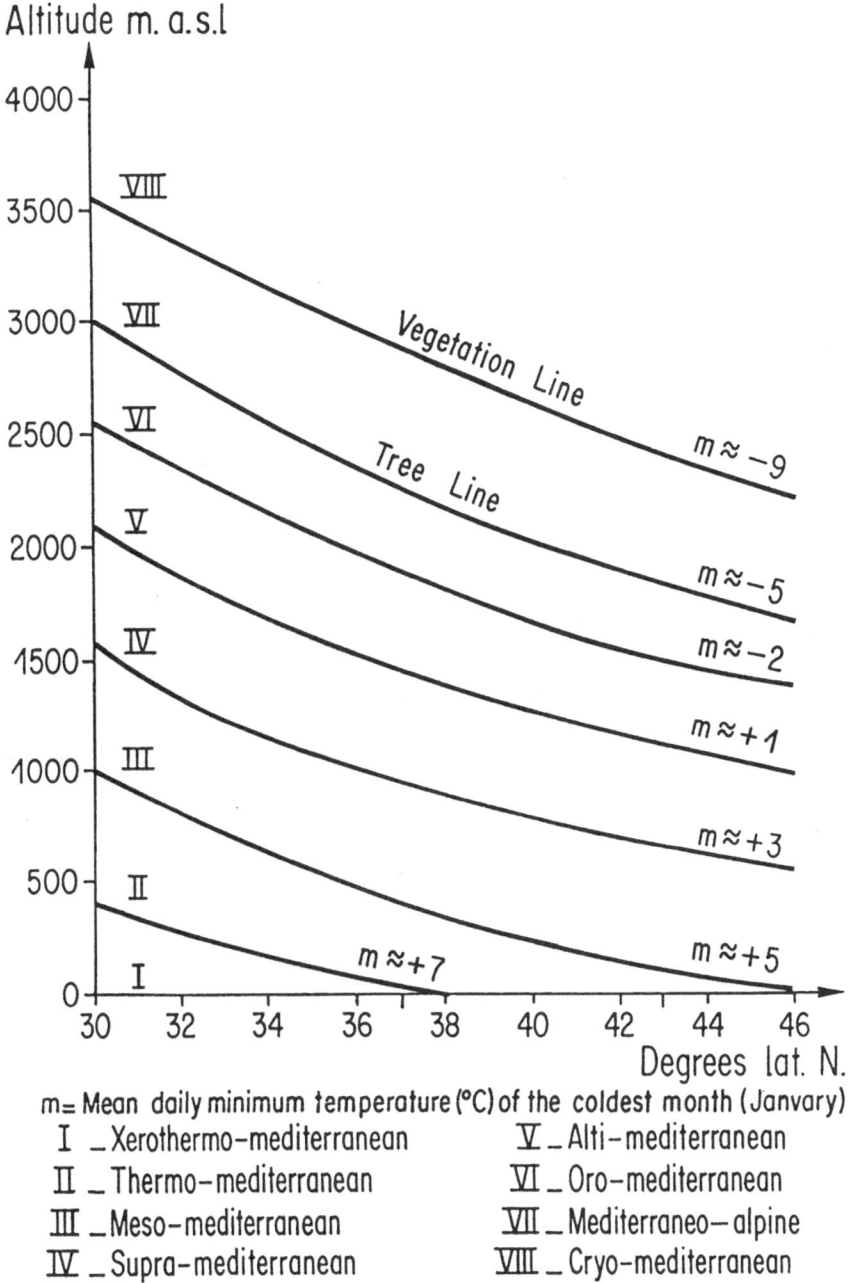

Figure 8. Altitudinal belts of natural vegetation types as a function of elevation and latitude (see also Table 4).

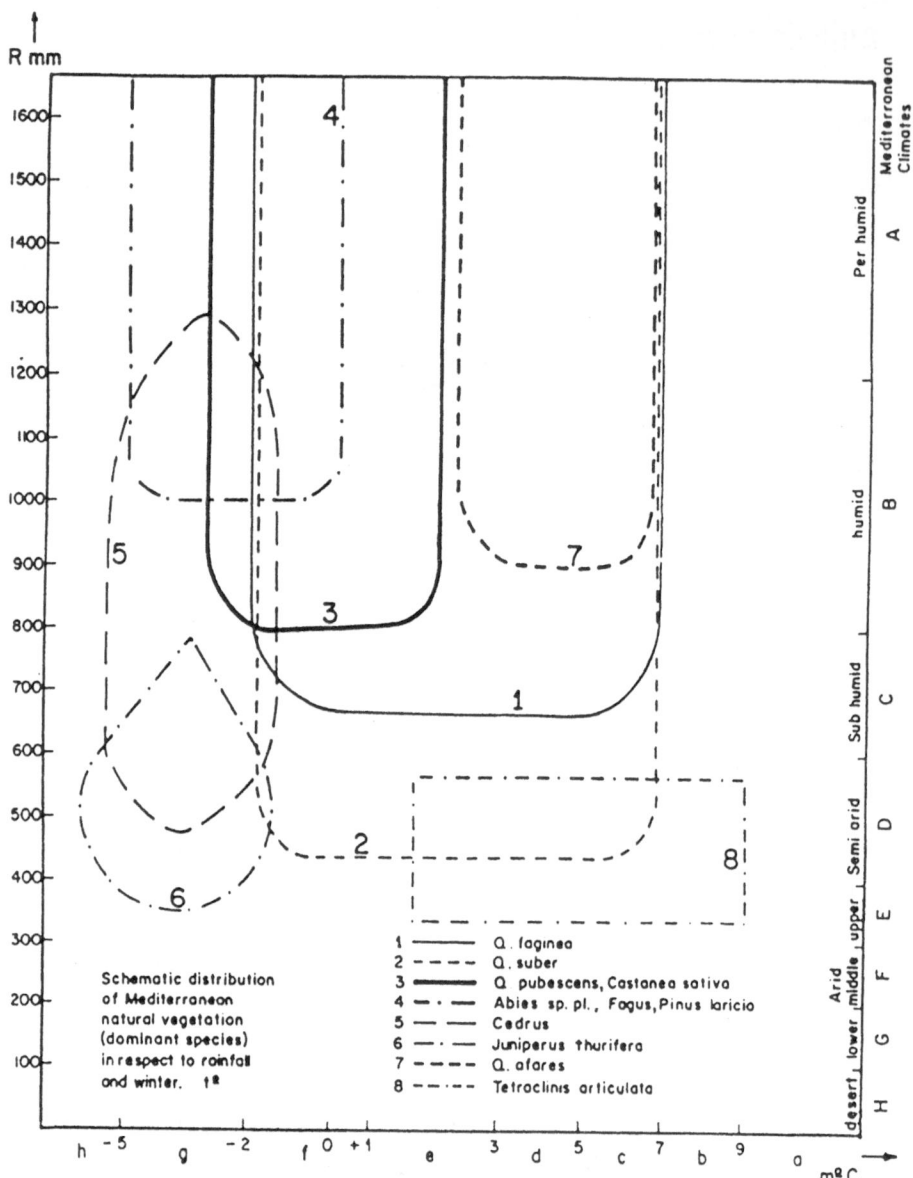

Figure 9. Distribution of main vegetation types as a function of mean annual rainfall and of winter cold stress in m (Le Houérou, 1973).

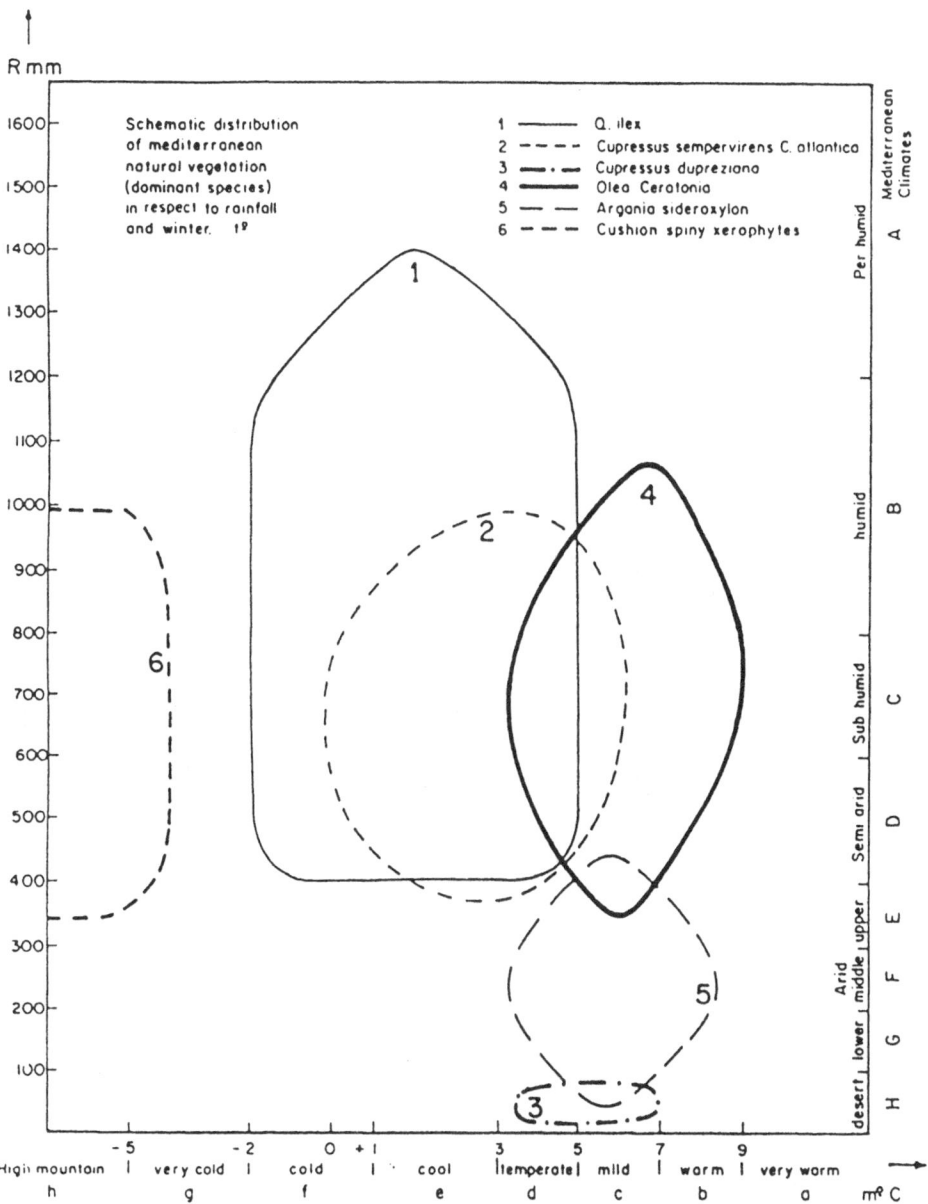

Figure 10. Distribution of vegetation as a function of P and m.

338

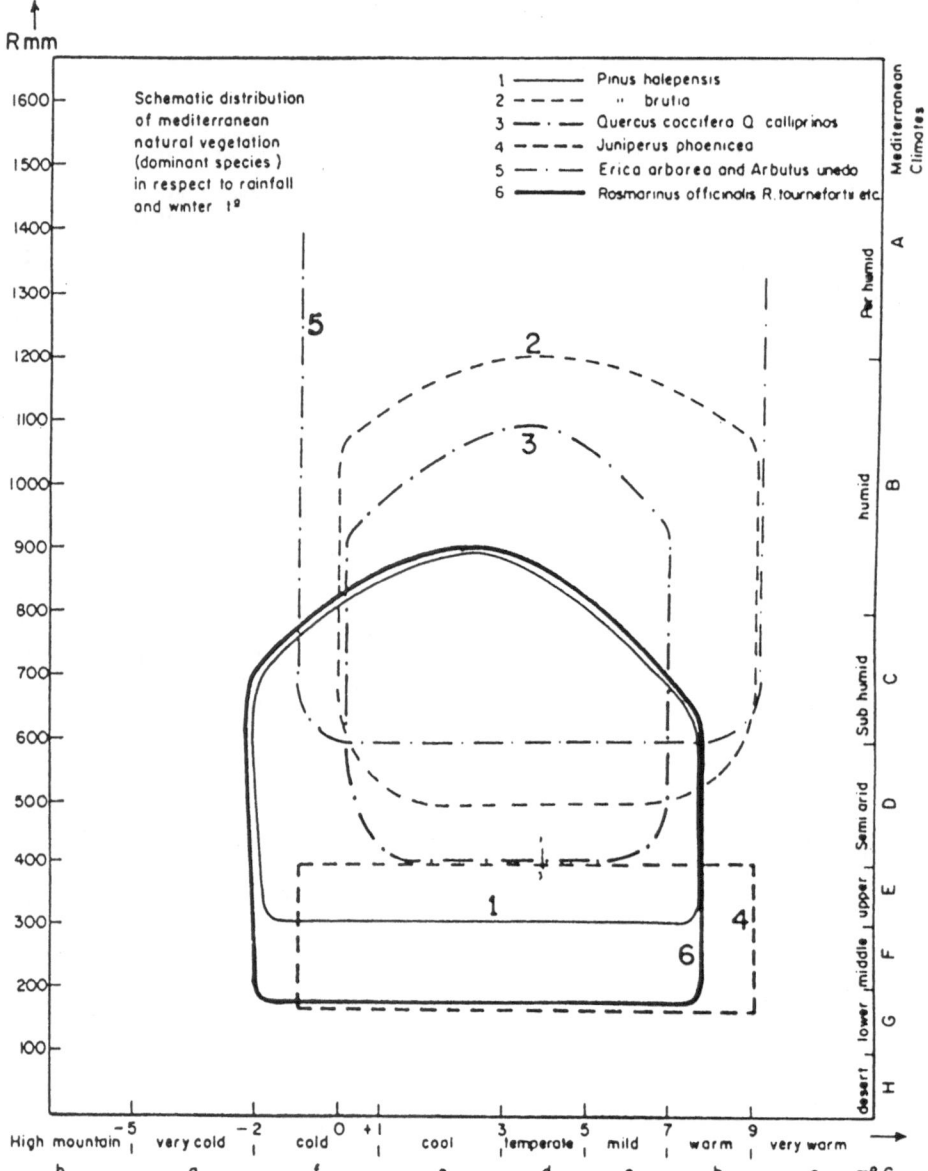

Figure 11. Distribution of vegetation as a function of P and m.

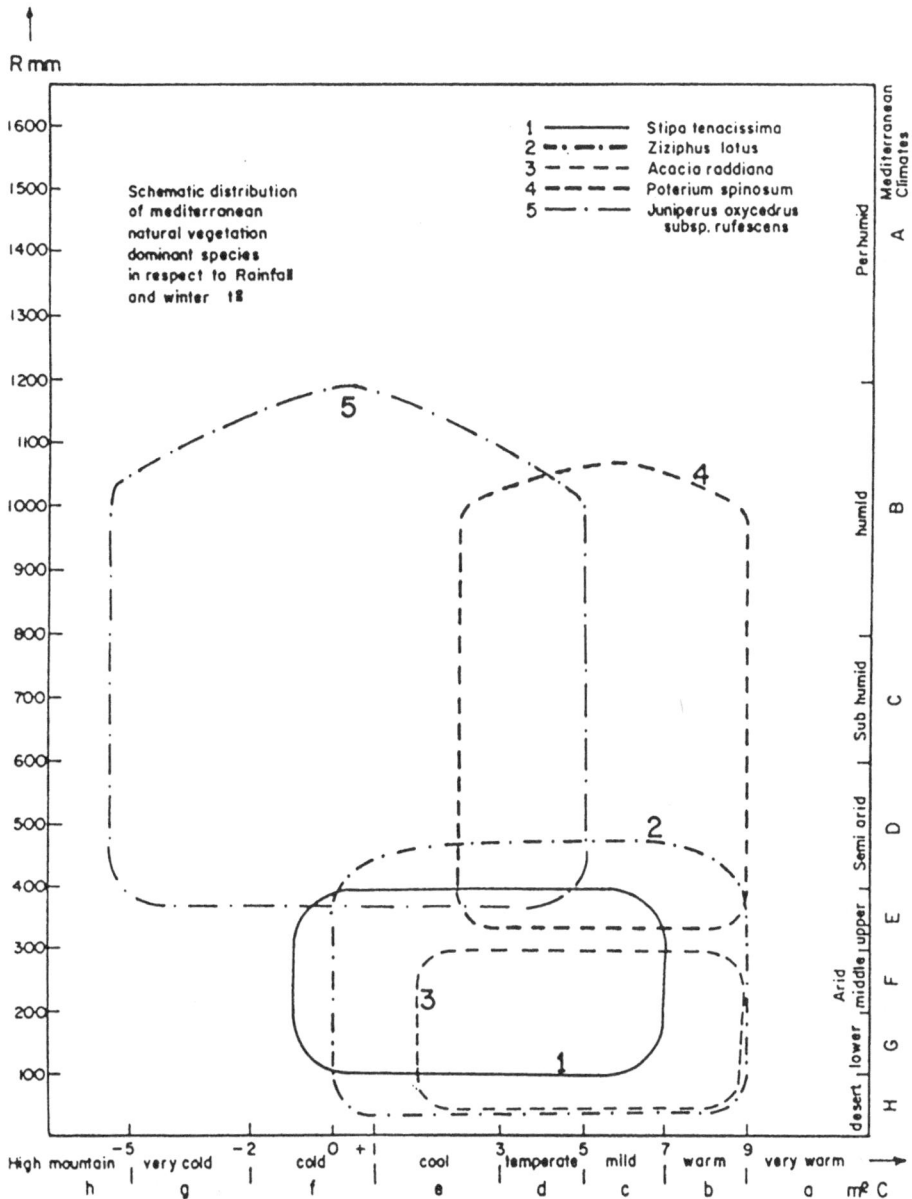

Figure 12. Distribution of vegetation as a function of P and m.

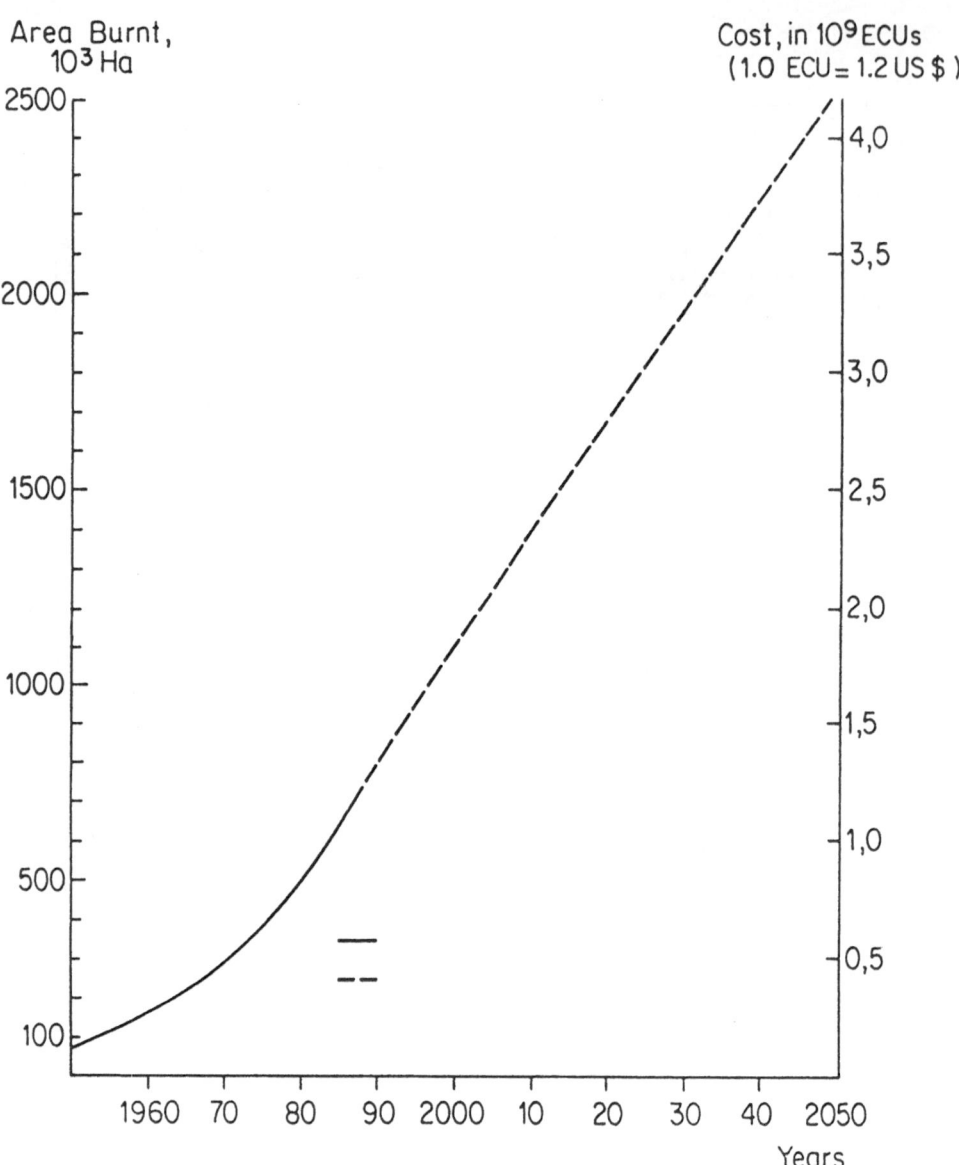

Figure 13. Evolution of the annual area of forest and shrubland burnt in the basin between 1960 and 1986.

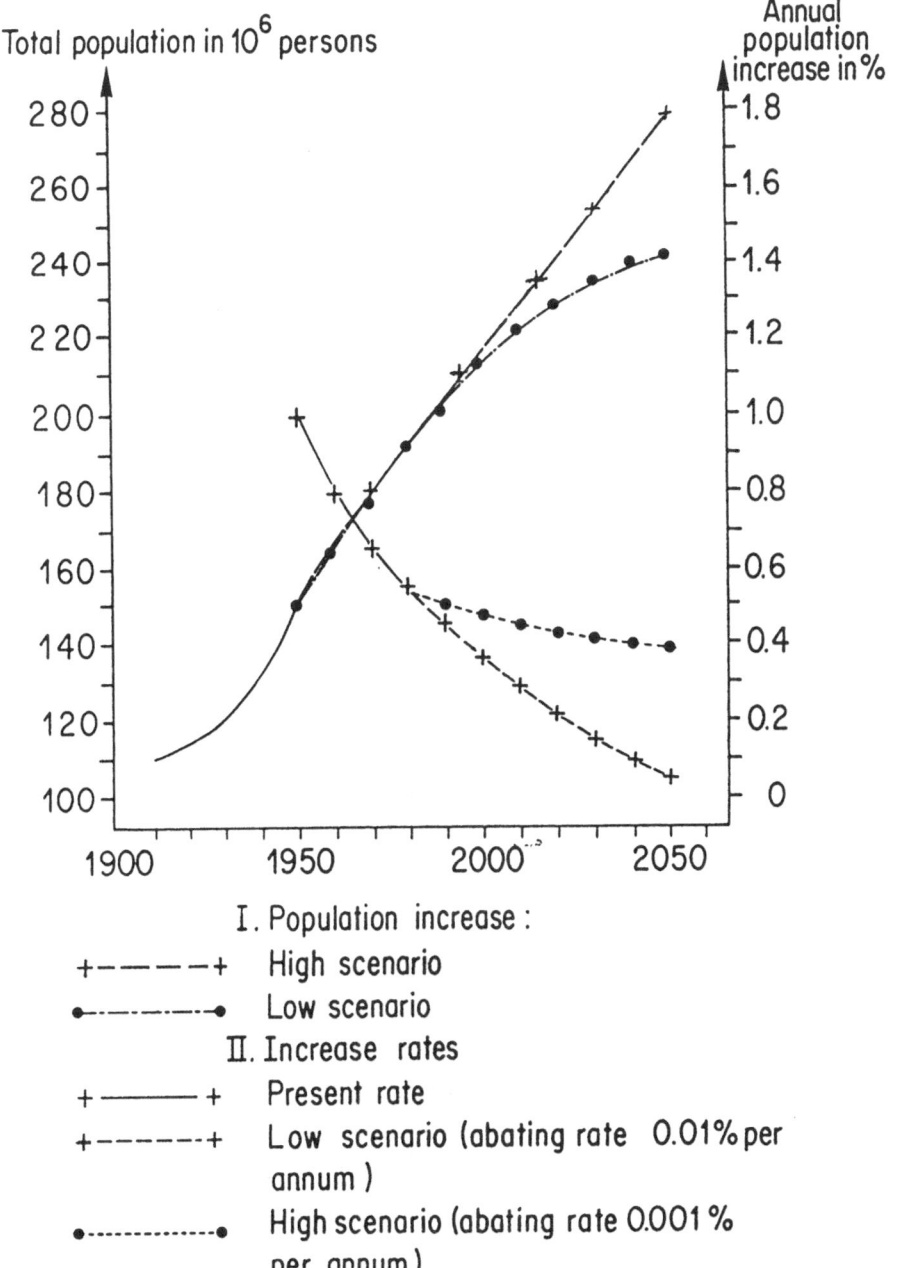

Figure 14. Evolution of the human population in the Euro-Mediterranean countries since the beginning of this century and projections to the year 2050.

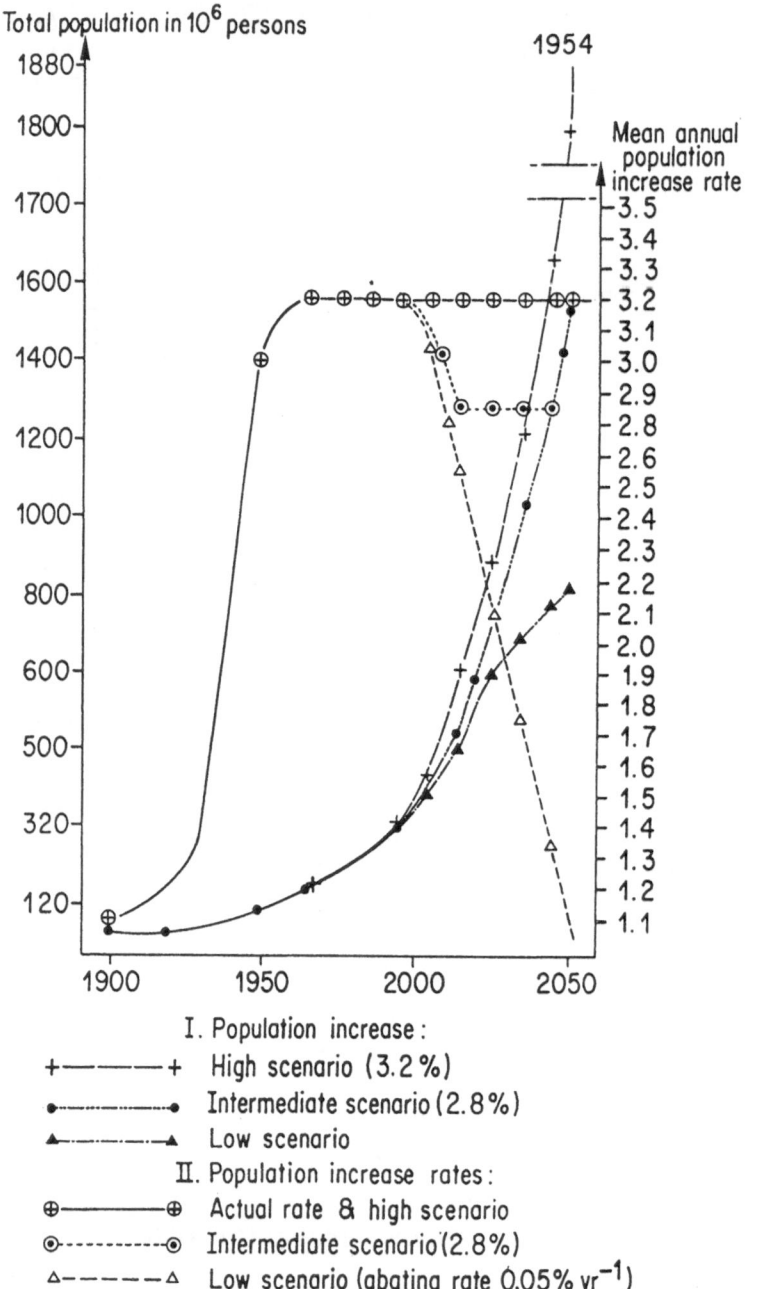

Figure 15. Evolution of the human population numbers in the Afro-Asian Mediterranean countries from 1900 to 1986 and projections to 2050 according to various growth rate scenarios.

TABLE 0. Surface area of the countries and provinces under Mediterranean climate in the Basin.

Area/Country	Percentage of country area with medit. climate	Surface area (10^3 km^2)	Area
Albania	18	5	Coastal plain
France	16	87	Coastal strip 50 km wide + Corsica
Greece	62	81	Except Macedonia
Italy	40	118	S.of Florence + coastal strip N.
Portugal	62	57	Southern 2/3
Spain	64	317	Except N.E.Corner & Pyrénées
Yugoslavia	10	25	Coastal strip & Adriatic Islands
EUROPE	9	690	
Algeria	50	1191	Except Central & South Sahara
Egypt	50	500	Except Central Sahara
Libye	50	880	Except Central Sahara
Morocco	100	620	
Tunisia	100	156	
AFRICA N.OF THE TROPIC OF CANCER	46	3347	
Cyprus	100	9	
Iran	80	1300	Except the Lut and Hyrcanian zones
Iraq	100	434	
Israel	100	20	
Jordany	100	97	
Lebanon	100	10	
Saudi Arabia & Emirates	50	1075	North of the Tropic of Cancer
Syria	100	144	
Turkey	30	231	Except the Euxinian Zone and interior mountains
ASIA S.OF 45° LAT.N. and W. 70° LONG.E.	45	3320	
Grand Total	-	7357	

TABLE 1. Bioclimatic zoning in the Mediterranean Basin (Le Houérou, 1975).

Main Zones (Water Stress)	Mean annual rainfall (mm)	P/PET ratios (Penman)	Sub-Zones based on Winter Cold Stress (m)							
			Very warm m>9	Warm 9>m>7	Mild 7>m>5	Tempe-rate 5>m>3	Cool 3>m>1	Cold 1>m>-2	Very cold -2>m>-5	Extrem. cold -5>m
Hyper Arid	100>P	0.05>R	+	+	+	+	+	+	-	-
Arid										
Lower	200>P>100	0.12>R>0.05	+	+	+	+	+	+	-	-
Middle	300>P>200	0.20>R>0.12	+	+	+	+	+	+	-	-
Upper	400>P>300	0.28>R>0.20	+	+	+	+	+	+	+	+
Semi Arid	600>P>400	0.43>R>0.28	+	+	+	+	+	+	+	+
Sub Humid	800>P>600	0.60>R>0.43	+	+	+	+	+	+	+	+
Humid	1200>P>800	0.90>R>0.60	+	+	+	+	+	+	+	-
Hyper Humid	P>1200	R>0.90	+	+	+	+	+	+	+	-

+ Combinaison present   - Combinaison absent

TABLE 2. Geographical location of Bioclimatic Zones in various Mediterranean countries.

Country	Hyper Arid	Arid Lower	Arid Middle	Upper	Semi Arid	Sub Humid	Humid	Hyper Humid	High Mountain
MEDITERRANEAN BASIN									
Portugal	-	-	-	-	+	+	+	+	(+)
Spain	-	(+)	+	+	+	+	+	+	+
France	-	-	-	-	(+)	+	+	+	+
Italy	-	-	-	(+)	+	+	+	+	+
Yugoslavia	-	-	-	-	(+)	+	+	+	+
Greece	-	-	-	(+)	+	+	+	+	+
Turkey	-	+	+	+	+	+	+	+	+
Syria	(+)	-	+	+	(+)	+	+	-	-
Lebanon	-	+	-	+	+	+	+	+	+
Israel	+	+	+	+	+	+	(+)	-	-
Jordan	+	+	+	+	+	+	(+)	-	(+)
Iran	+	+	+	+	+	+	+	-	+
Iraq	-	-	-	(+)	+	+	(+)	(+)	(+)
Cyprus	-	-	-	(+)	+	+	+	(+)	(+)
Malta	-	+	+	-	+	-	-	-	-
Egypt	+	+	+	+	+	-	-	-	-
Libya	+	+	+	+	+	(+)	-	-	-
Tunisia	+	+	+	+	+	+	+	(+)	+
Algeria	+	+	+	+	+	+	+	+	(+)
Morocco	+	+	+	+	+	+	+	+	+
OTHER MEDIT. ZONES									
South Africa	(+)	+	+	+	+	+	+	?	?
Australia	-	+	+	+	+	+	+	(+)	-
California	+	+	+	+	+	+	+	+	+
Mexico (Baja Calif.)	+	+	+	+	+	+	-	-	(+)
Chile	+	+	+	+	+	+	+	+	+

+ = Present ; - = Absent ; (+) = Pin-point ; ? = Dubious

TABLE 3. Vegetation zonation in Mediterranean mountains as a function of altitude and latitude.

Latitude (°N)	Mountains	Maximum altitude (m.a.s.l.) (rounded)	Country
30-32	Anti Atlas/Sarhro	2560	Morocco
28-35	Zagros	4550	Iran
31-32	High Atlas	4165	Morocco
32-33	Ksour Mounts	2240	Algeria
32-33	Liban	3090	Lebanon
32-34	Mid Atlas	3340	Morocco
35	Jebel Amour	1980	Algeria
35	Rif	2450	Morocco
35	Hodna & Aurès	2330	Algeria
35	Mount Ida	2500	Crete, Greece
35	Troodros	1950	Cyprus
35	Alborz	5600	Iran
35	Chambi	1540	Tunisia
35	Mount Alaoui	1560	Syria
36.5	Babor	2310	Algeria
34-38	Kurdistan	3610	Iran, Iraq, Turkey
37	Sierra Nevada	3480	Spain
37	Taygete	2410	Greece
37-38	Taurus	3920	Turkey
38	Etna	3260	Italy
38	Aspromonte	1960	Italy
38.5	Parnassus	2510	Greece
40	Olympus	2920	Greece
41	Sierra de Gredos	2590	Spain
42	" " Guadarrama	2470	Spain
42	Mte Terminillo	2210	Italy
42	Gran Sasso d'Italia	2910	Italy
43	Eastern Pyrénées	2300	Spain, France
44	Ventoux & S. Alps	2000	France
44.2	Aigoual (Cevennes)	1570	France
44.5	Mt Lozère (Cevennes)	1700	France

TABLE 4. Land-Use in the Northern Mediterranean Countries : Human population numbers and projections (in 10⁶ persons).

Country	1950	1960	1970	1980	1985	1990	2000	2010	2020	2030	2040	2050
Albania	1.2	1.6	2.1	2.7	2.8	3.0	3.3	3.6	3.9	4.2	4.5	4.7
France	41.7	45.7	50.8	53.9	55.2	56.5	59.3	62.9	66.0	68.6	70.7	72.8
Greece	7.6	8.2	8.8	9.6	10.0	10.4	11.2	12.0	12.7	13.5	14.2	14.8
Italy	46.8	49.5	53.7	56.4	57.1	57.8	59.0	60.2	63.4	64.7	66.0	67.3
Malta	0.3	0.3	0.3	0.4	0.4	0.4	0.4	0.4	0.4	0.4	0.5	0.5
Portugal	8.4	8.8	8.6	9.6	10.2	10.5	11.5	12.4	13.3	14.1	14.8	15.4
Spain	27.9	30.0	33.6	37.2	38.5	39.9	42.7	45.7	48.4	50.8	52.8	54.4
Yugoslavia	16.3	18.3	20.4	22.3	23.1	24.2	25.4	26.7	28.0	29.1	30.0	30.6
Total	150.2	162.4	169.8	192.1	197.3	202.7	212.8	223.9	236.1	245.4	253.5	260.5
Mean annual growth rate %		0.81	0.50	1.30	0.60	0.60	0.50	0.50	0.50	0.40	0.30	0.30

TABLE 5. Land-Use in the Southern Mediterranean Basin : Human population and projections, in 10⁶

Country	1900 (1)	1950	1960	1970	1980	1985	1990 (2)	2000	2010	2020	2030	2040	2050
Algeria	4.0	8.7	11.0	13.7	18.6	21.7	25.4	34.0	45.6	61.1	81.9	109.7	147.0
Egypt	8.0	20.0	25.9	33.3	42.0	46.9	52.4	65.0	80.6	99.9	123.9	153.6	190.5
Libya	0.7	1.0	1.2	2.0	3.0	3.6	5.2	6.3	9.4	14.1	21.2	31.8	47.7
Morocco	3.8	7.5	11.6	15.6	20.3	21.9	26.4	34.3	44.6	58.0	75.4	98.0	127.4
Tunisia	1.5	3.5	4.2	5.3	6.4	7.1	7.9	9.8	12.9	14.8	18.2	22.4	27.6
Cyprus	0.30	0.49	0.56	0.63	0.65	0.67	0.70	0.73	0.75	0.77	0.79	0.81	0.83
Iran	8.0	16.0	20.2	28.7	38.1	44.6	52.2	70.0	93.8	125.4	168.4	265.7	302.4
Iraq	2.6	5.2	7.1	9.7	13.1	15.9	17.7	23.9	32.3	43.6	58.9	79.5	107.3
Israel	0.3	1.4	2.1	2.9	3.9	4.2	5.3	7.1	9.6	13.8	17.5	23.6	31.3
Jordan	0.6	1.2	1.7	2.2	3.2	3.5	4.2	5.5	7.1	9.2	12.0	15.0	20.3
Lebanon	0.8	1.5	1.8	2.7	3.2	2.7	3.1	4.0	5.2	6.7	8.7	11.3	14.7
Saudi Arabia	1.7	3.2	4.0	5.3	8.4	11.5	16.1	22.5	31.5	44.1	61.7	86.4	121.0
Syria	1.9	3.4	4.7	6.1	8.6	10.5	12.8	17.9	25.1	35.1	49.2	68.9	96.4
Turkey	10.0	20.7	27.6	35.2	45.3	49.3	56.6	70.8	88.5	110.6	138.3	172.9	216.1
Total	44.2	93.8	123.7	163.3	214.8	244.1	286.0	371.8	486.0	636.7	836.1	1100.2	1451.0

Overall exponential annual growth rate : 2.8 % between 1950 and 1975 and between 1960 and 1985 ; Doubling period : 25 years.

(1) Estimate. (2) Projections from 1990 onward are made on the basis of an unabated growth rate identical to the 1950-1985 period ; all attempts to curb this rate have failed so far (Egypt, Tunisia).

348

TABLE 6. Land-Use in the Mediterranean Basin : Agricultural population in 10^3 and in % of total population.

Country	1950 nr	1950 %	1960 nr	1960 %	1970 nr	1970 %	1980 nr	1980 %	1985 nr	1985 %
Algeria	6500	75	7800	71	8279	60	9335	49	5720	26
Egypt	14200	71	14760	57	17848	54	21145	50	20407	43
Libya	800	80	720	60	634	32	467	16	519	14
Morocco	5550	74	8240	71	8601	57	10301	51	8965	40
Tunisia	2630	75	2860	68	2553	50	2578	41	2032	28
N. Africa	29680	73	34380	64	37915	55	43826	48	37643	36
Cyprus	290	60	310	55	233	39	212	34	153	23
Iran	9600	60	11510	57	12761	45	14319	38	13620	30
Iraq	3120	60	3550	50	4362	47	5260	40	3885	24
Israel	420	30	378	18	288	10	265	7	213	5
Jordan	600	50	560	33	776	34	836	26	272	7
Lebanon	900	60	990	55	487	20	263	10	304	11
Saudi Arabia	2650	80	2880	72	3794	66	5389	60	5136	43
Syria	2210	65	2585	55	3200	51	4265	47	2915	27
Turkey	16974	82	20980	76	23926	68	24641	58	24217	48
N.& M. East	36674	69	43743	63	49827	53	55448	44	50715	35
Albania	800	80	1200	80	1416	66	1647	60	1606	52
France	12000	30	10000	25	6961	14	4639	9	3310	6
Greece	4000	53	3800	50	4041	46	3574	37	2363	24
Italy	18000	44	20000	44	10077	19	6363	11	4318	8
Portugal	3000	40	3500	42	2875	33	2577	26	2108	21
Spain	11900	51	13200	48	8475	25	6185	17	4861	13
Yugoslavia	10606	76	12000	73	10145	50	8348	37	5493	24
Euro-Medit. Countries	60306	40	63700	39	34889	20	33333	17	24059	129

350

TABLE 7. Land-Use in the Southern Mediterranean Basin : Land Area, Forest and Shrublands.

Country	Land area	...Forest & Shrublands, 10^3 Ha...					
		1948/1952	1961/1965	1969/1971	1974/1976	1979/1981	1984/1986
Algeria	238174	3050	2549	2424	4122	4384	4385
Egypt	100145	2	2	2	2	2	2
Libya	175954	462	491	533	560	600	650
Morocco	44505 (62000)	5385	5302	5164	5182	5200	5200
Tunisia	16415	980	674	470	505	540	557
Cyprus	925	171	171	171	171	171	171
Iran	164800	18000 ?	18000 ?	18000 ?	18000 ?	18000 ?	18000 ?
Iraq	43492	1770	1953	1940	1930	1910	1900
Israel	2070	75	94	109	116	116	116
Jordan	9774	525	125	56	60	63	69
Lebanon	1040	92	94	95	90	85	80
Saudi Arabia	214969	400	1688	1630	1601	1200	1200
Syria	18518	432	446	468	445	466	516
Turkey	78058	10854 ?	20100	20170	20170	20199	20199
Total	1126334	42198	51689	51232	52954	52935	53044
Afforestation Rate % Land Area		4	5	5	5	5	5

TABLE 8. Evolution of Land-Use in the Mediterranean Basin between 1965 and 1984.

areas in 10³ Ha

Countries	Area	Forest & Shrubland 1965 Aff. Rate	1984 Aff. Rate	Change	Change %	Cropland 1965 Cult. Rate	1984 Cult. Rate	Change	Change %
Albania	2875	1266 44.0	1038 36.1	− 228	−18.0	501 17.4	713 24.8	+ 212	+42.3
France	54703	11905 21.8	14603 26.7	+2698	+22.7	20542 37.6	18812 34.4	−1730	− 8.4
Greece	13080	2545 19.5	2620 20.2	+ 75	+ 3.0	3854 29.5	3974 30.4	+ 120	+ 3.1
Italy	30123	5984 19.9	6410 21.3	+ 426	+ 7.1	5258 50.6	2233 40.6	−4125	−27.0
Portugal	9208	3165 34.4	3641 39.5	+ 476	+15.0	4370 47.5	3545 38.5	− 825	−18.9
Spain	50478	13160 26.1	15625 31.0	+2465	+18.7	20594 40.8	20540 40.7	− 54	− 0.3
Yugoslavia	25580	8744 34.2	9294 36.3	+ 550	+ 6.3	8266 32.3	7758 30.3	− 508	− 6.2
Subtotal Europe	186047	46769 25.1	53231 28.6	+6462	+13.8	73452 39.5	67575 36.3	−5877	− 8.0
Gran Total Whole Basin	978919	98355 10.0	106066 11.0	+7711	+ 8.0	148535 15.3	146103 14.9	−4432	− 3.0

Source : FAO Production Yearbooks.

TABLE 9 : Land-Use in Euro-Mediterranean Countries : Evolution and trends between 1965 and 1984 (areas in 10^3 Ha)

Country				1965			
	Land area	Arable	Cereals	Perman. Crops	Perman. Past.	Forest & Shrubl.	Misc.non Agric.
Albania	2740	497	321	55	736	1266	241
France	54475	21067	9160	1774	13221	11905	8382
Greece	13080	3800	1730	859	5100	2545	1635
Italy	29402	15454	6060	2763	5096	5984	2868
Portugal	9164	4332	1860	600	530	3165	1137
Spain	49978	20709	6830	4583	12300	13160	3809
Yugoslavia	25540	8349	5150	693	6472	8744	1975
Total	184379	74208	31161	10727	43455	46769	20047
				1984			
Albania	2740	713	360	121	400	1038	589
France	54563	18812	9707	1344	12385	14603	8763
Greece	13080	3974	1482	1026	5255	2620	1231
Italy	29402	12233	4810	3133	4930	6410	5829
Portugal	9164	3545	1124	585	530	3641	1448
Spain	49940	20540	7485	4920	10640	15625	3135
Yugoslavia	25540	7758	4230	740	6379	9294	2109
Total	184435	67575	29198	11369	40519	53231	23101
Difference Ha 1984-1965	+56	-6633	-1963	+1142	-2936	+6462	+3054
Difference % of 1965	--	-9	-6	+11	-7	+14	+15

Source : FAO, Production Yearbooks 1965 and 1985.

TABLE 10. Cereal cultivation in Euro-Mediterranean Countries ; Evolution and trend of cropped areas 1965-1984 (areas in 10^3 Ha)

Country	1965						
	Wheat	Rye	Barley	Oats	Maize	Rice & Misc.	Total
Albania	125	10	9	23	150	---	321
France	4520	221	2430	1070	871	170	9159
Greece	1258	16	188	120	144	8	1774
Italy	4288	48	197	367	1028	---	6056
Portugal	628	316	126	271	484	---	1960
Spain	3991	393	1374	502	478	6	6826
Yugoslavia	1683	146	405	321	2550	25	5152
Total Ha	16493	1150	4729	2674	5705	209	31149
Total %	53	4	15	9	18	1	100
	1984						
Albania	190	11	13	20	95	31	360
France	4832	90	2255	411	1858	261	9707
Greece	898	9	310	43	203	19	1482
Italy	3032	9	468	184	910	207	4810
Portugal	310	163	90	165	360	36	1124
Spain	2024	222	4155	465	516	103	7485
Yugoslavia	1348	45	264	151	2401	21	4230
Total Ha	12634	549	7555	1439	6343	678	29198
Total %	43	2	26	5	22	2	100
Difference Ha 1984-1965	-3859	-601	+2826	-1235	+638	+469	-1951
Difference % of 1965	-23	-52	+60	-46	+11	+224	-6

Source : FAO, Production Yearbooks 1965 and 1985.

TABLE 11a. Yields of main cereal crops : evolution and trends between 1965 and 1984, in Kg of grain per hectare.

| Country | ---------1965------- | | | ----------1984----------- | | |
	Wheat	Barley	Maize	Wheat	Barley	Maize
Albania	800	990	1030	2785	2756	4211
France	3270	3040	3980	6008	5066	6372
Greece	1650	1660	1730	1996	2106	8867
Italy	2290	1530	3230	2808	3479	6979
Portugal	980	570	950	1242	1033	1583
Spain	1120	1380	2390	2631	2570	6455
Yugoslavia	2060	1680	2320	3605	2667	4120

TABLE 11b. Yield increase in main cereal crops between 1965 and 1984, in Kg/ha and % of 1965.

| | Wheat | | Barley | | Maize | |
	Kg/ha	%	Kg/ha	%	Kg/ha	%
Albania	1985	248	1766	178	3181	309
France	2738	84	2026	67	2392	60
Greece	346	21	446	27	7137	413
Italy	528	23	1949	127	3749	116
Portugal	262	27	463	81	633	67
Spain	1511	135	1190	86	4065	170
Yugoslavia	1545	75	987	59	1800	78

TABLE 11c. Mean annual yield increase between 1965 and 1984, in Kg of grain per hectare and in % of 1965.

| | Wheat | | Barley | | Maize | |
	Kg/ha	%	Kg/ha	%	Kg/ha	%
Albania	99	15.5.	88	8.9	159	15.1
France	137	4.2	101	3.3	120	3.0
Greece	17	1.0	22	1.3	357	20.6
Italy	26	1.1	97	6.3	187	5.8
Portugal	13	1.3	23	4.0	32	3.3
Spain	76	6.7	60	4.3	203	8.5
Yugoslavia	77	3.7	49	2.9	90	3.9

TABLE 12. Land-Use in the Southern Mediterranean Basin : Cultivated Land (in 10^3 Hectares)

Country	1950	1960	1970	1980	1985
Algeria	6784	6820	6200	7500	7610
Egypt	2780	2610	2840	2450	2490
Libya	2509	1070	1725	1750	1790
Morocco	7858	8560	7500	8000	8400
Tunisia	4334	2970	4480	4700	4920
Cyprus	432	430	430	430	430
Iran	11593	16850	15700	13710	14830
Iraq	7496	5460	4990	5450	5450
Israel	996	410	410	410	420
Jordan	1140	890	370	400	420
Lebanon	296	280	325	310	300
Saudi Arabi	373	210	870	1110	1180
Syria	6130	6010	5910	5680	5620
Turkey	26384	25360	27380	28480	27540
Total	79105	77930	79130	80380	81400
Cultivated Rate % of Land Area	7	7	7	7	7

Change 1950 to 1985 : 81400 − 79105 = 2295 = 3 %

TABLE 13. Evolution of Land-Use in the Mediterranean Basin between 1965 and 1984.

areas in 10³ Ha

Countries	Area	Forest & Shrubland				Cropland			
		1965 Aff. Rate	1984 Aff. Rate	Change	Change %	1965 Cult. Rate	1984 Cult. Rate	Change	Change %
Algeria	238174	2549 1.1	4384 1.8	+1835	+72.0	6863 2.9	7744 3.2	+ 881	+12.8
Libya	175954	491 0.2	640 0.3	+ 149	+30.0	2509 1.4	2115 1.2	- 394	-15.7
Morocco	44655	5302 11.9	5200 11.6	- 102	- 1.9	7066 15.8	8331 18.7	+1265	+17.8
Tunisia	16361	674 4.1	555 3.4	- 119	-17.7	4406 26.9	4687 28.6	+ 281	+ 6.4
Iran	164800	18000 10.9	18020 10.9	+ 20	+ 0.1	15353 9.3	14840 9.0	- 528	- 3.4
Iraq	43492	1953 4.5	1900 4.4	- 53	- 2.7	4810 11.1	5450 12.5	+ 640	+13.3
Israel	2077	94 4.5	116 5.6	+ 22	+23.4	401 19.3	437 21.0	+ 36	+ 9.0
Jordan	9774	125 1.3	41 0.4	- 84	-67.2	391 4.0	415 4.2	+ 24	+ 6.1
Lebanon	1040	94 9.0	82 7.9	- 12	-12.8	276 26.5	298 28.7	+ 22	+ 8.0
Saudi Arabia	214969	1688 0.8	1200 0.6	- 488	-28.9	705 0.3	1156 0.5	+ 451	+64.0
Syria	18518	446 2.4	498 2.7	+ 52	+11.7	6523 35.2	5654 30.5	- 869	-13.3
Turkey	78058	20170 25.8	20199 25.9	+ 29	+ 0.1	25775 33.0	27411 35.1	+1636	+ 6.4
Subtotal Africa + Asia	792872	51586 6.5	52835 6.6	+1249	+ 2.4	75083 9.5	78528 9.9	+3445	+ 6.4

Source : FAO Production Yearbooks.

TABLE 14. Land-Use in the Mediterranean Basin : Irrigated Land
(in 10³ Ha).

Country	1950	1960	1970	1985	Differ. 1985-1970	Differ. % of 1970
Algeria	200	220	240	340	100	+ 42
Egypt	2500	2600	2840	2490	− 350	− 12
Libya	100	130	180	230	50	+ 28
Morocco	100	200	340	520	130	+ 53
Tunisia	40	70	90	220	130	+ 144
Northern Africa (intercountry averages)	2940	3220	3690	3800	110	+ 3
Cyprus	70	80	100	100	0	0
Iran	3800	4600	5200	5740	540	+ 10
Iraq	600	800	1480	1750	270	+ 19
Israel	100	140	170	270	100	+ 59
Jordan	50	50	30	40	10	+ 33
Lebanon	70	70	70	90	20	+ 29
Saudi Arabia	100	120	370	420	50	+ 14
Syria	300	400	450	650	200	+ 44
Turkey	1300	1700	1800	2150	350	+ 19
N. & M. East (intercountry averages)	6390	7960	9670	11210	1540	− 16
Albania	60	140	280	390	110	+ 39
France (x 0.8)	20	30	600	940	340	+ 57
Greece	340	410	730	1100	370	+ 51
Italy (x 0.8)	125	140	2050	2400	350	+ 17
Portugal	35	40	622	632	10	+ 2
Spain	130	210	2380	3220	840	+ 35
Yugoslavia (x 0.4)	40	60	50	70	20	+ 40
Medit. Europe	740	1020	6720	8750	2030	+ 30
Overall Medit. Basin	10070	12200	20080	23760	3680	+ 18

TABLE 15. Land-Use in the Southern Mediterranean Basin : Cereal cultivation (in 10³ Ha ; Yields in Kg/Ha).

Country	........Hectareage..........					..........Yields..........		
	1950	1960	1970	1980	1985	1969/1971	1979/1981	1984/1986
Algeria	2960	3160	3230	2970	2990	614	693	867
Egypt	1780	1870	1940	2010	1904	3847	4C52	4471
Libya	229	604	603	538	407	301	430	616
Morocco	4080	4180	4640	4410	4820	985	811	1145
Tunisia	1550	1550	1280	1420	1930	634	828	808
Cyprus	133	143	96	64	72	1109	1780	1610
Iran	3152	4480	7150	7800	8520	832	1149	1185
Iraq	2090	2500	3220	2160	2980	1079	832	1020
Israel	90	135	130	129	107	1445	1340	1679
Jordan	266	232	260	165	139	696	511	542
Lebanon	108	100	72	34	20	870	1307	1225
Saudi Arabia	94	160	190	388	630	1352	820	3355
Syria	1490	2220	2500	2640	2710	588	1156	899
Turkey	8250	12290	13240	13570	13750	1357	1860	1961
Total	26272	33624	38551	38298	40979	Mean 1122	1290	1527

Change   1950-1985 : 40979 - 26272 = + 14707 = + 56 %.
Increase 1970-1985 : 1527 - 1122 = + 405 = + 36 %.

TABLE 16. Land-Use in the Southern Mediterranean Basin : Cereal production (in 10³ metr.tons).

Country	1949/51	1959/61	1969/71	1979/81	1984/86	Change 1950-1985
Algeria	1053	1770	2060	2055	2920	+ 997 = + 51 %
Egypt	4101	6080	7470	8130	8860	+ 4759 = + 116 %
Libya	97	127	152	225	235	+ 138 = + 142 %
Morocco	2740	3150	4230	3580	5320	+ 2580 = + 94 %
Tunisia	690	603	615	1150	1270	+ 520 = + 75 %
Cyprus	98	151	165	114	113	+ 15 = + 15 %
Iran	3090	4560	5790	8950	10280	+ 7190 = + 233 %
Iraq	1410	1850	1970	1803	2204	+ 794 = + 56 %
Israel	75	216	200	239	175	+ 100 = + 133 %
Jordan	194	250	165	91	63	- 131 = - 68 %
Lebanon	99	92	62	41	29	- 70 = - 71 %
Saudi Arabia	106	230	255	300	1090	+ 984 = + 928 %
Syria	1200	1800	3300	3070	2390	+ 1190 = + 99 %
Turkey	9060	14830	15990	25230	27390	+18330 = + 202 %
Total	24913	35709	42424	54978	62339	+37426 = + 150 %

Production per inhabitant (kg/yr)

	261	289	260	256	255	- 6 = - 2 %

TABLE 17. Land-Use in the Southern Mediterranean Basin : Wheat cultivation (in 10³ Ha).

Country	1948/52	1961/65	1969/71	1974/76	1979/81	1984/86
Algeria	1597	1969	2214	2240	1943	1625
Egypt	605	557	551	583	577	500
Libya	124	149	163	210	251	270
Morocco	1287	1578	1952	1843	1673	1991
Tunisia	917	1002	908	967	887	824
Cyprus	75	69	74	64	15	12
Iran	2085	3580	5370	5840	5894	6156
Iraq	936	1595	1216	1513	1515	1045
Israel	35	58	111	100	96	82
Jordan	182	268	155	158	111	55
Lebanon	70	68	46	47	26	17
Saudi Arabia	36	92	57	73	71	519
Turkey	4770	7959	8732	9140	9265	9434
Total	13713	20340	22828	24385	26407	23687

Change 1950 to 1980 : 26407 - 13713 = + 12694 = + 93 % ; Annual increase : 2.3%

TABLE 18. Land-Use in the Southern Mediterranean Basin : Wheat production, in 10³ metr. tons.

Country	1948/52	1961/65	1969/71	1974/76	1979/81	1984/86	Prod. per inhabit. Kg/Yr	
							1948/52	1984/86
Algeria	996	1254	1359	1523	1270	1374	113	63
Egypt	1111	1459	1509	1960	1864	1872	54	42
Libya	11	37	41	82	125	174	11	48
Morocco	786	1336	1819	1872	1500	2717	86	124
Tunisia	452	446	520	867	837	855	129	121
Cyprus	48	63	75	75	20	16	97	25
Iran	1879	2873	3946	5438	6100	6430	122	144
Iraq	448	849	1080	1162	854	992	86	62
Israel	23	90	160	241	200	142	17	34
Jordan	128	180	127	120	67	47	101	13
Lebanon	51	64	39	60	32	22	34	9
Saudi Arabia	49	132	101	126	160	1846	15	160
Syria	761	1093	763	1657	1878	1584	221	151
Turkey	4770	8585	11423	14163	17058	17756	230	360
Total	11513	18461	22962	29346	31957	35827	120	147
Surface	13713	20340	22828	24385	26407	23687	---	---
Yield, Kg/Ha	840	910	1010	1200	1210	1510	---	---
Production per Inhabit. Kg/Yr	120	129	140	154	149	147	---	---

TABLE 19. Land-Use in the Southern Mediterranean Basin : Wheat Yields, in Kg/Ha.

Country	1948/52	1961/65	1966/70	1971/75	1976/80	1981/85
Algeria	620	640	560	600	650	660
Egypt	1840	2620	2580	3100	3670	3560
Libya	90	240	250	400	540	560
Morocco	610	850	860	970	970	860
Cyprus	640	910	1240	1210	1320	1330
Iran	900	800	910	890	1110	1000
Iraq	480	530	540	730	930	870
Israel	660	1540	1600	2360	1680	2030
Jordan	700	670	660	730	490	790
Lebanon	730	940	860	1090	1290	1230
Saudi Arabia	1370	1430	1460	1400	1360	2780
Syria	770	780	700	860	1010	1350
Turkey	1000	1080	1200	1400	1810	1890
Arithmetic Average	744	931	959	1124	1202	1351

TABLE 20. Land-Use in the Southern Mediterranean Basin : Sheep numbers, in 10^3 heads of mature animals.

Country	1947/52	1960/65	1966/70	1971/75	1976/80	1981/86
Algeria	4567	4622	6710	8246	11044	14285
Egypt	1254	1697	776	2024	1774	1982
Libya	1390	1376	1778	2697	5044	4974
Morocco	11249	10957	13785	16220	14755	13400
Tunisia	2462	3125	3433	3220	3929	5274
Cyprus	298	418	397	418	494	510
Iran	18000	30320	32500	35500	33260	34440
Iraq	9072	10245	11400	15610	11538	10210
Israel	53	190	196	189	230	247
Jordan	235	677	907	730	861	1041
Lebanon	55	200	208	237	254	139
Saudi Arabia	1098	2288	2940	3010	2990	3533
Syria	2968	4035	5930	5570	7577	11516
Turkey	24282	32863	35315	38534	43110	46227
Total	76983	103013	116275	132205	136860	147778

TABLE 21. Land-Use in the Southern Mediterranean Basin : Meat production, in 10^3 metr. tons.

Country	1948/52	1961/65	1969/71	1973/75	1979/81	1984/86	Prod. per inhabit. Kg/Yr	
							1948/52	1984/86
Algeria	56	84	108	120	173	204	6.3	9.3
Egypt	8	18	.46	46	128	154	8.0	42.7
Libya	115	156	201	212	288	316	12.5	14.4
Morocco	30	45	55	75	96	118	8.6	16.7
Tunisia	173	283	386	385	431	564	8.5	12.0
Cyprus	8	12	32	25	41	50	16.2	74.7
Iran	141	251	350	455	647	729	9.2	16.3
Iraq	46	110	117	170	168	262	8.9	16.5
Israel	2	83	121	166	187	197	4.0	46.5
Jordan	9	17	28	18	45	56	7.1	15.9
Lebanon	4	48	51	56	70	81	3.0	30.4
Saudi Arabia	48	61	70	88	171	210	13.9	20.0
Syria	13	44	51	82	157	396	4.1	34.3
Turkey	89	521	650	685	792	977	4.3	19.8
Total	742	1733	2266	2583	3394	4314	9.5	17.7
Prod.per Inh.	9.5	14.0	13.9	14.2	15.8	17.7		

Increase from 1962 to 1985 : 4314 - 1733 = 2581 = 150 % ; Exponential Annual Increase : 3.5 % .

TABLE 22. Land-Use in the Mediterranean Basin : Evolution of Gross Domestic Product Per Capita (GDPPC), in 1980 US $.

Country	1980	1987	Change in % of 1980
Algeria	980	1620	+ 65
Egypt	280	492	+ 76
Libya	6000	3400	− 47
Morocco	500	390	− 22
Tunisia	770	720	− 6
Northern Africa (intercountry averages)	1706	1325	− 22
Cyprus	1620	2600	+ 60
Iran	1930	2060	+ 7
Iraq	1400	1500	+ 7
Israel	2580	4000	+ 55
Jordan	630	730	+ 16
Lebanon	850	720	− 15
Saudi Arabia	4480	4020	− 10
Syria	800	960	+ 20
Turkey	770	780	+ 1
N. & M. East (intercountry averages)	1670	1483	− 11
Albania	530	530	0
France	6600	9300	+ 41
Greece	2550	2760	+ 8
Italy	3130	7800	+149
Portugal	1370	2040	+ 49
Spain	2880	4320	+ 50
Yugoslavia	1720	1050	− 39
Euro-Mediterranean (intercountry averages)	2680	4450	+ 48

TABLE 23. Evolution and projections of human population
(in 10^6 persons).

Time	1900	1925	1950	1975	2000	2025	2050
North	115	135	150	180	215	241	260
South	44	70	94	190	372	736	1450
Total	159	205	244	370	587	977	1710
N/S ratio	2.6	1.9	1.6	0.9	0.6	0.3	0.2

TABLE 24. Land-Use (10^6 Hectares).

Time	1965	1984	Difference	1984/1965
		North		
Forest & Shrubland	46.8	53.2	+ 6.4	1.14
Cropland	73.4	66.6	- 6.8	0.91
		South		
Forest & Shrubland	51.6	52.8	+ 1.2	1.02
Cropland	75.1	78.5	+ 3.4	1.05

367

TABLE 25. Land-Use : Cereals.

Time	1900	1950	1985	1985/1950	1985/1900
			North		
Area (10^6 ha)	34.3	31.0	29.2	0.94	0.85
Prod. (10^6 m.t.)	41.1	40.3	117.0	2.90	2.85
Yield (kg/ha/yr)	1200	1300	3780	2.91	3.15
Prod. per Capita	358	267	594	2.22	1.66
			South		
Area	–	26.3	41.0	1.56	–
Production	–	24.9	62.3	2.50	–
Yield	–	1222	1527	1.36	–
Prod. per Capita	–	261	255	0.98	–

TABLE 26. Sheep number and meat production.

Time	South 1950	1985	1985/1950
Sheep nrs (10^6 heads)	77	148	1.92
Meat prod. (10^3 m.t.)	742	4314	5.81
Meat prod.per Capita (Kg/yr)	9.5	17.7	1.86

TABLE 27. Gross Domestic Product (in 1980 US $) per Capita per year.

Area	Time	1980	1987	Difference %
North Africa		1706	1325	– 22
Near East		1670	1484	– 11
Mediterranean Europe		2680	4450	+ 48

AN INDICATION OF AN ABRUPT CLIMATIC CHANGE AS SEEN FROM THE RAINFALL DATA OF A MAURITANIAN STATION

GASTON R. DEMARÉE
Royal Meteorological Institute of Belgium
Ringlaan 3
B - 1180 Brussels, Belgium

ABSTRACT: This paper demonstrates the view that the Sahelian precipitation climate can be seen as an aperiodic but recurrent phenomenon alternating between two or more steady climate states. An "abrupt" climatic change is assessed through non-parametrical statistical tests applied to the annual, seasonal and monthly long-term precipitation time series of a Mauritanian station.

1. Recent Views on the Sahelian Drought

The Sahelian drought attracted, since its onset in 1968, wide-spread attention of the public, for its dramatic human consequences and the corresponding setting up of a vast international relief organization, as well as of the scientific world, emphasizing the value of a good data collection system and the need for new paths of research.

An early point of view of some scientists was that the drought could persist well into the next century (Winstanley, 1973), though others predicted the drought to be over soon and a return to "normal" conditions, whatever that might mean (Faure and Gac, 1981; Gac and Faure, 1987).

Landsberg (1975) states that this climatic event, how drastic its effect may be on population, animals and ecosystem, is an integral part of the climate of this area and should be seen as the result of a reversible climatic fluctuation. However, Landsberg reserves the term "climatic change" for a world-wide shift of climatic conditions to a new equilibrium position; the duration associated with it is about 104 to 106 years. At the same time, he states that the distribution of values of climatic elements has significantly changed. It is one of the findings of this paper, using some elaborated statistical tests, to show that a significant change in hydro-climatic parameters, evidencing a climatic change, took place in the second half of the sixties.

From the beginning of the 1980's many papers stressed, in one way or the other, the character of persistence, i.e. the continuance of the drought (Adefolalu, 1986; Beran and

369

R. Paepe et al. (eds.), Greenhouse Effect, Sea Level and Drought, 369–381.
© 1990 *Kluwer Academic Publishers.*

Rodier, 1985; Chouret et al., 1986; Gregory, 1982; Kerr, 1985; Lamb, 1982, 1983, 1985, 1988; Nicholson, 1983), but with reference to large regional variations in intensity, persistence and coverage (Ojo, 1987). Several papers reported that the application of regression techniques gave an indication of a significant downward trend for precipitation depths (Dennett et al., 1985; Hutchinson, 1985) while others compared average conditions in decadal, monthly and seasonal rainfall patterns and in rainy season characteristics of the drier and wetter periods (Gregory, 1983; Erpicum et al., 1988).

A new approach consists of looking at the long-term Sahelian precipitation series from a purely statistical point of view; it involves tests on the independence and on the stationariness of hydro-climatological time series.

Snijders (1983, 1986) studied the non-stationariness of the Burkina-Faso rainfall by producing a rainfall index compiled from daily data of 20 stations over the reference period 1923-1983. He showed that the rainfall exhibited a steadily decreasing trend over the period 1953-1983, and a single change-point around 1970. Furthermore, the annual values of the regional rainfall indicator remain stationary over the periods 1953-1969 and 1970-1983 separately.

Carbonnel & Hubert (1985), and later Hubert & Carbonnel (1987) pointed out the non-stationariness of West-African annual rainfall and indicated the years 1969-1970 as the most probable time over the whole Sudano-Sahelian area. Demaree and Chadilly (1988) showed significant changes in the mean, located in 1967-1978, for the annual precipitation amounts at Kaédi (Mauritania). Early results in this connection were obtained by Chadilly (1986) during a stay at the Royal Meteorological Institute of Belgium, and by Demaree (1987). Lately, West-African hydrological and hydrometric time series are viewed as alternating between two or more mean levels. The sudden shift between two levels defines an "abrupt" climatic change which is exemplary for the characteristic behaviour of the highly complex climatic system, with numerous non linear couplings (atmosphere, ocean, cryosphere, soil and biosphere), seasonly varying external forcings and phase transitions (Flohn, 1986). Carbonnel, Hubert and Chaouche (1987) determine transitions with a length of 10 to 15 years between "wet" and "dry" states, or vice versa, for 20th century Sahelian time series. They raise the question of increasing desertification and its connection with the steadily decreasing lower levels as well as with the impact of human activities in the Sahelian belt.

2. Droughts and Wet Periods

This section discusses droughts and wet periods as registered by the rainfall at Kaédi during the instrumental era. The station Kaédi is situated in the south-east of Mauritania, on the right bank of the Senegal river; its geographical coordinates are 16°08' N, 13°31' W and its altitude is 33 m. The land-use in the region is oriented towards agro-pastoralism. Regular daily observations started in October 1904 and constitute the longest available daily rainfall time series of Mauritania. An overview of the variation of the annual precipitation totals is presented in Figure 1.

The rainfall data are expressed in normalized departures of standard deviations from the long-term station mean. Such a representation is now commonly used to show temporal rainfall variability. A many-station index, named "standardized anomaly index", was computed by Ojo (1987) to bring out the interannual rainfall variability in West-Africa over the period of 1901-1985. It is clear that Figure 1 shows similar patterns as Ojo's graph, so that Kaédi can be considered representative for the precipitation climate of the West-Sahelian region in spite of the fact that some gaps in the record of the station and regional differences are known to exist.

The most striking feature is the rainfall below the mean over the entire period from 1968-1988, except for the year 1974 which is slightly above the mean. A decline of the rainfall started in the beginning of the 1960's, when nearly a decade of relatively wet years ended.

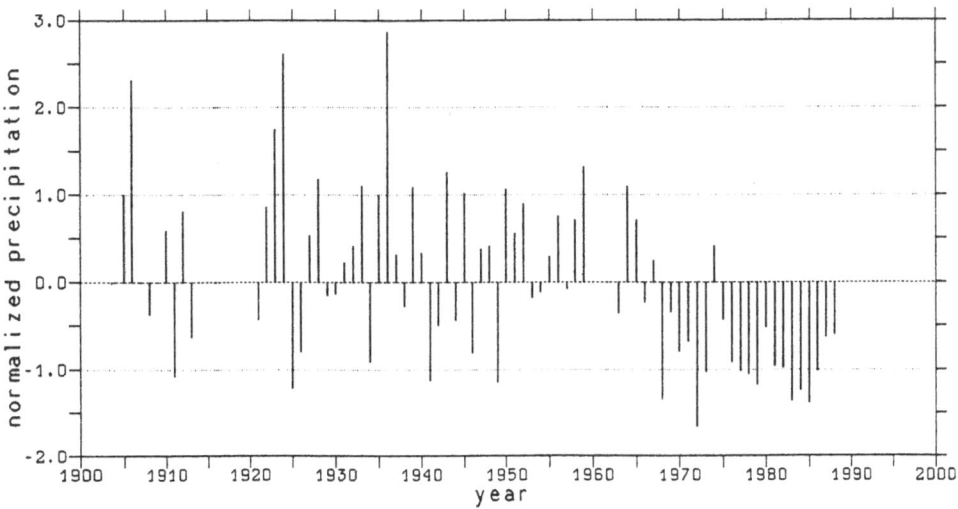

Figure 1. Annual normalized rainfall departures from the long-term mean at the station Kaédi, reference period 1904-1988.

The figure clearly shows the well-known "1913" and "1940" and the current "1968" droughts (Sircoulon, 1976; Grove, 1973; Paepe and Van Overloop, 1988); unfortunately, a gap in the record, presumably as a result of the First World War, does not allow further inferences about the first one mentioned here.

There is historical evidence of droughts and/or famine in Mauritania during the 19th century. "Les chroniques de Oualata" mention a severe drought in 1833/34 and several famine years, while "Les chroniques de Tichit" mention droughts in 1865/66, 1889/90, shortage of water in 1810/11, 1861/62, 1866/67, and excessive rains in 1893/94 (Tymowski,

1978). It is generally admitted that the information on the calamities reported by the author of the chronicles is correct, but also that some events in the early years of the 19th century may possibly have escaped his knowledge. However, no mention of prolonged periods of drought is retrieved; indeed, only isolated droughts or famine events are reported in the annuaries.

On the other hand, Daveau and Toupet (1963) and Toupet (1976) concluded from archaeological excavations that in medieval times rain-fed agriculture was practised on Mauritanian plateaus by the Gangara people. Such a method of farming needs an average annual rainfall of 450 mm, i.e. about 200 mm more than the actual average; thus a wetter climate prevailed in those times.

3. Statistical Tests

Figure 1 is also characterized by some alternation of drier and wetter years. Calculation of the serial correlation coefficient for the completed years 1921-1987 of annual rainfall totals leads to the acceptation of the null-hypothesis of no persistence (Sneyers, 1975). This finding corresponds to the one made by Krauss (1978), who characterized subtropical precipitation as highly variable with only a small degree of year-to-year persistence. On the contrary, Hare (1983) points to the existence of a marked trend for successive dry years to occur in a row, but this statement is probably based upon the ill-founded opinion that, by the time of the UNCOD Conference in 1977, the drought was over and that everything had returned to "normal". Deficient rainfall, however, continued into the 80's and even spread to East-Africa.

Persistent downward trends for the Sub-Saharan rainfall have been noted by Hare (1983), Hutchinson (1985), Dennett et al.(1985), and others. This section analyses monthly, seasonal and annual precipitation depths at the station Kaédi with the special object of detecting a hydro-climatic change through the statistical assessment of trends and/or abrupt change points.

In this context, the non parametric Mann-Kendall-Sneyers trend test, used in a progressive forward and backward way, allows the detection of the non random character of the time series. Interpretation of the time evolution of the test statistic permits to make deductions on the grouping of large or small values (Sneyers, 1975). In case of detection of an abrupt jump in the mean, the Pettitt (1979) non parametric test, allowing the detection of one single change-point at the most, is well suited (Buishand, 1982, 1984).

The analysis of the annual precipitation series is well documented in Demaree and Chadilly (1988). Their major finding was a highly significant decreasing trend which could be specified by the second test as a highly significant jump in the mean located in 1967-1968 and thus indicative of an abrupt climatic change. The jump in the mean, reaching 175.5 mm or 42.5%, separates the periods 1904-1967 and 1968-1988 with, respectively, a mean of 412.9 mm and 237.4 mm. Furthermore, a classic F-test shows that the variances over the two periods are significantly different at the 1% level.

The 1904-1988 rainfall series of Kaédi was used for the present study; the analysis being

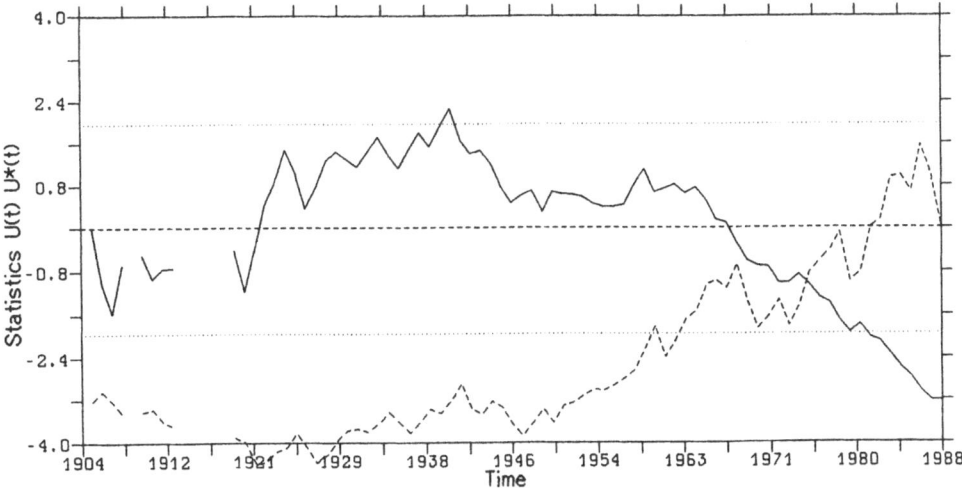

Figure 2. The Mann-Kendall-Sneyers sequential trend test applied in a forward (full line), and a backward (dashed line) way to the August rainfall depths at Kaédi.

(The horizontal dotted lines represent the critical values corresponding to the 5% significance level).

refined to monthly and seasonal aspects. In this paper "seasonal" stands for the April to October period which is generally considered to be the possible period with rain in the Sahel. In fact, at Kaédi, 90% of the annual precipitation falls within this period.

On the monthly level, a downward trend is found for all months of the rainy season but only the rainiest month, August, is significant at the 5% level. Figure 2 shows that the August depths had been decreasing since the early sixties and had reached the 5% level around 1980. It is the accumulation of the decreases over the successive rainy months that results in a highly significant downward trend for the seasonal and annual totals. Similarly, the Pettitt change point test localizes a highly significant jump in the mean of the August depths in the years 1964-1965, that means somewhat prior to the jump of the seasonal and annual totals. Figure 3 graphs the test statistics; the triangular form reaches its maximum in 1964, which corresponds to the change point. Stationariness over the separate periods 1904-1967 and 1968-1988 is assured.

The jump disclosed in the Kaédi monthly, seasonal, and annual precipitation series, splitting up the entire reference period into a "wetter" and a "drier" period, characterized by significantly different mean and variance, is indicative of the existence of multi-stable climate regimes. A stochastic climate model showing bistable behaviour has been introduced by C. Nicolis and G. Nicolis (1981); stochastic aspects of climatic transitions were studied by C. Nicolis (1982). Demaree and C. Nicolis (1990) proposed a bimodal stochastic model

374

to represent the Sahelian precipitation climate regime. Furthermore, by stochastic modelling of the **aperiodic** but **recurrent** behaviour of the transitions, results on the statistical predictability are obtained.

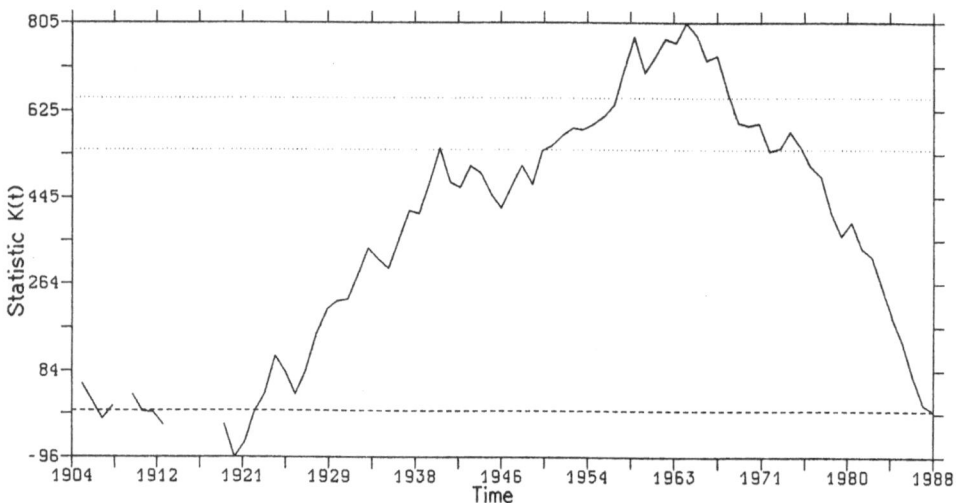

Figure 3. Pettitt's change-point test applied to the August rainfall depths at Kaédi.

(The horizontal dotted lines correspond to the approximate significance probabilities at the 5% and 1% level. The maximum, located in 1964, corresponds to the change point).

4. Measures of Central Tendency and of Variability for Precipitation Data

Katz and Glantz (1977) warn against the perception of what constitutes "normal" rainfall, which considers a below the long-term average to be abnormal or anomalous. For too long the notion of climatological normal or CLINO has been connoted with the normal distribution, which is determined unequivocally by its mean and standard deviation. This 19th century concept, based on the pioneering statistical work by Quetelet in Brussels, often remains maintained in climatology and is supported by the "hypothesis" of the nearly stationary character of most of the observed time series. This paper already showed that this property in particular does **not** hold for the time series under consideration. Katz and Glantz also state that the variability of Sahelian precipitation amounts is such that the mean does not represent "normal" rainfall well, and they suggest that the use of either the mode or the median may be preferable as a measure of a central tendency. Besides, measures of variability, other than the standard deviation, may be useful, e.g. rainfall deciles as proposed by Gibbs and Maher (1967), or quantiles as proposed by Katz and Glantz (1977), could be adequate for defining meteorological droughts.

In view of the detected non stationariness of the annual precipitation series of the Kaédi station, some statistical parameters, given in Table 1, are calculated over the two distinct reference periods corresponding to the "wetter" and the "drier" period respectively. The table stresses the large differences in the data over the two periods. As is well known for positively skewed distributions, the median is always smaller than the mean, which accompanies the fact that the number of samples, less than the mean, is greater than half the sample size. It is common knowledge that series of rainfall totals may deviate from a normal distribution; this is usually more the case for monthly series than for annual ones. Such deviations from the classic normality hypothesis are present in Table 1 and may eventually bias the application of parametric statistical tests. For that reason, distribution free tests were used.

Table 1. Measures of central tendency and of variability in August and in annual precipitation depths (0.1 mm) at Kaédi, Mauritania.

reference period	1904 - 1967		1968 - 1988	
	August	annual	August	annual
sample size	57	44	20	17
maximum	364.0	762.0	186.0	419.8
minimum	23.0	192.0	10.6	130.0
range	341.0	570.0	175.4	289.8
mean	160.1	412.9	85.1	239.2
median	157.4	411.9	74.4	225.2
standard deviation	73.1	136.1	46.9	67.8
coef. of variation	.46	.33	.55	.28
skewness	.46	.49	.53	1.04
upper quartile	210.8	507.4	115.6	281.5
lower quartile	99.2	307.0	57.5	193.3
interquartile range	111.6	200.4	58.1	88.2
upper decile	273.7	575.4	106.9	335.0
lower decile	79.5	222.6	14.2	162.2
number in sample less than mean	30	22	12	11

5. Frequency of Rainy Days

In this Section, one tries to find out if the changed rainfall characteristics, detected in the preceding Sections, are reflected in the number of rainy days with depth above some fixed thresholds. Thresholds of 0.1 mm, which is equal to the minimum measurable depth, of 10.0 mm, 20.0 mm and 40.0 mm (heavy rain) are defined.

Table 2 shows the average number of rainy days with precipitation equal to, or greater than the above mentioned thresholds over both periods. The hypothesis that both means are of the same population is tested by the classic Student's difference test. The hypothesis of equal means has to be rejected in all cases; at the level of significance .01 (**) for the thresholds 0.1 mm, 10.0 mm and 20.0 mm but only at the level .05 (*) for the heavy rains. Figure 4 shows the number of rainy days (rainfall equal to or greater than 0.1 mm) over the two periods; the significant difference in the means is illustrated by the levels of the dashed lines.

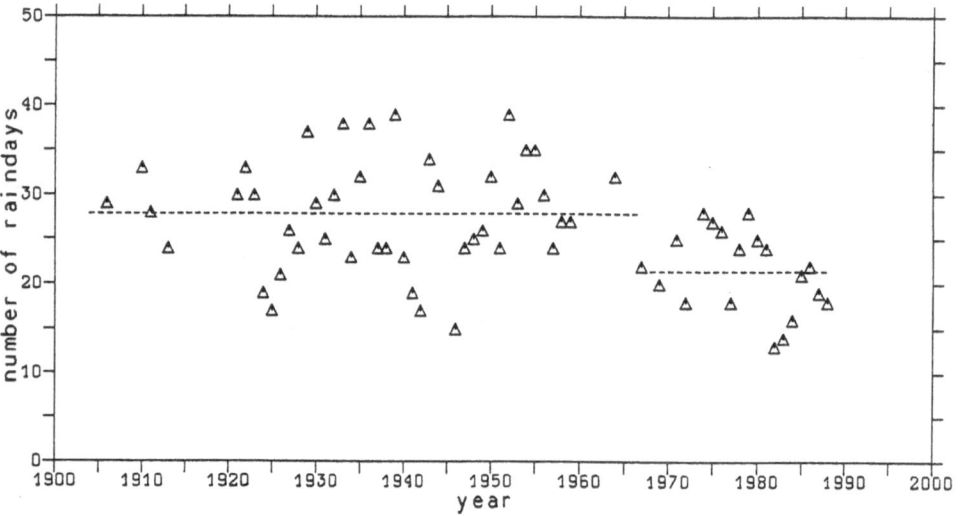

Figure 4. Number of rainy days (day sum equal to or greater than 0.1 mm) at Kaédi over the periods 1904-1967 and 1968-1988. The averages over both periods are represented by the dashed lines.

Table 2 shows the number of days with a day sum equal to or greater than the thresholds 0.1 mm, 10.0 mm, 20.0 mm and 40.0 mm over the "wet" period 1904-1967 and over the "dry" period 1968-1988 and the value of the statistic t for testing the equality of the mean.

Table 2.

Number of days with a day sum equal to or greater than the threshold

threshold	0.1 mm	10.0 mm	20.0 mm	40.0 mm
period 1904-1967 N1 = 44	27.8	13.4	6.9	2.2
period 1968-1988 N2 = 18	21.4	7.7	3.9	1.4
test statistic t	3.96(**)	5.70(**)	3.94(**)	2.08(*)

6. Conclusions

The findings of this paper add extra credence to the view that the western Sahel, in the second part of the sixties, underwent an abrupt hydro-climatic change in the sense of a transition from a "wetter" to a "drier" rainfall state. This is evidenced not only by the statistical examination of the annual rainfall series, but appears also clearly in the analyses of the monthly and seasonal data. The jump disclosed in the system is indicative of the existence of multiple stable climate regimes as a result of internal fluctuations or external random perturbations between two or more stable states of the precipitation climate (Demaree and Nicolis, 1990).

The rainfall pattern of the Kaédi station over the periods 1904-1967 and 1968-1988 not only differs by its monthly, seasonal, and annual totals but also by a highly significant modification of its intensity (number of events above fixed thresholds).

The large differences between the distinct periods bring forward the problem of assessment of climatological normals, their predictive value and their potential use in agro-meteorological decision-making (Todorov, 1985; Morel, 1986). Therefore, it is strongly suggested that the agricultural planning (e.g. irrigation) and the determination of long-term climatological means used for the design of hydraulic structures (e.g. dams), are based on the records of the current drier period.

More generally, the notion of climatological normals, consistent with the nineteenth century view of a gradual changing climatic behaviour, and based on relatively short and area limited records, should be abandoned since an increasing number of climatic changes do present themselves as "abrupt" (Flohn, 1986).

7. Acknowledgments

The author is indebted to the Data Bank Project, Programme Belgium-WMO, Brussels and to the Permanent Representative of Mauritania with WMO, Nouakchott, for having provided the data.

8. References

Adefolalu, D.O. (1986) "Further Aspects of Sahelian Drought as Evident From Rainfall Regime of Nigeria.", Archives for Meteorology, Geophysics, and Bioclimatology, Ser. B, 36, pp. 277-295.

Beran, M.A. & Rodier, J.A. (1985) *Hydrological aspects of droughts. A contribution to the IHP*, Unesco-WMO, Paris, 149 p.

Buishand, T.A. (1982) "Some methods for testing the homogeneity of rainfall records.", Journal of Hydrology, 58, pp. 11-27.

Buishand, T.A. (1984) "Tests for detecting a shift in the mean of hydrological time series.", Journal of Hydrology, 73, pp. 51-69.

Carbonnel, J.P. & Hubert, P. (1985) "Sur la sécheresse au Sahel d'Afrique de l'Ouest. Une Rupture climatique dans les séries pluviométriques du Burkina Faso (ex Haute-Volta)", C.R. Acad. Sc. Paris, t. 301, Série II, no. 13, pp. 941-944.

Carbonnel, J.P., Hubert, P. & Chaouche, A. (1987) "Sur l'évolution de la pluviométrie en Afrique de l'Ouest depuis le début du siècle.", C.R. Acad. Sci. Paris, t. 305, Série II, pp. 625-628.

Chadilly, M.S. (1986) *Rapport de Stage*, IRM, Uccle, 9 p. + tableaux et figures.

Chouret, A., Berthoult, C. & Pepin, Y. (1986) *Persistance de la sécheresse au Sahel; Etude de stations pluviométriques et hydrologiques de longue durée au Mali*, Observations de l'année 1985. Bamako, 81 p.

Daveau, S. & Toupet, C. (1963) "Anciens terroirs Gangara", Bulletin de l'I.F.A.N., T. XXV, sér. B, nos. 3-4, pp. 193-214.

Demaree, G. (1987) "Traitement des données pluviométriques de la station de Kaédi (Mauritanie).", Publications IRM, Misc. Série C, No. 24, 40 p.

Demaree, G. & Chadilly, M.S. (1988) "The Sahelian Drought(s) as Seen from the Rainfall Data of a Mauritanian Station.", in: Computer Methods and Water Resources, Vol. 3 Computational Hydrology. D. Ouazar, C.A. Brebbia and V. de Kosinsky (Eds.) Computational Mechanics Publications and Springer-Verlag, pp. 15-23.

Demaree, G. & Nicolis, C. (1990) "Onset of Sahelian Drought viewed as a Fluctuation

Induced Transition", Quarterly Journal of the Royal Meteorological Society (in press).

Dennett, M.D., Elston, J. & Rodgers, J.A. (1985) "A reappraisal of rainfall trends in the Sahel.", Journal of Climatology, 5, pp. 353-361.

Erpicum, M., Binard, M., Peters, J.P. & Alexandre, J. (1988) "Une méthode d'analyse des caractéristiques de la saison des pluies en région Sahélienne (exemples pris au Sénégal).", Actes des Journées de Climatologie, Mont-Rigi, Belgique, 5-7 novembre 1987. Université de Liège, Laboratoire de Géographie Physique, pp. 43-56.

Farmer, G. & Wigley, T.M.L. (1985) *Climatic Trends for Tropical Africa. Research Report for Overseas Development Administration*, Climatic Research Unit, University of East Anglia, Norwich, England, 136 p.

Faure, H. & Gac, J.Y. (1981) "Will the Sahelian drought end in 1985?", Nature, 291, pp. 475-478. Reply. Nature, 293, p. 414.

Flohn, H. (1986) "Singular Events and Catastrophes Now and in Climatic History.", Naturwissenschaften, 73, pp. 136-149.

Gac, J.Y. & Faure, H. (1987) "Le "vrai" retour à l'humide au Sahel est-il pour demain?", C.R. Acad. Sci. Paris, 305, Série II, pp. 777-781.

Gibbs, W.J. & Maher, J.V. (1967) "Rainfall deciles as drought indicators.", Bureau of Meteorology, Bulletin No. 48, Melbourne, 33p. + maps.

Glantz, M.H. & Katz, R.W. (1985) "Drought as a Constraint to Development In Sub-Saharan Africa.", Ambio, 6, pp. 334-339.

Glantz, M.H. & Katz, R.W. (1987) "African drought and its impacts: revived interest in a recurrent phenomenon.", Desertification Control Bulletin, 14, pp. 22-30.

Gregory, S. (1982) "Spatial Patterns of Sahelian Annual Rainfall, 1961-1980.", Arch. Met. Geoph. Biokl., Ser. B, 31, pp. 273-286.

Gregory, S. (1983) "A note on mean seasonal rainfall in the Sahel, 1931-60 and 1961-80.", Geography, 68, pp. 31-36.

Grove, A.T. (1973) "A note on the remarkably low rainfall of the Sudan zone in 1913.", Savanna, Vol. 2, No. 2, pp. 133-138.

Hare, F.K. (1983) *Climate and Desertification: A revised analysis*, WCP-44, WMO/UNEP, 149 p.

Hutchinson, P. (1985) "Rainfall analysis of the Sahelian drought in The Gambia.", Journal of Climatology, 5, pp. 665-672.

Hubert, P. & Carbonnel, J.P. (1987) "Approche statistique de l'aridification de l'Afrique de l'Ouest.", Journal of Hydrology, 95, pp. 165-183.

Katz, R.W. (1978) "Persistence of Subtropical African droughts.", Mon. Weather. Rev., 106, pp. 1017-1021.

Katz, R.W. & Glantz, M.H. (1977) "Rainfall Statistics, Droughts, and Desertification in the Sahel.", in Glantz, M.H. (ed.) *Desertification. Environmental Degradation in and around Arid Lands*, West-view Press, Boulder, Colorado, 346 p.

Kerr, R.A. (1982) "Fifteen Years of African Drought.", Science, 227, pp. 1453-1454.

Krauss, E.B. (1977) "Subtropical Droughts and Cross-Equatorial Energy Transports", Monthly Weather Review, 105, pp. 1009-1018.

Lamb, P.J. (1982) "Persistence of Subsaharan drought.", Nature, 299, pp. 46-48.

Lamb, P.J. (1983) "Sub-Saharan rainfall update for 1982: continued drought.", Journal of Climatology, Vol. 3, pp. 419-422.

Lamb, P.J. (1985) "Rainfall in Subsaharan West Africa during 1941-83.", Zeitschrift für Gletscherkunde und Glazialgeologie, 21, pp. 131-139.

Lamb, P.J. (1988) "Personal Communication."

Landsberg, H.E. (1975) "Sahel Drought: Change of Climate or Part of Climate?", Arch. Met. Geoph. Biokl., Ser. B, 23, pp. 193-200.

Morel, R. (1986) *Problèmes posés par les normes pluviométriques dans la région Sahélienne,* Colloque Int. Rév. Normes Hydrol. Suite aux Incidences de la Sécheresse. Com. Interafr. Etud. Hydraul., Ouagadougou, 15 p.

Nicholson, S.E. (1978) "Climatic variations in the Sahel and other African regions during the past five centuries.", Journal of Arid Environments, 1, pp. 3-24.

Nicholson, S.E. (1983) "Saharan climates in historic times.", in: The Sahara and the Nile, M.A.J. Williams and H. Faure (Eds.), Balkema, Rotterdam, pp. 173-200.

Nicholson, S.E. (1983) "Sub-Saharan Rainfall in the Years 1976-80: Evidence of Continued Drought.", Monthly Weather Review, 111, pp. 1646-1654.

Nicolis, C. and Nicolis, G. (1981) "Stochastic aspects of climatic transitions - Additive fluctuations", Tellus, 33, pp. 225-234.

Nicolis, C. (1982) "Stochastic aspects of climatic transitions-response to a periodic forcing.", Tellus, 34, pp. 1-9.

Obasi, G.O.P. (1984) *La sécheresse en Afrique,* PCM Bulletin, World Climate Programme, WMO, No. 6.

Ojo, O. (1987) "Rainfall trends in West Africa, 1901-1985.", in: The Influence of Climate Change and Climate Variability on the Hydrologic Regime and Water Resources. Proceedings of the Vancouver Symposium, August 1987, IAHS Publ. no. 168, pp. 37-43.

Paepe, R. & Van Overloop, E. (1988) "Archeo(logische)-Geologie of Geoarcheologie, een nieuwe transdisciplinaire wetenschap.", V.U.B. Magazine, 4, 14, pp. 18-23.

Pettitt, A.N. (1979) "A Non-parametric Approach to the Change-point Problem.", Appl. Statist., 28, 2, pp. 126-135.

Sircoulon, J. (1976) "Les données hydropluviométriques de la sécheresse récente en Afrique intertropicale. Comparaison avec les sécheresses "1913" et "1940".", Cahiers ORSTOM, Série Hydrologique, vol. XII, 2, pp. 75-174.

Sneyers, R. (1975) "Sur l'analyse statistique des séries d'observations.", Note Technique No. 143, OMM-No. 415, Genève, 192 p.

Snijders, T.A.B. (1983) "A study of the variability in space and time of Upper Volta rainfall.", II International Meeting on Statistical Climatology, September 26-30, 1983, Lisboa, 3.5.1-3.5.7.

Snijders, T.A.B. (1986) "Interstation Correlations and Non stationariness of Burkina Faso Rainfall", Journal of Climate and Applied Meteorology, 25, pp. 524-531.

Thirriot, C. and Fontin, M. (1988) "Simulation stochastique de la pluviométrie Sahélienne.", First International Conference on Africa, Computer Methods and Water Resources, Rabat, Morocco, March 14-18, 1988.

Todorov, A.V. (1985) "Sahel: The Changing Rainfall Regime and the "Normals" Used for its Assessment.", Journal of Climatology and Applied Meteorology. 24, pp. 97-107. Reply. 25, pp. 258-259.

Toupet, C. (1976) "L'évolution du climat de la Mauritanie du Moyen-Age jusqu'à nos jours.", in : La désertification au Sud du Sahara. Colloque de Nouakchott, Mauritanie, Décembre 1973. Les Nouvelles Editions Africaines, Dakar. 212 p.

Tymowski, M. (1978) "Famines et épidémies à Oualata et à Tichit au XIXè siècle.", Africana Bulletin, 27, pp. 35-53.

Winstanley, D. (1973) "Rainfall patterns and general atmospheric circulation.', Nature, 245, pp. 190-194.

PART IV: MANAGEMENT, TECHNIQUES AND CASE STUDIES

SECTION A: Sea Level

PRACTICAL PROBLEMS FOR COASTAL SUBMERGENCE IN THE LIGHT OF SECULAR TRENDS

P.A. KAPLIN
Geography Faculty
Moscow State University
Moscow 119899
USSR

ABSTRACT: A global tendency towards shore erosion is caused by both natural and anthropogenic processes, the main one being sea level rise. Reshaping of the coastal zone profile during sea level rise cannot be explained by the Bruun scheme alone. Caspian Sea evidence shows various types of coastal zone development depending primarily upon the inclination of submarine coastal profile. Different anthropogenic interference in the coastal zone are discussed. The author stresses an urgent need to create models of coastal management in specific natural and anthropogenic conditions.

In recent times there has been a major global tendency for shores to erode actively and for shorelines then to retreat, even where they have previously advanced. In recent decades shore erosion has gained strength and spread to many coastal areas around the World's oceans. As early as 1972, the Coastal Environment Commission of the International Geographical Union drew attention to this tendency and carried out special investigations of shoreline erosion (Bird, 1985). Over 15 years the Commission collected extensive information confirming that erosion was gaining strength and shorelines retreating, especially at depositional, i.e. previously advanced coasts. According to the Commission, more than 70% of the depositional coasts retreated landward 10 cm or more per year; nearly 20% of sand and gravel coasts retreated more than 1 m per year.

Shoreline erosion is most obvious in many densely populated regions of developed countries (Atlantic US coast, The Netherlands, Poland, Australia, USSR Black and Baltic Sea coasts) and appears to be a disastrous global phenomenon which destroys vast resources and even human beings.

There are several reasons for the global extent of shoreline erosion and their relative significance can change from time to time and from one place to another. General shoreline retreat is, on the one hand, a result of natural processes and, on the other, a consequence of the intensive economic development along coasts.

R. Paepe et al. (eds.), Greenhouse Effect, Sea Level and Drought, 385–393.
© 1990 *Kluwer Academic Publishers.*

The main natural causes of shoreline erosion are a rise of the world ocean level, formation of equilibrium coastal slope profiles, inundation and barrier development in many river mouths. The two last factors result in a sedimentary material deficiency for the coastal zone and a release of wave energy for shoreline erosion.

World ocean level rise has been traced by tide-gauge data (Klige, 1985; Pirazzoli, 1986; etc.). This process is going on at a rate of up to 1.5 mm per year since the last century and influences greatly the coastal zone evolution.

In view of the reshaping of the coastal zone profile and land erosion during the world ocean level rise, the mechanisms of these processes are being actively discussed. The scheme

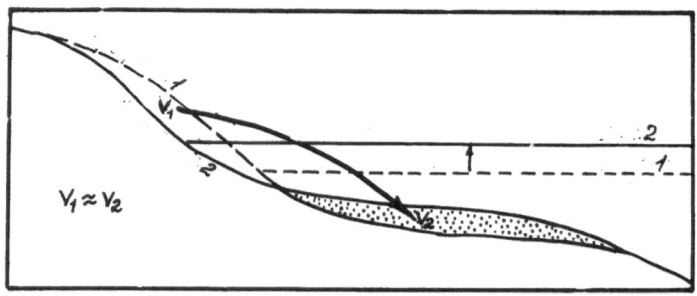

Figure 1. Coastal zone natural development during submergence according to Per Bruun (Bruun, 1962).

by Per Bruun (1962) is generally accepted. According to this, in the course of sea level rise, the upper part of the coastal slope and the above-water part of depositional features are eroded. Simultaneously with the erosion and horizontal retreat of the shoreline, sediments accumulate and the bottom surface is elevated in the lower part of the underwater coastal slope (Fig. 1). It is assumed that the volume of material eroded from the upper part of the coastal slope and that deposited on its lower part are equal.

In general, the Bruun scheme is supported by both experimental and natural data. But this scheme does not consider that during erosion of the frontal zone of depositional features some portion of the material in many cases would be washed over by waves and the body of the form would migrate towards the land (Zenkovich, 1967). Such a migration process has been carefully studied in the Chukchi Peninsula, Western Kamchatka, Sakhalin Island, etc. (Fig. 2).

The Caspian Sea evidence is of great interest for the studies of coastal zone deformations related to sea level changes. It is known that the Caspian Sea coasts from 1929 were exposed to a quick sea level fall to minus 29,2 m at the lowest point. The subsequent sea level rise of 1.2 m from 1978 to the present (i.e. 12 cm/year) is generally surprising though O.K. Leontyev supposed that the Caspian Sea level would rise at least up to 2000 because of Brückner's medium-scale climatic fluctuations.

Recent rise of the Caspian Sea level has a substantial complex influence on different economic factors. Many structures have been built on the surface of a terrace formed in the process of sea level fall since 1929. This terrace is now actively eroding. The rate of shoreline retreat in Dagestan (on the north-western coast) has reached 10 to 12 m per year. On the low-lying, gentle sloping shores, coastal ridges are breached, land behind them becomes inundated and shallow lagoons form.

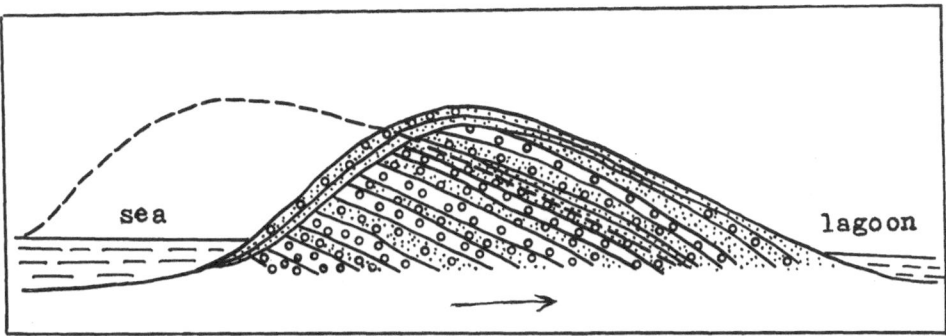

Figure 2. Coastal bar moved in the direction of land on the Chukchi Sea.

Coastal studies on the Caspian Sea and other USSR seas show that the Bruun scheme for submerging coasts is not universal (Leontyev, 1988). There are various types of coastal zone development under conditions of sea level rise. These types vary primarily with differences in inclination of the submarine coastal slope.

At the most gentle coasts (gradient ~ 0.0005), passive land flooding occurs without coastal zone reformation. These coasts are not subject to wave attack because waves lose their energy on the gentle submarine coastal slopes long before reaching the shoreline (Fig. 3a).

Where the gradient of submarine coastal slope is slightly more (~ 0.001), increasing sea level results in the formation of a coastal ridge in the wave-breaking zone at some distance from the shoreline. This ridge separates part of the water regime, forming a lagoon (Fig. 3b). This ridge turns into a barrier and is made of bottom material (at the Caspian Sea - mainly of shell material).

Steeper slopes (approx. 0.005) cause waves to break near the shoreline. Sediments on the submarine coastal slope undergo differentiation, and a significant portion of them is cast ashore, raising the beach ridge. Simultaneously sea water infiltrates across the ridge, while the lagoon receives freshwater from land. During the following sea level rise, the beach ridge changes into the bar/barrier form, with a shallow lagoon behind it, (Fig. 3c). However, where land behind the bar/barrier is not low-lying, a lagoon does not form. Instead, the beach ridge is raised and migrates in the direction of land. Essentially, this

388

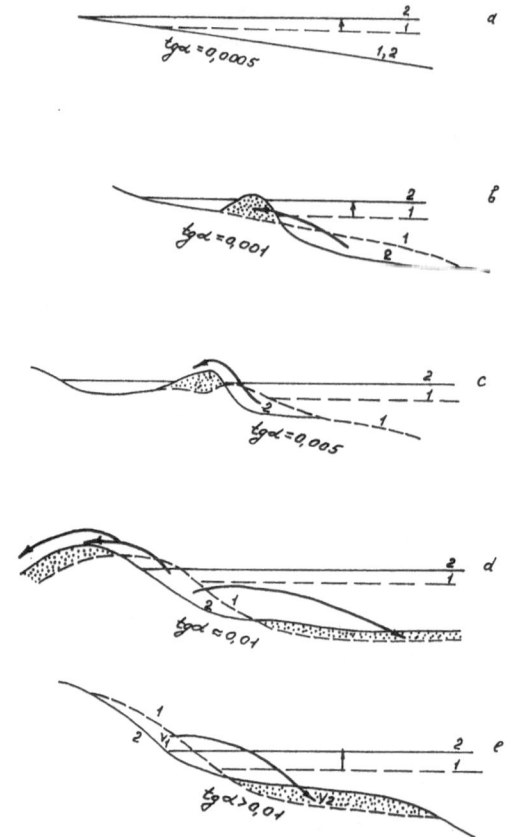

Figure 3. Different types of coastal zone natural development during sea level rise. Differences in schemes (a-e) are caused mainly by the gradient (tg α) of the primary submarine coastal slope.

(Dotted lines show profiles before sea level rise; solid lines, profiles after this rise).

is the next type of coastal zone development shown in Fig. 3d.

Coastal development, according to Per Bruun, occurs where the bottom gradient is high, 0.01 (Fig. 3e). There is much uncertainty, of course, in using inclination figures alone, because coastal development according to one or another scheme also depends upon other variables, in particular sediment quantity, particle size, and wave parameters.

Therefore the experience of studies on the Caspian coasts is very useful for the general analysis of coastal zone development during sea level rise. The Caspian Sea serves as a natural laboratory which can be used to simulate coastal processes during the projected world ocean level rise.

Depositional features are nourished not only by bottom sediments. Many depositional features and wide beaches are made of disintegrated rock material which migrates along coasts from river mouths and areas of bedrock outcrops. At the beginning of the post-glacial transgression, the increase of river discharges and the abundance of disintegrated rock material thus formed depositional features. In front of abrasion areas equilibrium underwater slope profiles were gradually formed by waves, so that sediment supply from these ares became exhausted.

In recent centuries, sediment nourishment from river mouths, previously the main suppliers of disintegrated material for the coastal zone, has decreased sharply. During the last stage of the post-glacial transgression the mouths of many rivers were flooded thus resulting in weakening erosion and decreasing supply of loose material from river basins. In addition, numerous small bays and estuaries formed in which river sedimentary load was deposited. On many coasts depositional features separated bays and bights from the sea, thus changing them into lagoons. Sediments from rivers and slopes of these water bodies do not reach the ocean coastal zone and are irretrievably lost in these enclosed and semi-enclose water

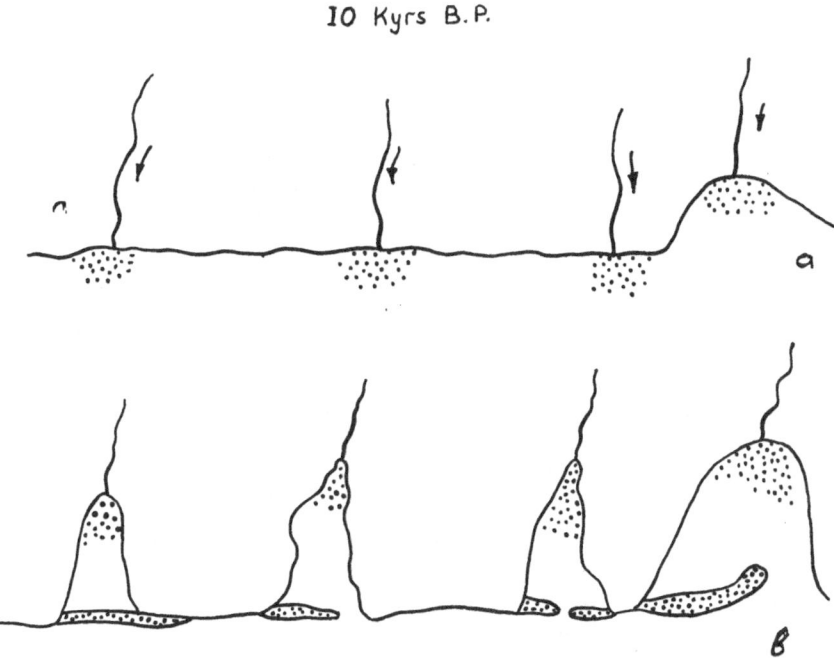

Figure 4. Changes of a coast with rivers supplying the coastal zone with disintegrated material: a - before the last stage of post-glacial transgression; b - at present when river mouths are blocked and sediments lost in lagoons.

areas (Fig. 4).

However, in natural conditions found up to the present in the world ocean coastal zone, depositional coasts would probably exist in an unstable equilibrium, even with the deficiency of sediments, being locally affected by erosion or accumulation. The coastal zone as a natural system is usually provided with some reserve of self-regulation and self-defence from degradational processes. However, the widespread technological interference in nature by man has apparently given rise to disruption of the natural equilibrium and has caused a progressively irreversible intensification of erosion processes.

From the late nineteenth century, there has been rapid economic development and use of coastal resources: ports were built, coastal areas were strengthened by various engineering constructions, embankments and water collectors were intensively built, etc. In most cases, technological interference did not take natural conditions into account and thus upset the existing equilibrium.

The most common type of interference by man into the natural coastal system is the extraction of sand and gravel for building. On many coasts in the recent past, this extraction has been massive. This is not surprising in view of the relative ease of obtaining construction material from the beach rather than from special quarries. For example, in the 1940-s to 1970-s some 30 million cubic meters of sand and gravel were extracted from the Georgian Black Sea beaches and delta areas for the construction of towns, resorts, roads, etc. It is nor surprising that this extraction has intensified a negative trend in the sedimentary budget of the coastal zone (Zerkovich and Schwarz, 1987).

Construction of jetties into the sea and other port facilities often greatly damages coastal zones. These engineering constructions make the alongshore migration of sedimentary materials more difficult, disrupt the integrity of natural coastal systems, and result in local but intensive erosion episodes. After the port construction in Poti (Georgia), a land strip 900 m wide was relatively rapidly eroded south of the port.

Anthropogenic activity has a great influence upon the amount and type of solids carried in river flow. According to some estimates, the total of such solids, carried into the world ocean coastal zone, is equal to 19.3 billion tons per year. A significant part of this material is kept in river mouths or passes irretrievably to great depths. Some 3 billion tons per year are pebble, gravel and sand sizes which accumulate in the coastal zone. At the same time, use of river runoff for irrigation and hydro-energy has sharply increased during the last several decades. As a result of dam construction for hydroelectric power, sedimentary material flow into the coastal zone from rivers has abruptly decreased. For example, after the construction of Aswan Dam, the Nile delta began to suffer from erosion at a speed of 40 m per year because of sedimentary deficiency.

In general, river runoff under technological activity has experienced a steady decline. According to predictions, continuation of this tendency will result in the reduction of river runoff by 50% by the year 2000; the solids carried in river flow will be reduced approximately by the same per cent. Thus, because of the decrease of alluvial supply for the coastal zone, the erosion of depositional features previously nourished by rivers, which has intensified in the last century, will increase further.

In the near future, as the sedimentary deficiency in the coastal zone increases, the global

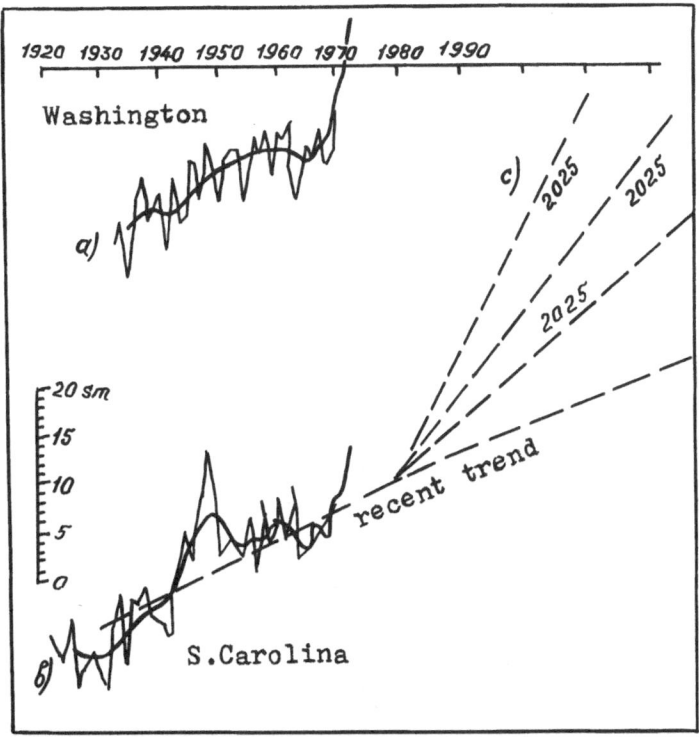

Figure 5. Ocean level changes: a - b - by tide-gauge data on U.S. Atlantic coasts; c - predicted sea level resulting from the "greenhouse effect" and global warming, with three scenarios considered (Barth & Titus, 1984).

tendency for shore destruction will be intensified. Firstly, human development on coasts especially those not previously touched by technological activity (for example of developing countries and the Arctic Ocean) will increase. Secondly, the rise of the world ocean level will continue and, according to predictions, may sharply accelerate in the near future. This rise of sea level is connected with the increase of CO_2 and some other gases in the atmosphere resulting in the so-called "greenhouse effect"; CO_2 accumulates in the atmosphere as a result of industrial burning of different fossil fuels (Fig. 5). According to data from the US National Academy of Sciences, doubling or even tripling of carbon dioxide, methane and some other gas concentrations in the atmosphere is expected by 2100. This might result in the warming of the earth's surface by 1.5 - 4.5°C. Such warming will cause glacier melting and, as a consequence, a rise of the world ocean level by 1.3 - 3.5 m (Barth and Titus, 1984).

Such a sea level rise in the next century can be regarded as a global disaster because vast densely populated coastal areas and many large cities will be inundated. Side by side with the global warming, the "greenhouse effect" will cause an increase in cloudiness, intensification of atmospheric circulation, rise of ocean storminess and strengthening of devastating coastal storms and typhoons. Even in the absence of greenhouse warming, shore erosion will continue to intensify in the near future because of continuing sea level rise, amplified by technological activity.

In view of these facts, shore protection is clearly a pressing need. In industrial practice, suitable engineering constructions have high priority for the protection of coasts. Revetments along coastal slopes are the most common and obvious means of protection. However, they increase wave activity at their toes which causes a fast pedestal destruction and sharp acceleration of shore erosion.

Groyne and block breakwaters built at angles to the coastline are more effective in shore protection. Together with breakwaters along the coastline, these constructions defend the shore only where they are built. But the construction of a groyne or breakwater series often causes an increase of erosion at the neighbouring coastal segments. Groynes, breakwaters and revetments can be undermined during storms, thus they are not durable constructions. In conditions of sedimentary deficiency, wave action on the hydrotechnical constructions causes an erosion of the underwater coastal slope in front of them. For example, the seabed in front of the constructions in Poti (the Black Sea coast) has deepened by 2 to 3 m. Increase in the gradient of the underwater coastal slope results in the sharp strengthening of wave attack. Deformation or destruction of engineering coastal works occurs suddenly under attack from medium waves. In general, experience demonstrates that "patching up" of coastal segments does not save the situation.

The above mentioned facts do not imply the uselessness and harmfulness of all hydrotechnical constructions. The success of many outstanding engineering constructions is beyond any doubt. Unfortunately, however, local successes of engineering solutions in coastal protection create an illusion that natural processes can be excluded from the analysis.

Engineering solutions should be included in the general scheme of measures for rational use and protection of coasts. They should cover the whole coastal zone influenced by related natural systems, should minimize breaches of natural conditions, and, in turn, should use natural conditions to the maximum to aid coastal self-defence.

There is now an urgent necessity to create and implement coastal management models. As one of these, we can consider a model of the sedimentary budget with artificial regulation of addition to and subtractions from it.

The best way to defend a coast from storm waves is to construct a beach in front of it to exclude wave energy. Thus more and more specialists conclude that the basis of coastal protection activity should be restoration of eroding beaches and the creation of new artificial ones. Depending on the specific conditions, different methods can be used to bring appropriate sedimentary material into the coastal zone: material can be unloaded from lorries at a specific point on the beach and then allowed independently to migrate alongshore to be distributed in the natural system; or material can be unloaded from barges onto the seabed to be brought ashore by wave action, thus forming a beach; or material can be transported from river mouths to the eroding coastal area by a pipeline, etc. These activities are carried out on a wide scale on the eastern Black Sea coasts (Zerkowich and Schwartz, 1987). On a regular basis, taking natural conditions into account, they introduce into the coastal zone considerable quantities of beach-forming material (sand, pebbles) from land quarries. For example, 510,000 cubic meters of sedimentary material were brought to the beach in the Gagra - Pitsunda area in 1982. The wide beach formed since not only protects dangerous areas but also sharply improves the state of the whole 22 km long coast.

Until the beginning of the 1980-s, 220 km of the total 312 km length of the Georgian coasts underwent destruction. The rate of erosion reached 16 m per year in some places. Paradoxically cases are on record to suggest that the worst situations have occurred where the coast had been protected by traditional engineering constructions.

During the last 5 years, significant improvement of the Georgian coastal zone has been accomplished, and at present, there are no heavily eroding coastal areas in the republic.Progress in Georgian coastal protection depends upon the basic principles of coastal natural process management keeping in mind all the peculiarities of natural systems and their interrelation.

Different geographical and geological conditions require, of course, different means of shore protection. In this connection there is an urgent need to develop and support scientific aspects of coastal studies as a specific discipline dedicated to the creation of models of natural coastal processes management in specific conditions. By coastal economic development and carrying out a policy of coastal management based on self-regulation and self-defence of coastal systems, we should be able to stop the destruction of coastal territories, which are exceptionally valuable, and even to create conditions for their successful development.

1.   References

Barth M. and J.D. Titus, Eds (1984) *Greenhouse Effect and Sea level Rise: A Challenge to our Generation*, Van Nostrand Reynhold Co., New York, 325 p.

Bird E.C.F. (1985) *Coastline Changes: A Global review*, J. Wiley and Sons. 219 p.

Bruun P. (1962) "Sea level rise as a cause of shore erosion.", J. Waterways and Harbour Div., 88, pp. 117-130

Klige R.K. (1985) *Izmeneniya Globalnogo Vobdoobmena*, Nauka, Moscow. 247 p. (Variations of Global Water Exchange, in Russian with English summary)

Leontyev A.K. (1988) "Problems of Caspian Sea Level and Stability of Caspian Coasts.", Moscow Univ. Herald. Geogr. ser., (1), pp. 14-21

Pirazzoli P.A. (1986) "Secular trends of relative sea level changes indicated by tide-gauge records", J. Coast. Res., S I - 1, pp. 1-26

Zenkovich V.P. (1967) *Processes of Coastline Development*, Wiley Intersci., N.Y. et al. 738 p.

Zenkovich V.P. and Schwartz M.L. (1987) "Protecting the Black Sea - Georgian S.S.R. gravel coast", J. Coast. Res., 3 (2), pp. 201-211

RISK ASSESSMENT AND CAUSES OF SUBSIDENCE AND INUNDATION ALONG THE TEXAS GULF COAST

JOHN M. SHARP, JR.
Department of Geological Sciences
University of Texas at Austin
Austin TX 78713-7909

STEVEN J. GERMIAT
Hart Crowser, Inc.
1910 Fairview Avenue East
Seattle WA 98102-3699

ABSTRACT: Analysis of tidal-gauge data from stations bordering the Gulf of Mexico indicate that the Texas Gulf Coast is sinking or submerging at a significant rate. This is attributed to a combination of eustatic sea-level rise and undifferentiated subsidence caused by man's withdrawal of pore fluids and by natural consolidation. The sinking is probably exacerbated by decreasing sediment input to the shore by reservoir impoundments. From 1958 to mid 1980's, submergence rates of 11.1 and 11.4 mm/yr are documented for station at Galveston Bay and Sabine Pass, respectively. The Galveston Bay station shows an increasing (second-order equation) rate over its period of record (1908-1986). This equation is almost identical to the Marine Board's (1987) middle estimate of sea-level rise created by the "greenhouse effect", Submergence at the mouth of the Rio Grande is 4.9 mm/yr. Data from relatively stable stations in Florida indicate that a "eustatic/tectonic" rise of approximately 2.4 mm/yr for the Gulf of Mexico. The remaining 2.5 to 9.0 mm/yr is thus due to combined, but undifferentiated, effects of petroleum production, groundwater pumpage, and the natural consolidation of low-permeability, clay-rich sediments. Groundwater withdrawals are well documented in the Houston area, but have been greatly reduced in the last 15 years (Houston has been switching to surface-water reservoirs for its major water supply). Groundwater pumpage is not everywhere a factor in the zones of high subsidence. Petroleum production has apparently created regional depressurization and concomitant subsidence along the uppermost Texas Coast.

A risk assessment couples these subsidence effects to three scenarios to future sea-level rise. Conservative analyses indicate that the upper Texas Coast will experience significant shoreline retreat due to undifferentiated erosion and submergence in the next 60 years. One complicating factor is the reduction of sediment input to the coastal areas - the Mississippi River, Brazos River, and Rio Grande now transport 50%, 10% and 5% of their predevelopment sediment loads, respectively,

R. Paepe et al. (eds.), Greenhouse Effect, Sea Level and Drought, 395–414.
© *1990 Kluwer Academic Publishers.*

to the Gulf. In addition, a Delphic analysis is presented which estimates the uncertainty of sea-level rise projections by the years of 2025, 2050, and 2100. Our data indicate that this and similar flat-lying coasts in the U.S. will experience major inundation problems in the near future. These have received minimal political or managerial considerations. Instead, large coastal development is underway.

1. Introduction

The recent Marine Board (1987) report, *Responding to Changes in Sea Level*, concluded that relative mean sea level is rising at the majority of the Earth's tidal gauging stations; that subsidence (and glacial rebound) are significant factors; that "the risk of accelerated mean sea level rise is sufficiently established to warrant" its consideration in land use planning and engineering design; and that beach erosion will also probably accelerate. The Marine Board also make several recommendations. These include:
1.   Consideration of accelerated sea-level rise effects in long-term planning and policy development.
2.   Decadal revisions of sea-level rise estimates; and
3.   Development of uncertainty analysis for sea-level (and subsidence) projections.
In this paper we evaluate the historical submergence of the northeast Texas Coast and its factors - eustatic and/or tectonic sea-level (ESL) rise, natural subsidence, and anthropogenic subsidence caused by extraction of subsurface fluids. We use three scenarios - baseline, low-rise, and high-rise - for future sea-level rise and couple these into projections for coastal retreat at 2050 from the combined effects of erosion, subsidence, and sea-level rise. Finally, we present the results of a preliminary "Delphic" analysis to provide an estimate of the uncertainty in ESL projections.

2. The Northeast Texas Coast

The shoreline along the Texas Gulf Coast has generally retreated over the historical record, resulting in a loss of coastal lands. We initiated our study after examining U.S. National Ocean Service tidal gauge station data (data at Sabine Pass and Pier 21, in Galveston, show high rates of relative sea-level rise - 11.4 and 11.1 mm/yr respectively - between 1958 and the mid 1980's), following a field examination of the area, and because of concerns of accelerated rates of sea-level rise.

The study area (Figure 1) comprises the northernmost portion of the Texas Gulf Coast, and extends from the Galveston Bay complex approximately 100 km northeastward to the Texas-Louisiana border (Sabine Pass), and approximately 30 km inland. Port Arthur is the only major city within the study area, although Galveston is immediately adjacent to the southwestern corner of the study area.

The area is susceptible to the loss of coastal lands; almost 50 percent of the area is below 1.5 m and 75 to 80 percent is below 3 m in elevation (Fisher and others, 1972, 1973). Like most areas along the Gulf Coast, the land surface dips gently toward the Gulf. The shore is characterized by a relatively steep and narrow beach zone, which grades inland

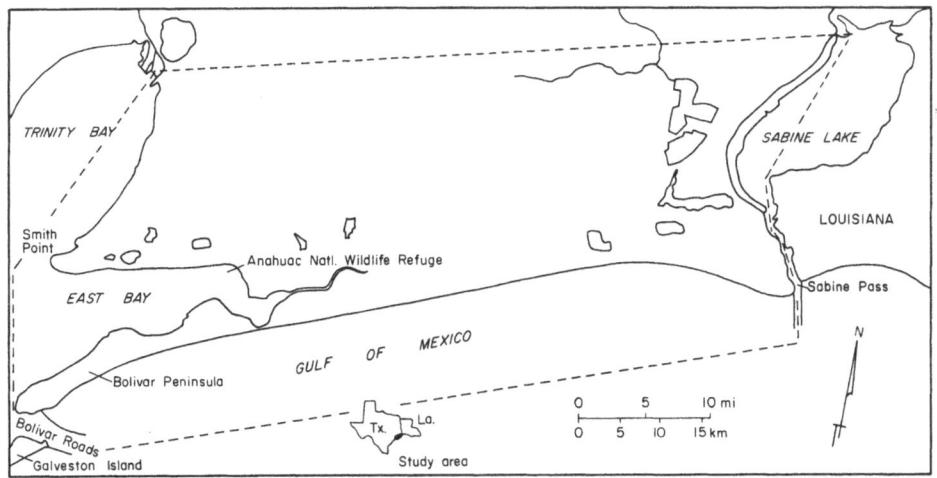

Figure 1. The study area encompassed by dashed line.

to coastal marshes. Consequently, even a small rise in sea level could cause a large landward displacement of the shoreline. Finally, the northeast Texas coast is sediment-starved, and there is little potential for coastal accretion. Major rivers bordering the study area, the Trinity and Sabine-Neches system, contribute little sediment to the shoreline because of reservoir development in the last 40 years and because sediment is trapped in Sabine Lake on the east and Trinity Bay on the west. Longshore transport is hindered by jetties at Sabine Pass and Bolivar Roads.

The geologic setting of the area is described in a number of reports including LeBlanc and Hodgson (1956), Nelson and Bray (1970), Wesselman and Aronow (1971), Jorgenson (1975), Kreitler and others (1977), Winkler (1979), Bentley (1980), Morton and Price (1987), Germiat (1988), and Sharp and others (1988). The interested reader is referred to these, among many reports, for geologic and hydrologic details. The area is similar to other prograding, submerging coasts with thick sedimentary deposits (Sharp and Domenico, 1976, Figure 2 therein). Many of these areas are prolific petroleum-producing provinces.

3. Historical Relative Sea-Level Rise

Relative sea-level (RSL) rise has two components: ESL rise and land-surface subsidence (vertical sinking of the land surface with respect to a stable datum). Sea level has fluctuated over geologic time as inferred by Vail and others (1977). The "Vail curves" cover all of Phanerozoic time but have poor definition for the late Holocene, the time period most critical to this study. Also well established are rising sea levels since the last peak of the Winconsin glaciation and the waning of sea-level rise to near present-day elevations by about 3,000 to 4,000 ybp, although Paine and others (1987) suggest more recent fluctuations of up

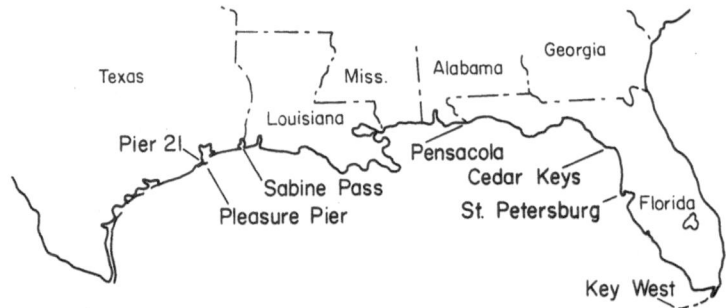

Figure 2. Tidal gauging stations in the Gulf of Mexico used in the study.

to a metre.

Analysis of tidal gauging records from the study area and the "stable" Florida Platform (Figure 2) are used to evaluate ESL rise and subsidence trends. Figure 3 shows dat for Texas stations; monthly averages of hourly tidal heights. Pier 21 provides the longest and most complete record, with data from 1908 to 1986 inclusive. Records for Sabine Pass and Pleasure Pier cover 1958 to 1983 and 1958 to 1986 respectively. They are less complete both in terms of period of record and number of missing data points within the record. Figure 4 shows similar data for the Florida gauges. Rates of sea-level rise are shown in Table 1.

3.1. EUSTATIC SEA-LEVEL RISE

The rate of RSL rise at Key West, Pensacola, and St. Petersburg is 2.4 mm/yr over the period of record (1913-1986, 1923-1986, and 1947-1986 respectively). The Cedar Keys station was not used because the possibility of localized land-surface movement and a gap in the data. Rates of RSL rise for the two baseline periods later used in coastal retreat projections (1930-1974, 1930-1984) are almost identical at the Pensacola and Key West stations (2.2 mm/yr). We considered this the baseline rate of "eustatic rise" for the Gulf of Mexico.

The assumption that tidal stations along the Florida coast are not subsiding may not be valid because worldwide tidal gauge indicate a global rise in sea level of between 1 and 1.5 mm/yr over the past century. Gornitz and others' (1982) value of 1.2 mm/yr is commonly cited in the recent literature (e.g., Revelle, 1983; Robin, 1986; Ramsey and Moslow, 1987; Marine Board, 1987). In a follow-up study, Gornitz and Lebedeff (1987) obtained values of 1.2 + 0.3 and 1.0 + 0.1 mm/yr by applying two different averaging methods to worldwide tidal gauge data for the last century. Thus, the value for eustatic rise in the Gulf of Mexico as determined from the Florida gauges is approximately twice the global rate. Fairbridge

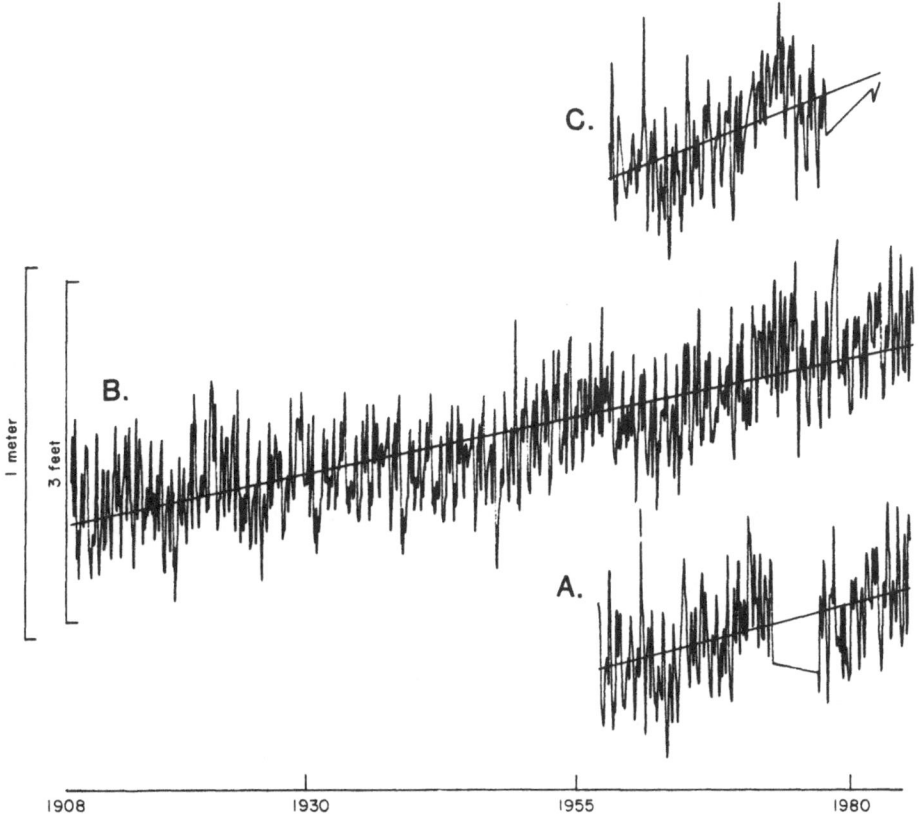

Figure 3. Tidal gauge data (monthly average tidal stages) for Texas stations: A, Pleasure Pier; B, Pier 21; C, Sabine Pass. Also shown are linear regression lines.

(1989, written comment) suggests that changes in oceanic (geostrophic) currents may be the cause of the rise. Alternatively, this anomaly may be created by tectonic plate deformation in the Gulf of Mexico beneath the very thick pile of Cenozoic sediments. Pirazzoli (1986) discusses the importance of local tectonic factors in analyzing local tidal gauge data. This tectonic deformation can be considered constant over the time periods of interest (to 2025, 2050, and 2100). The RSL rise over 2.4 mm/yr represents a valid estimate of eustatic/tectonic influences in the Gulf of Mexico during the past century.

In contrast, RSL has risen at rates of 6.3 mm/yr over the period 1908-1986 at Pier 21, 7.3 mm/yr at Pleasure Pier over the period 1958-1986, and 11.4 mm/yr at Sabine Pass over the period 1958-1983 (Table 1). The tidal record at Sabine Pass terminates in 1983, although the gauge is again (1988) in operation. The magnitude of the component of land-surface subsidence at the Pier 21 gauge is estimated by comparing its rate of RSL rise with

400

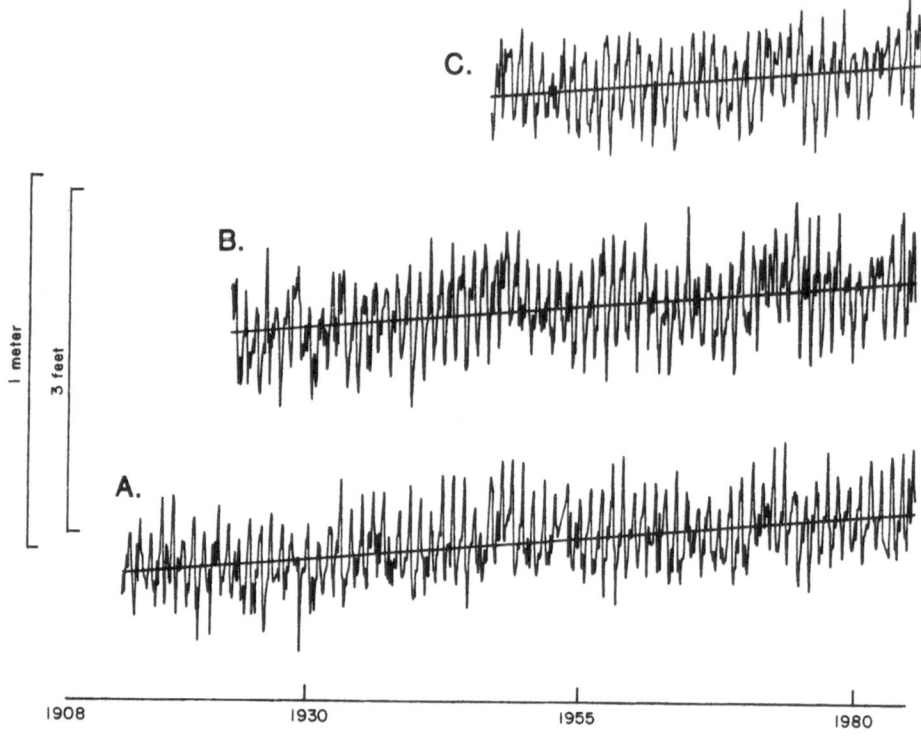

Figure 4. Tidal gauge data (monthly average tidal stages) for Florida stations: A, Key West; B, Pensacola; C, St. Petersburg. Also shown are linear regression lines.

the Florida rate (2.4 mm/yr), suggesting that approximately 60 percent of the observed RSL rise is accounted for by land-surface subsidence over the full period of record.

We also compared RSL at Sabine Pass and Pleasure Pier with Florida gauges for the same periods of record. For 1958-1983 Sabine Pass shows a rate of rise of 11.4 mm/yr, and for 1958-1986 RSL rose at a rate of 11.1 mm/yr at Pier 21 and 7.3 mm/yr at Pleasure Pier; the three Florida gauges registered rates of rise between 2.8 and 2.9 mm/yr. The reason is unknown for the rate difference or 3.8 mm/yr between Pier 21 on the landward side of Galveston Island and Pleasure Pier on the Gulfward side. Pier 21 is closer to the focus of groundwater pumping in metropolitan Houston (discussed below), but the distance is not great enough to account for the discrepancy in rates. Differential fault movement is a possible cause of the discrepancy.

These data suggest that land-surface subsidence accounts for approximately 75 percent of the observed RSL rise over the period 1958 to the mid 1980's at the Pier 21 and Sabine Pass stations, which bound the study area. Comparison of the ratio of land surface subsidence

Table 1. Relative sea-level rise (mm/yr) for various time periods for the seven gauging stations used in the study.

Station	Time of Series	Relative Sea-Level Rise (mm/yr)			
		Entire Series	1958-1986	1930-1974	1930-1982
Sabine Pass	1958 - 1983	11.4	11.4**	NA	NA
Pier 21	1908 - 1986	6.3	11.1	6.9	7.6
Pleasure Pier	1958 - 1986	7.3	7.3	NA	NA
Pensacola	1923 - 1986	2.4	2.8	2.2	2.2
Cedar Keys	1914 - 1986	1.8	1.7	NA	NA
St. Petersburg	1947 - 1986	2.4	2.8	NA	NA
Key West	1913 - 1986	2.4	2.9	2.1	2.2

Note. NA: indicates that the short length of series precludes regression for this period.
    **: Series ends in 1983

to eustatic rise at the Pier 21 station for the period 1908-1986 with the period 1958-1983 demonstrates that land-surface subsidence has been accelerating more rapidly than eustatic/tectonic rise over the period. This is supported by trend at the Sabine Pass station, although its record is not as long.

3.2. SUBSIDENCE

Consolidation of sediments is created as sediments dewater and effective stresses increase, causing a predominantly vertical shortening. This phenomenon has been extensively discussed. See, for instance, Ortenblad (1931), Gibson (1956), Skempton (1970), Sharp and Domenico (1976), and Gabrysch (1982, 1984). Three possible causes of subsidence in the study area are natural processes, groundwater pumpage, and extraction of oil and gas.

The sediments with the greatest capacity to undergo consolidation are clay-rich and shallowly buried. The Pleistocene Beaumond Formation has the greatest capacity for consolidation in the time spans under consideration, although the entire, thick Cenozoic sedimentary package is slowly consolidating (Sharp and Domenico, 1976; Sharp and others, 1988). Natural subsidence of Pleistocene and Holocene sediments in the area has received scant study. Table 2 lists the four available estimates of natural subsidence rates in the study area. The estimates could all be valid, because of their different locations and time intervals. Pier 21 and Sabine Pass gauges are sited near buried Pleistocene valleys with 30 to 40 metres in greater thickness of clay-rich sediments than the intervening areas (Kane, 1959; Nelson and Bray, 1970; Morton and Price, 1987), so that the higher estimate might be appropriate, but it is clear that natural consolidation is not solely responsible for the difference in RSL rise between the Texas and Florida gauges. This is emphasized by the recent acceleration of subsidence rates.

Subsidence created by pumping groundwater is well known along the Texas Coast, particularly in the greater Houston area. Based upon literature review and an analysis of benchmark relevelling data by Germiat (1988), we suggest that the westernmost portion of the study area has subsided somewhat in response to the groundwater pumpage in the Houston area.

Table 2. Estimates of natural subsidence rates for the Texas Gulf Coast.

Rate (mm/yr)	Reference	Basis for Calculation
0.1	Winker, 1979	Continuous tilting of Ingleside shoreline
0.15	Winker, 1979	Sedimentary cycle thicknesses
0.3	Bernard and LeBlanc, 1965	Present slope of Beaumont Formation surface
2.5	Morton, 1979	Subsidence of offshore Holocene strandline

Documented subsidence in Galveston from 1906 to 1978 was 4.1 mm/yr (Gabrysch, 1982). Therefore, some of the observed RSL rise at Pier 21 (6.3 mm/yr, 1908-1986; 11.1 mm/yr, 1958-1986) and Pleasure Pier (7.3 mm/yr, 1958-1986) must be related to groundwater pumpage in the Houston-Galveston area. This hypothesis is supported by the higher rate of rise at Pier 21 than at Pleasure Pier for the period 1958-1986 and the acceleration in RSL rise during the 1958-1986 epoch. This period corresponds to a 130% increase in pumpage from 1960 to 1978 in the Houston area.

The gauge at Sabine Pass shows a very similar acceleration in RSL rise, but there is little pumpage of groundwater in its vicinity. Surface water from the Trinity River and Neches River systems provide most of the water supply, although some groundwater is produced to the north of the study area (Wesselman and Aronow, 1971; Bonnet and Gabrysch, 1982). In most of the study area, water quality is poor and pumpage has been minimal. A review (Germiat, 1988) of all water level data for the study area demonstrated no water-level decline in excess of 11 m (36 ft.) within the area. Pumpage is limited to the upper Chicot aquifer. By comparison, maximum water-level declines in the Houston area were greater than 75 m (250 ft.) in the upper aquifer, and greater than 120 m (400 ft.) in the lower (Evangeline) aquifer. Furthermore, pumpage is restricted to a thin stratigraphic interval in the study area (75 m compared to 600 m) in the Houston area. Consequently, land-surface subsidence created by groundwater withdrawals is not significant within the study area, except in its westernmost fringes. This was also the consensus of Winslow and Wood (1959), Wesselman and Aronow (1971), and Bonnet and Gabrysch (1982).

Subsidence created by petroleum reservoir depressurization is a second anthropogenic mechanism. Such subsidence was first documented at the Goose Creek oil field by Pratt and Johnson (1926). Between 1917 and 1925 subsidence at the center of the field exceeded 1 m and affected an elliptical area of approximately 11 km². Normal faults as long as 0.2 km with displacements as great as 40 cm were recorded within the subsiding area. The subsidence caused permanent inundation of low-lying areas adjacent to the area of production. Both Kreitler (1977) and Holzer and Bluntzer (1984) concluded that the contribution of oil-and-gas withdrawal to localized land subsidence is minimal in the greater Houston area. Only 6 of 26 fields within the Houston area showed subsidence attributable to petroleum production.

On the other hand, Ratzlaff (1985) and Ewing (1985) presented examples of localized subsidence by hydrocarbon production within the study area. Between 1959 and 1977 subsidence of at least 0.33 m occurred in a 20 km² area in the Port Acres field (Figure 5), based on relevelling data and comparison of 1959 and 1977 topographic maps. The period of subsidence corresponded closely to the discovery and development of the Port Acres Gas Field in 1957 (Ratzlaff, 1982). Surface faulting in the High Island-Caplen area, where groundwater withdrawals are minimal, absent in 1930, was observed in 1982 aerial photos. Ewing (1985) suggested that the faulting is man-induced and that perhaps the faulting is caused by extraction of oil and gas from reservoirs with strong water drive.

The possibility of regional subsidence due to depressurization of hydrocarbon reservoirs is intriguing. In order to have significant land-surface subsidence near hydrocarbon reservoirs, four criteria must be met: 1) large reduction in reservoir pressure, 2) production from a large vertical interval, 3) poorly consolidated reservoir rock, and 4) shallow reservoir depth of burial. The fields within the study area, to differing degrees, appear to satisfy these four criteria. This area has supported extensive hydrocarbon production, but unfortunately historical data or reservoir pressures for oil-and-gas fields are scarce (Holzer, and Bluntzer, 1984). Only eight fields of those shown in Figure 5 have sufficient depressurization data to use Geertsma's (1973) simple analytical model. Table 3 summarizes predicted reservoir parameters and rates of subsidence over these fields. Geertsma's high and low estimates of uniaxial compaction coefficients for reservoir sandstones were used, corresponding to high and low estimates of subsidence. The calculations are for the reservoir only and not the adjacent clayey layers. In spite of the many assumptions required in such calculations, it is clear that extensive subsurface depressurization has occurred in oil-and-gas fields within the study area. The lack of documentation of significant localized subsidence, such as at Goose Creek or High Island, may represent a lack of observation and study or, perhaps, subsidence on a more regional scale. In any event, land-surface subsidence is occurring at rates which are difficult to explain solely by the combined effects of groundwater withdrawal and natural consolidation of the substratum. This implies that oil-and-gas production is a factor in the subsidence of the study area.

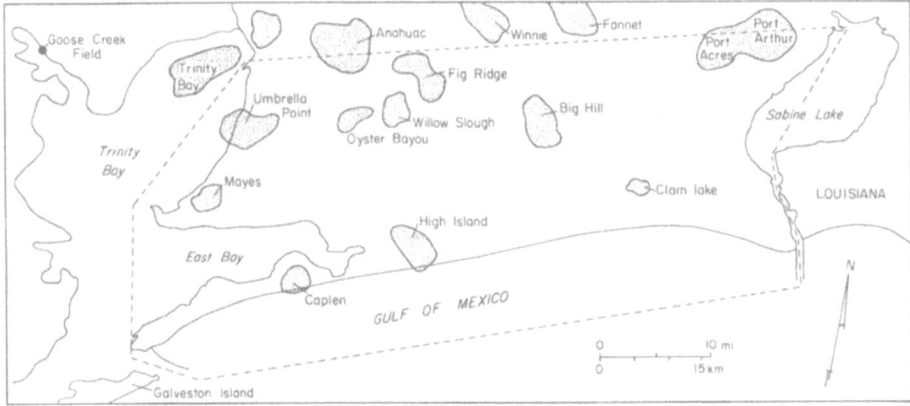

Figure 5. Oil and gas fields in the study area.

404

4. Projected Coastal Retreat to the Year 2050

In order to predict shoreline retreat for baseline, high-rise, and low-rise scenarios, as suggested by the Marine Board, we must consider how to relate RSL rise to coastal retreat. Previous studies (Fisher and others, 1972 and 1973; Morton, 1975; and Paine and Morton, 1986) provided both geological and historical shoreline-retreat data. We characterized the shoreline into four sediment types, shown in Figure 6; sand and shell beaches, marshes and coastal flats, clay bluffs,, and man-made shore. We then discretized the study-area coastline into segments, considering general uniformity in historical rates as well as shoreline orientation. Uniformity of shoreline type was shown to be a less important consideration. Baseline or historical movements were calculated for the period of 1930-1974 or 1930-1982, depending upon the data; data are shown in Table 4.

Table 3. Reservoir parameters, depressurization data, and simulated low and high subsidence for seven fields within or adjacent to the study area.

Field	D (m)	R (m)	h (m)	dP (kN/m2)	Avg. ø (-)	low Cm (m2/N)	high Cm (m2/N)	Low U (r=0) (cm)	Low U (r=R) (cm)	High U (r=0) (cm)	High U (r=R) (cm)
Anahuac	2030	2010	110	4964	0.28	4.2E-07	1.5E-06	5	3	16	10
Clam Lake	1590	800	470	1207	0.32	6.3E-07	2.1E-06	3	2	9	7
Fig Ridge	2590	1830	50	7722	0.27	1.2E-07	2.7E-07	1	0	1	1
Goose Creek	460	1680	30	7585	0.37	2.1E-06	6.0E-06	25	12	72	35
High Island	1680	950	90	19030	0.31	6.0E-07	9.0E-07	9	7	14	10
Oyster Bayou	2500	1390	50	5861	0.29	2.1E-07	3.0E-07	0	0	1	1
Trinity Bay	2440	2530	30	2275	0.32	3.0E-07	4.5E-07	0	0	1	0

Parameters are D, depth of burial; R, radius of ideal disc-shaped reservoir; H, reservoir thickness; dP, reservoir pressure decline; , porosity; Cm, uniaxial compaction coefficients; and U, estimated subsidence at $r = 0$ (center of reservoir) and $r = R$ (edge of reservoir).

The three scenarios for future land loss were developed:

Baseline scenario. This scenario is an extrapolation of historical shoreline movement to 2050. It is a conservative estimate of coastal land loss because it assumes no acceleration in the rate of RSL rise by the year 2050.

Low-rise and high rise scenarios. Nine projections of eustatic sea-level rise in response to climatic warming into the next century were taken from the combined efforts of the Carbon Dioxide Assessment Committee and Polar Research Board of the National Research Council (1983, 1985), the Environmental Protection Agency (Hoffman, 1986), and the Marine Board (MB, 1987) of the National Research Council. These are depicted in Figure 7.

Values for ESL rise at 10-year intervals between 2000 and 2100 were determined for each curve. At each 10-year increment, the mean and standard deviations (σ) of the nine values were calculated. The mean plus and minus one standard deviation calculated as a measure of the variance. The mean of the four high projections and the mean of the

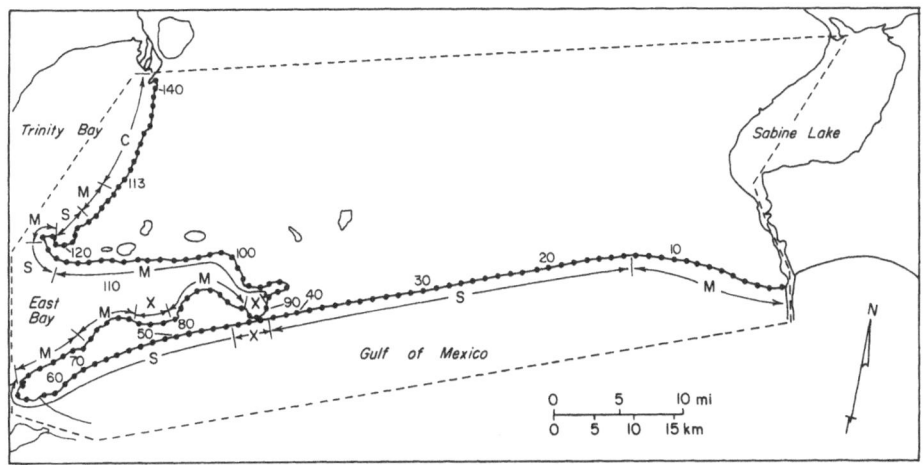

Figure 6. Shoreline segment numbers and sediment types: S, sand and shell beaches; M, marshes; C, clay bluffs; and X, man-made shorelines.

four low projections were also calculated. These data and the extrapolated baseline eustatic rise of 2.2 mm/yr are also displayed in Figure 7. The mean -1 σ and the mean of the four low projections (low-rise estimate) were virtually identical: this was also true for the mean +1 σ and the mean of the four high projections (high-rise estimate). Although there is no apparent statistical reason for this agreement, it implies that the envelope created by the curves represents a reasonable range of estimates for future eustatic rise, based on the best currently available projections. Although the synthesized low-rise and high-rise curves are non-linear, they approximate the linear trend to the year 2050. The mean of the nine projections for eustatic rise is the low-rise scenario (0.35 m at 2050); the mean plus one standard deviation (approximately equal to the mean of the four high projections) of eustatic rise is the high-rise scenario (0.57 m at 2050).

We calculated the component of subsidence at 2050 by subtracting the baseline eustatic rise from the baseline rate of RSL at Pier 21. The subsidence component was then added to the eustatic low-rise and high-rise estimates (Table 5). Multiplicative factors were derived by dividing these sums by the baseline RSL at 2050. Predicted shoreline retreats (as per the methodology suggested by Leatherman, 1984) are the baseline retreat values times this factor. Land loss (or gain) projections are thus the predicted shoreline retreat times the shoreline segment length. Shoreline retreats and land loss (or gain) predictions are listed in table 6.

5. Uncertainty in Sea-Level Rise Projections

We realise that there are considerable uncertainties in the analysis presented above. The

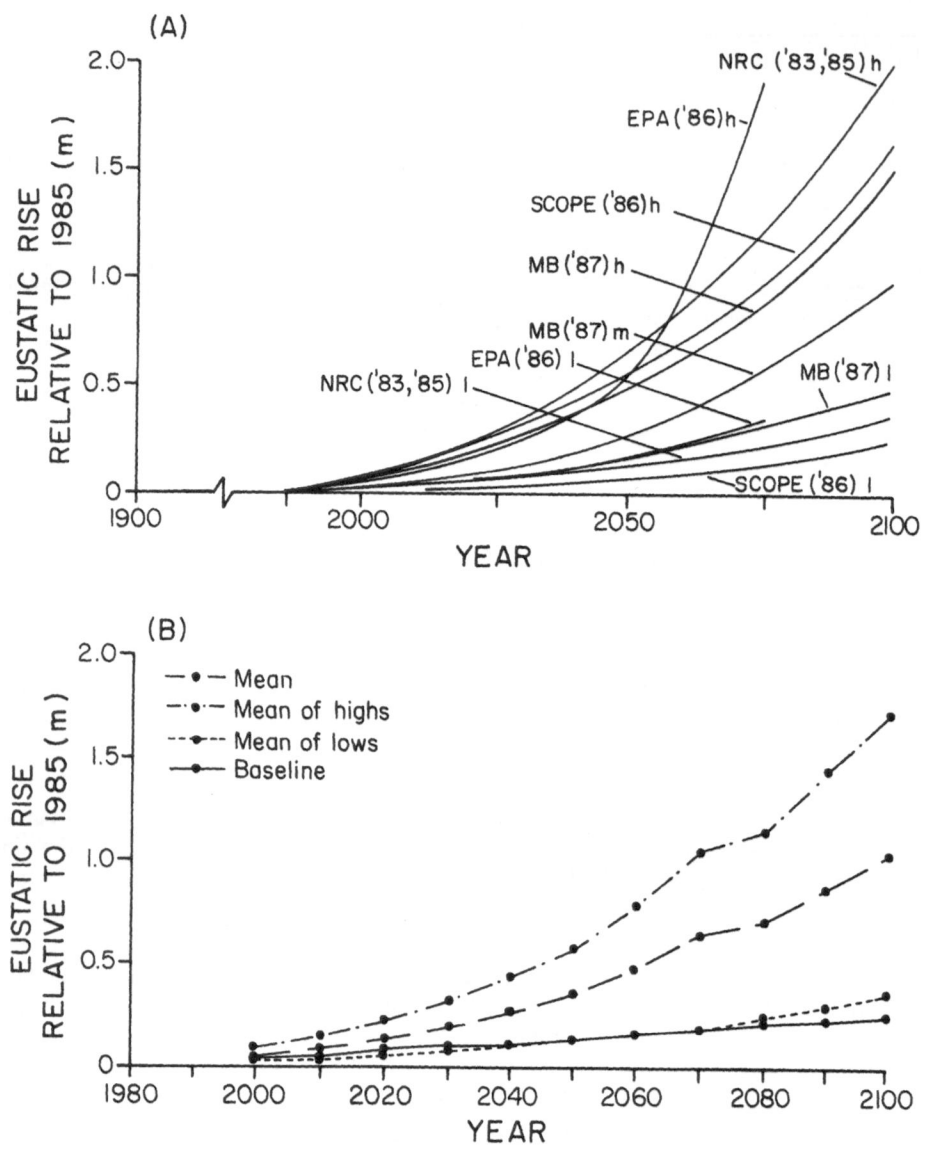

Figure 7. Eustatic sea-level rise projections (A) and baseline, total projections mean, and means of high and low projection means (B).

Table 4. Baseline rate of shoreline movement (mm/yr) for each of the ten shoreline segments.

Segment	Points	Average baseline movement (m)	Rate of baseline movement (m/yr)
1	1-13	-291	-7.0
2	20-42	-53	-1.2
3	51-62	+67	+1.6
4	63-64	-152	-2.9
5	73-87	-52	-1.0
6	90-95	-4	-0.1
7	96-110	-33	-0.6
8	111-117	-76	-1.5
9	118-128	-51	-1.0
10	129-142	-11	-0.2

Table 5. Estimated rates of eustatic sea-level rise, land-surface subsidence, and relative sea-level rise at 2050 for baseline, low-rise and high-rise scenarios.

	Segments 1-3	Segments 4-10
Baseline Period:	1930 to 1974	1930 to 1982
Rate of relative sea-level rise (mm/yr)	6.9	7.6
Rate of eustatic rise (mm/yr)	2.2	2.2
Land-surface subsidence rate(mm/yr)	4.7	5.4
Baseline scenario at 2050		
Eustatic rise (m)	0.14	0.14
Land-surface subsidence (m)	0.31	0.35
Relative sea-level rise (m)	0.45	0.49
Low-rise scenario at 2050		
Eustatic rise (m)	0.35	0.35
Land-surface subsidence (m)	0.31	0.35
Relative sea-level rise (m)	0.66	0.70
High-rise scenario at 2050		
Eustatic rise (m)	0.57	0.57
Land-surface subsidence (m)	0.31	0.35
Relative sea-level rise (m)	0.88	0.92

predicted increase of ESL rise created by the greenhouse effect is probably the major one. Quantification of this uncertainty was one of the Marine Board's (1987) recommendations. The use of baseline, low-rise, and high-rise scenarios above ignores their probabilities/uncertainties. We surmise that a small rise in the next century is highly likely, but not absolutely certain. Similarly, very high rises, of 10 metres for example, are highly unlikely, but not impossible. No data are available for standard statistical projection methods - the greenhouse effect has not yet set a historical trend in rising sea levels. Nevertheless, "subjective probability", or intuitive analysis are common in economic and social planning enterprises.

One way to attempt such an analysis is the Delphic approach, in which experts make the best guess on future outcome. We discussed this method with colleagues in statistics

Table 6. Shoreline retreat and loss (or gain) projections at 2050.

Shoreline segment:	1	2	3	4	5	6	7	8	9	10	NET
Length (km)*	21	55	18	4	18	9	24	10	16	18	193
Rate of movement (m/yr)											
Baseline	-7.0	-1.2	1.6	-2.9	-1.0	-0.1	-0.6	-1.5	-1.0	-0.2	
Low-rise	-10.5	-1.8	2.4	-4.1	-1.4	-0.1	-0.8	-2.1	-1.4	-0.3	
High-rise	-14.0	-2.4	3.2	-5.5	-1.9	-0.2	-1.1	-2.9	-1.9	-0.4	
Displacement by 2050 (m)											
Baseline	-455	-78	104	-189	-65	-7	-39	-98	-65	-13	
Low-rise	-683	-117	156	-264	-91	-9	-55	-137	-91	-18	
High-rise	-910	-156	208	-358	-124	-12	-74	-185	-124	-25	
Areal change (km2)											
Baseline	-9.6	-4.3	1.9	-0.8	-1.2	-0.1	-0.9	-1.0	-1.0	-0.2	-17.2
Low-rise	-14.3	-6.4	2.8	-1.1	-1.6	-0.1	-1.3	-1.4	-1.5	-0.3	-25.2
High-rise	-19.1	-8.6	3.7	-1.4	-2.2	-0.1	-1.8	-1.9	-2.0	-0.5	-33.8

* All segment lengths exclude the length of artificially-altered shoreline stretches.

and public administration, and the following approach was utilized. A questionnaire was prepared in the Appendix. This questionnaire hypothesized that the experts have intuitive normal of log normal uncertainty estimates of ESL rise.

Copies were sent to 16 scientists in the United States who have published estimates of sea-level rise and/or its effects. Fourteen responded and ten gave quantitative guesses. To be fair, we should state that several respondents considered the Delphic approach counterproductive; two suggested that their own deterministic but single-valued projections were more useful. One prepared a "true" Delphic response - several poems! Figure 8 represents Delphic data prepared as normal and log normal probability graphs of ESL rise. Shown on the charts are regression lines and, on the normal charts, values of ESL rise plus subsidence (the 9 mm/yr estimate from Sabine Pass and Pier 21 estimates). It is clear that experts have a probabilistic expectation of the risk of ESL rise. The log normal interpretation appears to be the better fit, especially at longer times.

As would be expected, the uncertainty range increases with longer projections and even encompasses a slight chance of sea-level fall by 2100. The consensus, however, appears to a reasonably good (50%) chance of a 1.5 m rise in ESL by 2100 and, for the northeast Texas Gulf Coast, a 2.5 m rise in RSL. We suggest that such estimates are valid planning criteria for evaluating coastal area development and management of coastal resources.

The statistical methodology for combining such subjective probability estimates with estimates from historical records is unknown. Nevertheless, these Delphic data represent the first estimate, to our knowledge, of ESL rise probability/uncertainty over the next century due to the greenhouse effect.

6. Conclusions

1. Analysis of tidal gauging station data, confirmed by benchmark relevelling data along the northeastern Texas Coast (Germiat, 1988), yields baseline rates of relative sea level (RSL) rise of 6.9 mm/yr for 1930 to 1974 and 7.6 mm/yr for 1930 to 1982 at Galveston's Pier 21. The most recent data yield higher rates of 11.1 mm/yr at Pier 21 and 11.4 mm/yr at Sabine Pass. Approximately 2.2 mm/yr is due to eustatic or "eustatic/tectonic" sea level (ESL) rise.

2. 4 to 9 mm/yr of recent subsidence in the study area is due to undifferentiated natural consolidation of clayey sediments and consolidation created by petroleum reservoir depressurization. Groundwater pumpage is a major factor only near the Houston/Galveston area.

3. Land loss projections for the year 2050 are developed by integrating future projections of RSL rise with empirical relations between RSL rise and shoreline movement. A range of RSL rise of 0.45 to 0.49 m by 2050 is obtained by projecting the baseline rates of RSL rise. Coupling baseline subsidence with projections of eustatic sea-level rise gives RSL rises of between 0.66 and 0.70 m by 2050 in the low-rise scenario, and between 0.88 and 0.92 m in the high-rise scenario.

4. Conservative estimates of shoreline retreat and land loss predict shoreline retreats of up to a kilometre by 2050. Using recent rates of RSL rise would yield higher estimates.

5. Future shoreline displacement within marshy segments may be greater than estimated since sea-level rise beyond a certain limit could cause inundation of huge areas if small steep berms are overtopped, allowing transgression across a gently sloping surface. Precise topographic mapping is needed to perform qualitative analysis of this process.

6. A Delphic approach to estimating ESL rise has been performed. Although preliminary, it may represent a method of providing valid uncertainty estimates. A log-normal probability distribution is inferred for 2100. Coupling such subjective uncertainty analyses to historical data sets would be an important procedure for integrated scientific and economic research.

7. Texas shoreline retreat and coastal land loss in the study area is historical fact. Certain factors were not addressed in this study, such as the long-term effects of diminished sediment input to the Gulf, the effects of coastal engineering and development, and precise geomorphologic responses to future patterns of subsidence and sea-level rise. However, we believe that our estimates of shoreline retreat and subsidence are as precise as possible given the uncertainties. Subsidence and the loss of coastal lands in Texas and similar coastal areas will continue at an accelerated rate. This must be considered in the management of coastal resources.

7. Acknowledgments

Acknowledgment is made to the donors of the Petroleum Research Fund, administered

410

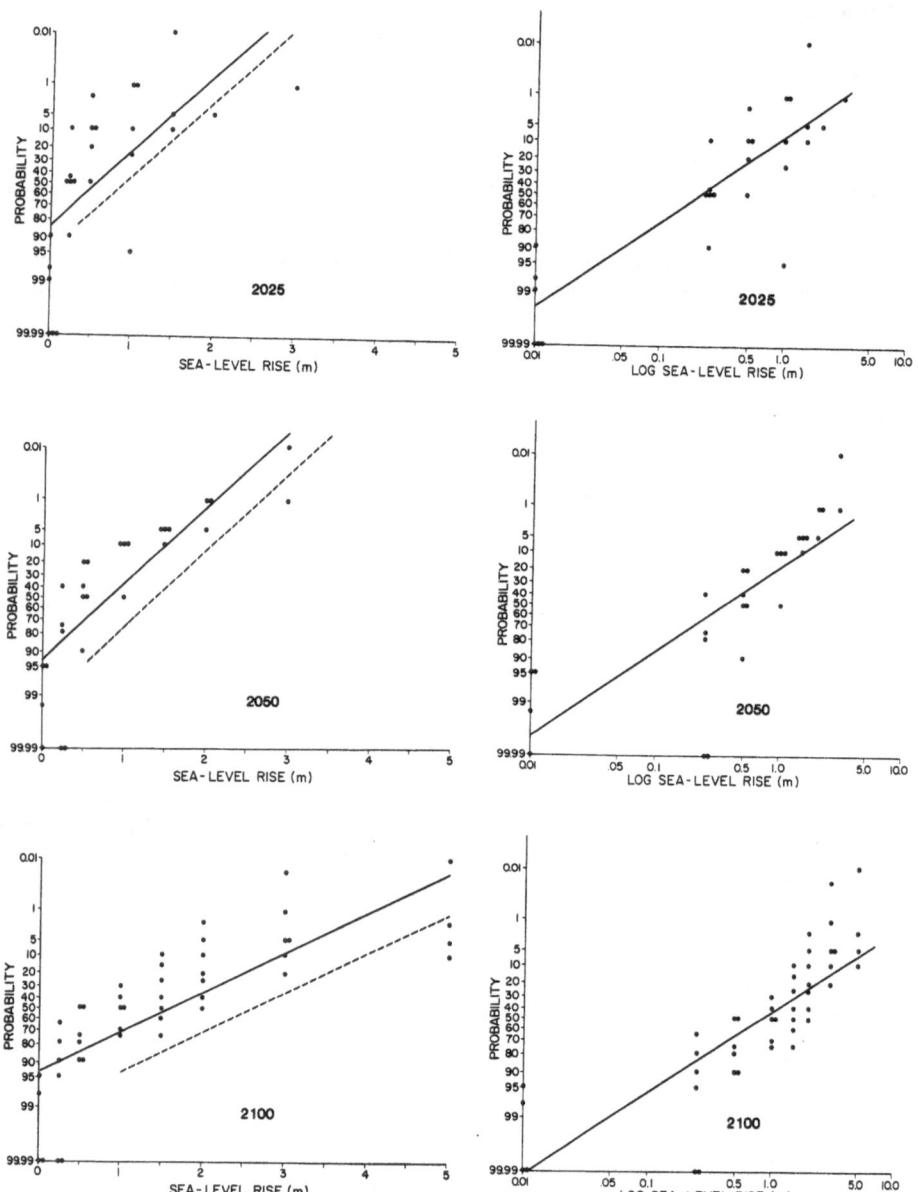

Figure 8. Eustatic sea-level (ESL) rise probabilities and log-normal probabilities compiled from expert projections. Values are given as 100% were plotted as 99.99%. ESL rise plus 9 mm/yr subsidence values are also on the probability curves.

by the American Chemical Society, for partial support for this research. Support for graduate research assistants was also provided by the Gulf Coast Association of Geological Societies and the University of Texas Research Institute. Manuscript preparation was funded by the Owen-Coates Fund of The University of Texas at Austin.

8. References

Bentley, M.E. (1980) "Hydrology of the Beaumond Formation (Pleistocene), Brazoria County, Texas", unpublished M.A. thesis, University of Texas at Austin.

Bernard, H.A., & LeBlanc, R.J. (1970) "Resume of Quaternary geology of the northwest Gulf of Mexico province", in *Recent Sediments in Southeast Texas*, University of Texas Bureau of Economic Geology Guidebook 11.

Bonnet, C.W., & Gabrysch, R.K. (1982) "Development of groundwater resources in Orange County, Texas, and adjacent area of Texas and Louisiana, 1971-80", U.S. Geological Survey Open File Report 82-330

Ewing, T.E. (1985) "Subsidence and surface faulting in the Houston-Galveston area, Texas - related to deep fluid withdrawal?", in Dorfman, M.H., and Morton, R.A. (eds.) *Geopressured-Geothermal Energy*, Proceedings of the 6th U.S. Gulf Coast Geopressured-Geothermal Energy Conference, pp. 289-298

Fisher, W.L., Brown, L.F., McGowen, J.H., & Groat, C.G. (1973) *Environmental Geologic Atlas of the Texas Coastal Zone - Beaumond-Port Arthur Area*, University of Texas Bureau of Economic Geology.

Fisher, W.L., McGowen, J.H., Brown, L.F., & Groat, C.G. (1972) *Environmental Geologic Atlas of the Texas Coastal Zone - Galveston-Houston Area*, University of Texas Bureau of Economic Geology.

Gabrysch, R.K. (1982) "Groundwater withdrawals and land surface subsidence in the Houston-Galveston region, Texas, 1906-1980", U.S. Geological Survey Open File Report 82-571

Gabrysch, R.K. (1984) "Case history No. 9.12 - the Houston-Galveston region, Texas, U.S.A.", in Poland, J.F. (ed.) *Guidebook to Studies of Land Subsidence Due to Groundwater Withdrawal*, pp. 253-262

Geertsma, J. (1973) "Land subsidence above compacting oil-and-gas reservoir", J. Petroleum Technology 25, pp. 734-744

Gibson, R.E. (1958) "The progress of consolidating in a clay layer increasing in thickness with time", Geotechnique 8, pp. 172-182

Germiat, S.J. (1988) "An assessment of future coastal land loss in Galveston, Chambers, and Jefferson counties, Texas", unpublished M.A. thesis, University of Texas at Austin

Gornitz, V., & Lebedeff, S. (1987) "Global sea-level changes during the past century", in Nummedal, D., et al. (eds.) *Sea-Level Fluctuations and Coastal Evolution*, Society of Economic Paleontologists and Mineralogists Special Publication No. 41, pp. 3-16

Gornitz, V., Lebedeff, S., & Hansen, J. (1982) "Global sea level trend in the past century", Science 215, pp. 1611-1614

Hoffman, J.S., Wells, J.B., & Titus, J.G. (1986) "Future global warming and sea level rise", in Sigbjarnarsson, T. (ed.) *Iceland Coastal and River Symposium*, National Energy Authority, Reykjavik, Iceland, pp. 245-266

Holzer, T.L., & Bluntzer, R.L. (1984) "Land subsidence near oil and gas fields, Houston, Texas", Groundwater 22, pp. 450-459

Jorgenson, D.G. (1975) "Analog-model studies of groundwater hydrology in the Houston district, Texas", Texas Department of Water Resources Report 190

Kane, H.E. (1959) "Late Quaternary geology of Sabine Lake and vicinity, Texas and Louisiana", Gulf Coast Association of Geological Societies Transactions 9, pp. 225-235

Kreitler, C.W. (1977) "Faulting and land subsidence from groundwater and hydrocarbon production, Houston-Galveston, Texas", in 2nd Internat. Land Subsidence Symposium Proceedings, Anaheim, California, 1976, International Association of Hydrological Sciences Publication No. 121, pp. 435-445

Kreitler, C.W., Guevara, E., Granata, K, & McKalips, D. (1977) "Hydrogeology of Gulf Coast aquifers, Houston-Galveston area, Texas", University of Texas Bureau of Economic Geology Geological Circular 77-4, pp. 72-89

Leatherman, S.P. (1984) "Coastal geomorphic responses to sea-level rise: Galveston Bay, Texas", in Barth, M.C., and Titus, J.G. (eds.) *Greenhouse Effect and Sea Level Rise*, Van Nostrand Reinhold, New York, pp. 151-178

LeBlanc, R.J., & Hodgson, W.D. (1959) "Origin and development of the Texas shoreline", Gulf Coast Association of Geological Societies Transactions 99, pp. 197-220

Marine Board, Committee of Engineering Implications of Changes in Relative Mean Sea Level, National Research Council (1987) *Responding to Changes in Sea Level*, National Academy Press, Washington, D.C.

Morton, R.A. (1975) "Shoreline changes between Sabine Pass and Bolivar Roads - an analysis of historical changes of the Texas Gulf shoreline", University of Texas Bureau of Economic Geology Geological Circular 75-6

Morton, R.A. (1979) "Temporal and spatial variations in shoreline changes and their implications, examples from the Texas Gulf Coast", J. Sedimentary Petrology 49, pp. 1101-1112

Morton R.A., & Price, W.A. (1987) "Late Quaternary sea-level fluctuation and sedimentary phases of the Texas coastal plain and shelf", in Nummedal, D., and others (eds.) *Sea Level Fluctuations and Coastal Evolution*, Society of Economic Paleontologists and Mineralogists Special Publication No. 41, pp. 181-196

National Research Council, Carbon Dioxide Assessment Committee (1983) *Changing Climate*, National Academy Press, Washington, D.C.

National Research Council, Polar Research Board (1985) *Glaciers, Ice Sheets, and Sea Level: Effects of a CO_2-Induced Climatic Change*, National Academy Press, Washington, D.C.

Nelson, H.F., & Bray, E.E. (1970) "Stratigraphy and history of the Holocene sediments in the Sabine-High Island area, Gulf of Mexico", in Morgan, J.P., and Shaver, R.H. (eds.) *Deltaic Sedimentation - Modern and Ancient*, Society of Economic Paleontologists and Mineralogists Special Publication No. 15, pp. 48-77

Ortenblad, Alberto (1931) "Mathematical theory of the process of consolidation of mud deposits", J. Mathematics and Physics 9, pp. 73-149

Paine, J.G., & Morton, R.A. (1986) "Historical shoreline changes in Trinity, Galveston, West, and East Bays, Texas Gulf Coast", University of Texas Bureau of Economic Geology Geological Circular 86-3

Paine, J.G., Prewitt, E.R., & Valastro, S., Jr. (1987) "Sea-level control of clay dune development at the Swan Lake site: evidence for a Holocene highstand?", Geological Society of America (South-Central Section), Abstracts with Programs 19, p. 175

Pirazzoli, P.A. (1986) "Secular trends of relative sea-level (RSL) changes indicated by tide-gauge records", J. Coastal Research 1, pp. 1-26

Pratt, W.E., & Johnson, D.W. (1926) "Local subsidence of the Cook Creek oil field", J. Geology 34, pp. 577-590

Ramsey, K.E., & Moslow, T.F. (1987) "A numerical analysis of subsidence and sea-level rise in Louisiana", in Coastal Sediments '87, American Society of Chemical Engineers, pp. 1673-1688.

Ratzlaff, K.W. (1982) "Land-surface subsidence in the Texas coastal region", Texas Department of Water Resources Report 272

Revelle, R.R. (1983) "Probable future changes in sea level from increased atmospheric CO_2", in Changing Climate, National Academy Press, Washington, D.C., pp. 433-448

Robin, G. de Q. (1986) "Changing the sea level", in Bolin, B., and others (eds.) Greenhouse Effect, Climate Change, and Ecosystems, Scientific Committee on Problems of the Environment (SCOPE) of the International Council of Scientific Unions, (ICSU), Chichester, John Wiley and Sons, Chichester, pp. 323-359

Sharp, J.M.,Jr., & Domenico, P.A. (1976) "Energy transport in thick sequences of compacting sediments", Geological Society of America Bulletin 87, pp. 390-400

Sharp, J.M., Jr., Galloway, W.E., Land, L.S., McBride, E.F., Blanchard, P.E., Bodner, D.P., Dutton, S.P., Farr, M.R., Gold, P.B., Jackson, T.J., Lundegard, P.D., Macpherson, G.L., & Milliken, K.L. (1988) "Diagenetic processes in northwest Gulf of Mexico sediments",Chapter 1 in Chilingarian, G.V., and Wolf, K.H. (eds.) Diagenesis 2, Elsevier Science Publ., New York, pp. 43-113

Skempton, A.W. (1970) "The consolidation of clays by gravitational compaction", Geological Society of London Quaternary Journal 125, pp. 373-411

Vail, P.R., Mitchum, R.M., & Thompson, S. (1977) "Seismic stratigraphy and global changes of sea level, part 4: global cycles of relative changes of sea level", in Seismic Stratigraphy - Applications to Hydrocarbons, American Association of Petroleum Geologists Memoir 26, pp. 83-97

Wesselman, J.B., & Aronow, S. (1971) "Groundwater resources of Chambers and Jefferson counties", Texas Water Development Board Report 133

Winkler, C.D. (1979) "Late Pleistocene fluvial-deltaic depositions of the Texas coastal plain and shelf", unpublished M.A. thesis, University of Texas at Austin.

Winslow, A.G., & Wood, L.A. (1959) "Relation of land subsidence to groundwater withdrawals in Upper Gulf Coast region, Texas", Mining Engineering 11, pp. 1030-1040

APPENDIX I. Sample Questionnaire on Future Eustatic Sea-Level Rise Probabilities

Please give your best estimate on the chances that the Greenhouse Effect will cause a given amount of sea-level rise (either in general or for the Gulf of Mexico):

I think there is a ____ % chance that mean sea level will rise above present level(s) at all by 2025.
I think there is a ____ % chance that mean sea level will rise above present level(s) by 0.25 m by 2025.
I think there is a ____ % chance that mean sea level will rise above present level(s) by 0.5 m by 2025.
I think there is a ____ % chance that mean sea level will rise above present level(s) by 1.0 m by 2025.
I think there is a ____ % chance that mean sea level will rise above present level(s) by 1.5 m by 2025.
I think there is a ____ % chance that mean sea level will rise above present level(s) by 2.0 m by 2025.
I think there is a ____ % chance that mean sea level will rise above present level(s) by 3.0 m by 2025.

I think there is a ____ % chance that mean sea level will rise above present level(s) at all by 2050.
I think there is a ____ % chance that mean sea level will rise above present level(s) by 0.25 m by 2050.
I think there is a ____ % chance that mean sea level will rise above present level(s) by 0.5 m by 2050.
I think there is a ____ % chance that mean sea level will rise above present level(s) by 1.0 m by 2050.
I think there is a ____ % chance that mean sea level will rise above present level(s) by 1.5 m by 2050.
I think there is a ____ % chance that mean sea level will rise above present level(s) by 2.0 m by 2050.
I think there is a ____ % chance that mean sea level will rise above present level(s) by 3.0 m by 2050.

I think there is a ____ % chance that mean sea level will rise above present level(s) at all by 2100.
I think there is a ____ % chance that mean sea level will rise above present level(s) by 0.25 m by 2100.
I think there is a ____ % chance that mean sea level will rise above present level(s) by 0.5 m by 2100.
I think there is a ____ % chance that mean sea level will rise above present level(s) by 1.0 m by 2100.
I think there is a ____ % chance that mean sea level will rise above present level(s) by 1.5 m by 2100.
I think there is a ____ % chance that mean sea level will rise above present level(s) by 2.0 m by 2100.
I think there is a ____ % chance that mean sea level will rise above present level(s) by 3.0 m by 2100.
I think there is a ____ % chance that mean sea level will rise above present level(s) by 5.0 m by 2100.

VULNERABILITY OF COASTAL LOWLANDS. A CASE STUDY OF LAND SUBSIDENCE IN SHANGHAI, P.R. CHINA

CÉCILE BAETEMAN
Belgian Geological Survey
Jennerstraat 13
1040 Brussels
and
Vrije Universiteit Brussels
Earth Technology Institute
Pleinlaan 2
1050 Brussels

ABSTRACT: Groundwater withdrawal caused drastic land subsidence in Shanghai, located in the Changjiang (Yangtze) coastal plain. Geological investigations, focusing on the need of a 3-dimensional approach, were carried out in the framework of the design of a mathematical model aiming at the management of land subsidence. The research was based on sedimentary environmental interpretation with special emphasis on the coastal processes involved.

The compaction of the sediments, due to groundwater withdrawal, is occurring mainly in the upper 70 m deposits, belonging to the Late Pleistocene and Holocene. The Upper Pleistocene deposits consist of a sequence showing the transition from an estuarine to a proper fluvial environment reflecting the end of a transgressive phase followed by a regressive phase. In the Holocene, although characterized by the post-glacial sea level rise, the coastal processes were dominated by a huge sediment supply from the Changjiang and Huanghe rivers. This resulted initially in a considerable salt marsh deposit followed by tidal flat deposits characterized by coastal progradation.

1. Introduction

Coastal lowlands are very vulnerable because these areas where land and sea meet are situated at, or even below, sea level. However, sea level is only one amongst a great number of elements determining whether an area is emerged or submerged. There is continuous lateral and vertical migration of the coastal environments in time and space depending on the net effect of influential processes acting on the coastal area. Kraft and Chrzastowski (1985) discussed extensively the process that may influence the coastal area resulting in coastal change. The authors concluded that a shoreline is stable only if there is a balance between

415

R. Paepe et al. (eds.), Greenhouse Effect, Sea Level and Drought, 415–426.

416

all forces and processes that tend to move the shoreline landward or seaward. These forces and processes may be of a climatic continental or marine origin but, and not at least, also be the result of human activity.

Besides, coastal lowlands, especially those along estuaries and deltas, are very vulnerable as most of them are very densely populated. For the last decades, this implied large investments in economy, industry and, in certain cases, tourism. It is obvious that these areas with high societal pressures require proper coastal management.

However, as coastal lowlands are the result of the combined actions of a large number of forces and processes, it is essential to know the geological history and understand the particular coastal processes involved before any management can be designed. An example of geological investigation, yielding the basic parameters for the design of a mathematical model for the management of land subsidence, will be discussed. The investigation has been carried out in the Changjiang (Yangtze) coastal plain (P.R. China) and, more particularly, in Shanghai (Monjoie, 1990).

The city of Shanghai with a population of more than 10 million is known to be in a critical situation as it belongs to the problematic group of sinking cities. Due to ground-

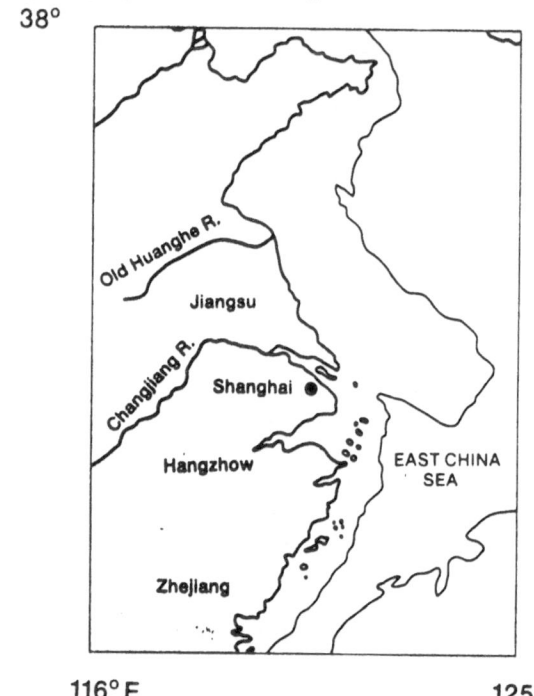

Figure 1. Location of the Changjiang estuary and the city of Shanghai.

water withdrawal, drastic land subsidence has occurred. Land subsidence has already been noticed in 1921, but in 1965 it reached a maximum of 2.63 m. The greatest subsidence was between 1956 and 1959 at an annual rate of 98 mm (Baeteman, 1989).

The city of Shanghai is situated in the vast, low lying coastal plain bordering the East China Sea in the East and characterized by the lower reach of the Changjiang (Yangtze) River (Fig. 1). The river debouches into the East China Sea just north-east of Shanghai.

The Chiangjiang coastal plain was formed and shaped during the Holocene under the influence of the rising sea level. But it is very well known that all over the world estuarine deposits rarely consist of a single fill sequence formed at only one high sea level stand. At least during the Pleistocene and Holocene, it is believed that the estuarine environment reoccupied the same places during various transgressions. From a geological point of view these coastal lowlands consist of a considerable sequence of unconsolidated sediments, sensitive to compaction.

The Changjiang coastal plain is pre-eminently a sea where at one time coastal processes dominated the situation and at another fluvial processes showed an overwhelming activity. The Changjiang river, however, has been the primary factor in the entire Late Quaternary history. The river forms an estuary which has been shifting back and forwards in time, leaving a complexity of deposits in the stratigraphical sequence.

2. Geological Investigation of Coastal Plain Deposits

It is not easy to intepret subsurface data of coastal plain deposits, investigated by means of drill-holes. Because of the similarity of the deposited lithofacies, the stratigraphy of each drill-hole is extremely complex.

A total understanding of changes in coastal and alluvial configurations, or rates of change in the position of coastal and fluvial sedimentary environments, must include an understanding of the vertical sequence of the area being studied. This third dimension to the sedimentary environments represents the element of time, and time is of overwhelming importance in understanding coastal and alluvial evolution (Kraft et al., 1985).

Inherent in the filling up system of coastal and alluvial plains are the frequent lateral and vertical changes in the sedimentary sequence. While studying coastal and fluvial sedimentary bodies or stratigraphic sequences, an important perspective to maintain is that the unit being examined is the result of a depositional environment that is not fixed in time and space. Therefore it is impossible to interpret one single boring and to recognize the units and their succession, because one boring only represents an isolated spot in a complex entity.

It is true that nowadays Quaternary Geology means a lot more than only establishing the boundary between units from subsurface data of one single spot. This approach does not reveal any information about the spatial distribution of deposits, which is essential for a fuller understanding of the development and evolution of the area through space and time.

Therefore it is imperative to establish the three-dimensional geometry of the deposits, i.e. the interrelationship between the different borings. Cross-sections must be made in detail.

The only possibility to correlate properly a series of boreholes implies the knowledge of the processes and mechanisms responsible for the deposition of the sediments and the formation of their facies, thus making an environmental interpretation. Although it is probably not true that each environment produces a unique sedimentary deposit, the basis of environmental interpretation rests on the assumption that particular environments generate deposits that bear the impress of environmental processes and conditions to a degree sufficient

418

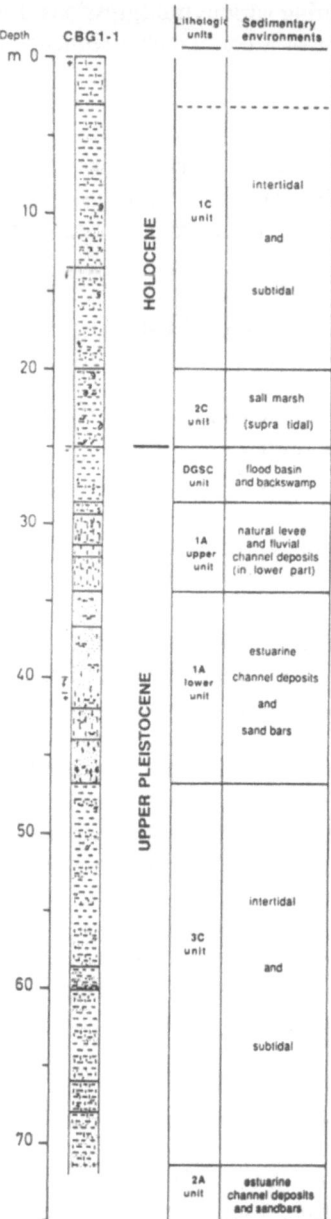

Figure 2. Lithological sequence of borehole CBG1 with indication of the units and related sedimentary environments.

to allow discrimination of the environment (Boggs, 1987).

In order to interpret facies of a depositional system which consists of an assemblage of interrelated environments and their associated processes, facies need to be analysed in terms of their relationship in space. Coastal plain systems as well as the rivers that feed them form an excellent example of an interrelated depositional continuum. Furthermore, it is often impossible to make a unique environmental interpretation on the basis of a single depositional facies. Therefore facies associations and sequences must be studied rather than individual facies. Facies associations are groups of facies that occur together and are genetically or environmentally related. The sequence in which they occur thus contributes as much information as the facies themselves (Boggs, 1987).

3. Geological Settings of the Shanghai Area

3.1. STRATIGRAPHICAL SEQUENCE AND GEOLOGICAL HISTORY

The compaction of the sediments due to groundwater withdrawal is mainly occurring in the upper 70 m, deposits belonging to the Late Pleistocene and Holocene. The Upper Quaternary deposits of Shanghai have mainly been regarded by the Chinese investigators in view of their geotechnical and hydrogeological properties. Hence the nomenclature of the Upper Quaternary of Shanghai characterized by units identified as "Compressible Layer", "Aquifer" and "Dark Green Stiff Clay". From a geological point of view the emphasis was restricted to the chronostratigraphy of some isolated borings leading to the subdivision of the Quaternary in Q_4^3, Q_4^2, Q_4^1, Q_3^2, Q_3^1, etc. (Q_4 for the Holocene, Q_3 for the Upper Pleistocene deposits). The identified "Compressible Layers" and "Aquifers" were labelled with this chronostratigraphy.

The present geological study was based on 42 boreholes to a depth of about 70 m, most of them previously carried out and having a simple description. Although the available data (literature and boreholes) were rather restricted, a stratigraphical sequence for the Upper Quaternary deposits could be put forward (Fig. 2).

The Upper Pleistocene deposits consist of a sequence showing the transition of an estuarine environment to a proper fluvial one. The estuarine environment is represented by a sandy deposit and forms what is usually named the Second Aquifer (2A). It is followed in the sequence by a dominantly clay deposit (3C) which is known to be sensitive to compaction. The latest unit of the estuarine environment is again formed by marine sand deposits (1A -unit, lower part) which gradually are replaced by fluvial sands and silts (1-unit, upper part) heralding the (last) fluvial phase of the Upper Pleistocene sequence. Both sand deposits are from the First Aquifer (1A).

Finally the silting up of the fluvial sequence is completed by the formation of a flood-basin clay. This clay deposit, known as dark green stiff clay (SGSC), represents one of the most interesting deposits in the area of Shanghai, since it is the only unit characterized by a high bearing capacity (for the upper 70m). All the evidence of the sedimentary and stratigraphical investigation yielded clearly that the clay has been deposited in a flood basin during a period of low sea-level.

The Upper Pleistocene was characterized by two main events. The beginning shows a transgressive phase whereby the area becomes occupied by an estuarine environment. In that evolution the decrease of the fluvial activity and the gradual overwhelming of the marine environment is clearly reflected. During that period the area obviously was characterized by (relatively) warm climatic conditions. The climatic optimum most probably coincides with the maximum of transgression.

In the literature it is often mentioned that a transgression occurred during the Late Pleistocene. Zhou Mulin et al. (1982) described a global transgression during an "interglacial (or interstade) of the Dali Glaciation" which he dated at 22900 - 32000 years ago, and in 1986 at 24000 y.B.P.

The transgressive phase resulted in the Shanghai area in the deposition of three distinct lithological units reflecting the vertical stratigraphical sequence of a transgressive estuarine environment: the 2C-unit, upper part, the 3C-unit and finally the 1A-unit, lower part. The absence of one of the units in the vertical sequence of a particular area should be explained as a prevalence of a subenvironment in that particular area during the evolution (or part of it), leading to a much better development of another lithologic unit.

The second main event during the Late Pleistocene consists of a regression. In the estuary the main influence decreased gradually while the fluvial activity became predominant. Finally the entire area was occupied by an alluvial valley, in which the river little by little arose the alluvial relief. Successive floods, each depositing a gradually outward-tapering layer of sediment, seem to have been capable over the years of slowly elevating the active channel by building up its more immediate surroundings. The process of elevating the channel cannot proceed indefinitely, however, for the more the channel and its borders are raised, the easier it is for the river to abandon its elevated position in order to construct a new course at a lower level in another location (Allen, 1985). In 1986, Nanson gave a rather different explanation about the process of abandoning a course in favour of a new one. He stated that overbank deposition gradually builds a flood plain of fine-textured alluvium over a period of hundreds or thousands of years, followed by catastrophic erosion by a single large flood, or a series of more moderate floods, which strips the flood plain to a basal lag deposit from which it slowly reforms. This periodic destruction appears due to the progressive development of large levee banks and flood-plain surfaces of highly variable relief. As the levees and flood plain grow, overbank flow is gradually displaced from the broad flood plain into the main channel and flood-plain back-channels, with a resulting concentration of erosional energy. Vertical-accretion flood plains at different stages of development result in a wide range of bank-full recurrence intervals, even along the same river.

Anyhow the process of abandoning an alluvial ridge in favour of constructing a new one is known as river avulsion and belongs to the most important of all fluvial processes.

Seaward of Shanghai the fluvial environment was very extensive as flood-basin clay (DGSC-unit) has been found in several locations of the (present) inner shelf (Milliman & Jing Qingming, 1985).

That period of regression and fluvial predominance coincided with cold climatic conditions leading to a low sea-level stand, and according to literature, most probably the lowest sea-level stand of Late Pleistocene.

In the Holocene the area changed again into an estuarine environment, represented by predominantly silty clayey sediments. Only two significant units can be recognized. The lower consists of mainly supratidal deposits and the upper represents the silts and clays deposited in the tidal flats of the estuary. Both deposits are very sensitive to compaction, hence are called the "Second Compressible layer" (2C) and the "First Compressible Layer" (1C), respectively.

Flood-basin clay from the Upper Pleistocene is directly overlain by fringing salt marsh deposits (2C-unit) representing the very first deposits owing to the postglacial sea-level rise as there is no evidence of basal peat (only in some exceptional cases). Basal peat reflecting the initial rising of the sea level and the amelioration of the climatic conditions after the glacial period, most probably could not come into being in the studied area due to the clayey substratum. Basal peat development may be expected on the sandy zones of the Upper Pleistocene fluvial deposits.

The salt marsh deposits already reflect the incoming marine influence and indicate that the area was changed again into an estuarine environment.

The general amelioration of climatic conditions resulted in an totally renewed activity of all sedimentary depositional and erosional processes, not in the least from the river itself.

This is indeed the reason why the deposits of the 2c-unit could reach such an important thickness. Normally, at a fast sea-level rise, salt marsh deposits very seldom reach an important thickness as the environment quickly submerges and changes to intertidal and subtidal, resulting in an onlap of tidal flat deposits on salt marsh deposits in the vertical sequence. Only when sea-level rise does not exceed the net rate of sedimentation, the salt marsh development can persist.

However, in the surveyed area the sea-level rise was very fast in the beginning of the Holocene, so it must be deduced that a huge supply of sediment was involved to keep up the salt marsh and prevent it from submergence during a rather long period. The huge supply of sediment was provided by the Changjiang and Huanghe rivers.

However, the salt marsh environment did not persist indefinitely and the sea-level rise finally became predominant over all other features. The salt marsh was pushed landward and in the studied area the marine influence became more and more pronounced, leading to the development of tidal flats in an estuarine environment, reflected in the deposits of the 1C-unit.

In the tidal flats first an intertidal subenvironment was generated, later evolving into a subtidal subenvironment.

The silts and clays comprising the flats from the deposits of the 1C-unit were supplied by the Huanghe and Changjiang rivers, and were largely dispersed to the flats by most probably tidal current littoral drift.

These subtidal deposits from the 1C-unit could have coincided with the maximum of the postglacial transgression often found in literature and situated in the early Atlantic period. There has been a seaward growth of the flats by mud accretion, also called a depositional regression. This was possible at the open coast because of high rates of mud supply and low wave energy. Such a depositional regression does not necessarily imply a lowering of sea level stand.

The depositional sequence as reflected by the 1C-unit represents a development of the mud flats accomplished during a period of relatively uniform conditions of high mud supply, very limited reworking (involving low wave energy) and continuous depositional regression or coastal progradation.

The process of coastal progradation is supported by the occurrence of sand ridges which, judging from 14-C dates of mollusc shells, accreted in a seaward direction (Chen Jiyu et al., 1985).

3.2. GEOMETRY OF THE LITHOLOGICAL UNITS

The 42 boreholes have been correlated in several N-S and W-E cross-sections covering the entire surveyed area. The correlations between the borings have been established on the basis of sedimentary characteristics and labelled with the distinguished lithological units. In the end two panel diagrams have been drawn in order to visualise the interrelationship between all the cross-sections and to give a general view of the spatial extension or geometry of the lithological units (Fig. 3 and Fig. 4).

One of the most striking features demonstrated by the panel diagrams is the occurrence of zones where the Upper Pleistocene deposits, more particular the compact flood-basin clay and partly the first aquifer are lacking (Fig. 3, cf. G41, J3, F7, G37, G35, G62). The absence of these deposits has a dual cause. The geometry of the clay deposit obviously shows several steps in the process of river avulsion. With the last (and lowest) step, the river did not yet have the opportunity to build up a new flood basin adjacent to its new position, so in this particular zone only predominantly sandy deposits from the channel itself are found.

The second origin is erosional rather than depositional. The beginning of the post-glacial period was characterized by a general amelioration of climatic conditions. This resulted in renewed activity of all sedimentary depositional and erosional processes. Especially the river itself became very active again because, as is known with large rivers, the channel regime is often controlled by the climatic conditions of the source area rather than those prevailing in the flood plain. This renewed activity of the river led to amongst others important erosion resulting in rather deep incisions in its own former flood plain. Although sea level rise was forthcoming, and the marine influence was extending more and more on the emerged offshore, the river eroded important parts of the upper Pleistocene deposits from its former flood plain. Some of these eroded deposits were found as channel lags at the base of the depression (reworked Pleistocene deposits, cf. borings F7, G37, G41).

The absence of flood basin clay in certain zones of the Shanghai area is of great importance to the hydrogeological conditions and must be surveyed in detail when evaluating the potential zones sensitive to subsidence. However, it was believed that the flood basin clay has had a rather regular distribution so that it has been sealing off the aquifers from the overlying compressible Holocene clay-layers, thereby preventing leakage and compaction.

Another important feature in the geological setting of the Shanghai area is the wedging

423

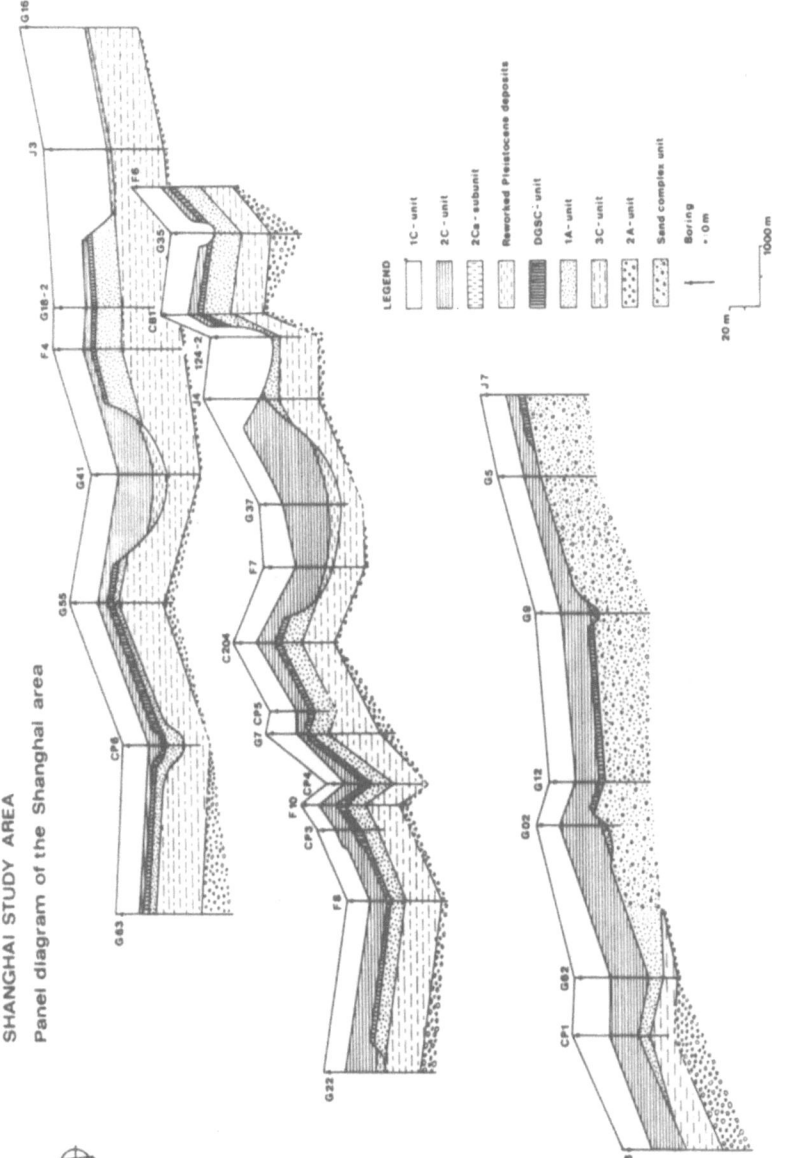

Figure 3. Panel diagram 1 of the Shanghai area.

424

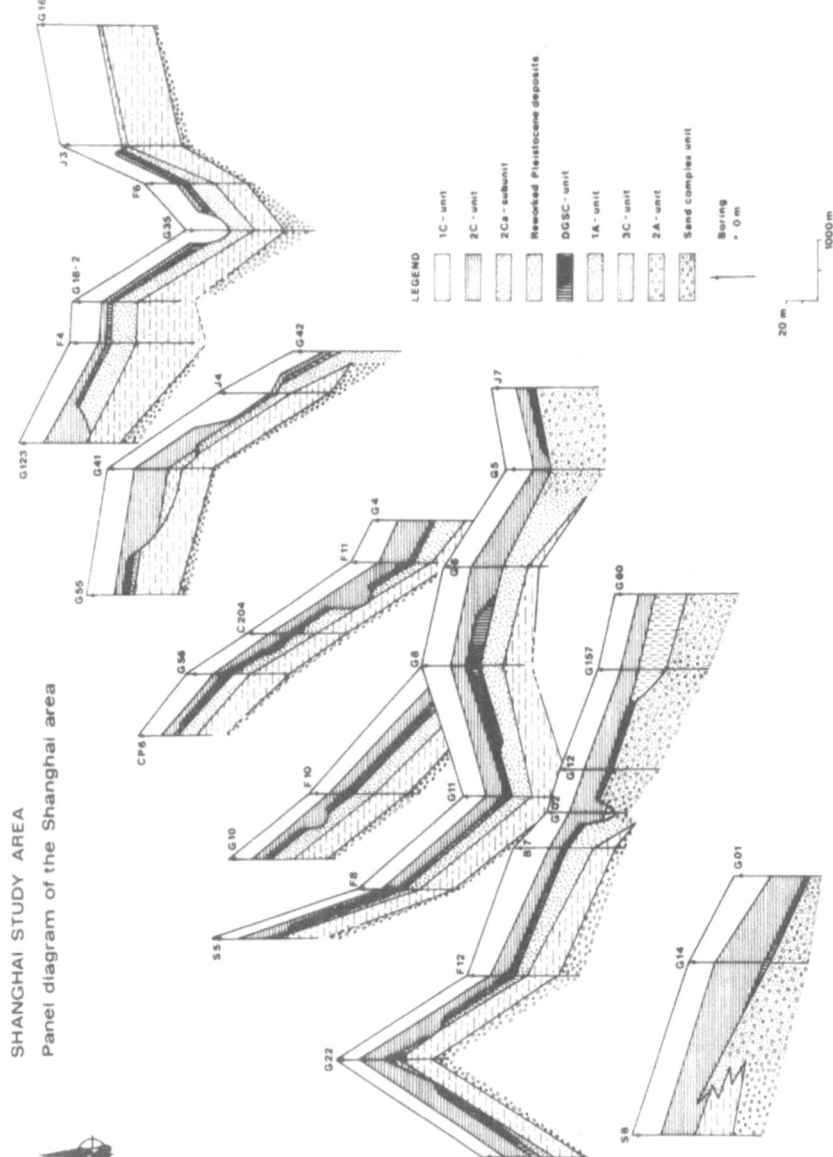

Figure 4. Panel diagram 2 of the Shanghai Area.

out of the Upper Pleistocene compressible layer (3C-unit) in southeastern direction and finally its complete absence in the southern part. In this area the channel activity from the estuary persisted during the entire period of the Upper Pleistocene transgressive phase and no mud flats could develop there. This resulted in an important accumulation of sand deposits, called the Sand Complex-unit. This is the case in the southernmost part of the city which, as a consequence, is less sensitive to land subsidence.

4. Concluding Remarks

Geological investigation based on sedimentary characteristics has proved to be very useful for the design of a mathematical model intending to manage man-induced land subsidence in Shanghai. Although the model was based on geological, hydrological and geotechnical parameters, geology yields one of the most essential elements.

The geological data, however, can only be of any use when the spatial distribution or geometry of the deposits has been established. Geological investigation restricted to the examination of one-dimensional data from the setting up of stratigraphical columns from some isolated boreholes is of no use in such a case. This does not imply that the stratigraphy is to be neglected. On the contrary, the stratigraphic sequence of the entire area under consideration, reflecting the geologic record of events, must be well identified.

The stratigraphic sequence and the geometry of the deposits can only be achieved once the local geological history and the coastal processes involved are well understood. Knowing the forces and processes contributing to the development and evolution of the coastal lowland, and considering their interrelationships, will enhance the resolution of treatment of subsurface data and hence increase the degree of accuracy of the interpretations.

5. References

Allen, J.R.L. (1985) *Principles of Physical Sedimentology*, G. Allen & Unwin, Boston, Sydney, 272 p.

Baeteman, C. (1989) "The Upper Quaternary Deposits of the Changjiang Coastal Plain - Shanghai area", Belgian Geological Survey Report, 155 p.

Boggs, S.Jr. (1987) *Principles of Sedimentology and Stratigraphy*, Merrill Publishing Company - Columbus, Toronto, London, Melbourne, 784 p.

Chen Jiyu, Zhu Huifang, Dong Yongfa & Sunjieming (1985) "Development of the Changjiang estuary and its submerged delta", Continental Shelf research, 4, 1/2, pp. 47-56

Kraft, J.C. & Chrazkowski, M.J. (1985) "Coastal Stratigraphic Sequences.", in: R.A. Davis Jr. (ed.) *Coastal Sedimentary Environments*. Springer Verlag - New york, Berlin, Heidelberg, Tokyo, pp. 625-659

Milliman, J.D. & Jin Qingming (1985) "Sediment dynamics of the Changjiang Estuary and the Adjacent East China Sea.", Continental Shelf Research, 4, 1/2, 251 p.

Monjoie, A. (1990) "Land subsidence in Shanghai." I.A.E.G. (in press)

Nanson, G.C. (1986) "Episodes of vertical accretion and catastrophic stripping: a model of disequilibrium flood-plain development", Geol. Soc. An. Bull., 97, pp. 1467-1475

Zhou Mulin & Liu You Rui (1982) "The Quaternary System of China.", in *Stratigraphy of China. An outline of the stratigraphy in China*. Chinese Academ. Geol. Science, Beijing, China. English Summary, pp. 442-445

Zhou Mulin (1986) "On China's Quaternary stratigraphy", in Hurford, A.J., Jager, E., and Ten Cate, J. (eds.) *CCOP Technical Publication* No. 16, Bangkok, pp. 5-22

THE IMPACT OF NEOTECTONICS WITH REGARD TO CANALS, PIPELINES, DAMS, OPEN RESERVOIRS, ETC. IN ACTIVE AREAS: THE CASE OF THE HELLENIC ARC

ILIAS MARIOLAKOS
Department of Geology
Section Dynamics, Tectonics, Applied Geology
University of Athens
Panepistimioupolis Zografou
15771 Athens, Greece

1. Introduction

Greece is a country in which more than 50% of all seismic energy of Europe is released. This is equivalent to 2% of the seismic energy of the world, whereas Greece covers only 0.09% of the world's surface. It has been estimated that in Greece an earthquake of M > 5R happens every 15 days. The reason is that Greece is situated at the front of a collision between two major plates: the African and the Euro-Asian.

Because of this geotectonic position a lot of earthquakes happen in the larger Aegean area, often followed by surface failures, faults, vertical land displacement (positive or negative) resulting in shoreline displacements.

These movements differ from place to place in both the same event and in different seismic events. In the case of the Alkyonide earthquakes for example (1981, eastern Corinthian Gulf), the maximum displacement of blocks, observed along a relatively short sector of the reactivated Kaparelli-Platees fault reaches about 1.60 m.

Because of its long recorded history Greece is one of the very few countries in the world, where such natural catastrophes have been recorded by ancient writers. It has been recorded for example, that the ancient city of Sparti (Lakonia, S. Peloponnisos) located in the center of a large neotectonic graben, was destroyed in 560 BC, in 492 BC (or 496 BC), in 464 BC, and in 412 BC. Especially with respect to the earthquake of 464 BC the ancient writers Strabo, Paunasias and Ploutarchos have described the total destruction of the city of Sparti, where only 5 houses remained upright and more than 20,000 people died. Fractures were observed on the land surface and during the earthquake of 492 BC the summit of mount Taygetos (the highest mountain of Peloponnisos) was torn off.

However, not only catastrophic events have been recorded by ancient writers. At that time a lot of constructions (buildings, temples) existed, which were vertically displaced,

427

R. Paepe et al. (eds.), Greenhouse Effect, Sea Level and Drought, 427–438.

so that at present they are found below the sea level.

Consequently engineers should, from the very beginning, take into account the dynamic behaviour of the area before planning large constructions such as canals, pipelines, dams, open reservoirs, etc. From my own experience I am not convinced that all of them are aware of the importance of these natural phenomena.

As far as sea level changes and the consequent shoreline displacement, combined with the worldwide eustatic movements are concerned, a lot of research has been done and some interesting papers have been published so far. However, as will be shown below Greece is not the most suitable area for such research because of its tectonic activity, expressed either by abrupt displacement or by very slow movements of the land, as proven by repeated levelling surveys which have been carried out in different areas as the Isthmus of Corinthos and in Kalamata.

Several field cases from Peloponnisos are described below, giving some simple data on the geometry and the kinematics of the faults induced by the earthquakes of 1981 (Corinthos - Beotia) and 1986 (Kalamata), without any dynamic interpretation which is, however, more complicated than is believed by many authors.

2. Area of Corinthos and Viotia (Earthquakes of 1981)

The eastern part of the Corinthian Gulf and its eastern prolongation were the epicentre of the earthquakes of February 24 (M = 6.7R), February 25 (M = 6.4R), and March 1981 (M = 6.4R). This is one of the most active areas in the Hellenic Arc.

The earthquakes damaged a very large area, mainly in the provinces of Corinthia, Viotia and Attiki. It is interesting to hear that the city of Corinthos has been destroyed more than 35 times during its long history.

From a neotectonic point of view the area affected by the earthquakes consists of a number of mega-horsts, such as the Gerania, Pateras and Kithaeron Mountains, separated by mega-graben structures e.g., the area of the Isthmus of Corinthos (SW of Gerania), the Magara basin between the Gerania and Pateras Mountains, the Aegosthena Bay and its landward prolongation, and the Thiva basin north of the Kithaeron Mountains.

Most of the Gulf of Corinthos is a neotectonic megastructure of the half-graben type, whereas the small Alkyonides Islands (one of the epicentres) consists of a horst emerging from the bottom of the eastern Corinthian Gulf.

From a kinematic and dynamic point of view, the neotectonic evolution of the above mentioned structures and their paleogeographic and morphotectonic evolutions is very complex.

2.1. THE ACTIVATED FAULTS

The earthquakes of February-March 1981 produced 4 main fault zones (Fig. 1). Two of them, namely the Platees-Kaparelli and Kalamaki fault zones are located north of the Kithaeron Mountains, whereas the other two, the Pissia and Schinos fault zones are north of the Gerania Mountains.

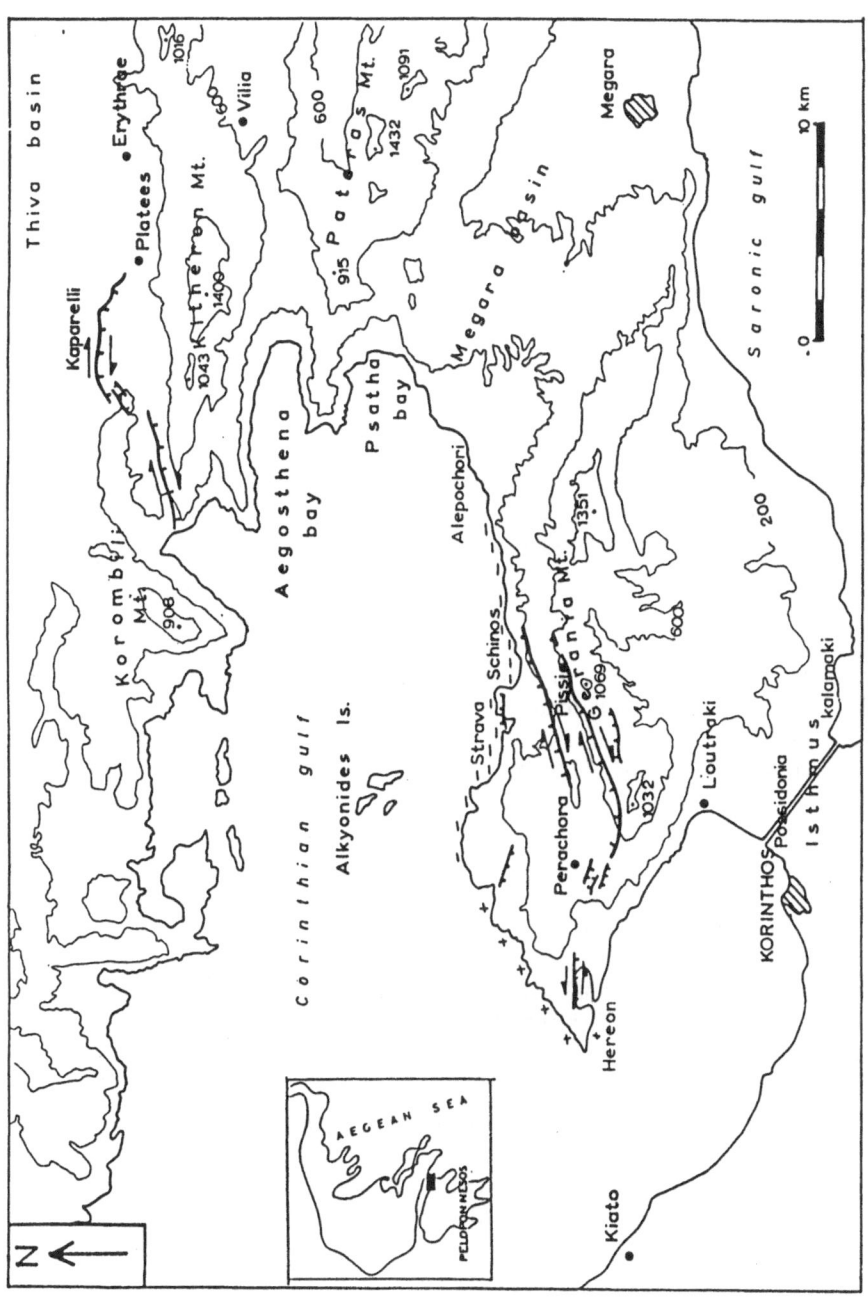

Figure 1. General map of the affected area showing the reactivated major fault zones and the vertical movements observed along the southern shoreline of the Perachora Peninsula.

Besides these two main fault zones, which can be considered first order faults a great number of smaller structures (2nd or 3rd order, etc.) were formed. These faults or fissures are diverse in size, shape and have different strikes.

There are several main characteristics which are common to the major fault zones. These are:

- they show a mean strike E-W;
- they follow older faults which means that older faults are reactivated;
- the fault surfaces are not plane but curved;
- they are not continuous;
- they show an "en echelon" arrangement;
- they are oblique slip-normal faults with right lateral movement. The ratio of the dip-slip component to the strike-slip component is about 7 : 3;
- the large number of microstructures which accompany the main fault zones are usually observed on the block that has moved downwards;
- although the mean strike of the reactivated faults is E-W, in some cases faults of another direction have been reactivated e.g., the Kaparelli-Platees fault. In this case two fault, with an E-W and NW-SE trend respectively, intersect;
- The older faults have not been reactivated at their whole length but only in part;
- the vertical displacement of the faulted blocks varies from place to place along the reactivated faults. The mean hight is 0.60 m, the maximum displacement, observed at the northern part of the Platees is 1.60 m;
- the maximum horizontal gap of the blocks is 80 cm, observed along the Platees branch of the Kaparelli-Platees fault zone.

2.2. OBSERVATION OF THE SHORELINES

The earthquakes of 1981 caused the subsidence of the land along the coast from west of Alepochori to as far as west of the Strava Bay (Fig. 1), where a very impressive subsidence of at least 50-60 cm was observed, which resulted in two weekend houses and a small church to be flooded by the sea. Before the earthquakes these buildings were located about 50 m from the shoreline.

Shoreline displacement at the coastal area around Corinthos is very common due to uplifting or subsiding of the land. These displacements have happened in different periods since ancient times, for example:

i.   The entrance of Lechaeon harbour (the harbour of ancient Corinthos, about 2000 years old) has been uplifted, as can be proved by the exposing of Lithofaga molluscs at a height of about 1.1 m above the present sea level.

ii. About 700 m east of the railway station of Corinthos, v. Freyberg (1973) reported rounded pottery sherds of undetermined age (Neolithic to Roman) cemented in a conglomerate from a marine terrace at a height of 1.30 m above the present sea level.

iii. At Possidonia at the western side of the canal exit at the Corinthian Gulf beach rocks cover the blocks used for the construction of the Diolkos (4th century BC). These beach rocks are found at a height of 0.75 m above the present sea level.

iv. Further north at Loutraki, probably the same beach rock structure is found at a height of 1.0 m (C.G. Higgins, in prep.).
v. At the eastern exit of the canal (the modern site of Kalamaki) submerged ruins have been identified as the ancient harbour of Schoenous.
vi. At the harbour of Kenchree (about 5 km south of Kalamaki), some submerged ruins have been found.
vii. At the Perachora Peninsula (Cape Haraeon), north of Corinthos, three old shorelines can be identified on a cliff above the present sea level, and some others are submerged.

3. The Isthmus of Corinthos

The Isthmus of Corinthos is a narrow land strip connecting Peloponnisos to the Hellenic mainland and separating the Corinthian Gulf from the Saronic Gulf. The neotectonic deformation of the area is very complex. As a macrostructure it is a neotectonic graben between two neotectonic megahorsts i.e. the Gerania Mountain (h > 1000 m) in the NE and the much lower Onia Mountain in the SW. Both mountains consist of alpine formations. The Isthmus area has been part of the Palaeo-Corinthian Gulf which has become land at least twice since the Pliocene, the second time during the Upper Pleistocene.

The construction of a 6.2 km long canal at the end of the 19th century has provided the opportunity to study in detail the late stages of the palaeogeographic and tectonic evolution of the area. The canal was dug more or less along the Diolkos, and ancient ramp used for vessel transportation connecting the Corinthian Gulf and the Saronic Gulf.

From this impressive section we see that the Isthmus consists of a succession of Pliocene marls (marine and fresh water), unconformably overlain by several, in cycles deposited near-shore Pleistocene sediments (mainly conglomerates and sandstones (Fig. 2a).

The Pliocene and Quaternary sediments are cut through by numerous, nearly E-W striking sub-vertical normal faults.

Philippson (1890) grouped them into two sets, namely the "Krommyonian system", with a moving down of the southern blocks and the "corinthian system", with a moving down of the northern blocks. Consequently the area of the Isthmus has been considered since to be a neotectonic horst created inside a bigger neotectonic macrograben. It is important to underline that the faults of the Krommyonian set are antithetic, whereas the ones of the Corinthian set are homothetic.

Most of the faults cross the Pleistocene deposits but some stop at the unconformity between the Pleistocene and the Pliocene deposits, while 2-3 faults stop inside the Pleistocene beds. Two of these faults were reactivated during the earthquakes of 1928 and another one (the first fault south of the national road bridge) was reactivated during the earthquakes of 1981. Displacement in this case was only 2 cm southward.

All this is evidence that the neotectonic activity is actually continuous and consequently the faults can be considered to be active, even from an engineering point of view. v. Freyberg (1973) proved, based on the construction of the strike contour map of the Pleistocene (Eutyrrhenium)/Pliocene unconformity surface, that the younger deformation of the Isthmus

432

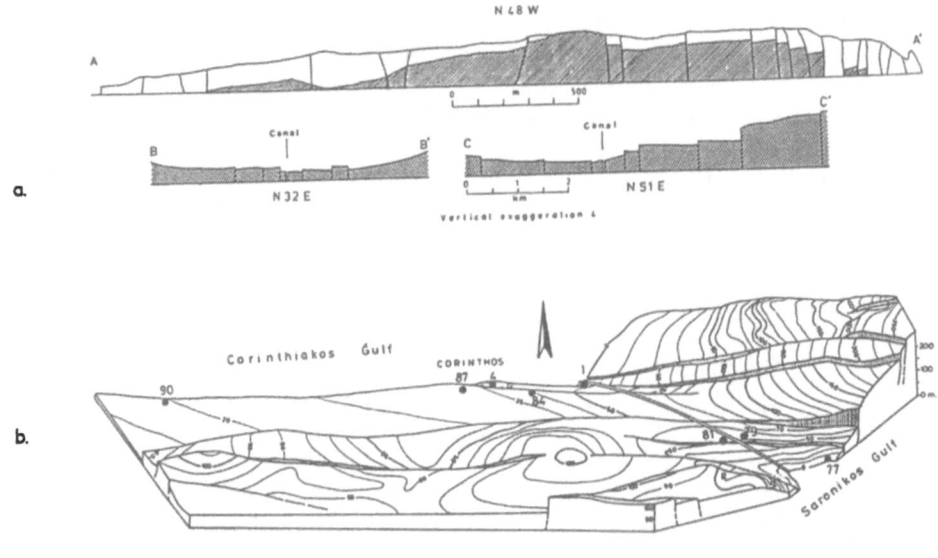

Figure 2. a. Map of the Isthmus of Corinthos showing the levelling benchmarks that have been used for the compilation of the curve of Figure 1b; b. Displacement, based on repeated levelling data. Benchmark 73 is considered invariable;

c. Height of the Euthyrrhenian (Fig. 3) along the levelling route.

land strip was caused by antitilted fault blocks due to a rotation of the whole area around and E-W striking horizontal axis (Fig. 2b). This new interpretations has two consequences:

i.   The normal faults are not the result of a regional stress field of extensional character but of a torsional one.

ii.   The area of the Isthmus of Corinthos does not represent a real tectonic horst. If the canal were constructed in another direction, the picture would be totally different, namely that of a graben (Fig. 2a, lower sections). Obviously it is neither; it is more complicated: it is a case of antitilted block faulting due to torsion of the whole area around a horizontal axis.

Taking into account the deformation of the area and its seismic activity, as was described above, as well as the submergence and uplifting of the land and the consequent shoreline displacement during the historical period and those caused by the earthquakes of 1981 it

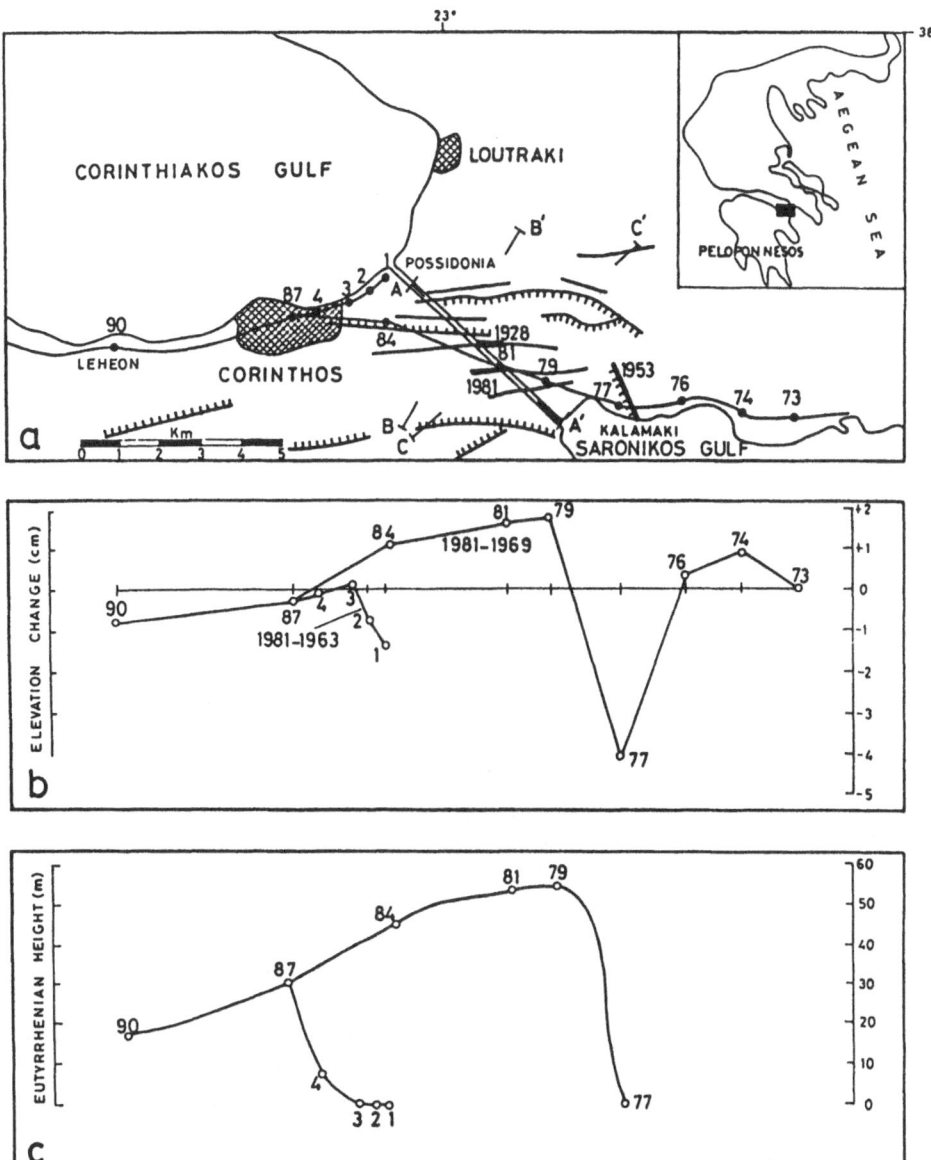

Figure 3. Perspective form of the contour map of the Euthyrrhenian layer surface (after v. Freyberg, 1973). The canal and some of the benchmarks are also shown.

is easy to realise its impact.

The question which is rising now is: can the present day kinematics be correlated with that of the Upper Pleistocene? In order to answer this question we have applied both geodetic and geological methods.

Geodetic methods. A levelling traverse crossing the Isthmus was carried out in 1969, whereas a minor one was carried out in 1963.

The same traverse was resurveyed after the 1981 earthquakes. Benchmarks common to both surveys are shown in Figure 3. The observed elevation changes have been used for the compilation of the displacement profile depicted in Figure 3b, which shows that the central part of the Isthmus (between benchmarks 79 and 84) appear uplifted relative to the near-coastal areas, whereas Possidonia shows a trend of subsidence relative to Corinthos.

Geological methods. In order to estimate the amount of displacement since the Euthyrrhenian, the geological profile of the unconformity surface between the Pliocene and the overlain Euthyrrhenian layer was constructed along the same route used for the levelling.

Figure 3 shows the change in height of this surface. The resemblance of the curves of Figures 3b and 3c, presumably showing the present day and the older (since the Euthyrrhenian) deformation of the Isthmus respectively, is probably more than coincidental and may reveal that the pattern of the crustal deformation has not changed since the Upper Pleistocene.

4. Messinia

4.1. THE NEOTECTONIC STRUCTURES

The province of Messinia (southwestern Peloponnisos) is another seismically active area of Greece. On September 13, 1986 an earthquake of M = 6.2R occurred, followed by a strong aftershock (m = 5.6R) on the 15th, which caused much damage in the city of Kalamata and in some villages to the east and northeast.

The major neotectonic macrostructure of the area is the Kalamata-Kyparissia graben. In its southern part it strikes N-S whereas further north it changes its strike to E-W (Fig. 4).

Inside this first order neotectonic megastructure a number of smaller of second order structures have been differentiated, in some cases with a totally different evolution.

The Kalamata-Kyparissia graben is extended between two major neotectonic megahorsts; the Taygetos Mountain to the east and the Kyparissia Mountain to the west, whereas between them a major fault zone and a great number of smaller neotectonic structures, 2nd and 3rd order horsts and grabens, exist.

The faults in the marginal fault zone are not continuous but they are interrupted and intersected by others which, although they belong to the same fault zone, have different strikes. As a matter of fact, they form conjugate sets of faults.

The main characteristics of the neotectonic faults are:

i. The density seems to be irregular in the major area and it is more or less independent

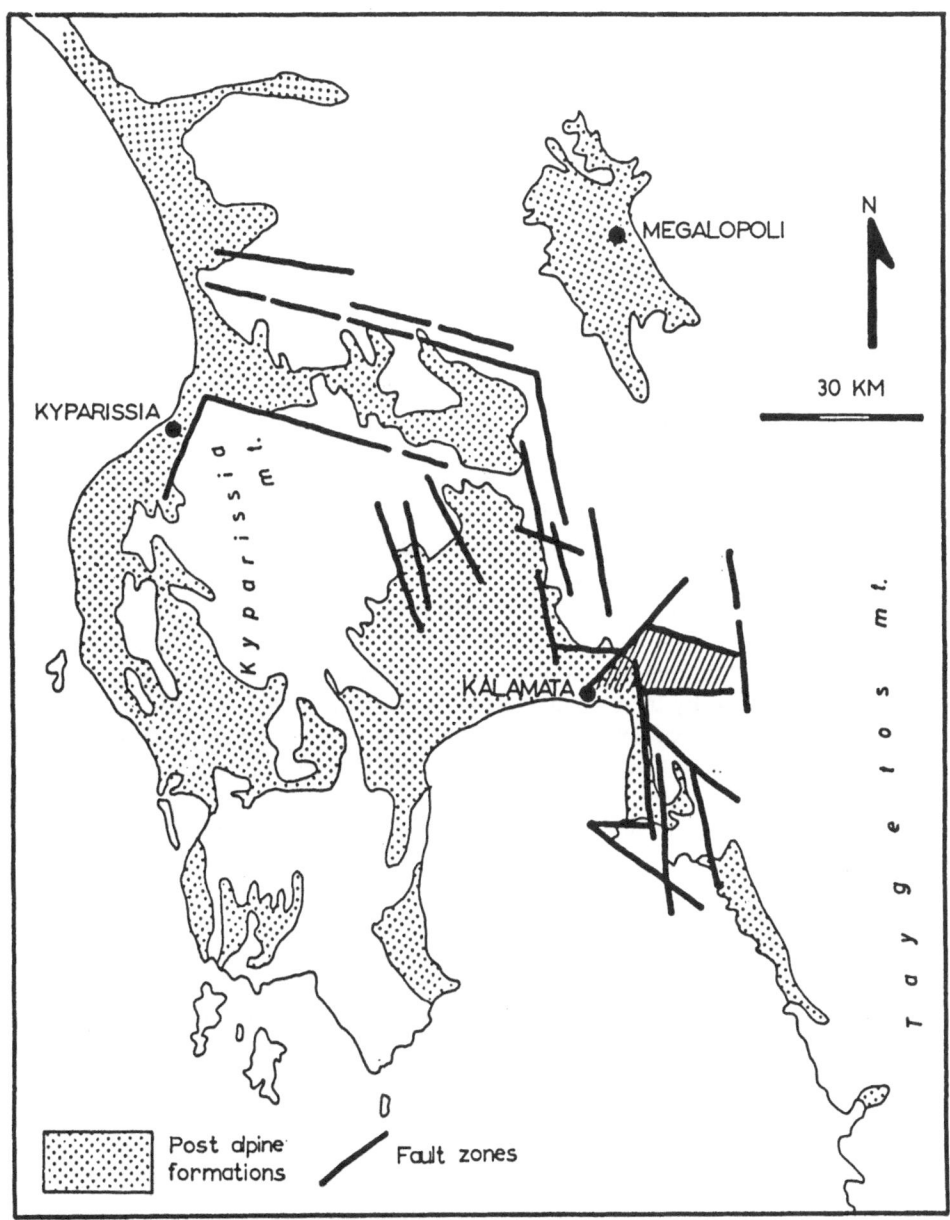

Figure 4. The main neotectonic structure of Messinia: the Kalamata-Kyparissia megagraben.

from the age of the strata e.g., the density of the faults in the Pleistocene deposits at the central part of the graben (north of the city of Messinia) is relatively higher than the one observed at the Neogene deposits, which outcrop at the eastern margin of the Kalamata-Kyparissia graben.

ii. The density of the neotectonic faults in the alpine neritic limestones is much higher than that of the Neogene deposits. But even inside the same formation e.g., the alpine neritic limestones, the fault density differs greatly from place to place.

iii. At the older fault surfaces successive slickenside generations have been observed. This is evidence of older reactivation of the same fault.

4.2. SEISMIC FAULTS AND SEISMIC FRACTURES

The Kalamata earthquake (9.13.'86), M = 6.2R, caused seismic faults, seismic fractures (without any displacement)(Fig. 5) and rock falls.

Concerning the seismic faults the following must be noted:

- They are mainly the result of the reactivation of older neotectonic faults. However, in one case in the area of the small village of Diasello, a totally new fault was formed.
- Faults occurred mainly during the first earthquake (M = 6.2) and only one (west of Elaiochori) during the second one.
- The reactivated faults strike in different directions.
- The throw of the seismic faults is generally small (max. = 20 cm) and of a normal character. It is interesting to note that the maximum throw has been observed at the seismic fault caused by the second and smaller earthquake (M = 5.6).
- Seismic faults have been observed in all kinds of formations, alpine and post-alpine.
- In many places, mainly due to the high relief, the faulting was accompanied by rock falls.

Concerning the seismic fractures (ruptures without visible throw) the following must be noted:

- Most of them are relatively small (4-5 m long), whereas some of them are very large)
- The seismic fractures form a zone or zones. The arrangement of the seismic fractures inside the zone is typically "en echelon". These fracture zones are in some places of a right-lateral and in others of a left-lateral character (Fig. 5).
- Seismic fractures have been created during both the first earthquake and the main aftershock. In some cases two separate fractures created by the first earthquake were linked by a new fracture created during the main aftershock.
- A number of seismic fractures did increase in width and length during the main aftershock.
- The traces of the seismic fractures on the land surface are not straight lines but angular, consisting of two sets of straight lines of different strike.
- The density of the fracture zones of large fractures varies from place to place. In one case the density is ten fracture zones on 100 m.

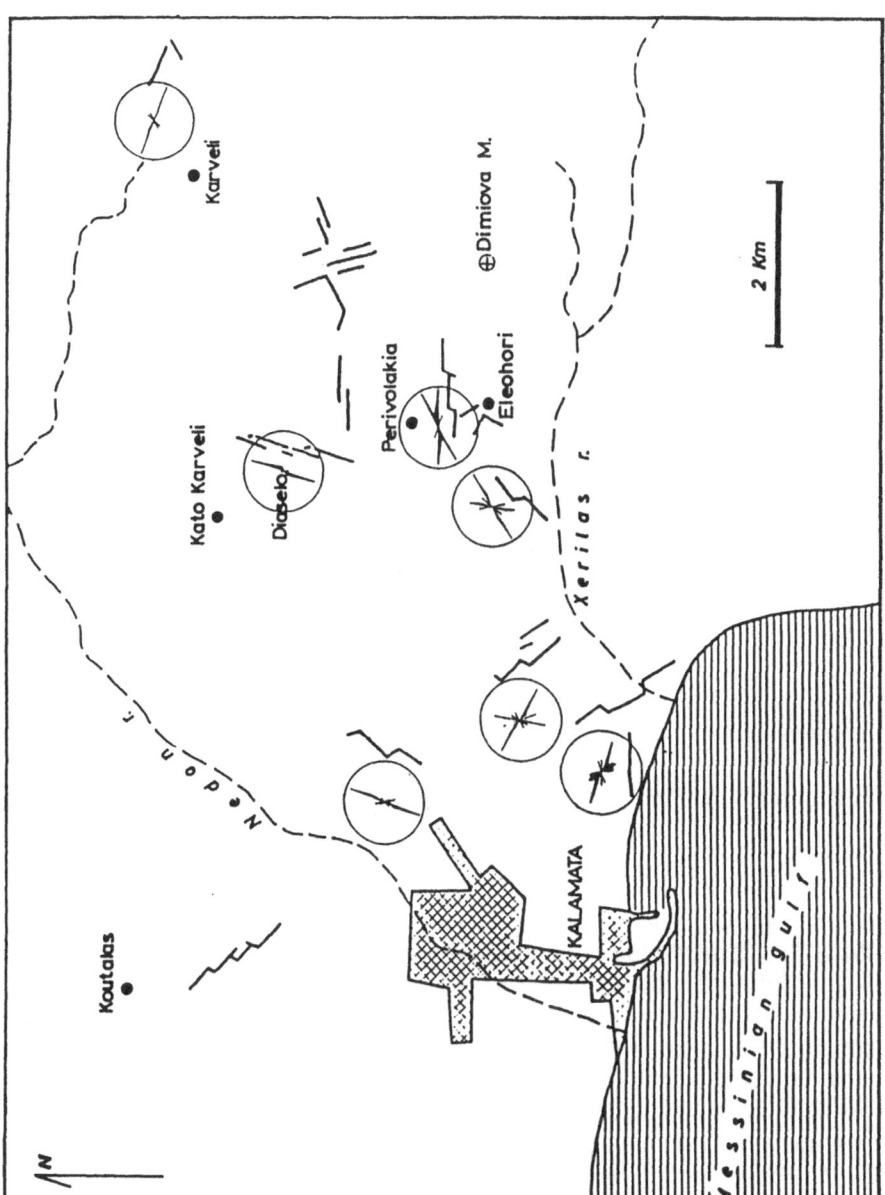

Figure 5. Seismic fracture zones and seismic faults caused by the earthquakes of September 1986 with fracture roses.

5. Conclusions

The presented data are taken from two relatively small, but representative areas of the Hellenic Arc. Their seismic activity allows us to come to the following conclusions.

- The density of the neotectonic faults, both inside the neotectonic macrograbens and at their margins is very high.
- The neotectonic faults or fault zones should be considered to be active and are very likely to be reactivated by earthquakes of M >5R.
- By reactivating older faults due to earthquakes of M >6.5R vertical displacement may reach 1.6 m.
- In coastal areas shoreline displacement connected with earthquakes is very common. The submergence or uplift of ancient buildings in the past, as well as that of recent earthquakes provides evidence that abrupt shoreline displacement should be considered a common phenomenon.
- The repeated levelling surveys in the area of the Isthmus have shown that the trend of the recent kinematics is the same as during the Upper Pleistocene. Consequently Greece and other general tectonically active areas are no suitable places for studying eustatic movements, as it is difficult to find stable areas.
- It is obvious that in tectonically active areas before planning big constructions such as dams, canals, pipelines, open reservoirs, etc., a very detailed neotectonic map should be prepared.

6. References

Freyberg, B. von (1973) "Geologie des Isthmus von Korinth.", Erlanger Geol. Abh. 95, pp. 1-83

Institute of Geology and Mineral Exploration (1986) *Geological Map of Greece: Kalamata sheet (1 : 50,000)*

Mariolakos, I., Papanikolaou, D., Symeonidis, N., Lekkas, S, Karotsieris, Z., & Sideris, Ch. (1981) "The deformation of the area around the eastern Corinthian Gulf affected by the earthquakes of February and March 1981.", Proceedings Int. Symp. Hellenic Arc and Trench, Athens, pp. 400-420

Mariolakos, I., & Stiros, S. (1987) "Quaternary deformation of the Isthmus and Gulf of Corinthos (Greece).", Geology 15, pp. 225-228

Mariolakos, I., Fountoulis, I., Logos, E., & Lozios, S. (1989) "Surface faulting caused by the Kalamata (Greece) earthquakes (9.13.86).", Tectonophysics, 163, pp. 197-203

Philippson, A. (1959) *Die Griechischen Landschaften.*, Band III, *Peloponnes, Teil 1: Die Osten und Norden der Halbinsel.*

EVOLUTION OF BARRIER ISLAND-LAGOON SYSTEMS FROM 200 KA AGO TO THE PRESENT IN THE LITTORAL ZONE OF ALICANTE (SPAIN). IMPACT OF A PROBABLE SEA LEVEL RISE

C. ZAZO, J.L. GOY, L. SOMOZA
Departamento de Geodinámica
Facultad de Geologia
Universidad Complutense
28040-Madrid
Spain

C.J. DABRIO
Departamento de Estratigrafía
Facultad de Geologia
Universidad Complutense
28040-Madrid
Spain

ABSTRACT: The distribution of barrier island-lagoon systems of the littoral zone of Alicante is controlled by tectonics. They occur in areas of subsidence to E-W oriented synclines and also to downthrown blocks of dextral faults oriented N 120°-140° E having some vertical movement. Separation of lagoons begun during the Middle Pleistocene but the Tyrrhenian beach ridges are the most prominent morphological feature. The subsiding trend of tectonic origin continued during Holocene times in the same areas, although progradation of beach systems was greatly reduced. In recent (historic) times, progradation has been further diminished due to building of dams along the Segura and Vinalopó rivers which are the main contributors of sediment to the coastal budget. Detailed morphological study of the beach ridges closing lagoons revealed a higher degree of subsidence in the area of Torrevieja since Tyrrhenian times to the present and a positive (although small) rate of progradation off the lagoons of Santa Pola and El Saladar.

1. Introduction and Geological Setting

In recent years, a growing interest in coastal zones and the foreseeable impacts of probable rise in sea level led to detailed morphological sedimentological and neotectonic studies aimed both at determining the changes that would be expected and designing effective strategies to minimize the resulting damage to natural ecosystems and human settlements. This

439

R. Paepe et al. (eds.), Greenhouse Effect, Sea Level and Drought, 439–446.
© 1990 *Kluwer Academic Publishers.*

is particularly true in the tectonically active littoral areas of Alicante Province (southeastern Iberian Peninsula) where large wild-life reservations (Lagunas de Torrevieja and La Mata, etc.) and resorts are found.

The study area is a part of the Elche basin (S.E. Alicante province) and it is located in the eastern Betic Cordillera. This area has been subjected to tectonic compression since the Miocene to date. During the Quaternary the direction of shortening was NNW-SSE (Montenat et al., 1987).

Intense neotectonic activity during the Quaternary, and specially in the Lower-Middle Pleistocene boundary, generated both new structures and the remobilization of older ones, all of which strongly influenced the present-day morphology of the littoral zones (Goy et al., 1987; Goy & Zazo, 1988).

This geodynamic framework favoured the development of lagoons along the coastal areas of southern Alicante province (Fig. 1) which were subsequently isolated from the sea by spit-bars formed under the prominent regional longshore drift with steady movement of sediment towards the south. Pleistocene changes of eustatic sea level also caused major modifications of the coastal morphology and large lateral shifting of the barrier island (spit-bar) and lagoon systems.

The aim of this paper is to show the main tectonic and sedimentary controls on the morphology of the coast and to report on the probable changes accompanying a foreseeable rise in sea level.

2. Geomorphological Features

The broad Elche basin is located adjacent to the Crevillente and Borbuño Ranges to the north (Fig. 1). The southern limit of the basin is marked by the fault-controlled present-day course of the Segura River. The basin fill consists of Neogene and Quaternary deposits. Quaternary stacked, alluvial-fan deposits derived from the mountains of Sierras de Crevillente and Borbuño descend to the south where the Laguna del Hondo and the lagoon of Santa Pola represent their associated playa-lake environments. More to the south, the asymmetric flood plain of the Segura River extends.

The occurrence of E-W oriented synclines favour the development of lowlands in the more depressed areas, which occur as coastal lagoons near the coast. The adjacent anticlines (Santa Pola, La Marina) provide rocky headlands suitable for spit-bars to attach. Sediment supplied by the River Vinalopó and Barranco de las Ovejas moved alongshore on a gently sloping shelf under a prominent drift and fed the spit-bars.

From the mouth of the Sagura River toward the south of the dominant tectonic features are dextral strike-slip faults trending N 120° E with some vertical component. They also induced the development of lowlands, coastal lagoons and spit-bars fed by the Segura River via longshore drift.

The tectonic activity increased during the transition from Lower to Middle Pleistocene as recorded also in many other coastal areas of the Eastern Betic Cordillera such as the Almeriá Province (Goy & Zazo, 1986). Plotting on a map the relative heights (Fig. 1)

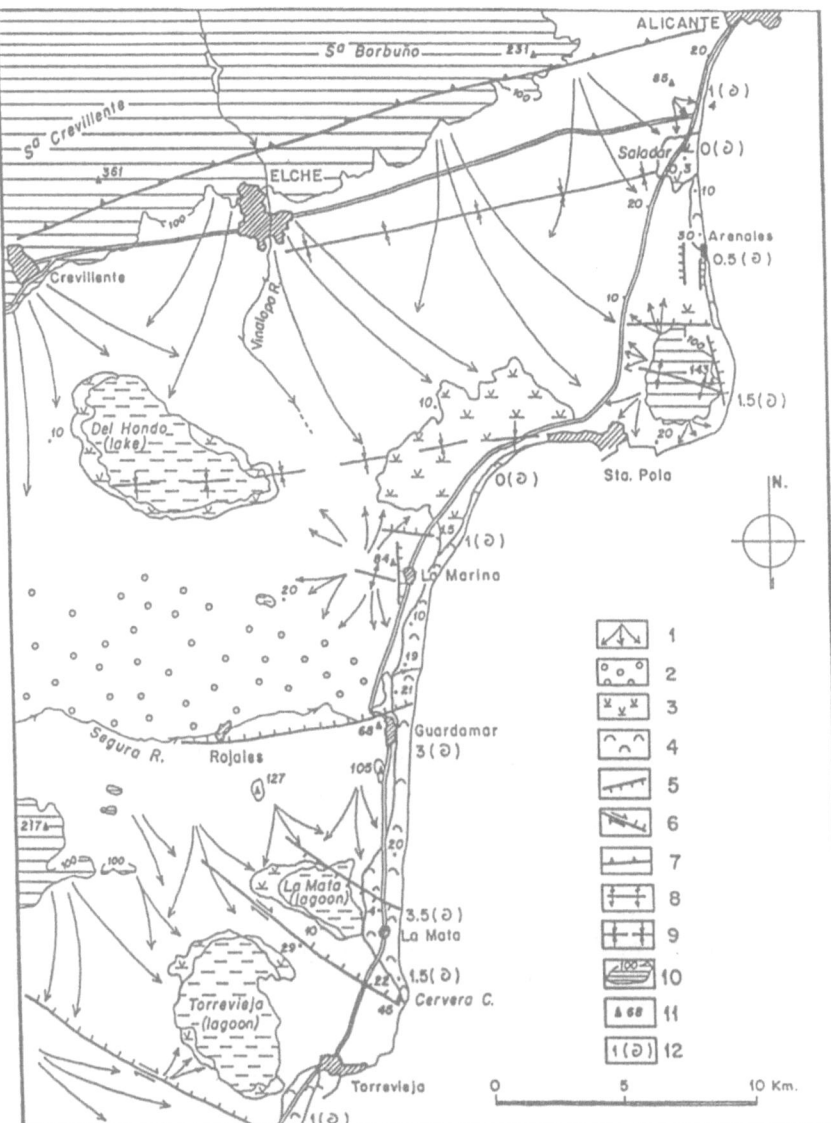

Figure 1. Map of the main geomorphological units.

- Legend: (1) alluvial fans and glacis, (2) alluvial plain, (3) marsh (marismas), (4) recent and ancient eolian dunes, (5) normal fault, (6) dextral strike-slip fault with vertical component, (7) overthrusting fault, (8) anticline, (9) syncline/inferred, (10) areas above +100 m, (11) topographic heights (metres), (12) topographic heights (in metres) of the Tyrrhenian III marine terraces aged ca. 100 Ka.

of the Tyrrhenian III episode dated as ca. 95 Ka ago (Th/U dating, Hillaire Marcel et al., 1986, Goy & Zazo, 1986) it becomes evident that there are areas where the terraces have been raised after deposition (Fig. 2) which coincide with the anticlinal axis (Santa Pola, La Marina) and uplifted fault blocks (Guardamar). The map also shows areas where present-day heights of terraces hardly surpass the mean sea level (Saladar, lagoon of Santa Pola, Torrevieja) corresponding to the syncline axis and downthrown fault blocks. The anomalous figures found in La Mata (Fig. 2) are explained as uplifting by minor faults.

Faults aligned N-S further control the present-day coastline and also the paleo-cliffs and marine terraces that record former positions of sea level.

3. Recent Seismotectonic Activity in Relation to Neotectonic Structures

As mentioned before the active tectonics in the area during the Quaternary is responsible for the coastal morphology and the topographical variability of the Tyrrhenian marine terraces (beach deposits bearing *Strombus bubonicus*), but also for the continued seismicity.

At present the seismicity in the area is among the highest in the Iberian Peninsula as was also the case in historic times. Estevez et al. (1987) mapped the distribution of earthquakes during the period 1396 and 1976 and showed that the area Guardamar-Torrevieja is located in the grade X isomaximum of seismic activity.

4. Dynamics and Evolution of the Barrier Island-Lagoon systems

It is known that La Mata lagoon formed as early as the Middle Pleistocene (Somoza, 1989) due to continued subsidence. It is likely that the dating is also valid for the rest of the coastal lagoons in the study area. In all cases barrier islands consist of several overlapping beach barrier sequences corresponding to repeated highstands with stationary or slightly falling sea level. The preserved depositional sequence of and individual beach was produced by the progradation of the foreshore and neighbouring subenvironments of spit-bars (Bardaji et al., 1987) growing towards the south. In Santa Pola, the lagoon was closed by a complex spit consisting of at least three overlapping Tyrrhenian spits, well exposed in La Marina quarry (Goy & Zazo, 1988).

Usually each coastal unit consists of beach and associated eolian dune deposits. The lateral and vertical associations of facies can be easily traced in each distinct unit. This scheme is widely recognized in all barrier island and lagoon systems along the coast of Alicante (Elche basin), but local tectonics may alter the number of units present.

In the last few years a drastic shortage of sediment input from rivers due to dams in Vinalopó and Segura Rivers coupled with systematic extraction of sand from the coastal eolian dunes (meant to transform beaches from gravelly into sandy) have dramatically altered the coastal budget of sand. Consequently, steady erosion is widely recognized, particularly from La Marina towards the south (Ayala et al., 1987): this segment of shoreline, with coastal retreat averaging 50-100 cm/yr, is classified as highly erosional.

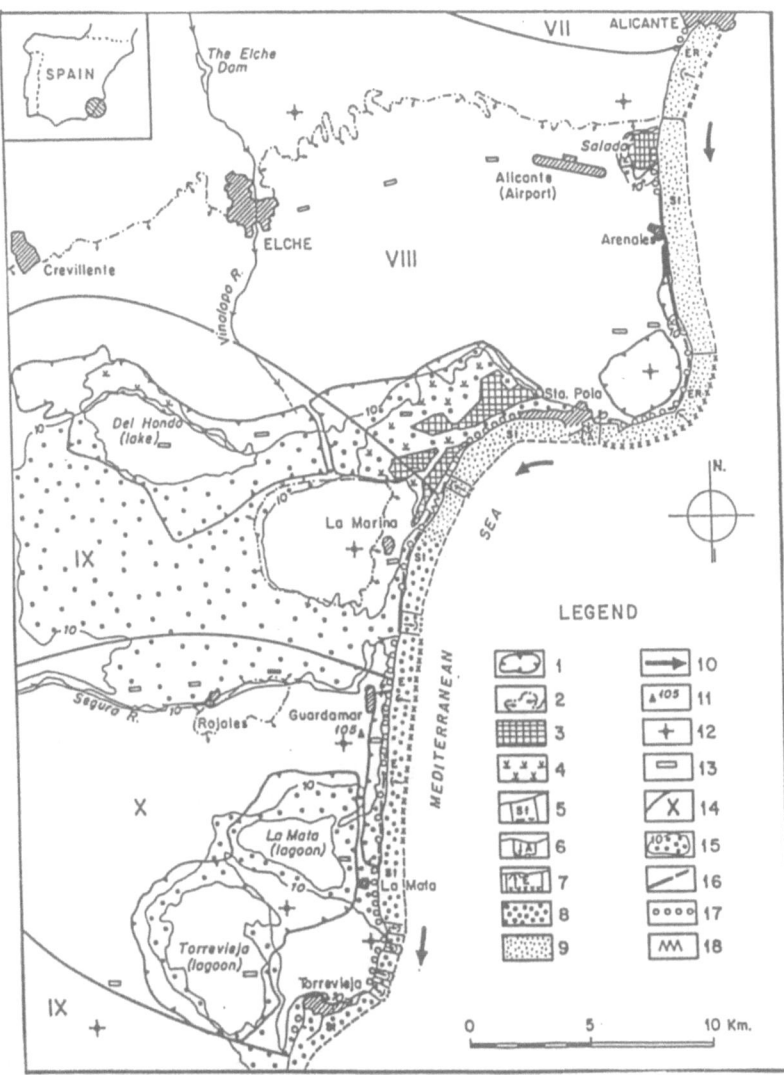

Figure 2. Map of main coastal uses and hazards (Alicante - Torrevieja, protected areas).

- Legend: (1) protected natural areas, (2) limits of the more important aquifers, (3) salt pan (salinas), (4) marsh (marismas), (5) stable coastline, (6) coastal accretion, (7) coastal erosion, (8) high coastal erosion, (9) low coastal erosion, (10) longshore drift, (11) height in metres, (12) areas with a tendency to rise, (13) areas with a tendency to subside, (14) isomaximum of seismic intensity (between 1396 and 1976), (15) areas below +1 m. Preferential coastal uses:, (16) preservation of nature and/or landscape, (17) fishing, (18) industry.

5. Areas to be Protected

In the coastal plains of Elche Basin these areas (Garcia Rodriguez et al., 1985) almost coincide with the spit-bar-lagoon systems (Fig. 2). Preservation of the associated wet, moist areas is vital for specific flora and fauna due to the arid to very arid nature of the area (average rainfall: 300 mm and aridity index almost 6).

Laguna del Hondo. A fresh-water lake, surrounded by palm trees, that form a part of the so-called wet areas of Alicante. It extends over a surface of 1,600 hectares with a mean height of about 2 to 5 m above sea level. The very high biological interest of the lake is based on the fact that it is the home of thousands of birds threatened by extinction as a result of the progressive reduction of these wet areas.

Eolian dunes of Guardamar-Elche. There are both mobile and fixed dunes that top the spits, covering an estimated surface of about 774 hectares. This area is covered with pine trees and bushes which must be preserved against the intense urban pressure.

Natural areas. The brackish Lagunas de la Mata (9 km^2) and Torrevieja (14 km^2), where large populations of *anatidae* and wading birds (including flamingo) rest and shelter, are of great importance for preservation. The salt pans (salinas) of Santa Pola, largely used by migratory birds, are included in this category as well.

6. Impact of a Rise in Sea Level

Most of the coastal areas of the study area lay below the + 10 m contour line (Fig. 2) and the lagoon zones in particular hardly rise above +2 m. In addition, recent neotectonics has tended to sink most of the coastal zones with only a few exceptions.

This means that even a small rise in sea level would produce:

(a) The drowning of the lagoons and spit-bars and the loss of the related natural areas which are of great interest for their ecological balance of flora and fauna.

(b) Increase of the already active coastal erosion triggered by the reduction of sediment and the peculiar petrological composition of the rocks exposed along the coastal zones (mostly non-compacted calcarenites and marls).

(c) Salinization of the aquifer of Vega del Segura-Campo de Cartagena, the largest in the Alicante Province which supports one of Spain's richest horticultural areas.

(d) Loss of one of the most intensively used recreational and touristic open spaces in Spain well known for mild temperatures (yearly average: 17-18°C).

7. Discussion

The study of sedimentary sequences and morphological disposition of the various barrier islands enclosing the lagoons of Santa Pola, La Mata and Torrevieja, together with Th/U measurements carried out on marine episodes (Goy et al., in litt.) reveals maximum subsidence in the area of Torrevieja due to movement of the dextral faults of San Miguel de Salinas

(Fig. 1). Consequently, only the eolian dunes of the Tyrrhenian (aged between 180 and 95 Ka, Hillaire Marcel et al., 1986) barrier islands are exposed above sea level. The rest of the Tyrrhenian beach deposits are drowned nowadays and they are found below sea level off the present-day lagoon of Torrevieja. Towards the south, close to the San Miguel de Salinas fault, the youngest Tyrrhenian marine episode (95 Ka) crops out 1 m above sea level (+1 m) associated with the coeval dune field (Fig. 1).

The subsiding tendency of the areas with Middle Pleistocene complexes of barrier island and lagoon persisted through the Tyrrhenian and Holocene cycles. This, coupled with a reduction of sediment input due to building dams in important estuaries (Segura and Vinalopó) implies that any rise in sea level (even of a small scale) would increase the already erosional trend of the coast. It is to be noted that coastal progradation (with very low values) only occurs off the lagoon of Santa Pola and El Saladar.

8. Acknowledgment

This is a contribution of IGCP 274: "Quaternary coastal evolution"

9. References

Ayala, F., Elizaga, E., Gonzalez de Vallejo, L., Duran, J.L., Beltran, F., Oliveros, M.A., Carbo, A., Guillamont, M.L., & Capote, R. (1988) "Impacto económico y social de los riesgos geológicos en España.", I.G.M.E., Serie Geol. Ambiental, 91 p.

Bardaji, T., Dabrio, C.J., Goy, J.L., Somoza, L., & Zazo, C. (1987) "Sedimentologic features related to Pleistocene sea level changes in the S.E. of Spain. In Zazo, C. (ed.) *Late Quaternary Sea-Level Changes in Spain*, Trab. sobre Neogeno-Cuaternario 10, pp. 79-93

Estevez, A., Pina, J.A., Auernheimer, C., & Montblanch, R. (1987) "El medio fisico de la franja litoral al sur de Alicante (Comunidad valenciana). Bases para la ordenación territorial", Comunicaciones III Reun. Nac. de Geologiá Ambiental y Ordenación del Territorio, Valencia, Vol. II, pp. 1393-1402

Garcia Rodriguez, J.J., Fresno, F., Moral, J., Mena, J.M., & Rey de la Rosa, J. (1985) *Mapa geocientifíco del medio natural. Provincia de Alicante. E: 1/100.000*, I.G.M.E., 100 p.

Goy, J.L., Zazo, C., Bardaji, T., & Somoza, L. (1987) "Tyrrhenian and Holocene levels disposition in the southeastern Spanish littorial, related to the Quaternary compression", Bull. INQUA Neotectonics Comm. 10, pp. 12-19

Goy, J.L., & Zazo, C. (1986) "Synthesis of the Quaternary in the Almeria littoral, neotectonic activity and its morphologic features, eastern Betics, Spain", Tectonophysics 130, pp. 259-270

Goy, J.L., & Zazo, C. (1988) "Sequences of Quaternary marine levels in Elche Basin (eastern Betic Cordellera, Spain).", Palaeogeography, Palaeoclimatology, Palaeoecology

68, pp. 301-310

Goy, J.L., Zazo, C., Bardaji, T., Somoza, L. Causse, C., & Hillaire Marcel, C. (in litt.) "Eléments d'une chronostratigraphie du Tyrrhénien des régions d'Alicante-Murcie, Sud-Est de l'Espagne", Geodin. Acta

Hillaire Marcel, C., Carro, O., Causse, C., Goy, J.L., & Zazo, C. (1986) "Th/U dating of Strombus bubonicus-bearing marine terraces in southeastern Spain", Geology 14, pp. 613-618

Montenat, C., Ott d'Estevou, P., & Masse, P. (1987) "Tectonic-sedimentary characters of the Betic Neogene basins evolving in a crustal transcurrent shear zone (S.E. Spain)", Bull. Centre Rech. Explor.-Prod. Elf Aquitaine 11, 1, pp. 1-22

M.O.P.U. (editor) (1976) *Plano indicativo de Usos del Dominio Litoral.*, Provincias de Valencia, Castellón y Alicante.

Somoza, L. (1989) *Estudio del Cuaternario litoral entre el Cabo de Palos y Guardamar (Murcia-Alicante). Las variaciones del nivel del mar en relación con el contexto geodinámico*, Ph. D., Universidad Complutense, Madrid, 352 p., 6 maps

AUSTRALIA'S HYDROLOGICAL STORAGE AND RISING SEA LEVEL HAZARD

C.D. OLLIER
University of New England
Armidale 2351
Australia

Abstract. Australia presents problems to climatic modellers: early climates were not related to latitude, and Quaternary climates cannot be explained by latitudinal shifts. Glacial periods were arid in the sense of mobile dunes, but pluvial in the sense of high lake levels. There is no convincing evidence of rising sea levels, and observed beach retreat can be explained by management. Groundwater is much used, but recharge is not generally useful because of great residence time. Schemes to divert water to dry areas are uneconomic, and the one major scheme of river reversal, the Snowy Mountain Scheme, has led to disastrous salinity problems in irrigation areas. There is no merit in filling the Lake Eyre depression. If it is ever considered desirable to increase water storage the best inducement in Australia would be to increase urban water supply.

1. The Australian scene

There is a widespread belief that sea level is rising, perhaps in response to climatic warming which in turn may be due to the greenhouse effect. If sea level does rise, many coastal areas will be at risk. They may be evacuated or defended by coastal engineering structures. A third possibility is that more water could be stored on land. This paper addresses these problems as they affect Australia.

Australia is the flattest of continents, and apart from Antarctica, the driest. Most of the centre of the continent is arid: the north is monsoonal and even in the temperate or Mediterranean climatic zones variability is great from year to year. It is a land of droughts and floods, which makes water management a major problem even in normal times. Rising sea levels only make water management more difficult.

1.1. PAST CLIMATES

Studies of seafloor spreading in the Southern Ocean, hotspot traces, and palaeomagnetism of Australian rocks all show that Australia has drifted from a position close to Antarctica

447

R. Paepe et al. (eds.), Greenhouse Effect, Sea Level and Drought, 447–455.
© 1990 *Kluwer Academic Publishers.*

to its present position in the past 55 million years. A possible explanation of past climates in Australia would be that as the continent drifted north it would pass through a series of climatic zones. This does not appear to have happened. Northern Australia did not drift through an arid belt, and has never been an arid region. The real story indicated mainly by palynology, is that as Australia drifted to its present position the climate remained warm and moist, and the modern type of global, zonal climate set in later. Dryness started to appear in the Upper Tertiary, but real aridity with sand dunes and salt lakes did not appear until the Quaternary. In climatic modelling it cannot be assumed that climate was very different in the past. In Australia the Permian glaciation was followed by the Mesozoic warmth without a significant change of latitude (Ollier, 1986). Another climatic lesson from Australia relates to the last glacial. Many models of climatic change have simple shifts of climatic belts, but in the last glaciation in Australia the continent was arid in the sense that vegetation was reduced and sand dunes were active, but pluvial in the sense that lake levels were high and rivers were large (Bowler, 1986). The pressure gradient between Australia and the Southern Ocean was steepened, and winds were stronger than any blowing today at any latitude.

In the past few decades there is no firm evidence of a general increase in temperature in Australia.

1.2. SEA LEVEL CHANGES

Sea level changes are recorded throughout earth history, but their interpretation has been very much controlled by ruling paradigms of geology. Plutonist and Neptunists, fixists and continental drifters, and in recent years plate tectonic interpreters have all had their day. Chappell (1987 a) has produced a recent review of ocean volume change and the history of sea water, but the topic still has many controversial aspects. For instance, most workers assume that the volume of sea water is almost constant, but Carey (1988) claims that new sea water is produced as ocean surfaces expand by seafloor spreading.

World schemes of changing sea-level have been produced, especially those of Vail and his colleagues (e.g. Haq et al., 1986). They are inevitably controversial in detail, but major features seem to be well-established. One recurrent theme is that sea level seems to fall rapidly on many occasions, and slowly rise again to high levels. For glacial periods this could be explained by rapid onset of glaciation followed by slow melting, but the same pattern is found in times like the Cretaceous when the earth was free of ice. Carey (1988) explains: "The cycles reflect the balance of total seawater volume against the total capacity of the ocean basins, as each increased with time." The present rapid rise in sea level, through trivial on a global scale, is contrary to sea level behaviour through most of geological time.

Changes in sea level are recorded in Australia in varying degrees of detail. The evolution of the Australian coast is reviewed by Jenkin (1984). The Cretaceous transgression flooded the already flat continent, dividing it into three large islands. The Eocene transgression is preserved by marine incursions up valleys in Western Australia, now about 300 m above present sea level. In South Australia the Eocene transgression is marked by coastal beach ridges and associated aeolian dunes at about 150 m (Benbow, 1988).

On a more modern scale, sea level was low during the last ice age, and rose later only to fall in the last few thousand years. Late Quaternary sea level changes have been reviewed by Chappell (1987b). Basically, he seems to find a fall of about 2 metres in the past 6000 years, but there is evidence of genuine local variation. Other detailed analyses are by Hopley (1987), Aubrey and Emery (1986) and Bryant et al. (in press). Bird (1988) notes that Australia has very few reliable long-term gauge records, but analysis of data from Sydney indicated that mean sea level rose about 7 cm between 1841 and 1972 (Hamon, 1987). Bryant (1988) points out that climatic variables have such complex links with sea level that sea level predictions become uncertain.

1.3. COASTAL PROBLEMS

The Australian coastline, about 33000 km long, consists of about 56 % beaches backed by dunes (Bird, 1988). Most beaches have sand from more than one source. Many Australian beaches reveal sequences of cut and fill, but in recent decades the erosional system has predominated on Australian beaches (Bird, 1988: see also Thom, 1974; Chapman et al. 1982; Bird, 1985). Ninety Mile Beach in Victoria is typical, but it is backed by a series of successively formed parallel dune ridges indicating a previous history of coastline progradation, so the onset of erosion is reversal of earlier conditions. Sectors of coast with sustained beach accretion during the past century are few and far between: Bird (1988) gives examples, but notes that recession of beach fringed coastlines has been much more widespread. He lists 14 possible causes of beach erosion, one of which is rising sea level. Studies of coastline movements by Gordon (1988) of 32 beaches along the New South Wales coast show that "most beaches show a shoreline recession trend over the past 50 years equal to or greater than 0.2 m per year, $\pm$ 0.1.m/year".

A strong possibility on many Australian beaches is that beach loss follows badly managed coastal development, and the widespread construction of jetties, groynes and other engineering structures has provided a plethora of examples of beach destruction.

Another problem is that the supply of sand to the coastal zone has been first increased (by extensive land clearance) and then reduced (by construction of dams). Any proposals to build further dams will inevitably also reduce the sand supply to beaches. Artificial beach nourishment from offshore is being tried with varying degrees of success.

There are some parts of Australia where coastal protection is unlikely to be successful if sea level rises. The most vulnerable is probably the low coast of the Gulf Country around the Gulf of Carpentaria, where the low gradient, monsoonal climate, and high tidal range already combine to flood large areas every wet season.

The cities will have their own problems, and in Australia all capital cities except Canberra are on the coast. Structural engineering solutions will no doubt be economically possible because of the density of the population, but even so the large populations living close to sea level will probably suffer.

The management of sea level changes have scarcely been considered yet in Australia, apart from several papers in Pearman (1988). The Australian Water Research Advisory Council was set up to advise the Federal Government on all aspects of water research, and in its 1988 Report no mention is made of possible sea level rises. Short (1989) is optimistic enough to think that, if suitable coastal research is carried out "The management of Australia's coast can therefore proceed with a reasonable degree of confidence that Greenhouse is being accommodated, and more importantly our coast is being safeguarded if this two stage response is incorporated in all planning."

1.4. AUSTRALIA'S WATER RESOURCES

Mean annual rainfall and runoff across Australia are meagre in comparison with most other land masses (Pigram, 1988). These mean figures are themselves misleading, because of great variability in space and time - Australia is a land of flood and drought. Development of water resources has been an article of faith for generations, and until recently the construction of water storages was easy to justify to a water sensitive population.
Per head of population Australia has a relatively large amount of water storage, as water supply for towns, for irrigation, and for hydroelectricity.
A summary by state is as follows (in Gigalitres):

New South Wales	25.000
Victoria	11.000
Queensland	8.000
Western Australia	7.000
Tasmania	23.000
Total	64.000

This is equivalent to 64 km^3, which is about one fiftieth of all the water stored on land in the world. More dams are still under construction and projected, but it is unlikely that this figure will ever be doubled.

Some storages such as the Snowy Mountain Hydroelectric Scheme are multi-purpose (including irrigation, recreation and tourism), even though the purposes do not appear to have been thought out at the start of the scheme.

1.5. GROUNDWATER

Groundwater occurs in most parts of Australia and makes up about 14% of total annual water use, 2.5 million megalitres (2.5 km^3). About 60 % of Australia is totally dependent on groundwater, and even Perth may draw 50% of its supplies from groundwater in a dry year.

By far the greatest groundwater storage is the Great Artesian Basin (Habermehl, 1980). This vast basin of 1.7 million km^2 or 20 % of Australia yields about 200 million m^3 per year. (This is rather less than the annual groundwater yield from the Burdekin delta in Queensland, and about 10 % of Warragamaba Dam which supplies Sydney. If all the water were suitable it would supply Sydney for only about 100 days (Warner, 1986). Nevertheless it is important to Queensland to support stock.

But despite exploitation and waste, the Great Artesian Basin is still almost full. Some conservationists are concerned that industrial developments such as that at Olympic Dam (Roxby Downs) in South Australia may make serious inroads into the water storage of the Great Artesian Basin. However the total use, for industrial processing and the town supply is only 7 Megalitres per day or 81 litres per second.

This may be compared with flowing artesian waterbores. There are about 10 bores with a flow of 100 litres per second, comparable to the total use of Olympic Dam. These free flowing bores have 95% wastage, which does not seem to concern the conservationists. The total estimated flow from all bores in the Great Artesian Basin is 1200 Megalitres per day.

In parts of Western Australia groundwater recharge is happening following the removal of forest and planting of wheat. The trees transpired much more water than wheat does, and the change has resulted in a rise of the water table. Unfortunately the rise brings salt with it, and salinity is creeping steadily up the sides of valleys. There is no scope here for any more recharge. Re-afforestation may be effective, but of course land operators want profits in their life time.

At present re-charge of aquifers on a large scale is restricted to mining towns such as those of the Pilbara region.

1.6. IRRIGATION

Of Australia's water resources, 74 % goes to irrigation, 18 % to urban/industrial uses, and 8 % to other rural uses (Crabb, 1983). Most of the irrigation is very uneconomic (Davidson, 1969).

It is also clear that some irrigation projects are actively destroying the environment by salinity. The Murray Basin is the most extreme example. Over-irrigation has raised the water table until it intersects the ground surface, with resultant salinity. An area of 1 Million hectares is salinised at the surface, and over 140,000 hectares the water table is less than 2 m deep. Over two-thirds of the basin the water table is rising at 20 cm/year, as it has been for about 100 years. About 100,000 hectares are lost each year, and many millions of dollars are being spent on studying, ameliorating and (hopefully) curing the problem. But since the basic cause is irrigation agriculture, a sacred cow, there is little chance of a solution without major changes in political attitudes, and meanwhile public perception of land destruction means an unfavourable attitude to more dams or more irrigation. The Murray Basin region already loses about $200 million each year in agricultural revenue through salinity problems.

Some schemes such as the Ord River Scheme based on the artificial storage of Lake Argyle with a capacity of 5.8 million Megalitres, and designed to open up development in the north, have been very slow to prove worthwhile, and the new Burdekin Scheme in Queensland is also facing problems. With growing public awareness of the lack of economic sense in some irrigation schemes, and a strongly growing urge to conserve environments where they are not already destroyed, the Australian public is generally against more dam building and water storage.

1.7. THE LAKE EYRE LOWLANDS

Newman and Fairbridge (1986) have suggested that filling of the earth's large natural depressions may be one way of alleviating rising sea-level. Australia has just one major depression, the Lake Eyre basin. This has an area of about 100 000 km^2 and is about 11 m below sea level. Various schemes have been developed in the past to flood this area. It has a volume which is insignificant compared with other major depressions in the world, and with its shallowness would simply be an evaporating pan. Lake Eyre filled in the 1950's, but evaporated within about 2 years, and had no effect on rainfall in the surrounding areas. Even Moondarra Dam on the Leichhardt River serving Mount Isa, in a monsoonal climate, has evaporation losses about twice the assured annual supply for consumptive use.

Some old ideas concerned the harnessing of eastern rivers and diverting them to the inland. This was the Bradshaw plan, enunciated in the 1930's by Dr J.J. Bradshaw. The dams that Bradshaw envisaged on the Burdekin and Tully rivers have been built, or are being built. But the rivers are still flowing east, where all the water is needed for the most rapidly growing population in Australia. Indeed the latest plan for the Tully scheme will take water that flowed west and transfer it to the east. There are now no arguments for diverting water from the east to the west, and the Snowy Mountain Scheme will probably be the last attempt at this kind of manipulation.

1.8. POSSIBLE SOURCES OF WATER

One of the possible aims of any world-wide water management scheme would be to take sea water from ocean storages on to land. But in Australia salinisation is the enemy, and the need is for freshwater. The answer seems to lie in adding desalinated water to the continent. Desalination off seawater would provide and expensive additional supply of freshwater to coastal cities, but (given almost instant return) would have no effect on sea level. Addition of freshwater to irrigation areas is likely to increase salinity problems in most areas. Addition to groundwater storages is completely uneconomic except for mining towns, and it would generally be better to provide water direct to the town than to groundwater. For instance, there is scope for recharge in the Great Artesian Basin, but only of fresh water. If freshwater is to be manufactured by de-salting, it would be better to provide it direct to Roxby Downs, or to some other city water supply, because the residence time for water in the aquifer, between recharge in the Eastern Highlands and discharge at bores or mound springs is about 2 million years (Habermehl, 1980).

Alternatively freshwater could be derived from elsewhere. One possibility is to use icebergs. The details have been summarised by Quilty (1988). A medium-large berg of 30 megatonnes has a volume of 0.04 km^3. Such a berg would keep Perth going for 7-8 weeks in summer. It would be worth $14 million in Perth, $18 million in Adelaide. There are technical problems in delivering the ice, mainly to prevent it dissolving (not melting) before it reaches Australia.

Perhaps the bergs could be wrapped in plastic. It is also possible that the temperature differential of berg and sea water could be used to propel the bergs.

If the use of icebergs becomes feasible another problem will arise: over-use of icebergs could affect ocean currents and so cause further far-reaching changes.

1.9. POPULATION GROWTH IN AUSTRALIA

One possible strategy to counter rising sea-level is to have more water stored on land, but there is general "anti-dam" attitude amongst many Australians. It would be very difficult to persuade them to build dams simply because of a suspected, but so far undemonstrated, rise in sea level. Any dam building in the near future must be for more immediate aims. Increased irrigation seems to be an unlikely aim, as does further development of hydro-electricity. The one good reason to store more water immediately is for urban water supply.

Australia is generally considered an underpopulated country, and many visionaries have dreamed of large, even huge populations occupying the continent. In 1988 Dr Alan Reynolds, an economic adviser to President Reagan, told us to embark on a crash program to build the population to around 60 million. Much higher estimates have been made for the numbers that could live in Australia (regardless of the standard of living) but most fail to understand the limitations of the continent, of which water is paramount.

"A consultancy study prepared for the Fitzgerald Inquiry reports recent research suggesting a capacity of between 25 and 50 million people. It is argued though that "the upper limit could be met only through massive investment in water-resource development in northern Australia and Tasmania and redistribution to centres of population in mainland southern or eastern Australia, or by a major population redistribution." (McCracken, 1989). In brief, the greatest constraint on population growth, and of maintenance of present living standards in Australia is water supply. The major cities are already feeling effects of shortages. Informed opinion in Australia is currently against more water storages for environmental reasons, but when the economic disadvantages of "no dams" becomes plain, opinion will probably change. For political reasons the best argument for more storages will be for urban supply, and we can expect an increase in storage capacity. Unfortunately the physical configuration of Australia combined with its climate makes it a poor option for onland water storage.

2. References

Australian Water Research Advisory Council (1988) *The First Two Years*, Australian Government Publishing Service, Canberra

Aubrey, D.G. and Emery, K.O. (1986) "Australia - an unstable platform for tide-gauge measurements of changing sea levels.", J. Geology, 94, pp. 699-712

Benbow, M.C. (1988) "The Ooldea Range, and Eocene coastal dune on the northeast margin of the Nullarbor Plain, Australia.", International Geographical Union, 26th Congress,

454

Sydney. Abstracts, Vol. 1, 40

Bird, E.C.F. (1985) *Coastline Changes*, Wiley Interscience, Chichester

Bird, E.C.F. (1988) "The future of the beaches", in Heathcote, 1988, pp. 163-177

Bowler, J.M. (1986) "Quaternary Landform Evolution", in Jeans, D.N. (ed.) *Australia - A geography, vol. 1: The Natural Environment*, Sydney University Press, pp. 117-147

Bryant, E. (1988) "Sea level variability and its impact within the greenhouse scenario", in Pearman, 1988, pp. 135-146

Bryant, E.A., Roy, P.S. & Thom, B.G. "Australia - an unstable platform for tide-gauge measurements of changing sea levels.", discussion, in press, J. Geology

Carey, S.W. (1988) *Theories of the Earth and Universe*, Stanford University Press, Stanford

Chapman, D.M., Geary, M., Roy, P.S. & Thom, B.G. (1982) *Coastal evolution and coastal erosion in New South Wales*, Coastal Council of New South Wales, Sydney

Chappell, J. (1987 a) "Ocean volume change and the history of sea water", in Devoy, 1987, pp. 33-56

Chappell, J. (1987 b) "Late Quaternary sea level changes in the Australian region", in Tooley and Shennan, 1987, pp. 296-331

Crabb, P. (1983) "The water resources of New South Wales: an overview", in Crabb, P., Rich, D.C. and Riley, S.J. (eds.) *Water Resources Conference*, Geographical Society N.S.W., Conf. Papers No. 4, pp. 6-20

Davidson, B.R. (1969) *Australia Wet or Dry*, Melbourne University Press, Melbourne

Devoy, R.J.N. (ed.) (1987) *Sea Surface Studies*. Cross Helm, London

Gordon, A.D. (1988) "A tentative but tantalizing link between sea level rise and coastal recession in New South Wales", in Pearman, 1988, pp. 121-134

Habermehl, M.A. (1980) "The Great Artesian Basin", BMR J. Geol. Geophys., 5, pp. 9-38

Hamon, B.V. (1987) *A century of tide records: Sydney (1886-1986)*, Technical Report No. 7, Flinders Institute for Atmosphere and Marine Science

Haq, B.U., Hardenbol, J. & Vail, P.R. (1987) "Chronology of fluctuating sea levels since the Triassic", Science, 235, pp. 1156-66

Heathcote, R.L. (1988) *The Australian Experience*, Longman Cheshire, Melbourne

Hopley, D. (1987) "Holocene sea level changes in Australasia and the southern Pacific", in Devoy, 1987, pp. 375-408

Jenkin, J.J. (1984) "Evolution of the Australian coast and continental margin", in Thom, B.G. (ed.) *Coastal Geomorphology in Australia*, Academic Press, London

McCracken, K. (1989) "50...150...200 Million Australians?", Geography Bulletin, 21, pp. 3-35

Newman, W.S. & Fairbridge, R.W. (1986) "The management of sea level rise", Nature, 320, pp. 319-321

Ollier, C.D. (1986) "Early landform evolution", in Jeans, D.N. (ed.) *Australia - A Geography, vol. 1: The Natural Environment*, Sydney University Press, pp. 97-116

Pearman, G.I. (ed.) (1988) *Greenhouse - Planning for climate change*, Brill, Leiden

Pigram, J.J. (1988) "The taming of the waters", in Heathcote, 1988, pp. 151-162

Quilty, P. (1988) "Formulating the future", Australian Natural History, 22, pp. 116-118

Short, A.D. (1989) "Coastal implications of a greenhouse sea level rise, Abstract.", Institute of Australian Geographers Conference, Adelaide

Thom, B.G. (1974) "Coastal erosion in eastern Australia", Search 5, pp. 198-209

Tooley, M.J. and Shennan, I. (ed.) (1987) *Sea level changes*, Basil Blackwell, Oxford

Warner, R.F. (1986) "Hydrology", in Jeans, D.N. (ed.) *Australia - A Geography, vol. 1: The Natural Environment*, Sydney University Press, pp. 49-79

PART IV: MANAGEMENT, TECHNIQUES AND CASE STUDIES

SECTION B: Greenhouse

POSSIBLE EFFECTS OF MAN ON THE CARBON CYCLE IN THE PAST AND IN THE FUTURE

H. FAURE, L. FAURE-DENARD
LGQ-CNRS, University Luminy, Case 907
F-13288 Marseille, Cedex 9, France

R.W. FAIRBRIDGE
Columbia University and NASA - GISS
2880 Broadway, New York 10025, USA

ABSTRACT: On a global scale, vegetation belts are the natural expression of the response of major ecosystems to climatic parameters. Environmental conditions are mainly controlled by air temperature and water availability. For Example, in herbaceous savannas and steppes a direct linear to exponential relation exists between total vegetation weight (biomass in dry matter or in carbon) and rainfall (Le Houerou, 1990). In high latitudes or high altitudes temperature is the limiting factor.

Mankind's increasing population has grossly modified the global environment by the progressive replacement of forests by grasslands and cultivated areas. Today 33 to 47% of the Holocene phytomass (275 to 490 Gt of carbon)(Gt = Gigaton = 10^{15} g) is burnt or decayed (Faure et al., 1989) and passed as carbon dioxide into (or through) the atmosphere. An unknown mass of soil carbon is oxidized each year in what the American poet Robert Frost described as "the slow smokeless burning of decay". The representative curve of this global change (forest loss) is parallel to the curve of the population growth. (SCOPE 13, 1979). Since the acceleration of the industrial revolution (about 1850) the action of man on the atmosphere has also dramatically increased by the burning of about 175 Gt of fossil fuels (up to 1987). By these cumulative and accelerating actions, mankind has very greatly increased the supply of carbon dioxide to the atmosphere (SCOPE 16, 1981).

As a counter-effect, so far very minor, recent efforts have been made to stimulate the withdrawal of CO_2 from the atmosphere (Grantham, 1990), such as by reforestation as opposed to deforestation, by soil building or by acceleration of carbon storage by sedimentation. In the quest of food and energy, mankind has behaved as an oxidising agent, first on the biosphere, then on the pedosphere and finally on the lithosphere.

During the same interval of time, somewhat over 100 years, the action of Mankind on the hydrologic cycle has been comparatively limited in terms of global consequences. Major

R. Paepe et al. (eds.), Greenhouse Effect, Sea Level and Drought, 459–462.
© 1990 *Kluwer Academic Publishers.*

460

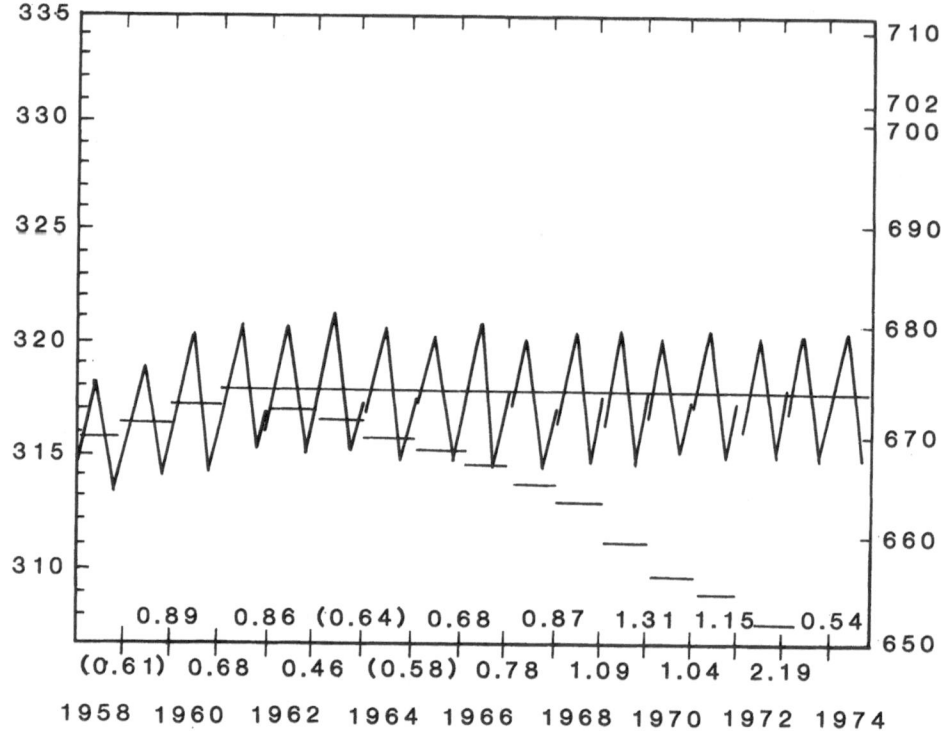

Figure 1. A natural and theoretical model for carbon dioxide control of the atmosphere by acting on natural sinks and sources with the measures proposed in the text.

On the left, values for the global atmosphere are in part per million CO_2 by volume (ppmv), and on the right in Gigatons (Gt) of carbon. On the bottom, the time scale is in years A.D. (with the recorded annual increase in ppmv above the line of years).

The graph shows the two main variations of atmospheric carbon dioxide: the seasonal variation and the effective annual increase in ppmv above. The first may reach 15 Gt amplitude as measured by Keeling et al. (1985), and the second is about 2 to 3 Gt y^{-1}.

The natural variations are shown for 4 years on the left (from measures by Keeling, for 1958-1962). From 1962 onward the theoretical model shows the effect of acting on the natural seasonal sources by stopping the return to the atmosphere of a part of the seasonal flux (as proposed in the text, point 1 to 6).

The natural annual flux of carbon exchanged between the phytosphere and the atmosphere is about 110 Gt (photosynthesis, respiration). That exchanged by the ocean/atmosphere system is about 100 Gt (Bolin et al., 1979, 1981). A one percent action on the global natural fluxes may be more feasible than to oblige mankind to reduce its fossil fuel consumption to 50%.

dams, canals and irrigation schemes, even when they are locally spectacular, only achieve a minor change in the planetary asymmetric distribution of water. Equatorial zones remain over-watered while giant tropical and subtropical deserts are little modified by a management of latitudinal shifts in the global water budget.

For the future, it might be feasible (theoretically, if not practically) for Mankind to reverse some of these former inadvertent global "experiments" by voluntary scientifically guided action. To this end, the following ideas could be investigated as to their potential action on the carbon cycle:

1. Increased geohydrologic management can provide for the irrigation of extensive areas of deserts and semi-arid regions, leading to widespread reforestation. Each million square kilometres reforested can absorb 10 Gt of carbon.

2. Vegetation (e.g., from mangrove or salt marsh species) can be submerged and buried in marine gulfs, sebkhas and depressions, and may contribute to an artificial sink of atmospheric CO_2. For example, four cubic kilometres (4 km^3) of vegetation, thus artificially fossilized, would subtract one Gt of carbon from the global cycle.

3. Water table control of major continental depressions (e.g., Chad, Sudan, Mali) may be used for another carbon sink by development and management of artificial peat-lands and shore reforestation.

4. Nutrient stimulation of calcareous algae and coral growth would contribute to increase the carbonate reservoir of calcareous sediments.

5. This nutrient enhancement could be achieved in combination with an artificially engineered upwelling, thus forming a "biological pump". This upwelling would introduce essential metabolic nutrients (phosphates, nitrates, etc.).

6. A "salt pump" mechanism that depended on sea-ice formation could be accelerated by the breaking-up of sea ice. This would enhance the formation of new sea ice and increase the sink of new cold salt water with its carbon dioxide.

Eventually, Mankind should be in a position to know enough about the carbon biogeochemical cycle to recommend which of the various natural carbon dioxide pumps would be best to stimulate. Sinks and sources may than be controlled in a harmonious way to be able to protect our global environment in the next millennium.

1. References

Bolin, B. (ed.)(1981) "Carbon cycle modelling", SCOPE 16, J. Wiley, p. 390

Bolin, B., Degens, E.T., Kempe, S., & Ketner, P. (eds.)(1979) "The global carbon cycle", SCOPE 13, J. Wiley, p. 491

Faure, H., Fabre, M, Faure-Denard, L., Lezine, A.M., & Petit-Maire, N. (1989) "Une estimation de la biomasse globale à 18.000 ans BP.", Colloque *Biogéographie, Environment, Aménagement*, Paris, juin 1988, AFGP

Grantham, R. (1990) "Approaches to correcting the global greenhouse drift by managing tropical ecosystems.", *Tropical Ecology*, vol. 30 - n° 2, pp. 1-12

Keeling, C.D., Whorf, T.P., Wong, C.S., & Bellagay, R.D. (1985) "The concentration of atmospheric carbon dioxide at ocean weather station P from 1969-1981", J. Geophys. Res. 90, 10, 511-10, p. 528

Le Houerou, H.-N. (1989) *The grazing land ecosystems of the African Sahel.*, Ecological Studies, vol. 75, 282 p., 114 Figs., Springer-Verlag

THE RELATIONSHIP OF INQUA TO THE GLOBAL CHANGE PROGRAM AND OTHER INTERNATIONAL GROUPS

NAT RUTTER
Department of Geology
University of Alberta
Edmonton, Alberta
Canada, T6G 2E3

ABSTRACT: The ICSU-IGBP: A study of global change, is in its incipient stage, but will probably last for over 10 years. The emphasis will be on the total earth system, concentrating on the various interacting factors and processes that can cause change. The development and testing of models will be an important component. INQUA and Quaternarists enter the picture by supplying information of past global changes in order to test cause an effect hypothesis and to help evaluate interactive earth system processes. Numerical data from time slices and time series of the relatively recent past on a worldwide scale will be particularly important for the core program. However, changes that have taken place during the last glacial cycle, not detectable in the short term, must be considered. Much of this information is available now or is being acquired in existing projects, but has to be organized and presented in a form easy to utilize by others. New programs should extend our present knowledge in traditional areas, particularly in geographic areas where little information is available. Key archives include ice cores, tree rings, lacustrine sediments, coral deposits, loess, paleosols, ocean cores, pollen, historical records and glacial deposits. To ensure input of past global changes is included in the core program, the IGBP Special Committee has established a Scientific Steering Committee on Global Change of the Past. The Committee has suggested the following programs consisting of two streams. One focuses on the last 2,000 years, and the other concentrates on coarser temporal resolution of climatic changes in earth history through the last two glacial cycles. However, the Special Committee emphasizes that the core program will not be all-encompassing or an "umbrella" program for all aspects of global change. On the contrary, organizers encourage other organizations and individuals to initiate and carry out their own programs. INQUA, besides participating in the core program, is participating in global change programs of the IUGS. Here, other dimensions of global change will be considered. Important projects recently identified at an IUGS workshop include: 1) past global reservoirs of carbon and their change with time; 2) global sea level change and impacts; 3) abrupt changes, extreme events and short-term fluctuations in continental records; 4) pilot projects on land-sea correlation; and 5) long continental records. In addition, INQUA, through its Intercongress Committee on Global Change will coordinate and identify projects of its own commissions, sub-commissions and working groups that are of particular interest and importance to the global change program. The Intercongress Committee may also initiate its own

R. Paepe et al. (eds.), Greenhouse Effect, Sea Level and Drought, 463–481.
© 1990 *Kluwer Academic Publishers.*

projects taking advantage of our worldwide network of Quaternary scientists. The past global reservoir of carbon has been suggested as a project we should undertake. It is apparent, therefore, that INQUA and Quaternary scientists in general have an important role to play in the global change program if it is to be a success. This could be our "finest hour".

1. Introduction

The Special Committee for the International Geosphere-Biosphere Program: A Study of Global Change (IGBP) was appointed by the International Council of Scientific Unions (ICSU) Executive Board in January 1987 in response to the decision by the ICSU General Assembly (September 1986) to launch an IGBP. The Special Committee will develop a coherent interdisciplinary program to study the linkages between earth system components (biota, land, sea and air). The overall objective of the IGBP is:

> To describe and understand the interactive physical, chemical and biological processes that regulate the total Earth System, the unique environment that it provides for life, the changes that are occurring in this system, and the manner in which they are influenced by human actions (Figure 1).

The initial emphasis of the IGBP will be to increase our understanding of key elemental cycles among the terrestrial, marine and atmospheric systems. The interactive effects between climate and biosphere will also be a major focus of the IGBP. The IGBP will pay particular attention to processes important on time scales of decades to centuries and spatial scales of regional to global dimensions. The Program will concentrate on those interactions which are most sensitive to human perturbation with the ultimate goal of developing a procedure to predict future changes in our global life support system.

1.1. THEMES AND PROJECTS COMPONENTS

Four underlying themes form the basis for the development of the IGBP:
> Documenting and predicting global change
> Observing and understanding of dominant forcing functions
> Understanding transient phenomena
> Assessing global change effects on our resource base

The development of the IGBP will depend on three major program components:
> Modelling of earth system processes on a regional to global scale;
> Observation and monitoring of key phenomena across the globe;
> Experimental research into processes which have relevance to global-scale cycles.

These components are interconnected in that model development will depend on experimental results whereas validation of model predictions will rely on observations on key processes and properties of the global system. Each of these components will contribute to the understanding of the cause-and-effect relationships of global change.

Simulation models are critical to the success of the IGBP. Modelling will be used to

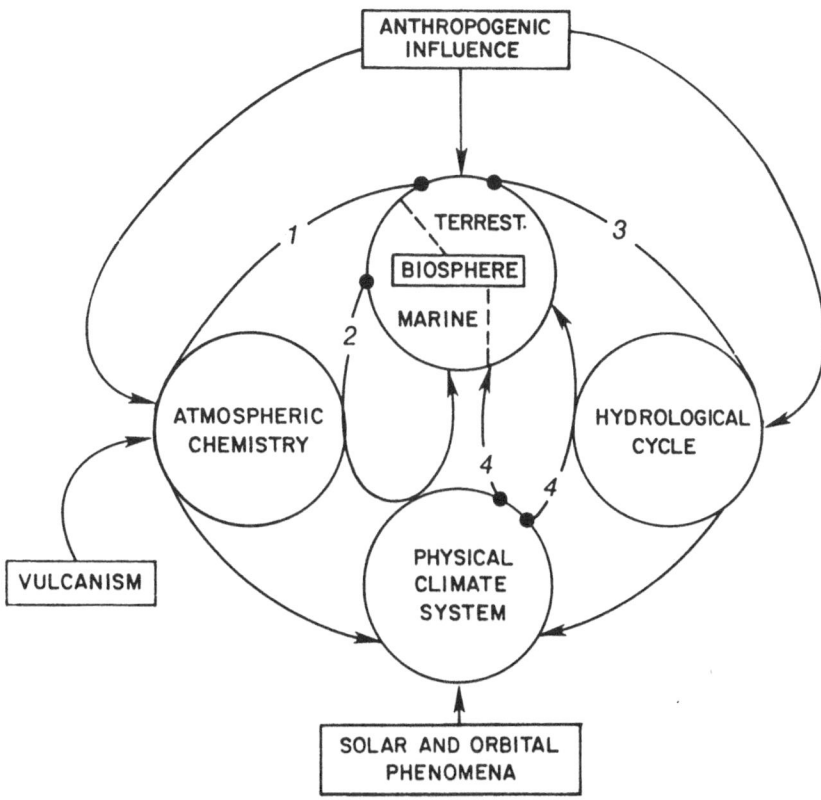

Figure 1. The various forcing factors affecting the earth's systems.

te·t hypotheses and resolve the problems of transcending temporal and spatial scales. One goal will be to develop a mechanistically based simulation model which will represent the interactions between those earth system components which affect climate and thus predict the consequences of global change.

A world-wide network of Geosphere-Biosphere Observatories is essential to the investigation of global change problems. The observatories will serve as benchmark locations for specific ecosystems, ground-truth sites for remotely-sensed observations, validation sites for simulation models, and locations for experimental studies. The long-term records from the observatories may provide evidence of global change effects.

Much of the understanding of linkage between the different components of the Earth system will be elucidated through experimental studies. These studies will be designed to test cause-and-effect relationships between human activities and climate change, to determine the coupling of the biota, land, sea, and air to global change, and resolve problems of spatial and temporal scaling.

1.2. THE PROGRAM'S ORGANIZATION

The initial design of the Program will be governed by five Coordinating Panels and two Working Groups. Research objectives and projects will be identified by the Coordinating Panels. Assessment of current status of knowledge and future prospects for Program activities in areas of interest to IGBP will be conducted by Working Groups
The Coordinating Panels are:

      Terrestrial Biosphere-Atmospheric Chemistry
      Marine Biosphere-Atmospheric Interactions
      Biosphere Aspects of the Hydrological Cycle
      Climate Change Effects on Terrestrial Ecosystems
      Global Analysis, Interpretation of Modelling

The Working Groups are:

      Data and Information Systems
      Geo-Biosphere Observatories

The material presented above was published with some modifications from the first issue of Global Change News prepared by the IGBP Secretariat in 1988. There is no question of the complex task ahead. To what degree the objectives will be met is unknown. The program will require a massive effort and probably will last over ten years.

The next phase of the program, currently being undertaken, will result in recommendation for a number of IGBP core projects. Recognition of problem areas resulted in the rapid initiation of the following international research projects: the role of biota in the cycles of chemicals in the atmosphere which give rise to the greenhouse gas effect; the role of oceanic organisms in the global carbon dioxide cycle; the role of land plants in the exchange of energy and moisture between the land and the atmosphere; and a coordinated effort to recover information from natural archives that will illuminate connections between atmospheric composition, global temperature, ice extent, solar history, and the distribution of land and oceanic organisms.

The core projects will be coordinated by the Special Committee who will sometimes work in connection with other bodies, for example, the International Global Atmospheric Program (IGACP) and the Joint Global Flux Study (JGOFS). There are literally hundreds of international projects that have a bearing on global change which will have to be considered, although many may not require a formal designation by the Special Committee.

It was made clear early on that this "core" program is not an umbrella for all global change projects. On the contrary, other initiatives are welcome.

It is recognized by the Special Committee that an understanding of global changes of the past is necessary if we are to understand the causes, mechanisms and manifestations of future global changes. To this end, the Special Committee has appointed a Scientific Steering Committee on Global Changes of the Past (SSC), (an outgrowth of an earlier Working Group on Techniques of Paleo-Environmental Research) under the chair of Hans Oeschger. Members of the Committee until 1992 are:

      H. Oeschger, Switzerland, Chairman, (ice cores)

J. Eddy, USA, (historical records)

A.L. Berger, Belgium, (paleo-models)

B. Berglund, Sweden, (Quaternary geology)

B. Frenzel, FRG, (palynology)

N. Shackleton, U.K., (ocean sediments)

T. Lui, China, (loess)

J. Jouzel, France (on behalf of C. Lorius),(ice cores)

C. Pfister, Switzerland, (historical records)

N. Rutter, Canada, (Quaternary geology)

J. Imbrie, USA (paleoclimate)

F. Schweingruber, Switzerland, (tree rings)

M. Stuiver, USA, (geochemistry)

A.A. Velichko, USSR, (paleogeography)

G. Bjorklund, Sweden, (IGBP Secretariat)

The terms of reference for the Scientific Steering Committee (SSC) include:

1)   Assessing the possible contributions of planned or existing national and international efforts, including those of, for instance, the International Union for Quaternary Research (INQUA), as they pertain to the underlying themes and objectives of the IGBP;

2)   developing, within one year of appointment, plans for an initial multi-technique core IGBP project for coordinated field activities focused on the IGBP themes and objectives; and

3)   initiating, within two years of appointment, pilot segments of the project.

The SSC held its first meeting in Bern, Switzerland on April 4-7, 1989 to define a core IGBP program. The decision of the group was to develop a strong program of two distinct but related streams, aimed at establishing a vastly-improved record of global variations in climate, biochemistry, and biomass and the interactions that link these fundamental Earth system parameters.

STREAM 1 of the proposed core program will focus on the last 2000 years of the earth history, with the goal of reconstructing the detailed history of climatic and environmental change for the entire globe for the period since 2000 years BP, with temporal resolution that is at least annual and ideally seasonal. The period encompasses major climatic shifts between such features as the Medieval Warm Epoch, the Little Ice Age, and the general warming that has characterized global climate in the last one or two centuries. This is the only period for which written history is available, and the full period in which human impacts have become a recognized force in fixing the global environment.

The purposes of STREAM 1 are:

a)   To provide a baseline of natural change against which human impacts can be measured;

b) To establish a record against which environmental signals from the more distant past can be calibrated and quantified;

c) To illuminate the connections and phase-relationships between biogeochemical and climate changes in the most recent and most accessible period of Earth history;

d) To provide a data base for testing numerical models of climate and environmental processes.

STREAM 2 will concentrate, with coarser temporal resolution, on dominant changes in Earth history through the last two glacial cycles; a period fixed by the anticipated results from planned deep ice-coring programs in Greenland and Antarctica.

The goal is to reconstruct a continuous history of climatic and environmental change through the time of the most recent glacial-interglacial cycles (the last 250,000 to 300,000 years) in order to improve our understanding of the sequence of events that control epochs of major climatic and environmental change. Coordinated studies will focus initially on agreed-upon time slices and spatial regions.

Two types of reconstructions are envisioned:

A. The elucidation of long term variations, such as the transitions from interglacial to glacial periods, as well as periods of "abrupt" climate changes and "short term" fluctuations;

B. A more detailed examination of variations during the Holocene (last 10,000 years) and the Last Interglacial (about 120,000 years BP), to assemble regional and global environmental data from many sources.

The first step towards the goal for STREAM 2 will be to collect existing data sets that bear in significant environmental changes of both short and long term during this period, and put them in a form that allows intercomparison and study. As in STREAM 1 this initial reconnaissance of what is now available will identify needs and data gaps.

A particular need in STREAM 2 is that of establishing a uniform global chronology of events, and the improvement of methods for establishing exact temporal fiducia. In addition to information on temperature, precipitation, and atmospheric composition, there is a need for more information on organic composition, vegetation description and extent, and global biomass.

Preliminary steps to meet the goals of STREAM 1 and STREAM 2 will require efforts to establish calibration and end-to-end data analysis procedures that will make intercomparison possible between various data sets, and to establish on agreed upon chronology of climate and earth system change during this period.

New data will need to be collected to "fill the gaps", including data from both long- and short-term records. Both field and laboratory data will need to be assembled and interpreted, and methods developed for cross-correlating data that have been collected at different sites, by different methods, and from different archives.

The collected data will be used as inputs for environmental models. These include time dependent global models to better understand the dynamic behaviour of the environmental system, and coupled ocean and biosphere models that describe regional differences in the environment.

There is a need to organize teams to collect and archive new data describing the Earth system and to stimulate the establishment and improvement of new laboratories and archival centres that serve these ends.

There is an equivalent need to increase the educational opportunities that deal with the recovery and interpretation of environmental data from the past, and in this way to increase the number of scientists and technicians that work on recovering and interpreting environmental data of the past.

Efforts within the two streams, though focused on the last 2000 and on the last 300,000 years, will utilize data on Earth history from a much longer span to put these most recent epochs in the context of significant changes of longer term.

The SSC's program has sufficient scope to allow us almost unrestricted development of new and innovative projects that would contribute to a better understanding of the Earth systems' processes during these important time intervals. In other words, other groups, such as INQUA or IUGS can easily develop specific programs that can fit neatly into the broad program outlined above.

2. INQUA and the IGBP

Quaternarists and therefore, INQUA enter the picture by supplying information of past global changes in order to test the hypotheses of cause and effect and to help evaluate interactive earth system processes. Numerical data from time slices of the relatively recent past on a world wide scale will be particularly important for the core program. However, changes that have taken place during the last glacial cycle, not detectable on the short term must be considered. Much of this information is available now or is being acquired in existing projects, but must be organized and presented in a form easily utilized by others. Sources of information that can be derived from key archives are summarized in Tables 1 and 2 (modified from IGBP Report No. 4, "Global Changes of the Past", 1989).

Traditionally INQUA, through its worldwide network of scientists on various Commissions (see appendix), Subcommissions and Working Groups, has concentrated on the collection and interpretation of a variety of proxy data in order to present the condition of the earth during specific time intervals. Closely allied to this is our development of better and more reliable dating methods. These will no doubt continue to our most important activities. To this end, however, and to ensure maximum input into the global change program, we must work more closely with climatologists and modellers. They need numerical data in addition to our interpretation of past conditions. In turn, we will gain a better insight into forcing mechanisms causing change, a better picture of worldwide conditions during specific time intervals, and in general a better understanding of the total earth system. Workshops should be encouraged, dealing with general circulation and simulation models of interaction between the various spheres such as climatic change and vegetation dynamics, atmosphere-ocean interaction, orbital variations, and growth of ice sheets. In other words, it is time we broadened our outlook but not at the expense of our traditional well established activities.

INQUA as an organization representing Quaternary scientists from around the world has the primary responsibility to ensure that our science is being fully utilized in the IGBP. Our network of scientists and organization of commissions, subcommissions and working groups has been in place for over 50 years and is recognized by others as the primary organization representing Quaternary scientists. As an affiliate of the International Union of Geological Sciences (IUGS) we were recently called upon to plan and participate in an IUGS workshop at Interlaken, Switzerland (April, 1989) to determine what role the IUGS should play in the IGBP. Four working groups were organized: WG-1, Ocean record of past global change; WG-2, Terrestrial record of past global changes; WG-3, Anthropogenically induced global change; WG-4, Interaction of biodiversity reduction and global environment.

The one closest to our work was WG-2, Terrestrial record of past global changes. The group consisted of 12 scientists, most highly involved in INQUA activities. After much thought and deliberation they generated several broad projects that they thought should be undertaken in order to make the program a success.

2.1. RATIONALE FOR PROPOSED PROJECTS (MODIFIED FROM RUTTER ET AL., IN PRESS).

Quaternary geologic data from the continents can contribute in two principal ways to addressing the problems that have been identified by the SSC - Global Changes of the Past Program.

A. There is a need to document the state of the lithosphere, pedosphere, cryosphere, hydrosphere, and biosphere for particular times in the past when environmental conditions differed from today. This is required especially in relation to attempts to model past global conditions, in some cases to provide input to models and in others as a means of validating model simulations.

B. Assessments of the dynamics of lithospheric, pedospheric, criospheric, hydrospheric and biospheric responses to environmental changes are required. Rates of change, responses to "extreme events", threshold responses and the phase relationships between different components of the global system must all be evaluated, as must be the magnitudes of responses to perturbation. The extent and sign of any feedbacks between components of the system, and their impact upon the dynamics of response, must also be investigated. These data are fundamental to the assessment of the rates of many of the changes predicted for the future, as well as to evaluating their likely magnitude and consequent impact upon mankind.

Both forms of contribution are of vital importance, but the projects that they demand are to a great extent qualitatively different in character. Many data exist that can be synthesized in order to document the past status of components of the global system, whereas rather few detailed studies, independently controlled by absolute dating, have been made of the dynamics of these components' responses to environmental change.

Projects can be grouped into these two programs according to the nature of the contribution that they will primarily make. These groupings reflect also a degree of parallelism of methodology and/or of use of the same geological samples amongst the projects in each program. Within the two programs the proposed projects are intended only as examples or in some cases pilot projects, and are not intended to represent an exhaustive list of what we believe can or should be done within each program.

PROGRAM A. Documenting past states of the Global System

Projects within this program should identify common aims relating to:
- the time slices (or time windows) to be documented,
- the forms of cartographic (or profile) presentation to be used, and
- the variables to be evaluated and the units in which they will be measured.

Initial work in these projects will synthesize existing data and can be regarded as an imperative for the rapid delivery of Quaternary geologic data in a form which is relevant to the Global Change program as well as being readily communicated both to scientists in other disciplines and to non-scientists. Data synthesis should result in well-documented data bases both of raw data and of derived data that are mapped.

This initial work will also help identify spatial and temporal gaps in the existing data. Current projects that will generate relevant data should be identified and further projects to collect data that will fill gaps and/or improve resolution should follow the initial data syntheses.

PROGRAM B. Assessing the dynamics of responses to environmental change

Projects within this program share requirements for high-quality stratigraphic sampling and for extensive absolute dating of materials that will often lie beyond the range of ^{14}C dating. Many of the projects would use the common resource of the lacustrine record, although others will require samples from aeolian, fluvial, or peat deposits. In all cases there is a need for high quality cores that are adequately documented and archived and for a central database so as to facilitate and encourage multidisciplinary study of their sediments.

Particular emphasis should be given to obtaining long and continuous stratigraphic records that document one of more cycles of glaciation and that also have the potential for correlating with atmospheric and marine records. Shorter records will also be of value in examining the dynamics of responses since the last glacial and will be of enhanced value when their sediments display annual laminations. Wherever possible the investigation of these records should be multidisciplinary and include examinations of aspects of the sediments that reflect both internal responses of the system in which the sediment accumulated (e.g., sedimentology, geochemistry, isotope chemistry, magneto-stratigraphy and biostratigraphy of organisms living within the system) and external responses occurring on the landscape or in the region around the site of deposition (e.g., pollen stratigraphy, tephras, aeolian and/or fluvial particles). Whenever appropriate or possible, the sites chosen for investigation should also serve to fill spatial or temporal gaps identified by projects within Program A.

Projects within this program should commence as soon as possible by may not deliver initial products until somewhat later than those in Program A.

<u>Examples of Outline Project Descriptions</u>

PROGRAM A

2.2. PROGRAM ON PAST GLOBAL RESERVOIRS OF CARBON AND THEIR CHANGE WITH TIME

<u>Problem Statement</u>: Changes in terrestrial landscapes, irrespective whether triggered by climatic perturbations or other causes, need to be validated by quantitative data, for which carbon content in major ecosystems is one of the best parameters. Proxy data of past carbon reservoirs in the continent should be summarized and interpreted on a set paleomaps of their content in vegetation, soils, lakes, etc. for defined times slices, on a scale of ca. 1:15 million or larger for selected continents and regions where better data are available.

<u>Objectives</u>: - To contribute to the knowledge if the past carbon cycle and its rate of change and to quantify such data.
- To determine the most sensitive regions which are subject to rapid C reservoir variations.
- To serve as a database for modelling of fluxes and perturbations of C with time and especially for comparison with present and predicted geochemical cycles and models which are already available.

PROGRAM B

2.3. PROJECT ON GLOBAL SEA LEVEL CHANGE AND IMPACTS

<u>Problem Statement</u>: Global sea level rise has been identified as a potentially important impact of "greenhouse" warming. However, estimates of the amount of sea level rise to be expected over the next 50-100 years are notoriously uncertain. It should therefore be a priority to accurately monitor the component of sea level rise that is of climatic (as opposed to tectonic of glacial isostatic) origin and to assess the impacts that might be expected on coastal communities, having identified those that are especially at risk. Accurately separating the climatic component of sea level rise will require application of modern space-based geodetic techniques to remove the tectonic component and the use of models of glacial isostasy to separate this additional influence from the tide gauge records.

<u>Objectives</u>: The objectives of the project are to bring together the expertise of geomorphologists, geophysicists, climatologists, oceanographers, geographers, and coastal engineers, so as to:

- Accurately quantify the sea level variations that have occurred over the last two glacial cycles, where possible, focusing especially upon the most recent 20,000 years of earth history.
- To establish local late Quaternary and late Holocene geological trends in vertical movement to which rates of present and future sea level variations may be compared.

- To separate, where possible, the different contributions to the local sea level trends associated with tectonic, glacial, isostatic, climatic and local anthropogenic influences, etc.
- To support the development of geophysical models of the glacial isostatic adjustment process, of atmospheric and oceanographic models of climate variability and of models of storm surges, etc.
- To identify the coastal locations that would be most vulnerable to future sea level rise and storm surges, and to specify the range of consequences (flooding, wetland degradation, salt intrusions, etc.) likely to occur at each location.

2.4. PROJECT ON ABRUPT CHANGES, EXTREME EVENTS AND SHORT-TERM FLUCTUATIONS IN CONTINENTAL RECORDS

Problem Statement: Although abrupt changes and extreme events have played, and will increasingly play, an important role in the well-being of the world's people - probably extracting a heavy toll on life and property - we have failed to consult the geological record for help in understanding such events. From the geological archives we may be able to predict the likelihood that a certain extreme event will occur, estimate its magnitude, establish its recurrence rate, and measure its spatial and temporal impact. It is important to investigate factors leading to and initiating abrupt changes, extreme events and short term fluctuations in recent earth history. The response to a natural (or man-induced) event may or may not be reversible once a threshold value is reached, and may or may not initiate future environmental changes. For example, the abrupt introduction of overflow from glacial Lake Agassiz 11,000 years ago, first altered river and lake systems carrying this water, and subsequently may have affected waters of the Gulf Stream which, in turn, may have caused global cooling (i.e. the Younger Dryas). Warmer conditions only returned concomitant with the end of the Lake Agassiz influx. Lacustrine, mire, aeolian, fluvial and colluvial sediments in which chronological indicators provide good time resolution offer the best proxy records to elucidate past events of this nature.

Objectives: - To identify short-term, high-magnitude effects (e.g. biotic of sediment anomalies) and their pattern of recurrence through time.
- To identify causal mechanisms (e.g., flood, drought, cold, earth movement, volcanic event) and areas of greatest vulnerability.
- To seek links with forcing mechanisms (e.g., solar variation, orbital parameters, aerosols, degree and shift in seasonality, precipitations character).
- To identify initiating factors and threshold conditions and to determine whether preparatory or compounding events/conditions were involved (e.g., preceding drought,

low flood incidence, high lake level).
- To select case histories to demonstrate the impact that such events may have on mankind.
- To assess the relative importance of human influence versus natural change and examine recovery processes following abrupt events.
- To establish a record of late Quaternary extreme events and short-term fluctuations on the baseline of natural environmental change in order to formulate predictive models.

2.5. PROJECT ON LAND-SEA CORRELATION

Problem Statement: There has been little correlation between terrestrial and marine events even though some excellent records from each are known. A correlation between land and sea archives is one way to understand regional-scale terrestrial geosphere-biosphere feedbacks in terms of the globally-integrated changes of the oceans and atmosphere. Nature has already carried out numerous short-term experiments with our environmental systems. Predictive models of anthropogenic global change cannot yet provide adequate regional-scale information for formulating response strategies. The high-resolution geological record provides an important opportunity to understand responses on a regional scale. It is thus necessary to improve our ability to correlate terrestrial and marine records and to link these with the atmospheric history held in ice cores. A pilot project is proposed for the climatic sphere of influence around central and southern Europe, the circum-Mediterranean rim, northern Africa and the eastern Atlantic Ocean. This program can take advantage of the high density of potentially correlative time-series elements available in this area, and link with ongoing projects.

Objectives: - To identify and document the high-resolution glacial/interglacial sequences in central and southern Europe and correlate these with marine isotope stages in adjacent oceans.
- To determine phase relationships between land and sea events and among environmental signals held in archive components.
- To identify and quantify continental signals in marine records.
- To lay the groundwork for input for modelling experiments.

2.6. PROJECT ON LONG CONTINENTAL RECORDS

Problem Statement: Long, continuous continental records that extend back over two interglacial periods are lacking in many parts of the world, although deposits of loess or alluvium and especially lacustrine sediment suitable for such investigations are widespread. Records from such continental deposits are necessary in order to compare responses on the continents to those of the oceans, and to match the high-resolution records of the atmospheric environment provided by the ice cores. The continental records will reveal the shifting long-term patterns of the various components of the biosphere and geosphere, and so enable the nature of these systems' response to long-term environmental change to be inferred. Long records will also increase the possibility of examining mechanisms of rapid change that did not occur during post glacial times by means of high resolution investigations of appropriate intervals in the pre-Holocene record.

Objectives: - To determine the time-series of changes, both internal and external to the site of deposition, that reflect responses to changes in the atmospheric environment.
- To complete a series of transects of such long records across the major zonal climate regions and the longitudinal extent of the continents, and to promote improved spatial and temporal extent of long continental records.
- To enable global mapping of data for time-series during "the last interglacial" or earlier.

The proposed projects fit conveniently into the broad program outlined by the Scientific Steering Committee. Therefore, these projects should contribute directly to the core program. It should be emphasized, however, that other projects suggested by other working groups at this Workshop may or may not fit into the core program. They will, however, contribute to a better understanding of global change. Further, specific projects, say by national groups or scientific associations, can be designed to meet the broad objectives of these projects.

The above discussion relates to one example of INQUA's contributions in planning a joint global change program. Even though we are an affiliate of IUGS, we may be called upon to aid outside groups needing our expertise. In the meantime, we have our own Intercongress Committee on Global Change with H. Faure as President. We see the need of INQUA having its own projects. At the present time the Intercongress Committee is formulating appropriate programs that take advantage of our expertise in our various Commissions, Subcommissions and Working Groups. One that has been suggested and has been endorsed by the INQUA Executive is the project on Past Global reservoirs of Carbon and presented at the Interlaken meeting. This is an ideal INQUA project because it draws on many areas of our expertise and provides numerical data to modellers. In addition to our own projects, the Intercongress Committee will suggest, identify and encourage INQUA Commission projects that are important to the Global Change Program.

3. Funding and the Individual Scientist

The individual scientist's funding and participation in specific projects of the Program has not been addressed in this paper because it is not altogether clear at this point.

In my opinion, the national IGBP committees will be the backbone of the Program. They will propose and identify projects that contribute to the core program. The committees will probably have to raise project money from the traditional funding agencies. The national committees in addition, must coordinate the efforts within the country. The big question will be how much "new" money will be available for global change projects? Another question, how much money will be spent on projects concerning global changes in the past, our primary interest here? Outside the national committees, there are literally hundreds of organizations, both worldwide and regional, that will contribute to the Program. INQUA is one of them. Funding for "world" projects will come from a variety of sources. UNESCO may be a principal source. No matter what, there will be competition between various groups, and only projects judged to be important to the overall goals of the Program will be funded.

As far as the individual Quaternary scientist is concerned, I do not see much change in an individuals current effort. He or she will continue to contribute to an agency's goal which in turn may contribute to a national global change projects which eventually will feed it into the core program (see Table 3). This does not eliminate an individual from being a part of planning, initiating and contributing to a national or "world" project. After all, the entire Global Change program is designed by grass roots scientists. In most cases, an individual may see his or her contribution form a part of a larger effort in keeping the goals of the Global Change program.

4. Conclusions

The organizers of IGBP: A Study of Global Change have recognized the importance of understanding past global changes in order to better predict and understand future changes in the earth's condition. The Scientific Steering Committee on Global Changes of the Past is charged, among other things, with ensuring that relevant paleo information is utilized in order to fulfil the requirements of the overall objectives of the Global Change program. INQUA and Quaternarists in general, have an obligation and opportunity to demonstrate the value of our science to the understanding of the total earth system and man's interaction with it.

The program is still in its incipient stage. The organization and overall objectives are fairly well in place, whereas other aspects, especially funding, are yet to be worked out. Broad programs for the Global Changes of the Past component are, or are being, formulated. Specific projects that meet the goals of the broad programs, are in various stages of development by various national groups and scientific organizations. The question of funding as well as the enormous task of coordinating and transferring pertinent information into the overall objectives are still, in most cases, unanswered.

5. Selected Bibliography

Oeschger, A., & Eddy, J.A. (1989) "Global changes in the past", Report No. 6, IGBP, *A Study of Global Change*, 39 p.

Rutter, Nat, Ammann, B., Faure, H., Huntley, B., Kelts, K., Peltier, W.R., Pirazzoli, P.A., Schlüchter, C., Schnack, E., Starkel, L., Teller, J., & Yaalon, D.H. (in press) *Global Change and the Terrestrial Record, Global and Planetary Change.*

Scientific Advisory Council for the IGBP (1989) *The IGBP: A Study of Global Change*, A Report from the First Meeting of the Scientific Advisory Council for the IGBP, Report No. 7, 1 and 2, 341 p.

Special Committee for the IGBP (1988) "IGBP: a study of global change, a plan for action.", Report No. 4, *IGBP: A Study of Global Change*, 200 p.

6. Appendix

INQUA COMMISSIONS

C-1   Commission on Stratigraphy
      President: Dr. M.N. Alekseev (USSR)

C-2   Commission on Genesis and Lithology of Quaternary Deposits
      President: Dr. J. Lundqvist (Sweden)

C-3   Commission on Quaternary Shorelines
      President: Dr. S. Jelgersma (The Netherlands)

C-4   Commission on Loess
      President: Prof. M. Pecsi (Hungary)

C-6   Commission on Palaeopedology
      President: Dr. J. Catt (U.K.)

C-7   Commission on Neotectonics
      President: Dr. N.A. Mörner (Sweden)

C-8   Commission for the Study of the Holocene
      President: Dr. B. Ammann (Switzerland)

C-10 Commission for the Quaternary of Africa
      President: Dr. Abdoulaye Faye (Senegal)

C-11 Commission on Paleogeographic Atlas of the Quaternary

President: Dr A.A. Velichko (USSR)

C-12   Commission on the Paleoecology of Early Man
President: Dr. H. Muller-Beck (GFR)

C-13   Commission on Paleoclimate
President: Dr. A. Berger (Belgium)

C-14   Commission on Applied Quaternary Studies
President: Prof. G. Luttig (FRG)

C-15   Commission on the Quaternary of South America
President: Dr. J.O. Rabassa (Argentina)

Intercongress Committee on Global Change
President: Prof. H. Faure (France)

Intercongress Committee on Tephra Studies
President: Dr. J.A. Westgate (Canada)

Table 1. Characteristics of natural archives.

Archive	Temporal precision	Extent (yrs)	Derived parameters							
Tree-rings	yr/season	10^4	T	H	C_A	B	V	M	L	S
Lake sediments	yr	$10^4 - 10^6$	T		C_A	B	V	M		
Polar ice cores	yr	10^5	T	H	C_A	B	V	M		S
Mid-latitude ice cores	yr	10^3	T	H	C_A	B	V	M		S
Coral deposits	yr	10^5	T	H	C_W	B		M	L	
Loess	10 yr	10^6	T	H	C_S	B		M		
Ocean cores	100 yr	10^7	T		C_W	B		M	L	
Pollen	10 yr	10^6	T	H	C_S	B				
Paleosols	100 yr	10^6	T	H	C_S		V	M		
Sedimentary rock	2 yr	10^7	T	H	C_S		V	M	L	
Historical Records	day/hr	10^3	T	H		B	V	M	L	S

T = temperature
H = humidity or precipitation
C = chemical composition of the air (A), water (W) or soils (S)
B = information on biomass, as in pollen samples
V = volcanic eruptions
M = geomagnetic field
L = sea level
S = solar activity

Table 2. Information derived from natural archives.

	ICE					SEDIMENTS/PEAT				TREE-RINGS	
	²H	^{18}O (H₂O)	Dust/Chem	Gases	Radio Isotopes	Comp.	Faunal	Plants	Isotopes	Structure	Isotopes
SUN			x		XX						XX
ATMOSPHERE											
Aerosols		x	XX		XX	XX			XX		
Circulation			XX	XX	XX	x			XX		XX
Gas composition			XX	XX					XX	x	XX
Humidity	XX	XX	XX	x	XX	x			x	XX	X
Temperature	XX	XX	XX	x	XX				XX	XX	X
OCEAN											
Circulation	XX		XX	XX	XX	XX	XX		XX	XX	XX
Bio-pump			x	XX	x	XX	XX		XX		X
Sea-level	XX		x	X		XX	XX		XX		X
Sea-ice	X		x	X		XX	XX		XX		X
BIOSPHERE											
Vegetation			XX	XX	x	x	XX	XX		XX	XX
LITHOSPHERE											
Volcanoes/erosion			XX	x	x	XX				x	
BIOCHEMICAL CYCLES											
Carbon, N, S			XX	XX	XX	x		x	XX	XX	XX
ICE SHEETS											
Mass, height, sea ice	XX		XX	x	XX	XX			XX	XX	XX
ANTHROSPHERE	X		XX	XX	XX	XX	x	x		x	XX
DATING	XX		XX	XX	XX	XX	XX	XX	XX	XX	XX

Table 3. Possible path of an individual scientist's contribution to the Global Change program.

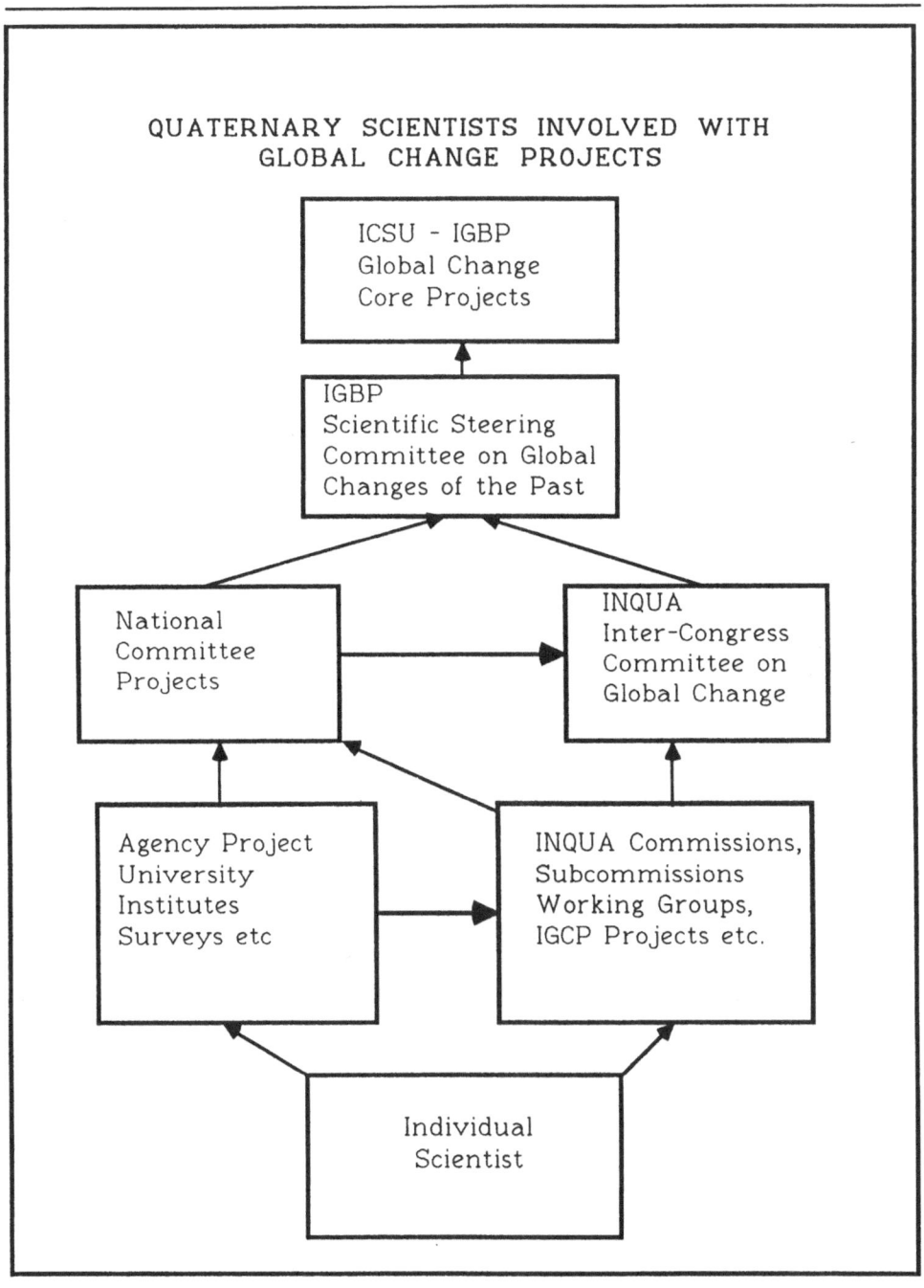

PART IV: MANAGEMENT, TECHNIQUES AND CASE STUDIES

SECTION C: Management

ESTIMATION OF NATURAL AND MAN-INDUCED GROUNDWATER RECHARGE

E. ROMIJN
Consulting Hydrogeologist
Mariënbergweg 1
6862 ZL Oosterbeek
The Netherlands

ABSTRACT: In order to investigate large scale artificial recharge of groundwater in arid and semi-arid regions both technical problems related to infiltration capacity and environmental side-effects caused by large scale infiltration have to be studied. Great advantage can be taken from recent studies of natural and man-induced recharge in arid and semi-arid regions as presented for example at the Nato workshop held in Turkey in 1987. Scarcity of data, however, is a commonly encountered problem. One way suggested to solve this problem is Issar's method of comparable hydrogeological provinces. Others are regionalisation techniques in combination with remote sensing.

1. Introduction

In March 1987 a NATO advanced research workshop on Estimation of Natural Recharge of Groundwater was held in Antalya, Turkey (Simmers, 1988). It was decided that the target would be the world's arid and semi-arid zones, where need for reliable estimates of groundwater recharge is greatest. Results were summarised in a manual of practice on recharge estimation to be published by IAH as a contribution to Unesco's IHP (Lerner et al, 1989). Some of the results are also presented in this paper.

Unesco (1979) defines arid zones by a precipitation (p)/potential evapotranspiration (etp) index, in which both are mean annual values and etp is calculated by the Penman formula. Arid implies when $0.03 < p/etp < 0.20$ and semi- arid that $0.20 < p/etp < 0.50$. In semi-arid regions p varies between 300-400 mm and 700-800 mm in summer rainfall regimes and between 200-250 mm and 450-500 mm for winter rainfall areas.

The above study contributes to investigations into large scale artificial recharge of groundwater in semi-arid regions when information on the hydrological cycle of (semi-)arid regions under natural conditions is available, thereby forming the basis for environmental impact studies. Three domains, the (deeper) lithosphere, the atmosphere and the zone of

485

R. Paepe et al. (eds.), Greenhouse Effect, Sea Level and Drought, 485–493.
© 1990 *Kluwer Academic Publishers.*

human occupation, strongly influence the soil-water-vegetation complex in which infiltration and recharge processes take place. The deeper lithosphere is still the "hardware" which does not undergo drastic changes under human influence. However human impact on the upper parts of the lithosphere (the soil) is dramatic, both in a direct way by landuse, pollution, civil engineering works etc and indirectly through influence on climatic factors i.e. the radiation/heat balance, atmospheric circulation and vapour, gas and dust content of the atmosphere. Large scale groundwater recharge schemes will not only change the soil-water-vegetation complex but also influence the regional climate because large infiltration reservoirs will cool air masses and increase air humidity. Large scale groundwater recharge will also create new groundwater flow systems, eventually resulting in more shallow groundwater tables and seepage zones. From irrigation projects in (semi-)arid regions related salinization problems are well known, as are parasitic diseases connected to these environments. Direct (subsurface) artificial groundwater recharge would be the best choice but is only possible in highly permeable rocks, i.e. fissured or karstic rocks or mountain front alluvial fans. However, a major problem concerns how to collect enough data for preparing sound feasibility and impact studies for such widespread engineering schemes.

2. Space-Time Extrapolation of Data

Most arid and semi-arid zones are regions of scarce hydrogeological data due to sparse habitation and randomness of climatic events.

The "regionalisation" problem of how to extrapolate point measurements urgently requires solutions. Fieldwork and aerospace surveys should indicate more or less homogeneous zones which allow extrapolation of local data. This mapping problem applies to all three domains i.e. hydrogeology, climate and human occupation which, through the soil-vegetation complex, determine infiltration and recharge rates. The last 30 years' aerial photo-interpretation has been supplemented with powerful remote sensing techniques (Tiros since 1960, Noaa since 1970, Landsat since 1972, Spot 1986). However, spatial (G) and temporal (T) resolutions conflict: Landsat has G of 30 m and T of 18 days, Noaa T of 12 h but G 1-4 km, the geostationary Goes-Meteosat system half-hour T but only direct subsatellite G of the order of 1 km. Spot's off-nadir technique allows for high G of 10-25 m and for T of a few days. Still lacking is active radar on space platforms (ERS 1990?) which is of particular relevance for soil moisture survey. Passive radiometers on space platforms have a G of about 20 km only.

Hydrogeological zoning can be derived from landforms (e.g dunes), erosional characteristics, geological setting (folds, faults, fractures, outcrops, volcanic and karstic features), drainage patterns, classification of crops and vegetation, and biomass determination with the vegetation index NVI. The aim is to infer aquifer systems and groundwater flow systems (Larsson, 1984; LaMoreaux et al, 1984). Remote sensing techniques have not yet contributed to estimates of recharge by calculating evapotranspiration in (semi-)arid zones (De Bruin, 1988). De Bruin argues that by measurement of cloud temperature (Meteosat) rainfall estimates may be obtained comparable in accuracy to existing raingauge networks. This would allow extrapolation of runoff and recharge data with the help of remotely sensed timeseries of convective storms, which means temporal extrapolation.

A further development involves geographical information systems (GIS) for remotely sensed data handling, some of which are available also for PCs (Meijerink et al, 1988). These will allow more objective mapping and zoning procedures by comparison with the hand made maps of 25 years ago.

3. Methods for Recharge Estimation

The working group on recharge estimation (Lerner et al, 1989) distinguished five sets of methods for recharge estimation:
a) direct measurements
b) water balance methods
c) Darcian approaches
d) tracer techniques
e) empirical methods.

Their possible application depends on factors such as conceptual realism, sensitivity of the method for assessing errors and for variability of data both in a spatial and temporal sense and costs. Suitability of the methods also depends on different situations of groundwater recharge, e.g. precipitation recharge (P), river recharge (R), irrigation losses (I), urban recharge (U) or recharge over the whole catchment area (C), including interaquifer flow (A).

a) Direct measurements incorporate lysimeters (a1) and seepage meters placed under water (a2). They measure the flow through a parcel of the soil. Lysimeters are expensive, give only "point" measurements and are more suitable for humid climates. Seepage meters are easy to handle and also produce point values.

b) Water balance methods are not expensive provided precipitation data, data for the calculation of evapotranspiration and riverflow measurements are available. They calculate recharge as the difference between inputs (rainfall) and outputs (evapotranspiration, surface runoff etc.), so their accuracy diminishes when inputs and outputs approach each other, e.g. under arid circumstances. They can be applied to soil parcels (b1, soil moisture budgets), to parts of river channels or canals (b2) and to whole catchment areas (b4). The more sophisticated hydraulic approach uses flood routing models (b3). Another method uses fluctuations of the groundwater table to calculate groundwater stored by recharge, taking into account other inflow and outflow estimates (b5).

c) Darcian methods calculate flows based on Darcy's law and the equation for mass conservation. However, measurements of hydraulic heads, hydraulic conductivities, moisture contents and storativities are needed which leads to low accuracy when data is limited. Calculations for vertical flow (c1) whether saturated or unsaturated give only point values. Regional values may be obtained with numerical models for groundwater flow coupled with surface water models and models for evapotranspiration, provided that enough accurate data on aquifer properties and heads are available (c2). For many "simple" cases analytical solutions can be used. Inverse modelling (c3) calculates aquifer properties from groundwater heads and estimated recharge, the latter being a "by-product" which is very sensitive to data errors. There is also the problem of non-uniqueness of solutions from inverse techniques. Modelling together with sensitivity analysis has the advantage of giving an indication on data needed and uncertainties in calculated results.

d) Tracer techniques are amongst the most useful techniques for recharge estimation, especially in arid zones. Signature methods use dating or labelling of water parcels and throughput methods involve a mass balance of the tracer. Contamination and flow through preferred pathways complicate calculations. Environmental signature methods (d1) using T or carbon-14 provide data on water movement and therefore on recharge. Applied tracers (T, etc.) can be used to label the unsaturated flow. Others, like bromide-82 and iodide-131, may be used for determining flowrates in the saturated zone with borehole dilution techniques (c3). Non-evaporating environmental tracers (e.g. Cl) may be used to estimate evapotranspiration and hence recharge by measuring increased concentrations in groundwater.

e) Empirical methods are often used for regions with scarce data. For precipitation-recharge, recharge (= r) may be formulated as $r = f(p)$, where f is not only a function of soil and vegetation characteristics but also of climatic data, e.g. of precipitation itself. Formulae with threshold values b (mm) are used for regions or even for whole catchment areas: they are of the type $r = a(p-b)$. In semi-arid regions Goldschmidt & Jacobs used for example $r = 0.86(p-360)$ and Bredenkamp $r = 0.30(p-313)$, both on a yearly basis for limestone/dolostone terrains (Lerner et al, 1989). Due to the stochastic character of rain storms some recharge of groundwater will take place even under semi-arid situations, which means that threshold values do not exist or depend on the rainfall regime. Recharge may also be derived from river flow rate, sometimes with the aid of groundwater table response functions. Seepage losses from canals may be estimated from canal water depth, discharge, soil coefficients and the like.

These methods are roughly evaluated in Table 1. For estimating P (precipitation recharge) methods a1, b1, d1 and possibly c1 are the most suitable for humid zones and d1, d2, and possibly c1 for more arid zones. River recharge (R) should be estimated by methods b2 & b3, irrigation losses (I) by a2, b1, b2 and regional estimates (C) could be based on b4, d1, and d2. For first approximations method e1 is useful and for detailed studies d3.

4. Some Results from Case Studies of Semi-Arid Zones

Issar & Passchier (in Lerner et al, 1989) summarised a number of studies on recharge estimation and grouped them together according to hydrogeological provinces. They distinguished as provinces Riverbed basins (R), Mountain front basins (M), Sands (S), Sandstones (ST), Limestones and Dolostones (L), Chalk (C), Plateau basalts (P), Vesuvian type basalts (V), and Granitic (incl. acid metamorphic) rocks (G).

R show localised recharge through infiltration into gravel layers. Recharge depends on aquifer properties, depth of groundwater table and channel characteristics (flow rate, stream flow velocity, suspended load etc.).

Table 1. Evaluation of different methods for recharge estimation

method	judgement (+ or -) concept	accuracy	costs	applicability (X) P	R	I	U	C	A
a1 lysimeter	+	+(*)	-	X					
a2 seepage meter	+	+	+		X	X			
b1 soil moisture budget	+	+(*)	+	X	X	X			
b2 balance of river channel & canal	+	+	+		X	X			
b3 flood routing	+	+	+		X				
b4 balance of catchment or region	+	+/-	+/-				X	X	
b5 water table response	-	-	+		X			X	
c1 Darcian vertical flow calculation	+	+/-	-	X	X	X			X
c2 numerical modelling & analytical solutions	+	-	+/-	X	X	X			X
c3 inverse modelling	+	-	+/-					X	
d1 environmental signature tracers	+	+	+	X	X			X	
d2 environmental throughput tracers	+/-	+(**)	+	X	X	X		X	
d3 applied tracers	+	+/-	+/-	X	X	X		X	
e1 empirical methods	-	-	+						
e2 watershed (lumped) models	-	-	+	X	X			X	

(*) for arid zones (-); (**) for humid zones (-)
P = precipitation recharge; R = river recharge; I = irrigation losses;

M   are also recharged locally from rivers and subsurface inflow, both coming from adjacent mountains. Recharge depends on topography, aquifer properties, depth of groundwater table, and type of precipitation. Winter precipitation is more effective than summer precipitation because of spatial and temporal spread and lower evaporation losses.

S   show direct recharge, depending in a non-linear way on precipitation intensity and duration. Winter precipitation is usually more effective than summer precipitation. Influence of vegetation on recharge is by interception of rainfall and different rates of evapotranspiration. Rootholes may create preferred pathways for infiltration.

ST   have direct recharge along outcrops or at unconfined regions. They often contain fossil water recharged during more humid geological ages.

L,   with their characteristic solution processes (karstification) and often without any surface runoff, show high recharge percentages. Often localised recharge takes place (e.g. in swallow holes). Vegetation is an important factor both for evapotranspiration and karstification. C have typical double porosity characteristics caused by solution along fissure patterns and show high recharge charcteristics.

P   have a very low primary permeability. For recharge the fractures are of importance. Weathered parts form aquifers although soils developed from weathered zones may contain very impermeable clays.

V   are very inhomogeneous layered rocks often crossed by impermeable dikes and with perched aquifers stored above each other.

G   recharge also depends on weathered zones and fracturing, just like in case P.

Table 2. Recharge as % of precipitation. Data from case studies by Lerner et al (1989). See also appendix

Hydrogeological areas	Precipitation (mm); between () nr of case studies			
	< 200	200 - 400	400 - 600	600 - 800
Sands (S) Dunes	1-11(4)	1-19(4) 60(1)	1-12(2) 65(1)	15(1)
Limestone-dolostone (L)	1-27(3)	1-41(2)	3-60(6)	13-53(4)
Plateau basalts (P)				8-9(1)
Granitic rocks (G)	1-15(2)	1(1)	1-13(1)	3-21(2)

Table 2 gives recharge values for direct precipitation recharge as % of precipitation. Locally recharged riverbed and mountain front basins are excluded from the table as well as large confined sandstone aquifers. High recharge factors for L are typical; dunes (S) show extremely high recharge factors. Below 600 mm precipitation recharge may be as low as 1 % per year in all the hydrogeological areas.

5. References

Bruin, H.A.R. de (1988) "Evaporation in arid and semi-arid regions", *Estimation of natural groundwater recharge*, NATO ASI Series C Vol. 222, Reidel, Dordrecht.

LaMoreaux, P.E., Wilson, B.M., & Memon, B.A. (1984) *Guide to the hydrology of carbonate rocks*, Unesco, Paris.

Larsson, I. (1984) *Ground water in hard rocks*, Unesco, Paris.

Lerner, D., Issar, A., & Simmers, I. (1989) *Estimation of natural groundwater recharge*, Heise, Hannover.

UNESCO (1979) *Map of the world distribution of arid regions*, MAB Techn. Notes 7, Paris.

Meijerink, A.M.J., Valenzuela, C.R., & Stewart, A. (1988) *ILWIS, the integrated land and watershed management information system*, ITC publ. 7, Enschede.

Simmers, I. (1988) *Estimation of natural groundwater recharge*, NATO ASI Series C Vol. 222, Reidel, Dordrecht.

6. Appendix

Here results by Lerner et al (1989) are summarised.

Hydrogeological area		Method	Region	Climate	Precip. mm/a	Infil. mm/a	Recharge mm/a	Recharge % of precip.
1. RIVERBED BASINS								
Hillel & Tadmor	1962	b1, b2	Negev, ISR	arid	50-150		300-800	
2. MOUNTAIN FRONT BASINS								
Gupta & Sharma	1984	d3	Sabarmati, IND	sub-humid summer monsoon	800			15
3. SANDS								
Dincer et al	1974	d1	Saudi Arabia	arid	50-100		2.3	3
Caro & Eagelson	1981		id	"	80		6	7.5
id			id	"	130		14	11
Edmunds et al	1988	d2	Khartoum, SUD	"	180		1	0.5
Phillips et al	1984	d1	New Mexico, USA	"	200	2.5		1
Stephens & Knowlton	1986	c1, c2	id	"	200		37	19
Issar et al	1985	c1, c2	coast. dune, ISR	"	200			55-60
Sharma & Gupta	1985	d3	Thar, IND	"	150-300			6-14
Verhagen et al	1979	d1	Kalahari, BOT	semi-arid	330	12-88		
Allison & Hughes	1983	d1	AUS	"	335		3-4	1
Edmunds & Walton	1980	d2	Cyprus	"	420		50	12
Foster et al	1982	d1, d2	Kalahari, BOT	"	450		0-5	0-1
Issar et al	1985	e1	coast. dune, ISR	"	500			65
Sharma	1988		Swan coast, AUS	sub-humid	775			15

4.LIMESTONE, DOLOSTONE

Hydrogeological area		Method	Region	Climate	Precip. mm/a	Infil. mm/a	Recharge mm/a	Recharge % of precip.
Wright et al	1982	b4, d2	Bahrain	arid	73		5-20	7-27
Lloyd et al	1987	b4	Qatar	"	75			10-12
Issar et al	1985	b4	Avdat, ISR	"	100			1-2
Burdon	1961	b4	Damascus, SYR	"	262			41
Commander	1983	d2	Fortescue, AUS	"	350			1
Tixeront et al	1951	b4	Tunisia	semi-arid	300-780			0-53
Schoeller	1948	b4	id	"	348-676			35-90
Smit	1978	b4	Ghaap, RSA	"	445			3
Bolelli	1951	b4	Morocco	"	500			14
id		b4	id	"	600			20
Bredenkamp	1988	b4, b5, e1	Transvaal, RSA	"	560-590		65-85	12-14
Bredenkamp et al	1970,1974	b4, d1, e1	Pretoria, RSA	"	560		17.7	3
Mero	1958	b4	ISR	"	600			53
Fleisher	1981	b4	Transvaal, RSA	sub-humid	590-700			13-27
Shachori et al	1965	b1, b4	Carmel, ISR	"	672		318	47

5.PLATEAU BASALTS

Hydrogeological area		Method	Region	Climate	Precip. mm/a	Infil. mm/a	Recharge mm/a	Recharge % of precip.
Athavale et al	1988	d3	Deccan Trap, IND	semi-arid	612		46	7.5
id		d3	id	"	652		56	8.6

6.ACID CRYSTALLINE ROCKS

Hydrogeological area		Method	Region	Climate	Precip. mm/a	Infil. mm/a	Recharge mm/a	Recharge % of precip.
Issar & Gilad	1982	b4, d1	Sinai, EGYPT	arid	50			15
Allan & Davidson	1982		Western AUS	"	<300			0.1-0.5
Athavale & Rangarajan	1988	d3	Southern IND	semi-arid	390-615			1-13
Sukhija & Rao	1983	b4, d1	Vedavati, IND	sub-humid	616			13-21
Thiery	1988	b1, e2	Ouagadougou, BUR	"	690		23-45	3-7

DEVELOPMENT AND MANAGEMENT OF FOSSIL GROUNDWATER RESOURCES FOR PURPOSES OF DROUGHT MITIGATION

KEN W.F. HOWARD
Groundwater Research Group
Scarborough Campus
University of Toronto
Scarborough, Ontario
M1C 1A4 Canada

ABSTRACT: Immense volumes of fossil groundwater underlying one third of North Africa and the Arabian Peninsula, provide the key to the mitigation of drought and alleviation of famine. However, given that these groundwaters are finite and essentially non-renewable, the development of an appropriate aquifer management strategy represents a difficult challenge, particularly as this strategy must reject the traditional and conservative "safe yield" approach to groundwater development, and recognize groundwater mining as a viable alternative. It is concluded that groundwater mining is an acceptable approach provided, i) it is positively planned and realistically evaluated in advance, ii) close control over groundwater production is exercised, and iii) there is a clear and feasible plan for alternative water supplies when the groundwater resources are exhausted. It is also important that the development program is kept flexible, thus allowing refinements to be incorporated as development proceeds and data on the aquifer's response to pumping become available. This "operational approach" to management policy development would likely involve staged groundwater production, intense monitoring and a continual upgrading of resource and economic assessments.

1. Introduction

The key to alleviating drought and famine lies locked in vast epicontinental basins which underlie one third (6.5×10^6 km^2) of North Africa and the Arabian peninsula. These basins are noted for their extensive tabular sandstone and carbonate aquifers, Cambrian to Tertiary in age, which contain immense reserves of good quality groundwater. Perhaps ironically, these reserves often lie within just a few hundred metres of the parched desert soils.

Burdon (1977, 1982) has estimated that the eleven principal groundwater basins of the Saharan and Arabian regions (Figure 1 and Table 1) store as much as 80,000 km^3 (8×10^{13} m^3) of water, representing approximately 2% of the world's actively circulating groundwater. The origin of this water is the subject of continuing controversy (Lloyd and

495

R. Paepe et al. (eds.), Greenhouse Effect, Sea Level and Drought, 495–512.
© 1990 *Kluwer Academic Publishers.*

Farag, 1978; Edmunds and Wright, 1979; Bakiewicz et al., 1982; Wright et al., 1982; Lloyd and Miles, 1986). Model studies and the local presence of tritium suggest that the aquifers are receiving some replenishment. However, radio-carbon isotope data confirm that most of the stored water is "fossil water" which recharged the system during a wetter climatic regime some thousands of years before present. For practical resource development purposes, most major arid-zone groundwater projects can be considered reliant on groundwater reserves that are non-renewable.

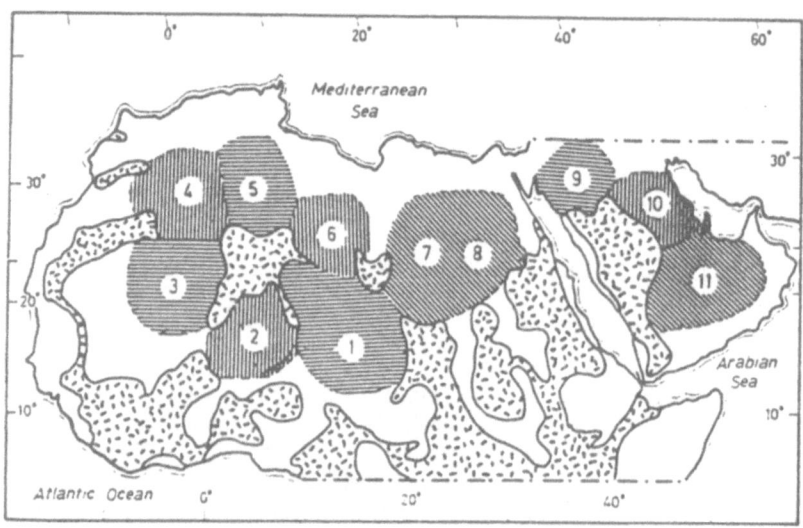

Figure 1. Major groundwater basins of North Africa and the Arabian Peninsula (from Burdon, 1977).

The potential for utilization of the large groundwater basins has been recognized in countries such as Libya, Egypt, Israel and Saudi Arabia where major resource development and irrigation projects have been launched. In Libya, for example, the first phase of the "Great Man-Made River Project" (Pearce, 1984) will extract 23 m^3 per second from the Sarir Basin and convey it 1500 km to the coast where it will be used for water supply and irrigation of 1,800 km^2 of agricultural land. At an estimated cost of $ 4 billion, the project is proceeding despite warnings from critics that recharge to the system, mainly from the mountains of Tibesti and Ennedi, over the border in Chad (Figure 2), may be as little as 5 m^3 per second.

While the Libyan project clearly illustrates the potential rewards of groundwater resource exploitation, it also highlights some of the complexities of the issue. It demonstrates, for example, the difficulties of resource assessment and raises the thorny question of alternative water supplies when the groundwater resources have been exhausted.

Recognizing these types of problem, this paper reviews some of the more important principles of groundwater development and management as they relate to the particular circumstances affecting drought prone regions of the world. It examines the difficulties of resource assessment and the concept of "safe" or "sustained" yield. It also reviews various approaches to resource management and in particular to the potential for resource augmentation by artificial recharge.

Table 1. Areal extent and groundwater storage capacity of the eleven major groundwater basins of North Africa and the Arabian Peninsula (after Burdon, 1977).

Groundwater Basins	Area km^2	Capacity km^3
SAHARA		
1. Chad	1,100,000	14,000
2. Niger	525,000	7,200
3. Tanezrouft	450,000	2,000
4. Western Erg	250,000	4,200
5. Eastern Erg	375,000	6,400
6. Fessan	450,000	4,800
7. Kufra	350,000	3,400
8. Western Desert	1,300,000	18,000
SAHARAN SUB-TOTAL	4,800,000	60,000
ARABIA		
9. Nafud	350,000	4,000
10. Riyadh	150,000	1,500
11. Rub al Khali	1,200,000	14,500
ARABIAN SUB-TOTAL	1,700,000	20,000
TOTAL for 11 Basins	6,500,000	80,000

2. Groundwater Development Using the Principle of Safe Yield

For generations, many groundwater scientists have wrestled with the complex and continuous issue of utilizing non-renewable groundwater resources. As recognized by Walton (1970), "There is a real problem in development and management in arid regions: should the water be extracted for maximum benefit of the present generation as minerals and other non-renewable resources are mined, or should the pumping be limited to the negligible quantity that can be supported perennially?".

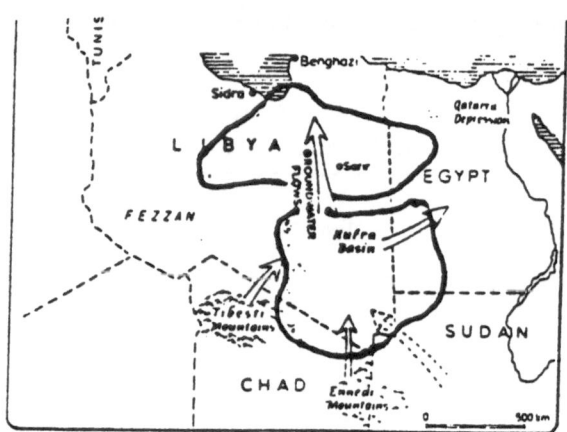

Figure 2. Groundwater sources for the Libyan "Great Man-Made River Project" (after Pearce, 1984).

2.1. DEFINING SAFE YIELD

Early attempts to define the amount of water that could be exploited from a groundwater reservoir invoked the term "safe yield". This term was originally used by Lee (1915) to describe the amount of water that could be pumped from an aquifer "regularly and permanently without dangerous depletion of the storage reserve". In 1923, Meinzer modified Lee's definition by expressing the term "dangerous depletion" more explicitly in terms of economic considerations. Meinzer defined safe yield as "the rate at which water can be withdrawn from an aquifer for human use without depleting the supply to the extent that withdrawal at this rate is no longer economically feasible", a definition which remains well entrenched in the hydrogeologic literature.

Since the early work of Lee and Meinzer, the term "safe yield" has undergone considerable evolution. Conkling (1946) expanded the definition, describing safe yield as an annual rate of extraction which i) does not exceed the average annual rate of recharge, ii) does not lower the water table so the permissible cost of pumping is exceeded, and iii) does not the water level so as to permit intrusion of water of undesirable quality. Banks (1953) added a fourth condition to provide protection for water rights. In 1961, the Committee on Groundwater of the American Society of Civil Engineers tried to remove some of the ambiguities of meaning that had developed by introducing and defining such terms as "maximum sustained yield", "permissive sustained yield", " maximum mining yield", and "permissive mining yield". However, these terms have failed to gain wide acceptance, and most groundwater scientists and engineers tend to favour the term safe yield, even though they define it in many various ways. Bouwer (1978), for example, describes safe yield as equal to the average replenishment rate of the aquifer, defining it as the rate at which groundwater can be withdrawn, without causing long-term decline of the water table or piezometric surface. Fetter (1988), on the other hand, proposes that safe yield be defined as "the amount of naturally occurring water that can be withdrawn from an aquifer on a sustained basis, economically, and legally, without impairing the native groundwater quality or creating an undesirable effect such as environmental damage".

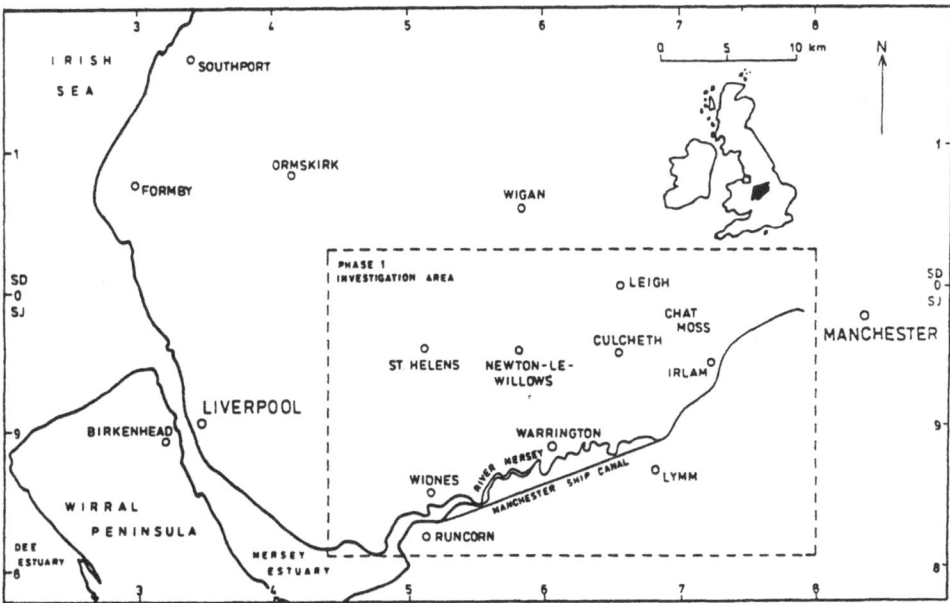

Figure 3. Location map of the Permo-Triassic Sandstone aquifer model study (Howard, 1987).

2.2. CRITICISMS OF THE SAFE YIELD APPROACH

Unfortunately, the debate over the definition of safe yield has diverted attention from a more fundamental question; that is, whether safe yield, according to any of these definitions, is necessarily an appropriate principle for determining groundwater management policy. In the first place, many will argue that its primary goal of protecting or "sustaining" the viability of the groundwater resource for all time fails to recognize that controlled long-term depletion of the resource, also known as groundwater mining, may be considerably more acceptable from a socioeconomic standpoint. The second problem with safe yield stems from the common but false assumption that "safe" and responsible use of water entails rates of production that do not exceed annual rates of aquifer replenishment and do not cause a long-term decline of water level. In the first case, it must be recognized that *any* extraction of groundwater is potentially undesirable since it will ultimately produce a depletion of natural outflows including spring discharge and baseflow contributions to rivers and streams. If annual groundwater production equals annual average recharge and the groundwater water levels are to be maintained, natural outflows would cease completely. Furthermore, the avoidance of long-term groundwater level decline is hardly commendable when this is achieved as a consequence of stimulating additional inflows from adjacent aquifers, lakes or even the sea.

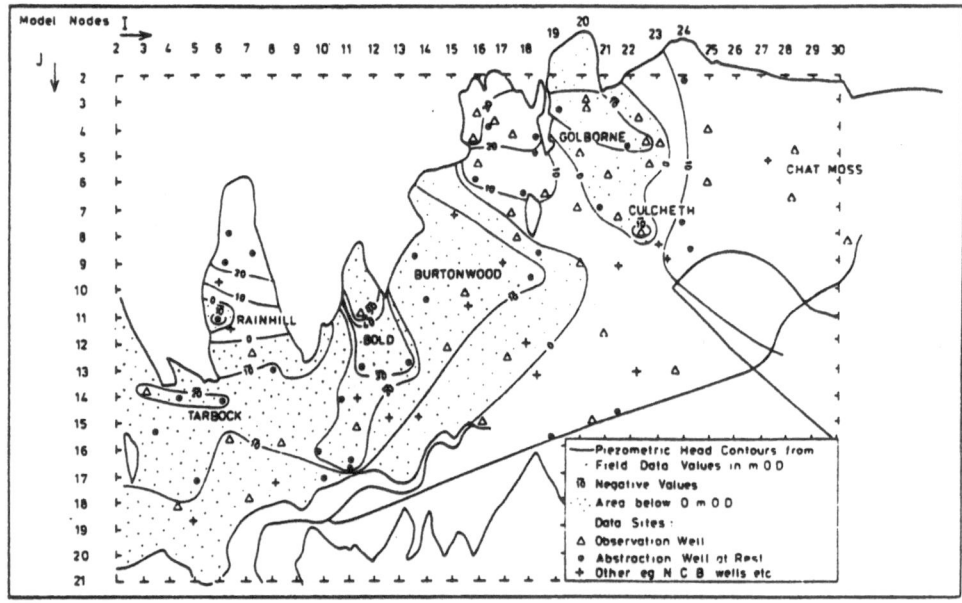

Figure 4. Potentiometric contours for the Permo-Triassic Sandstone aquifer in 1979, showing an extensive area with water levels considerably below sea level (0 m O.D.).

These and other criticisms of the safe yield principle are illustrated by a model study of the Permo-Triassic sandstone aquifer of North West England (University of Birmingham, 1981; Howard, 1987). For over 140 years this aquifer has been the major source of water for the region between Liverpool and Manchester (Figure 3). Production began in the mid-19th century and, despite a gradual long-term decrease in water level, was increased steadily until 1970 when it reached a maximum of 160×10^3 m³/d. Subsequently, production was decreased following an accelerating decline of ground-water levels (Bow et al., 1969) and a notable deterioration of water quality, particularly in wells bordering the coast. The imposed reductions in fresh ground-water production slowed the rate of decline of the potentiometric heads, but did little to improve water quality. By 1979 (Figure 4), ground-water levels throughout much of the area remained considerably below sea level.

In response to future management concerns, a comprehensive hydrogeological study of the area was undertaken which included development of a two-dimensional numerical flow model, designed to represent aquifer behaviour throughout its 140-year history (Howard, 1987). The primary purpose of the model was to allow various future management options for the aquifer to be evaluated. However, more interesting is the 140-year flow balance shown in Figure 5 which provides an incisive view of the aquifer's long-term behaviour and the highly interactive nature of the component flows and their response to pumping over time. In the Figure, six different quantities are plotted, the tow showing least variation appearing above the other four. This is done for ease of reading the diagram and has no other significance.

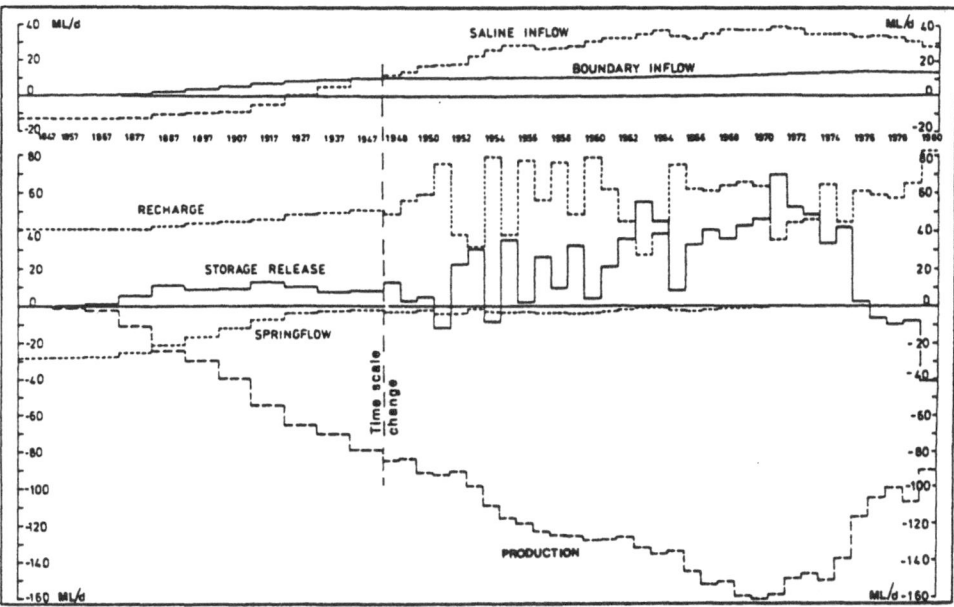

Figure 5. Model flow balance for the period 1847 to 1980.

The model water balance shows that, prior to development, 70% of the annual discharge of 40 x 10³ m³/d discharged naturally as springflow where it was able to supplement surface water supplies. The remaining 30% discharged to the estuary (represented as a negative saline inflow). By 1907, groundwater production had grown to 40 x 10³ m³/d (equal to the annual approximate recharge and thus according to Bouwer (1978) and others, equal to the aquifer's "safe yield") springflows had reduced to less than half of their original value and over 25% of the water was being drawn from storage, thus continuing a water level decline that had begun within the first ten years of pumping. Due to the distribution of the production wells at this time, the increased productivity did little to alleviate losses of fresh groundwater to the estuary which reduced from an original 11 x 10³ m³/d to 7 x 10^3 m³/d.

The behaviour of the aquifer changes significantly over the next 40 years (to 1947). This period sees production doubled to 80 x 10^3 m³/d (approximately double the annual rate of recharge), an increase that is met almost entirely by a cessation of discharge to the estuary followed by an inflow of seawater. Only minor changes to other flows occur during this period, the most important being an increase in recharge resulting from elevated vertical hydraulic gradients. There is also a slight increase in the volume of water induced across low permeability fault boundaries to the aquifer (boundary inflow in Figure 5). It is apparent that sea-water intrusion had become a crucial feature in the continuing ability of the aquifer to supply fresh water at production wells.

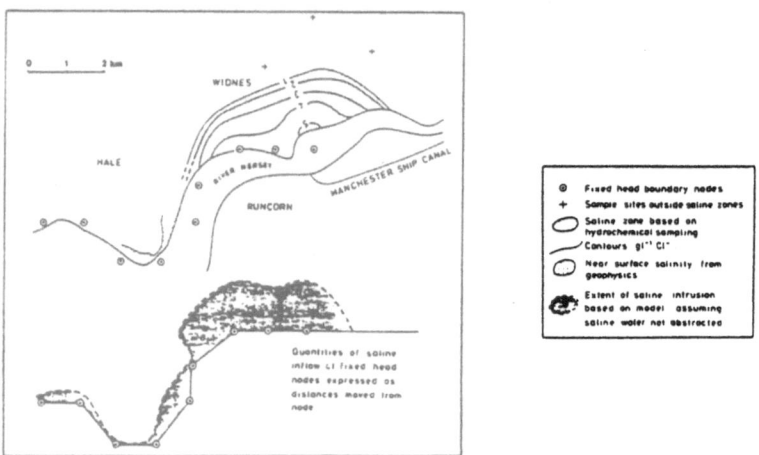

Figure 6. Comparison of ground-water salinity variations with model predictions of saline inflow.

The importance of the sea-water inflow component becomes even more evident during the period 1948 to 1978. By 1970, production of fresh, uncontaminated water had reached 160 x 10^3 m³/d, nearly three times the average rate of discharge. Over the same period saline inflows increased to 40 x 10^3 m³/d, even though as indicated by Figure 6, the *maximum* inland movement of saline water was no more than 2 km.

The Permo-Triassic aquifer model illustrates how aquifer management relies not only on recognition and quantification of the various contributory flow components, but more importantly on an understanding of the interdependence of one flow component on another and how this interdependence varies with time and rates of groundwater renewal. It is interesting to note that had the safe yield principle been invoked, whereby groundwater withdrawals would have been such as to avoid groundwater level decline, depletion of springflow or intrusion of sea water, production would have ceased during its early phase of development. It is speculated that the region owes its economic prosperity to overexploitation of the aquifer. It is this prosperity which is now able to make available to the region alternative fresh water supplies.

In fairness, it should be recognized that the groundwater management program adopted for the Permo-Triassic sandstone of north-west England was neither positively planned nor carefully evaluated prior to implementation. The development program simply evolved over history, with changes being made continuously to balance demand with the economic and environmental consequences of the declining water levels. It was not until the aquifer was thoroughly investigated in the late 1970's (University of Birmingham, 1981) that the merits of the development program were established.

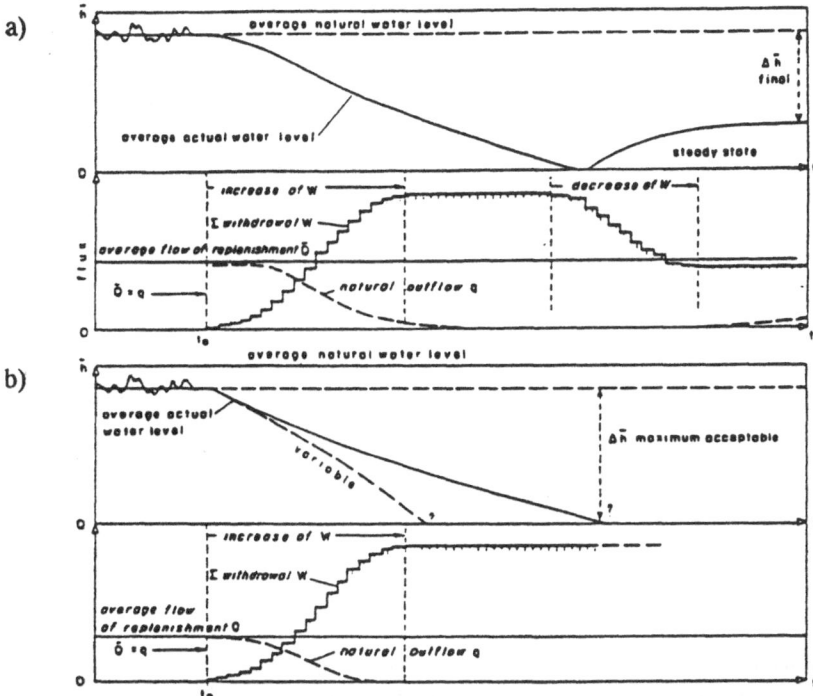

Figure 7. Aquifer response for groundwater mining by a) alternating phases of excessive withdrawal with phases of reduced production, and b) long-term, uninterrupted overexploitation. (After Margat and Saad, 1983).

Table 2. Examples of groundwater mining (after Margat and Saad, 1983).

Country	Aquifer	Average recharge (10^6 m³/year)	Total Volume withdrawn (10^6 m³)	Proportion of volume withdrawn corresponding to a storage depletion (%)	Maximum water level drop (m)	Time period
Australia	Great Artesian Basin	800 (initial)	35,100	70[a]	120 (in 83 years)	1880-1973
Algeria/Tunisia	Intercalary Continental of northern Sahara	270	1,360	85[b]	29 (in 14 years)	1950-1970
U.S.A., Arizona	Quaternary alluvium aquifers	370	225,000 annual 6600	90[b] 94	70 (in 60 years)	1920-1980
U.S.A., Texas	Ogallala High Plains	60	annual 5500 annual 9900	98.9 99.4	20 (1949-1960)	1968-1974
France	Lower Trias sandstones	50	6,000 to 700	50	44 (in 76 years)	1900-1975
	Lorraine		annual 164	70		1978
Spain	Limestone of Crevillente	10	annual 40 (present day)	75	200 (in 15 years)	1967-1982
Spain	Tenerife Island (Canaries)	310 of which 70 by irrigation	annual 210	57[b]	10 m/year (present day)	1980
Iran	Plain of Ghazvin	281	annual 377	31[b]	--	1965
Iran	Plain of Varamin	264 (with irrigation)	annual 252	12[b]	--	1973

[a]Ratio: decrease in storage/total volume withdrawn
[b]Taking into consideration a residual outflow from natural outlets

There are many other examples, where disregard for the basic principle of safe yield has achieved similar, if not greater socio-economic benefits (Table 2). While it is likely that the majority of these examples were born out of ignorance of the hydrogeologic system rather than a deliberate, carefully assessed venture to develop a finite resource (Mandel, 1979) they clearly represent an alternative approach to groundwater development which may be viable, at least in principle, in other situations.

3. Alternative Approaches to Aquifer Management

To many groundwater scientists and engineers, the safe yield concept continues to provide an easily implemented, relatively secure set of guidelines for resource development policy decisions. Once it can be accepted that safe yield, in its conventionally recognized form, is an outmoded concept and should simply be an option rather than a basic, driving principle, the scope for alternative approaches to aquifer management becomes limitless.

Among the strongest proponents for alternative approaches to groundwater resource management are Ambroggi (1966), Shata (1982), Margat and Saad (1983) and Issar (1985). They recognize that water has value only by virtue of use, and that the safe yield approach has little or no relevance in regions of the world underlain by vast reserves of non-renewable fossil groundwater. Instead, they propose alternative development schemes which are more in line with the "optimal yield" approach described by Domenico (1972). Unlike "safe yield", and aquifer's "optimal yield"

1) considers the hydrogeology of the entire basin and gives due regard for the importance of groundwater-surface interaction;

2) recognizes that the temporal and spatial distribution of groundwater production are as important considerations as the total yield produced;

3) is based primarily on socioeconomic considerations, considering *all* costs and returns associated with a particular use rate;

4) is not a fixed yield but may be varied to reflect anticipated changes in such factors as social objectives, economic growth and the cost of alternative water supplies.

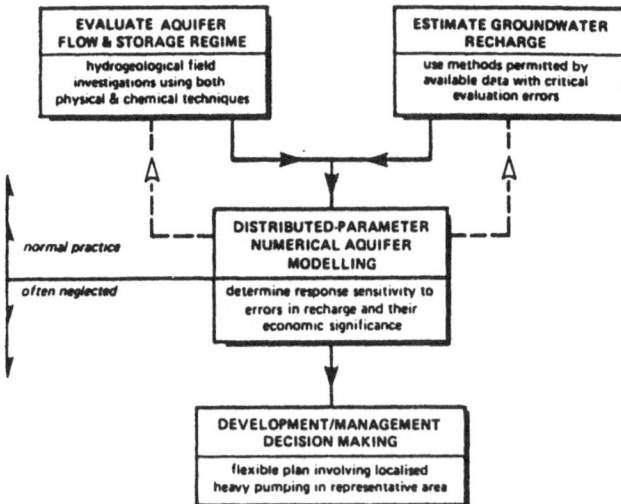

Figure 8. Organizational plan for evaluating groundwater management options (after Foster, 1988).

In considering concepts for utilization of essentially non-renewable groundwater resources in regional development. two basic strategies can be identified (Mandel and Shiftan, 1982; Margat and Saad, 1982). In the first (Figure 7a), phases of controlled pumping at rates considerably in excess of annual recharge, are alternated with phases of reduced production. This approach draws on the groundwater reserve, but minimizes the long-term impact on groundwater levels, and creates temporary storage which can be utilized by artificial recharge during infrequent storm events. The short-term excessive depletion of water levels may also maximize aquifer inflows by increasing recharge through enhanced vertical hydraulic gradients (Howard, 1987) or by inducing the release of water from aquitards.

The second basic strategy is more straightforward in that it involves planned depletion of groundwater storage through long-term, uninterrupted overexploitation. As shown in Figure 7b, this approach will ultimately cause a permanent loss of natural outflow and provides for a continuous decline of water level. Overexploitation or "mining" can be an important tool for facilitating economic growth and allowing postponement of investment in dams, long distance transfers and desalination plants etc. (Margat and Saad, 1983). It can prove to be a particularly sound approach if positively planned and realistically evaluated in advance, if close control over groundwater abstraction is exercised and if there is a clear and feasible plan for alternative water supplies when the groundwater resources are exhausted.

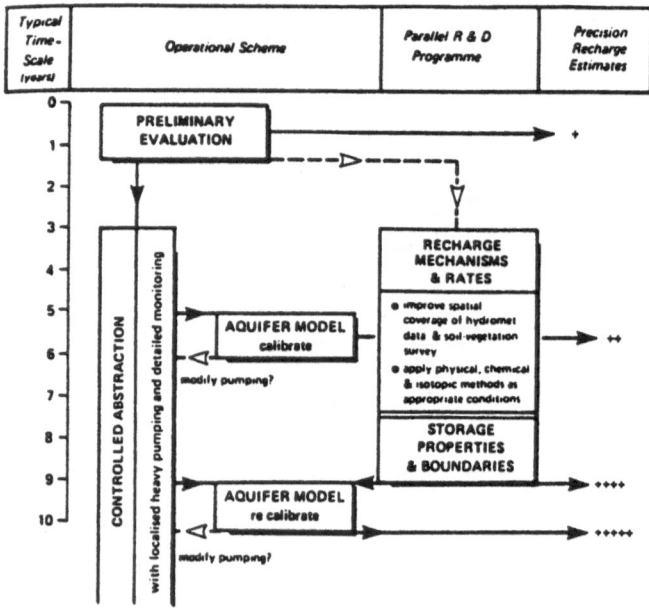

Figure 9. Operational approach to the development of groundwater management policy (after Foster, 1988).

A third possible management strategy recognizes the interdependency between the ground and surface water resources of a basin and optimizes their development through conjunctive use (Paling, 1984). While this type of approach is strongly recommended where appropriate, its application is often limited in arid and drought prone areas where surface water resources are infrequent and unreliable.

4. Practical Considerations

4.1. SELECTION OF MANAGEMENT STRATEGY

In most studies, the appropriate management strategy is normally selected with the aid of a numerical computer model. This model allows the historical behaviour of the groundwater flow regime to be simulated. Subsequently it can be used to predict the likely response of the aquifer under various conditions of recharge and production. By incorporating economic factors into the model, the cost associated with possible alternative management scenarios can be evaluated.

Theoretically, it should be possible to use this type of model in conjunction with linear and nonlinear programming to determine the optimal management strategy (Li et al., 1987). However, experience over the past 20 years has shown that the rigorous application of optimization techniques to hydrologic basins is an impracticable task. As explained by Kazmann (1979) the numerous variables combined with the difficulty of predicting changes in contributory factors such as the cost of oil, developments in technology and the effects of political unrest tend to make the result of optimizing techniques obsolete, well before they can be implemented.

The problem of selecting an appropriate aquifer management strategy is particularly difficult in arid environments where reliable hydrogeologic data are rarely available (Lloyd, 1986) and groundwater recharge estimates are subject to considerable uncertainty and large error. Given these constraints Foster (1988) recommends that the aquifer management strategy be kept flexible, thus allowing refinements to be made during implementation as more data become available (Figure 8). This "operational approach" to management policy selection might include

1) installation of borehole pumps and pipeline distribution systems suitable for operation over wide ranges of pressure and flow rate;
2) conservative spacing of production wells to avoid risk of well interference and yet permit possible construction of additional boreholes at intermediate locations, should the need arise;
3) staged development with localized heavy production and intensive monitoring to provide calibration data for the numerical model;
4) continually upgraded resource evaluation and economic assessment.

Figure 10 gives an example of staged groundwater development for the southern desert of Jordan (Howard Humphries Ltd., 1982; Lloyd, 1986). It illustrates how predictions

508

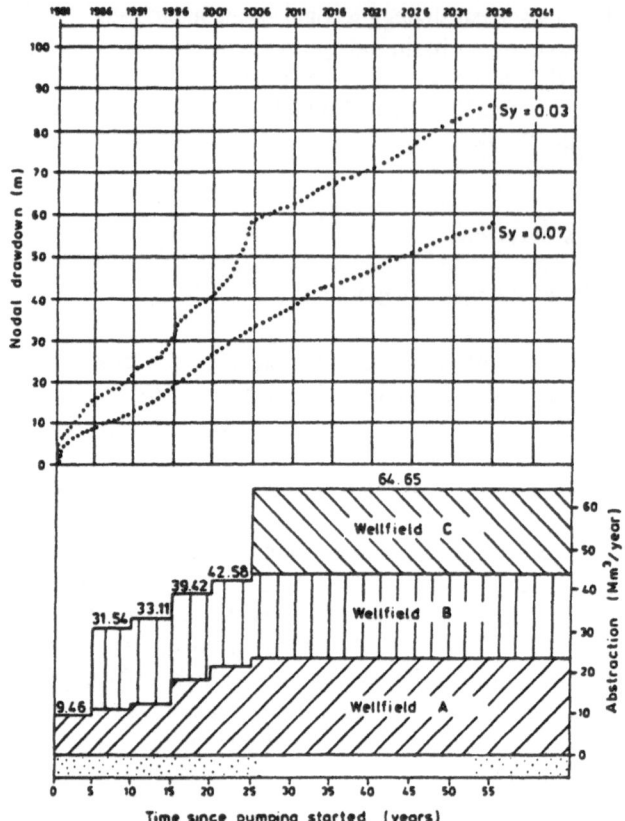

Figure 10. An example of staged groundwater development as proposed for the southern desert of Jordan (Howard Humphries Ltd., 1982, after Lloyd, 1986).

of the ling-term decline in potentiometric head depend on an ability to obtain a reliable estimate of specific yield (Sy).

4.2. GROUNDWATER MINING - PLANNING FOR THE FUTURE

While groundwater mining through planned and controlled overdevelopment can provide an effective option for the management of groundwater resources, it carries with it the responsibility for devising a plan to deal with the period immediately prior to and following total exhaustion of the groundwater resource. This plan may include a phased reduction of extraction to a level commensurate with steady state equilibrium, or it may require complete cessation of production. More usually, however, the post-development plan will provide for phased augmentation of the groundwater supply with surface water imports.

For example, in the western U.S., the Central Arizona Project (CAP) plans to augment dwindling groundwater supplies for Tucson and Phoenix by diverting nearly 2×10^9 m^3 (2 km^3) of water per year from the Colorado River and raise it by over 800 m along a series of canals, 540 km long. Similar surface water transfers are under development in California. In Egypt, where the development of the "New Valley" in the western Desert relies on the mining of fossil reserves in the Nubian Sandstone, post-development plans include the possibility of conveying water from the Aswan reservoir on the Nile.

In many respects, the planning and costing of possible post-development schemes deserve similar consideration to that devoted to the development phase itself. This is particularly true when the scheme includes provision for some form of artificial recharge. In recent years, artificial recharge of aquifers for storage and groundwater supply augmentation has generated considerable interest (Jacenkow, 1984; Asano, 1986; Li et al., 1987). Compared to conventional storage systems, located at or above ground level, aquifers provide large volume storage at low cost, afford reasonable protection from pollution, and minimally impact upon land-use. Aquifers are also less prone to yield depletion through drought and, under favourable conditions, can provide inexpensive distribution systems. In arid environments, the sub-surface approach to water storage holds particular appeal since it eliminates evaporative losses of between 1000 and 3000 mm per annum, and is unaffected by silting, a problem which often causes a drastic reduction in the storage capacities of surface reservoirs.

In practice, efficient and cost-effective use of artificial recharge can be seriously infringed by a series of hydrogeologic constraints, all of which require careful evaluation. Four are particularly important.

4.2.1. *Infiltration Capacity.* There is little value in accessing large volumes of surface water for purposes of artificial recharge if this water cannot be efficiently transferred to the sub-surface reservoir. Unconfined aquifers comprising coarse grained, unconsolidated sediments represent the ideal sub-surface reservoirs since they provide high storage and are readily recharged by water spreading, recharge basins and irrigation techniques. Experience has shown, however, that infiltration rates can decline significantly with time as a result of clogging by the fine grained sediments and microbial growths. Air entrapment between the wetting front and water table, and the swelling of clay colloids in the aquifer matrix also retard recharge rates. When confined aquifers are recharged, injection wells must be used. Injection wells require frequent maintenance since they are particularly prone to clogging.

4.2.2. *Water Chemistry.* It is important that the chemistry of the recharged water be compatible with the native groundwater. The first concern is the risk of water quality deterioration through cross-contamination. However, often more important is the possibility of geochemical reactions which can reduce intergranular permeability through the production or precipitation of certain mineral species. Typical problems include the precipitation of dissolved iron by oxygenation, and the swelling of montmorillonite clays through Na^+ - Ca^{++} exchange.

4.2.3. *Regional Water Balance.* A common mistake is to assume that water diverted for the purpose of artificial recharge would otherwise have gone to waste. It should be recognized that wet season run-off may be an important source of water for local rivers; it can also be an extremely important source of indirect recharge for aquifers, particularly in arid areas (Lloyd, 1986). Also, some studies have shown that reduction of evaporation can lead to lower rainfall, in which case artificial recharge of surface water may have implications on future resource availability.

4.2.4. *Aquifer Water Balance.* To be worthwhile, hydrogeologic conditions must be such that most of the artificially recharged water can be intercepted and recovered for use. If augmented recharge simply leads to increased discharge at springs and increased losses by deep percolation to underlying aquifer systems, few socioeconomic benefits will be derived.

It seems clear that while artificial recharge and sub-surface water storage provide an attractive means of solving some water supply problems, each application demands critical evaluation. It is essential that the aquifer system be clearly defined, that hydrogeologic parameters of storativity and transmissivity be determined, and that all inflows and outflows to and from the system be quantified. It is only with these data, incorporated into a transient flow numerical model, that it becomes possible to reliably evaluate the feasibility of the scheme.

5. Conclusions

One third of North Africa and the Arabian Peninsula is underlain by vast reservoirs of fossil groundwater. Developed and managed responsibly, these resources provide the key to the mitigation of drought and the alleviation of famine. They can also provide a catalyst for industrial development and economic prosperity.

Unfortunately, recognizing that the fossil waters are finite and essentially non-renewable, the development of an appropriate management strategy represents a difficult challenge. First and foremost, this strategy must reject the traditional and conservative "safe yield" approach to groundwater management, and recognize groundwater mining through planned and controlled development as a viable alternative. Secondly, the strategy must incorporate an appropriate long-term plan for provision of alternative water supplies when the groundwater resources are exhausted.

Theoretically, it should be possible to select an optimal management strategy by using a numerical aquifer flow model and applying routine optimization techniques. Experience has shown, however, that an inability to predict the long-term behaviour of important variables makes rigorous application of optimization procedures a meaningless task. In practice, the ideal course of action os to implement a flexible management program which can be refined as development proceeds and data on the aquifer's response to pumping become available. This "operational approach" to management policy development would likely involve staged resource development, intensive monitoring and a continual upgrading of resource and economic assessments.

6. References

Ambroggi, R.P. (1966) "Water under the Sahara", Sci. Am. 214, pp. 21-29

Asano, T. (1986) *Artificial Recharge of Groundwater*, Butterworth Publishers, 767 p.

Bakiewicz, W., Milne, D.M., & Noori (1982) "Hydrogeology of the Umm Er Radhuma Aquifer, Saudi Arabia, with reference to fossil gradients", Quaternary Journal of Engineering Geology, 15, pp. 105-126

Banks, H.O. (1953) "Utilization of underground storage reservoirs", Transactions, American Society of Civil Engineers, 118, pp. 220-234

Bouwer, H. (1978) *Groundwater Hydrology*, McGraw-Hill Inc., 480 p.

Bow, C.J., Howell, F.T., Payne, C.J., & Thompson, P.J. (1969) "The lowering of the water table in the Permo-Triassic rocks of south Lancashire", Water and Water Eng. 74, pp. 461-463

Burdon, D.J. (1977) "Flow of fossil groundwater", Quarterly Journal of Engineering Geology, 10, pp. 97-124

Burdon, D.J. (1982) "Hydrological conditions in the middle east", Quarterly Journal of Engineering Geology 15, pp. 71-82

Conkling, H. (1946) "Utilization of ground-water storage in stream system development", Transactions, American Society of Civil Engineers, 3, pp. 275-305

Domenico, P.A. (1972) *Concept and Models in Groundwater Hydrology*, McGraw-Hill

Edmunds, W.M., & Wright, E.P. (1979) "Groundwater recharge and paleoclimate in the Kufra and Sirte basins, Libya", Journal of Hydrology, 40, pp. 215-241

Fetter, C.W. (1988) *Applied Hydrogeology*, Merrill Publishing Company, 2nd ed., 592 p.

Foster, S.S.D. (1988) "Quantification of groundwater recharge in arid regions: A practical view for resource development and management.", in Simmers, I. (ed.) *Estimation of Natural Groundwater Recharge*, D. Reidel Publishing Company, pp. 323-338

Howard Humphries Ltd. (1982) *Modelling of the Disi Sandstone Aquifer, Southern Desert of Jordan*, Report of the Water Supply Corporation, Hashemite Kingdom of Jordan, 44 p.

Howard, K.W.F. (1987) "Beneficial aspects of sea water intrusions", Ground Water, 25, pp. 398-399

Issar, A. (1985) "Fossil water under the Sinai-Negev Peninsula", Scientific American, 243, pp. 104-128

Jacenkow, O.B. (1984) "artificial recharge of groundwater resources in semi-arid regions", presented at *The Proceedings of the Harare Symposium*, IAHS Publ. No. 144, pp. 111-119

Kazmann, R.G. (1979) "Groundwater: new directions", J. Hydrol. 43, pp. 555-569

Lee, C.H. (1915) "The determination of safe yield of underground reservoirs of closed basin type", Transactions, American Society of Civil Engineers, 78, pp. 148-151

Li, C., Bahr, J.M., Reichard, E.G., Butler, J.J.Jr., & Remson, I. (1987) "Optimal siting of artificial recharge: an analysis of the objective functions", Ground Water, 25, pp. 141-150

Lloyd, J.W., & Farag, H.M. (1978) "Fossil ground-water gradients in arid regional

512

sedimentary basins", Ground Water, 16, pp 388-393

Lloyd, J.W., & Miles, J.C. (1986) "An examination of the mechanisms controlling groundwater gradients in hyper-arid regional sedimentary basins", Water Resources Association, 22, pp. 471-478

Mandel, S. (1979) "Problems of large-scale groundwater development", J. Hydrol., 43, pp. 439-443

Mandel, S., & Shiftan, Z. (1981) *Groundwater Resources*, Academic Press, Inc., 269 p.

Margat J., & Saad, K.F. (1983) "Concepts for the utilization of non-renewable groundwater resources in regional development", presented at The United Nations Natural Resources Forum

Meinzer, O.E. (1923) *Outline of Ground-Water Hydrology, with Definitions*, U.S. Geological Survey Water-Supply Paper 494

Paling, W.A.J. (1984) "Optimization of conjunctive use of groundwater and surface water in the Vaal basin", presented at *The Proceedings of the Harare Symposium*, IAHS Publ. No. 144, pp. 121-128

Pearce (1984) "Why Gadaffi's well may run dry", New Scientist, 4

Shata, A.A. (1982) "Hydrogeology of the great Nubian Sandstone basin, Egypt", Quarterly Journal of Engineering Geology, 15, pp. 127-133

University of Birmingham (1981) *Saline Groundwater Study, Phase I, Lower Mersey Basin*, Final Report to the North West Water Authority, 81 p.

Walton, W.C. (1972) *Groundwater Resource Evaluation*, McGraw-Hill, pp. 606-607

Wright, E.P., Benfield, A.C., Edmunds, W.H., & Kitching, R. (1982) "Hydrogeology of the Kufra and Sirte Basins, eastern Libya", Quarterly Journal of Engineering Geology, 15, pp. 83-103

SMALL DAMS AND SUBSURFACE DAMS IN ARID AND SEMI-ARID AREAS: THE EXAMPLE OF SAHARAN AND SAHELIAN AFRICA

R. GUIRAUD
Department of Geology
Faculty of Science
33 Rue Louis Pasteur
84000 Avignon
France

ABSTRACT: The principle of using small dams, sand dams and subsurface dams are described along with their advantages for improving water resources and quality in Saharan and Sahelian Africa. Examples are provided. Multiplication of these types of dams worldwide would help mitigate drought and decrease the amount of water flowing to the oceans.

1. Introduction

The hydroclimatic framework of arid and semi-arid areas is characterized by:
- low precipitation, concentrated in short periods;
- periodical river floods lasting for some hours, days or at times weeks;
- a high evaporation rate.

In this context it is important to try to retain as much water as possible in valleys, on the surface or, better still, in alluvial deposits.

Three main concepts of damming can help us to attain our objective:
- surface dams;
- sand dams, or sandy flats or reaches;
- subsurface dams.

Sometimes it is possible to combine two kinds of dam. The choice of the type of dam is controlled by the geological, hydrological and topographical conditions prevailing in the area. This paper will only concern small-scale rural projects, which can be realized frequently and very easily, with few resources and with local labour.

513

R. Paepe et al. (eds.), Greenhouse Effect, Sea Level and Drought, 513–522.
© 1990 *Kluwer Academic Publishers.*

514

2. Small Surface Dams

Small surface dams are very common, all over the world. In Sahelian Africa thousands of them have been built:
- in Burkina-Faso (formerly Upper Volta), where about one thousand have been built during the last 20 years;
- in the Canary Islands, and in particular in Fuerteventura where about two thousand small earth dams were built during the last four years.

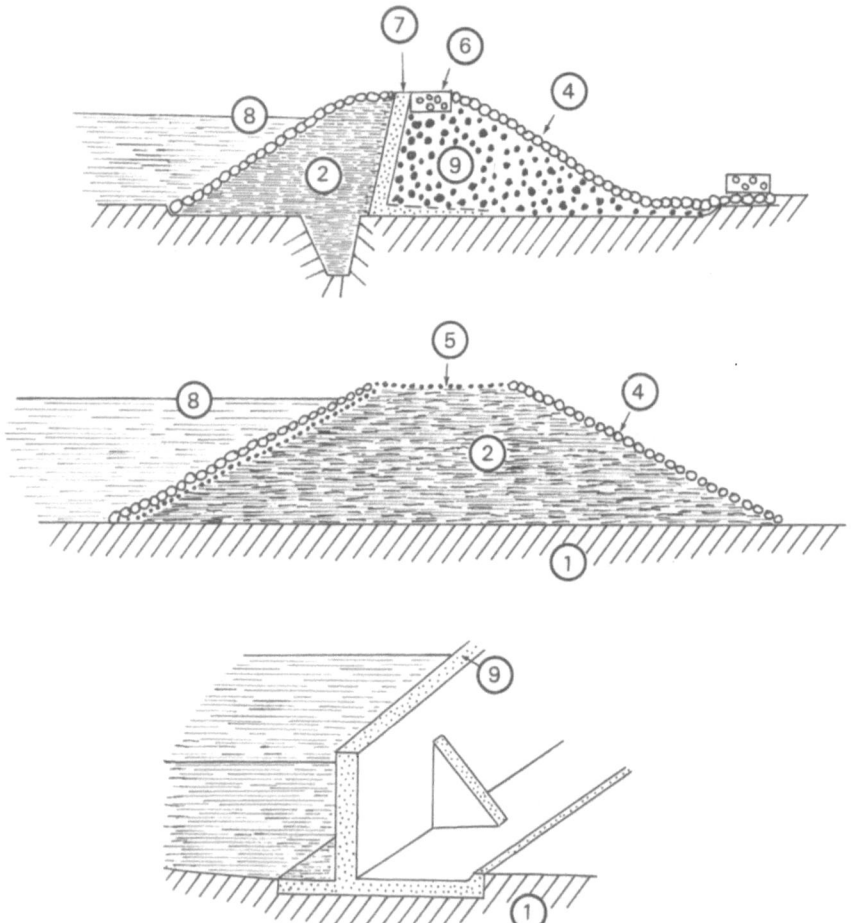

Figure 1. Examples of small surface dams: 1, bedrock; 2, compacted earth (clay dominant); 3, boulders; 4, blocks; 5, gravel; 6, gabion; 7, filter-drain; 8, high water level; 9, concrete.

The sites ought to have, as far as possible, impervious basement but if such is not the case, water will seep and can be exploited downstream of the dam, by a well.

The dams can be built with different techniques and materials: earth, boulders, blocks, gabions, concrete, plastic veils. Figure 1 shows some examples. General experience has shown that it is to be recommended that a dam should be built to be overflowed. Cost can be very low, for instance when a bulldozer can be used.

Finally, we note here some water resource developments which exhibit more or less the same type as surface dams:
- terraces, commonly built with blocks and boulders in mountainous regions (Canary Islands, Cape Verde Archipelago, North Africa, northern Cameroon, etc.));
- "gabias", a sort of semi-circular dam constructed in the Canary Islands in order to divert a part of the high water floods and thereby retain both water and sediments.

3. Sand Dams

3.1. HYDROLOGICAL AND GEOLOGICAL FRAMEWORK

The use of such dams has been well described by Classen (1980), following studies on the basement areas of eastern and southern Africa.

In regions where the dominant rocks are granites or gneisses, erosion produces large quantities of sand in addition to silt and clay. Under conditions of high intensity rain storms, usually of short duration, the volume and velocity of runoff are high and have the power to move the eroded materials rapidly into the valleys. The finer particles of clay and silt are carried further downstream, leaving clean sand behind. Sand, being the heaviest, is the first material to settle as soon as floods recede. The sand deposits occur mainly upstream of barriers of unweathered and non-eroded rocks, thereby forming sand reaches (Fig. 2a).

After the surface flood has stopped, subsurface water flow continues through the sand. A rock barrier across a river, acting as a dam, will bring the water to the surface for a short distance after which it will once more disappear into the sand downstream.

Such sites are often encountered in the Precambrian regions of Africa, South America and India. In Sahelian Africa many examples are found, for instance in Burkina Faso, western Niger, northern Cameroon and south-western Chad.

3.2. CONSTRUCTION OF ARTIFICIAL SAND DAMS

Small dams can be constructed in order to increase the volume of a sandy aquifer: preferably in association with a rock barrier (Fig. 2b) but also feasible without a rock barrier.

As sand transported by a flood is abundant, the reservoir is susceptible to rapid filling by sand, perhaps even after the first rainy season if the dam is not high enough. Therefore the dam must be capable of resisting the passage of flood water over its crest.

That is why it should be built, as far as possible, with resistant materials. Appropriate techniques can employ:

- thick masonry walls;
- concrete walls;
- gabions;
- plastic bags filled with clay, earth of sand.

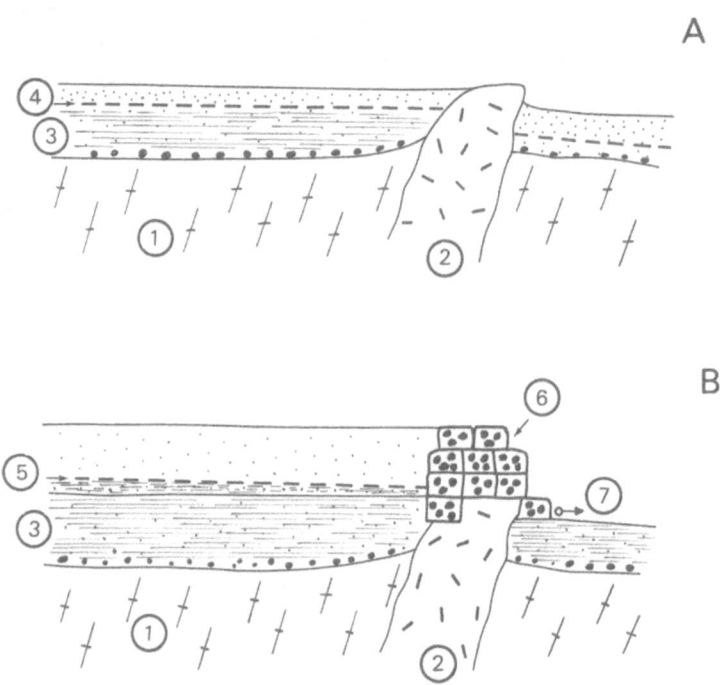

Figure 2. A, sand reach; B, sand dam. 1, tight bedrock; 2, rock barrier; 3, underflow; 4, original water table; 5, raised water table; 6, gabions; 7, spring.

The main advantage of the last two - gabions and bags - is that the dam can be raised as the reservoir is filled by sand. A watertight veil can also be made immediately upstream of the dam, with clay or PVC.

3.3. SOME EXAMPLES

Examples of sand dams are known in northern Cameroon (in the Mandara Mountains), in south-western Chad, in north-eastern Somalia and in Botswana.

4. Subsurface Dams

4.1. THE PRINCIPLE

The aim of such dams consists of raising the water table of an alluvial aquifer. In fact, one often notices the presence of subsurface flows in alluvial valleys in arid and semi-arid areas; these subsurface flows are evidenced by the existence of springs or swamps, and/or trees. However, as the water table of the alluvial aquifer tends to fall appreciably by the end of the dry season, it is important to make an effort to raise it or, at least, to stabilize it in order to facilitate rural water development.

To achieve this objective, one could construct a subsurface dam. Figure 3 illustrates the principle of such dams.

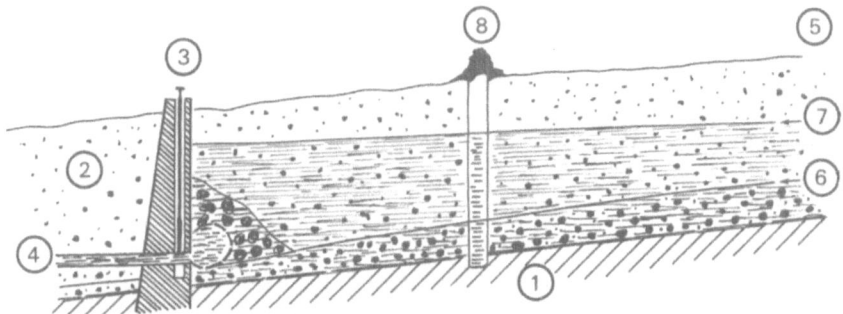

Figure 3. Cross-section showing the principle of a subsurface dam. 1, tight bedrock;, 2, alluvial deposits; 3, dam; 4, optional water-catchment; 5, sense of flow; 6, initial water table; 7, raised water table; 8, perennial well.

4.2. THE GEOLOGICAL FRAMEWORK

Potential sites must exhibit an alluvial aquifer and an impermeable bedrock. Such a situation prevails in many regions.

Alluvium can be found almost everywhere. This is linked with the lowering of the sea level which happened about 16,000 to 20,000 years B.P., associated with the last main glaciation. The sea level previously reached approximately -120 m, and the rivers were overdeepened. Then, the sea level rose to its present level: this was accompanied by large alluvial infilling of rivers valleys.

Tight bedrock can arise from different kinds of rock:
- clays, marls or shales in sedimentary basins;
- granito-gneissic basement;
- and, sometimes, volcanic or volcano-sedimentary rocks.

518

E

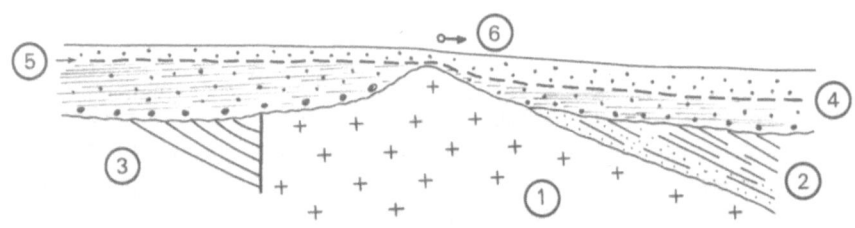

Figure 4. The natural subsurface dam of Wadi Nakheil, 2 km west from Quseir (Egypt). 1, Precambrian basement; 2, Cretaceous; 3, Tertiary; 4, alluvial deposits; 5, water table; 6, springs.

Natural subsurface dams are sometimes encountered, linked to the presence of rocky thresholds or sills. Here, one can observe natural springs or ponds located upstream or downstream from the threshold. Many examples have been described in northern Cameroon (Tillement, 1971), in eastern Chad (Schneider, 1989) and in south-eastern Egypt (Guiraud, 1987). Figure 4 represents the example of Wadi Nakheil, in the Egyptian eastern Desert.

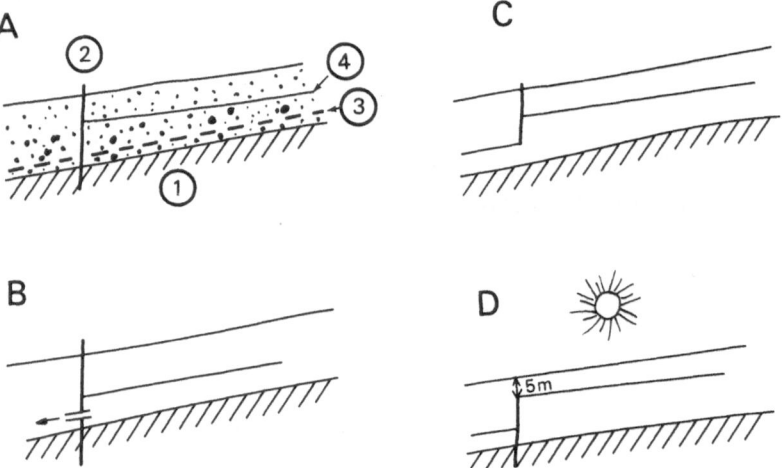

Figure 5. Different types of subsurface dam. 1, tight bedrock; 2, dam; 3, initial water table; 4, raised water table.

4.3. CONSTRUCTION OF SUBSURFACE DAMS

Various techniques can be considered, including:
- ditches filled with compacted clay;
- ditches covered by a watertight plastic veil and filled with compacted clay;
- ditches and built walls;
- interlinking piles;
- watertight screens obtained by drilling and cementations;
- bins;
- etc.

There are numerous options related to the local conditions and the means at one's disposal. Some dams can be very cheap when it is possible to use a bulldozer and/or local labour.

According to the hydrogeological and hydroclimatic framework - discharge of the aquifer, evaporation rate, etc. - different types of subsurface dam can be considered (Fig. 5).

4.4. THE ADVANTAGES OF SUBSURFACE DAMS

Advantages include:
- constitution of a groundwater reserve, which may be important in some cases;
- raising of the water table;
- formation of perennial water points and possibility of permanent settlement;
- elimination of surface water loss due to evaporation with a parallel reduction of salt concentration (where the dam does not reach the surface in arid and hot areas);
- great reduction of human and animal pollution;
- the cost of construction can be low, with the possibility of using local labour;
- maintenance cost is minimal;
- definitive type of work (generally);
- recovery of part of flood water, if the dam is surmounted by a threshold.

4.5. SOME EXAMPLES

Examples of subsurface dams can be found in several countries from Sahelian or Saharan Africa:
- Cape Verde Archipelago, where some hundreds of dams have been built during the past ten years (Fig. 6);
- south-western Mauritania;
- Niger, west and south of the Aïr Mountains;
- north-eastern Sudan;
- northern Somalia;
- Kenya;
- Botswana, in the arid zone of southern Africa.

Some older constructions are also known in Morocco (Sous Valley), Algeria (Laghouat), Tunisia (Kairouan), southern France, etc.

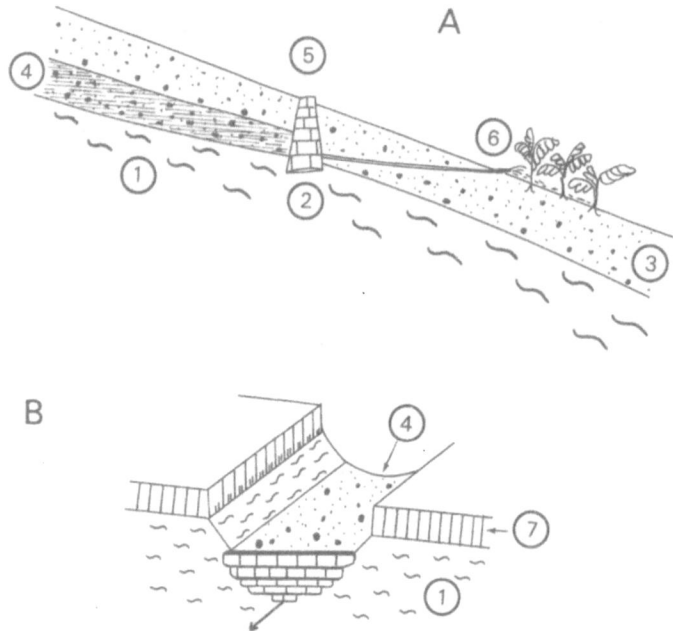

Figure 6. Example of subsurface dam built in the Cape Verde Archipelago (after Andreini and Bourguet, 1984). A, cross-section; B, bloc diagram.

(1, tight volcanic bedrock; 2, natural threshold; 3, alluvial deposits; 4, aquifer; 5, dam; 6, irrigation; 7, basalts).

5.   Conclusions

In Sahelian Africa, the geological and hydrogeological conditions often favour the construction of small-scale surface and subsurface dams. This is, sometimes, also the case in the Sahara. Regions with mountainous reliefs, or with immediate surroundings in which the hydrographic network functions for few days or weeks in a year are the most suitable; there the possibilities for aquifer recharge exist. The coarse alluvium, steep slopes and gorges in the valleys commonly associated with these regions give rise to ideal conditions for subsurface dam site locations.

The following are examples of these favourable regions: the archipelagos of Cape Verde and the Canaries, the south of Mauritania and the surrounding regions of Mali, some valleys in the Hoggar, the Andrar Iforas, Aïr and Tibesti, the Damagaram, northern Cameroon and southern Chad, the Ouaddai and the Darfur, the Red Sea hills, the regions close to the Gulf of Aden and some valleys in the east African Rift. Elsewhere, especially in the regions where sedimentary basins have clayey-marl substrata, suitable sites could also be found locally after a rapid hydrogeologic investigation.

Mixed surface, and subsurface dams would often permit more rapid recharge during high waters and reduce losses due to evaporation in flood-plains.

In summary, taking into account both requirement and potential, it appears urgent to launch a campaign for the construction of small dams in all arid and semi-arid regions and more particularly in the Sahel. It would then be much easier to maintain rural settlement. Besides, multiplication of these types of dam worldwide would help to decrease the amount of water flowing to the oceans, which would contribute to the stabilization of sea level.

6.   References

Andreini, J.C., & Bourguet, L. (1984) "Les barrages souterrains. Mise en oeuvre au Cap Vert. Conditions d'applications au Sahel", Hydrogéol.-Géol. de l'Ingénieur 4, pp. 393-399

B.C.E.O.M. (1978) *Les Barrages Souterrains*, Minist. Fr. Coop., Paris

Biswas, A.K. (1984) "Climate and water resources", in *Climate and Development*, Tycooly Intern. Publ., Ltd., Dublin, Natural Res. and the Environm., Ser. 13, pp. 96-116

Classen, C.A. (1980) *Small Scale Rural Water Supplies*, Report Min. Agric., Gov. Botswana, Gaborone

Cone, L.W., & Lipscomb, J.F. (1972) *The History of Kenya Agriculture*, Univ. Press of Africa, Nairobi

Faillace, C. (1987) "Results of a country-wide groundwater quality study in Somalia", Geosom 87, Somali Nat. Univ., Mogadishu, Abstracts.

Gautier, M. (1952) "Le Barrage de Tatjmout", in *La Géologie et les Problèmes de l'Eau en Algérie I*, 19e Congr. Géol. Int., Alger 1952, pp. 270-275

Guiraud, R. (1981) "Données Géologiques et hydrogéologiques en faveur de la réalisation de barrages d'inféroflux dans le Sahel", Bull. Liaison C.I.E.H., Ouagadougou, 45, pp. 16-19

Guiraud, R. (1987) "Les barrage d'inféroflux: l'intérêt de la réalisation de tels ouvrages dans le bassin du Nil et dans les régions voisines", Int. Symp. Nile Basin, Cairo, 1-7 March 1987, Comm. 53

Guiraud, R. (1988) "L'hydrogéologie de l'Afrique", J. African Earth Sci., 7, 3, pp. 519-543

Guiraud, R. (1989) "Les barrages d'inféroflux. Leur intérêt pour l'Afrique saharienne et sahélienne", in Demissie, M, and Stout, G.E. (eds.) *The Sahel Forum on the State-of-the-Art of Hydrology and Hydrogeology in the Arid and Semi-Arid Areas of Africa*, Ouagadougou, Burkina-Faso, in press

Joseph, A., & Diluca, C. (1986) "Les barrages souterrains en zone sahélienne. Présentation d'un projet d'aménagement de barrage souterrain dans l'Aïr au Niger", Bull. Liaison C.I.E.H., Ouagadougou 66, pp. 27-36

Margat, J. (1985) "Hydrologie et ressources en eau des zones arides", Bull. Soc. Géol. France 7, pp. 1009-1020

Rodier, J.A. (1964) "Régimes hydrologiques de l'Afrique noire à l'Ouest du Congo",

522

ORSTOM, Paris

Rodier, J.A. (1975) "Evaluation de l'écoulement annuel dans le Sahel tropical africain", Trav. et Doc. ORSTOM 46

Rodier, J.A. (1982) "Evaluation of annual runoff in tropical African Sahel", Trav. et Doc. ORSTOM 145

Sandford, K.S. (1935) "Sources of water in the north-western Sudan", The Geographical Journal 85, 5, pp. 412-431

Seguis, L. (1986) *Recherche, pour le Sahel, d'une Fonction de Production Journalière (Lame Précipitée - Lame Ecoulée) et sa Régionalisation*, Thesis U.S.T.L. Montpellier

Tillement, B. (1971) "Hydrogéologie du Nord-Cameroun", Bull. Dir. Mines Géol. Cameroun, 6

Schneider, J.L. (1978) *Géologie et Hydrogéologie de la République du Chad*, Thesis Avignon

UNESCO (1978) *World Water Balance Resources of the Earth*, Studies and Reports in Hydrology, 25, Paris

STUDY OF THE CHAD BASIN WATER SUPPLY BY CATCHMENT CARRIED OUT IN THE CONGO-ZAIRE BASIN

G. MOGUEDET
Labo. Géologie, 2 B^d Lavoiseur
49045 Angers Cd, France
P. GIRESSE
Labo. Sédimento, A^e Villeneuve
66000 Perpignan, France
and
KINGA-MOUZEO
Labo. Géologie, BP 69, Université de Brazzaville
Congo

ABSTRACT: The Chad basin is submitted to a persistent drought. Some think that one means of countering this problem would be to catch the Congo-Zaire basin waters, which are often in excess. The characteristics of the relevant rivers, Congo, Oubangui and Chari, are presented and a 1200 km long route is proposed from the catchment area on the Congo River as far as Lake Chad. Some consequences are discussed.

1. Introduction

The Chad basin constitutes a large endorheic system of 1.5 million km^2 located in the middle of the African continent, between 5 and 25° N, and 7 and 25° E (Fig. 1). Lake chad, which occupies a small part of this basin, has known varying fortunes during past millennia (Servant-Vildary, 1979; Maley, 1977, 1981; Rognon, 1979; Servant, 1983). Over that period this exclusive continental environment has experienced wide climatic variations and the lake's surface has strongly fluctuated.

Submitted to a persistent pluviometric deficit since the end of the 1960's, the area is marked by drought effects so that Lake Chad's surface is diminishing in a disquieting way. Following a maximum of 25,000 km^2 in 1963 this surface has since been in continuous recession. In 1984 it covered only 5,000 km^2, reduced by a further half by 1986.

To fight against this phenomenon we have planned to capture waters in the Congolese basin, where these waters are sometimes in excess. This option seems to be logical because

R. Paepe et al. (eds.), Greenhouse Effect, Sea Level and Drought, 523–539.
© 1990 *Kluwer Academic Publishers.*

it appears that the Oubangui had supplied the Chad basin before being caught by one of the Congo's tributaries.

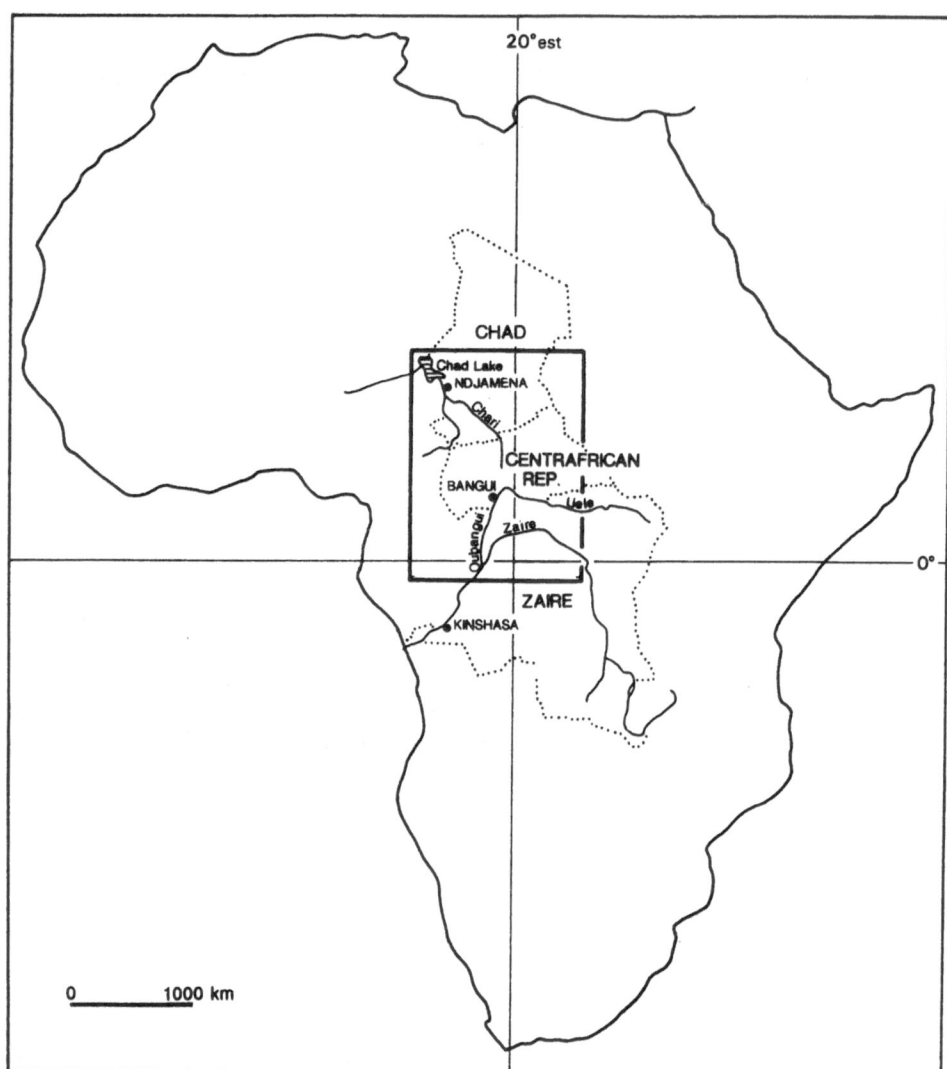

Figure 1. Lake Chad, Chari and Congo basins.

525

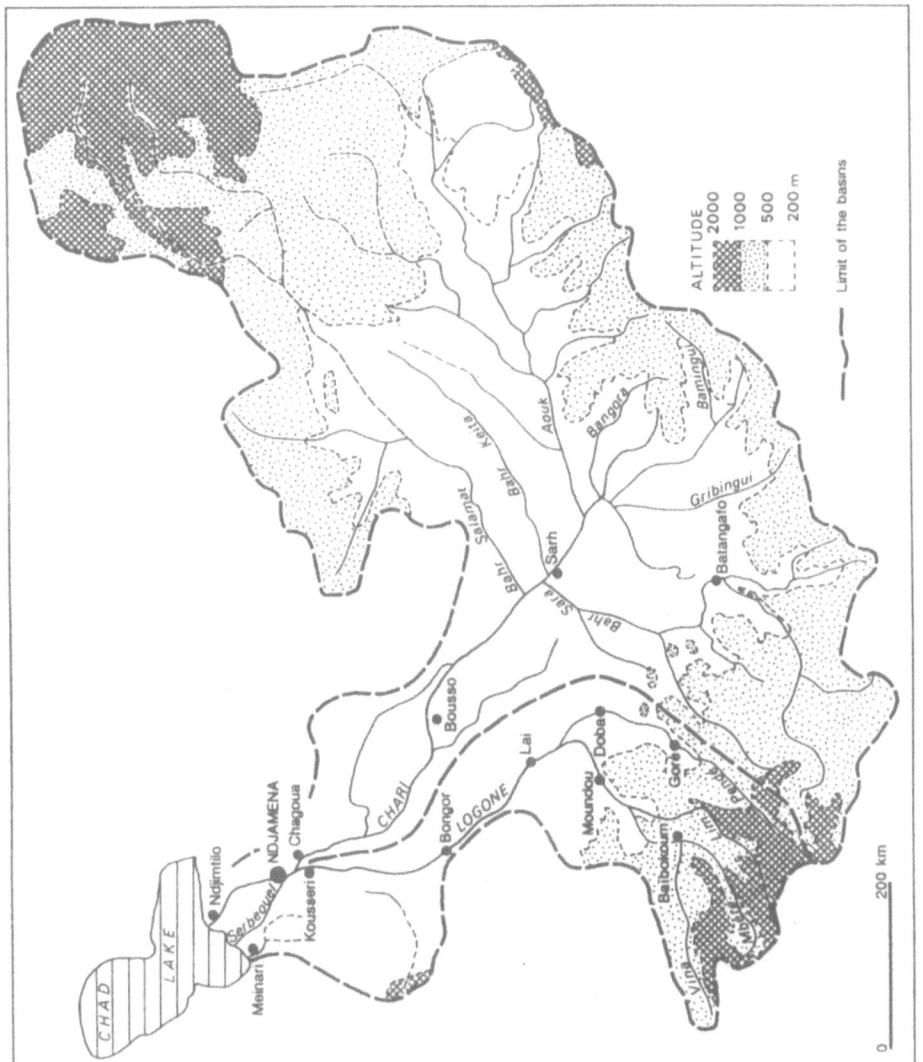

Figure 2. The Chari basin.

2. Flow Conditions

Before going slightly beyond the fifth parallel at Djoukou, the Oubangui has a northwesterly direction Fig. 3). It then turns to the west and next, after Possel, where the confluence with the Kemo is located, it has a sharp bend and takes a southwesterly direction after Bangui.

This sudden change in direction of the flow has suggested the hypothesis that the Oubangui-Uele flowed towards the Chari and then into Lake Chad and made the Belgian geographer Wauters think, at the beginning of the century, that the Oubangui had been captured by a small tributary of the Mpoko across the Bangui-Zongo canyon. To strengthen this hypothesis Borgniez (1935), quoted by Boulvert (1987), mentioned the following facts.

- existence of the bend near Fort-Possel.
- existence of a depression between Fort-Possel and Bouca. "The bottom of this depression rose of course by about 100 metres."
- existence of a wide alluvium bed north-east of Fort-Possel which is a former river arm filled in by sandy deposits. Some have seen in it traces of a former central lake named Fiba of Liba.
- near Bangui the subsidence seems to have affected a number of small streams.
- in the karstic plains of the Ombella and the Mondjo the drainage is so poor that it is sometimes difficult to determine which way the water is flowing.

So, if we attempt to catch the Congo-Zaire basin waters to supply the Chad basin, we would merely reproduce a phenomenon that existed before recent tectonic events probably caused this reversal.

3. Characteristics of the Rivers (Tab. 1)

3.1. LAKE CHAD AND THE CHARI BASIN (FIG. 2)

From time to time Lake Chad has occupied all or part of an exclusive hollow. When this hollow is filled in, the lake is like a little inland sea, as was the case at the end of the last century and also in 1963-1964. Then it reached the maximum height of 283 m, above which the waters flow out towards the Bahr El Gazal (Carmouze, 1976). At other times, for example between 1904 and 1907, 1941 and 1946, and since 1973, its surface has diminished significantly. Then the lake begins to divide up into smaller bodies when the height of the waters falls to 280 m (Leveque, 1987).

The lake's surface changes of course with the climatic conditions. The interannual average of the precipitation, calculated between 1954 and 1972 is 305 mm and the evaporation 2110 mm (Carmouze, op. cit.). As a matter of fact, due to the low depth of the lake, which is 3.8 m on average, even weak climatic variations can change the lake's surface in an important way.

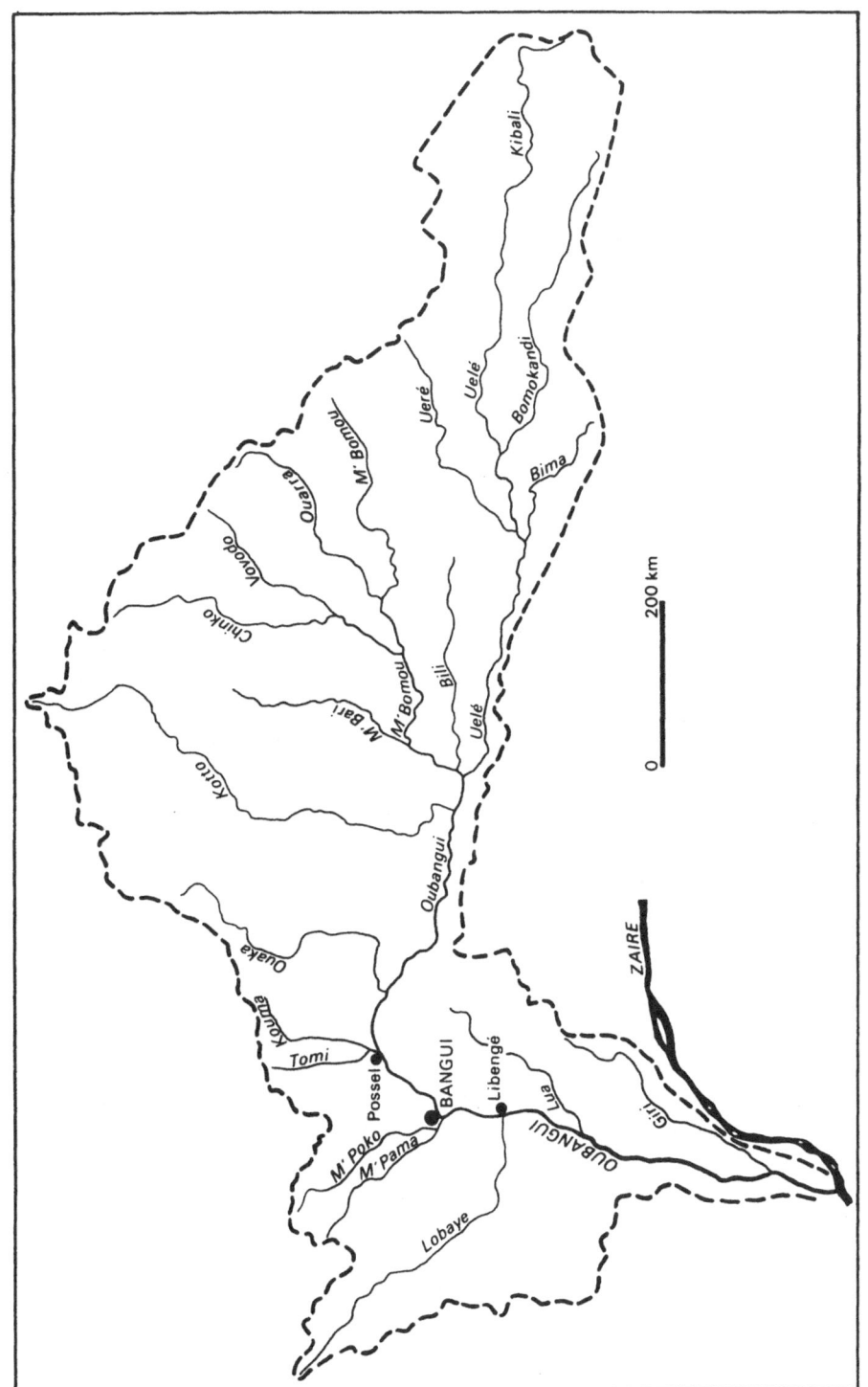

Figure 3. The Oubangui Basin (after Thiébaux, 1987).

Table 1. Compared characteristics of the rivers.

	CHARI (1)	OUBANGUI (2)	CONGO-ZAIRE (3)
Basin's surface (km^2)	500,000	479,000	3,700,000
Discharge (m^3/s^{-1})	1,200	4,080	42,000
Spec. module ($l/s^{-1}/km^{-2}$)	2.4	8.7	11.4
Spec. mech. erosion ($t/km^{-2}/y^{-1}$)	3.0	11.5	10.0
Suspension load (mg/l^{-1}) (4)	40.0	42.5	27.0
Total yearly load (10^6 t/y^{-1})	1.5	5.5	36.0

(1) from Boulvert (1987)
(2) from Thiébaux (1987)
(3) from Moguedet (1988)
(4) from Gac (1980)

The fluvial supply for the lake is composed of the waters of the Komadougou Yobe, the El Baid, the Serbeouel, which is in fact an arm of the Chari, and above all of the Chari itself, the flood of which occurs in October and November. The Chari basin, the surface of which exceeds 500,000 km^2, occupies only 25% of the whole surface of the Chad basin. The lower Chari is supplied by the Fhari proper and by the Logone. The upper Chari is constituted by the joining of the oriental Chari, to which the Gribingui, the Bamingui and the Aouk flow, and of the occidental Chari. This one is called Ouham upstream and Bahr Sara downstream. It often happens in Africa that rivers change their names.

The Logone is constituted by the joining of the oriental Logone or Pende and the occidental Logone or Mbere.

Before the recent drought the Chari and the Logone yielded at lake Chad an output of more than 1000 m^3/s. The average yearly discharge in Ndjamena was then 1183 m^3/s, which represents an additional yearly quantity of water of 40 x 10^9 m^3/s (Billon et al., 1974). Its run-off coefficient is 8.5% and its run-off deficit is 400 x 10^9 m^3/s. Boulvert (1987) suggested, however, that the discharge of the Chari and the Bousso is 935 m^3/s, and that of the Logone at Lai is 541 m^3/s. Therefore about 1500 m^3/s are shed annually by the Chari into Lake Chad. The average low levels, according to Boulvert (op. cit.), are about 200 m^3/s, whereas the flood exceeds 5000 m^3/s.

In 1984 and 1985, the basin was cruelly hit by the drought. In 1984 the yield of the Chari has decreased in a catastrophic way an the deficit in Ndjamena reached 82% of the interannual value (Sircoulon, 1984). In spring 1985 the discharge diminished yet again reaching 7 m^3/s at Kousseri and Chagoua, located respectively on the Logone and the Chari, just above Ndjamena. This is the lowest water level observed since the beginning of the gauging (Boulvert, 1987). The Chari's waters are acid with PH varying between 6.2 and 6.6, and the dissolved load between 47 and 74 mg/l (Gac, 1980). The waters, which contain an average of 31.4 mg/l of HCO_3-, 4 mg/l of Ca^{++} and 2.92 mg/l of Na^+ calculated from the data of Carmouze (1976), characteristic of the calcium bicarbonate water group and of the sodium chloride subgroup, like 33% of the surface waters of the world (Meybeck, 1979). The other cations, K^+ and Mg^{++}, have concentrations of about 1.8 mg/l, whereas silica reaches almost 20 mg/l. Therefore 1.36 Mt of HCO_3 and nearly 1 Mt of SiO_2 are poured yearly into lake Chad. There the concentration of the dissolved elements increases as we move from the delta (Carmouze, op. cit.). This is multiplied by 2 in the southern part of the lake and by 4 in the northern, but the relative chemical composition of the waters is altered.

The suspended matter in the Chari is estimated to be about 40 mg/l (Gac, op. cit.). It is composed mainly of clay minerals among which kaolinite is predominant. Moreover there is also a lot of quartz silt accompanied by illite, goethite, and traces of chlorite and interlayered clays. The total amount of solid material brought into Lake Chad, calculated on the basis of 1183 m^3/s, is 1.5 Mt/y and the specific mechanical erosion of the whole of Chari's basin is 3 t/km^2/y.

3.2. THE OUBANGUI

The Oubangui flows from the junction of two rivers, the Uele and the Mbomou. The entire Oubangui-Uele system is over 2400 km long (Fig. 3). The Uele originates between the Congo and the Nile with a regime that is clearly equatorial and its low level discharge can be three times as high as that of the Mbomou. The Uele's average discharge is from 1500 to 1600 m^3/s and the Mbomou's 1350 m^3/s (Boulvert, 1987).

The interannual flow of the Oubangui at Bangui is 4080 m^3/s with an average low level of 849 m^3/s and an average flood of 9260 m^3/s (Thiébaux, 1987). The highest and lowest recorded discharges were 15800 m^3/s in October 1916 and 315 m3/s in March 1985 (Fig. 4). The flood occurs in October and November and between August and November its discharge is between 6000 and 12000 m^3/s. The interannual value is about one tenth of the Congo's at Brazzaville, but more then three times that of the Chari.

The Oubangui upstream of Bangui has a basin covering a surface area of 479,000 km^2. The average rainfall is 1491 mm and therefore the specific discharge is 8.5 l/s/km^2, which corresponds to a run-off coefficient of 17.9% (Thiébaux, op. cit).

The concentration of dissolved elements varies between 29 and 76 mg/l, and the suspended matter between 35 and 52 mg/l. For the latter, the average quoted for a short period between July and October 1987 is 42 mg/l, so that the specific mechanical erosion can be estimated to be about 11.5 t/km^2/y.

3.3. THE CONGO-ZAIRE RIVER

3.3.1. *Hydrological Characteristics.* With a surface of 3.7 million km^2, the Congo basin is second only to the Amazon. It spreads over 9 countries and covers part of western Africa's high lands and the Congolese central depression. It is 4370 km long and rises at an altitude of 1600 m.

530

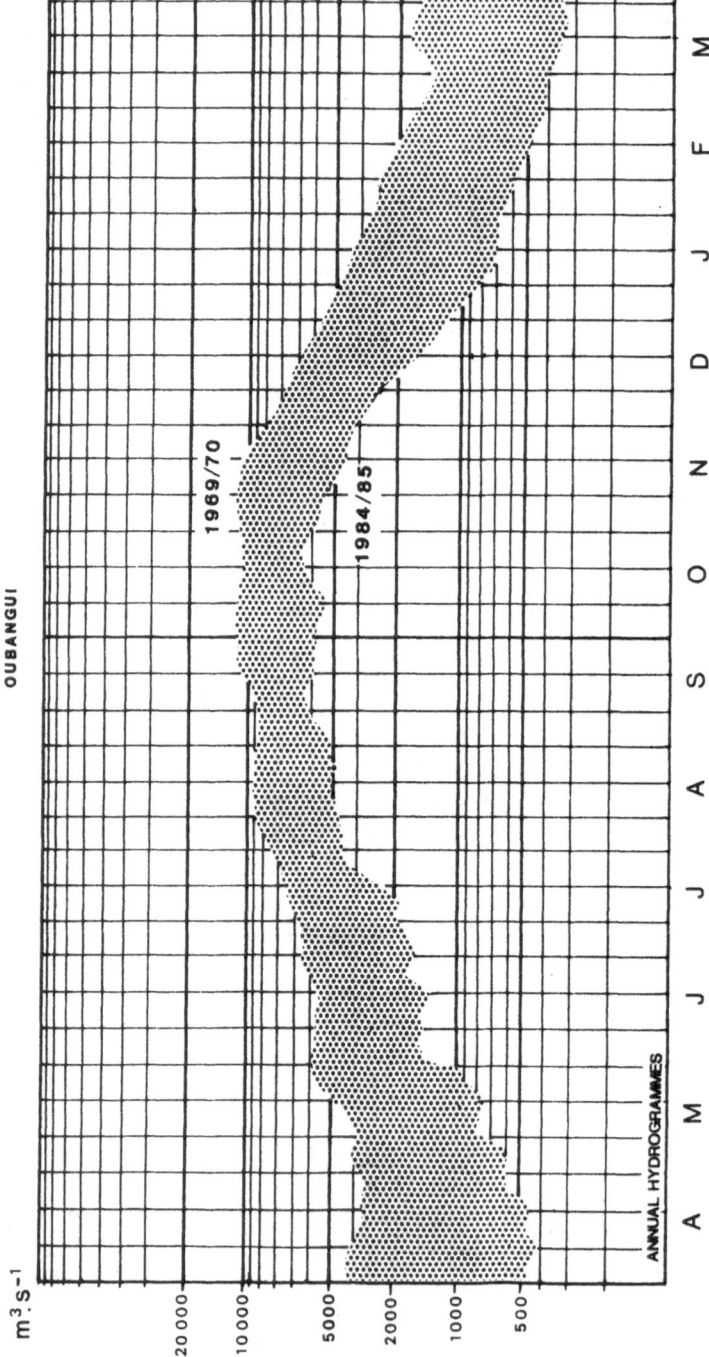

Figure 4. Variation of the discharge between a humid year (1969-1970) and a dry year (1984-1985) in m³/s⁻¹ (after Thiébaux, 1987).

It is characterized by a gentle incline, approximately 0.033%, which is unstable as is the case with most of its tributaries. The river crosses a series of low areas, forming lakes, pools and flooded plains when in spate. Hence at the junction with the Oubangui and Sangha, floods cause the formation of a huge lake of more than a hundred thousand km^2 (Fig. 5). The Congo bore can flow up the Oubangui for several hundred kilometres and for a month one can observe the inversion of the relative heights of the river level between upstream and downstream. The passage from one low area to another is through rapids and canyons,

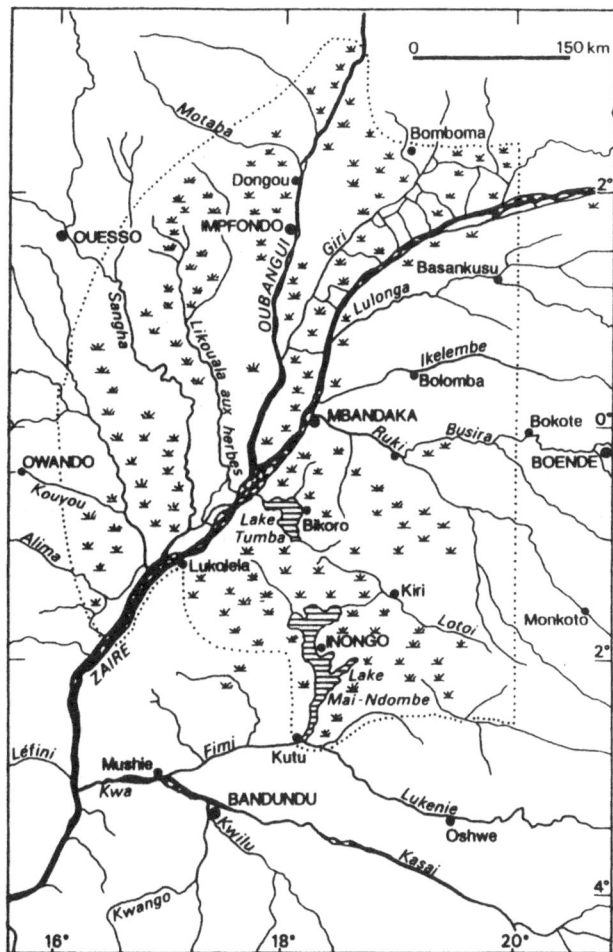

Figure 5. The floodable plain at the confluence of the Congo and the Oubangui (after Compere and Symoens, 1987).

the presence of which results from new motions of existing faults.

Its basin occupies both sides of the Equator and it relatively stable. The stream regimes (Fig. 6) are the hydrological reflection of climatic patterns, with one to two months delay. In the north, the rainy season is from July to October and the rivers, in particular the Oubangui are in spate from September to November. In the southern hemisphere, the rainfall regime is more complex. In the eastern part of the basin, most of the rainfall occurs between April and June, whereas in the western part the rainy season extends from October to May.

Before joining the Oubangui the Congo, which has flowed along the Equator for a thousand km, has a very regular regime resulting from the basic flow built up during its equatorial passage. An important peak in October and November, due to additional waters coming from flooding of its main tributaries in the northern hemisphere appears after junction with the Oubangui.

In Brazzaville variations in the discharge appear to be simple at first sight, but on closer scrutiny they prove to be particularly complex and correspond to the arrival of waters from various origins. The first flooding takes place in April and May and the second, which is more important, occurs in November and December. Records since 1902 in Kinshasa and since 1941 in Brazzaville show 85% chance of maximum flooding in December. High water levels are on either side of two low water periods, the first being centred round the month of March and the second, which corresponds to the lowest water level of all, occurs in August. In Brazzaville, the average yearly discharge is 42000 m^3/s and 60000 m^3/s in spate.

The highest recorded flood is thought to have been more than 80000 m^3/s in December 1961, but it may have been more than 80000 m^3/s. On the other hand, the lowest recorded flow was 21400 m^3/s on July the 20/21, 1905, according to Van Ganse (1959). Between Brazzaville and the ocean, the Congo receives and additional 2% water and the average discharge at the mouth is 42800 m^3/s, thus discharging yearly into the ocean an average of 3.7 x 10^9 m^3. This constitutes 4% of global fluvial discharge into the world's oceans, and 60% of the total fresh water into the Guinea Gulf.

If we take 1500 mm as the average yearly rainfall over the entire basin, the total yearly rainfall is 5560 x 10^9 m^3. Only 24% of the rainfall waters go into run-off waters. This percentage, which seems to be low, can be attributed to the importance of evapo-transpiration, the basin being to a large extent covered by forest.

The average high water-low water ratio, of about two, gives the Congo its equatorial regime, which in reality does not accurately reflect the complexity of its hydraulic regime. Its high regularity results from the combination of several factors, namely:

1) the consistency and abundance of equatorial rainfall
2) the distribution of the basin on both sides of the Equator, the flooding periods of one hemisphere making up the low water periods of the other
3) the existence of lakes and pools in the basin area which act as spillways absorbing the overflow and releasing it when water level is low.

3.3.2. *Variations in discharge since the beginning of the gauging.* When we observe the variation in discharge since the beginning of the gauging we notice a sudden change in

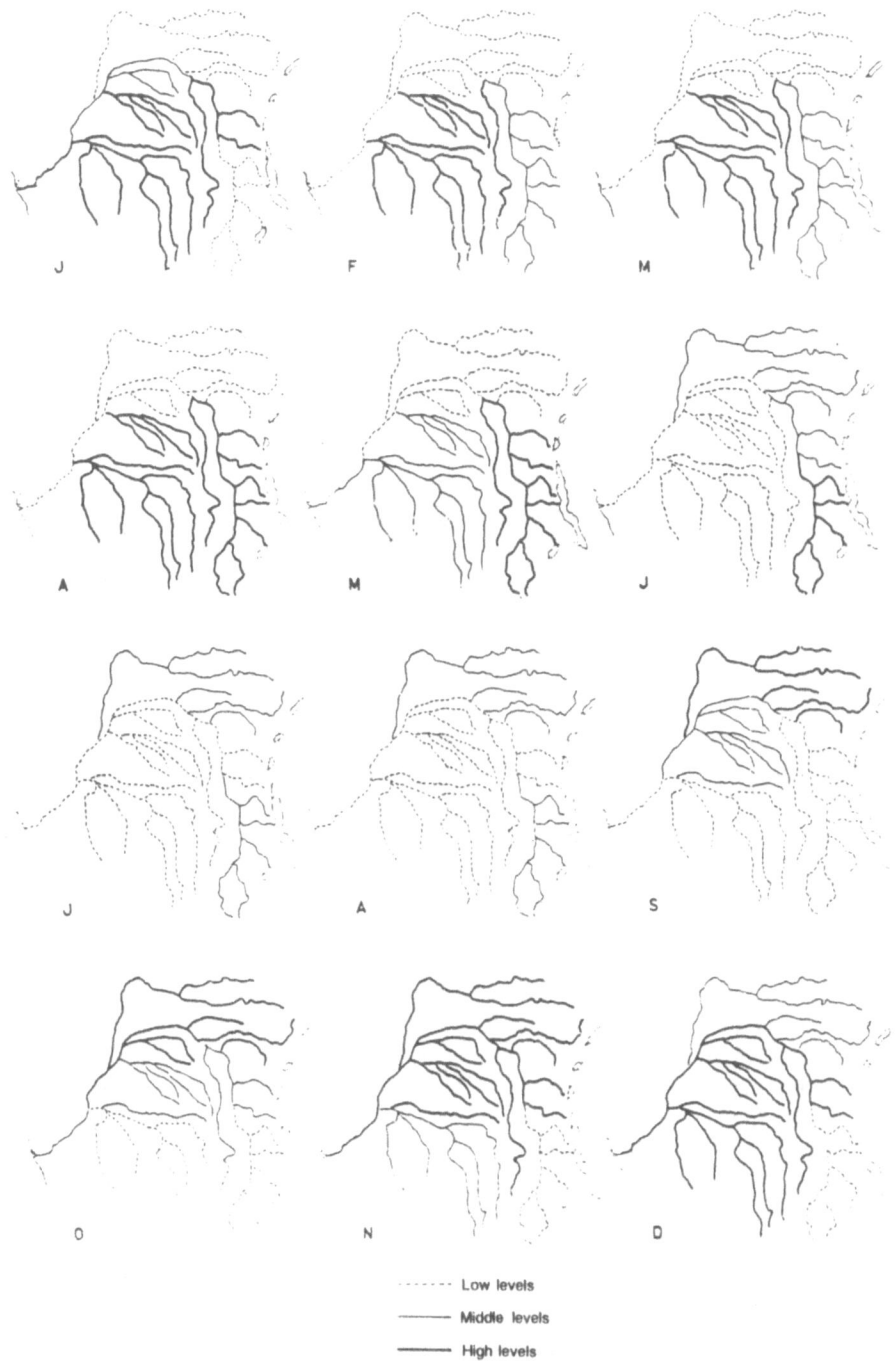

Figure 6. Maps of the mean monthly hydrometric river levels of the Congo basin (after Bultot, 1959).

the amplitude of flooding after 1960. Until then the variation in the average yearly discharge seemed to be shrinking so that extremes were getting nearer and nearer the average figure obtained over the last 60 years, which is 41000 m³/s.

In 1961 a value of 56000 m³/s, for a major flooding of over 75000 m³/s (and maybe more than 80000 m³/s) may seem quite unexpected. Frequency analysis based on data prior to 1961 leads us to expect, according to ORSTOM, a return period of such flooding of 100,000 years.

Between 1959 and 1970 all the yearly flows and floodings were above average. The observation period prior to 1960 was too short and the sampling insufficient. The yearly average discharge, integrated into the general results obtained between 1902 and 1984, is therefore inconsistent with the other data. Studies made by ORSTOM at Inga for the building of a dam on the Congo have shown that we could expect the Congo to increase to 90000 m³/s for very exceptional flooding, which corresponds to a yearly average discharge of over 60000 m³/s.

The general evolution of the module according to the weather is mirrored by climate evolution in central Africa since the turn of the century. We can observe that between 1960 and 1970 there has been in central Africa a rise in humidity comparable to that at the end of the last century which was reported by chroniclers of the time.

The Congo discharge at the confluence of the Oubangui has not been appraised very precisely. In Brazzaville it is 42000 m³/s as an average but, before that, it receives the waters of its most important tributary, the Kasai, the interannual module of which is about 10000 m³/s so that the discharge of the Zaire-Congo confluence is about 30000 m³/s

3.3.3. *Physical characteristics of the Congo's waters.* It is very important indeed to know the characteristics of the waters we want to discharge into the Chad basin. At Brazzaville (Moguedet, 1988) the PH varies between 6 and 7.7, with an average of 6.94. The mean conductivity is 32.65 μS/cm, which corresponds to an average concentration of dissolved elements of 30 mg/l, including 10 mg/l of dissolved silica. Conductivity and PH increase when the water level is low. The PH then becomes basic, while the conductivity is often over 50 μS/cm. As water levels start rising, PH and conductivity progressively go down until reaching the lowest values recorded at the end of flooding. Conductivity then approaches 25 μS/cm and PH becomes acidic.

When the water level is low the basicity results from an important concentration of bicarbonate in the water, the total amount of dissolved elements varying very slightly. When there is flooding the PH decreases because of this dilution but also because of the presence of organic matter and therefore of dissolved carbon, which is washed out of humus during the rainy season.

Although the mineralization of the river appears to be dominated by a phenomenon of dilution-concentration governed by the river discharge, other factors also intervene. Indeed, the highest flood levels correspond above all to rain water running off an already leaching soil, whereas at the beginning and end of flooding especially there is infiltration water rich in dissolved salts. Although its waters have a low mineralization rate, the Congo is one of the ocean's main suppliers of dissolved salts.

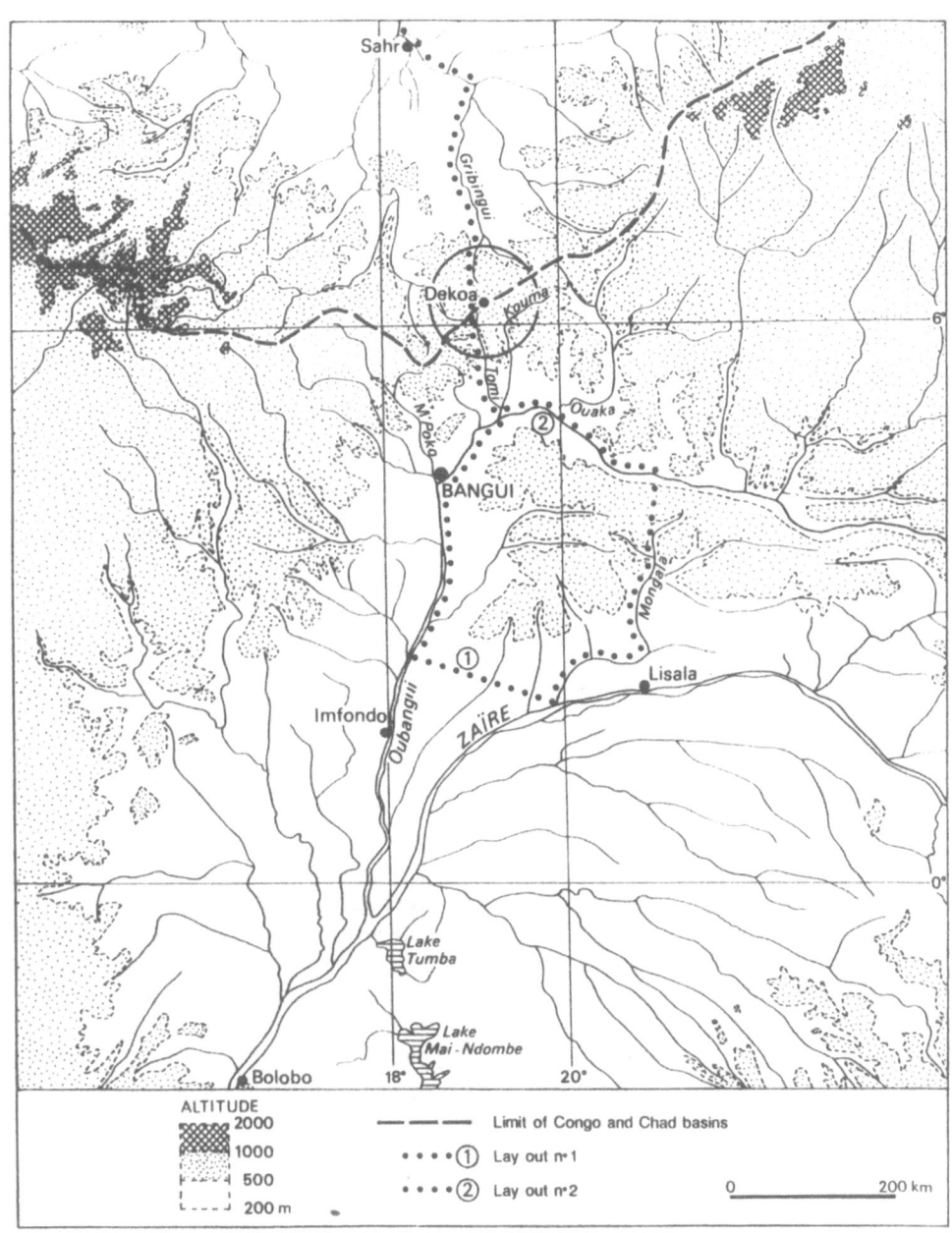

Figure 7. The interfluve between the Congo basin and the Chad basin. The proposed layouts.

536

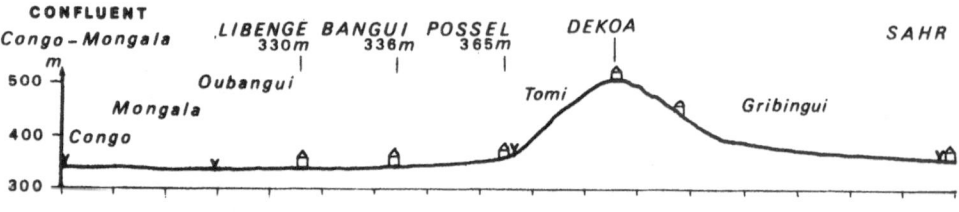

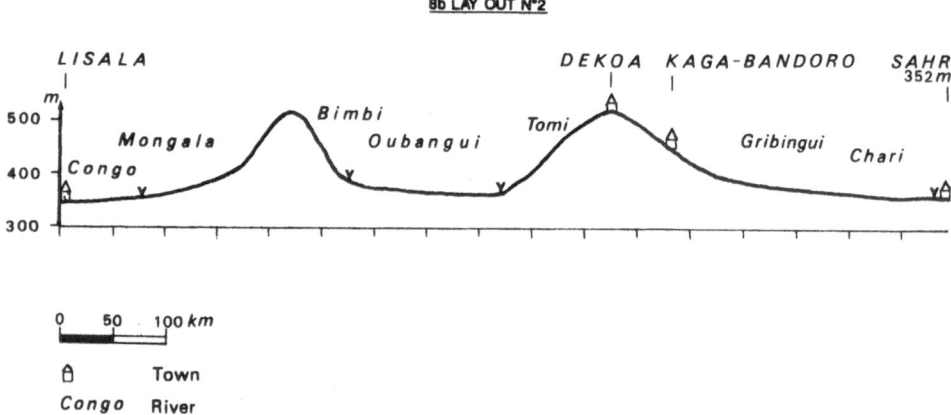

Figure 8. The layout propositions.

The amounts of suspended matter, 27 mg/l on average, are equally unimportant, especially because there is a generally low gradient of the river and its tributaries. The nature of the transported matter is, on the other hand, closely linked to the climatic environment and is thus completely characteristic of the wet tropical zone. It is composed of mainly kaolinic clay and quartz silt. Iron hydroxides, in association with organic matter especially in the form of ferrohumic compounds (Kinga-Mouzeo, 1986), account for approximately 3 mg/l, about one sixth of which corresponds to microparticles less than 0.45 μm (Meybeck, 1979).

The total amount of suspended matter delivered to the ocean is about 36 Mt/y.

4. Proposed Configuration

Because of the low discharge recorded on the Oubangui in March 1985 (315 m³/s), it is thought better to capture the Congo's waters rather than those of the Oubangui. The best place for the capture seems to be at the Congo-Mongala confluence which is the nearest to Lake Chad. This is at an altitude of 340 m. The boundary begins by crossing the flood plain to join the Oubangui, south of Libenge, at an altitude of about 330 m (Figs. 7 & 8a). It then goes up the Oubangui beyond Bangui where the river is at an altitude of 336 m. It next reaches Possel at the confluence with the Kemo at which point the altitude is 355 m. To that point the slope has been 0.08 m/km.The boundary then follows the portage roads which was used at the beginning of the colonisation. It goes on up the Kemo and then the Tomi, the average slope of which is 0.97 m/km. It then crosses the crestline which is a little above 500 m and reaches the Chad basin down the Gribingui, the average slope of which is 0.5%. The town of Sahr on the eastern Chari is situated at an altitude of 352 m at about 550 km from Lake Chad.

Another option (Fig. 8b) is to go up the Mongala, a Congo tributary, and join the Oubangui upstream of Kouango. The advantage of this option is that we go down the Oubangui for about 150 km. The drawback is that we have a second crestline situated at an altitude of about 510 or 520 m between the Congo and the Oubangui basins

5. Conclusions

It seems reasonable in view of the important water deficit of the Sahelian area, to attempt a water transfer from the Congo-Zaire basin to the Chad basin. Because of the low levels seen on the Oubangui in March 1985 and also because most of the waters of the Congo come from the southern hemisphere, where the rainy season occurs at the time of the dry season north of the Equator, it is preferable to aim at capturing the Congo waters. If 1% of the Congo's discharge were sufficient to relieve the water deficit in the Chad basin, its capture would not have important consequences for the Congo basin. In fact, it may even lead to advantages such as diminishing the surface of the flood plain, when the Congo is in spate.

The project may also be justified by the fact that large sums of money are already invested every year for mitigation of the drought, not always with success. Such a project would not only be carried out for the mitigation of drought but above all to enable the countries involved to achieve important development projects.

The project seems to be technically feasible. It would not be the first time that water transfer has been carried out, even if not on this scale. According to Llamas (1983, 1986) it must turn, however, into hydroschizophrenia. Accurate studies must be carried out to discover:
- The water needs in the Chad basin. It is certainly important to keep up the lake level and to maintain irrigation around the lake. The water supply of remote areas could also be considered.
- The water resources of the Chad basin and particularly the underground waters, which can be used to fight against the water deficit.

- The eventual effects on the climate, although they would probably be negligible even if evaporation were to increase.
- The consequences on the hydrochemical evolution of Lake Chad, the waters of which do not have the same chemical composition as those of the Congolese basin.

It is only after these studies have been done that we shall see if such a project is truly feasible.

6. References

Billon, B., Guiscaffre, J., Herbaut, J., & Oberlin, G. (1974) *Le bassin du fleuve Chari*, Monog. Hydrol. N° 2, ORSTOM, 450 p.

Boulvert, Y. (1987) *Carte oro-hydrografique de la République Centrafricaine, feuille Ouest - feuille Est, à 1/1000000*, Ed. de l'ORSTOM, coll. Notice explicative n° 106, Paris

Bultot, F. (1959) "Sur le régime des rivières du bassin congolais", Bull. Séances Ac. Roy. Sciences colon. belges, Bruxelles, pp. 442-456

Carmouze, J.P. (1976) *La régulation hydrogéochimique du lac Tchad*, Trav. et Doc. de l'ORSTOM n° 58, Paris, 418 p.

Gac, J.Y. (1980) *Géochimie du bassin du lac Tchad*, Trav. et Doc. de l'ORSTOM n° 123, Paris, 251 p.

Compere, P., & Symoens, J.J. (1987) "Le bassin du Zaïre", in Burgis, M.J., and Symoens, J.J. (eds.) *African Wetlands and Shallow Water Bodies*, Directory, Ed. ORSTOM, Paris, pp. 401-456

Kinga-Mouzeo (1986) *Transport particulaire actuel du fleuve Congo et de quelques affluents; enregistrement quaternaire dans l'éventail détrique profond (sédimentologie, minéralogie et géochimie)*, Thèse Univ. Perpignan, 251 p.

Leveque, Ch. (1987) "Le bassin tshadien", Ch. Leveque coordinateur, in Burgis, M.J., and Symoens, J.J. (eds.) *African Wetlands and Shallow Water Bodies*, Directory, Ed. ORSTOM, Paris, pp. 233-277

Llamas, M.R. (1983) "The role of groundwater in Spain's water resources policy", Proceedings of the Fourteenth Biennial Conference on Groundwater, California Water Resources Center, University of California, Davis, Report No. 56

Llamas, M.R. (1986) "Influence de certains facteurs physiques et admininstratifs sur la définition d'une politique des ressources en eau. Cas de l'Espagne", Eau et Développement, n° hors série, Revue maroccaine de l'eau, Rabat

Maley, J. (1977) "Pollen analysis and paleoclimatology of the last 12 millennia of the Chad region (Africa)", in *Palaeoecology of Africa*, AA Balkema Ed., Rotterdam, Vol 11, pp. 79-81

Maley, J. (1981) *Etudes palynologiques dans le bassin du Tchad et paleoclinatologie de l'Afrique Nord Tropicale de 30000 ans à l'époque actuelle*, Trav. et Doc de l'ORSTOM, Thèse Doc. Etat Montpellier, 586 p.

Meybeck, M. (1978) "Dissolved elemental contents in the Zaire (Congo) River", Nether. J. Sea Res., Vol. 12 (3/4), pp. 293-295

Meybeck, M. (1979) Concentrations des eaux fluviales en éléments majeurs et apports en solution aux océans.", Rev. Géol. dyn. et Géogr. phys., Vol. 21, Fasc. 3, pp. 215-246

Moguedet, G. (1988) *Les relations entre le fleuve Congo et la sédimentation récente sur la marge continentale entre l'embouchure et le Sud du Gabon. Etude hydrologique, sédimentologique et géochimique*, Thèse Etat, Univ. Angers, 335 p.

Rognon, P. (1979) "Mécanismes climatiques actuels et paleoclimats au Sahara", in *Palaeoecology of Africa*, AA Balkema Ed., Rotterdam, Vol. 11, pp. 1-12

Servant, M. (1983) *Séquences continentales et variations climatiques: évolution du Bassin du Tchad au Cenozoïque supérieur*, Trav. et Doc. de l'ORSTOM n° 159, Paris, 573 p.

Servant-Vildary, S. (1979) "Paleoclimnologie des lacs du bassin tchadien au Quaternaire récent", *Palaeoecology of Africa*, AA Balkema Ed., Rotterdam, Vol. 11, pp. 65-78

Sircoulon, J. (1984) "La sécheresse en Afrique de l'Ouest. Comparaison des années 1982-1984 avec les années 1972-1973", Cah. ORSTOM, série Hydrol., Vol. XXI, n° 4, pp. 75-87

Thiébaux, J.P. (1987) *PIRAT-Opération Grands Bassins Intertropicaux: Transports de matières sur l'Oubangui à Bangui. Premiers résultats (1986-1987)*, ORSTOM-INSU, 60 p., multigr.

Van Ganse, J. (1959) "Les débits du fleuve Congo à Léopoldville et à Inga", Acad. Roy. Sc. Col. Belge, Bull. Séances, n° V, 3, pp. 737-763

Yayer, J. (1951) "Caractéristiques hydrographiques de l'Oubangui", Mém. Inst. Roy. Col. Belge, Bull. des Séances, pp. 803-835

GROUNDWATER RECHARGE AND STORAGE BY DAMMING A SMALL EPHEMERAL STREAM

PER WEDEL & RODNEY L. STEVENS
Department of Geology
Chalmers University of Technology
and University of Göteborg (Gothenburg)
S-412 96 Göteborg, Sweden

ABSTRACT: This project deals with the additional storage capacity of the surrounding sediment and bedrock along a stream system. The cost-benefit ratio of a dam depends largely on topographical and lithological conditions. With narrow and steep valleys the fractured bedrock and leakage from a reservoir is a crucial consideration. Sedimentological investigations along a small ephemeral stream in south-east Cyprus have revealed that there is storage capacity in the former stream channels but, more importantly, these sediments function as a buffer against irregular discharge impulses and aid infiltration into the bedrock. Dams might be constructed along recent stream channel. Leakage from the valley stream sediment to the underlying limestone would recharge the bedrock aquifer.

1. Introduction

The benefit of a dam depends upon several factors of geological and hydrological nature. A dam site must generally be positioned where the underlying strata are impermeable enough to prevent loss of the stored water. A minor amount of leakage can be allowed. Furthermore, sites which are favourable from a topographical point of view might, nevertheless, be abandoned when the elevation of the reservoir is of importance as it is when considering irrigation or hydroelectrical power programmes.

In this paper a concept is outlined for storing water in a more general way whereby near-surface leakage and topographical requirements are less restrictive to the reservoir location.

The presence of permeable strata under a dam might, in fact, be favourably utilized. The controlled or intentional leakage of water to the subsurface surroundings offers the following advantages:

1) More water than just the volume of the reservoir can be stored. This is especially important if there is more water in the stream discharge than can be stored during

541

R. Paepe et al. (eds.), Greenhouse Effect, Sea Level and Drought, 541–551.

the rainy season in the reservoir since the store in the adjacent rock and soil can also be utilized.

2) The water stored in the surrounding rock and unconsolidated sediments will, to a large extent, be shielded from evaporation.

3) The water table in communication with the dam will rise.

The artificial recharge obtained will offer a groundwater with generally improved chemical and bacterial characteristics compared with surface water. On the other hand, there is some risk for deterioration of the chemical quality with time due to reaction with salt and pore water within the beds through which the water percolates.

As an example of how such a water scheme might be protected we have used an ephemeral stream in Southeast Cyprus where we have also studied sedimentological and stratigraphical features. There are no plans to carry out this damming in the actual valley. There has been no investigation of the physical properties of the bedrock at the projected dam sites in this study since the purpose has primarily been to illustrate the infiltration and artificial groundwater-recharge technique. However, this valley is a suitable example of the general principles and a good result is anticipated if certain rainfall and up-stream conditions could be met. The scope of the NATO Advanced Research Workshop "Geohydrological Management of Sea Level and Mitigation of Drought" is also very well illustrated by this project, even if the effect upon sea level would be negligible.

2. Environmental Setting

The valley area under investigation is hilly and has a flat coastal plain downstream. The climate of the area is eastern Mediterranean. Annual mean precipitation is 500 mm, with considerable variation from year to year. Several hydroelectrical and irrigation dams have been considered in the region to the West as well as various domestic water-supply schemes. However, the cost-benefit ratios are usually low because of topographical and lithological conditions. Due to the narrow and steep valleys the planned dams would have limited volume. The bedrock is fractured and leakage from the reservoir is considered to be a crucial consideration with negative implications in connection with conventional dam exploitation but with positive possibilities regarding groundwater recharge, as explained in this report.

3. Sedimentological Investigation

The stream banks have been investigated and the Quaternary stratigraphy recorded (see Fig. 1). The stream has eroded into the local bedrock comprised of pillow lava and flaggy limestone. The clastic material has been alternately deposited and reworked by current action. Multiple sequences reflect the shifting balance between constructional and erosional phases. Along most of the reaches terraces have fine-grained intercalations due to overbank deposition. Also, low-energy sediments have accumulated in former oxbows (e.g. Fig. 1). Alluvial fans have built out from the valley sides and increasingly interfinger with stream deposits in the valley's central portion.

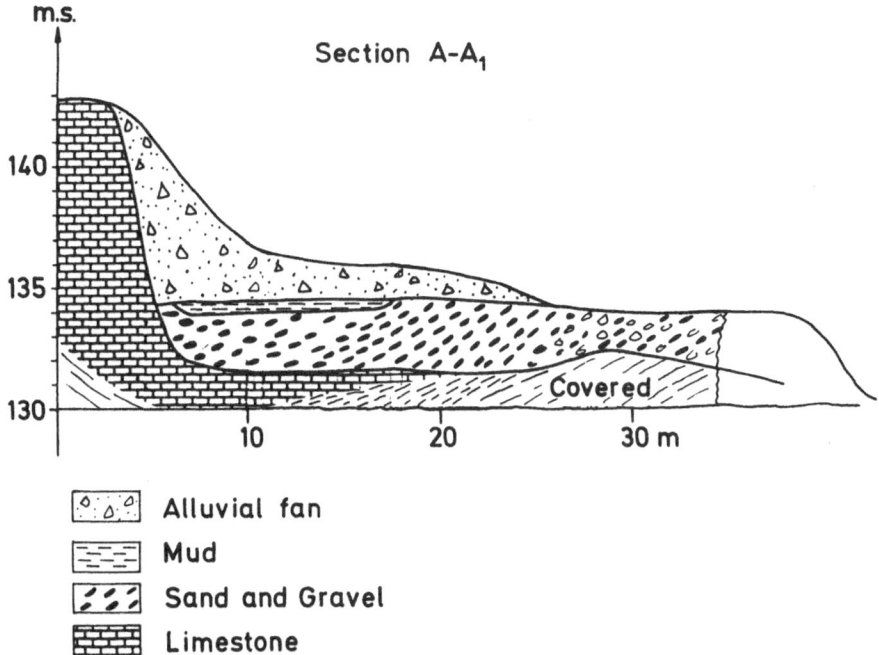

Figure 1. Section A-A of an example of an ephemeral stream valley, as given in Fig. 2.

(The initial erosion of the limestone has created a relic valley floor a few metres above the present stream base. The aggradation prior to the latest erosive phase deposited coarse stream-channel material as well as fine-textured sediment from low energy subenvironments. Uppermost, there is alluvial-fan material).

The contrasting petrology of the local bedrock formations aids the identification of the beds of different genesis. In the mid and lower stretches of the stream the fluvial deposits have a mixture of dark-coloured clasts from the basic igneous rocks from the distal, upland source area and light-coloured limestone from the adjacent bedrock hills through which the stream also erodes. The alluvial-fan material is entirely derived from the surrounding limestone hills and. therefore, lacks basic igneous clasts except in rare examples of mixing by landslides and soil-creep over former stream banks.

Clast morphology provides further support for the visual interpretation of provenance and, thereby, sedimentological origin. Those from the upland area are well rounded due to their non-stratified, basic-igneous source and greater distance of transport along the stream. The debris from the limestone area is very angular and platy in contrast. The sedimentary structures are similarly indicative but not always so unambiguous.

4. Valley Development

The object of this report does not justify an extensive palaeoenvironmental analysis. However, to evaluate the storage capacity of the stream banks it is necessary to understand the general lithostratigraphy and its background.

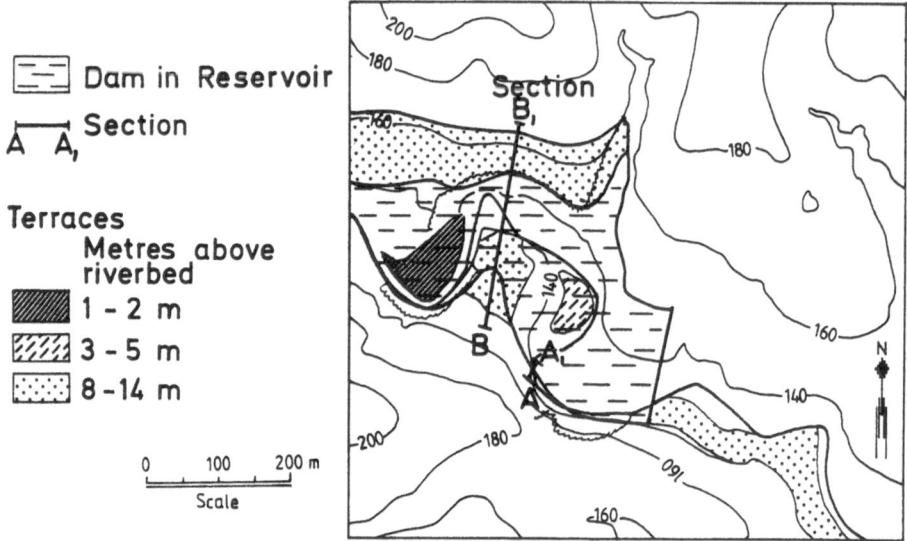

Figure 2. An example of a dam site in an ephemeral-stream valley. Sections A-A, and B-B are shown in Figures 1 and 3, respectively.

Initially, the stream erosion incised into the bedrock and developed a valley which was considerably broader than today's reaches. The present stream thalweg has been observed at several localities to be 3-4 m below the former valley floor suggesting that the most recent fluvial phase has been primarily erosive. Several relic channels in the limestone surface indicate stream meandering. The former channel dimensions have at certain times been of about the same magnitude as today (Fig. 2). This does not necessarily mean that the hydrogeological situation or the climate was the same as today, since the environmental evolution has been influenced by various factors which we hope to define better with additional work.

After the initial erosional phase a period of aggradation took place, perhaps alternating with times of revived erosion. The fluvial gravel along the stream was eventually built up to at least 14 m above the present stream bottom, although there are even higher terraces which are interpreted to be due to stream erosion (Bagnall, 1960). It has not, however, been shown how they fit into the valley evolution of aggradational and erosional sequences.

Along the slopes there are fans with debris from the adjacent hill (Figs. 1 and 3). This debris is most often matrix supported. Generally, the clast-supported portions have a substantial fine-grained matrix which makes them essentially impermeable, as are the matrix-supported portions. There are, however, also some beds of permeable clast-supported debris within the alluvial-fan sequences. On the whole, the hydraulic conductivity and storage capacity of these strata must be regarded as too low to be of any importance. This is in contrast to the high porosity and permeability of most alluvial fans because of the carbonate lithology which is very easily broken down to form a fine-grained matrix. Where the valley is wider and along the stream where the slopes are only a few degrees, there is commonly colluvium of local origin on top of the stream beds. Also, there is probably a substantial contribution to the soil from wind transportation.

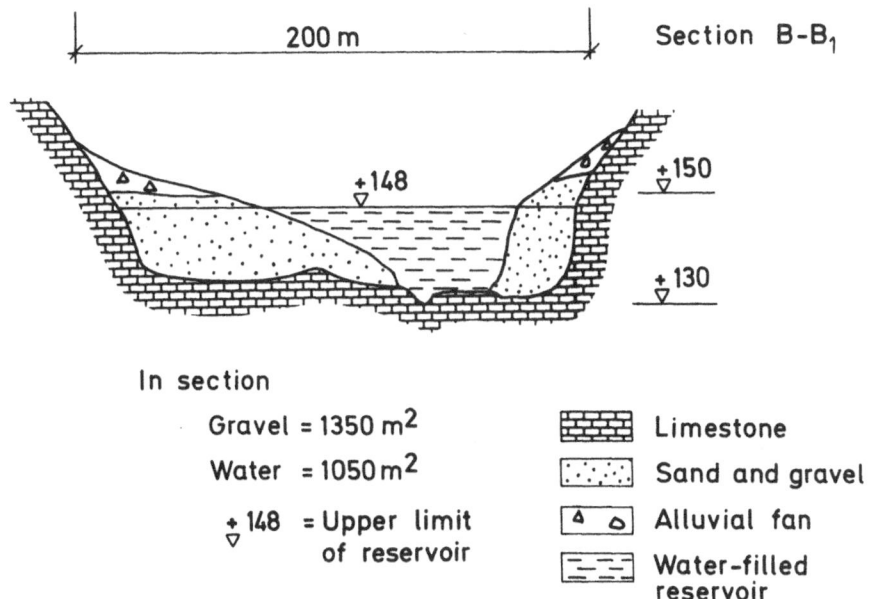

Figure 3. Valley cross section B-B (Fig. 1) with a hypothetical damming. A substantial amount of water will be stored in the sediments for further percolation into the bedrock.

5. Groundwater Recharge

Subsurface storage of water is influenced by many factors. There must be sediment of rock which can serve as an aquifer and there must be a mechanism to convey the water from the surface to the store. When dams are used for catching water for this purpose there is also the problem of sedimentation by impermeable silt and clay in the reservoir.

Three different types of aquifer can be distinguished from a general point of view. These are:

- aquifers in unconsolidated strata,
- aquifers in fractured bedrock,
- aquifers in karstic bedrock.

As can be seen from Table 1, rock porosity varies over a great range. In particular, the difference between fractured and unfractured rock can vary over several orders of magnitude (Liedholm, 1987).

Table 1. Porosity ranges for selected geological materials (after Knutsson and Morfeldt, 1978).

Material	Porosity due to pores	Additional porosity sources
Gravel	30 - 40	
Coarse sand	30 - 40	
Medium-fine sand	30 - 35	
Silt	40 - 50	Occasionally fractures
Clay till	45 - 55	Local fractures and lenses
Limestone, dolomite	1 - 50	Karstic structures and fractures
Coarse-medium textured sandstone	<20	Fractures
Fine-textured sandstone	<10	Fractures
Claystone - Siltstone	-	Fractures
Basalt	-	Fractures
Porphyry	-	Fractures
Granite, gneiss and others	-	Fractures

Hydraulic conductivity shows even greater variability with respect to bedrock fracturing (Table 2). The bedrock within the study area is marl flagstone. Along the valleys it appears moderately fractured except in a few distinct portions where only a few fractures oblique to the bedding were visible in the escarpments. No karstic structures have been observed.

The fluvial sands and gravels overlying the limestone have a high hydraulic conductivity due to their porosity. However, the hydraulic conductivity of finer grain sizes varies considerably so that the interbedding of silt and mud is of crucial importance. It can be concluded for practical considerations that silt and finer material have such a low hydraulic conductivity that they prevent percolation of water into the subsurface at artificial groundwater-recharge sites. In contrast, sand and gravel consistently permit infiltration. Natural sediment is often, of course, a mixture of grain-size fractions. From a hydraulic point of view the smaller fractions are decisively important in hydraulics since they infill the voids between the larger particles. In a stream valley the gravelly sand of former stream-channel beds can be considered to be good conveyer with sufficient hydraulic conductivity to accept rapid infiltration from ephemeral-stream flow.

Although these beds are volumetrically not sufficient for substantial aquifer capacity, they do extend beyond the area of the surface reservoir. Furthermore, the sand and gravel often have a direct contact with the fractured bedrock in the area so that they could serve as a buffer for initial infiltration while groundwater is simultaneously but more slowly transferred to the much larger aquifer of the fractured bedrock. The matrix-supported alluvial-fan deposits, on the other hand, are aquicludes which strongly reduce water percolation from a reservoir to the subsurface. The long-term leakage from such beds to an aquifer can, however, be of considerable importance.

Table 2. A comparison of permeability and representative aquifer materials. Calculated in m/s (after U.S. Dept. of the Interior, 1981).

PERMEABILITY				
10^{-2}	10^{-4}	10^{-6}	10^{-8}	10^{-10} m/s
Relative Permeability				
VERY HIGH	HIGH	MODERATE	LOW	VERY LOW
REPRESENTATIVE MATERIALS				
Clean gravel	- Clean sand sand and gravel	- Fine sand	- Silt, clay, mixtures of sand, silt and clay	-Massive clay
Vesicular and ascoriaceous basalt and cavernous limestone and dolomite	Clean sandstone and fractured igneous and metamorphic rocks	Laminated sandstone, shale, mudstone		Massive igneous and metamorphic rocks

6. Damming and Stratigraphy

In a stream valley, as the one described above, the coarse fluvial sediment on top of the bedrock serves as a filter for the rock aquifer or, more generally, for any potential groundwater storage (Fig. 4). When the stream is dammed the water in the reservoir will infiltrate into the bedrock both beneath the recent channel and into outcrops adjacent to the fluvial sediments of the stream banks. In the reservoir there will be rapid sedimentation and the bottom will soon be clogged and impervious whereas the banks, mostly fluvial gravel and sand in their lower portion, will also be flooded but will remain open to infiltration much longer. Water will percolate into the permeable beds which will convey it to the fractures in the bedrock.

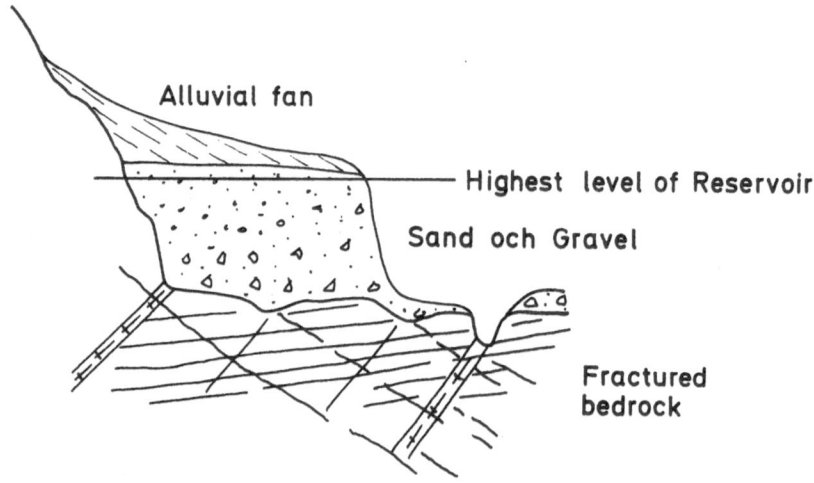

Figure 4. The coarse-grained sediments overlying fractured rock will facilitate infiltration into the bedrock. The sand and gravel bed will help distribute the water to fractures beneath.

Recharge characteristics depend on several factors considered below.
1) The pressure head, determined by the water depth in the reservoir and its variability over time. In areas such as that of the present investigation the precipitation is highly variable and the reservoir will be unlikely to be filled to its capacity every year. After the peak flow the head will decrease due to infiltration and evaporation. The flow into the banks will, to some degree, support the banks from erosion.
2) The hydraulic conductivity of the unconsolidated beds. This parameters is, of course, most important but it is also variable in these deposits due to the intercalations. The capacity of the stream-channel sediments with their high porosity will most likely not be a limiting factor within the system.
3) The hydraulic conductivity of the bedrock. Without further, detailed study the bedrock conditions are essentially unpredictable as is suggested from Tables 1 and 2. The frequen-

cy of fractures along the valleys is much higher than in the surrounding hills. In fact, the location of the valleys has, to a large extent, been determined by this anisotropic property of the bedrock. Bedrock areas with high frequencies of open fractures and overlain by sand and gravel beds favour high infiltration rates into the bedrock. There are numerous documented examples in which accidental lowering of the water table has occurred in residential areas when fractured bedrock zones have been drained during tunnelling of excavations (Knutsson & Morfeldt, 1978). Also, the capacity of high production from wells in crystalline bedrock demonstrates the importance of fractures.

4) Sedimentation of silt onto the surface of the sand and gravel beds. Most reservoirs, especially in arid regimes, are filled in with silt and clay. If this continues at an appreciable rate, the total reservoir capacity of the dam can be restricted to the surface volume. This may be a major limiting factor of many systems including the projected site in this study.

5) Clogging due to bacterial activity in the surface beds. The bottom of an artificial-recharge reservoir is always clogged by silt and bacterial activity. Continued effectiveness usually requires the top layer frequently to be removed. In these small reservoirs the walls of the channel must be cleared if silt and bacterial growth are clogging the sand and gravel. This is, however, a minor problem compared to the sedimentation on the bottom of the channel. The relatively steep slopes of the stream banks may be important in limiting silt sedimentation and clogging.

7. Dam Siting

The site where the dam is built must be chosen where the stratigraphy is favourable, but topographical and stability considerations cannot be ignored. The topographical requirements are the same as for any dam and the cost-benefit ratio is dependent upon local conditions. The stability problem must always be evaluated very carefully concerning dam construction. The efficiency of the reservoir as an infiltration and recharge area and the conditions for stability are, unfortunately, contradicting. The danger of peripheral leakage and erosion which could undermine dam construction are reduced it the bedrock is impermeable, but this is less advantageous for groundwater-recharge purposes. If grouting is required the dam construction will be more complicated and likely to be technically infeasible in many rural areas of developing countries. The present study area lies close to several towns and this represents no complication.

The project results can be best illustrated with reference to the chosen example. As can be seen from Figure 3, the cross sections over the dam and reservoir reveal a wide area of coarse fluvial sediments where the water easily can penetrate into the bedrock along the bottom and the valley side. Sand and gravel cover the entire valley upstream and downstream from this section, as suggested by eroded sections in the same gravel body. Sedimentation of silt from the reservoir will not easily build up thick layers that can prevent infiltration on these vertical sides. The interface between this coarse sediment and the bedrock of the former, wider valley has a dimension which is many times larger than the present channel bottom. The water percolating through the entire sediment volume will infiltrate into the bedrock, primarily via fractures and bedding. The frequency and width of these openings will essentially determine the infiltration capacity of the bedrock. The frequency of fractures and their degree of openness for specific sites can be measured. Subsequently, the average values can be estimated for particular rock types from local experience . After drilling and hydrogeological tests the hydraulic properties of the rock can be calculated with a known degree of certainty. This is, however, not necessary for small dams if proper field work is done by the geologists prior to the choice of dam sites.

8.   Summary

The described example of an ephemeral stream has revealed that there can be a substantial recharge of groundwater by means of artificial infiltration. The efficiency in this case is largely dependent on the presence of highly permeable sand and gravel from the former stream channels and an underlying fractured bedrock. With suitable dam sites and up-stream water management, substantial amounts of water might be artificially recharged (Fig. 5).

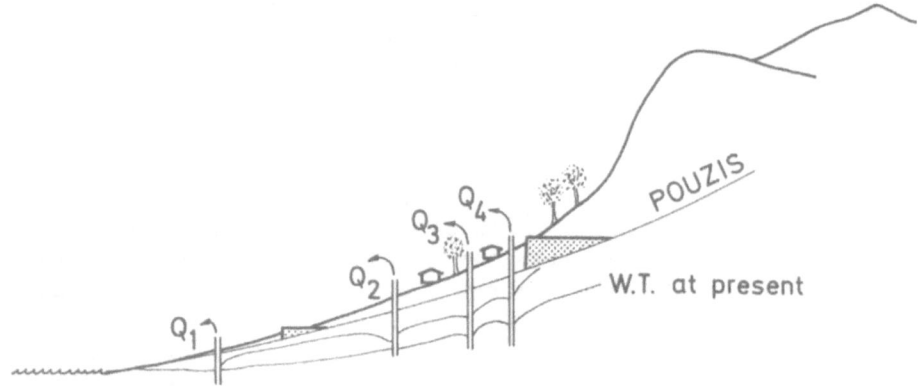

Figure 5. A hypothetical section along the stream after recharge dams have been constructed. W.T. = Groundwater table.

9. Acknowledgments

We want to express our sincere gratitude to the authorities of Cyprus, personified in Mrs. Milicori at the Ministry of Defence, who made it possible for us to carry out field courses in Cyprus for several years, and to our colleagues, Dr. Afrodisis at the Geological Survey and Dr. Grivas and Dr. Makrides at the Soil Survey Department, who guided us and introduced the geology and soil development of the area in a most interesting way. The students participating in the field courses have worked with diligence and their results have been significantly helpful.

10. References

Bagnall, P.S. (1960) *The Geology and Mineral Resources of the Pano Lefkara-Larnaca Area*, Geol. Survey Dpt. Cyprus Mem. No. 5, Government Printing Office, Cyprus

Knutsson, G., & Morfeldt, C.-O. (1978) *Vatten i jord och berg*, Ingenjörsförlaget AB, Stockholm.

Liedholm, M. (1987) Swedish Hard Rock Laboratory, Progress Report 25 8707 SKB, Regional Well Data Analysis.

U.S. Department of the Interior (1981) *Ground Water Manual*, Water Resources Technical Publication, Water Power and Resources Service, United States Government Printing Office, Denver

SEA LEVEL RISE AND ARTIFICIAL GROUNDWATER RECHARGE. A STUDY ON THE FEASIBILITY OF GEOHYDROLOGIC MANAGEMENT

ALBERT J. ROEBERT
Deputy director of the Municipal Waterworks of Amsterdam
Condensatorweg 54
1014 AX Amsterdam
The Netherlands

ABSTRACT: Intercepting 1% of the global rainfall of a one year period and storing it by artificial recharge in the subsurface of semi-arid areas could lead to a drop in sea level of less than 15 mm. Artificial recharge for this purpose is not the solution to the problem of sea level rise. "Geohydrologic management", as defined by Fairbridge is not deemed feasible. Artificial recharge of 1% of the global annual rainfall means the equivalent of 110 times the artificial recharge in the State of California, U.S.A. A calculation is presented on the dimensions of a pilot project on the same scale as the artificial recharge in California. This pilot project is situated SW of Lake Chad in Nigeria. Artificial recharge, though perhaps not on that scale, is certainly feasible if intended to provide water for development. It is not recommended to start construction of an artificial recharge project at its full capacity. Generally, a phased growth of capacity is essential for the sound development of artificial recharge projects.

1. Change in the World Water Balance

The question raised in this workshop has been described by Fairbridge in this Volume: "Water deficiency versus water excess: global management potential". It is well known that there are water-deficient regions where people face the hazards of catastrophic droughts and, on the other hand, water-excess areas, where people suffer from catastrophic flooding and storms.

In this workshop the feasibility of intercepting an amount of water equal to 1-2% of the annual global rainfall and storing it underground in areas suffering water shortages has been studied. The catchment of river discharge and its storage underground could interfere with the global water balance and change it. The present equilibrium in the world water balance could change in such a way that a larger part of global water is stored in the pores of the upper layers of the earth crust and less water in the oceans. This change in water

R. Paepe et al. (eds.), Greenhouse Effect, Sea Level and Drought, 553–564.
© 1990 *Kluwer Academic Publishers.*

balance has been defined by Fairbridge as *Geohydrologic Management*. The general idea is that this storage, resulting in a lower general sea level, would diminish dangers to flood prone river valleys and coastal regions. The amount of water involved, i.e. 1-2% of the global rainfall can be estimated at 5180 km³ (1%). The workshop studied the feasibility of recovering this amount of water into present or potential aquifers.

In my contribution to the workshop I presented the geohydrologic component, including:
- a summary of the water movements on earth in relation to the aim of the project (section 2);
- an explanation of the technique of artificial recharge of aquifers (section 3);
- two examples of large-scale artificial recharge in the U.S.A. and The Netherlands (section 4);
- a description of the prospect of a large project in a semi-arid region;
- a conclusion on the feasibility of Geohydrologic Management, and some statements and recommendations.

2. Water Movements on Earth and the Aims of the Project

In order to study the feasibility of the idea of Geohydrologic Management some figures on the world water balance are presented in Table 1. The project aims at 1% of global rainfall. As can be calculated from Table 1 that equals 5180 km³ of water. This is the additional amount to be stored underground on top of the present storage. It is also the amount that should be withdrawn from the oceans in order to lower the ocean level. The lowering of sea level if the said amount is removed from the oceans is easily calculated from Figure 1. The con-

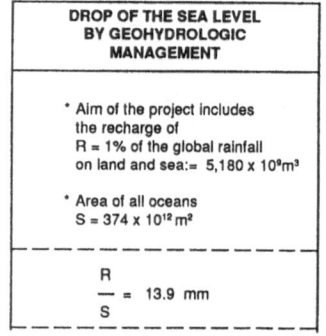

Figure 1. Drop of the sea level by geohydrologic management.

clusion of Figure 1 is that the change in the world water balance after the transfer of 1% of global precipitation would be a decrease of ocean level of only 13.9 mm.

Geohydrologic Management does not offer a solution for the indicated problem of increasing floods in low coastal areas.

In this paper the concept of artificial recharge on a large scale will be worked out in detail. Thus can be explained what would be the dimensions of a recharge operation including 1% of global precipitation. Artificial recharge on the mentioned scale, as a solution for the problem of sea level rise, is not feasible. However, the principle of artificial recharge on a smaller scale is very promising for the improvement of the water supply of arid and semi-arid areas.

Table 1. Global rainfall and evaporation (average over a long period)(after Huisman and Olsthoorn, 1983).

ON LAND

Land area	136 million km^2
Precipitation	790 mm/y
Evaporation	460 mm/y
Net evaporation	340 mm/y
Net evaporation i.e. available for surface and subsurface waters and equal to total discharge to the sea	46×10^{12} m^3/yr

ON SEA

Sea area	374 million km^2
Precipitation	1100 mm/y
Evaporation	1220 mm/y
Net evaporation	120 mm/y
Net evaporation i.e. vapour transport from the oceans to the continents	46×10^{12} m^3/yr

3. The Technique of Artificial Recharge

Artificial recharge means essentially introducing water into a pervious geologic formation through works designed:
- to maintain the infiltration rate,
- to increase the watered area and
- to increase the period of infiltration (Richter et al., 1959).

These works can be basins, modification of river streambeds, pits, ditches or furrows, injection wells and flooding of additional area along a river. A lot of literature has been written on this subject (Bize et al., 1972; Schmidt, 1982; Johnson and Finlayson, 1989).

Each artificial recharge project is different which results in an endless list of case-histories, each elucidating some, but never all aspects of artificial recharge. These can be compared in order to improve on the planning of a next project.

Nowadays changes in the quality of recharged water, changes in the subsoil and environmental aspects of recharging (pretreated) riverwater, alien to the aquifer, are emphasized. The theoretical aspects of the essential geohydrologic calculations are described by Huisman and Olsthoorn (1983) but for the present purpose it seems sufficient to mention only the most essential aspects of artificial recharge as far as they are needed to understand a project.

The present project of Geohydrologic Management aims at the recharge of water underground in order to store more water in the continental subsurface systems. Therefore, the storage aspect of recharging surface water underground is essential. However, there are many other aims. For the sake of clearness they will be mentioned briefly. Bize et al. (1972) clearly distinguish several aims for the application of artificial recharge:
a.   modification of the water quality;
b.   restoration of a disturbed equilibrium or protection against various disturbances, e.g. preventing or stopping the intrusion of saline groundwater in aquifers or reducing subsidence due to dewatering;
c.   increase of the capacity of underground resources and/or optimation of the recovery regime by storage.

A cumulation of aims occurs in nearly all the projects. In the case of Geohydrologic Management is, according to Fairbridge, the first part of c the most essential one.

A second way to classify artificial recharge projects is the following:
1.   Californian type of recharge: large-scale projects essentially meant for regional use. Recharge of this type can be found in California and near the North Sea coast of The Netherlands.
2.   European type of recharge: smaller scale facilities in the field of drinking-water supply, mainly intended to purify river waters and as a way to increase recovery from aquifers, e.g. in Germany, France and Switzerland.

It holds true after all that there is no general formula for the application of artificial recharge of aquifers. The construction must be modified according to local circumstances.

This workshop includes the study of artificial recharge on a global scale, so the description of the Californian type of artificial recharge may give the best view of the dimensions and the feasibility.

4.   Some Examples of Artificial Recharge

Artificial recharge for purposes of Geohydrologic Management includes recharge operations on a scale that surely exceeds all previous projects. Therefore it is advisable to focus on the larger projects in the world. The author does not intend to present an accurate and complete description of the projects nor to mention all important projects. His only aim is to present the reader a view on some of the larger facilities of artificial recharge in the world.

Most installations for artificial recharge have some elements in common. These are:
- artificial recharge nearly always starts after a period of recovery of large amounts of

pure groundwater from the aquifer;
- a gradual growth is more likely than immediate production at the full capacity.
The advantage of the first element is that the geohydrologic situation will be quite well
known before starting the recharge operation resulting from the experience gained during
the period of abstraction. A good knowledge of the situation is essential to predict the
hydrologic behaviour of the aquifer to be recharged.

4.1. CALIFORNIA STATE (RHONE AND MORGAN, 1989)

Area of the State	: 409,000 km^2
Inhabitants	: 27,000,000
Start of the artificial recharge operation	: 1895
Area involved	: . x 1000 acres of ponding basins
	. x 1000 miles of stream channels and unlined channels.
Number of artificial recharge installations	: an inquiry in 1958 (Richter, 1959) showed 276 installations.
Number of agencies running artificial recharge operations	: 65 agencies (1988).
Total artificial recharge in 1988	: 2,200 million m^3/y
Infiltration rate (basin recharge)	: 0.15-1.20 m/day
Transfer of recharge water	: by several pipeline systems over large distances, some more than 1000 miles.

4.2. THE COASTAL AREA OF THE NETHERLANDS (PROVINCES OF NOORD- AND ZUIDHOLLAND (Peters, 1989))

Area of the two provinces	: 6780 km^2
Inhabitants	: 5,800,000
Start of artificial recharge operations	: 1940
Area involved	: 800 acres spreading basins
Number of recharge installations	: 10
Total artificial recharge in 1988	: 165 million m^3/y
Infiltration rate (basin recharge)	: 0.1-0.5 m/day
Transfer of recharge water	: by several pipeline systems
Distance between the river and the discharge area	: 50-75 km.

4.3. EXPERIENCES FROM THE BIGGER ARTIFICIAL RECHARGE PROJECTS

- Artificial groundwater recharge projects are always subject to a slow evolution rather than a rapid development. The first stage in most cases is a normal groundwater development scheme, during which the constitution of the subsoil and the hydrological situation become clearer. That knowledge is essential for a fruitful start of an artificial recharge project.

- 1 % of annual global rainfall equals 5180 km^3 i.e. about 2355 times the yearly artificial recharge operation of the Californian water industry.

It is not necessary to recharge the mentioned 5180 km^3 of water in one year. This storage should be built up in a long range of years. A calculation is presented below for a filling period of the subsurface of 20 years.

5. Pilot Recharge Project for Global Geohydrologic Management

5.1. ENGINEERS' DREAMS

Before starting calculations the author wishes to state that prof. Fairbridge's idea should be considered without being hampered by the magnitude of the figures and the cost involved. Only the question: what is technically feasible. The author found several similar ideas in a book of Ley (1955) *Engineers' Dreams* and he took this book as a guide in the world between fantasy and technology. In section 5.3 one of these dreams is recalled in relation to a pilot project for Geohydrologic Management.

5.2. PILOT PROJECT FOR GLOBAL GEOHYDROLOGIC MANAGEMENT

At first one is shocked by the magnitude of the figures. Therefore the whole idea should be divided into a number of reasonably sized projects. Only then the feasibility of a recharge operation on such a large scale can be investigated. Then the real magnitude of the operation will also become clear. The author made several assumptions and calculations for a recharge project on a large scale. The first step is to calculate the amount of water that can be stored in the pores of a dry underground aquifer.

I started by postulating a circular project area with a diameter of 100 km, i.e. 7850 km^3.

```
 _____
|  ARTIFICIAL RECHARGE PILOT PROJECT  |
|                                     |
| Circular Area    R = 50 km          |
| Surface          S = 7850 km²       |
| Dry aquifer      d = 30 m           |
| Porosity         p = 0.2            |
|                                     |
| Content equals:  S x d x p = 47 km³ |
 - - - - - - - - - - - - - - - - - - -
```

Figure 2. Increase of storage.

```
 _____
|  ARTIFICIAL RECHARGE PILOT PROJECT  |
|                                     |
| Ultimate content         :  47   km³ |
| Filling of the aquifer in :  20 years|
| Q                         = 2.35 km³/a|
|                           = 2,350,000,000 m³ /a|
|                           = 6,500,000 m³ /day|
|                           = 75 m³/sec|
 - - - - - - - - - - - - - - - - - - -
```

Figure 3. Continuous water flow to the project.

What are the specifications of this circle/cylinder in order to make artificial recharge possible? The following assumptions have been made for a calculation to give an idea of the dimensions:

- porous and dry material on the surface and subsurface to allow infiltration; this material forms the potential aquifer;
- as a desired thickness of this pervious formation on the surface a minimum of 150 m is taken;
- an aquiferous part of 100 m;
- a dry part of 50 m;
- a possible rise of the groundwater level of 30 m;
- a porosity minimum of 0.2
- reasonable original groundwater quality in order to allow re-use of the water;
- not too many intercalated aquitards;
- a reasonable infiltration rate on the surface;
- the whole area must form - or be a part of - a basin so that it is able to hold water, preferably with natural boundary conditions.

Figure 2 shows the calculation of the amount of water that can be stored with the given assumptions. The additional storage in the cylinder amounts to 47 km^3 of water. This amount will be gradually introduced into the aquifer in a period that is called the filling period of the basin. After the period the outflow of the basin will be enlarged with the yearly influx. In reality the outflow of the groundwater basin will start earlier. Thus the filling will take longer than 20 years. In Figure 3 the flux of water to that assumed project has been calculated, allowing for a filling period of 20 years. The annual quantity of water involved has been calculated to be 2.35 km^3.

It will be shown below that the setup of this cylinder-project is of the same size as the total artificial recharge activity in the State of California.

5.3. NUMBER OF PROJECTS NECESSARY FOR GEOHYDROLOGIC MANAGEMENT

On the basis of the figures of section 5.2. the number of the above mentioned projects that together would store (after 20 years of filling) 1 % of global precipitation can be calculated. The number of the assumed projects is 110.

To get another impression of the project it is a meaningful exercise to compare this continuous influx of 75 m^3/sec. into one "cylinder" with the supply of recharged water of the Californian water industry. This amount of 2.2 km^3 equals the continuous influx of the pilot project. The conclusion is that Geohydrologic Management up to the scale of 1 % of global precipitation includes 110 artificial recharge and storage projects that each has the size of the total recharge capacity of the Californian water industry without any water recovery during at least 20 years.

5.4. LOCATION OF A PILOT PROJECT

In search of areas that could fulfil the requirements mentioned in section 5.2. the author's attention was drawn to the Chad area.

The first indication to study that area was the place where the conference took place. That place urged to look at the nearby dry Sahara region.

The second indication was one of the chapters in Ley's book on central Africa (1955). He recalled a project of Sörgel (Fig. 4), proposing to place a dam in the Congo River and to create a large lake in central Africa, called Lake Congo and another lake, to be filled by the Oubangui-branch of the Congo, called Lake Chad, being much larger than the former and present lake of that name. The project also mentioned an outlet river to the Mediterranean.

The third indication was a hydrogeological map of the lake Chad area, prepared by UNESCO Water Science Division in Paris.

An ideal geologic profile of the area SW of the present Lake Chad is shown in Figure 5.

Figure 4. The lake project of Herman Sörgel after Ley (1955).

Sadowa-Gajigana Area		
	thickness	lithology
layer 1	100 m	more or less clayed sand
layer 2	250 m	impervious clay
layer 3	50　m	sand, artesian aquifer

Figure 5. Ideal geologic profile, region SW of Lake Chad.

Based on this knowledge the thickness of the dry part of the superficial aquifer of a circular area is presented in Figure 6. Some further calculations, assuming a transmissivity of 100 m^2/day, showed that under the given assumptions the recharge of the said amount of water could be feasible, if there is really a need to do so. A very important additional advantage is that in the deeper formations an artesian aquifer is present. One of the major lessons to be learned from already realized projects (see section 4.3) is related to the start of such projects: the first operations should be the recovery of the available groundwater and only afterwards artificial recharge of surface water is to be executed.

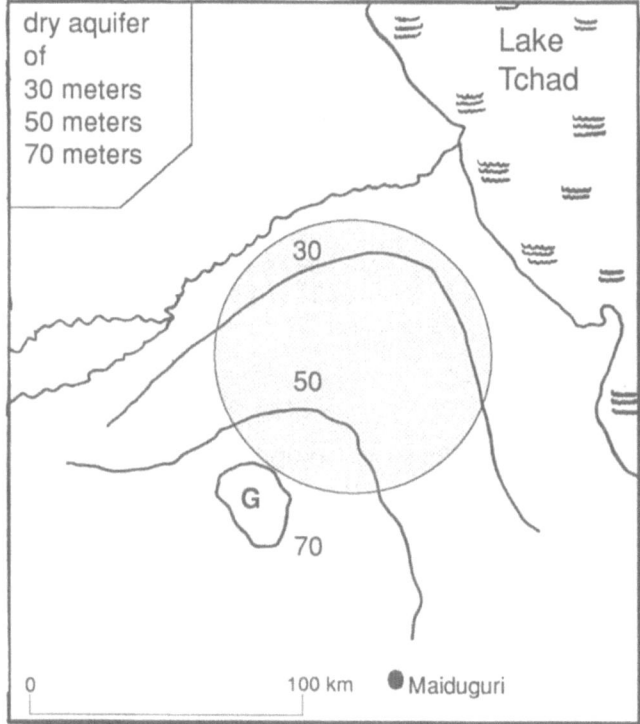

Figure 6. Artificial recharge pilot project.

During the Workshop in Fuerteventura this idea was discussed with scientists more specialized than the author in the local geologic and economic situation of the area. They appreciated the conception and that gave the impetus to put all this on paper, open to further discussion.

6. Conclusion, Statements and a Recommendation

The idea of prof. Fairbridge holds two major components. They are:
- controlling sea level rise;
- by recharging potential aquifers with river water that otherwise would flow to the sea.

This idea became known as Geohydrologic Management. In section 2 the conclusion was reached that the result of recharging river water in aquifers (or reservoirs) to an amount of 1% of global precipitation would only lead to a decrease in ocean level of less then 15 mm. In section 5 the conclusion was reached that it is possible to recharge aquifers with large amounts of water. In section 5.3 a comparison has been made between the recharge of 1% of global precipitation and the size of the Californian recharge installations. It became quite clear that the aim of artificial recharge operations should not be Geohydrologic Management but the development of the arid regions of the world. The feasibility of projects on such a large scale should be elaborated one after the other and could not be predicted in advance. In California and The Netherlands the recharge of large quantities of water have been considered feasible. The distance over which the recharge water is to be transported may be long, but even that did not stand in the way of implementation.

6.1. STATEMENT 1

Controlling sea level rise by storing water underground is not feasible because:
- even intercepting an amount of water equal to 1% of annual global precipitation and storing it in dry aquifers would only lead to a drop of sea level of 10-15 mm.

6.2. STATEMENT 2

In arid and semi-arid areas the storage of riverwater in the subsoil by artificial recharge is highly recommended to:
- bridge the gap between wet and dry periods;
- increase the yield of an aquifer;
- enlarge the evapotranspiration rate
- enhance the water quality;
- provide water for development.

6.3. RECOMMENDATION FOR A POSSIBLE RESEARCH PROJECT

Implement a research project for water recovery and artificial recharge to be developed in several phases near Lake Chad. The scale would be much smaller than that of the pilot project described in section 5. The capacity would be no more than a few million m^3/y.

Objectives of the research project:
a. a step towards continuous supply of water for development;
b. education and on the spot training directed towards groundwater management techniques. Possible side effects of the recovery should be taken care of.

Phases:
Phase 1: making an assessment of the available groundwater and developing the recovery of the groundwater present as a start of operations in the proposed area (see phase 2).
Phase 2: recharging the partly dry aquifer SW of Lake Chad (Nigeria) with Komadugu-river water. This phase should start on a small scale. An adequate project management must be available.
Phase 3: transporting river water over a long distance across the watershed by artificial means (see Pool in this Volume).

564

7. References

Bize, J., Bourguet, L., & Lemoine, J. (1972) *L'alimentation artificielle des nappes souterraines*, Masson & Cie., Ed., Paris

Huisman, L., & van Olsthoorn, T.N. (1983) *Artificial Groundwater Recharge*, Pitman Books, London

Johnson, A.I., & Finlayson, D.J. (eds.)(1989) *Artificial Recharge of Ground Water*, American Society of Civil Engineers, New York

Ley, W. (1955) *Engineers' Dreams*, The Viking Press, New York

Madrid, C. (1989) "Artificial Ground Water Recharge in Southern California", in Johnson, A.I., and Finlayson, D.J. (eds.) *Artificial Recharge of Ground Water*, American Society of Civil Engineers, New York, pp. 378-384

Peters, J.H. (1989) "Artificial recharge and water supply in The Netherlands", in Johnson, A.I., and Finlayson, D.J. (eds.) *Artificial Recharge of Ground Water*, American Society of Civil Engineers, New York, pp. 132-144

Richter, R.C., & Chun, R.Y.D. (1959) "Artificial recharge in California", Transactions of the American Society of Civil Engineers, 3274, pp. 742-766
Also published in December 1959 in the Journal of the Irrigation and Drainage Division, Proceedings Paper 2281

Rhone, R.A., & Morgan, H.V. (1989) "Artificial recharge with imported waters in Central and West Coast basins of Los Angeles County", in Johnson, A.I., and Finlayson, D.J. (eds.) *Artificial Recharge of Ground Water*, American Society of Civil Engineers, New York, pp. 405-415

Schmidt, K.H. (1982) *Artificial Groundwater Recharge*, Bulletin 11-14 DVWK, German Association for Water Resources and Land Improvement, Verlag Paul Parey, Hamburg/Berlin

CLIMATIC CHANGES IN THE LEVANT AND THE POSSIBILITY OF THEIR MITIGATION

ARIE S. ISSAR
The Jacob Blaustein Institute for Desert Research
Ben-Gurion University of the Negev
Sede Boqer Campus 84993, Israel

ABSTRACT: Variations in the composition of oxygen 18 and carbon 13 in core samples taken from the Sea of Galilee and speleothems from caves in Northern Israel show that during historical times the region went through dry and wet periods. Each period lasted from decades to centuries. Similar changes can be expected in the future.

In order to prepare for such changes so as not to entirely dependent upon the randomness of the climatic regime, it is recommended that use could be made of water from fossil aquifers together with desalinated seawater for domestic use, part of which could be reused as treated sewage water. These sources could be utilized in addition to conventional surface and groundwater systems. The long-term plan for utilization of these resources should be calculated according to the ratio between storage and long-term expected demand, not to mention economic issues forecast within the framework of future socio-economic scenarios.

1. Introduction

The availability of water resources is a function of precipitation, which is a function of the climatic regime. This regime is stochastic. One has a better chance of giving a more correct answer when prediction data are collected for long periods. The methodology for forecasting the probability of events in a stochastic system is empirical, i.e. data from the past are processed in order to detect a pattern of changes, without the need for understanding the mechanism which causes the changes.

The requirements for processing stochastic series are long-term observations. The reliable forecasting period is a function of the range backwards. As in most regions, meteorological observations are restricted to a few decades and are very seldom recorded for more than a century. "Safe" predictions are also for this time range. This is mainly the case for developed countries, where the duration of observations and, thus, prediction is sufficient for normal time ranges. When it comes to Third World countries, planning has to be carried out more on the basis of correlation with other regions and, many times, on the basis of synthetic

565

R. Paepe et al. (eds.), Greenhouse Effect, Sea Level and Drought, 565–574.
© 1990 *Kluwer Academic Publishers.*

data based on statistical processing of the short-term data available. This approach is also used in developed countries when the range for forecasting, based on time series analysis of the observed past, is not sufficient. This approach will, however, fail in cases of climatic changes beyond the range of time that was covered by observations, especially if such processes are of a catastrophic nature, namely that their range (i.e. higher or lower levels of precipitation) is beyond the range of the absorption capacity for the planned system. The system may then collapse. The simplest example would be for a dam that is planned to withstand a certain maximum magnitude of flood water. A more complex example would be for an entire economical system to be based on a certain predicted water supply level. The problem of predicting the impact of a climatic change on a hydrological system, in order to plan accordingly, becomes even more acute when countries on the margins of the arid belts are involved. The reason is that fluctuations are more severe and when precipitation is below a certain minimum threshold, irreversible environmental and socio-economic processes may begin.

A response to these dangers and problems could be a combination of three approaches. In the first approach synthetic climatic curves would be constructed on a multidisciplinary data basis and not solely from meteorological records. These include botanical data, such as tree rings and pollen assemblages in the sediments of marshes and soil profiles, as well as data derived from historical and archaeological research. Although such are not directly quantitative, they can be incorporated into the models as an indication of what changes can occur and the extent such catastrophes might reach.

The second approach would try to understand the atmospheric systems which cause climatic changes and try to predict their likelihood and dimensions; that is, an effort to open, as much as possible, the "black-box" or boxes which are behind the stochasticity of the climate. For this purpose the data obtained from the first approach are important and may serve to calibrate the climatic model when it is run backwards. This calibration is necessary due to the fact that these models are very complex and incorporate many assumptions, which can be tested by observations derived from historical data.

The third approach is practical. It takes into consideration the fact that many observations show our globe to be undergoing a gradual warming process. Whether this change is due to the greenhouse effect of other reasons, man must prepare in order to face this change and do all that is possible to mitigate its effects. This can be achieved by making the water supply system more "deterministic", by trying to become less dependent on the random character of nature. This aims to widen the safety margins of supply systems. In such cases groundwater aquifers, if available, would play an important role in the water supply system. The reason for the propriety of groundwater for this purpose is that it flows through porous media. One can regard the minute rock particles as natural retarding dams which reduce the velocity as well as lengthen the flow path of the underground water. This effect is cumulative and, under certain geological conditions, may be responsible for the storage of vast quantities of water recharged over many millennia. Yet, incorporation of these resources needs to be carried out in the framework of a long-term master plan in order to guarantee the quantities needed at the appropriate time and as economically as possible.

For this reason the following questions must be answered:

1) How will future climatic change regimes affect the water resources in each region? It is clear that while some region may become drier other may become more humid.

2) What is the potential storage capacity of groundwater resources and how can they fit into future water supply systems in order to answer local demand, as well as into the framework of a regional system?

These questions will be answered in the following article by using Israel as a case study. This country is situated on the border of a desert and during the past, present and in the future influenced by long and short-term climatic fluctuations. In this context what has already been done as well as what should be done in order to overcome problems of water deficiency are discussed.

2. Changes in the Future Learned from Changes in the Past

Israel lies between two climatic zones: The Mediterranean zone and the Sahara zone. The Mediterranean zone is part of the Westerlies belt, and the Sahara zone is the northern part of the Inter-Tropical Convergence zone (ITCZ). The Sahara zone affects the entire Middle East during the summer season when its influence spreads northward. During winter the dry Sahara zone, characterized by high barometric pressure, moves southward so that the region is influenced by cyclonic cold fronts from the west and northwest which bring in moist air that has accumulated over the Atlantic Ocean and the Mediterranean Sea. The southward movement of these fronts and thus the number of rain storms reaching the region varies from year to year. In years when a belt of high pressure remains over the area rain storms are less abundant resulting in droughts.

Two main factors, namely the rate of movement of the high pressure belt of the Sahara and the changes of the trajectories of marine storms, affect the mean annual quantity of rain throughout Israel. Scarcity and high variance of rains are more evident the further one travels to the South due to the configuration of the coastline of the southeastern edge of the Mediterranean Sea. The southern part of Israel, the Negev, lies outside the main path of rainstorms coming in from the northwest: hence its aridity (Fig. 1).

Rain usually falls between November and March. Rain storms may be preceded by a barometric high over the desert area. In this case dry and sometimes hot air brings dust from the desert which

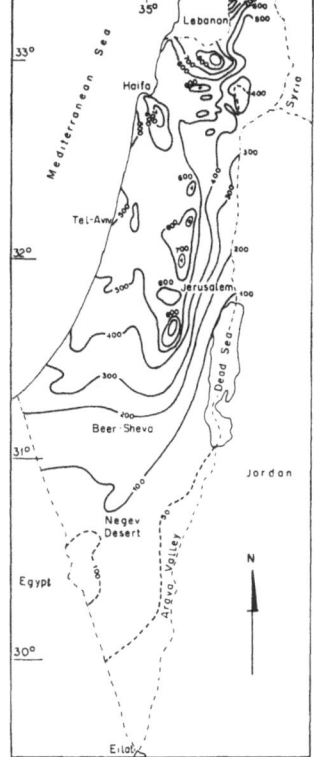

Figure 1. Map of rainfall in Israel.

travels towards the barometric low. In the autumn and spring, when dust storms are most abundant the hot, dry spell or "Khamsin" may break with a heavy rain storm. Most of the duststorms are connected with barometric lows and come from the west. Some of them travel over northern Africa and dust is brought from the Sahara. In some special climatic conditions, which were dominant presumably during glacial periods, the cold front entered the region over North Africa. This is the reason for the meteorological conditions that cause frequent heavy dust storms to be followed by rain storms. This also caused large quantities of dust to be washed out of the atmosphere of transported by flood water and deposited as Pluvio-Eolian deposits, namely loess. Many of these storms, which occurred during the autumn or even summer reached land via the northern coast of Africa and blew over salt and gypsum marshes (sabkhas) that were created by the regression of the sea due to glaciation. This caused the air to become enriched with salt and gypsum, characterizing the loess layers and the palaeowater which recharged the aquifers of the Nubian sandstone found under the Sinai and Negev deserts.

The period of massive deposition of these layers ended at about 15,000 years B.P. arising from an abrupt change from a humid to a more arid climate at the end of the Pleistocene. It was followed by a period of sand deposition brought to shore by winds. The sand dunes covering the loess layers advanced to a distance of about 40 km from the sea shore covering everything in their path and blocked stream channels which previously flowed to the sea. Two main sand layers can be distinguished: a more extensive one, which began its penetration towards the end of the last glacial period, namely at about 15,000 years ago, and continued its inland drift throughout prehistorical times; and a younger one, which covered the western stretches of the coastal plain. The sands penetrating during dry periods settled during the more humid periods. The settling down of the sand dunes is distinguished by the layer of fine reddish silts which accumulated on the surface. The reddish colour was derived from the decomposition of minerals containing iron (Tsoar, 1976). The young layer of sand that covered only the coastal area is still mobile. Its began its penetration at the end of the Byzantine period, the beginning of the Moslem conquest ca 1,500 B.P., which was most probably a function of an aridification phase (Issar and Tsoar, 1987; Issar et al, 1987). The sands originated in Nubia whence they were transported by the Nile river and brought to the shores of Israel and Sinai by counter-clock-wise currents (Emery and Neev, 1960; Goldsmith and Golik, 1980). The terraces of the Nile river show that its tributaries in Ethiopia enjoyed a humid climate at the end of the Pleistocene due to the northern migration of the monsoonal belt in the wake of the migration to the North of the desert belt. This explains the sudden change in the loess deposits to sand deposits in the western Negev and northern Sinai, when larger quantities of sand from Nubia were dumped into the Mediterranean Sea.

Another characterizing feature of the periods which experienced climatic changes are the rises and later the fall in the levels of the Dead Sea. During the last glacial period this lake extended from the Sea of Galilee in the North to about 30 km South of the present shores to form the palaeo-Lake Lisan (Begin et al., 1974). Its level reached an altitude of -200 metres below MSL compared to the present level of -430 metres. The water of the lake was brackish. The sediments are known as the "Lisan Formation" and they consist of varves of marl, chalk, gypsum, and sands.

Changes in the oxygen and carbon isotope compositions of the carbonates deposited at the bottom of the Sea of Galilee coincide with variations in the pollen assemblages of these deposits (Issar and Tsoar, 1987; Issar et al., 1987), and have shown that climatic changes also occurred during the last 7000 years. This conclusion has been confirmed by changes observed in the isotopic composition of samples from speleothems taken from caves in the Upper Galilee of Israel (Geyh et al., manuscript). A good correlation can be found between these isotopic data and changes in the level of the Dead Sea and Nile river (Fig. 2). All this is in compliance with the palaeoclimatic model suggested for the transition period between the Pleistocene and Holocene period, as discussed above.

The isotopic and sedimentary evidence for climatic changes can be correlated with changes in the settling and abandonment of settlements in the more arid parts of the country. This is especially clear in the plain of Beer Sheva, situated on the northern border of the Negev desert (Fig. 1). During the Chalcolithic period a humid and colder climate prevailed, as can be seen from the Sea of Galilee and the speleothem samples taken from caves in the Upper Galilee, which exhibit lower values (Fig. 2). Some settlements dating from this period have been mapped in the plain of Beer Sheva. (Govrin, in press). Many of these settlements were located in the proximity of riverbeds. The most famous is Beer Sheva (Perrot, 1968), which most probably received its water supply from shallow wells in the riverbed.

Towards the end of the Chalcolithic period and at the beginning of the early Bronze age, a climatic change took place which caused a major decline in the number of settlements. The warmer period which followed, as can be seen from the isotopic composition, caused the abandonment of all the settlements in this region (Govrin, in press). During the Middle Bronze Age the area was practically devoid of settlements. At about 2000 B.C. aridification reached a critical phase throughout the Levant. Marine archaeological investigations (Raban, 1987) along the coast of Israel show that the inlets of the river became silted up, which indicates that there was an increase in the supply of sand from the Nile river. A new settlement phase started at the beginning of the Iron Age (ca 1000 B.C.), when the Negev was part of the Kingdom of Judea. Sometime during the 6th or 7th century B.C. aridification phase is indicated by the isotope curves. From about 300 B.C. to about 600 A.D. a colder, more humid phase affected the entire region. This can be seen in the depletion of the isotopes in the core sample taken from the Sea of Galilee and the speleothem samples, as well as in the rise in the level of the Dead Sea, which during part of this period reached a maximum level of about 70 metres above the present level. At the same time the Nile river was low. A short, drier interval occurred about 200 A.D. The humid period was also a time in which many settlements all over the Negev were established by the Nabatean people and continued during the Roman and Byzantine periods. Many cities and towns were built. The cities were very rich and surrounded by terraced valleys irrigated by water harvesting techniques (Evenari et al., 1971).

Beginning at about 600 A.D. a gradual abandonment of the settlements is in evidence from archaeological excavations. At the same time the isotope curves show a warming trend which peaked at about 700 A.D. The Arab conquest took place during this period

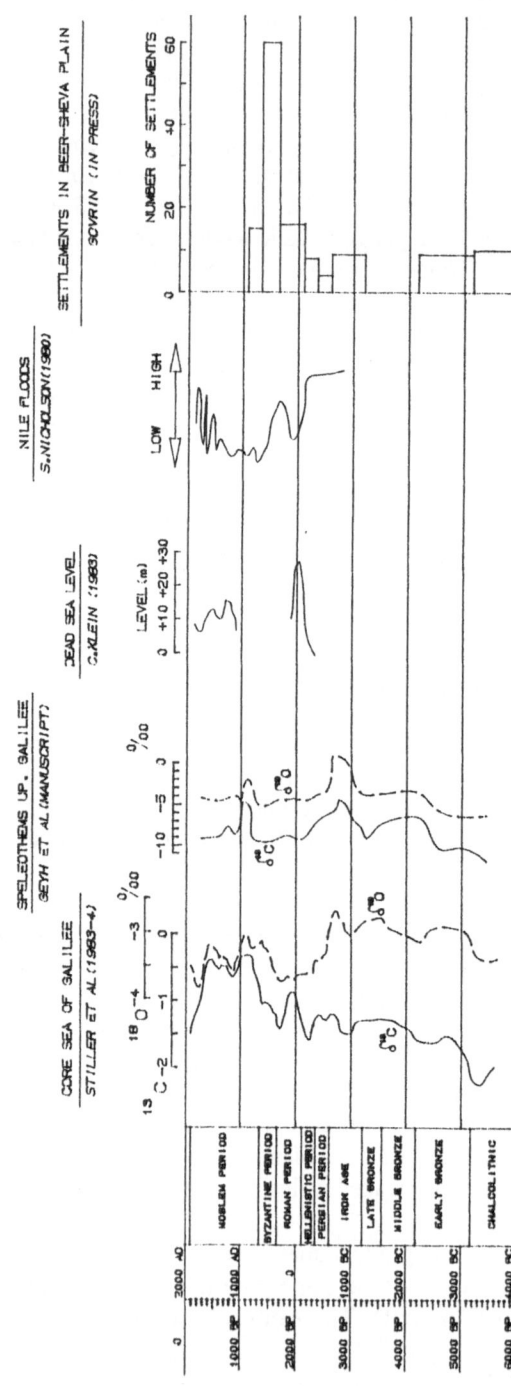

Figure 2. Correlation between isotope composition of sediments in core samples taken from the Sea of Galilee and speleothems in caves of Galilee, and the levels of the Dead Sea and Nile river.

and, after the resettling of some of the desert towns, they were again totally abandoned. From this period a invasion of sand dunes to the coastal plain of Israel began. This is explained by a rise in the level of the Nile river, which began at about 300 A.D. Duringthat period a short, warmer phase occurred causing a rise in the level of the sea. A more humid climate started at about 1300 A.D. and finished at about 1500 A.D. This corresponds to the time of the Crusaders or the Little Ice Age. The isotope curve for the core samples taken from the Sea of Galilee shows a cold phase during the last century and a warming trend since then. There is support of a warming trend from the meteorological data collected in Jerusalem since the mid-19th century. Striem (1985) processed these data and reported that the winter temperatures rose by more than 1°C, while summer temperatures remained steady. He also found that the mean monthly barometric pressure rose by about 1 mb, and that the annual rainfall decreased from about 600 mm in the second half of the 19 century to about 500 mm in the first half of this century. An increase in the amount of rainfall has taken place since 1985.

3.   Groundwater Storage as a Major Factor to Mitigate Drought

Living on the margins of a desert, the inhabitants of Israel have had to face the problems of ensuring their water supply during dry seasons and years since the early stages of history. This has been done by relying on the groundwater supply from either springs or wells. At present the main operational aquifers are the following (Nativ and Issar, 1988):
1.   The limestone aquifer of Mount Hermon (mainly of Jurassic age). Springs feeding the Jordan emerge from this aquifer. The water flows to the Sea of Galilee, which functions as a storage lake. The average annual pumpage rate is about half a billion m^3.
2.   The limestone-dolomite aquifers of the northern and central parts of the country (mainly of Middle Cretaceous age). The water from this aquifer supplies about a third of a billion m^3 each year.
3.   The calcareous sandstone aquifer of the coastal plain (mainly of Plio-Pleistocene age). This water supplies another 430 million m^3.
4.   The Nubian sandstone aquifer underlying the southern arid part of the country (mainly of Lower Cretaceous age). The water from this aquifer is fossil. Annual pumpage is about 30 million m^3.

These aquifers, except for the last, are integrated into one system by the National Water Carrier.

The engineering requirements for this integration which started to materialize in the early fifties were two-fold: a) to transfer water from the humid north, where water is ample but land is limited, to the dry plains of the south, where water is scarce but land is ample; and b) to mitigate seasonal and annual dry spells.

In order to carry out the National Water Supply Plan (NWSP), a special water law was passed declaring that all water resources are national property, while guaranteeing the existing rights of users. The importance of such a law for regional planning on a national basis cannot be overstated. It is important to point out that the law was prepared after the Master Plan was drafted. In other words, the law was written in order to enable the execution of the NWSP.

The law and plan have put emphasis on the exploitation, management and distribution of water form a quantitative point of view, and less on qualitative aspects. This has proven to be the weak point of the system, and has brought about a deterioration in the quality of water, mainly in coastal areas (Nativ and Issar, 1988). Another negative aspect of this plan was that it guaranteed rigid water quotas to the farmers, who were the main users. In 1985 agriculture used 1.4 m^3 per year, while domestic consumption and industrial use was only 0.5 m^3. Thus, when several dry years followed each other, the aquifers were over-pumped, which caused a general decline in the water table and penetration of sea water took place. This called for a revised National Water Supply Plan, which has recently been prepared. The main outline for this plan is a reduction in the over-pumping of aquifers in the north and central parts of the country by approximately 25%, and compensating the agricultural demand by using treated sewage water for irrigation purposes after it has through the process of recharge into the groundwater. Recharge of the depleted aquifers by water being pumped from the Sea of Galilee is also planned. Seawater desalination is not foreseen as a solution unless there is a breakthrough in desalination methods, which would reduce the present cost by a third. On the other hand, the development of efficient methods for water use in agriculture has already played, and is going to continue to play, an important role in the water balance of the country.

Another resource available for future development is the brackish water found in various aquifers. This water is used to grow special salt-tolerant crops. This is practised in many instances through the use of special irrigation methods such as drip irrigation. The total annual amount of brackish water (1,000 to 6,000 ppm TDS), forecast for future use is about 0.15 billion m^3.

The total amount of water forecast for development by the end of this century is about 2.1 billion m^3. Yet, this plan does not take into consideration a severe climatic change which would affect the present hydrological balance by an appreciable amount. The questions are: a) what might be the order of magnitude for such a change; and b) what can be done to mitigate the influence of such a change?

To assess the scale it is suggested that w look at historical records and estimate the magnitudes for the changes which occurred in the past. The most prominent aridification change took place during the Byzantine period, whereas the most humid phase took place during the Roman period (Fig. 2). Unfortunately there are no up-to-date records for the lower level of the Dead Sea which might give an idea of the impact of a negative change in precipitation. On the other hand, there is a possibility of estimating the influence of a positive change. This can be done by calculating the inflow, and thus the precipitation needed to raise the water level of the Dead Sea by about 70 metres during a period of maximum change. This lasted for about 70 years during the Roman period. The increase calculated was about 40% (Klein, 1982). It is suggested that such a change is an approximation.

With regard to the second question, three directions that may be pursued as follows:

1) To utilize more extensively the Nubian sandstone aquifer containing fossil water. Calculations carried out by the author have shown that about 0.2 billion m^3 can be drawn from this aquifer on a long-term basis. This water can be pumped either from wells or specially constructed galleries along the western margins of the Arava, to be pumped northwards. Brought to the Beer Sheva area the cost of one m^3 is estimated to reach 0.30 U.S.$.

2) The desalination of seawater along the shore, especially in connection with electrical power plants. The cost for this water is estimated to be in the range of 0.60 to 1.00 U.S.$. This water would mainly be for domestic use.

3) About 60% of the desalinated water can be reused for irrigation as reclaimed sewage water. The cost of desalination of seawater and the treatment of sewage water for irrigation can be shared by urban and agricultural users of this water.

In conclusion, a climatic change can be mitigated by using a combination of mined fossil water, desalinated water for domestic use and its reuse as treated sewage water for irrigation purposes.

4. References

Begin, Z.B., Ehrlich, A., & Nathan, Y. (1974) *Lake Lisan, the Pleistocene Precursor of the Dead Sea*, Geological Survey, Ministry of Commerce and Industry, State of Israel, Bull. No. 63, 30 p.

Emery, K.O., & Neev, D. (1960) *Mediterranean Beaches of Israel*, Geological Survey, Israel, Bull. No. 26

Evenari, M., Shannan, L., & Tadmor, N. (1971) *The Negev: The Challenge of a Desert*, Harvard University Press

Geyh, M.A., Wakshal, E., & Franke, H.W. (manuscript) *Carbon-14, 13 and Oxygen-18 data of Speleothems from Upper Galilee*

Goldsmith, V., & Golik, A. (1980) "Sediment transport model of the southeastern Mediterranean coast", Marine Geol. 37, pp. 147-175

Govrin, Y. (in press) *Map of Molada (139) 14-07*, Published by the Department of Antiques and Museums, the Archaeological Survey of Israel, (in Hebrew with English Abstract)

Govrin, Y. (in press) "Settlement patterns in the Northeast Negev, in the fourth-third millennium B.C.", in Orion, E., and Cohen, R. (eds.) *Archaeology of Nomads in the Southwest Deserts of Asia*, (in Hebrew)

Issar, A.S., & Bruins, H.J. (1983) "Special climatological conditions in the deserts of Sinai and the Negev during the latest Pleistocene", Paleo 3, Elsevier Science Publishers, Amsterdam, 43, pp. 63-72

Issar, A.S., Tsoar, H., Gilad, I., & Zangvil, A. (1987) "A palaeoclimatic model to explain depositional environments the Late Pleistocene in the Negev", in Berkofski, L., and Wurtele, M.G. (eds.) *Progress in Desert Research*, Rowman & Littlefield Pub, pp. 302-310

Issar, A.S., & Tsoar, H. (1987) "Who is to blame for the desertification of the Negev?",

574

Proc. IAHS Symp., Vancouver, Canada, IAHS Publ. No. 168, pp. 577-583

Katsnelson, J (1970) "Frequency of dust storms ad Beer Sheva", Israel J. of Earth Sciences 19, pp. 69-77

Klein, C. (1982) "Morphological evidence of lake level changes: western shore of the Dead Sea", Isr. J. of Earth Sciences 31, pp. 67-99

Nativ, R., & Issar, A.S. (1988) "The problems of an over-developed water system: the Israeli case", Water Quality Bulletin 13 (14), pp. 126-131

Nicholson, S.E. (1980) "Saharan climates in historic times", in Williams, M.A.J., and Faure, H. (eds.) *The Sahara and the Nile*, A.A. Balkema, Rotterdam, pp. 173-200

Perrot, J. (1968) "La Préhistoire Palestinienne", *un Supplement au Dictionnaire de la Bible* VIII, pp. 416-438

Raban, A. (1987) "Alternated river courses during the Bronze Age along the Israeli coastline", Colloque Internationeaux C.N.R.S. *Déplacements des lignes de rivages en Méditerranée*, Ed. du C.N.R.S., Paris, pp. 173-189

Stiller, M., Ehrlich, A., Pollinger, U., Baruch, U., & Kaufman, A. (1984) "The Late Holocene sediments of Lake Kinneret (Israel) - multidisciplinary study of a 5 m core", in *Geological Survey of Israel*, Ministry of Energy and Infrastructure, Jerusalem, Israel.

Striem, H.L. (1985) "Quantitative and qualitative aspects of the recent climatic fluctuations", Israel J. of Earth Sciences 34, pp. 47-48

Tsoar, H. (1976) "Characterization of sand dune environments by their grain-size, mineralogy and surface texture", in Amiran, D.H.K., and Ben-Arieh, Y (eds.) *Geology of Israel*, IGU, Jerusalem

Williams, M.A.J., & Adamson, D.A. (1980) "Late Quaternary depositional history of the Blue and White Nile rivers is central Sudan", in Williams, M.A.J., and Faure, H. (eds.) *The Sahara and the Nile*, A.A. Balkema, Rotterdam, pp. 281-304

Yaalon, D.H., & Ganor, E. (1975) "Rate of aeolian dust accretion in the Mediterranean and desert fringe environments of Israel", 9th Int. Congr. of Sedimentology, Nice, 2, pp. 169-174

THE LEGACY OF SAHELIAN MANAGEMENT: 1965-1988

G.L.WELLS
School of Geography
Oxford University
Mansfield Road
Oxford OX1 3TB
England

K.BURKE
Department of Geosciences
University of Houston
Houston, Texas 77204-5503
USA

ABSTRACT. The Sahel undergoes episodes of extreme climatic change on timescales ranging from millennia to decades. Following a period of high rainfall in the 1950s, plans were initiated in the 1960s for a number of large-scale irrigation schemes based upon the assumption that climatic conditions of the 1950s and 60s were normal for the region and would persist into the future. Many of these projects were completed in the 1970s and early 1980s during a time of prolonged, intense drought. A review of NASA orbital photography and field reports from the Sahel illustrates the mixed record of attempts to manage regional hydrology in the face of changing environmental conditions. In contemplating the development of new water management projects designed to assuage the problems brought about by drought and the growing regional population, the various fates of these earlier attempts to control the changeable Sahelian environment should be recalled. Small-scale water management schemes may prove to be more resilient to climatic change in the Sahel than traditional large-scale engineering projects sponsored by Western nations and favoured by the central governments of Sahelian nations.

1. Introduction

In the 1960s at the beginning of the post-colonial era, the Sahelian nations of former French West Africa and Nigeria, Gambia and Sudan continued to pursue courses of economic development that emphasized the export of raw materials to European markets. With the

575

R. Paepe et al. (eds.), Greenhouse Effect, Sea Level and Drought, 575–592.
© 1990 *Kluwer Academic Publishers.*

exception of Nigeria with its enormous petroleum reserves, nations in the Sahel turned to the expansion of cash-crop monoculture as the primary source of capital with which to construct modern societies. Subsistence agriculture has given way to widespread groundnut, rice, wheat and cotton farming and cattle ranching, as governments encourage the production of export commodities. Whereas the Green Revolution of the mid-1960s swept through developing countries of eastern Asia and Latin America to eliminate chronic famines and, in some cases, to create grain surpluses for profitable export, the introduction of advanced farming technology and high- yielding cereal varieties has had little impact upon the Sahel. Reasons for this failure are many and include the lack of coordinated, sustained investment, the decline of international commodity prices, administrative mismanagement by central governments and, in several instances, internal conflicts involving neighbouring populations, but of all the possible explanations for the limited development of regional agriculture, one cause appears of central importance—the absence of a reliable source of water.

Yet in the Sahel there are projects where large investments have been made in systems designed to provide water for agriculture on the scale envisioned for the transformation of Sahelian economies by an African Green Revolution. In this chapter, we examine the performance of several of the most notable water management schemes set against the background of regional climatic change. The protracted drought in the Sahel beginning in 1968 and lasting at least through the mid- 1980s with a brief remission in the mid-1970s severely disrupted plans to base future economic growth upon an expanded agricultural export sector. In fact, by contrast with conditions 25 years ago when western African nations were largely self-sufficient in food production, during the early 1980s the volume of food imports rose at an annual rate of $>4.0\%$, while domestic production fell in most areas (Grove 1986). With populations growing at estimated annual rates ranging from 2.5% to 3.4% (World Bank 1987), Sahelian countries confront a steady decline of per capita food production, if present trends persist. In the face of this prospect, it has become essential to identify the factors which have led to the failure of large-scale irrigation projects and to determine which water management systems offer the best hope for long-term success.

2.   A Background of Climatic Change

The landscape of the Sahel has been shaped by a succession of radically different regional climates over the course of the late Quaternary. At the peak of the last ice age (18,000 years ago), an increased poleward temperature gradient created by the presence of continental ice sheets and expanded pack ice cover strengthened the zonal circulation across the northern hemisphere (Wells 1983) bringing the westerly airstream into northern Africa (Nicholson 1982; Street-Perrott and Roberts 1983). Meridional circulation weakened and was displaced southward, causing precipitation to decline over the region of the modern Sahel (Kutzbach and Street-Perrott 1985). During this period of late-glacial hyperaridity, sand dunes formed across the Sahel. Longitudinal dunes blocked the course of the Senegal River south of Kaedi (Michel 1980) and the course of the Niger River north of Mopti near Lac Debo (Grove and Warren 1968). In the Chad Basin, the ancestral Lake Chad vanished, and crescentic dunes migrated over the desiccated lake floor (Servant and Servant-Vildary 1980). In central Sudan, the Sudd and Marchar marshlands dried, and sand dunes formed along the White Nile and Blue Nile floodplains and throughout the Qoz of Kordofan (Warren 1970; Adamson et al. 1985). Archaeological investigations (Petit- Maire and Riser 1983; Wendorf et al. 1985) report no evidence of human occupation of oases in the southern Sahara. In the region of the modern Sahel, human habitation was restricted to a few, isolated areas located primarily in southern Nubia (Butzer 1980). The lack of archaeological finds in the Sahel combined with geomorphological evidence for the creation of desert landforms lead to the conclusion that during the period 18,000 to 12,500 years BP the southern boundary of the Sahara lay along latitude 14°N in the western Sahel and 11°N in the Chad Basin and Sudan. This position corresponds closely to the southernmost limit of modern Sahelian grassland and mixed acacia woodland. It represents a southward shift of the vegetation belt by 450- 600 km. The southern extent of Sahelian vegetation during the glacial maximum is not known in detail, but may have locally reached the Gulf of Guinea.

From the period 12,500 to 4500 years BP, intensified meridional circulation led to much greater monsoonal precipitation over the Sahel and Sahara (COHMAP Project Members 1988). Many large lakes and marshes developed within Saharan basins during the late Pleistocene to middle Holocene pluvial episode. At 22.5°N in the Taoudenni area of northern Mali, freshwater lakes existed from 8300 to 6700 years BP (Fabre and Petit-Maire 1988). During the same period of the early Holocene, Lake Mega-Chad, a great inland sea of 320,000 km^3, covered much of the Chad Basin with a shoreline closure stretching from Bongor to Faya Largeau (Grove and Warren 1968; Servant and Servant-Vildary 1980). In central Sudan, the regional water table rose ponding freshwater lakes between Pleistocene sand ridges. In northern Sudan, Wadi Howar, currently a dry river bed choked by eolian sediments, connected drainage from the Jebel Marra region to the Nile (Pachur and Kropelin 1987), and a freshwater oasis surrounded by savanna grassland existed at Selima near 21.5°N from 9500 to 4500 years BP (Haynes 1982; Ritchie and Haynes 1987). Early to middle Holocene archaeological sites are scattered across the central Sahara and supply evidence of widespread human occupation of a savanna landscape with an abundance of large fauna, including the lion, giraffe, elephant, hippopotamus and rhinoceros. An exact position for the northernmost limit of savanna vegetation during the Holocene remains speculative, but it must have been at least 23°N in the western Sahel and 21.5°N in Sudan.

2.1. HISTORICAL CLIMATIC FLUCTUATIONS

Following the last Saharan pluvial phase (which ended 4500 years BP), Neolithic settlements in the Sahara were abandoned as the regional climate returned to hyperarid conditions. Contrary to some popular beliefs, the last 4000 years have been characterized neither by climatic stability with the desert and savanna zones remaining in essentially the same positions nor by progressive aridity causing an inexorable spread of desert. Hassan (1981) has shown large fluctuations in Nile discharge recorded by the Nilometer over a nearly 1300-year period. The Nile flood maxima are related to monsoonal precipitation over Ethiopia and Sudan, whereas flood minima are connected to rainfall over the equatorial region drained by the White Nile. Major peaks in Nile floods occurred in the periods 1070-1180 and 1350-1470, and minor highs were recorded during 1725-1800 and 1830-70. Similar episodes of increased monsoonal rainfall are displayed by changes in the level of Lake Chad as determined by radiocarbon dates obtained from shoreline sediments (Maley 1973, 1976). Highstands of the lake are inferred for periods circa 1330-80, 1470-1540, 1600-1710, 1750-1810 and 1860-90. Though Lake Chad receives a large proportion of its input from a drainage system with its source in the Adamawa region to the south of the Sahel, studies of changes in the interannual distribution of African rainfall have shown that precipitation over the two regions is strongly correlated (Nicholson 1981; Nicholson and Chervin 1983).

Historical records confirm relatively wet intervals interspersed with severe droughts in the Sahel during the past 400 years. According to Nicholson (1980; 1981), the sixteenth and seventeenth centuries experienced generally greater rainfall when compared to the twentieth century. The eighteenth century was similar to the present with episodes of extended, severe drought, and the nineteenth century was wetter, especially in its latter half. Between 200 to 350 years ago, local chronicles and the journals of travellers noted a longer rainfall season, the presence of vegetation, large game animals, floods, permanent water bodies and freshwater wells in areas of the northern Sahel virtually uninhabited during the twentieth century. Dense stands of coastal mangroves and gallery forest stood along the banks of the Senegal River according to reports by Chambonneau in 1677 and Adamson in the 1750s (Nicholson 1981). In the 1790s, Browne (1799) encountered woodland in the vicinity of the Bahr el Ghazal in an area of central Chad close to the northern margin of twentieth century savanna grassland and wrote of a verdant landscape in northern Darfur near Kutum now on the edge of the Sahara. In the 1780s or 90s, the Bahr el Ghazal valley may have been sufficiently flooded to allow boat travel northward from Lake Chad to southern Borkou (Nicholson 1981).

Together with the evidence for moister conditions than those of the present century, historical records reveal the occurrence of severe droughts in the 1680s and 1710s (Nicholson 1980) and the periods 1738-56 and 1828-39 (Nicholson 1981). It was during the latter drought period that Charles Darwin made his famous observation of a Saharan/Sahelian dustfall over Sao Tiago in the Cape Verde Islands in January 1832. Among the surprising aspects Darwin discovered in his dust samples were siliceous plant fragments of more than 67 different west African species (Darwin 1839; 1846), an indication that drought stress had led to the deterioration of vegetation cover which was transported by harmattan winds into the Atlantic by deflation of the dried plant residue.

In the twentieth century, meteorological records reveal major droughts in the Sahel in the 1910s, reaching an extreme in 1913-14, the 1940s, particularly 1941-42 and 1947-48, and 1968-87, with 1970-73, 1977 and 1982-85 representing the driest years (Nicholson 1980, 1981; Lamb 1985). In contrast, the 1950s were exceptionally wet in the Sahel with precipitation 15-20% above the 1931-60 mean annual rainfall. Stark differences in decadal trends of precipitation can be observed when comparisons are made between the period 1950-59 and 1970-84. Mean annual rainfall during the 1970-84 Sahelian drought period declined 50% at Dakar, 39% at Timbuktu, 54% at Agadez and 35% at Khartoum compared to the 1950s (Nicholson 1989, Table 1). The amplitude of differences in annual rainfall

LATITUDINAL FLUCTUATION OF AFRICAN RAINFALL

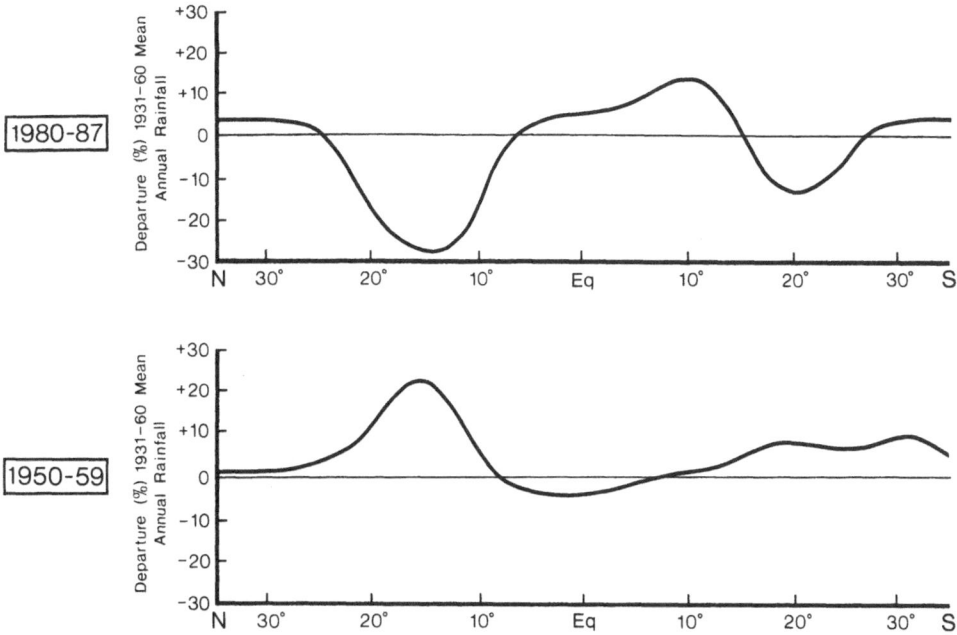

Figure 1. Changes in the latitudinal distribution of African rainfall on a decadal timescale are reflected by the records of 170 meteorological stations located 36°N-34°S.

and the latitudinal distribution of precipitation over the African continent have been studied to identify recurrent patterns of wet and dry years for particular regions (Nicholson and Chervin 1983; Nicholson 1986). One recurrent pattern of the 1950s was unusually wet conditions in the Sahel and drier than average conditions in equatorial Africa (e.g., Nicholson's (1986) Continental Anomaly Type 3 in 1952-56 and 1958). On the other hand, drought years in the Sahel (e.g., 1931, 1947, 1968, 1970) have often been associated with unusually wet years in equatorial Africa.

Using the precipitation records of the British Meteorological Office and World Meteorological Organization for 170 inland African stations from 36°N to 34°S, comparisons can be drawn between the latitudinal fluctuation of African rainfall during the periods 1950-59 and 1980-87 (Figure 1). Though no attempt has been made to weight station means by their longitudinal zones, and changes of rainfall over the central Sahara are highly inferential because of the lack of data collection, the latitudinal precipitation gradient reflects the decadal trends of precipitation anomalies of the types discussed by Nicholson and Chervin (1983) and Nicholson (1986). The 1950s were typified by high rainfall over the Sahel and Kalahari and below average rainfall near the Equator. Through 1987, the 1980s were marked by rainfall deficits in the Sahel and Kalahari and higher than average rainfall over the equatorial region. The recent latitudinal fluctuations of the African hydrologic budget can be seen in the levels of lakes contained within closed drainage basins of the Sahel and equatorial region. From 1966 to 1985, the surface area of Lake Chad fell from about 22,000 km² to 1689 km² (Grove 1985; Mohler and Amsbury 1988). By contrast, from the early 1970s to 1985, the surface area of Lake Rukwa (8°S) increased by approximately 30%. A nearly parallel record of regional rainfall and hydrologic trends is presented by Lamb's (1985) normalized rainfall departures for 20 western Sahelian stations which in the period 1968-85 failed to achieve the long-term mean annual precipitation in every year except 1969 and the discharge of the White Nile at Mogren, Sudan, that for the same period exceeded the long-term (1912-85) mean discharge in every year from 1968 to 1982 (Sir M. MacDonald and Partners 1988). The increased flow of the White Nile created by higher equatorial rainfall during the past two decades has led to the expansion of permanent swamp in the Bahr el Jebel floodplain of the Sudd from Bor to Malakal from 2700 km² in 1952 to 16,200 km² in 1980, as detected by aerial photographic surveys, satellite image analysis and field mapping (Howell et al. 1988, p. 44).

After a decrease of intensity in 1986-87, the long Sahelian drought may have been broken in 1988 by a return to rainfall close to the 1931-60 annual mean. For Niger, January-August 1988 precipitation was only 3% below the 1931-60 mean, while stations in Senegal received rainfall 8% below the long-term mean. In Sudan, remarkable rains came in July and August leading to totals well above the long-term annual mean for stations in the central and northern parts of the country. On 4 August 1988, Khartoum recorded a torrential 200.5 mm (Hulme and Trilsbach 1989); the 1931-60 mean annual rainfall for Khartoum is only 172 mm! Whether this year of good rains in the Sahel signals an end to drought or only an interlude in a longer period of regional aridity, such as occurred in 1974-76, will be determined by the character of regional rainfall as we enter the next decade.

2.2. CHANGE IN THE POSITION OF THE SAHEL/SAHARA MARGIN: A CONSEQUENCE OF MAN OR NATURE?

In the 1970s, particularly in the years following the United Nations Conference to Combat Desertification held in Nairobi in 1977, much of what has been stated in both professional journals and the popular media about recent desertification in the Sahel has placed major emphasis upon human degradation of the environment. As Wade (1975, p. 235) expressed the position, "the primary cause of the desertification is man, and the desert in the Sahel is not so much a natural expansion of the Sahara but is being formed in situ under the impact of human activity." News reports often present an image of the Sahara relentlessly advancing southward by 10s of kilometres per year under the stimulus of overgrazing by the herds of nomadic pastoralists and poor land management by sedentary farmers. In opposition to these views is the position that the retreat of Sahelian vegetation during the 1968-87 drought has been largely the result of natural changes caused by years of low rainfall and only in small areas abetted by human activities. To determine which position reflects the reality of environmental change in the Sahel requires an examination of the location and rate of movement of the savanna margin in response to population pressure and rainfall patterns over the past 40 years.

A regional analysis of the historical extent of savanna vegetation in the Sahel can be done with the aid of aerial photographic surveys and orbital remote sensing techniques. The first comprehensive aerial photography of the region was collected in the 1940s and 1950s by trimetrogon surveys conducted by French and British colonial governments and the U. S. Air Force. These photographs were used to prepare the IGN series topographic maps and other specialized maps which showed the extent of Sahelian vegetation (e.g., Raisz 1952). Additional aerial photographic surveys often using colour and colour infrared films were made in the late 1950s and early 1960s by private French and British firms and the U. S. Air Force for the Agency for International Development (USAID). The products

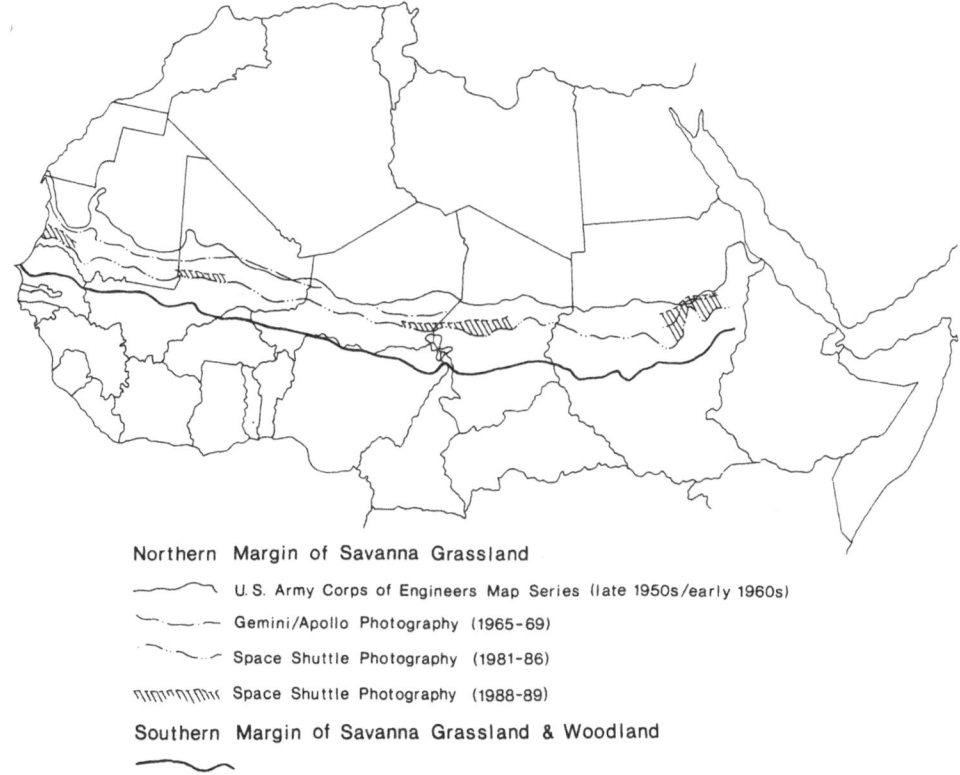

Northern Margin of Savanna Grassland

———⌐ U.S. Army Corps of Engineers Map Series (late 1950s/early 1960s)

⌐·—·— Gemini/Apollo Photography (1965-69)

⌐·⌐·· Space Shuttle Photography (1981-86)

ʍʍʍʍ Space Shuttle Photography (1988-89)

Southern Margin of Savanna Grassland & Woodland

⌐⌐⌐

Figure 2. The movement of the northern margin of the Sahelian grassland belt over the past three decades.

of these later surveys and, in some areas, photographs dating back to the early 1950s were used to create a series of 1:2,000,000 physiographic maps for the U. S. Army Corps of Engineers. The maps identified the location of grassland, tree savanna and forest in the Sahel. Although the Army Corps of Engineers map series was first published in 1967, the maps record the state of Sahelian vegetation as it appeared in the late 1950s and early 1960s. A general coincidence can be seen in the vegetation boundaries of the Army Corps of Engineers map series and those noted on the earlier trimetrogon survey maps (Raisz 1952). The northern limit of Sahelian grassland during the late 1950s and early 1960s reached above 18°N in the western Sahel and 16°N in Sudan (Figure 2). Photographs taken by NASA astronauts during the Gemini and Apollo space missions from 1965 to 1969 show the status of the savanna grassland in the western Sahel and Sudan in the period just prior to the onset of severe drought. With the exception of small areas in northeastern Mali and in Sudan, the northernmost extent of Sahelian grassland lay south of the boundary represented on the Army Corps of Engineers map series (Figure 2).

Since the first launch of the Space Shuttle in 1981, NASA astronauts have regularly photographed the Sahel for evidence of environmental change. By January 1986, the northernmost limit of savanna grassland had retreated to 16.5°N in the western Sahel and 14.5°N in Sudan (Figure 2). However the change in the position of the grassland margin of > 150 km in some areas since the Gemini and Apollo photography did not occur gradually over the 20-year period. Early Landsat images of the Sahel acquired in 1972-75 and orbital photographs made during the three NASA Skylab missions in 1973-74 show the northern grassland margin > 100 km south of its 1965-69 position. Between late 1972 and January 1974, the northern half of Lake Chad dried, and many small lake basins in Mali, Niger and Chad became desiccated. From this sequence of events and the relative stability of the grassland margin as photographed in 1981-86, it appears that the southward shift of the savanna grassland limit took place rapidly in the period from 1969 to 1972. Support for this conclusion can be drawn from the analysis of regional dust storm frequencies. For instance, Nouakchott experienced only 2 dust storm days in 1968, but had 22 occurrences in 1972 and 65 days with dust storms in 1974 (Middleton 1985). Such an exponential increase in the frequency of dust storms can happen only during a period of rapid environmental deterioration. Over the entire Sahelian zone from the early 1960s until 1986, approximately 900,000 km^2 of former savanna grassland were displaced by desert or became severely degraded.

A sudden northward advance in the position of the savanna grassland margin was photographed by NASA astronauts during Space Shuttle missions in 1988-89. By October 1988, the line of green grassland in western Mauritania had moved 110 km north of its location in January 1986 (Figure 3), and the change persisted into the dry season, when the area was again photographed in March 1989. Major northward advances in the extent of savanna vegetation also were recorded in areas of eastern Mali, Niger, Chad and Sudan (Figure 2). Near the Bahr el Ghazal, small lake basins, which had been entirely dry during 1981-86, once again held water as did many shallow basins in central Sudan. The expanse of open water in Lake Chad continued to increase, but remained only about one- quarter of the area attained in the mid-1960s. Throughout the Sahel, the landscape seen from orbit

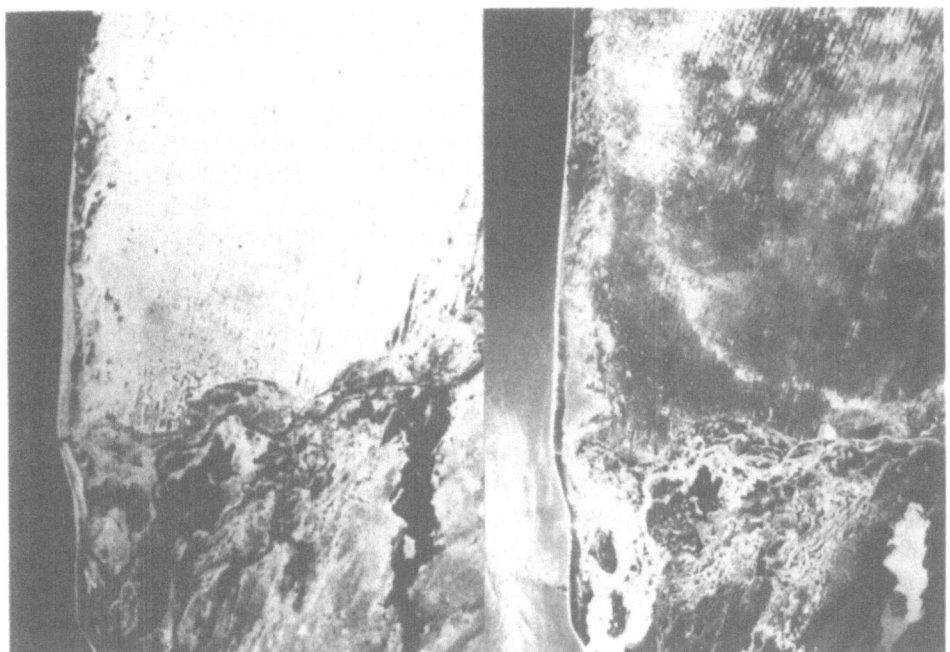

Figure 3. Rapid changes in the density of the green vegetation between 10 November 1984 (left) and 30 September 1988 (right) are recorded by orbital photographs of coastal Senegal and Mauritania near the mouth of the Senegal River.

was remarkably more green than during any period since the late 1960s.

From the preceding discussion of vegetation changes through time in the Sahel, it may be concluded that the Sahara has not advanced consistently at the rate of "100 kilometres a year (Wright 1985)", though movement both southward and northward of several 10s of kilometres may occur in the course of a single year. Likewise, blanket statements invoking anthropogenic causes, such as the one by Le Houerou (1977, p. 28) that "expansion of deserts along their edges is due only to human action . . . ." are clearly refuted by evidence from the Sahel. To the contrary, grassland savanna forms a highly dynamic environment that is rarely stable for long periods. Its margin undergoes broad-scale changes linked to natural fluctuations in the distribution of regional rainfall.

In the relatively wet 1950s, Sahelian grassland achieved a greatly extended northern boundary. By the mid-1960s, several drier years prior to the beginning of drought in 1968 (Lamb 1985) caused a substantial southward retreat of savanna from its most extreme locations. The harsh early years of the 1968-87 drought produced a rapid contraction of the grassland belt before 1974 that remained in essentially the same location through the mid-1980s. With higher precipitation in 1986-87 and a return to rainfall near long- term norms in 1988, the grassland limit advanced northward by > 100 km in some areas.

The consequences of local desertification by human actions, particularly the removal of savanna woodland for fuel, should not be ignored, but these effects have not been great in terms of the total regional area. Village barren haloes in Mali, Senegal and Mauritania identified by MacLeod et al. (1977) using orbital photographs taken during the 1973-74 Skylab missions were not appreciably larger when photographed by Space Shuttle astronauts a decade later. A similar retrospective analysis in central Kordofan (Hellden 1984) using aerial photography from 1962 and Landsat images collected during the Sahel drought found no systematic growth of cleared areas surrounding villages. The well-documented cases of man-induced land degradation, such as along the Israel/Egypt border (Otterman 1974), Angola/Namibia border (Wells 1989), Republic of South Africa/Lesotho border (Rapp 1975) and Ekrafane Ranch within Niger (Wade 1975; Freden 1976; MacLeod et al. 1977), are limited in area and stem from a complex interaction of different pressures upon the environment involving questionable political policies, local land management practices, regional climatic gradients and climatic change. For the Sahel as a whole, the greatest changes in the position of savanna grassland have occurred in the least populated regions (e.g., eastern Mauritania and northwestern Mali), whereas the areas of large and increasing populations along the Senegal and Niger valleys and in central Kordofan have demonstrated much less change in their vegetative cover. Natural causes, not human actions, have been responsible for the widespread desertification of the Sahel and its revegetation during both the geological and recent past.

3.   Exemplary Fates of Large-Scale Water Management Schemes

The changeable environment of the Sahel presents special challenges to the construction of water management systems to sustain agricultural irrigation. Unfortunately for Sahelian nations, in many cases these challenges were not fully recognized until after the installation of large, costly projects. This circumstance arose primarily because the hydrological conditions of the 1950s and early 1960s created by ample regional rainfall were viewed as `normal' for the Sahel, and the drought beginning in 1968 was regarded as a temporary distortion of the climatic norm. A detailed review of the historical climate undermines this conception. Droughts with durations of 10 and even 20 years are part of the natural climatic fluctuation of the Sahel and are no more unrepresentative of long-term precipitation patterns than the high rainfall totals recorded in the 1950s and previous wetter periods. The degree of interannual variability in Sahelian rainfall (Katz and Glantz 1977) may lead to a conflict between the perception of long-term average rainfall as `normal' and the reality of frequent periods of precipitation well above and below the historical mean. Projects designed to operate only when hydrological conditions are near the long-term mean face great problems when rains fail for several years in succession.

Just as orbital remote sensing techniques have been useful in mapping the extent of Sahelian vegetation, similar methods may be applied to monitor various water management schemes in regions where surface data are difficult to obtain. Orbital photography can be used to detect the area of open water in surface storage systems, the seasonal flood pulses along river channels, the presence of water in canals and the greenness of cultivated fields over the course of several years. In the following sections, we evaluate the performance of some of the principal water management schemes in the Sahel.

3.1. LARGE-SCALE WATER MANAGEMENT SCHEMES

3.1.1. *Two Upstream Diversion Projects: Niono and Lac Maga.* The irrigation schemes

Figure 4. Location of large-scale Sahelian water management schemes.

of Niono in Mali and Lac Maga in Nigeria (Figure 4) divert water from rivers with headwaters in the region to the south of the Sahel. The Niono project is located 40 km north of the Niger in the vicinity of Segou near the northern terminus of the Canal du Sahel and an adjacent canal constructed in the late 1970s. The project is the first major irrigation scheme to tap the Niger along its upstream course through the central Sahel. Developed in two sections beginning in the 1960s, the irrigation scheme involves approximately 6000 ha laid out in a classic gravity-fed trellis pattern of conduits to supply rectangular fields. During the period of intense drought in 1983-85, photographs by Space Shuttle astronauts showed the scheme to be a green polygonal oasis set in the middle of parched savanna. In the worst phase of drought in early 1985, when water levels of the Niger in western Mali became so low that even downstream travel by pirogue became difficult, agricultural fields in the Niono scheme appeared to maintain their productivity throughout the dry season. The relative success of irrigation at Niono has probably been achieved at some expense to traditional floodplain farming along the Niger in the region north of the canal diversions, but to our knowledge no study of this possible consequence has been undertaken. For efficiency in sustaining and expanding its area of productive fields during the drought period, the Niono scheme has no parallel amongst large Sahelian water management schemes.

The Lac Maga irrigation scheme became operational in 1979 following the construction of a dam near Pouss, Nigeria, designed to capture the seasonal flood pulse down western distributaries of the Logone River. It irrigates approximately 8000 ha for rice cultivation in a grid pattern of fields immediately north of the dam site. Although orbital photographs made in October 1984 after the disastrous failure of summers rains revealed a 55% decline in the area of open water held in the reservoir, irrigated fields were maintained in production throughout the mid-1980s. Seasonal monitoring of regional water transport using thermal data from Meteosat has been done by Rosema and Fiselier (1988). Their work indicates profound changes in the distribution of the Logone flood surge during the 1980s, with Lac Maga capturing a substantial proportion of the seasonal floods in 1983 and 1985. Changes in the distribution of seasonal floodwaters must have had an impact upon the northern Logone plain in Nigeria and Cameroon and upon downstream floodplain farmers who are not participants in the Lac Maga scheme. As with the Niono project, irrigation at Lac Maga can be regarded as a success measured in terms of its having maintained agricultural productivity during the severe drought, but this success may be qualified by the disruption caused to traditional agriculture located downriver. Both the Niono and Lac Maga schemes owe a significant part of their successes to favourable sites along the upstream segments of major rivers which enter the Sahel.

3.1.2. *Surface Storage: Bakolori Dam.* Plans for the Bakolori Dam on the Sokoto River in northwestern Nigeria (Figure 4) began in the mid-1960s, and construction was completed in 1978 at a cost of 350 million Naira (Bird 1984). The irrigation project associated with the dam came into use in the early 1980s with a goal to irrigate 27,000 ha for rice cultivation (Adams 1986). Through the mid- 1980s, the scheme failed for several reasons to bring more than a small fraction of the planned field area into production. Discharge of the Sokoto River reached such low levels during the early years of the project that sustained irrigation of the entire scheme was not feasible. Social problems also intervened. The expropriation of land from floodplain farmers without adequate compensation and delays in construction of the irrigation system led to attempts to plant crops on areas of the scheme without the consent of its management. A blockade of the project was broken by police action in April 1980 at the cost of many lives and much civilian property (Adams 1984). The failure to make a smooth transition from the construction stage to initial operation caused at least a temporary loss of cooperation with the farming communities which were to be among the project's beneficiaries. Given these results, Bird (1984) has made a strong case that the Bakolori scheme actually reduced productive capacity in the area by its inundation of 7200 ha of arable land along with forfeiture of an additional 20,000 ha of downstream floodplain farmland. Agricultural production statistics confirm that the area produced less food in the years after completion of the Bakolori Dam.

3.1.3. *Abstraction from a Lake: South Chad Irrigation Project.* Following the success of a small pilot project in the 1960s, the installation of a major irrigation scheme in northern Borno State, Nigeria, was completed in 1979 to draft water from Lake Chad via a canal constructed to Kirenowa (Figure 4). The South Chad Irrigation Project cost more than $400 million and was intended to irrigate 67,000 ha for the production of rice, wheat and cotton (Kolawole 1987). Only about 7500 ha were irrigated in the initial years of the scheme because of the low level of Lake Chad at the time of the project's completion. Rice farming was curtailed from the beginning, and irrigation stopped entirely in 1983, when the water reaching the intake canal fell below the minimum abstraction level (279.9 m) required for siphons at the Kirenowa pumping station. As the water level continued to drop to below 277 m in 1984, the irrigation scheme disintegrated and approximately 60,000 people abandoned the area (Kolawole 1987). Many of these refugees moved onto the floor of Lake Chad to grow maize and cowpeas in areas which were submerged by metres of water in the 1970s. Kolawole (1988) has traced the establishment of many new communities on the lake floor, where about 25,000 people live in 49 villages within his study area. Most of these villages were founded in 1984-85. Although farming of the lake floor produced good crops in the early years after settlement, the new communities remained vulnerable to a rapid rise in lake level after the increased rainfall recorded during 1986-88 in the Chari/Logone watershed. In 1988, the maximum seasonal discharge of the Chari at Ndjamena was four times greater than its 1984 flow and reached a value near its long-term median discharge (Terry Evans 1989, written communication). Many of the lake-floor villages in Kolawole's study area were submerged by the rise of lake level to approximately 279.5 m in October 1988 (Are Kolawole 1989, written communication). Even with a continued trend toward the expansion of Lake Chad, failure to maintain the pumps and conduits of the South Chad Irrigation Project during the period of its abandonment by local farmers may delay an efficient return to crop production. Some irrigation has started once more within the South Chad Irrigation Project, and it stands to benefit if there is a sustained trend toward higher lake levels.

3.1.4. *Regional Transfer: Jonglei Canal.* Whereas environmental conditions in the Sahel during the past decade could not have been made much worse for the prospects of the South Chad Irrigation Project given its design limitations, changes in the volume of equatorial rainfall created the best circumstances in which to attempt the transfer of water from the expanding southern Sudd swampland to the drought- stricken areas of central Sudan. The Jonglei Canal project (Figure 4) dates back to proposals made in the 1920s and postwar studies completed in the 1940s and early 1950s (Howell et al. 1988). A decision to construct the canal was made in 1974, and excavation began in 1978. The scale of the planned canal is colossal with a length of 360 km dug to depths of 4-8 metres across a 50 m width designed to transport an average discharge of 25 million m³ per day (Howell et al. 1988, p. 49). According to estimates made in the early 1980s, the $175 million cost of the Jonglei Canal would be rapidly offset by a gain of 252,000 ha of irrigated farmland in northern Sudan.

After excavation across 240 km to the south of the White Nile near Malakal, construction was halted in November 1983 (Howell et al. 1988). The rising tide of civil war swept over the area as the Sudan Peoples Liberation Army fought a guerrilla campaign to wrest control of southern Sudan from Islamic rule by the northern elite. The rebel soldiers have received considerable support from the rural population in the Sudd that tends to perceive the Jonglei Canal as beneficial only to interests in central and northern Sudan, while its construction threatens to disrupt the traditional movement of livestock to seasonal pastures and to place limits upon fishing. Whether this costly project will ever be completed depends upon the outcome of current hostilities in Sudan.

3.2. SMALL-SCALE WATER MANAGEMENT SCHEMES

3.2.1. Boreholes. Several thousand waterwells have been drilled across the Sahel with the greatest concentrations occurring in southern Mauritania, Senegal, Niger and northern Nigeria. It not possible to be as unequivocal in judgment about the success of such an enterprise as in the performance evaluation of the large-scale projects. Many of the boreholes have undoubtedly brought benefits and today provide reliable sources of potable water for numerous villages and towns. Some irrigation of farmland has been attempted with water extracted from local aquifers, but the chief agricultural use of boreholes has been for the supply of cattle tanks. During the severe drought years of the early-1970s, this use was a decidedly mixed blessing for the Sahel and was noted by many observers (Glantz 1976; Le Houerou 1977). In several areas, especially in southern Mauritania and Niger, cattle became concentrated around boreholes. When the drought intensified, well-watered livestock ate all available forage and then died of starvation. Poor management of livestock at the boreholes merely postponed the reduction of herds during severe drought and led to examples of "borehole-made-desert" in some areas (Le Houerou 1977, p. 28).

The technology to drill for water is widely available and can be applied to many areas in the Sahel where shallow aquifers exist, such as in the Chad Basin. Still, the advisability of drilling waterwells on an increased scale is debatable. Many of the wells drilled to date have been financed by the relief efforts of donor groups and nations, and the economics of borehole irrigation for agricultural development have not yet been proven in the Sahel. The initial drilling costs of $20-30 thousand for a shallow well are substantial for Sahelian villages in a region where a credit system has not been fully established. Recurrent costs of well maintenance are an additional burden.

3.2.2. Small Dams and Subsurface Sand Dams. In many instances modern technology can contribute to the improvement of well-established methods of Sahelian agriculture. As Adams (1986) has observed for Nigeria, over 1 million ha are devoted to traditional methods of irrigated agriculture, but less than 31,000 ha have been brought into production using formal irrigation schemes. Many steps such as the improvement of small-scale shadoof irrigation of river floodplains by the use of better pumping systems, the introduction of fertilizers, pesticides and herbicides and the cultivation of new strains of high-yielding cereals can be introduced at a fraction of the cost required to construct large-scale water management schemes. Small surface dams built at the proper sites and subsurface sand dams ponding seasonal floods along ephemeral drainages offer other effective strategies to increase arable land around rural population nodes of 100-200 people. Small dam projects utilize local labour, are maintained by the farmers and their families and can be installed for much less expenditure than other irrigation technologies (see Guiraud, this volume). Although these advancements upon traditional farming practices are not immune to collapse during the worst drought phases, a local failure would not suddenly displace thousands of farmers, as happened with the South Chad Irrigation Project. The cultural adjustments to drought traditionally undertaken by Sahelian farmers are more in keeping with the operation of these small-scale water management schemes than with the larger projects.

4. Conclusions

The water management schemes which encountered the greatest difficulties during the period 1965-88 were those planned in the 1960s after a period of unusually high rainfall. When these "normal" conditions ended in the late 1960s, the largest of the new projects were in many instances the most vulnerable to environmental change, and some schemes such as the South Chad Irrigation Project disintegrated completely during the worst phase of the recent drought. Despite their obvious limitations, plans for expensive, large-scale schemes such as the Manantali Dam on the Bafing River in western Mali have continued to dominate the development of Sahelian watersheds. Whereas the economic pressure to create an agricultural export sector and the ability for central governments to promote their control through regional water management will continue to make large irrigation schemes attractive, small-scale projects offer the prospect of a more immediate contribution to subsistence agriculture and a more flexible response to future episodes of environmental change.

The Sahelian climate changes rapidly. To improve the lives of those living in the Sahel requires the development and deployment of technologies resilient to climatic change.

5. References

Adams, W.M. (1984) "Irrigation as hazard: Farmers' responses to the introduction of irrigation in Sokoto, Nigeria", in Adams, W.M. and Grove, A.T. (eds.) *Irrigation in Tropical Africa: Problems and Problem-Solving*, (Cambridge African Monographs 3), African Studies Centre, Cambridge, pp. 121- 130

Adams, W.M. (1986) "Traditional agriculture and water use in the Sokoto valley, Nigeria" Geographical Journal 152, pp. 30-43

Adamson, D.A., Gasse, F., Street, F.A. & Williams, M.A.J. (1980) "Late Quaternary history of the Nile", *Nature* 288, pp. 50-55

Bird, A.C. (1984) "The land issue in large scale irrigation projects: Some problems from northern Nigeria" in Adams, W.M. and Grove, A.T. (eds.) *Irrigation in Tropical Africa: Problems and Problem-Solving*, (Cambridge African Monographs 3), African Studies Centre, Cambridge, pp. 75- 85

Browne, W.G. (1799) *Travels in Africa, Egypt and Syria*, Cadell and Davies, London.

Butzer, K.W. (1980) "Pleistocene history of the Nile Valley in Egypt and Lower Nubia", in Williams, M.A.J. and Faure, H. (eds.) *The Sahara and the Nile*, Balkema, Rotterdam, pp. 253-280

COHMAP Project Members (1988) "Climatic changes of the last 18,000 years: Observations and model simulations", Science 241, pp. 1043-1052

Darwin, C. (1839) *The Voyage of the Beagle*, Colburn, London

Darwin, C. (1846) "An account of the fine dust which often falls on vessels in the Atlantic Ocean", Quarterly Journal of the Geological Society of London 2, pp. 26-30

Fabre, J. and Petit-Maire, N. (1988) "Holocene climatic evolution at 22-23° N from two palaeolakes in the Taoudenni area (northern Mali)", Palaeogeography, Palaeoclimatology, Palaeoecology 65, pp. 133-148

Freden, S.C. (1976) "Survey for the Landsat program" in Short, N.M., Lowman, P.D.Jr., Freden, S.C. and Finch, W.A.Jr. (eds.) *Mission to Earth: Landsat Views the World*, NASA SP- 360, Washington, pp. 1-26

Glantz, M.H. (1976) "Nine fallacies of natural disaster", in Glantz, M.H. (ed.) *The Politics of Natural Disaster: The Case of the Sahel Drought*, Praeger, New York, pp. 3-24

Grove, A.T. (1985) "The physical evolution of the river basins", in Grove, A.T. (ed.) *The Niger and Its Neighbours*, Balkema, Rotterdam, pp. 21-76

Grove, A.T. (1986) "The state of Africa in the 1980s", Geographical Journal 152, pp. 193-203

Grove, A.T. & Warren, A. (1968) "Quaternary landforms and climate on the south side of the Sahara", Geographical Journal 134, pp. 194-208

Haynes, C.V.Jr. (1982) "Great Sand Sea and Selima Sand Sheet, eastern Sahara: Geochronology and desertification", Science 217, pp. 629-633

Hassan, F.A. (1981) "Historical Nile floods and their implications for climatic change", Science 212, pp. 1142-1145

Howell, P., Lock, M. & Cobb, S. (1988) *The Jonglei Canal: Impact and Opportunity*, Cambridge University Press, Cambridge

Hulme, M. and Trilsbach, A. (1989) "The August 1988 storm over Khartoum: Its climatology and impact", Weather 44, 82-90

Katz, R.W. & Glantz, M.H. (1977) "Rainfall statistics, droughts and desertification in the Sahel", in Glantz, M.H. (ed.) *Desertification*, Westview, Boulder, pp. 81-102

Kolawole, A. (1987) "Environmental change and the South Chad Irrigation Project (Nigeria)", Journal of Arid Environments 13, 169-176

Kolawole, A. (1988) "Cultivation of the floor of Lake Chad: A response to environmental hazard in eastern Borno, Nigeria", Geographical Journal 154, 243-250

Kutzbach, J.E. & Street-Perrott, F.A. (1985) "Milankovitch forcing of fluctuations in the level of tropical lakes from 18 to 0 kyr BP", *Nature* 317, pp. 130-134

Lamb, P.J. (1985) "Rainfall in subsaharan west Africa during 1941-83" Zeitschrift fur Gletscherkunde und Glazialgeologie 21, pp. 131-139

Le Houerou, H.N. (1977) "The nature and causes of desertization", in Glantz, M.H. (ed.) *Desertification*, Westview, Boulder, pp. 17-38

MacLeod, N.H., Schubert, J.S. & Anaejionu, P. (1977) "Report on the Skylab 4 African drought and arid lands experiment" in Wilmarth, V.R., Kaltenbach, J.L. and Lenoir, W.B. (eds.) *Skylab Explores the Earth*, NASA SP-380, Washington, pp. 263-285

Maley, J. (1973) "Mecanisme de changements climatiques aux basses latitudes", Palaeogeography, Palaeoclimatology, Palaeoecology 14, pp. 193-227

Maley, J. (1976) "Les variations du lac Tchad dupuis un millenire: Consequences paleoclimatiques", Palaeoecology of Africa 9, pp. 44-47

Michel, P. (1980) "The southwestern Sahara margin: Sediments and climatic change during the recent Quaternary", Palaeoecology of Africa 12, pp. 297-306

Middleton, N.J. (1985) "Effect of drought on dust production in the Sahel", *Nature* 316, pp. 431-434

Mohler, R.R.J. & Amsbury, D.L. (1988) "Extension of a drought monitoring and vegetation methodology to the western Sahel", Geocarto International 3(4), 29-36

Nicholson, S.E. (1980) "Saharan climates in historic times", in Williams M.A.J. and Faure, H. (eds.) *The Sahara and the Nile*, Balkema, Rotterdam, pp. 173-200

Nicholson, S.E. (1981) "Saharan climates in historic times" in Allen J.A. (ed.) *The Sahara - Ecological Change and Early Economic History*, Menas, London., pp. 35-59

Nicholson, S.E. (1982) "Pleistocene and Holocene climates in Africa", *Nature* 296, p. 779

Nicholson, S.E. (1986) "The spatial coherence of African rainfall anomalies: Interhemispheric teleconnections", Journal of Climate and Applied Meteorology 25, pp. 1365-1381

Nicholson, S.E. (1989) "Long-term changes in African rainfall", Weather 44, pp. 46-56

Nicholson, S.E. & Chervin, R.M. (1983) "Recent rainfall fluctuations in Africa", in Street-Perrott, F.A., Beran, M. and Ratcliffe, R. (eds.) *Variations in the Global Water Budget*, D. Reidel, Dordrecht, pp. 221-238

Otterman, J. (1974) "Baring high-albedo soils by overgrazing: A hypothesized desertification mechanism", Science 196, pp. 531- 533

Pachur, H.-J. & Kropelin, S. (1987) "Wadi Howar: Paleoclimatic evidence from an extinct river system in the southeastern Sahara", Science 237, pp. 298-300

592

Petit-Maire, N. and Riser, J. (eds.) *Sahara ou Sahel? Quaternaire Recent du Bassin de Taoudenni (Mali)*, Lamy, Marseille

Raisz, E. (1952) *Landform Map of North Africa*, Published privately, Boston

Rapp, A. (1975) "Soil erosion and sedimentation in Tanzania and Lesotho", Ambio 4, pp. 154-163

Ritchie, J.C. & Haynes, C.V. (1987) "Holocene vegetation zonation in the eastern Sahara", *Nature* 330, 645-647

Rosema, A. & Fiselier, J.L. (1988) "Meteosat Based Evapotranspiration and Thermal Inertia Mapping for Monitoring Transgressions in the Lake Chad Region and Niger Delta", EARS, Ltd., Delft

Servant, M. & Servant-Vildary, S. (1980) "L'environnement quaternaire du bassin du Tchad", in Williams, M.A.J. and Faure, H. (eds) *The Sahara and the Nile*, Balkema, Rotterdam, pp. 133-162

Sir M. MacDonald & Partners (1988) *Rehabilitation and Improvement of Water Delivery Systems in Old Lands* (Project Nr.EGY/85/012), Final Report to the International Bank for Reconstruction and Development, World Bank, Washington

Street-Perrott, F.A. & Roberts, N. (1983) "Fluctuations in closed-basin lakes as an indicator of past atmospheric circulation patterns", in Street-Perrott, F.A., Beran, M. and Ratcliffe, R. (eds.) *Variations in the Global Water Budget*, D. Reidel, Dordrecht, pp. 331-345

Wade, N. (1974) "Sahelian drought: No victory for Western aid", Science 185, pp. 234-237

Warren, A. (1970) "Dune trends and their implications in the central Sudan", Zeitschrift fur Geomorphologie Supplementband 20, pp. 154-180

Wells, G.L. (1983) "Late-glacial circulation over central North America revealed by aeolian features", in Street- Perrott, F.A., Beran, M. and Ratcliffe, R. (eds.) *Variations in the Global Water Budget*, D. Reidel, Dordrecht, pp. 317-330

Wells, G.L. (1989) "Observing Earth's environment from space", in Friday, L.F. and Laskey, R.A. (eds) *The Fragile Environment*, Cambridge University Press, Cambridge, pp. 148- 186

Wendorf, F., Close, A.E. & Schild, R. (1985) "Prehistoric settlements in the Nubian Desert", American Scientist 73, pp. 132-141

World Bank (1987) *World Development Report 1987*, Oxford University Press, Oxford

Wright, P. (1985) "Will Africa ever feed itself?", London Times, May 15, 1985, p. 12

6. Acknowledgments

Discussions with Bill Adams, Terry Evans, Dick Grove, Rene Guiraud, Are Kolawole and Michael Lock contributed to the ideas contained in this manuscript.

SALINE GROUNDWATER IN THE CANARY ISLANDS (SPAIN) RESULTING FROM ARIDITY

E. CUSTODIO
International Course on Groundwater Hydrology
Department of Ground Engineering
Polytechnic University of Catalonia
Jordi Girona Salgado, 31
08034 Barcelona, Spain

KEYWORDS/ABSTRACT: Brackish groundwater/ climatic effect/ arid areas/ Canary Islands. Groundwater recharge in arid lands is only a very small fraction of rainfall. its salinity is high due to the concentration of rain-contributed and airborne salts by evapotranspiration in the upper layer of the ground. This is enhanced in windy coastal areas and islands due to the important contribution of airborne marine salts. There brackish to salty groundwater is commonly found. Examples from the Canary Islands are presented and discussed, and possible explanations are outlined. Local solutions for water supply are briefly presented.

1. Introduction

Arid and semi-arid areas, especially in intertropical and Mediterranean climates, not only suffer from shortage of renewable water resources, but also from excessive groundwater salinity due to evaporative concentration of salts in recharge water. In many instances climatic, topographic and soil conditions are excellent for human settlements, but fresh-water resources become a limiting factor for development. The limitation is especially serious in areas far from rivers of without thick pervious sedimentary deposits containing fresh-water recharged in other areas or during previous more humid climate. This is even more serious in small islands since the only water resources available are those existing on, or that can be generated inside the island itself. In coastal areas and small islands the water salinity situation is worsened by the effect of high atmospheric salinity due to the proximity to the sea, especially in windy areas. In such instances groundwater recharge, besides being scarce, is too salty to be used to supply most water needs. This situation is considered taking the Canary Islands as an example. The origin of the salinity is discussed taking into account chemical and environmental isotopic data. Some alternative solutions are briefly reviewed.

593

R. Paepe et al. (eds.), Greenhouse Effect, Sea Level and Drought, 593–618.

2. Groundwater Salinization Processes

All the following comments and discussion refer to groundwater recharge and aquifer water in areas in which there is no direct contamination by sea water penetrating the ground or existing in it. A detailed description of these processes can be found elsewhere (Custodio and Bruggeman, 1987; Custodio and Llamas, sect. 13, 1976)). Hence only salinization processes due to aridity will be considered.

The sources of salts in soil water are:

R_P.- salts contributed from precipitation (rain, dew, snow)(ML^{-3})

R_R.- salts contributed from inflowing surface and subsurface water (ML^{-3})

S_R.- salts contributed by weathering of soil rock $(ML^{-2}T^{-1})$

S_A.- salts deposited on the soil by dry fallout $(ML^{-2}T^{-1})$

S_P.- salts contributed by plant cover decay $(ML^{-2}T^{-1})$

S_W.- salts contributed by wind-borne solids and dry fallout $(ML^{-2}T^{-1})$

The outflow of salts depends on the following factors:

O_G.- salts carried away with groundwater recharge (ML^{-3})

O_R.- salts carried away with outflowing surface and subsurface water (ML^{-3})

O_W.- salts removed by wind $(ML^{-2}T^{-1})$

O_F.- salts taken away to plants or incorporated into them $(ML^{-2}T^{-1})$

Q_R.- salts precipitated into the voids of the soil and rock $(ML^{-2}T^{-1})$

It is assumed that only conservative salts (dissolved in ionic form) are considered; that is to say, salts that are largely non-sorbed and do not change their chemical stat in the soil, such as Cl^-, Ca^{++}, Na^+, ... easily change their concentration by gas inflow or outflow or by redox processes, and then the balance is more complicated. Also it is assumed that evaporation in the soil and plant transpiration do not take away salts. This is true for conservative salts, but not for volatile substances in equilibrium with gases, such as NH_4^+ and HCO_3^-.

Considering a unit of surface area and time, the balance of conservative salts can be written as:

$$(R_P P + R_R Q_1 + S_R + S_A + S_P + S_W) - (O_G R + O_R Q_0 + O_W + O_P + O_R)$$

$$= \int_0^{z_0} dz\ d(\theta c)/dt$$

in which:

P	=	precipitation depth (LT^{-1})
R	=	groundwater recharge (LT^{-1})
Q_1; Q_0	=	surface and subsurface runoff; [1]inflow, [0]outflow (LT^{-1})
z	=	vertical coordinate (down from the soil surface)(L)
z_0	=	depth of soil subjected to evapotranspiration (limit below which only downward flow exists)(L)
t	=	time (T)

θ = soil humidity (adim.)

c = salt concentration in soil water (ML^{-3})

The water balance leads to:

$$(P + Q_1) - (R + Q_0 + E) = \int_0^{z_0} dz \, d\theta/dt$$

in which E = actual evapotranspiration.

In most areas, especially in dry areas, some simplifications can be introduced. Let us assume steady state for all the variables; that is to say, there is no trend in them and data used are long-term averages in an area devoid of significant human activities capable of producing a transient situation. Then the second term is zero. Let us also assume that plant cover does not change ($S_P = O_P$). Then:

$$R_P P + R_R Q_1 + S_R + S_A + S_W = O_G R + O_R Q_O + O_W + O_R$$

$$P + Q_1 = R + Q_0 + E$$

When surface and subsurface runoff is not significant and there is no net salt precipitation into the soil, as is the case for chloride in most circumstances except for very extreme situations, and the soil weathering does not contribute salts (as is true in most cases for chloride):

$$R_P P + S_A + S_W = O_G R + Q_W$$

$$P = R + E$$

then:

$$O_G = \frac{R_P P + S_A + S_W - O_W}{R} = R_P \frac{P}{R} + \frac{S_A + S_W - O_W}{R}$$

This means that if $S_A + S_W - O_W$ is small the salinity of recharge for a conservative non-precipitable solute such as chloride, O_G depends directly on rainfall salinity enlarged by a factor P/R, that depends predominantly, though not exclusively, on climate.

This factor P/R may exhibit values greater than 100 in arid climates. The actual value has to be calculated for each circumstance taking into account daily rainfall distribution, daily evapotranspiration (dominated by evaporation if canopy cover is scarce) and soil and root conditions. Changes in soil conditions in the same area easily produce significant changes in the results. The role of macropores, soil cracks and other open spaces is important and difficult to take into account properly.

To know R_P, repeated samples of rain are needed in order to get a representative weighted means. One or a few measurements are not enough, although they may indicate the order of magnitude for qualitative assessments. The accurate measurements of R_F needs sampling rain-gauges with their surfaces free of exotic salts, protected against dry fallout and wind-borne salts. In some instances a lid is used with a sensor that opens the gauge only when it is raining. Protection from evaporation until the sample is taken away is not strictly needed if rainfall depth is determined from the sample volume and only conservative salts are considered. Otherwise evaporation has to be avoided by means of good isolation and early collection of the sample, or by the use of floating paraffin oil. This is also needed when environmentally stable isotopes of water are to be measured, since evaporation produces isotopic fractionation.

In most practical cases such measurements are difficult to carry out. Normal rain-gauges or tanks with only a partial lid to reduce evaporation are commonly used. They collect dry fallout and wind-transported solids as well as rain. If the deposition rate is not greatly changed by the sampling device itself and all soluble salts collected are incorporated into the water sample, the result is indicative of $R_P P + S_A + S_W$ and then $R_P'P'$ can be approximately substituted for the addition, R_F' and P' being respectively the salt concentration and rainfall depth obtained from the collected sample. This sample, if not evaporated, is also valid for water O^{18} and H^2 isotopic analyses.

The isotopic composition of water is an important tool for understanding the process of evaporation of rainwater in the soil. Under conditions of good vegetation cover, transpiration dominates over evaporation and there is no important isotopic fractionation. The isotopic composition of recharge water is the weighted mean of rainfall (see Fontes, 1980; Gat and Gonfiantini, 1981).

But under arid climate the vegetation cover is scarce and discontinuous. In such cases evaporation dominates over transpiration and isotopic fractionation of soil water is the rule. This is enhanced in bare rocks and when vapour transport is significant in the process of evaporation. Then recharge water is isotopically heavier than the corresponding rain. O^{18} is relatively more enriched than H^2 (Levin et al., 1980), so that the representative points in a δO^{18} - δH^2 graph plot on a line with a slope clearly less than the meteoric water slope, the value of which is 8. Sometimes it is even less than the slope obtained from evaporating surface water, generally between 4.5 and 6.5. Dinçer (1980) reports values as low as 2 in the Sahara and Saudi Arabia for sand moisture in desert areas.

3. General Characteristics of the Canary Islands

The following discussion refers to conditions observed in the arid areas of the Canary Islands. Existing data are far from sufficient for a thorough discussion but they are enough to establish the climatic effect on groundwater salinity.

The Canary Islands, the SW extreme of Europe and a part of Spain, are located between 100 to 500 km west of the African coast of the Sahara. They lay between the latitudes 27°37' N to 29°25' N and the longitudes 13°20' W to 18°10' W (Fig. 1). The archipelago is formed by seven inhabited main islands. The most relevant data are as follows, from

East to West:

Island	Area km^2	Altitude m	Form	Other
Lanzarote	835	670	Elongated	Extensive recent volcanic outflows. Includes three "isletas" and several "islotes".
Fuerteventura	1729	807	Elongated	Dissected massif plus peripheral "flat" lands. Includes the "isleta" of Lobos.
Gran Canaria	1558	1950	Circular	Big, deep radial gullies.
Tenerife	2036	3718	Triangular	Guitar shaped, with a big collapsed caldera and a very high central volcano.
Gomera	378	1484	Circular	Big, deep radial gullies.
La Palma	706	2426	Triangular	Deep erosional caldera and very steep slopes.
Hierro	278	1501	Triangular	High sea cliffs.
TOTAL	7520			

Steep land slopes are common, up to 10-20%, sometimes mor than 50%, and almost vertical cliffs are frequent. Flat-lands are scarce and many of them are coastal platforms.

The islands are wholly volcanic. There is a "basal complex" of basic plutonic rocks and submarine volcanic outflows and pyroclastics, densely injected by dykes, that outcrops in Fuerteventura, Gomera and La Palma, but probably also underlies the other islands. On top of it, from the Miocene until the present, a thick pile of subaerial volcanics (lavas, pyroclastics, dykes, ignimbrites) form the present islands, starting with up to 1000 m thick fissural basalts. Later eruptions ejected more differentiated volcanics and finally alkaline basalts dominate.

Total population is close to 1.5 million concentrated in the two main islands, Gran Canaria and Tenerife. The main economical activities are services, tourism, irrigated agriculture and fisheries, with some basic industries.

Total water demand is about 430×10^6 m^3/year, mainly groundwater but surface reservoirs, desalination and reuse also play an important role. Further details can be found in Custodio

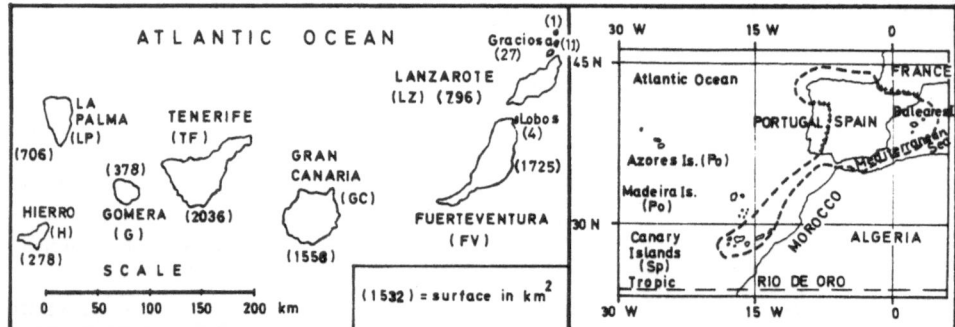

Figure 1. The archipelago of the Canary Islands (Islas Canarias) and its situation at the SW corner of Europe. The figures indicate the surface of the islands in km².

et al. (1989) and in SPA-15 (1975).

The general hydrogeological structure of the islands can be sketched as a core of low-permeability old volcanics with successive covers of younger, more pervious ones (see Custodio, 1985, 1989). Wide variations can be found from one area to another.

Under the existing permeability distribution, even with small recharge, high ground-water domes can form inside or lining the island cores, where they exist. But most of the recharge, if there is enough, flows downslope inside the younger volcanics high and medium altitudes, where recharge is maximum. Discharge is produce where the cover is interrupted, producing springs and seeps or, when the conductive blanket extends down to the shore or coastal pervious formations, groundwater is directly discharged into the sea.

Many of the thick, pervious materials at high altitudes are dry since a thin bottom-saturated thickness is enough to transmit the recharge under the prevailing high groundwater gradients, sometimes exceeding 0.1. Then, up to many tens and even hundreds of metres of unsaturated materials can be found, although the intercalation of low-permeability formations (red backed soils, altered ash deposits, dense flows, etc.) may produce local, small perched aquifers and the corresponding small springs and seeps that frequently have a clearly seasonal discharge. They may exist in wet as well as dry areas.

4. Rainfall and Groundwater Recharge

The Canaries are situated in the dry belt off the Sahara desert and thus regionally the annual rainfall is less than 150 mm or even less than 100 mm. This is the case in the "low islands" and in the coastal areas, especially in the South. But even in these conditions, intense down-pours can be expected from time to time. Temperature is ameliorated by the effect of the sea, the cold sea water up-wellings in the area and the humid trade winds. Mean monthly values vary between 17 and 25.5°C in coastal areas. In such conditions potential evapotranspiration values up to 1500 mm/year in sunny areas can be expected, with peak values exceeding 10 mm/day.

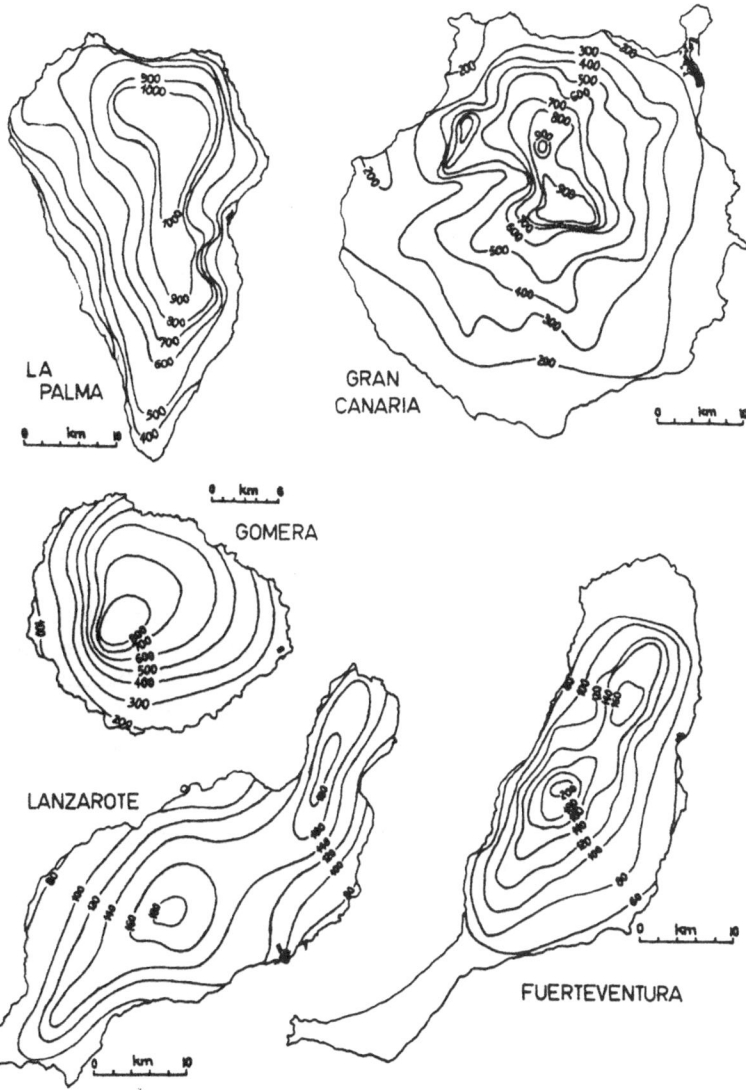

Figure 2. Mean rainfall of some of the islands in mm/year. Two wet (La Palma & Gomera), one intermediate (Gran Canaria) and the two driest islands (Lanzarote & Fuerteventura) are shown. Data from SPA-15 (1975).

In the high islands the situation changes dramatically since they produce the uplift of the humid trade winds, thus increasing the rainfall and cloudiness. Precipitation may attain 1000 mm/year on the northern slopes. Temperatures are colder than on the coast, although freezing conditions are absent. Yearly evapotranspiration is down to less than 1000 mm/year. Figure 2 shows the mean rainfall in the two driest islands, Lanzarote and Gran Canaria, and in the two wettest, La Palma and Gomera. The orographic effect clearly shows up, although the Figure does not include altitude contours.

There is a clear Mediterranean-type seasonal distribution of rainfall, with wet autumn and spring time and dry summer and also a part of the winter. In the driest areas a large part of yearly rainfall concentrates in a few intense downpours that produce flooding problems from time to time. Variations in yearly rainfall are remarkable with clear sequences of dry and humid years. Recent dry conditions in the Saharan belt have been felt clearly in the Canaries.

Dew deposition and fog interception by tree leaves are important for the vegetation in mid-northern slopes, where forested belts of pine trees and lauri-silva still exist. Dew is also noticeable in some dry areas.

Surface runoff is highly variable, depending on local conditions of soil, geology and slope. Some bare areas of flat-lying young volcanics are devoid of significant surface runoff, even under intense downpours, whilst others, corresponding to old acidic volcanics and basal-complex materials, behave as though impervious after the thin soil zone is water saturated. Most of the deeply cut gullies (barrancos) are dry and only carry water during and shortly after exceptional rain storms.

Related to the spatial variability of rainfall there is a wide range of mean groundwater recharge values, from up to 300 mm/year in the wet areas down to less than 1 mm/year in the driest. These values are only rough estimates from soil water balances. Great changes can be expected from one year to another depending not only on total rainfall but also on precipitation distribution. As is the rule in arid lands, most of the meagre recharge in Lanzarote and Fuerteventura and in the dry southern areas of the other islands corresponds to exceptionally wet periods in wet years, the groundwater recharge produced by small rains being negligible. Dew condensation and fog interception in trees is probably unimportant for groundwater recharge on a regional basis in wet areas and the effect is unknown in dry areas.

Existing data on chemical rainfall composition is still scarce and inaccurate, but indicates the general behaviour. In all cases rainfall samples include dry fallout.

The precipitation falling on the island of Hierro was studied in 1981 and 1988 (Soler, 1989), sampling the numerous rainwater cisterns there. Data have to be used with care since there are possible chemical changes in some cisterns due to evaporation, pollution of water collecting surfaces and refilling with water carried from other localities. Nevertheless the main trend can be seen. Since this is the westernmost island, far from concentrated human settlements and with the only thermoelectric plan in the western extreme, man-produced sulphate can be ruled out in most cases. Also agricultural practices are limited to the northern coastal platform. The chloride content is scarcely dependent on altitude in that windy island except for the low coastal areas, especially in the north (Fig. 3). Normal values vary between

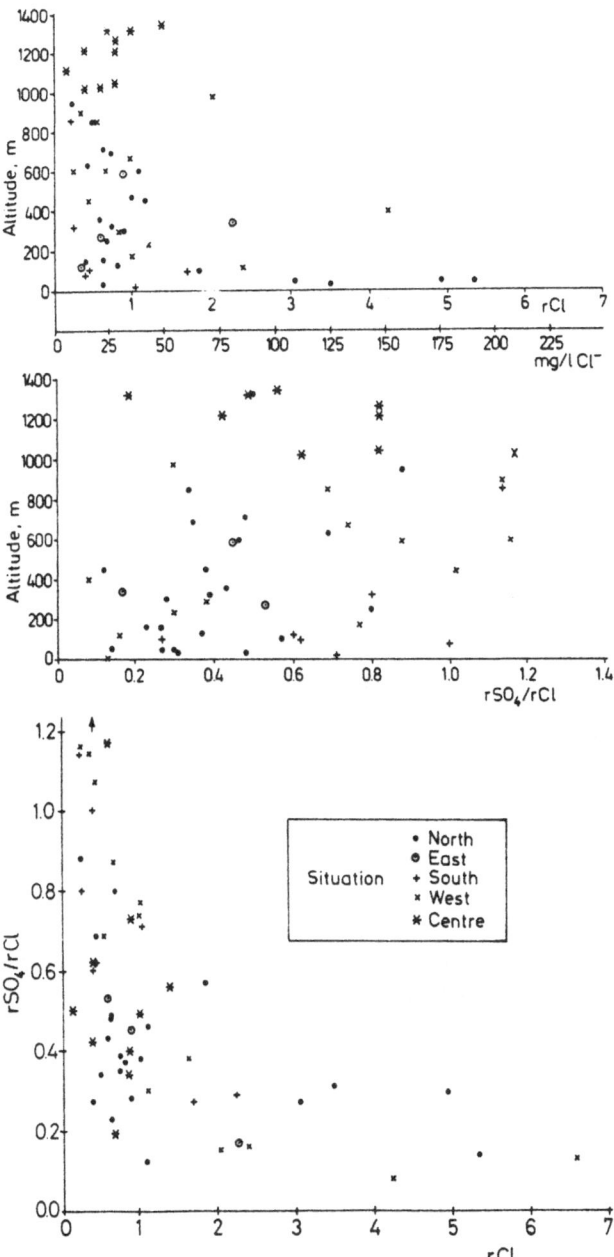

Figure 3. rCl and rSO$_4$/rCl (r = meq/l) values for cistern collected rainwater at the island of Hierro (data taken from Soler, 1989).

10 and 35 mg/l Cl⁻ but up to 200 mg/l are possible on the north coast. The rSO_4/rCl ratio does not show any altitudinal trend; it ranges from 0.1 to 1.2, but most of the values rangebetween 0.2 and 0.9. It seems that they are systematically greater than the value for seawater (0.10 to 0.11), although the influence of chemical inaccuracies has to be studied in more detail. The plot of rSO_4/rCl vs. rCl (Fig. 3) shows an inverse relationship. For rCl less than 1 the rSO_4/rCl ratio covers the full range. For higher values there is a clear tendency to smaller values of the ratio, down to about 0.2. It seems that the more saline rainwater is close to the seawater ratio, although still higher, and that the less saline rainwater indicates very variable increases in SO_4^-.

In Gran Canaria, at 900 m altitude on the windward slopes was found between 7 and 14 mg/l Cl⁻ (SPA-15, 1975) Recently a repeated sampling of a N-S line of rain-gauges has yielded the results in Figure 4 (Gasparinin et al., 1989). In Figure 4, the map shows the chloride content (mg/l) in wells that penetrate a few tens of metres below the regional water table. Samples have been protected from evaporation using floating paraffin oil. On the northern (windward) slope there is a slight decrease of Cl⁻ with altitude, from about 20 mg/l near the coast to 6 mg/l at the top. The rSO_4/rCl (r = meq/l) ratio varies between 1.4 and 0.2. On the southern (leeward) slope there is a clear increase in salts near the dry coastal plain, up to 100 mg/Cl. The rSO_4/rCl ratio is around 0.5 to 0.7. Data from rain cisterns in Lanzarote give values of about 80 mg/l Cl⁻ and rSO_4/rCl of 0.4 to 0.15 Custodio, 1974 b).

The relatively high values of the rSO_4/rCl ratio relative to that of seawater (0.10 to 0.11) are not well understood. In some cases SO_4^- derived from fossil fuel combustion can be adduced, but in many cases this is unlikely. It has been suggested that this is the result of wind-transported dust coming from the Sahara desert. There are no measurements to evaluate the importance of this effect but its likelihood is doubtful due to the low frequency of this type of event (once or twice a year). A few preliminary determinations of dust taken at Lanzarote (Custodio, 1974 b) show a soluble fraction with rSO_4/rCl ratio between 0.4 and 0.2. This dust contains 1 to 5% by weight of Cl, which constitutes a significant but not dominant contribution to rainfall salinity, even in dry areas. Further studies are needed to obtain valid conclusions.

There is the possibility of chemical fractionation during air-borne salinity formation over the sea. due to the ready formation of gypsum micro-crystals from partially evaporated sea droplets. But local studies about this are unknown. This effect is expected to increase, if it is real, in windward-oriented areas, especially when rough seas dominate.

5. Sources of Salinity

Subaerial volcanic rocks have a very low content of Cl (Custodio, 1978). This is confirmed by chemical analyses of well and spring water from well known areas, even where there is uptake of volcanic CO_2 from cooling deep magma chambers and the consequent deep weathering of rocks (Custodio, 1978, 1988). The same can be said for S, but there are exceptions for the sulphate contribution. This may be increased near areas with concentrated

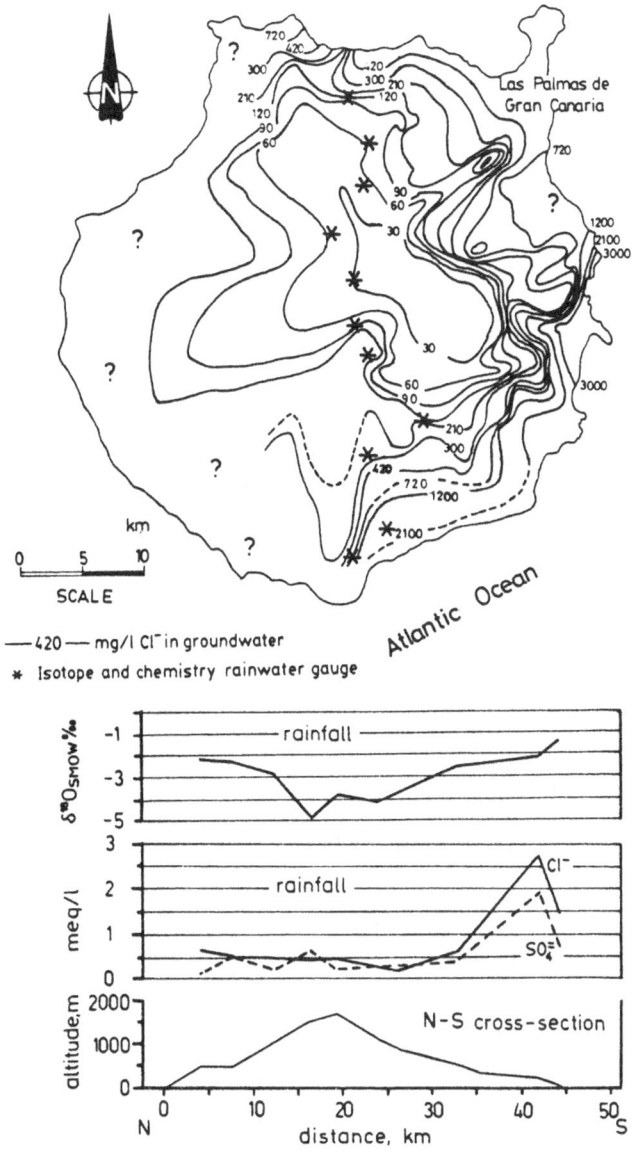

Figure 4. Rainfall depth (main yearly values) and mean weighted Cl⁻ and SO_4^- content in mg/l in rainfall incorporating dry fallout, and oxygen-18 in rainfall water (δO^{18} SMOW), in a N-S profile.

volcanic emissions, representing recent eruption sites. Also there is the possibility of SO_4^- increase from oxidation of sulphides in old submarine volcanics, which formed from seawater In this case water outflows devoid of oxygen are given off sometimes with high iron contents. Basal complexes have plenty of these submarine volcanics, now emerged.

Large altitudinal changes can be expected in volcanic islands. In Gran Canaria submarine lava flows, probably of Pliocene age, are observed at about 100 m above present sea level. A deep bore in Lanzarote found subaerial basaltic flows of Oligocene age between 700 and 800 m below present sea level, covered with 60 m of submarine lavas and pyroclastics and above them new subaerial volcanics (Sánchez-Guzmán and Abad, 1986). These altitudinal changes mean that areas presently above sea level were under seawater some time ago so that some relict salinity from seawater can be found in poorly drained and very low permeability formations.

The old marine water is diluted by salt diffusion into continental water after saline water in the more pervious formations is flushed out. Diffusion is a very slow process and its effect is easily erased in small aquifers, as in the Canaries. The detailed mathematical analysis is outside the scope of this paper (for more details see van Genuchten and Alves, 1982; Javandel et al., 1984; Davidson and Airey, 1982).

When two semi-infinite bodies of static water of different salinity, separated initially by a sharp interface, are allowed to start a diffusive process, the concentration can be described by:

$$(C - C_2)/(C_0 - C_2) = 0.5 \text{ erfc } [x(4D_m t)^{-1/2}]$$

in which
C = concentration at distance x from the interface and at time t after the start of the process.
C_0 = initial concentration of the saline water body.
C_1 = initial concentration of the freshwater body.
D_m = molecular diffusity.

$$\text{erfc} = \text{complementary error function} = 1 - \frac{2}{\sqrt{\pi}} \int_0^z \exp(-z^2) \, dz$$

A volcanic aquifer is really a medium of two separate porosities (Custodio, 1989) but when long times are involved in low permeability formations, the behaviour is like a porous one. The value of D_m to be used is that of water molecular diffusity for the solute of interest, increased by the tortuosity effect of the porous medium. For $D_m \sim 0.01$ m^2/day the following values are obtained for x in metres and t in years

$x/\sqrt{t}$	0.001	0.01	0.1	1.0
$(C - C_1)/(C - C_0)$		0.99	0.89	0.180.00

To get significant saline increases at some tens of metres from the initial interface, centuries must elapse and even then groundwater flow cannot be disregarded. Real concentrations are smaller in short flow systems.

Let us consider a diffusive process of solute into a semi-infinite space A from another semi-infinite space B at constant salinity. There is no flow and only diffusion takes place from B into A, initially devoid of solute. Concentration C in space A after a time t at distance x from the interface AB is given approximately by the aforementioned error function (the factor 0.5 disappears due to the constant concentration at semispace B):

$$C/C_0 = \text{erfc } [x(4D_m t)^{-1/2}]$$

For $z < 0.1$ (large time and short distance) erfc $\sim 2z/\sqrt{\pi}$. Then for two solutes 1 and 2:

$$\frac{(C/C_0)_1}{(C/C_0)_2} \sim \frac{D_{m2}}{D_{m1}} = \frac{D_1}{D_2}$$

in which D refers to open water if tortuosity effects can be assumed to be the same for two different solutes.

For a not very concentrated solution consisting on Na_2SO_4 and NaCl, at 25° the handbook of Chemistry and Physics gives $D_1 = 1.12 \times 10^{-5}$ cm^{2}s^{-1} and $D_2 = 1.54 \times 10^{-5}$ cm^{2}s^{-1}. Then:

$$(rSO_4/rCl)_1 \sim 1.17 \ (rSO_4/rCl)_2$$

for a mixture of $MgSO_4$ (D = 0.71 cm^{2}s^{-1}) and NaCl this becomes:

$$(rSO_4/rCl)_1 \sim 1.47 \ (rSO_4/rCl)_2$$

This means that diffusive processes from a seawater body into a freshwater body increases the rSO$_4$/rCl ratio, but under normal circumstances it cannot be increased beyond 0.15 when semispace B contains seawater. For shorter times and longer distances there is a smaller increase.

6. Saline Groundwater in the Amurga Massif

The Amurga massif forms the southeast of Gran Canaria island, limited by the sea and two deep gullies (barrancos), with the apex at 1122 m of altitude (Fig. 5, after Gasparini et al., 1987). It is composed of phonolites and phonolitic ignimbrites from an area not far from the peak. They are assumed to rest on Miocene basalts at unknown depth. These basalts are below sea level near the coastal area. The massif has not been covered by more recent volcanic extrusions due to its elevation relative to more recent volcanic centres.

A series of drilled wells have been constructed recently at 3 to 5 km from the coast. There is no other source of groundwater information in the massif except for a few

606

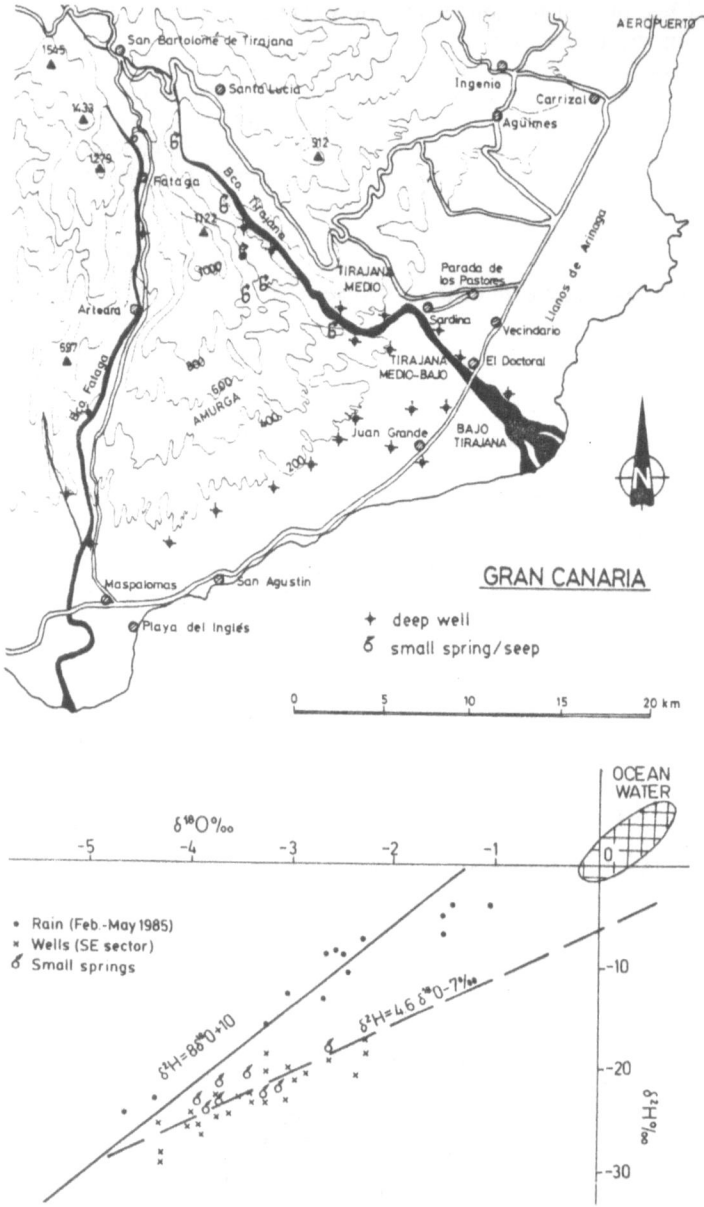

Figure 5. The Amurga phonolitic massif in SE Gran Canaria, and the δO^{18} vs. δH^2 content in rain and groundwater (deep wells and some small springs). Only rains for the period February-May are shown.

insignificant springs. The wells penetrate below sea level and the piezometric levels are up to 20 m above it. The theoretical Ghijben-Herzberg freshwater-saltwater interface is always well below the bottom of the wells.

Groundwater is brackish with about 1.5 g/l of Cl^-. It seems that there are no important vertical salinity changes. The rNa/rCl (r = meq/l) ratio is about 0.9 to 1.0, with rSO_4/rCl ~ 0.2. This water is desalinated in a large electrodialysis plant.

Modern seawater intrusion can be ruled out as the origin of groundwater salinity due to existing hydrodynamic conditions and the lack of known saline stratification. Also. it seems that relict seawater is not an explanation for Amurga since the rSO_4/rCl ratio is too high and the rNa/rCl ratio does not agree with the expected cation exchange effect when freshwater is replacing saltwater in a material with non-negligible exchange capacity (see Custodio and Bruggeman, 1987). The age of the water, as noted below, is young relative to the rate of possible altitudinal changes. Also it has to be ruled out that the salts come from the rock itself as an infinite time of contact will not yield the observed salinity.

Diffusion from deep seated saline water seems very unlikely since the observed thick layer of nearly homogeneous brackish water (more than 50 m in some instances) is difficult to explain by this process. Also the rSO_4/rCl ratio is too high.

The only remaining explanation for groundwater salinity is a climatic effect. Rainfall varies from about 100 mm/year at the coast up to about 300 mm/year at the peak. Expected groundwater recharge under dominantly bare, rocky soil, with scattered euphorbia-type bushes and cactuses with deep roots, may vary between less than 1 mm/year up to 20 mm/year. Surface runoff is only a small fraction of rainfall. With a combined rainfall salinity between 20 and 50 mg/l in the assumed recharge area, recharge water salinity varies between 200 and 500 mg/l in spite of the relatively high elevation. This agrees with the above. In other areas of the island with a higher rainfall the chloride content is 30 to 60 mg/l, clearly different from that observed in the isolated massif.

The environmental isotopic composition of rainfall in Gran canaria follows the typical oceanic line represented by $\delta H^2 = 8\delta O^{18} + 10\%.$, for δ values in %. relative to the SMOW standard. This is valid also for the rains in the dry area, except at summer time, but their contribution to groundwater recharge is negligible.

The Amurga area groundwaters plot on a different line (Fig. 5) of the equation (Gasparini et al, 1987, 1989) $\delta H^2 = 4.6\delta O^{18} - 7\%.$. The slope is typical of an evaporative process that happens only in soils devoid of a continuous vegetation cover, as in the present situation. Conditions are not as extreme as those found in dune belts in deserts. This agrees with the evaporative concentration as the source of salinity. The groundwater line is well below the seawater point, which excludes seawater as the origin of salinity. Also rock salt dissolution is excluded due to the relatively large range of values along the groundwater line.

The intersection of the groundwater line with the meteoric line corresponds to a δO^{18} content of $-5.0 \pm 0.8\%$. According to the altitudinal profile of Figure 4 and the altitudinal regression line of δO^{18} vs. elevation (h) (Gonfiantini, 1974; Gonfiantini et al., 1976):

$$\delta O^{18} = - (1.43 \pm 0.15) \times 10^{-3} h(m) - (2.91 \pm 0.20) \%.,$$

the recharge area corresponds to an elevation of 700 to 1950 m. Since the maximum elevation of the Amurga itself is 1122 m, the area of recharge thus defined is about a quarter of the total area. It corresponds to the triangle above 700 m. The result is a mixture of water recharged at different altitudes weighted by the corresponding surface area.

C^{14} determinations have been corrected using C^{13} values in dissolved inorganic carbon to compensate for possible dilution by inert volcanic CO_2. The four corrected ages for the Amurga wells vary between 7400 and 11600 years (Gasparini et al., 1989). Three ages are between 10100 and 11600 years, a surprisingly small spread, but consistent with the chemical uniformity of the waters. Since well altitude is less than 200 m, considerably below the above-determined recharge area, a compound piston and exponential flow model seems reasonable for that case (see Zuber, 1986). The groundwater exchange time can be taken as of 10000 to 12000 years, including the transit time through the thick unsaturated zone. If the piston flow model is admissible for such a fissured and porous medium, the transit time through the unsaturated zone of thickness H, with a steady content of mobile water m_r and a recharge R, is $t_R = H\ m_r/R$. Assuming values of $H = 300$ to 500 m, $m_r = 0.02$ to 0.03 and $R = 1$ to 10 mm/year results in $t_R = 600$ to 15000 years. This means that it is possible that under some circumstances the exchange time deduced above can be explained almost entirely by the unsaturated flow duration.

Let us assume that Amurga is a circular sector of radius (up to the wells) r_p and angle α (radians), with the top at the peak; r_0 is the radius to the edge of the recharge area, receiving a mean recharge R, that arrives at the saturated zone delayed by $t_R = H\ m_r/R$. If r is the radius of a section between r_0 and r_p, H_0 the saturated thickness (assumed constant) and m the kinematic porosity, the conservation of groundwater flow Q, implies:

$$Q = \frac{\alpha}{2} r_0^2 R = r_0\ \alpha\ H_0\ m\ v_0 = r\ \alpha\ H_0\ m\ v = r_p\ \alpha\ H_0\ m\ v_p$$

in which v is groundwater intergranular velocity at a given distance, assuming uniform flow along the section. Then:

$$v = \frac{r_0^2\ R}{2H_0 mr}\ \frac{1}{}$$

The flow time from r_0 to r_p is:

$$t = \int_{r_0}^{r_p} \frac{dr}{v} = \frac{2H_0\ m}{r_0^2\ R} \int_{r_0}^{r_p} r\ dr = \frac{H_0\ m}{r_0^2\ R} (r_p^2 - r_0^2) = t_t - \frac{H\ m_r}{R}$$

in which t_t = turnover time, deduced from corrected radiocarbon data.

$$R = \frac{1}{t_t} [H_0m (-1 + (r_p/r_0)^2) + H\, m_r]$$

Using the following values, obtained from existing knowledge of the islands (Custodio, 1985) and local conditions as deduced before:

r_0 = 10000 m H_0 = 200 to 400 m   m = 0.03 to 0.04
r_p = 20000 m   H = 300 to 500 m m_r = 0.02 to 0.03
    t_t = 10000 to 12000 years

The result is R = 2 to 6 mm/year, within the range of recharge values expected. Thus, the observed salinity can be explained by a climatic effect notwithstanding the fact that 10000 years ago climatic conditions were not the same as today. This is no reason to invalidate the climatic impact on salinity, but the recharge values deduced are not necessarily those prevailing at present. It has also been assumed that the system is under steady state conditions. Although long transients from former climatic conditions cannot be ruled out without more data, it seems unlikely that they are important at the present time given the small area concerned.

Consider a groundwater system consisting of an aquifer between a water divide and a drainage line parallel to it. According to well known aquifer behaviour properties (Custodio and Llamas, 1976, Chap. 9.17; Lloyd and Miles, 1986), head perturbation disappears after a time $t_d = \alpha\, L^2 S/T$ in which $\alpha \approx 2.5$, L is the distance between the water divide or aquifer limit and the drainage line, S is the storage coefficient and T the aquifer transmissivity, which is assumed constant. This is for linear flow, but for radial flow it is also a first approximation. In Amurga, for L $\approx$ 10 km, S $\approx$ 0.01 to 0.02 (drainable porosity) and T $\approx$ 50 to 200 m²/day, this results in t_d being less than 250 years. In that case residual head changes from past climatic conditions are unlikely.

The chemical and isotopic study of groundwater is appropriate since chemical values present a longer memory than hydrodynamic ones.

7. Saline Groundwater in the Famara Massif

The Famara massif forms the northern part of Lanzarote Island, with cliffs at the West side and more gentle slopes towards the East. It constitutes a thick pile of tabular basalts of Miocene age, gently dipping towards the East, and includes the injection dykes and the associated ejecta close to the fractured emission points.

Groundwater is tapped by a few galleries drilled from the West coast and shallow wells penetrating perched aquifers concentrated in some flat areas. Studies were carried out early in the seventies and included intensive drilling, head measurement and chemical sampling (Custodio, 1974 a, 1974 b) but no environmental isotopic studies are available. Groundwater is dominantly brackish.

There is a groundwater dome (see Fig. 6, after Custodio, 1974 a, 1978) below the highlands (altitude about 600 m) as the result of the small recharge derived from rainfall that barely reaches 200 mm/year, plus some dew deposition, the importance of which for increasing groundwater recharge is unknown. The groundwater body behaviour is conventional (Custodio, 1974 a) in spite of the existence of a variable combination of lava flows, dykes, pyroclastic materials and red layers. Soil water balances show a mean recharge from less than 1 mm/year at the eastern coastal plains up to 10 mm/year at the top. These are an indication only of the order of magnitude.

The salinity of the regional groundwater body increases with depth and where rock permeability is very small. It ranges from about 550 mg/l Cl⁻ up to 2500 mg/l Cl⁻in the more saline bore-holes inside the galleries (Fig. 6). Depending on the area a background salinity of 500 to 1000 mg/l may be assumed, which according to the rainfall salinity data given corresponds to about 7.5 to 20 mm/year of mean recharge, assuming that runoff is negligible. These are quite large values. A possible explanation is the frequency of fog and dew in the area.

The increases in observed salinity need another explanation. The rSO_4/rCl ratio is about 0.14, the less saline waters presenting a ratio slightly higher than the more saline ones. All waters are chemically similar but the rNa/rCl ratio ranges from 1.2 in the less saline waters (about 550 mg/l Cl⁻) to 0.8 in the more saline ones (up to 2800 mg/l Cl⁻), very close to that in marine water (0.82).

Although more erratic, the same pattern is true for the small water seeps in Famara (salinity between 300 and 600 mg/l Cl⁻). The same holds for the shallow wells in the perched aquifers (1000 to 1700 mg/l Cl⁻), as well as in the wells at low altitude in the Southwest area, although in that case the salinity range is greater (1000 to 2100 and up to 4000 mg/l Cl⁻). Other small seeps scattered throughout the island at relatively high altitude, some at the foot of young volcanic cones, with a range in salinity from 50 to 140 mg/l Cl⁻, exhibit rNa/rCl commonly about 1.2 and rSO_4/rCl typically around 0.4.

The increase in water salinity in the galleries seems to be related to seawater contamination, but it cannot be explained as an unfinished leaching process of materials saturated with seawater a long time ago, if it is assumed that the island was then at a lower altitude relative to sea level. This leaching process implies an increase in the rNa/rCl ratio through cation exchange but that is not apparent. The reverse seems, in fact, to be the case.

A recharge of about 10 mm/year, an active saturated thickness of at least 200 m and a kinematic porosity of about 0.06 (the value deduced for the Miocene basalts) suggest a mean exchange time higher than 1200 years. In practice it is much higher for the deeper levels. This allows one to assume that the downward increasing salinity may be the effect of very slow salt diffusion from below. A very thick continental water-marine water mixing zone can therefore be expected. It is thicker in the areas where the rock is less permeable, since continental water flow decreased there. The disruption in the salinity stratification seen in Figure 7 for gallery III is the result of water extraction around the gallery bottom through the crown of bore-holes drilled there. The preferential orientation of vertical fissures parallel to the coast line allows for relatively high yields in the boreholes in spite of the very low permeability in the direction normal to the coast line. The ascent of saline water, invading rock volumes previously saturated with fresher water produces a decrease in the

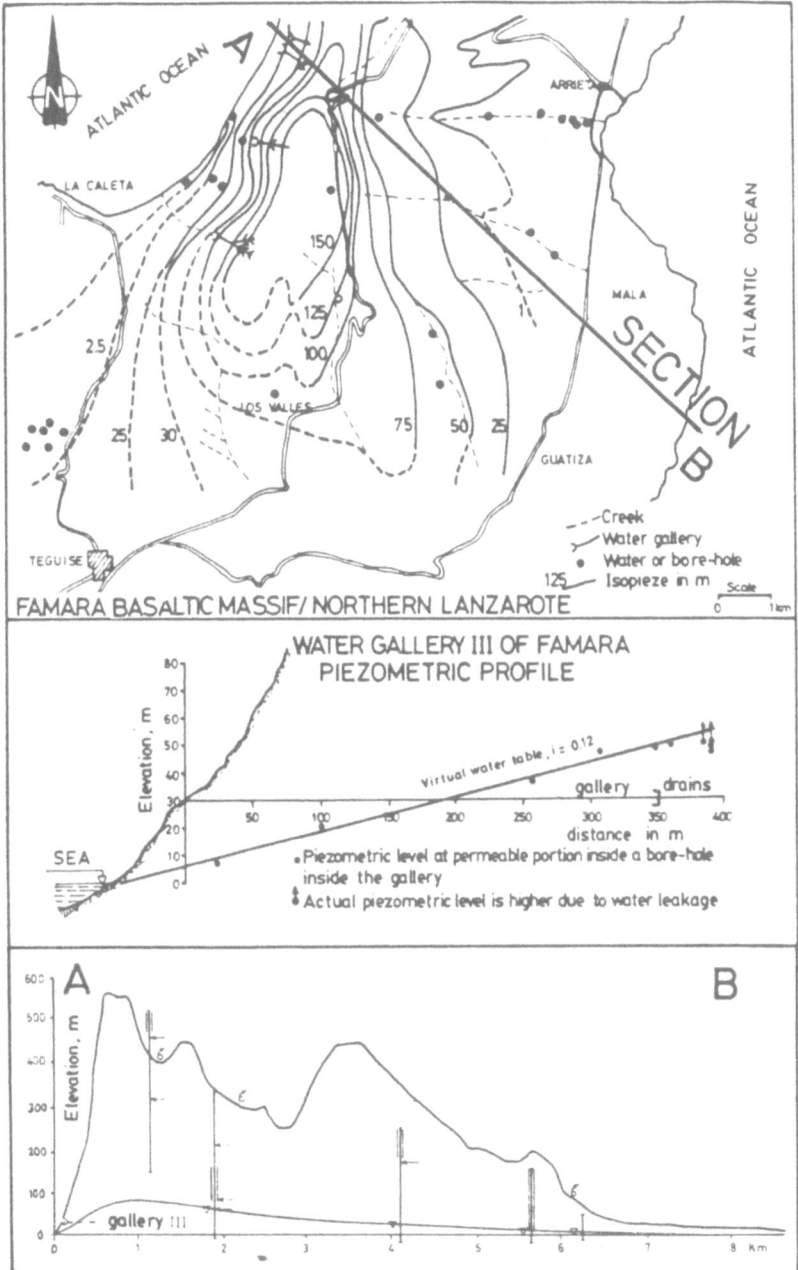

Figure 6. The Famara massif. The regional groundwater table, the piezometric profile along one of the water tunnels (gallery III) and a cross-section of the massif showing the large thickness of the unsaturated zone are shown.

rNa/rCl ratio, as the data seem to show. Calculations are needed to substantiate this. The saltwater body is probably under non-equilibrium conditions and this requires further explanation.

Famara's environmental aridity seems to be softened by the frequency of dew and fog in the highland, but even so groundwater recharge is saline. This salinity increases with depth in the regional groundwater body, probably due to diffusion from deep-seated trapped saline water of marine origin, not responding to Ghijben-Herzberg saltwater wedge conditions.

8. Saline Groundwater in Gomera Island

The island of Gomera, although small, exhibits a mean rainfall varying between barely 200 mm/year in the southern coastal area up to about 900 mm/year in the central highlands as shown in Figure 2. There are no chemical analyses of rainfall. Groundwater chemical information is good and will be completed shortly with environmental isotopic analyses.

The chloride content in groundwater decreases clearly from the centre (30 to 50 mg/l Cl⁻) towards the coast where values up to 350 mg/l Cl⁻ are common in wells and springs not influenced by seawater encroachment. The values are smaller (70 to 140 mg/l) on the dry southern coast the on the much wetter northern coast. This is explained by the prevalence of winds from the sea in the northern area. These winds contribute more atmospheric salinity than in the South.

Most low-salinity waters exhibit a rSO_4/rCl ratio close to that of marine water, but the more saline groundwaters at low altitudes, especially along the northern coast, exhibit much higher ratios, up to 2.5, generally associated with relatively high total inorganic carbon contents and high pH. The latter is due to the intense rock weathering in a system open to soil CO^2. Since the rock is practically devoid of Cl, there is also a concentration by evapotranspiration of rainfall and airborne salts in groundwater recharge.

In this area combustion or other man-made or natural sources of S are unlikely. Agriculture may be a possible source of SO_4^- but this is also unlikely. In the possible areas of recharge of the springs showing high rSO_4/rCl ratios there are only small agricultural plots, under dry farming and without intensive use of fertilizers and soil improvers.

A possible explanation for the high rSO_4/rCl ratios is fractionation of marine salts but this has not been checked in the field. It cannot be excluded that some of the dissolved S comes from oxidation of sulphides from submarine volcanics in the basal complex where there is deep groundwater circulation, as also happens inside the Caldera de Taburiente on the island of La Palma. In other places this origin for SO_4^- is unlikely.

9. Saline groundwater in Fuerteventura

Fuerteventura is the most arid of the Canary Islands. The orientation and its altitude, although important, do not produce a sufficient orographic effect to increase the rainfall significantly.

613

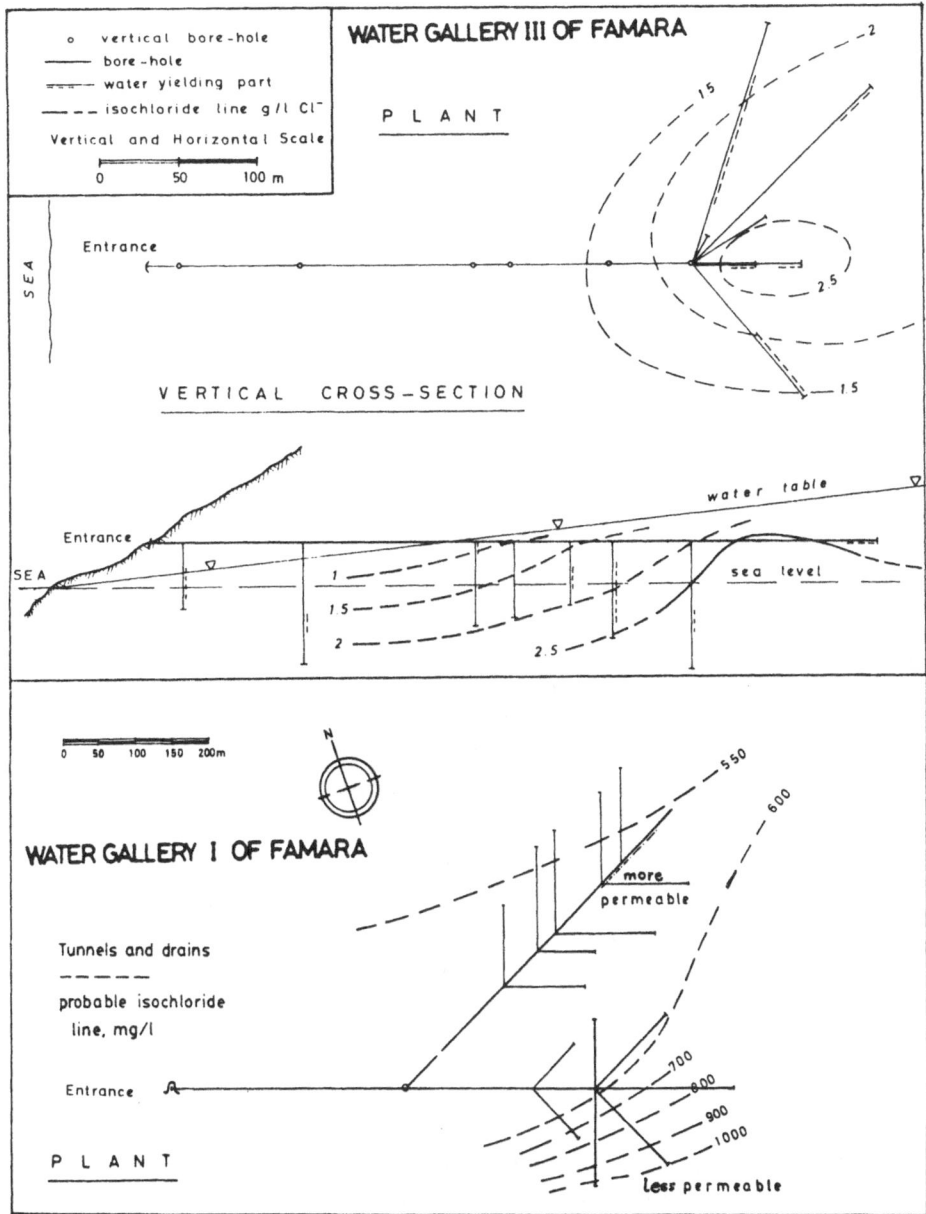

Figure 7. Salinity distribution along two of the galleries in the Famara massif, in mg/l Cl⁻. Data come from samples taken at water inlets and drains inside the gallery. They correspond to known pervious tracts.

Mean rainfall varies barely, 60 mm/year at the coast and 200 mm/year in a small area at the centre (Fig. 2). The conditions are very close to those existing in Lanzarote, with which it has many similarities. Useful information comes from the SPA-15 (1975) report and other related studies. Groundwater samples are mainly from shallow wells.

The approximate range of chloride content in groundwater uncontaminated by marine water is 200 to 5000 mg/l Cl^-, with low values along the central axis and increasing towards the coast. The lowest values do not correspond with the maximum altitude. This may be interpreted as the effect of these elevations being towards the windward side. The leeward side, at the island centre, still receives some rainfall but it is less affected by sea spray and air-borne salinity.

The rSO_4/rCl ratio exhibits a rather wide range (0.13 to 0.40) that does not permit easy interpretation. Analytical accuracy is rather low. Also some wells are just in farmed areas or in the bottom of creeks that contain lands levelled for cultivation. As happens in Lanzarote, the less saline groundwaters have a rNa/rCl ratio greater than unity and those with high salinity a ratio less than unity. The salinity seems to be of purely climatic origin and the salts are mainly marine, transported through the atmosphere.

10. Conclusions

Existing data point to climatic impact as the cause of brackish water found in the Canary Islands except where there is direct influence of modern or old seawater (as in deep wells in East and Southeast Gran Canaria, North of Amurga and perhaps in Famara). The climate also affects groundwater coming from more rainy areas in the interior of the wet islands due to the increasing area for the receipt of wind-transported salts, that are leached towards the saturated zone by the small recharge existing there.

The rSO_4/rCl ratios do not coincide with that of seawater, especially near the coast. This has not been studied in detail, but chemical fractionation of airborne marine salinity is a possibility to be considered. Diffusion of marine salts from below cannot produce the ratios observed.

In some areas salinity of climatic origin increases downwards as it also does in very low permeability formations due to diffusion of deep-seated saline water of marine origin, probably under non-equilibrium conditions.

11. Comments on Water Supply Alternatives in Areas with Brackish Groundwater

The aridity of two of the Canary Islands and the southern part of the others, jointly with the contribution of marine salts through the atmosphere, create the conditions for the existence of brackish groundwater. This is the only local water resource in many areas. Most of the major ions are in concentrations higher than the admissible limits in drinking water and even for many household uses and for crop and garden irrigation.

Common alternatives practised used in the Canaries are:

a) importing of water from other areas of the island, when they exist and it is administratively and economically feasible. Imported water can be used directly or after mixing it with local brackish water to increase total availability. The same can be applied to the alternatives that follow.

b) obtaining additional surface water resources from the establishment of stormwater storage reservoirs, if there is enough runoff and sufficient sites. This includes both small and large rainwater cisterns. In many cases these additional resources are too small or too expensive. To reduce evaporation in warm and windy areas, covers of different kinds can be used over basins, or water can be infiltrated into the ground to be recovered later, generally with some mixing with brackish groundwater.

c) importing water by tanker-ship from another island or from the continent. This is an expensive solution, used only for emergencies. Systematic use requires, in addition to the tanker, a loading station and an unloading station, each with a reservoir plus a pipeline to the harbour or to a buoy offshore. There is also the problem of the reliance on another administration or country for security of the supply, and this may be socially and politically undesirable.

d) encouraging efficient use of water, generally introducing rapidly increasing water prices for increased volumes per person, house or building, or introducing incentives for water saving, mainly by tax reductions or cheap loans for new investments. Pipe-leak detection and repair is very important in urban areas. Where agricultural water demand is significant or dominant, improved irrigation techniques are needed, but this is not a long-term solution. Efficient use of irrigation water produces small but very saline return flows that in most cases add to natural recharge, thus impairing the aquifer salinity after some time. Drip irrigation is an effective method since it reduces non-useful evaporation from the soil, but other irrigation methods are also efficient if there is a soil cover (mulch) to reduce evaporation. For saving water it is much more effective to reduce the irrigated area by shifting to other more labour-intensive crops and producing a higher income.

e) introducing water recycling in industry and sewage water reuse. After adequate treatment urban sewage can be used for irrigation of crops, gardens and public parks and for street cleaning. The feasibility depends on the original quality of supply water, the avoidance of noxious and difficult-to-treat pollutants and the efficiency of the sewer system. Reuse is valid for green areas in towns and tourist centres and where irrigated agricultural areas are close to the treatment plant.

f) installing desalination plants where the water use can support the cost of the process: urban demand can generally meet it in tourist and residential areas, and also for special agriculture for which feed water is brackish. Local brackish water can have good potential for reverse osmosis and in some cases electrodialysis potential for desalination. If the flow and steady chemical composition of water can be assured and adequate chemical pretreatment is applied to avoid early encrustation. Seawater desalination by reverse osmosis or distillation (multi-stage flash and multi-effect for large plants; vapour compression for medium and small plants is becoming competitive in some areas. The water produced can be supplied directly of mixed with local brackish water.

12. Acknowledgments

Data used here has come from studies done by or for the Public Administration of the Canaries, or from several research projects carried out jointly with the Polytechnic University of Catalonia (Department of Ground Engineering - ETSICCP) and the International Course on Groundwater Hydrology. Part of the research has been financed by the Spain-US Joint Committee for Scientific and Technical Cooperation. Also some data has been generated from studies carried out jointly with the Université de Paris Sud and financed by CGS and ELMASA.

13. References

Custodio, E. (1974 a) "Flujo de aguas subterráneas y existencia de un nivel de saturación en las formaciones volcánicas de la Isla de Lanzarote (Islas Canarias, España).", Simposio Internacional sobre Hidrologia de Terrenos Volcánicos, Arrecife de Lanzarote, Gobierno de Canarias-Cedex, Madrid/Las Palmas, 1987, Vol. 1, pp. 185-215

Custodio, E. (1974 b) "Contribuciones al conocimiento geohydroquímico de la Isla de Lanzarote (Islas Canarias, España)", Simposio Internacional sobre Hidrologia de Terrenos Volcánicos, Arrecife de Lanzarote, 1974, Gobierno Autónomo de Canarias-Cedex, Madrid/Las Palmas, 1987, Vol. 1, pp. 463-510

Custodio, E. (1978) *Geohidrogeologia de terrenos e islas volcánicas*, CEDEX-Centro de Estudios Hidrográficos and Instituto de Hidrología, Madrid, Pub. 128, 303 p.

Custodio, E. (1985) "Low permeability volcanics in the Canary Islands (Spain)", *Hydrogeology of Rocks of Low Permeability*, Tucson, Arizona, International Association of Hydrogeologists, pp. 553-554

Custodio, E. (1988) "Hydrochemistry of Tenerife Island", Revista Española de Hidrogeología, Associación Española de Hidrología Subterránea, 3, pp. 1-19

Custodio, E. (1989) "Groundwater characteristics and problems in volcanic rock terrains", *Isotope Techniques in the Study of the Hydrology of Fractured and Fissured Rocks*, Proc. Advisory Group Meeting, STI/PUB/790, Vienna 1986, Inter. Atomic Energy Agency (in press)

Custodio, E., & Llamas, M.R. (1976) *Hidrología Subterránea*, Ediciones Omega, Barcelona, 2 vols., 2350 p.

Custodio, E., Bruggeman, G. (1987) *Groundwater Problems in Coastal Areas*, Studies and Reports in Hydrology 45, UNESCO, Paris, 596 p.

Custodio, E., Jiménez, J., Núñez, J.A., Puga, L., & Braojos, J.J. (1989) "Hydrology of the Canary Islands (Spain)", *Guide on the Hydrology of Small Islands*, Studies and Reports in Hydrology, UNESCO, Paris, (in press).
Also in Hidrología y Recursos Hydráulicos, Vol. XIV, Assoc. Española de Hidrología Subterránea, Madrid, (in press)

Davidson, M.R., Airy, P.L. (1982) "The effect of dispersion on the establishment of a paleoclimatic record from groundwater", Journal of Hydrology, 58 (112), pp. 131-147

Dinçer, T. (1980) "Use of environmental isotopes in arid-zone hydrology", *Arid Zone Hydrology*, Inter. Atomic Energy Agency, Vienna, pp. 23-30

Fontes, J.Ch. (1980) "Environmental isotopes in groundwater hydrology", in Fritz, P., and Fontes, J.Ch. (eds.) *Handbook of Environmental Isotope Geochemistry*, Elsevier, Vol.1, pp. 75-140

Gasparini, A., Fontes, J.Ch., Custodio, E., Jiménez, J., & Núñez, J.A. (1987) "Primeros datos sobre las características químicas e isotópicas del aqua subterránea del macizo fonolítico de Amurga, Gran Canaria", Hidrogeología y Recursos Hydráulicos, Associación Española de Hidrología Subterránea, Vol. XI, pp. 281-298

Gasparini, A., Fontes, J.Ch., Custodio, E., Jiménez, J., & Núñez, J.A. (1989) "Example d'étude géochimique et isotopique de circulations aquifères en terrain volcanique sous climat semi-aride (Amurga, Gran Canaria, Îles Canaries).", Journal of Hydrology, (in press)

Gat, S.R., & Gonfiantini, R. (1981) *Stables Isotope Hydrology: Deuterium and Oxygen-18 in the Water Cycle*, Inter. Atomic Energy Agency, Technical Report Series 210, 339 p.

Gonfiantini, R. (1974) "Environmental isotope investigation in Canary Islands groundwater", Simposio Internacional sobre Hidrogeología de Terrenos Volcánicos, Gobierno de Canarias-Cedex, Madrid/Las Palmas, 1987, Vol. II, pp. 617-659

Gonfiantini, R., Gallo, G., Payne, B.R., & Taylor, C.B. (1978) "Environmental isotopes and hydrochemistry in groundwater of Gran Canaria", *Interpretation of Environmental Isotope and Hydrochemical Data in Groundwater Hydrology*, Inter. Atomic Energy Agency, Vienna, pp. 159-170

Javandel, I., Doughty, C., & Tsang, C.F. (1984) *Groundwater Transport: Handbook of Mathematical Models*, American Geophysical Union, Water Res. Monograph 10, 228 p.

Levin, M., Gat, S.R., & Issar, A. (1980) "Precipitation, flood- and groundwaters of the Negev highlands: an isotopic study of desert hydrology", *Arid Zone Hydrology*, Inter. Atomic Energy Agency, Vienna, pp. 3-22

Lloyd, J.W., & Miles, J.C. (1986) "An examination of the mechanisms controlling groundwater gradients in hyper-arid regional sedimentary basins", Water Res. Bull., 22 (3), pp. 471-478

Sánchez-Guzmán, J., & Abad, J. (1986) "Sondeo geotérmico Lanzarote-1: significado geológico y geotérmico", *Física de los Fenómenos Volcánicos*, Anales de Física-B, Madrid, 82, pp. 102-109

SPA-15 (1975) *Estudio científico de los recursos de aqua den las Islas Canarias*, Dirección General de Obras Hidráulicas-UNESCO, Madrid, 3 vols. + 2 vols. of plates

Soler, C. (1989) "Estudio hidrológico del Hierro", Plan Hidrológico de Canarias, Dirección General de Aguas, Santa Cruz de Tenerife (internal report)

Van Genuchten, M.Th., & Alves, W.J. (1982) *Analytical Solutions of the One-Dimensional Convective-Dispersive Solute Transport Equation*, U.S. Dept. of Agriculture, Tech. Bull.

618

1661, 149 p.

Zuber, A. (1986) "Mathematical models for the interpretation of environmental radioisotopes in groundwater systems", *Handbook of Environmental Isotope Techniques*, Vol. 2: Terrestrial Environment B., Elsevier, pp. 1-59

PART IV: MANAGEMENT, TECHNIQUES AND CASE STUDIES

SECTION D: Techniques

GEOLOGICAL RESOURCES FOR CONSTRUCTION WITH LOW ENERGY AND LOW TEMPERATURE INPUT

GEORGES A. PATFOORT
Vrije Universiteit Brussel (V.U.B.)
Dept. Constructions
Pleinlaan 2
1050 Brussels, Belgium

ABSTRACT: Through chemical polymerization reactions, minerals in clay can be hardened and transformed into useful construction materials. The mineral polymers are produced at atmospheric pressure and low temperature with a minimum of energy input. They can be used as structural materials with attractive properties in construction, building, hydraulic works, dykes and other applications.

1.   Environmental Requirements for Engineering

Geologists, climatologists, hydrologists, ecologists and other scientists are more and more concerned about the environment. They are preoccupied about protection from anthropogenic and natural events and they try to foresee, anticipate and estimate short-term and long-term changes, and even make provisions for possible errors. They think about cataclysms such as floods, inundations, hurricanes, typhoons, earthquakes, volcanic eruptions and about environmental changes such as the greenhouse effect, ozone concentration, coastal changes, climatic fluctuations and feedbacks, air and water pollution, deforestation, aridity, erosion, sea level rise, desertification, water management, and also in general about the supply, depletions and limits of resources and perspectives concerning energy problem.

The engineer has to look for solutions and to develop appropriate materials, processes, designs, and finished products. Problem solving techniques have to be used to adapt the products to the ever changing ecological environment. The information he gets is difficult to integrate: the daily news is tendentious, and disaster prevention is problematical. A method for assessing long-term evolution prospects has to be developed from global dynamic system analysis. The lack of appropriate information in data banks has to be complemented and then processed and developed further as part of a synergetic strategy.

Material engineering in this context has not been, strictly speaking, successful. For example,

621

R. Paepe et al. (eds.), Greenhouse Effect, Sea Level and Drought, 621–624.
© 1990 *Kluwer Academic Publishers.*

the most important construction materials: cement and reinforced concrete are still not resistant to acid rain and corrosion; building construction with brick walls is not even earthquake resistant due to the limited tensile and flexural strength and toughness; window panes and glass bottles are dangerous; the weight of a glass bottle is often more than that of its content.

Mitigation of disasters and the rethinking of materials and product design, considering resources and energy, is urgent. Indeed, every material and process uses an input of energy: more than 40% of our energy is used in industrial production.

We urgently need construction materials produced by easy processing methods and with the following properties: low density, incombustibility and fire resistance, resistance to chemicals, resistance to high temperature, high toughness, low energy input and of course low price, available in very large and inexhaustible quantities and ecologically attractive.

In the future very large quantities of such construction materials will be necessary, to maintain our ecological system. One should consider housing, infra-structures for houses, sewerage, dykes, dams, hydraulic works, soil and road stabilization, erosion problems, safe storage of polluted materials and so on. Such materials have not previously existed.

2. Inadequate Material Properties

The engineer might consider, among others, three classes of materials: metals, ceramics and organic polymers. Nearly all materials are of geological origin, except polymers which are from plants and animals. To extract, transport and transform these materials, significant amounts of energy are needed: steel production absorbs 30-40 MJ/kg, cement 6-8 MJ/kg, bricks 3-5 MJ/kg, and organic polymers (plastics) 70-100 MJ/kg.

Some raw materials are only available in limited quantities in the earth's crust whereas other are available nearly everywhere and have been exploited since prehistoric times yet have experienced the least technological development.

The earth's crust contains 46% oxygen, 27% silicium and 7% aluminium as the most abundant, by comparison with copper 0.007%, nickel 0.008%, silver 0.00001%. Fortunately these three important elements are spread all over the world and are incorporated in rocks, clay and sand deposits. Our ancestors used them as bricks and ceramics. Later one they were used in cement and porcelain. Primitive civilisations dried and hardened their materials in the sum. At present these materials are sintered or baked through energy-intensive processes. It is important to establish whether there are other possibilities of transforming the basic elements oxygen, silicium and aluminium to valuable construction materials with a minimum of energy consumption.

Sometimes our vocabulary and semantics restrict out thinking patterns. It became customary to use the word "polymerisation" to describe the process of forming large molecules from smaller units, and to use it in organic polymer technology: one is polymerising polyethylene out of the ethylene monomer or polystyrene out of the styrene monomer. Notwithstanding the fact that most solid materials are built up by atoms linked together by primary chemical bonds as ate polymers, it is only in some highly technological publications that the word polymerisation is used in this context for inorganic materials. In fabrication of molecular sieves in sol-gel processes in making sulphur concrete one is (unintentionally) using polymerisation technology.

For many years, research has been going on to make linear and other macromolecules with phosphorus, silicium, germanium and others analogous to organic polymer chemistry. Beyond the field of silicones this research has limited industrial application for various reasons as: unprocessability, chemical instability, high production costs, insufficient molecular mass, etc.

On the other hand combinations of oxygen, silicium and aluminium, such as Al_2O_3 and SiO_2, in various proportions and with different reagents, are able to form larger molecular structures with metal oxides or metalloids in alkaline as well as in acid environments. Usually the bond is realised by the aluminium atom. In aluminosilicates the aluminium is in sixfold coordination, the material becomes extremely reactive.

Fortunately some of the reactions occur at atmospheric pressure and low temperature obtainable with a solar collector. The most important properties of these mineral polymers are absolute fire resistance, thermal and chemical resistance and non conductivity. These are all basic properties for construction.

3. Mineral Polymers (MIP)

At the construction department of the Vrije Universiteit Brussel (V.U.B.) in Belgium, the study of these mineral polymerisation reactions, mostly unknown, and the use of the polymerised materials in general construction application, has been the subject of ongoing research for several years. The class of materials is called "mineral polymers" (MIP).

Many mineral materials with different formulations can be obtained by chemical polymerisation or polycondensation of different oxides from the elements H, Li, Na, K, Al, Si, Mg, Ca, P,..... It is even possible to seek sources of Al_2O_3 and SiO_2 that are ecologically benign, such as fly ash, furnace slag and a range of industrial waste materials.

The properties of mineral polymers (MIP) can be compared to those of organic polymers and ceramic materials. They have similar properties to ceramics, porcelain pottery, glass or cement but they are not obtained by fusion, sintering or firing at high temperature: a temperature of below 100°C is sufficient to complete the reaction and to obtain the final properties.

Raw materials can be produced as semi-dry powders and processed by pressing into finished products. Other formulations produce raw material mixes as pastes and permit pressure forming, extrusion and injection moulding. The raw material can also be obtained as a liquid and all processes usually applied for composites with organic matrices are possible. It follows that fibres and fillers can be incorporated in the mineral matrix to achieve a low density. The matrix can even be foamed.

Materials can be obtained with unconfined compression strengths up to 150 N/mm^2. Therefore properties can be adapted to needs and toughness can be increased to obtain composite mineral polymers with important tensile and bending characteristics and impact

624

resistance.

Present experience suggests that in many countries, especially tropical countries, a range of large exploitable deposits can be found.

Any mineral polymer can be fabricated with a high amount of additives such as sand and waste to reduce the price.

4. Future Prospects and Conclusions

At present two completely different fields are being exploited at the V.U.B.:
- low price building materials with emphasis on incombustibility, low energy input and toughness with respect to earthquake resistance.
- high technology materials with emphasis on strength, toughness and high temperature resistance.

However, bearing in mind the novelty of the technology and the unusual interactions between such different fields of science as geology, chemistry and structural engineering, a well-versed multidisciplinary team is a necessity for a successful outcome.

5. References

Aveston, J., Mercer, R.A., & Sillwood, J.M. (1975) *The Mechanisms of Fibre Reinforcement of Cement and Concrete*, Part I, National Physical Laboratory Publ. No. SI 90/11/98, January 1975

Davidovits, J., & Lavau, J. (1976) "Solid phase synthesis of a mineral blockpolymer by low temperature polycondensation of alumino-silicate polymers", IUPAC International Symposium on Macromolecules, Royal Institute of Technology, Stockholm, Sweden

SITE-INVESTIGATION IN SEMI-ARID AND ARID REGIONS FOR RECONNAIS-SANCE AND ENGINEERING-GEOLOGICAL PURPOSES

M.A.POOL
Vrije Universiteit Brussel
Earth Technology Institute
Pleinlaan 2
1050 Brussel

ABSTRACT. Electro-magnetic investigation techniques have been successfully used in assessing soil/rock interfaces for pipeline construction in Saudi Arabia and Argentina.

Next to refraction-seismics the penetrating Georadar could provide excellent data on the lateral continuation of rockhead, saline groundwater level and sometimes soil/soil discontinuities.

With the detection of North African palaeodrainage systems some years ago from space using the SIR A/B (Shuttle Imaging Radar) it was thought that a mayor break-through was achieved in non destructive profiling from space. Yet there are still some obstacles to be "removed" before this becomes reality. Some of these obstacles are mentioned in this article with emphasis on radar techniques, both spatial and terrestrial.

1. Introduction

During recent decades many water transport systems such as canals and pipelines have been constructed to carry water from source areas to consumers. Source areas may be surficial catchment areas in mountainous regions with sufficient rainfall, aquifers or even desalination plants along coastlines. This paper deals with the geotechnical aspects of pipeline transport systems and the use of Georadar as a tool in the geotechnical survey.

During the many shallow surveys that were carried out it became obvious that the benefit of the Georadar-system depends to a large extend on the moisture content of the subsurface. This means that arid regions are favoured in terms of penetration-depth and interpretation of data. The greatest advantage of the system is the continuity of the profile that can be surveyed, even from a helicopter. At the end of this article a rough cost-benefit analysis for a pipeline construction to Lake Chad is proposed.

R. Paepe et al. (eds.), Greenhouse Effect, Sea Level and Drought, 625–635.
© 1990 *Kluwer Academic Publishers.*

2. The Concept of Georadar-profiling

The name of Georadar is a general denomination from the original name E.S.P., meaning Electro-Magnetic Subsurface Profiling. As the words indicate the principle is detecting and ranging of subsurface objects and interfaces by E.M.-waves. The equipment functions as a radar system, radiating repetitive short time duration electro-magnetic pulses into the ground. These pulses are partly reflected by subsurface discontinuities, received, processed and analyzed.

The discontinuities may consist of different soil transitions, soil- rock transitions, watertables and buried objects like boulders, cables and pipes. Penetration depth is dependent on the type of EM-pulse and the electro-magnetic parameters of the subsurface. Other techniques using EM-pulses exist nowadays, but their principle is based more on induction than reflection. Georadar techniques are a little like seismic reflection techniques but, due to a totally different energy-content of the input wave, the acquired data are distinctive.

Seismic waves can be generated at the surface or near surface using explosives or mechanically induced vibrations. They are detected by geophones stuck into the ground. Seismic frequencies are in general in the range of a few hundred Hz (elastic waves). The E.M.- waves, as used with the E.S.P., have frequencies in the low V.H.F.- range, ranging from 10 to 1000 MHz.

The fact that the subsurface- properties generally allow the low-frequency components to propagate to the greatest depth is an important factor in choosing the type of survey-equipment. However, the higher frequency components give rise to higher resolutions or the ability to discriminate between closely spaced interfaces and targets.

3. Limitations of Georadar use

The success of a Georadar survey or E.S.P. is generally determined by subsurface conditions, data processing techniques and the right choice of equipment and survey parameters. This paper will not deal with all factors in detail but intends to summarise the advantages of E.S.P. in arid and semi-arid regions. The complexity of E.M.-waves while penetrating the subsurface and, for example, oscillations in the reflection of the pulses make it necessary to go into the E.M.-wave theory in some detail.

In general E.M.-waves consist of two systems of transverse waves that propagate through a medium with a velocity of 3×10^8 m/sec. or less. In a loss-free dielectric medium (space) the speed is that of light. In other media the velocity is governed by the so-called dielectric constant (ε). The wavelength (λ) is related to the velocity and the frequency of the pulse. This wavelength together with the so called conductivity determines the actual penetration depth of the E.M.-wave in the subsurface medium. While travelling through a medium the E.M.-wave will loose energy (converted to heat) and will be reduced in propagation velocity. This energy loss or attenuation depends on the conductivity (σ) of the subsurface.

To show the advantages in arid zones, some electromagnetic parameters are listed in Table 1. From this table it can be seen that any salt in solution gives rise to very high

Table 1. Approximate Electromagnetic Parameters of Typical Rocks and Soils at a frequency of 100 MHz., and Normal Temperature

Material	Conductivity (σ, in mho/m)	Dielectric constant (ϵ_r)	Attenuation dB/m	Velocity cm/nsec (approximate)
Air	0	1	0	30
Fresh water	10^{-3}	81	0.18	3.3
Sea water	4	81	330	3.3
Sandy soil (dry)	1.5×10^{-4}	3	0.14	17
Sandy soil (wet)	7×10^{-3}	25	2.3	6.0
Clayey soil(dry)	2.5×10^{-4}	3	0.28	17
Clayey soil(wet)	5×10^{-2}	15	20	7.8
Granite (dry)	10^{-8}	5	10	13
Granite (wet)	10^{-3}	7	0.6	11
Basalt (wet)	10^{-2}	8	5.6	11
Shale (wet)	10^{-1}	7	45	11
Sandstone (wet)	4×10^{-2}	6	24	12
Limestone (wet)	2.5×10^{-2}	8	14	11

Table 2. Approximate Possible Georadar Penetration Depths.

Type of antenna	ϵ_r/ϵ_0	Soil type (low σ)	Penetration depth
900 Mhz	9	wet sand	1 m
300 Mhz	9	wet sand	5 m
80 Mhz	9	wet sand	15 m
900 Mhz	3	dry sand	40 m
80 Mhz	3	dry sand	100 m
80 Mhz	81	saline wet sand	0 m
80 Mhz	3.5	dry clay	70 m

conductivities, relatively low velocities and large energy absorption with no penetration to deeper strata, only reflection.

4. Profiling Depth

From Table 1 some considerations of the maximum penetration depth in different subsurface conditions can be distilled immediately. As this depth is dependent on the subsurface conditions in terms of its conductivity and dielectric constant on one hand and the input frequency on the other (hence the wavelength of the radar pulse), the right choice of frequency is important.

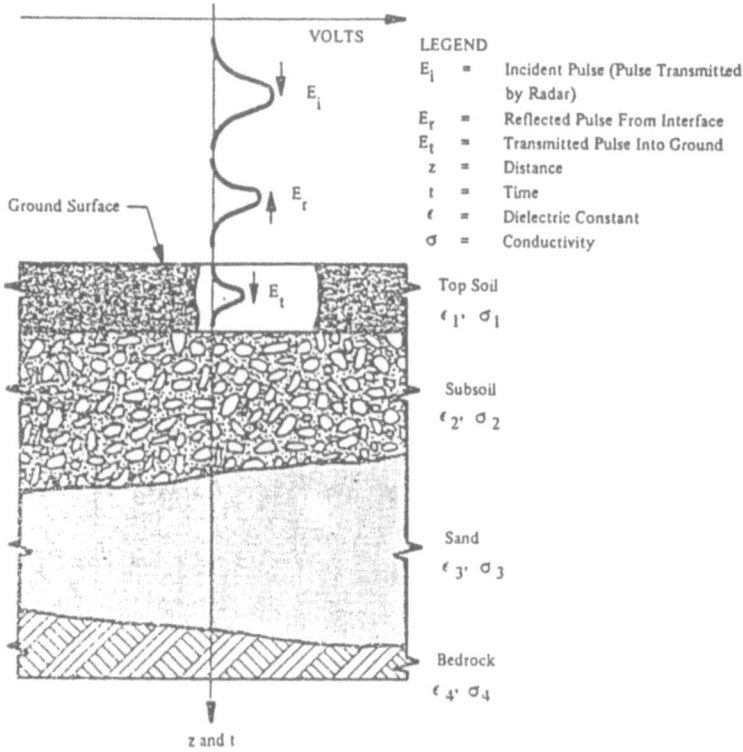

Figure 1. General Principles of (Pulse) Georadar.

The most common and available antennae have central frequencies of 80, 120, 300 and 900 Mhz. The lowest frequency will penetrate to the greatest depth; the highest will have the best resolution. From experience it appears that penetration depths, as listed in Table 2, can be obtained.

Fig.1 shows clearly that Georadar profiling is based on a reflection technique. The E.S.P.-pulse transmitted into the earth is composed of a wide frequency spectrum. The system is therefore called a broadband video pulse radar. Not all frequencies around the transmitted central frequency of the antenna will be reflected at a boundary or discontinuity; special conditions in terms of coefficients (Fresnel's and Snell's laws) should be fulfilled.

Fig.2 gives a visual explanation of how obtained data are displayed on screen and/or printout. These data can be displayed in real time on a graphic recorder, identical to those used in marine bottom and sub-bottom profiling. Data can also be directly digitised as to process them efficiently by e.g. a personal- or a minicomputer (Fig. 3).

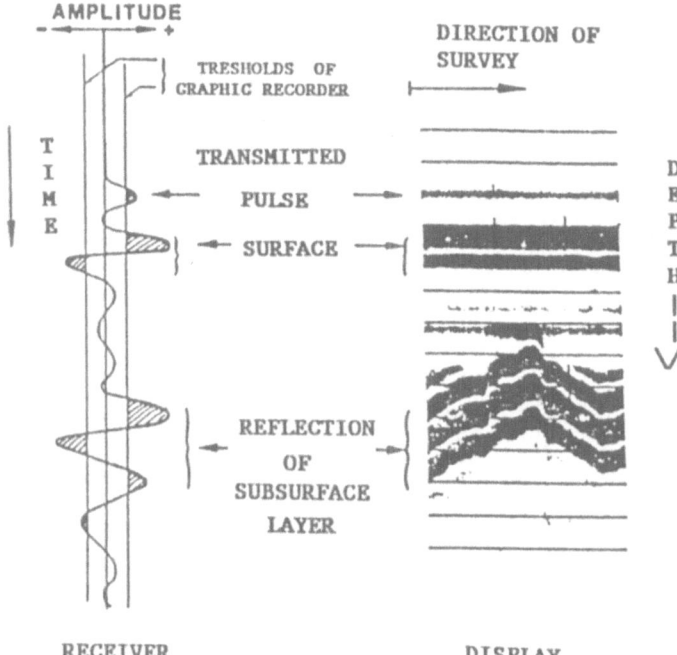

Figure 2. Typical Display of Reflections on a Graphic Recorder.

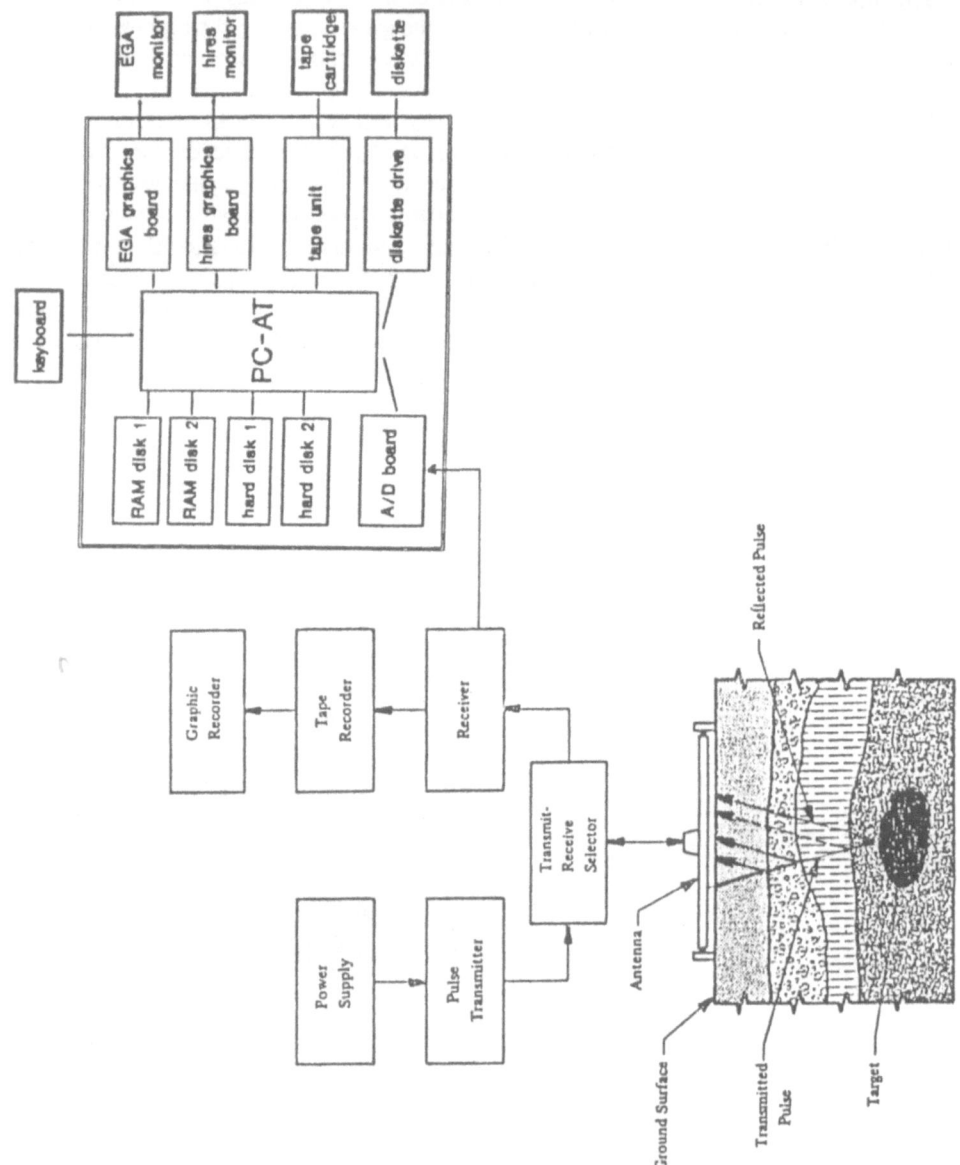

Figure 3. A Possible Hardware Configuration for Georadar Data Processing.

Table 3. Some Comparative Values and Characteristics of Both Systems.

	SIR A/B ($\pm$ 1300 MHz)	Georadar (100 MHz)
Incident angle (Θ) at subsurface contact.	$\pm$ 35°- 50°	conical "beam" normal incidence
Wavelength λ (in space)	24 cm	$\pm$ 300 cm
Wavelength λ' in dry sand ($\epsilon_r = 3.5$)	13 cm	$\pm$ 160 cm
Effective penetration: (or skin depth δ) *, at incident angle Θ and wavelength λ'	1-6 m	12-70 m (at normal incidence)
Signal scattering	important	important
Boundary conditions: Microrelief and geo- metric setting	extremely important	moderately important

*)Skin depth (δ): The depth below surface where the field amplitude ("strength of the signal") is reduced to $1/e$ ($=0.37$) of its original value at the surface, and the power (proportional to the square of the field amplitude) reduced to $1/e^2$ ($=0.135$) of the transmitted power at the surface.

N.B. *If a reflector exists at that depth the magnitude of the reflected wave will be further reduced in travelling upwards and arrive at the surface with a maximum of only $1/e^2$ ($=0.018$) of its original value.*

Advantages of Georadar in Arid Conditions:
1) Continuous profiling.
2) Direct continuous display of boundary reflections.
3) Groundwater table can be traced and lateral followed.
4) Any salt in solution gives no further penetration: so detectable.
5) Groundsurvey up to 15 km/hr, helicopter survey even faster.

Disadvantages:
1) Limited attainable depth.

2) Difficult interpretation of data without
   any reference borehole or other on the spot measurement.
3) Depth scale of reflections must be calculated from
   the time scale, which is actually displayed.

5. The Shuttle Imaging Radar (SIR-A/B) and the Georadar, a comparison

At approximately the same time responses from both Georadar and SIR-A were recorded at and above the Arabian Peninsula. Both projects were totally independent from one another but served the same goal: the investigation of subsurface conditions by EM-waves. Unfortunately the SIR-A track did not cover the Georadar test site so that results could not be compared directly. Table 3 shows a comparative example of some of the main characteristics of both systems. Nevertheless it may be useful to compare both systems to identify their characteristics for complementary uses in arid regions.

6. Conclusions

In general it should be noted that the above mentioned SIR A/B configurations are not suitable for penetrating the subsurface. Only in extremely dry conditions may a few meters be penetrated and reflections recorded. Nevertheless, if the signal field strengths are increased, the penetrating properties of these microwaves become more suitable for semi-arid and arid regions. However one might perhaps need nuclear power to transmit signals from which penetration (i.e. comparable to Georadar) can be obtained.

Georadar (with frequencies ranging from 60 - 900 Mhz) has produced very encouraging results both in penetrating soil/rock layers and discrimination of hidden geometrical structures. The accessability of the terrain for vehicles may cause difficulties in obtaining good profiles. The helicopter mounted version of the system is available and has given good results, even in wet areas.

As mentioned above it is recommended at this stage that research on the responses of both systems should be further compared to determine the rate of complementary use.

7. References

Berlin, G.L., Tarabzouni, M.A., Al-Nasser, A.H., Sheikho K.M. & Larson, R.W., "Sir-B Subsurface Imaging of a Sand-Buried Landscape: Al-Labbah Plateau, Saudi Arabia. IEEE Trans. Geosci., Remote Sensing, Vol. GE-24, no.4, July 1986.

Farr, T.G., Elachi, C., Hartl Ph., & Chowdhury, K., "Microwave Penetration and Attenuation in Desert Soil: A Field Experiment with the Shuttle Imaging Radar" IEEE Trans. Geosci. Remote Sensing Vol. GE-24, no.4 July 1986.

Hippel, A.R. von (1954) *Dielectric Materials and Applications*, The M.I.T.-Press, Cambridge, Massachusetts.

Lord, A.E., Koerner R.M., & Reif, J.S., "Determination of Attenuation and Penetration Depth of Microwaves in Soil", Geotech. Testing Journal, ASTM, 2, 1979.

McCauley, J.F., Schaber, G.G., Breed, C.S., Grolier, M.J., Haynes, C.V., Issawi, B., Elachi, C., & Blom, R. (1982) "Subsurface valleys and Geoarcheology of the Eastern Sahara revealed by Shuttle Radar," Science, vol. 218 no. 4576, pp. 1004-1020.

McCauley, J.F., Schaber, G.G., Breed, C.S., Grolier, M.J., Haynes, C.V., Issawi, B., Elachi, C., McHugh, W.P., & El Kilani, A., "Palaeodrainages of the Eastern Sahara-The Radar Rivers Revisited" IEEE Trans. Geosci. Remote Sensing Vol. GE-24, no.4 July 1986.

Morey, R.M. (1974) "Continuous Subsurface Profiling by Impulse Radar", Proc. Subs. Expl. for Undergr. Excav. and Heavy Constr. American Society of Civil Engineers, New York.

Pool, M.A., "Geotechnical Investigations for Pipeline Construction"
Arab Oil and Gas, Vol 20. no 2. Paris 1984.

Pool, M.A., "Geotechnical Aspects of the Centro-Oeste Pipeline Construction in Argentina. Proc. of the 5th Int. Congress of the Int. Ass. of Eng. Geol. Buenos Aires, 1986..

Schaber, G.G., Breed, C.S., McCauley, J.F., & Olhoeft, G.R., "Shuttle Imaging Radar:Physical controls on Signal Penetration and Subsurface Scattering in the Eastern Sahara" IEEE Trans. Geosci. Remote Sensing Vol. GE-24, no.4 July 1986.

A1. ANNEX: A PIPELINE PROPOSAL TO RECHARGE LAKE CHAD

Due to the fact that the author has been to be involved with pipeline construction during the period he worked for a pipeline contractor, a rough cost estimate for water transport systems is presented below. As an example the recharge operation to the aquifers near Lake Chad is chosen.

Source area is the Kainji Reservoir (Nigeria), receiving its water from the Niger river. The distance along the most suitable route is approximately 1000 km. Avoiding the highest elevations, the main difference in height will be of the order of 600 m. This proposal is based on experience with a big project in Saudi Arabia and the figures as presented in Table A1 should be adjusted for local conditions, both financially and environmentally. Figure A1 shows the proposed pipeline route. As the water level in the Kainji Reservoir recently became very low this project does not seem quite feasible, but should be regarded as a cost example.

Table A1. Cost indication for a water pipe line system to Lake chad

Pipeline Nigeria	Costs (in 1982 $):
	Pipes:
Length: 2 x 1000 km	500 US$/m: 1000 x 10^6 $
Estimated throughput: 2 x 450,000 m^3/day Working pressure: 60 bar.	Installation: 600 x 10^6 $
Ductile iron pipe: steel grade X52-X60	
Pipe diameter: 60" (153 cm)	
Wall thickness of pipe: 16 mm	
PE-coating	
2 x 5 pumpstations (45 MW)	Pumpstations: Installation: 450 x 10^6 $
Total output: 450 MW	Rate/MW ?
Estimated height-difference: 600 m	$\dfrac{\qquad\qquad +}{}$ Subtotal: 2050 x 10^6 $ Contingencies: 10% 200 x 10^6 $ + Total: 2250 x 10^6 $

Financing (7% for 15 years)
= ± 60% of total . . . . . .   1350 x 10^6 $
Maintenance per year . . . .   45 x 10^6 $

Costs per year . . . .   285 x 10^6 $
Cost per m^3 . . . .   ± 0.80 $/$m^3$

635

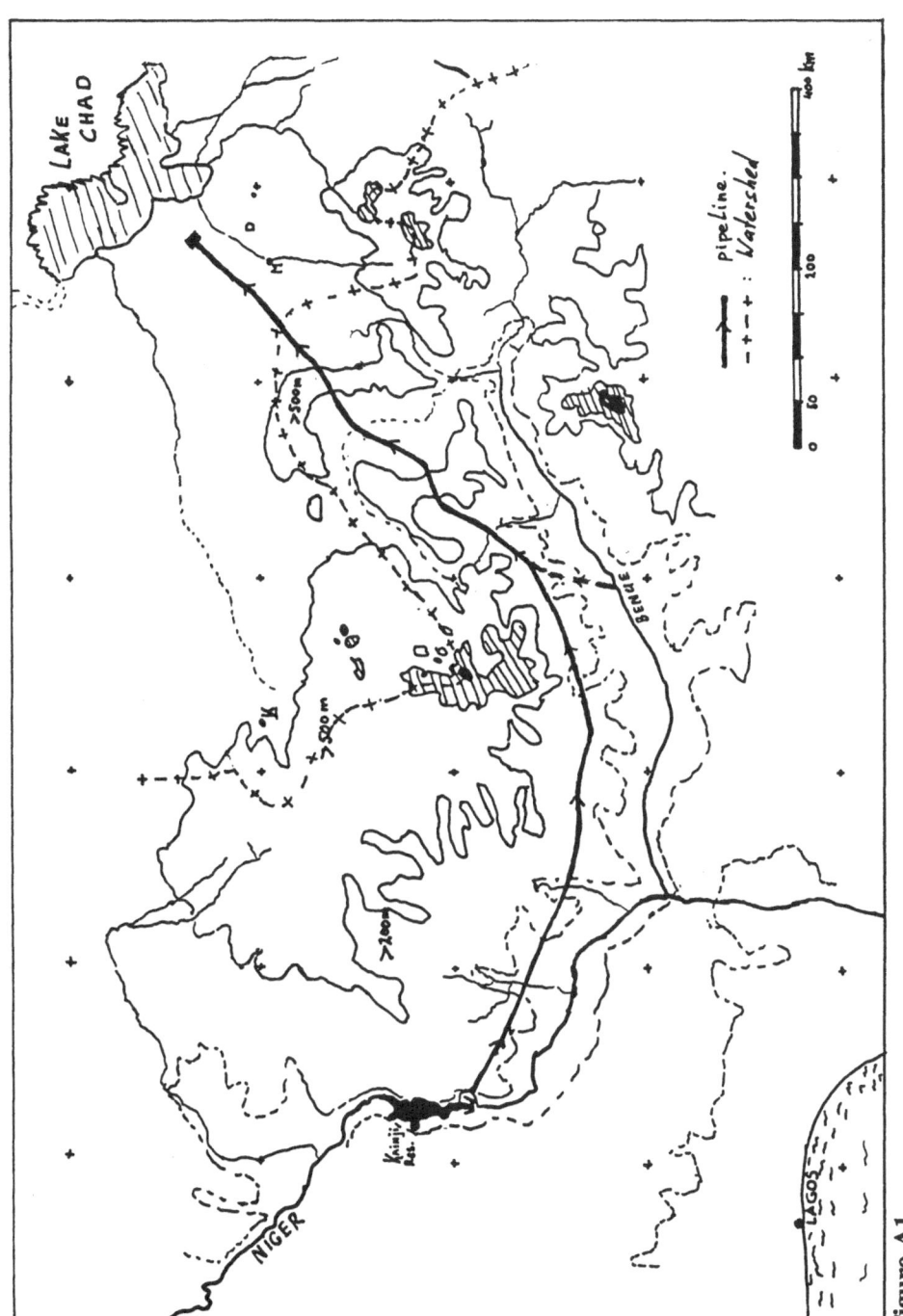

Figure A1.

POSSIBLE ENGINEERING SOLUTIONS AND A CASE STUDY

C.P. DE MEYER
President Chief Executive Officer of HAECON N.V.,Belgium
Lecturer Vrije Universiteit Brussel (VUB)
Belgium

ABSTRACT: Throughout the ages the seashore worldwide has exerted a continuous attraction on settlers. Why? Originally because of food supplies and transportation possibilities, and more recently because of its high potential for socio-economic developments.

As the burden on coastal settlements gets heavier due to population concentration and competition for space and the coast is relentlessly battered as a consequence of the steady sea level rise, all possible solutions should be taken into account to prevent disasters.

Theoretically, sea level rise could be controlled by reducing the amount of water returning from the continent to the ocean.

Construction of large reservoirs, artificially achieved using dams or natural depressions, leads to a conceivable redistribution in time and location of the water on the continents.

Indeed, many projects of global scale have been considered! Several such "engineers' dreams" are not science fiction. One of them is a water-transfer scheme from the Zaire River Basin to Lake Chad.

1. Introduction

The level of Lake Chad is currently dropping at a rate of approximately 30 to 50 cm a year, so that an important polderization program along the shore is running out of water. As a consequence the four bordering countries are dramatically over-pumping fossil water, and thus the possibility to recharge water to the Lake and/or to underground geological structures was considered.

On the other hand, the Zaire River discharges approximately 42,000 m^3/sec in the ocean, as it, and its affluents, drain a 3,800,000 km^2 basin. The Zaire River to Lake Chad water-transfer project proposes to tap about 350 m^3/sec in a location where the system intake would disturb neither fluvial transportation nor the feeding of the Inga Barrage retaining lake. The aqueduct would transport water over a theoretical distance of about 1200 km.

637

R. Paepe et al. (eds.), Greenhouse Effect, Sea Level and Drought, 637–640.
© 1990 *Kluwer Academic Publishers.*

Is this unrealistic, an engineer's dream, or academic science fiction? It may have been for a very long time but such projects are now being carried out. Pipelines have been laid under the harsh climatic conditions of the Arctic, water ducts have been installed in the American west, and plans for an aqueduct in South Africa (The Lesotho Highlands Water Scheme), which have been off and on the shelf for years, are not only seriously considered, but construction procedures have been launched.

2. The Lesotho Highlands Water Scheme

This multipurpose scheme would allow, through a system of dams and tunnels, the export of large quantities of water and the generation of hydro-electricity for local use. As water resources exceed all projected local water demands, no major issues arise here. However, the need for hydroelectricity generation to satisfy local demand increases. By 1995 South Africa's industrial hub will need more water than the Vaal River basin can supply. Though the RSA could cover the expected shortage from sources within its territory, for instance through the proposed Orange-Vaal Transfer System (OVTS) scheme, the long distances and high lifts make it an expensive and vulnerable construction. A proposal to study water transfer from within Lesotho to the Vaal system by gravity has been made to Lesotho.

First studied in the 50's, again in the 60's, and now in the 80's, a steadily larger scheme was considered, as water requirements in South Africa grew. The Lesotho and South Africa joint technical committee feasibility report (1979) recommended first the construction of a 35 m^3/sec water transfer scheme, later of 70 m^3/sec, requiring the phased construction of four dams (initially at Katse, later at Mohale, Mashai and Tsoelike), about 100 km of water transfer tunnels and a hydropower generation component. Stage 1 studies identified an optimum project layout and concluded that there were no unsolvable environmental, socio-economic or legal difficulties. The two governments agreed in 1984 to proceed with Stage 2 of the feasibility studies completed in April 1986, regarding an export of 70 m^3/sec. Scheduled to begin in 1990, phase 1A would provide a continuous supply to the RSA of some 17 m^3/sec of water from the Katse reservoir and a transfer tunnel to a hydropower complex, a tailpond, and a delivery tunnel for discharge into the Ash River in the RSA, from where water would then flow to the Vaal Dam, into the main reservoir of the Johannesburg-Pretoria area.

3. African Water Transfer

This impressive undertaking can be compared to some other huge projects in Africa. The "Great Artificial Stream" was started four years ago under Libyan sponsorship. It foresees a subterranean conduit with 4 m diameter that will tap underground water layers over 1,200 km and be the spine of 4,000 km pipe-network. Some 4 to 6 million m^3 of water per day would be available. Professor Hense, however, commenting on his undertaking, considers it environmentally unwise and ultimately nefarious. Such would not be the case for a 6,000 km trans-Sahara canal linking Nouakchott, at the Atlantic coast, and the Bay of Saloum at the border between Egypt and Libya; this scheme encompasses numerous side schemes providing energy, fresh water and desert recuperation; its sociological effects would be very far reaching.

4.   Zaire River Water for Lake Chad

Other water transfer projects are occasionally taken into consideration. The idea is basically the same whether one considers the Transacqua proposal of Bonifica (Italy) or the ASWA proposal of HAECON (Belgium); the differences lie in scope, itinerary and auxiliary uses.

Water from the Zaire River could be led to the drought-stricken areas of the Sahel and could alleviate the lack of water threatening the polderization schemes around Lake Chad. Water needs are urgent here and a HAECON preliminary study was based upon delivery of up to 350 m^3/sec. Of several routes considered, linking the Ubangi River is the most direct one. Open-channel and closed conduit transportation appears to be the most appropriate method.

The matter of water supply to the Sahel is one of some urgency. Climatic studies have shown that rainfall is steadily decreasing and drought will worsen in the near future. "Dry" and "humid" years group themselves into periods of unequal length, but no significant periodicity has been established. While this poses serious problems for farmers, some of these have to cope with another consequence. For several decades Sahel farmers have developed polder schemes in the flood areas around Lake Chad; the steady lowering of the lake's water level leaves the polders dry and farmers abandon them to create new ones. If, by supplying water the level were raised again, the abandoned polders could be integrated anew in the agricultural system. If some stability of level could be attained, polderization would no longer have to be limited to the traditional, may we say "artisanal" type, but modern, industrial polders could be established.

4.1. THE ZAIRE-CHAD AQUEDUCT

For decades the South Africa-Lesotho project has been put on and off the shelf; it has been decried as an impossible undertaking. It is now in the construction stage. Similarly, a water pipeline bringing water from a powerful river to a drought stricken area is often decried as an utopia. Environmental considerations as well as engineering problems have been brought forward to block even a thorough assessment. It has been shown that physical and sociological environmental effects can be overcome in South Africa; why should this not be possible in Central Africa? Similarly, as a whole, the project is technically feasible and a transport of 350 m^3/sec can be considered. A canalization-closed conduit approach seems indicated. While a choice of conduit diameter cannot be made on data presently at hand, a pipeline with a diameter of 4.6 m would require 2.5 kwh/m^3 based upon a pump output of 60%. Using 21 such conduits, the transport of 350 m^3/sec from Bangui to Lake

Chad requires 1.2 kwh/m^3. Evidently, to remain economically realistic this energy must be sought from within the hydro-electrical possibilities of the Zaire basin. Only a hydraulic axis study of the conduit will determine whether river beds can be used for part of the journey. The Chari River comes to mind. Provided slope and evaporation requirements are met an open air duct could be used if a depth of 3 to 5 m is assured. One should however keep in mind that pipeline transport eliminates evaporation and infiltration problems and substantially reduces health impact. While a larger number of conduits provides advantages it causes load losses.

The Zaire River constitutes by its discharge the second largest source of fresh water to reach the ocean. Its mean discharge varies from 40,000 to 45,000 m^3/sec; the lowest discharge rate (in August) is about 25,000 m^3/sec and the highest is 80,000 m^3/sec (in December). The Zaire River straddles the Equator. This provides some advantages: the mean discharge of the river is second only to the Amazon River.

The conduit itinerary is selected to keep terrain accidents to be crossed to a minimum. There is of course an alternative solution consisting in blasting tunnels through the mountains which divide the two hydrographic basins, as is being done in the South Africa-Lesotho project, or to opt for a mixed approach of pumping and tunnels. However, tunnelling appears only an advantageous alternative if the intake site is the Ubangi River. Preliminary studies do not consider it a good solution as water is tapped at Bangui. A Ubangi route and intake has been considered but according to some hydrologists pumping at its confluent is not advisable because it would remove too large a percentage of water when the water level is low. However, at Bangui pumping influence could be spread over a large area of marshes. Taking 350 m^3/sec in the northern area of the basin, directly from the Zaire River poses no problem at all. This adds some kilometres to the scheme, but at first sight no other solution is hydrologically or environmentally acceptable. However, an intake to be studied in the Bumba-Lisala-Mobeka region can be suggested. Pumping directly from the Zaire River also reduces sediment problems; transport of sediments create friction, wear and tear both in conduits and related installations.

5. Conclusion

Environmentally acceptable water transfer schemes may become necessary, with certain reserves, rather than remain engineers dreams.

PART V: PANEL DISCUSSIONS, CONCLUSIONS AND RECOMMENDATIONS

NATO - ADVANCED RESEARCH WORKSHOP on

Geohydrological Management of Sea Level and Mitigation of Drought

Fuerteventura, Canary Islands, March 1-7, 1989

Conclusions and Recommendations
Reports of Sessions and Panel Discussions

by R. Paepe, Director of the Workshop
with the collaboration of E. Van Overloop and I. Demolin

643

R. Paepe et al. (eds.), Greenhouse Effect, Sea Level and Drought, 643–689.
© 1990 *Kluwer Academic Publishers.*

1. Session I: Feasibility

1.1. CONCLUSIONS AND RECOMMENDATIONS

a. Sea level. There is a perceived rise of sea level in many parts of the Globe, the steric part of which is unquestioned and locally the tectonic, but a number of points remains unresolved.

b. Tide Gauges. The observed rise of sea level from tide gauge readings is not only eustatic but includes tectonic and oceanographic components.

c. Tide gauges should be installed in "white areas" to get more data in future, and the use of satellites and laser techniques should be taken into account.

d. CO2-Forest. Because of their role in the CO_2 cycle, tropical forests should be protected.

e. It is hoped that further research will clarify the importance of the ocean in the greenhouse gas cycle.

1.2. SESSION REPORT [by S. Jelgersma].

During this session five lectures were presented:

Rampino presented a discussion of all greenhouse gasses involved in the so-called Greenhouse Effect and the reductions of stratospheric ozone:

Increasing greenhouse effect form increases in carbon dioxide, CFC's, methane and tropospheric ozone. The recognized and proposed interactions on marine primary productivity on global hydrology are extensively discussed.

Blumenthal presented models that can be used to predict the effects of changes in West African surface specific humidity on the sea surface temperature (SST) patterns of the tropical Atlantic.

In relation to future sea level rise due to the greenhouse effect Jelgersma presented the results of the latest study by Wigley of the Climate Research Unit of the University of East Anglia at Norwich (U.K.). The best guess is a rise of sea level of 10-35 cm by the year 2040.

The climatic signal by that time is thought to be represented by
1. an observed relative rise of sea level of 10-15 cm by tide-gauge reading
2. the retreat of mountain glaciers.

Pirazzoli discussed point 1 and suggested that most of the observed relative sea level rise would be due to tectonic movements.

Jelgersma discussed a glacier-temperature model by Verlemans of mountain glaciers. His conclusion was that half of the retreat is due to the 0.5 °C temperature rise during the last 100 years and the other half to changes in the radiation balance (dust veil).

Climatic changes during the last glaciation on the extension of tropical African forest were discussed by Leroux. An important conclusion is that during the last glacial maximum this forest was only present in three refuges: Liberia, Cameroon and Zaire.

Climatic changes in the geological past, especially those observed in the studies of deep-sea cores, were discussed by Jelgersma. Especially studies of the late glacial and the Holocene indicate that changes in temperature do not respond in a gradual smooth way but in sharp jumps.

The geological record suggests a beginning of the next Ice Age. How the predicted greenhouse warming will interact with this natural trend is not known.

1.3. PANEL DISCUSSION REPORT   [by M. Leroux].

The first session is devoted to the atmospheric, oceanic an climatic response to greenhouse and feedback effects.

Jelgersma makes a statement of the knowledge in that field, and wonders how the warming caused by the greenhouse that is foreseen will interfere with the "natural" tendency towards a cooling.

Leroux reminds us that the tropical forest assures its own protection, but that its maintenance is related to a fragile equilibrium. Pirazzoli asks if the sea level is really rising, but cannot assert it definitely, admitting however the possibility of a rise of the sea level during the next century. Rampino evokes the climatic interaction between the greenhouse effect, the stratospheric ozone and the global hydrology and considers that several parameters may be affected, causing noticeably a global warming of the climate and a rise of the sea level.

Before launching the discussion, Leroux wishes to stress the fact that a tendency of the temperature established at a global scale (Jones et al.) is not necessarily the best reference for the analysis of meteorological consequences. The equatorial belt records but weak variations: The thermic changes which are the most strongly felt concern the polar regions, which are meteorologically the most important. One needs to know the nature of this tendency: What is the contribution of each season (the winter being maybe more important in the middle latitudes as opposed to the monsoon regions).

Latitudes and seasons both generate different atmospheric responses, in particular by their influence on the general circulation:
a. A cold situation in the polar regions corresponds to a fast general circulation, but
b. a warm situation in the polar regions corresponds to a slow general circulation, the speed of the fluxes (temperate as well as tropical) being a function of the frequency and of the power of the transport of cold polar air (or, in other words, of the gradient of temperature and pressure between the poles and the equator).

Berger states that the time and space scale must be clearly defined, abrupt changes on a local scale being difficult to observe on a global scale.

Issar observes that the shifting in latitude of the climatic belts produces effects inverse to the latitudes, an increasing of mediterranean precipitation going along with a decrease in the sahelian region and vice versa.

Regarding the tropical forest, Wells suggests that the Amazonian forest retreated like the African forest to "refuges" during the maximum glaciation. Llamas suggests that these positions ought to be protected absolutely against deforestation. Fairbridge asks that an inventory of these refuges is made. Leroux thinks that the real problem is not located there, but at the existing capital should be protected.

Da Cunha asks if the reforestation represents a real solution for the CO_2 absorbtion. Leroux thinks that the forest may be overestimated (compared to the role of the ocean), that it is not realistic (particularly from an economic point of view) to reconstitute the forest, but that it is necessary to determine better associations between forest/agricultural zones, whose thermal behaviour allows the forest to protect itself against the invasion of continental "alizés" (easterlies) with their strong evaporation capacity.

At the end of the discussion, a lot of questions remain unanswered, in particular the link between the evolution of the temperature, the rising of the sea level and the increase of CO_2.

2. Session II: Distribution of Rainfall and Modelling Predictions

2.1. CONCLUSIONS AND RECOMMENDATIONS

a. Droughts are natural, quasi periodic, and may be eventually predicted.
b. A wide variety of options are available for transfer of water from rainfall areas to subsurface aquifer storage.

2.2. SESSION AND PANEL DISCUSSION REPORT [by R.W. Fairbridge]

In his keynote address, the meeting president (RWF) urged that the fatalistic "Act-of-God" attitude to climatic and other planetary hazards should be abandoned. Instead, trained scientists ought to insist on a deterministic philosophy. Failure to recognize this principle by many modellers resulted in naive, unrealistic and misleading forecasts.

As concerns sea level, the consensus favours Pirazzoli, who claims that so many tide stations are subject to neotectonic departures that no systematic pattern is identifiable. The assumption that mean sea level is rising (at about 1.2 mm/a) was based on the false premise that a filtered series of long-time records would indicate a global trend. But tide stations are selectively biassed towards delta and estuary sites, both affected by geological subsidence, about which we can do nothing. Two crucial tests could be applied:

1. To a Central Pacific atoll, far removed from plate boundaries, glacio-isostatic warping, sedimentation, etc.; thus, e.g. Tonk Atoll shows El Nino fluctuations but no general trend over a half century.
2. To a Baltic tide gauge series, also far removed from plate boundaries and sedimentation, as well as from geostrophic currents, but undergoing isostatic rebound for which a correction factor can be used, e.g. Hanko in Finland (100 yr), or Stockholm in Sweden (200 years long). No departure is observed except for what is appropriate only to steric expansion (warmer water) in the 20th century. Thus, as presently observed, no global effort can control sea level. Individual area submergence relates to neotectonics or current

velocity (Coriolis effect), and only local action can mitigate those effects.

Concerning the drought hazard, this is an area where hydrogeologists can clearly play a decisive role.

First, we recognize that droughts are natural, cyclical events which we may eventually be able to predict using astronomical analysis. Second, a wide variety of options are available for transfer of water from high rainfall areas to subsurface aquifer storage, or for a very small scale (but multiple) storage in existing ephemeral channels.

Howard outlined the principles of management of water resources. He pointed out that the idea of artificial recharge of aquifers was perfectly feasible but that it was not just as simple as filling a tank and emptying it in time of need. Complex systems of withdrawal would be required with large cones developing.

The next paper was by Wells, who displayed dramatic satellite pictures of the total aridity of such areas as Lake Chad and Khartoum in the decades up to 1987 when heavy rains occurred, greening the desert over a wide area. This effectively disproved the rather commonly accepted "sociological theory" (that desertification was caused by mankind's lack of skill, etc., etc.).

Kukla indicated that GCMs currently employed were more successful with temperature than precipitation prediction. Some of the different versions are different in larger terms than obtained for the anticipated doubled CO_2 warming. The GISS model also failed to replicate conditions in the last interglacial and the onset of glaciation. Evidently there is need for substantial improvement.

Hugues and Faure, jointly with Fairbridge, submitted an abstract but due to Faure's illness it was postponed.

3. Session III: Glacier Nourishment, Ocean Control and Fluviatile System in response to climatic periodicities

3.1. CONCLUSIONS AND RECOMMENDATIONS

a. Climate is continuously changing at all time and space scales. In the absence of any man's influence on climate, the astronomical theory which is confirmed by many geological data is particularly useful for predicting a global glacio-eustatic sea level drop which will ultimately reach 100 m within the next 60,000 years.

b. There is no conclusive evidence of a eustatic sea level rise taking place at current time.

c. There was an increase of the global continental runoff from 1910 to 1975, which can be correlated with the global increase of temperature during the same period. On a regional or on a continental scale shifts, lags and oppositions are observed.

d. Climate changes on the geological and historical time scale should continue to be investigated and modelled to provide a background of natural climatic variability within which man's impact could be more easily evaluated.

e.   Satellite monitoring of sea level elevation including departures from the normal geoid should be intensified.

f.   Changes in the ocean floor topography and the influence of the pressure distribution over the oceans should be taken into consideration.

g.   Global and regional hydrological balance should be conducted together with analysis of cyclicity at different space scales.

h.   The volume of dry sediments available in arid countries capable of storing water away from evaporation should be measured carefully.

i.   Continuous observations of hydrology and water quality fluctuations of the large rivers and streams flowing close to arid areas possibly influenced by drought should be made.

3.2. SESSION REPORT [by A. Berger]

The session on Glacier Nourishment, Ocean Control and Fluviatile systems started with a prediction of glacio-eustatic sea level change at the astronomical time scale (Berger). For the next 60,000 years, assuming no human interference or any other forcing besides the astronomical one, climate will progressively deteriorate to reach glacial conditions, similar to 18,000 YBP. In these conditions, the mean cooling rate should be 0.01 °C per century and the corresponding glacio-eustatic sea level drop would be 1.5 mm per year.

At this palaeoclimatic time scale, there are clear cyclicities in geological data. Petit-Maire has shown that the extension and shrinking of the arid tropical belt is clearly related to the Milankovitch periodicities. Particularly during the last humid optimum, the extension and volume of surface water were very important in presently arid areas (Petit-Maire). The large depressions inundated in the Holocene (as the Tarfaya area, South Morocco and the Taousse Defile, Mali) could be flooded again now through simple engineering. Paepe showed 100 kyr and 400 kyr periods in palaeosols and river deposits in various parts of the globe. For example, at the beginning of the last interglacial-glacial cycle, soil developed albeit after the maximum of the marine transgression. The stratigraphy of the deposits of the Ruzizi Plain suggests also several palaeoclimatic fluctuations from wetter to drier climates (Ilunga).

There were two papers on global water cycle fluctuations. Tardy demonstrated that there was an increase of the global continental runoff from 1910 to 1975 which can be correlated with the global increase of temperature during the same period. On a regional scale, shifts and lags can be observed from south to north and to a larger extent from east to west. Klige showed that during this century sea level rise followed the Earth's warming by approximately 20 years. The main sea level rise was 1.5 mm/year (corresponding to an increase of the oceanic volume by 540 km^3/year).

At the "present time", sea level is rising along 62.5 % of the coasts and is falling along 12.6 %. Climate warming induces an increase of water exchange intensity by 4 %, the continents becoming drier and glaciers melting.

One paper dealt with the impact of a probable sea level rise on the local scale (littoral of Alicante, Zazo). In the Elche Basin (Alicante), tectonic activity and sedimentation indeed favoured the development of lagoons which were isolated from the sea by split bars. These are particularly sensitive regions to any future sea level rise.

Another example of non eustatic sea level change from the Mediterranean was given by Goldsmith. Monthly mean sea levels determined from tide gauges that local neotectonics and meteorological oceanic effects (which can amount to a 15 - 20 cm range) mask eustatic effects there. These last factors related to local seasonality must therefore be taken into consideration in any proposal for geohydrologic management.

3.3. PANEL DISCUSSION REPORT [by Y. Tardy]

Two major points have been discussed:
1. The sea level rise question, related to glaciation-deglaciation, greenhouse gases and tectonics.
2. Components and effects of cyclic phenomena at the Earth's surface.

1.   The Sea Level Rise Question
It seems that the large majority of participants have accepted the general conclusion that at present there is no conclusive evidence of the eustatic sea level rise taking place at the current time (i.e. the last fifty years). There are some regions where sea level seems to rise and some others around which it decreases. The difficulty comes from taking into account the local tectonics subsidence which has been shown to be very active everywhere.

The second point of the general discussion focused on possible causes of sea level rise. It seems clear we generally admit that the increase of CO_2 in the atmosphere is due to human industrial activity. It is also obviously admitted that CO_2 is a greenhouse gas. However, it is accepted that we should not unduly focus on CO_2 only and forget for example H_2O vapour which is 75 % responsible for atmospheric temperature control. At any rate, no one was convinced that the observed increase of CO_2 at the present time has already caused the an observable sea level rise.

Continuing with the question of present sea level fluctuations one may perhaps recommend the continuation of satellite monitoring of sea level elevation and changes of departures from the normal geoid. One also recommends taking into consideration changes in ocean floor topography and the influence of the pressure distribution over the oceans.

The third debate was devoted to Berger's modelling based on the Milankovitch' cyclicity calibrated on the observed fluctuations during glacial times. Clearly people admitted the prediction of a probable increase of global ice volume over the next 60,000 years. The subsequent eustatic sea level drop could compensate, at least partly, for the expected sea level rise due to the greenhouse effect.

2.   Components and Effects of Cyclic Phenomena at the Earth Surface
Examination of soil and river cyclicity clearly shows that these phenomena are complex

and have to be treated at different time scales and at different space scales.

Global runoff has been fluctuating but has also increased this last century. Regarding the objectives of this conference one must point out that this recently observed apparent excess of water going into the oceans is probably due to an excess of water evaporated from the oceans. However, nothing is sure and we must recommend that the monitoring of the hydrological balance of the ocean be carefully performed. The climate of the world is indeed is not restricted to the climate of the continents but includes also that of the oceans.

However, this global increase is not observed over all the different continents considered separately: Since the beginning of this century, some of them became more humid, while some became more arid. The same kind of discrepancy is also observable at a regional scale within a given continent and at a basin scale within a given continent etc. For example an increase of humidity in the Sahel is sometimes compensated by a decrease of humidity close to the equator and an increase of humidity over Western Africa is generally compensated by a decrease of humidity over Brasil etc.

Thus, a great majority of people have been aware that fluctuations observable in small regions are higher in intensity than those observable over large regions, at a continental scale and *a fortiori* at a global scale. If cyclic factors may be recognised through cyclic regional or global fluctuations we emphasize here that their effect may be different, shifted or in opposition, from time to time and from place to place. No global regional hydrological balance should be conducted without analysis of cyclicity at different space scales.

4. Session IV: Deserts and Drought

4.1. CONCLUSIONS AND RECOMMENDATIONS

a. The ongoing desert encroachment is due to a complex interaction of climatological changes and human impact. Depending on the sites the latter varies from weak to very important.

b. When introducing new technologies or setting up important projects, the changeable environmental components must be assessed properly and effectively controlled. Even then, storage for the possible bad times must be respected.

c. More and better forecasting of environmental disasters such as drought and wet spells are urgently needed. Hence more research should be done.

4.2. SESSION AND PANEL DISCUSSION REPORT [by M. De Boodt]

During the afternoon session "Desert and Drought",, three papers were presented. In the keynote lecture "Desertification or Desert Reclamation" the question was raised whether desertification was mainly due to a change in climate or to human behaviour. Whereas in the last decade(s) it was mainly human beings that were thought to have caused the desert encroachment, it is now agreed that the interaction "human behaviour change on climate" should be considered as the major reason for the ongoing desertification. Of course it is not possible to indicate how much each is responsible for what is happening now but it is certain that a 90:10 ratio blaming mankind is not correct.

In the keynote speech, modern methods which are at the same time economically justified were considered to lead to more than 50 or 80 % more efficiency in water use by comparison with traditional overland flood irrigation.

The impact of historical climatic changes on the availability of water resources in the Mediterranean region was discussed showing that fossil water is present in very large quantities guaranteeing a supply for a few hundreds of years when careful use is made of it and reserves against catastrophes are put aside.

A less optimistic note was heard when the legacy of the Sahelian water management (1965 - 1988) was discussed. Although the majority of the cited examples were poor, a few could be mentioned, for example the large rice irrigation scheme at Niono, considered a successful operation, could go on through the worst phase of the current drought. To avoid misfortunes in the future when huge projects are planned the environmentally changeable Sahelian environment should be controlled first of all.

When the "Basic Issue on the Concepts of Drought" were discussed, the major point was that we need more and better forecasting of drought spells although everyone agreed that this is a difficult task as these phenomena are considered to be (apparently) stochastic.

In summary, a general agreement was reached at the panel discussion that:
1. The ongoing desert encroachment is due to a complex interaction of (apparently) stochastic climate changes and human behaviour, the latter certainly not being negligible.
2. When introducing new technologies or setting up important projects the changeable Sahelian environment must be better known and eventually controlled. The storage principle must be respected.
3. More and better forecasting of environmental disasters such as droughts and wet spells is badly needed. Hence more research should be done on this (apparently) stochastic phenomenon.

5. Session V: Environmental Changes and Hazards

5.1. CONCLUSIONS AND RECOMMENDATIONS

a. It is recommended that better-evaluation of the geological record and climatic change predictions on a regional and global scale should be made in order to explain changes and recommend priorities of research.

b. More research on the fundamentals of known geological hazards and "sudden events" should be undertaken. Otherwise we will be known only as "journalists".

c. Anthropogenic reaction to change and the impact of change on mankind must be thoroughly investigated.

5.2. SESSION REPORT [by F. Ayala Garcedo]

Ayala analysed the role of "Hazards in Climate and Sea Level Changes". Hazards analysed were volcanoes, earthquakes and tsumanis, coastal subsidence and potential big extraterrestrial impacts. Amongst anthropogenetic factors analysed briefly were erosion, dams, and anthropogenic dust.

Le Houerou analysed the probable effect of a 3 °C increase from the greenhouse effect over the Mediterranean basin. The effects would be significant in marginal regions (climate or soil conditions) because of a higher evapotranspiration. Areas for cultivation of olives, citruses and vegetables might expand. Despite this, Le Houerou thinks demographic explosion would be a more important and would increase detritic sedimentation from the South.

Mariolakos analysed the impact of seismicity and neotectonics in Greece, where up to 50 % of the seismic energy of Europe is released. During major earthquakes, displacement of shorelines takes place.

Liu Tungsheng analysed loess deposits from China (440,000 km²). They are a fragile environment in the face of climatic changes. The loess column has a cyclic nature, with a 37 loess-palaeosol sequences occurring in Baoji (2.5 million years span). Its study may therefore be interesting for palaeoclimatic change.

Rutter emphasised the importance to Quaternarists if the ICSU International Geosphere-Biosphere Programme (IGBP). He described ice cores, tree rings etc. as key archives and exposed New Programme as the core programme. IGBP has established a Steering Committee on Global Change of the Past.

5.3. PANEL DISCUSSION REPORT   [by N. Rutter]

The papers presented in Session V "Environmental Changes and Hazards" centered to a large extent on four categories:
1. Geological changes such as isostasy and erosion, earthquake phenomena, and volcanism;
2. climatic changes caused by volcanic gases and dust, and airborne loess;
3. anthropogenic influences such as man's activity and populous growth on the biosphere-lithosphere and;
4. organizational aspects of the ICSU-IGBP Global Change Programme and the need for scientific Quaternarists.
During the discussion period, several diverse subjects were discussed, both within and outside the framework of the papers presented. Important topics included:
a. *Prediction of change.* It was brought out by several that more and better data are needed in order actually to predict geological and climatic change. There is confusion of cause and effect. If we consider the "Big G" (gravity) model perhaps many changes that occur in the Earth's system could be explained by one phenomenon - such as earthquakes followed by drought. However, not everyone accepts the "Big G" theory. We do know natural hazards as times are related to climatic variations that show periodicity. We

also have the ability to present many kinds of periodic changes, and have the statistics on the probability of other changes.

We should consider what happens when these changes occur together. However, it was suggested that we commonly have difficulty in predicting change of a simple model over a short term.

The suggested recommendation was to evaluate our record of success in predicting geological and climatic change on a regional and global scale with a view to explaining these changes and recommending priorities of research.

b. *Data, communication and modelling.* There is a real concern that high quality, standardized numerical data on a world-wide basis is lacking. If any input into the simulation models is to credible we must have more and better data. Modellers commonly fit their data to prove their models; some blame us for not supplying reliable field data. It is incumbent upon us to influence climatic modellers and ensure that their models are as accurate as possible. It was suggested that a databank be established or at least that a real effort should be made to standardize data and work out ways to evaluate the accuracy of data. Many agree that earth scientists are reluctant to give up their basic data, let alone prepare it in standardized form. However, we must try to establish some sort of homogeneity. In addition we must communicate better with scientists of other disciplines that can help with our work. It was noted that there was no representation at this meeting from such related groups as the International Tsumani Association.

The suggested recommendation was to survey existing data bank systems such as Goddard in New York with the intention of proposing the establishment of a world data bank by an appropriate authority to ensure the accessability of high quality, standardized data.

c. *Neglected Research.* Several people suggested that more research be directed to the study of ignimbrite volcanoes and their effect on climate, their distribution and their frequency. In addition, more research on the fundamentals of known geological hazards and "sudden effects" should be undertaken or else we will be known only as "journalists".

Finally, anthropogenic reaction to change, and the impact of change on mankind must be more thoroughly investigated.

Suggested recommendation: To encourage research in the "neglected research" mentioned above.

6. Session VI: Continental Water: Flux, Stock, Recharge, and Exchange

6.1. CONCLUSIONS AND RECOMMENDATIONS

a. The control of sea level rise by current and foreseeable technologies for storing water underground is not economically feasible, at least not for the next few decades.

b. In arid and semi-arid areas the storage of river water in the subsoil (by artificial recharge) is highly recommended and economically feasible on a broad scale.

c. Better control on the water demand and better management of water resources is needed in order to reduce the drastic effect of drought.
d. *Recommendation 1: To establish an international arid region coordination center for applied research and development.*
   The University of Ghent is willing to make a start with this center and will provide the necessary facilities.
e. *Recommendation 2: To implement a pilot project in an arid region (e.g. close to Lake Chad in several phases for recovery of artesian groundwater and for artificial groundwater recharge.*

6.2. SESSION REPORT [BY E. ROMIJN]

During the session seven speakers covered different topics such as estimation of natural recharge, water conservation techniques including artificial recharge of groundwater, hydrology of Lake Chad, mathematical groundwater modelling in West Africa and salination problems in the Canary Islands. Wedel, Guiraud and Kogbe stressed the importance of small scale and partly buried dams to diminish runoff and to store riverwater underground, thus protecting the water from evaporation. Kogbe developed a draft management scheme for the Lake Chad basin aiming at improving the hydrological potential of the basin with the help of earth dams, polders, drainage systems and import of water taking into account ecological conditions.

Kogbe noted that he is in favour of taking some 10 % of the flow of the Ubangi-river to the lake area. A warning came from the audience to consider the evaporation from an open channel viz. the river.

Surface and subsurface dams were explained by Guiraud. A number of figures illustrated the method.

Of special interest were:
a. A complete list of advantages of subsurface dams;
b. subsurface dams relating to permanent settlement of people;
c. three methods of constructing of subsurface dams;
d. 500 dams in Cap Verde;
e. 2000 dams in Burkina Faso constructed in 20 years;
f. dams in Mauretania, Kenya, Botswana, Somalia;
g. low cost of subsurface dams, e.g. $ 1.000 for a dam for a village of 100 inhabitants, whereas a borehole can be twenty times more expensive.

Roebert tried to evaluate the consequences of storing 1% of world precipitation underground, which is about 4250 km³/a, and lowering the sea level about 11 mm in total.. Water supply, including artificial recharge in California, is less than 1 km³/a, so storing 1 % of global rainfall would mean 5000 California projects.

A pilot-project in the Lake Chad area could only store 45 km³ underground which is only 1 % of the objective. We should investigate whether there are 100 of these locations available on earth. On the other hand, such an amount of water is not needed for water supply. As the result would be only a very minimal lowering of sea level it is deemed to be not a realistic objective.

Dieng showed how groundwater systems can be modelled in arid zones. Custodio studied the origin of salinity of groundwater in some areas in the Canary Islands, caused by evaporation and the contribution of airborne salts.

Romijn summarized data from case studies of natural recharge as a percentage of precipitation. Due to the stochastic character of rainfall, the recharge in arid zones varies between 1-25 % of precipitation, the higher values found in limestone/dolostone terrains. In semi-arid zones natural recharge can be as high as 50 % in coastal dunes and limestone/dolomite. Moreover, huge amounts of fossil water in sandstone basins are available in North Africa. It can be concluded that efficient groundwater management should be of first priority to mitigate droughts, whereas hydrogeologic management of sea level is not feasible for the moment.

6.3. PANEL DISCUSSION REPORT [by A.J. ROEBERT]

6.3.1. *Storage in Reservoirs*. Pirazzoli raised a question on storage in reservoirs as mentioned by several authors e.g. Fairbridge and Klige. The chairman took this as an example to show the difficulty of filling a general and centralized databank. The data prepared by Pirazzoli are presented as Annex 1 (6.3.3.).

The cited literature indicates that sea level is decreasing at a certain rate per year due to new surface reservoirs. The result of the discussion - easy in this conference because several authors were represented - was that the intention of the authors was to indicate the total available storage at a certain time and not extra storage each year although that was mentioned in the publication in very prestigious journals.

6.3.2. *Dams*. Camfield gave his comments on the construction of surface and subsurface dams; he stressed the need to care about the interests of the downstream locations. Wells added that the distance between villages in former times was fixed in relation to the available quantity of water. Guiraud answered that he agrees with the principle but that the risks are minor because:
a. Low density of population, villages are at a distance of 5-15 km;
b. many tributaries reach the stream in between villages and
c. dams are mostly pervious.

6.3.3. *Statements and recommendations*. A number of statements and recommendations were discussed, see below.

Annex 1

Author	Period of time	Subsurface considered	Reservoir water	Balance storage
Newman & Fairbridge, 1986	1832-1982	disregarded	- 0.75	(- 0.75)
Robin, 1986	1900-1975	+ 0.07	- 0.19	- 0.12
Gormitz & Lebedeff, 1987	past century	+ 1.30	- 1.50	- 0.20
Klige & Dobrovolski, 1988	1900-1975	+ 0.38	- 0.20	+ 0.18

Note: For comparison, the recent lowering of the Caspian Sea level corresponds to a global sea level rise of about 0.1 mm/yr from 1920 to 1960.

STATEMENT 1:
The control of sea level rise by current and foreseeable technologies for storing water underground is not economically feasible, at least not for the next few decades because even interception of 1% of global precipitation leads to a drop of sea level of only 10-15 mm.

Note: 1% of global precipitation is 4250 km^3/a. That amount is 2000 times as large as the total artificial groundwater-recharge in California (U.S.)

STATEMENT 2:
In arid and semi-arid areas the storage of riverwater in the subsoil (by artificial recharge) is highly recommended and economically feasible on a broad scale in order to:
- bridge the gap between wet and dry periods;
- enlarge the yield of an aquifer;
- reduce the evapotranspiration rate;
- enhance the water quality.

STATEMENT 3:
Better control of water demand and better management of water resources is needed in order to reduce the drastic effect of drought.

RECOMMENDATION 1:
To establish an international arid area coordination center for applied research and development to foster the application of suitable techniques for water management and to increase the efficiency of information exchange.
* The University of Ghent is willing to make a start with this center and will provide the necessary facilities.
The aim of this center will be to provide better training, research and development and

it will probably be sponsored by EEC, World Bank and/or other possible sponsors.

One of the first tasks should be to investigate the already existing institutions to seek cooperation, e.g. with the UNESCO/IUGS-project 252 and to organise a workshop on the subjects.

The evaluation is recommended of the volume of dry sediments/rocks in (semi-)arid zones in which we are able to store water.

An aim in the long run will be to establish a really International Arid Area R & D Institute.

RECOMMENDATION 2:

To implement a pilot project for recovery and artificial recharge in several phases.

Objectives:

Phase 1   Make an assessment of available groundwater and develop it to commence operations in the proposed area, see phase 2.

Phase 2   Recharge the partly dry aquifer SW of Lake Chad (Nigeria) with river (Komadugu)-water. This phase should start on a small scale but with adequate project management.

Phase 3   Transport of riverwater from long distance across the watershed by artificial means.

7.   Session VII: Impact on Coastal Lowlands, Geological Land-Use Planning Techniques

7.1. CONCLUSIONS AND RECOMMENDATIONS

STATEMENT:

Submergence and erosion of coastal areas is clearly a significant problem, regardless of future eustatic sea level rise; there are a number of land-use planning and engineering techniques which might be used in mitigation.

RECOMMENDATION

There are recommendations included for data collection, analysis, and immediate political and scientific actions.

Finally we must consider future long-term trends, how to cope with them on an international basis and integrate scientific with economic and social studies.

7.2. SESSION AND PANEL DISCUSSION REPORT [BY JOHN M. SHARP, JR.)

This session had three formal talks and one brief report, followed by vigorous discussion. The first talk (by Sharp) dealt with a risk analysis and causes of subsidence along a portion of the Texas Gulf Coast. The second (by Patfoort) dealt with new materials for use in coastal engineering. The third (by Baeteman) concerned subsidence in Shanghai, P.R.C., plus some general comments. Finally a paper by Faure was read by Petit-Maire.

In the following panel discussion, the following points were raised:

1   Rising relative sea levels are already a major problem around the world. Actions and further research are required.
2   The effects of future eustatic sea level rise are uncertain, but rises up to 1 m. in the next 100 years are not beyond possibility.
3   Jelgersma presented a list of short-term and long-term recommendations (provided by an earlier European meeting), which will be incorporated with modifications in this report.
4   The use of dredged material (if non toxic) for beach replenishments encouraged, but it is realised that this may be only a short-term solution. It might provide "breathing space" until long-term solutions are found.
5   The effect of major storms or other meteorological or climatological events must be considered in coastal management. Fairbridge provided a list of major historical storm events. These major events are important in coastal responses.
6   Several raised the question of how we can influence political and socio-economic decisions. Among the solutions:
    a.   a follow up NATO conference with input from economists and social scientists to provide further recommendations
    b.   a global system for evaluating coastal problems and political/engineering responses
    c.   compensation for developing countries in response to problems of rising sea level
    d.   a system of carefully located tidal gauges is suggested to provide the most accurate possible estimate of eustatic sea level rise, not influenced by tectonic, subsidence, oceanographic, or meteorological processes. We should look carefully at sea level rise estimates and calculate isostasy, tectonic, and subsidence effects and how we select the sites
    e.   the use of satellite data for evaluating coastal hazards and both eustatic and relative sea level rise appears promising
    f.   we should not resist abandoning (giving back) land to the sea, when appropriate.

8.   Session VIII: Engineering Solutions

8.1. CONCLUSIONS AND RECOMMENDATIONS

STATEMENT:
We should consider prevention first, mitigation if prevention is not possible, and adjustment to the situation if neither prevention nor mitigation is possible.

RECOMMENDATIONS
1 A policy of developing non-fossil fuel energy should be encouraged.
2 Reductions in energy-use through improved utilisation and technological advances should be encouraged.

3 Dat should be gathered carefully on the possible adverse effects of any large-scale water transfer and irrigation plans.

8.2. SESSION REPORT [BY C. DE MEYER)

A case history in the Lesotho province was presented to show what enormous water diverting and collecting schemes are possible. Many of these have since been constructed, including tunnels through hard rock. With reference to this project De Meyer presented a pilot study, including some figures on costs, and topographic heights of the route for a similar pipeline between the Zaire-river and Lake Chad basin.

Pool talked about Georadar, its advantages and disadvantages, and also presented a pipeline-scheme, this time between the Niger and Lake Chad. Kogbe remarked that the latter was rather useless as the water level of the Niger had been too low lately, even for electricity generation purposes.

Ollier showed us some excellent examples of beach accretion and decretion mechanisms, sometimes introduced by man, in Australia. He also spoke about problems with recharging aquifers in Northern Australia, as estimated by Habermehl.

Kaplin showed some fine slides of beach erosion and subsequent reinstatement by man's dumping of "natural size" material from adjacent hills into the sea and waiting for the next big storm to do the rest. This was done at 3 to 4 m. below sea level so that waves and hydrodynamic forces could transport and remove all pebbly material. There was also a brief discussion about the Kara Bogar, which was sealed off by a dam and then later reopened to the sea, so that it could serve as an evaporation basin.

8.3. PANEL DISCUSSION REPORT [BY F.E. CAMFIELD]

Engineering solutions can either be immediate responses to a problem, or long-term projects to provide relief from a problem over a long period of time. An example of an immediate response was the sill placed in the lower Mississippi River in 1988 to stop salt water intrusion during the 1988 drought. These immediate responses were generally not discussed during the workshop.

Long-term projects fall into three categories, in which:

1 We can accept the problem, and adjust available resources to counteract it;
2 We can mitigate the problem. Types of mitigation suggested have ranged from planting forests to utilize CO_2, to diverting water to irrigate drought stricken areas;
3 We can prevent the problem. This involves technology policy issues and population control policies. Policies may include moving from fossil fuels to nuclear power to reduce emissions and policies to reduce energy use.

It was suggested that we should probably look at these in reverse order, looking at prevention first, mitigation if prevention is impossible, and adjusting to the problem if mitigation is not possible. It was noted that there is a full range of non-fossil fuels including hydroelectricity, geothermal or wind energy, and-so-forth.

Also, recent developments in heat pumps could greatly reduce energy consumption if they were widely adopted.

Additional improvements in insulation of buildings and fuel efficiency in vehicles would also reduce energy use.

It was noted that diverting surface water to mitigate droughts may not be practical for a number of reasons. These are large-scale (high-cost) projects, it takes a long time fully to implement a project, they frequently cause political and social problems, they have long periods of underuse, and they add to the debt of the host areas.

If groundwater aquifers are available it may be better to develop them. Wells can be drilled as they are needed, using local labour. The cost of water from an existing aquifer will generally be less than the cost of water from a large-scale water diversion project; investment is more gradual and the water is used more efficiently and economically.

It was also noted that many project studies are not based on recent and reliable field data. It is extremely important that people proposing projects visit the project sites and develop sufficient data for preparing alternative plans.

Solutions for relative sea level rise will vary greatly from area to area. Relative sea level rise is much greater in some areas than others, and the cause of local relative sea level rise and the characteristics of the area will influence any recommended solution.

9. Session IX: Case Studies and Cost Benefit Ratio

9.1. CONCLUSIONS AND RECOMMENDATIONS

1. Cost-Benefit analysis is not strictly feasible for many suggested schemes because the costs are borne by a different community from that gaining the benefits.
2. "Small Scale Schemes" of water management are preferable in many ways to large capital-intensive ones.
3. The real costs of large schemes should include costs of prevention of salination.

9.2. SESSION AND PANEL DISCUSSION REPORT [BY C. OLLIER]

In this session it was decided unanimously to omit discussion of sea level management, as existing knowledge is too meagre, and concentrate on aspects of mitigation and drought.

As usual, discussion concentrated on Africa and ignored places like Pakistan, China, Chile, etc. It was pointed out that in Australia the cost (of irrigation) is borne by the public but the benefits are taken by the irrigators. The "user pays" principle is largely avoided. It is difficult, perhaps even undesirable, to talk of cost-benefit ratio when the benefits are not taken by the people who pay the cost.

Guiraud gave detailed figures on small dams and bores in Africa. The cost (generally tens of thousands of dollars) was fairly small, the benefit obvious, and the time of implementation short. The cost was paid from outside; this seems an affective form of aid.

Issar suggested that assessments of costs should be standardized, and recommended that they should refer always to the cost of water (dollars per cubic metre).

Ollier pointed out that in irrigation schemes it is necessary not only to add water, but also to remove it to prevent salinisation. The cost of removal is often the most expensive part of the operation but is usually ignored and not costed at all.

Kogbe suggested that precise costing is not possible but that we could distinguish two categories:
1. cheap, small scale
2. capital-intensive schemes

10. Suggested Conclusions

1. All participants agreed that the continuing CO_2 rise has a warming effect with impact on (1) sea level rise (due to expansion of warming sea water and melting of land ice), and (2) precipitation regimes.
2. Sea level rise and precipitation shifts have recently been observed in many parts of the world. Most of these shifts are in qualitative agreement with the projections of numerical climate models projecting the impact of increased CO_2.
3. In order to increase the reliability of the estimates of future sea level rise and precipitation changes, the CO_2 component of the changes has to be isolated and the principal CO_2-unrelated causes identified and quantified.

   The participants discussed in detail these issues (as summarized in other sections of this report) and concluded that:
   - at present there is no possibility of reliably separating the CO_2-related component from the natural components of the observed changes,
   - efforts should concentrate on improved understanding of various causes of sea level rise and precipitation shifts, and
   - a reliable network of long term observation sites should be selected and maintained.
4. Finally it is concluded that with the information available at present a major sea level rise along densely inhabitated coasts is much more likely than a sea level fall as is the increased frequency of droughts in areas of marginal precipitation. Contingency plans should be urgently developed to prepare for such situations.

11. Comments on Session I

11.1.   PIRAZZOLI'S PAPER

11.1.1. *A. Berger:* The CO_2-T relationship is probably frequency dependent. For the last 150,000 years, at least for the low frequency part of the record, CO_2 concentration changes seem to be a consequence of climatic changes. As far as man's activities are concerned, CO_2 is clearly part of "external" forcing, so that a response from the climate system is expected.

According to Hansen and Lebedeff, the changes in temperature from 1880 to 1985 is maximum in high polar latitudes. Where have you seen that maximum warming was in temperate latitudes?

Pirazzoli: My statement that the recent warming was not consistent with greenhouse-induced models was referring to the period since 1980, not since 1880. The reference, if I have a good memory, is a paper published in episodes (March?) in 1988, quoting Jones.

11.1.2. *R. Paepe:* I believe you cannot simply compare temperature variation with sea-level changes since the latter changes are responding to many factors not related to climate.

Pirazzoli: Yes, but the many factors related to local or regional factors may not be predominant on a global scale. An increase in temperature, on the other hand, would induce global sea-level changes, through the melting of continental ice and steric effects in the ocean.

11.2.   RAMPINO'S PAPER

11.2.1. *M. De Boodt:* Is the influence of DMS as a parameter to form cloud condensation nuclei (CCN) not overemphasised in your paper? There are so many other sources to form CCN's. Has an estimate been made of the part played by DMS in this matter?

Rampino: A recent paper by Schwartz in *Nature* has considered the question of anthropogenic sulphur as CCN in the northern hemisphere. He finds, I believe, that DMS can account for no more than 15% of marine CCN, but the issue is not settled. Water vapour availability is also important in cloud formation, so difference between the two hemispheres should be expected. Charlson of the University of Washington is replying to the criticism in a forthcoming issue of *Nature*.

11.2.2. *Y. Tardy: Dr.* Rampino has shown a graph relating CH_4 production and soil temperature. Does he mean that the greenhouse effect related to methane can be locally registered?

Rampino: The point is that the warming of the climate will cause a secondary effect of releasing methane by increased soil bacterial activity and breakdown of methane hydrates. The additional methane released will add to the global greenhouse.

11.2.3. *A. Issar:* Were there any experiments done on a small scale in order to get an answer to key problems; for example, the influence of UV change on plankton migration?

Rampino: Yes. Our plan is to use small-scale experimental results to parameterize various relationships for incorporation into more complex models. A literature exists (very detailed) on experimental results of UVB on Phytoplankton (e.g. Calkins and Blakefield, 1986 in UNEP/EPA Report,Effects of Changes in Stratospheric Ozone on Climate, pp. 211-235, or Calkin's book "The Role of UV Radiation in Marine Geosystems", New York Plenum, 1982).

11.2.4. *E. Romijn:* Is there any influence of DMS on ozone?

Rampino: I don't believe so. DMS is confined to the troposphere where it forms H_2SO_4. This would rain in a few days/week and not reach the stratosphere.

11.2.5. *A. Berger:* Two questions related to sea-ice in climatic models:
- Which is the season in which the response of the climate system is the largest in your CO_2 increase experiment?

Rampino: In our model, changes in late fall and early spring are the largest. This is related to the disappearance of ice cover during a period it would normally not disappear.
- Do you have brine rejection modelled in your experiment?

Rampino: Our 2-D model experiments do include brine rejections. We have done a number of seasonal experiments , so I am not sure what our latest model shows. Results will be presented at Spring AGU, and are *in press* by Hoffert. I will ask him to send Dr. Berger a preprint of the work.

11.2.6. *P.E. Camfield:* This question relates to the increase in stratospheric water and cloud cover. How would this affect polar precipitation? It would appear that over a period of time this would have some effect. If such an effect occurred, there would be an influence on polar ice cover which in turn would affect sea level change (global sea level).

Rampino: The increased water vapour from methane destruction in the troposphere is negligible. However, increased water vapour in the dry stratosphere is significant. It provides for a possible increase in ice crystals on whose surfaces the various heterogeneous reactions that destroy ozone may take place. An increase in "polar stratosphere clouds" could cause additional ozone destruction but would have no detectable effect on precipitation.

12. Comments on Session II

12.1.   BURKE'S [WELLS'] PAPER

12.1.1. *N. Petit-Maire:* Plotting together data for "African Palaeolakes" (i.e. about 1000 ^{14}C ages) is a temptation for all of us. However, each lacustrine situation is related to different parameters (altitude, longitude, oceanic proximity, tectonics, surface or aquifer feeding, etc.) and must be considered individually before making any real *climatic* conclusions. This gives nice graphs but dangerous interpretations.

Wells: A well stated caution: lakes may rise or fall for a variety of reasons. Changes in meteoric precipitation, evaporation (by changes in wind velocity and/or cloudiness), episodic communication of water from neighbouring basins, fluctuations of regional groundwater and many other factor make each closed basin a special case. The value of the Oxford global data base for Quaternary lake level status is that it allows broad regional trends in the climates of the past to be better visualized upon a regional, continental, or even a global basis. To achieve this goal, a great deal of specific information about a basin must be generalized, so drawing specific, detailed climatic conclusions for individual areas based upon a representation of the Oxford lake level date can be risky.

12.2.   KUKLA'S PAPER

12.2.1. *B. Denness:* George Kukla has pointed out various deficiencies of current GCM's with regard to forecasting climate patterns in a CO_2-rich world; Rhodes Fairbridge has admitted to being controversial in claiming that the climate will ultimately be predictable.

The writer would like to suggest that until GCM's build in an element for consideration of the possibility that Gravity (G) may be a variable instead of the constant (Universal Gravitational "Constant") that it normally thought to be they cannot take account of such changes should they be occurring as Dirac postulated some 50 years ago. The consequence would be that, no matter how otherwise all-embracing the models they could never predict accurately the climate at different levels of G.

Kukla gave an example of a failed hindcast for the onset of the late Pleistocene glaciation, when G was -according to the writer's deterministic geophysical model, expanded in this volume and elsewhere- considerably different from that today. It may not be the fault of the GCM's or even the modellers that failure is so common but rather that the input data is incomplete in more ways than currently considered. Perhaps thought should be given not only to air/sea interaction, cloud cover, etc., but also to a time dependent G variation.

12.2.2. *A. Berger:* This is a type of model which provides the climate in equilibrium with the prescribed boundary conditions.

There is another kind of model which gives the transient response of the climate system to a forcing, astronomical forcing for example. The latter simulate the time-dependent behaviour of climate including the generation of the boundary conditions themselves. In this way, they take into account the memory of the system, which can not be done explicitly with an equilibrium snapshot experiment.

12.3.   BLUMENTHAL'S PAPER

12.3.1. *R. W. Fairbridge:* The modelling may be helpful for trying to understand a palaeontological problem in the Last Interglacial period. A distinctive coastal fauna of invertebrate organisms known today in the Gulf of Guinea and as far north as Senegal, typified by a large gastropod *Strombus bubonius*, requires a non-seasonal climate and a water temperature in the range of 24 to 28°C. This assemblage appeared suddenly in the Canary Islands and penetrated the Mediterranean from west to east, but only during the late Pleistocene. Its presence could not be explained if the presently cold Canaries Current was effective then in its modern conformation. Dr. Joachim Meco, a local palaeontologist, proposes an unusual in-phase alignment of the Milankovitch orbital elements to explain it. The writer (RWF) noted a paper by Schell some thirty years ago that correlated upwelling and cold SST along the Benguella Current coast of SW Africa with the solar cycle, whereas a weakened trade-wind mechanism could trigger Blumenthal's mechanism, to create an "Atlantic El-Nino" effect bringing a tongue of warm water as far north as Gibraltar.

Blumenthal: Dr. Fairbridge is quite right in saying that the mechanism presented could greatly contribute to oceanic surface warming along the African coast. The effect would be particularly dramatic if the winds reverse relative to the coast, i.e. if the winds change from blowing dry air over the ocean (which leads to a strong latent heat cooling) to winds blowing moist air from the ocean to land (which leads to relatively little latent heat cooling), the SST would increase to maintain the heat flux balance. Such a change in winds would lead to coastal warming though other mechanisms as well. In particular, if the change in winds stops the upwelling and/or alters the Canaries Current there would also be a strong warming along the West African coast. I suspect that the latter set of effects would be stronger than the air humidity effect, though the speculation could be verified by utilizing an ocean model. Connections with a solar cycle or the Milankovitch cycle are less clear, since the effects of these cycles on wind direction are not simple.

12.3.2. *N. Petit-Maire:* At 6000-4000 BP a warm Guinean/Senegalese Mollusc fauna[*] reached up to Cape Juby (28° N) along the occidental Sahara-Morocco coast. Its ecological requirements now limit it south of 17° N/21° N. This resulted in the possibility for the larvae to migrate northwards (i.e. weakening of the Canaries Current) and to proliferate along the coast (i.e. reduction of upwelling) during the last climatic optimum and persist during the regressive stage (maximum of the transgressions dated 5500 BP at Nouakchott). [*]12 species

12.3.3. *M. Rampino:* Do you have any published material on this or occurrence of warm water fauna in previous interglacials? Perhaps we could collaborate, we have a model at New York University which produces *shut off* of upwelling at interglacial peaks and in mid-Holocene time.
  Blumenthal: I too am interested in more information about warm water/cold water fauna alternation. It is very helpful to be able to check models such as the one I presented against climates distinctly different from the present.

12.4. PANEL DISCUSSIONS

12.4.1. *K. Howard:* In response to Fairbridge's insistence that groundwater storage if wet season derived surface water is an efficient process, there is a considerable difference between pumping water out of a storage tank and pumping water from an aquifer. The former can be drained entirely - the water level being lowered across the whole tank instantaneously. The latter must be pumped using wells; pumping must induce flow to these wells, by developing cones of depression. Drainage of the system will be incomplete and take considerable time.

12.4.2. *A. Issar:* I think that the question of cost-benefit is important, because there will always be more projects than money and the question will be of priority, namely which projects can be financed and which not.

12.4.3. *M.A. Pool:* A new desalination plant in Saudi Arabia in 1982 provided enough water for Riyadh-capital. Costs were not really a point for this project. But environmental problems not foreseen before, like where to put waste water came along with this new sufficient water supply.

12.4.4. *A.J. Roebert:* 1-2% of the Global rainfall is a very high quantity of water and if one is able to intercept that water and make it infiltrate in the subsoil aquifer, the influence on the sea level is only about 11.4 mm. That is not very much. My suggestions:
1. leave sea level out of the discussion
2. climatologists may concentrate on change of the climate in which areas become wet
3. leave cost out of the discussion at present; it is too early.

12.4.5. *E. Romijn:* Dike building in Holland was certainly based on cost/benefit calculation. After the storm surge in 1953 the design level for sea dikes has been based on a probability of 1% in 100 years (or return period of 1 in 10,000 years).

12.4.6. *G.L. Wells:* Regarding short-term responses to rising water levels: politicians in the state of Utah were placed in an almost impossible position in the early to mid 1980's. The Great Salt Lake was rising rapidly and threatening valuable farm land and residential areas. Climatologists could not tell the politicians whether the rising levels would continue, yet they had to act or risk disgrace for failure to make countermeasures. After installing huge pumps to export the excess water to the western Bomeville Desert, the lake ceased rising, not because of the water extraction but because unusually high snowfall in the Great Basin ended. A huge capital project was installed to no immediate effect.

12.4.7. *M. Leroux:* This session is concerned with the distribution of rainfall and its prediction, and especially the inability to predict. I wish to emphasize that the present evolution of rainfall in northern and southern tropical Africa is exactly the inverse of the evolution of temperature, as seen globally by the Jones' warming trend.

Will a CO_2 doubling signify a drought doubling or is it necessary to search for the parameter, or temperature trend to which the rainfall evolution is really connected.

12.4.8. *J.M. Sharp:* Three points we need to consider in this conference are:
1. What is the feasibility of geohydrological management?
2. Can we afford management?
3. Are there environmental effects, not anticipated, which the management schemes would bring about?
We need to get a better understanding of the uncertainty of sea level rise due to climatic changes. These data or predictions are needed for the consideration of management plans.

12.4.9. *G. Kukla:* How can one assess the accuracy of predictions? Can we do it at all?

12.4.10. *F.E. Camfield:* Politics relating to flood control need to be revised. Development of flood prone areas (flood plains) results in flood control projects, designed to carry off the water as rapidly as possible. These projects may be in relatively arid regions, resulting in water being wasted down concrete lined channels instead of spreading it on the ground to recharge groundwater. Many of these areas are subjected to infrequent flash flooding rather than steady flows of water.

On the question of cost, recent experience in the Great Lakes (United States, Canada) indicates that the public along the shorelines are willing to pay for protective measures against rising water levels. It is necessary, however, to spend the money effectively.

In response to other discussion, the United States Army Corps of Engineers is considering the politics relating to disposal or use of material dredged from navigation channels. In particular they are looking at the beneficial use of the material for shoreline restoration

12.4.11. *V. Goldsmith:* Many of the problems of the danger to coastal resources from rising sea level can be ameliorated by better management of our coastal resources, e.g.,
1. New Orleans would be dry if not for river diversion.
2. Sand dredged from inlets and harbours can be placed back on the beaches (nourishment). This can be encouraged by (a) redesigning dredges; (b) economic incentives to do this rather than having sand dumped in the sea.

13. Comments on Session III

13.1.   BERGER'S PAPER

13.1.1. *K. Howard:* You indicate that your climate models must incorporate the memory of the system. We find the same problem with groundwater models. I am interested to know what starting point you use for your models, e.g., how far you go back and what are the starting conditions? These must have an important influence on how the model behaves.

Berger: Our model is indeed very sensitive to initial boundary conditions. The integration was started 125,000 years ago, a time at which we may suppose the Greenland ice sheet was 2/3 the present-day value and the overall climate over the Earth was similar to that of today. The problem of longer simulation is, to find a date for which geologists can provide adequate and accurate information.

13.1.2. *F.J. Ayala:* Do you think isostasy has played any role in glacial-interglacial cycles?

Berger: The isostatic uplift of continents and sinking of oceans has been taken into account in models of the ice-sheets and of the ocean for sea level rise computations.

13.2.   TARDY'S PAPER

13.2.1. *C. Baeteman:* When you make interpretations of wet and dry phases in climate considering large rivers which cross nearly entire continents, crossing different climatic belts, for which area exactly are you drawing a conclusion?

Tardy: I have shown comparisons between San Francisco-Missouri-North Saskatchewan which occupies a smaller area than "nearly entire continent". Neither do the Loire, Elba, Ural and Yenissei occupy the whole of Europe or the whole of Asia. It is a question of scale and information is not concentrated in only one region. The rainfall over a basin of 3 km^2 is also highly heterogeneous. Information, to be complete, has to be compared at both small and large scales.

In natural systems logic or truth is no more reached in small scale than in large scale observations. It is not because a system is small that you will find in it better accuracy or more precise information; generations of people called scientists were unproductive for centuries through looking only at one scale phenomena. Religions are not only available for individuals but also for populations of very different individuals.

By showing the shifts between fluctuations I hope only to show you (if you can see it) that there are shifts. This is what Krisnamnity has called teleconnected propagations.

13.3.   PAEPE'S PAPER

13.3.1. *A. Issar:* It will be interesting to correlate the Holocene curve from Greece with the isotope curve of the Sea of Galilee. This will enable us to discern climatic changes and tectonic changes.

Paepe: The study of Greece will be completed in the future by archaeological site study of the coastal and river valley areas. Dating will be mainly based on archaeology and fluctuations of climate studied by Van Overloop on the basis of stable isotopes and pollen. Correlation with Mesopotamia has been done.

13.4.   MOGUEDET'S PAPER

13.4.1. *C.A. Kogbe:* The estimates of discharge of the Congo and Ubangi rivers are impressive and commendable. It is, however, quite premature to conclude that diversion of a small amount of water to the Chari river will have a negative effect on the environmental conditions of the Gulf of Guinea as a different and less negative conclusion can be derived from the same data after calculating an optimum amount of water that such a diversion will require.

Moguedet: We must be careful with such a project and we must study precisely the prior conditions (and the consequences).

13.5.   PANEL DISCUSSION

P. Pirazzoli: The answer to the question asked by Tardy: "Should we avoid measuring sea level changes in subsidence areas?" can only be negative. No area on the Earth's surface can be considered as absolutely stable. Furthermore, for geological reasons, human induced factors, and ocean dynamic effects (in the case of a warming climate), submergence is more frequent than emergence in coastal areas. All vertical measurements made from the Earth's surface should therefore be interpreted as essentially relative. With this in mind, sea level measurements can be very useful in most places (subsiding as well as uplifting)

C. Baeteman: Considering sea level changes and tide gauges, subsiding areas should not be neglected or rejected. These data and observations should be regarded critically with respect to all other processes influencing the sea level. The vulnerability of these areas and their dense populations can for sea level observations be better developed in order to recognise or detect any signal of a coming sea level rise.

One should not force any immediate conclusions about how to consider sea level data from subsiding areas before related problems are explained later.

C. Ollier: The Cail curves go back to Cambrian times, and have certain very distinctive features, such as very rapid fall in sea level and slow rise. This is found in non-glacial periods, and so is essentially tectonic, not climatic. Possible explanations are professed in a recent publication: Palaeogeography, Sea Level & Climate - Bureau of Mineral Resources Record 1988/42, Canberra.

E. Romijn: As to "absolute" measurement or sea level rise I agree with the proposal to refer to satellite measurements albeit that for the moment they are not accurate enough. This however is a matter of further development.

J.M. Sharp: Is it possible that a regularly spaced distribution of tidal gauging stations throughout the world would provide a decent estimate of eustatic sea level rise? If so, I propose this as one of our recommendations.

M. Leroux: It is impossible to compare (or contrast) on a global scale the evolution of African and American climates. In the distant past, as in recent decades, the whole tropical area has shown the same evolution (on a global scale), this zone (including tropical America, Africa and Asia) having the same general aerological conditions (structural conditions of rain-making), these conditions having evolved in the same manner.

V. Goldsmith: Although tidal data indicate a relative rise in sea level, we do not know the cause. It may mean that this indicates that these areas of tide gauges are subsiding tectonically (e.g. river delta's), or are subject to changes in winds (onshore/offshore), or currents (e.g. Gulf Stream), and may have nothing to do with the rise of CO_2 or the so-called "Greenhouse" effect.

G. Kukla: A serious problem with sea level surface topography is that the departures from geoid may not be stable in time. Available satellite monitoring is not long enough to clarify the problem.

Recommendation: continue satellite monitoring of sea level elevation.

B. Denness: In response to the question whether there has been a sea level rise over the last century, Klige and Pirazzoli have demonstrated that overall there has been a rise but during the last 50 years there have merely been fluctuations about a stable mean. As we are concerned with the likelihood that sea level will rise in response to a temperature rise caused by the greenhouse effect this is exactly what we should expect. It is a fact that the global temperature has risen over the last century but has merely fluctuated over the last 50 years - almost exactly in step with sea level change.

I have provided a deterministic model that combines natural and manmade temperature changes and matches very closely the measured global temperature changes during the last century. That model takes into account the steadily rising greenhouse temperature and adds to it the natural change embodied in resonances from the long term Milankovitch cycles and beyond: those natural changes have experienced fluctuations on a decreasing mean over the last 50 years, having increased before that.

Of paramount interest is the future . Both the natural and manmade temperatures from the above model are scheduled to increase during the 1990's and beyond. Evidence for the 1980's (after the model was developed) illustrates that this is already underway, as forecast by the model. It is logical to suppose sea level will respond similarly upwards in the future too.

Most of the discussion session was devoted to the diversion (a complete red herring) that uncertainties of sea level rise over the last 50 years appear to introduce similar uncertainties regarding the validity of the greenhouse effect. The above summary indicates that the introduction of uncertainty is totally unnecessary. If leading scientists allow themselves the luxury of this type of debate it is hardly surprising that it takes so long for politicians and planners to listen to them.

M. Leroux: Among the variables relating to the level of the sea we must consider the atmospheric pressure, its regional (or zonal) distribution and the variability in space or time of this distribution (which forces differences of several cm in sea level).

14. Comments on Session IV

14.1.   M. DE BOODT'S PAPER

14.1.1. *C.A. Kogbe:* Desertification in the Sahel belt and Southern fringe of the Sahara is not only due to human action but to inevitable climatic changes.

De Boodt: Yes, we agree that desertification is a complex phenomenon in which both climatic changes and man's interference are important. To make things clearer it is more appropriate to speak about: *aridity or aridification* when the climatical influence (mainly drought) is predominant and about *desertification* when man is the major cause of the desert encroachment.

14.2.   A. ISSAR'S PAPER

14.2.1. *E. Van Overloop:* How far does ancient literature on "natural hazards" like for instance famines, diseases, drying up of wells, wars, migrations, contribute to knowledge about climatic changes?

Can archaeological discoveries on filled up wells, and abandoned cities be used for the same purpose?

Issar: Yes, they have to be examined carefully to avoid local anomalies. For instance, we found out that the cleaning of ancient wells can show whether there was a change in the groundwater table. (Personally I think that even a careful study of the Bible can give us insight on climatic change).

14.3.   G. WELLS' PAPER

14.3.1. *C.A. Kogbe:*
1.   Statistics on demography or agricultural productivity in African states can be very misleading as recent censuses have been quite unreliable. This is due to poor accessibility to rural areas as well as other political factors.
2.   Large scale technological solutions to problems, such as the construction of dams, often disrupt the traditional approach to agriculture with disastrous consequences. What is needed is *relevant* technology NOT technology transfer. Traditional methods can be improved upon and made to be more effective.

Wells:
1.   No doubt it is true and makes relief operations and planning for the future development more difficult. Better information must be made available.
2.   I agree entirely. More *relevant* and *resilient* technology needs to be applied.

14.3.2. *M. Leroux:* The problem of respective responsibilities of nature and man in the desertification is already an "old" problem. Everybody knows that the decreasing of rainfall is commanded by aerological factors (and especially by the diminution in the latitudinal extensions of the tropical zone, the Sahelian area of Africa being particularly affected, like all the tropical margins. Man acts on the desertification by destroying the vegetal cover, but he is not responsible for the rain decrease.

Wells: Human actions may accelerate the removal of vegetation cover during drought (and even non-drought) conditions. These effects are extremely limited in areal extent and include the classic village barren "haloes", created by the removal of fuel wood for cooking and overgrazed areas surrounding boreholes, where livestock have been concentrated during severe drought. The cumulative extent of these effects forms a minor fraction of the total area of drought desiccated vegetation and exposed soil surfaces created by natural processes.

14.3.3. *N. Petit-Maire:* "Is the recent Sahel belt retreat southward anthropic or natural?" asked Dr. Kogbe. The answer is both. Since 6,000 BP, the Sahel has retreated c. 600 km southwards, the natural (astronomical?) trend is probably going on in parallel with deforestation and overgrazing.

The Sahara-Sahel changes in the last 18,000 years were *not* due to any anthropic factors and they were huge. Nowadays something like 90% are natural (wind force intensity and frequency increase is not man made) and 10% human. We could cope with this 10%!

Wells: I am not sure we have the means to place absolute percentages on the relative contributions of natural and anthropogenic causes of desertification, but the latter can be shown to be very small. Late Quaternary shifts of the vegetation zone southward and northward across vast areas occurred without human intervention.

What we are now beginning to realize is that the Sahelian-Saharan margin may shift by 100-200 km north and south on a decadal timescale and that advances and retreats of 105 km can occur in a single year, all quite apart from man's influence.

14.3.4. *A. Berger:*

1.   How are the mean precipitations computed for the different latitudes?

2. How do you explain such a population growth during such adverse conditions in some Sahel countries?

Wells:

1. The historical mean precipitation (1931-1960) serves as a baseline for the precipitation departures. A total of 75 stations are used in the meridional transect from 35° N to 30° S. Precipitation over the Sahara is highly generalized and no attempt has been made to weight the rainfall data from different stations occurring within the same latitudinal band. Despite these limitations, the technique has produced curves of precipitation distribution and amplitude very similar to those discussed by Leroux in his presentation.

2. African population as a whole is believed to have increased by 40 percent during the past 20 years. This figure is dominated by the populations of Nigeria, Kenya, Ethiopia and Egypt with their steep (>3.0%/year) rate of growth.

   The Sahelian population has also increased during this period, but not at the same rate. The growth of urban areas in the Sahel and maintenance of the agrarian population by food relief during famine and improved medical care has allowed an expansion of population despite the adverse climatic conditions of the past two decades.

14.3.5. G. Kukla: Why can you not get the same (and even better) results with *regular landsat*, TM, spot, channel 1, 2 and 3 of A VHRR, DMSP (and so on)?

Wells: Comparable results can be generated for the monitoring of the Sahelian-Saharan margin using various orbital radiometer systems. However, only manned orbital photography has an image archive extending back to 1965, *before* the onset of the prolonged Sahelian drought. NOAA Polar Orbiter A VHRR and its allied sensor system on the DMSP series yield images that must be carefully corrected for the effects of atmospheric turbidity. Once these corrections have been made and confidence established in the product, the NDVI or other CHANNEL 1 and 2 ratios and combinations can be represented as images of the status of the Sahelian vegetation zone in a timely and cost-effective manner. The early landsat MSS suffered from a lack of gain control over high albedo surfaces. Landsat TM is much superior, but also much more expensive to collect and process. TM images may provide the means for point-sampling A VHRR products at higher spatial resolution over critical areas. SPOT image products are 10-100 times too expensive to be of any benefit in a Sahelian monitoring program, unless the French government defrays virtually all costs.

14.3.6. M.R. Llamas: Which is the use of groundwater in the Lake Chad region?

Wells: At the present time, groundwater extraction by boreholes serves a very limited area of the Chad Basin and is limited to a few municipal water supplies and livestock concentrations. Agriculture has traditionally depended upon Lake Chad itself and more recently, upon diversions and surface storage of basin drainages, as in the South Chad Irrigation Project, Baga Polder and Lake Maga schemes.

These engineering projects have proved to be vulnerable to total collapse during prolonged drought. A thoughtful policy of groundwater extraction and distribution may be beneficial to the recovery and future development of the region.

14.4.   PANEL DISCUSSION

B. Denness: Dr. Kukla suggests that climate change, at least the natural element of it, is stochastic. As evidence he cites the droughts in the US mid-west in the 1930's and late 1950's which appear to have no relation to commonly accepted frequencies.

In fact both of the above droughts correspond to high points in the deterministic gravity variation model I have described to this workshop: the model shows not other significant "highs" this century. However, it does forecast another for the early 1990's towards which the model is already accelerating. With this in mind I published the forecast of the US mid-west droughts beginning in the late 1980's. This was reiterated in public in Colombia S.A. in april 1988: during the summer of 1988 the US mid-west experienced a drought that cost \$5 billion of Federal relief to resolve. That is evidence of determinism against a stochastic process.

G. Wells: I would appreciate hearing any objective evidence for the broad-scale desertification of the Sahel by human actions.

The margin of the Sahel along the southern Sahara is a very dynamic boundary. Between 1968 and 1970, the zone of desiccated vegetation and denuded soil surfaces moved southward 100 to 175 km. The southward advance was most dramatic *in the least populated areas of the Sahel*. It is highly unlikely that the activities of nomadic pastoralists in these areas of northwestern Mali and eastern Mauritania with a historically low population could have stripped the surface of vegetation over an area of $> 200,000$ km^2 in only 2 years. Moreover, in 1988, the zone of green vegetation from coastal Mauritania to central Sudan advanced northward 50 to 100 km in a single year. As there has been no concerted policy of revegetation of the northern Sahel by human intervention, it is not possible to explain this sudden, northward advance by any other means than a return to higher precipitation over the drought region. From studies of both the past and present climate, one may conclude that the margin of the Sahel along the Sahara would change by 100's of km northward and southward even in the complete absence of human occupation of the region.

H.N. Le Houerou: I have been studying desertification over the past 35 years on both sides of the Sahara. There is much indisputable evidence of strong human interference. Drought of course helps!

The albedo effect has nothing to do with it; the low albedo of the Assouan Lake has not increased rainfall. Rainfall in Assouan was 1 mm/year before the lake and it is still the same 25 years later!

Y. Tardy: Human deterioration may be due to climatic change because populations do not move together with the climate limits. At a given time they can be in equilibrium and do not cause any irreversible damage. However, when climate changes the population is in disequilibrium and causes damage to the environment. But this is a climatic effect.

A. Berger: In response to Wells and Le Houerou: there are complex interactions between the climate system and man's impact. For example, the Charney mechanism tends to show that deforestation in a sensitive climate zone, like the Sahel, would create a climate situation more favourable to desertification. Do we have any observation to support such theory?

G. Wells: [Charney hypothesis & drought amplification mentioned by Berger] Regarding the drought in the Sahel, the biogeophysical feedback mechanism proposed by Charney in the mid-1970's has not been confirmed by remote sensing techniques or field inspection. Charney's model required a doubling of the surface albedo of the Sahel to a value approaching that of mobile Saharan sand dunes.

Recent work by Courel and Rasool has shown that this increase of regional surface albedo has not occurred despite the southward extension of the Sahelian-Saharan margin during the past two decades.

My own albedo maps of the western Sahel generated from NOAA Polar Orbiter A VHRR data show that the Sahelian albedo of southern Mauritania along the Senegal border is at most 0.21 (measured in April 1985), while the barren sands of northern Mauritania have an albedo of 0.44. A modification of the Charney hypothesis has been proposed by Wells (1989) and Wells & Middleton (1988).

The presence of brightly-reflective terrigenous aerosols over the western Sahel during the precipitation season may serve to limit the northward advance of monsoonal moisture by pranoting regional subsidence above the lower troposphere. This "airborne" increase of regional albedo approaches the Charney model, where radiometer measurements have shown dust palls to possess albedos of 0.34-0.38. (Wells, G (1989)"The Fragile Environment.", Cambridge University Press *and* Wells, G. & Middleton, N.J. (1988)"The alternation of land surface cover across the western Sahel recorded by orbital photography 1965-1986.", International Land Surface Climatology Project Second Scientific Results Meeting, April 25-29, 1988.)

M. Leroux: It is necessary to be precise about what we are really speaking: rainfall or desertification.

If it is rainfall, I wish to emphasize that rainfall responds to three simultaneous conditions:
- precipitable water,
- the ascent of air,
- and aerological structural conditions favouring the ascent (that is to say without subsidence, or without air stratification or wind shearline).

In these conditions, how can man have influence?

H.N. Le Houerou: The failure of the Green Belt in Northern Africa is not necessarily a human failure as it depends on local conditions of soil, runoff and rainfall. In good local conditions, one planting only may be successful.

N. Petit-Maire: Nomadic populations with their cattle accelerate anthropic desertification which adds to the natural aridification trend. Governments should be advised against it before it is too late.

Man may impede the normal biological regeneration at the end of a dry cycle.

15. Comments on Session V

15.1.   PANEL DISCUSSION

M. Rampino: For the last ten years we have been working on building up a data base of information on volcanic eruptions for modelling studies at the Goddard Institute for Space Studies. We have used historic eye-witness accounts, information from ice cores, tree-rings, deep-sea cores, astronomical observations, etc. The most important information is that of volume of erupted products and chemical composition (sulphur volatiles) on the magmas. These data have been used in model experiments.

It must be remembered that volcanic eruptions affect the atmosphere only by $< 0.5°C$ for 2-3 years since this is the residence time of aerosols in the stratosphere. The experiments and actual events like El Chichón show that volcanoes, even quite large ones (Tambora 1815, for example) can only temporarily delay the Greenhouse warming (look at 1982-1983 effects of El Chichón).

As far as very large ignimbrite eruptions are concerned, these are usually quite poor in sulphur volatiles, and also very rare.

P. Pirazzoli: More complete data banks would certainly be most useful. However, earth scientists usually prefer not spending much time to make their own data available to data banks. In the field of sea level research, this has been experienced by IGCP projects 61 and 200. which have attempted to create a data bank of radio-carbon dated sea level indicators, but obtained only very limited success.

G. Wells: A comment testing Pirac's hypothesis of the decay of the gravitanional constant: since 1969 NASA has repeatedly performed laser reflection interferometry of the Earth/Moon system using mirrors placed upon the Moon by astronauts. The reflection surveys can determine the baseline distance to within a fraction of one cm. If Dirac's hypothesis were correct, a change of > 3 cm should have been observed by now. No change has been detected.

[For Nat Rutter] "Sudden events" and their propagation through the Earth system need to be investigated as a part of the Core Program of the INQUA Global Change initiative. For instance, recent isotopic dating of the Toba eruption of > 1000 km^3 has been placed at 75 Ky B.P.. Any student of the cycle of ice age formation and retreat should be curious about the coincidence of the Toba event and the period of rapid ice accumulation in the Northern Hemisphere. Does it augment Milankovitch orbital forcing? Is not this the kind of special event that needs study?

F.J. Alaya: Ignimbrite eruptions do not have an influence on climate proportional to tthis volume because the plinian part is at the beginning, being converted later to a pyroclastic flow with minor influence.

A possible mechanism to explain the influence on climate is heating of the stratosphere by plinian volcanic eruptions, thereby causing deviation of jet-streams (proven after the El Chichón eruption 1982) and anomalous atmospheric circulation beneath in the troposphere.

G. Wells: Refering to Dr. Garcedo's comment upon the effects of an ignimbrite eruption: our models for the climatic consequences of an ignimbrite eruption have been developed for relatively small events in the historical record (e.g., Katmai in 1912). To my knowledge, no one has simulated the possible effects of an explosive caldera eruption of the scale of the Toba eruption 75 Ky ago. On the Altiplanoof Chile, Bolivia, Argentina and Peru, more than 20 post-mid Miocene explosion calderas with massive (500-1000 km^3) ignimbrite blankets have recently been discovered. The altitude of these events (> 3800 m), their latitudinal position, geochemistry and Plinian column mixing characteristics may have special and unrealized consequences for Global Change.

F.E. Camfield: Groups studying earthquakes along ocean arcs with respect to tsumani generation have studied probabilities of earthquakes for purposes of coastal zone planning, and also to consider potential (probable) hazards to existing projects. Extensive studies have been carried out in that regard.. These investigators are generally associated with the Tsumani Commission of ICSU.

On modelling, the Corps of Engineers recently had a meeting where modelling was discussed. The conclusion was that models are not giving good results because of the lack of field data to verify them. Emphasis has been placed on obtaining supercomputers (which will provide bad results faster), when we really need the data to verify models. Also, good quality control is needed on the data that is gathered.

F.J. Ayala: Physical knowledge of Plate Tectonics and other phenomena related to earthquakes must be well known prior to applying statistics to find cycles.

A. Issar: We have today quite a lot of data on periodic changes in world climate due to astronomical factors. We know something about the chances of earthquakes, sunspots, volcanic eruptions, etc. What is needed now is to put this information together and find out what will be the triggering influence or the accumulative influence of random or stochastic phenomena on periodical ones.

B. Denness: Ayala-Garcedo quite correctly introduced a range of geophysical phenomena - volcanoes, earthquakes and so on - that may have an influence on climate change: he also pointed out that such change takes place on a variety of scales over all periods of time from the geological timescale to a matter of only a few years.

It is suggested that, while correlations can indeed be shown between volcanic or seismic events and climate change, there is no reason whatsoever to suppose that the correlation represents cause-and-effect. In fact, one might consider that all these geophysical phenomena, including climate change, may be responding to the influence of change in a further mor fundamental geophysical phenomenon, the Universal Gravitational "Constant" (G), as proposed by the great theoretical physicist Dirac in 1938 and more recently quantified to some degree by Lyttleton in relation to mountain building.

If G changes so the distance between astronomical bodies changes: as G increases the Earth approaches the sun and becomes hotter and vice versa - hence the relation between G and climate.

Similarly as G changes so the rotational velocity of the Earth changes with the consequence that tectonic plates, under the influence of the changing centrifugal force thus induced, adjust their position with respect to one another and in so doing cause earthquakes and volcanity - hence the relation between G and volcanics and earthquakes.

Therefore, if we have a model to describe the variation of G on all timescales, we can expect it to describer simultaneous changes in climate, earthquakes, etc. That model was prepared by the writer in 1980 and has been tested in forecast since then.

Its origin, construction and explanation have occupied at least 10 substantial publications since that time. It has been used, for example, to forecast successfully dramatically increasing rainfall in Dubai (Middle East) and Colombia (South America) and drought in the USA. It has also successfully forecast a Richter Magnitude 5.9 earthquake in Colombia and matches in hindcast the distribution in time and space of the 200-plus Magnitude 6-plus twentieth century earthquakes in China. Countless other examples both of hindcast matching and of practical forecasting are also available.

The gravity model, with its forecast of natural climate change, has been combined with the greenhouse effect temperature response to match very closely the measured changes in global temperature this century - including the 9 years since the model was developed. Last year's droughts in the USA were correctly forecast in 1983 with the additional advice that they will return frequently and deepen over the next decade.

Repercussions of these droughts (which are essentially the product of natural climate change, merely aggravated by the greenhouse effect) in the main food-producing part of the world (accentuated by parallel drying in other important food-producing parts of the world) for US and hence global economy are clear and were published several years ago. The G-model (via global temperature, regional drought and its economic implications) forecast an intense economic depression for the 1990's. It is suggested that this would already have begun but for the misleading effects of Reaganomics that has allowed the US to build a still-growing national debt of about 3 times its national turnover - a feature that would not be entertained for an individual and whose brittleness will be exposed by further droughts.

In summary, indeed geophysical phenomena are relevant to climate change on all timescales, but please do not let us get unnecessarily confused by the minutiae of their apparent interactions when there is already a model that combines them and leads to a practical (and tested) forecasting procedure.

R. Paepe: This gathering is lacking representatives from such associations as the International Tsumani Ass., the Earthquake Prediction Research Ass., etc. This group of NATO should recommend greater communication amongst scientists in these various fields and stimulate research in broader fields of related sciences.

The need of fundamental research on Natural Hazards is necessary because otherwise the study of Natural Hazards remains at a journalistic level.

Natural hazards show definite trends of periodicities which hazards are interrelated and also related to climate.

A.J. Roebert: I support the idea of Prof. Paepe for better communication among scientists directed towards a world wide data base as intended. I will present an example of evaluation of data in the panel discussion of Session VI.

B. Denness: In response to the quest for a model which covers the geological and historical time scales and yet also enables an interpretation of use to forecasting events over the next few years, it is suggested that the writer's model of G variation (with its various geophysical, climatic and socio-economic spin-offs) can accomplish this. It has been tested for several years with striking results. It is suggested that it would be unjustifiably cynical not to take that model seriously: either we accept the possibility of determinism being real, in which case such a model must exist, or we do not. A deterministic model should convince a politician; a probalistic model is easier to discount.

R.W. Fairbridge: Data collection and exchange is essential, but its acquisition is fraught with difficulty. Each item of data is often collected by a different person in a different place. How can we assure the synthesist of homogeneity and quality control?

Concerning geologically related catastrophic events, little provision appears in the IGCP agenda for their study. Earlier Ayala Garcedo noted the more than 1 cm (not mm) thickness of ash over two-thirds of the United States. A few years ago (1980? in *Pal. Pal. Pal.*) a Japanese researcher, Tawai, showed historical evidence of a 30 m tsumani on average once a century on the coast of Japan. No one would survive such events.

On the subject of ignimbrites I, and several colleagues present today, examined magnificent examples on the west coast of Fuerteventura. I also noted that about 25% of the volcanic cones on the island are Holocene. Catastrophic eruption may occur any time.

16. Comments on Session VI

16.1.   E. ROMIJN'S PAPER

16.1.1. *J.M. Sharp:* How much would temperature variability affect estimates of natural recharge?
Romijn & Issar: It would affect these estimates. Also, the intensity of a rain storm would be a major factor.

16.2.   C.A. KOGBE'S PAPER

16.2.1. *R.W. Fairbridge:* What percentage of Oubangi River discharge would be needed?

Open channels are not favoured because of blockage to wildlife migration, farmers or their stocks; and because of pollution.

16.2.2. *panel discussion*

E. Custodio: Big water management schemes are very difficult to carry out. Political, social, and environmental problems are very difficult to overcome and deep sentiments of people are difficult to change. Small schemes are always preferable and much more possible. Spain, the most arid country in Europe has a lot of experience to be taken into account. Plans made 20 years ago have not been possible to realise. In two years time in the Canaries an IAH meeting on groundwater over-exploitation that will be a good opportunity to discuss this in depth.

N. Petit-Maire: In addition to Custodio's comment: Yes, underground water is international. Beware!
   The Libyan project for an artificially induced river through aquifer pumping worries Egyptian and Sudanese hydrogeologists: the western oasis in Egypt would go dry and id. in NW Sudan.

K. Howard: [Comment re: Recommendation No. 2 - Artificial Recharge Scheme/Pilot Project. (follow up to Issar)] While I support the proposed recommendation, we must not lose sight of the fact that of the order of 60,000 km^3 of water - fossil water - are currently available under the North African continent. This water is surely the first choice in our efforts to mitigate drought. Artificial recharge is an important, supplementary approach to the problem but not the *only* approach.

M. De Boodt: Remark on statement No. 1. I propose to add: "at the time being" or "for the time being" or "at least for the next decade(s)"

N. Petit-Maire: The UNESCO/IUGS Project IGCP 252 is on "Past and Future Evolution of Deserts". It has at present 350 members from 46 countries. About a third of them are very active. One could extend this existing international structure into a Research or Applied Research Center at:
1.   the European level
2.   the international level
Proposal for Recommendation: creation of an International Center for Arid Areas Studies. (Multidisciplinary research)

F.E. Camfield: Would building surface and subsurface dams cause any problem by preventing water from getting to families/villages at downstream locations?

R. Guiraud: The risk of drying wells providing groundwater for villages located downstream

from small underground dams is really minor, for three reasons:
- the density of the population is generally very low and the villages are distant, 5 to 15 km from each other;
- there are many tributaries reaching the river between two villages;
- the dams are scarcely ever impervious.

But, of course, one must be careful in the desert, for example when there is an oasis downstream of the planned dam.

N. Rutter: Before an International Recent Research Institute or coordinating body is established, it is necessary to survey what existing institutes or organizations are already in existence which could advise on the best way to establish an international organization and what group would be the best to organize and operate it.

R. Guiraud: This concerns the "Pilot Project for Artificial Recharge in NE Nigeria":
1. I think there is not enough water in the Komadougou River, which has been quite dry in its lowest part during recent years.
2   I agree with Prof. Issar: drilling a deep artesian aquifer would be cheaper than bringing water with very long pipes and/or canals.

R. Klige: Different estimates of sea level change due to human management of water stocks result from difference methodologies.

Nowadays a steadily growing water volume is impounded in artificial reservoirs. There is an increasing total volume of world reservoirs (changes approximate to 100-120 km^3 per year) and an increasing minimal volume (dead) world reservoirs (changes approximate to 70-75 km^3 per year). We must take for estimates of sea level change due to human management of water stocks only the minimal increased volume (dead) world reservoir (70-75 km^3). This is an equivalent to -0.2 mm per year sea level change.

In general ocean volume increases as a result of climate warming. Continents become dryer and glaciers decrease. This process is intensified by thermal expansion of the ocean waters (0.3-0.4 mm/year).

Geological processes changing the Earth's surface exert an influence on natural processes controlling water. Geological studies demonstrate that topographical contrasts coincide with hydrosphere volume increases. This conclusion can be well traced by the analysis of ocean level fluctuations during the last 300,000-500,000 years. Its fall rate is equal to 0.3 mm per year.

A. Issar: I suggest the consideration of an International University specializing in arid-zones *research and development*, namely *applied research* and *teaching*.

I also suggest that before considering recharge of the Chad Basin we should investigate the possibility of large scale pumpage on the artesian aquifer.

R.W. Fairbridge: The Fairbridge/Newman data (published in *Nature*, 1986) was based on the latest accumulated statistics of major dams (worldwide), which represents an annual elevation of the annual holding of the millions of small agricultural dams, recharge, swimming pools, etc. I have no knowledge of the origin of the Gornitz-Lebedef figure (but I will ask them).

International Arid Area Research and Development Board would be a better name for a hydrologically oriented research organization. I am not sure if it should have a fixed centre, secretariat, etc., etc., *or* whether it should be like a dependency of I.C.S.U. that simply meets at regular intervals.

J.M. Sharp: The use of small-scale artificial recharge projects, especially in arid and semi-arid zones, should be encouraged.

F.E. Camfield: [Response to Roebert] It is not correct to say that small changes in sea level are not significant. The retreat (or advance) of shoreline as a ratio of sea level change may be 100 to 1. Billions of dollars worth of property may be affected by what appears to be a relatively small change in sea level. A one metre rise in sea level could be catastrophic. (Also a one metre drop could make many ports in-operative).

M. De Boodt: When speaking about establishing an international Desert Research Institute, we think this is already too much and perhaps not realistic! On the other hand, there are international organizations which are willing to call together at regular intervals meetings or even to establish a permanent secretariat to exchange result. There are many institutes now already working on "Research and Development" in arid and semi-arid zones! They like to share and exchange ideas.

Y. Tardy: Recommendation: evaluate the volume of dry sediments in which we are able to store water on continents away from evaporation. If we store water in lakes in arid zones water will go back to the ocean by evaporation.

A. Issar: We can say that most of the differences in the percentage of recharge is due to differences in the regime of the storms. Thus, for instance, in sand dunes all water from storms below a threshold of 3 mm will be lost to recharge (the salts will remain and will be added by a flushing storm to groundwater).

17. Comments on Session VII

17.1.   J.M. SHARP'S PAPER

17.1.1. *G. Wells:* How does one do effective risk assessment for low frequency but high magnitude meteorological events superposed upon coastal subsidence and CO_2-created sea level rise?

17.1.2. *V. Goldsmith:* The phenomenon of subsidence/sea level rise (SLR) has been well

known for a long time at Houston.
a.   What plans are being made to deal with this?
b.   If Houston and Venice van serve as "role models", since they have the most extreme SLR, how can we scientists, who are concerned with possible rises, best communicate to decision makers?

Sharp:
   a.   Very few. Some coastal wetlands have been made into wildlife reserves and the Houston-Galveston Subsidence District is working to reduce subsidence from groundwater pumpage.
   b.   A very good question. The answer is, I think, continued publicity and redoubled efforts to communicate the risks.

17.2.   G. PATFOORT'S PAPER

17.2.1. *E. Romijn:*
   1.   Could you give a cost estimate of the new mineral polymers?
   2.   What kind of catalysts are used for polymerisation?

Patfoort: Mineral polymers are a group of new materials wit very different properties and different formulations. So we have to compare materials that exist in organic polymers. So it is impossible to give any answer, without specification of each material, on the price of catalysts.

For example: lateric soil has been polymerised with the use of a catalyst made from waste of plants without economic value. The reaction can take place at low temperature with solar energy. No other material or energy is necessary but in this case the material is not of high quality.

17.3.   C. BAETEMAN'S PAPER

17.3.1. *S. Jelgersma:* I have tried to indicate the serious effect of man-induced extraction of groundwater in deltaic areas. Man-induced ground surface lowering due to over-exploitation of groundwater is emphasised to the audience.
The City of Bangkok experience at the moment a surface lowering of 12 cm per year. In Taipeh (Taiwan) a surface lowering of 2 metres has occurred during the past 20 years due to groundwater exploitation. This man-induced land subsidence is a more serious feature than the future sea level rise in those areas.

17.3.2. *P. Pirazzoli:* Are the aquifers exploited in Shanghai made of fossil water? This could increase subsidence rates.
Baeteman: Yes, this actually is the main reason of the land subsidence. These subsurface deposits were never dewatered in a natural way.

17.4.   PANEL DISCUSSION

G. Wells: [Discussion about the impact upon decision makers (re.: Environmental Atlas

Series of the Texas Coastal Zone)] These coastal zone environmental surveys have had mixed success. For popular initiatives, such as the preservation of coastal wetlands important to regional waterfowl, the reports have been of substantial assistance to the maintenance of these areas through public policy. However, for the regulation of development of the barrier islands, economic incentives for rapid development exceed the long-term benefits of preservation.

E. Romijn: [to G. Wells] Are satellite measurements of any help, now or in the future (radar or laser) for eustatic sea level rise measurement?

R. W. Fairbridge: Concerning episodicity of coastal disasters mentioned by Wells, I submit a list of coastal floods and dike breaks over the last 1000 years, based in part on work of Wood (book "Tidal Dynamics"). All breaks, except military man-made ones, correspond to Perigee-Syzysy Spring tides (often plus (Perihelion influence). Thus they are astronomically predictable. Warnings can be issued, and appropriate preparations can be made.

One questioner asked if there was clustering on this list? Answer: yes, the concentration in the late 1900's matches that of the 14th century. Both are peaks of long-term tide cycles.

P. Pirazzoli: In order to improve our knowledge of present sea level movements and of their various components (eustasy, compactions, subsidence (natural and/or man-induced), isostasy, active tectonics, ocean dynamics, etc.), I suggest that a recommendation is prepared encouraging new local relative sea level studies, not only with tide-gauges but also with geological investigations and levellings, at different time scales adapted to the components to be identified. This would facilitate the identification of the areas which are at risk, or which would be at risk according to various scenarios of possible sea level rise, and the type of risk.

R. W. Fairbridge:
1. The International Association of Oceanography has a MSL commission, but we should recognize and publicize the need for isolated island stations both for MSL and weather.
2. Pirazzoli recommended that Pleistocene and Holocene terraces should everywhere be measured in relation to tide gauge trends in order to evaluate submergence or emergence. Also we should try to remove interannual climate effects (El Nino, etc.).

S. Jelgersma: Tooley asks me to tell Fairbridge that the following floods in the North Sea Basin should be added: 1792, 1916, 1977 and 1983.

A. Issar: It is recommended that the next meeting should include also economists. Especially when it comes to reconsidering the substitution of fossil fuel by nuclear power.

S. Jelgersma: The impact of a future sea level rise on the coastal lowlands of Europe was discussed during the European Workshop on Bioclimatic and Interrelated Land Use Changes.

C. Baeteman: [Comment on Faure's paper] Some proposals for the protection of the global environment rather contradict the general goal, e.g. the submergence of salt marshes and mangrove areas in order to reduce CO_2 emission, is just inviting the sea to invade the land. Salt marshes and mangroves are known to be a "buffer-zone" protecting the back-land from flooding by the sea. Furthermore, if the existing salt marshes are submerged they are going to develop again at the newly established high water line. The result would be loss of land and a greater risk of sea flooding.

With regard to subsidence and coastal retreat, the least harmful method is beach nourishment (a relatively cheap method). It is only valid for a short time. However, it gives time to investigate long-lasting engineering solutions (it is known that most engineering solutions to keep out the sea induce more destruction than protection). One should consider areas that have to be given up (back to the sea) in order to save and protect other areas (industrial and densely populated).

B. Denness: The imminent submergence of the Maldives has been mentioned earlier: an excellent challenge to the planning of land-use in a coastal environment. Failure to solve it will result in the demise of a nation, or at least its homelands and it should not be expected to face that fate alone.

However, man must be realistic. Economic factors cannot be ignored on the scale that will be involved in the event that sea level rises worldwide (for greenhouse or natural reasons, or both). Sacrifices will have to be made; consciences will have to be addressed. Compensation planning should be commenced now in parallel with barricade planning around areas of heavier economic investment in nations with more muscle.

The world can no longer accommodate easily the traditional reaction of man to sea level rise, i.e. migration. Now we have national boundaries and sovereign states that did not exist at the time of the migrations that must have taken place when sea level rose about 100 metres some 17,000-12,000 years BP at the end of the last ice age.

One thinks, for example, of the many seamounts in Polynesia that are now within 100 metres of lea level. Thor Heyerdahl demonstrated that South American indian could have travelled by raft from Chile to Easter Island (westwards) and on to Polynesia on a raft like the Kontiki. Had he continued he would have demonstrated that Polynesians could equally well have travelled to South America on the return part of the South Pacific gyre (eastwards). That leg of the journey makes more sense: Polynesians had more reason to make the trip - their homelands were drowning. Surely it is no coincidence that the earliest settlements in South America are dated at about 17,000 B.P. Should this hypothesis be correct it must be acknowledged that it is a tribute to the power of alcohol and a consequent perception shared with my friend Tomás Shuk in Colombia.

N. Rutter: How successful has the "Texas Environmental Atlas" been? Are politicians using them? Do they need help with interpretation? Were the efforts worth the money?

G. Patfoort: It has been emphasized that the resolutions of this workshop should be transmitted with its most important findings to political important authorities, people with decision-making power and to the media. Our resolutions are very many and important but they would lose their impact in detail. It is necessary to present, for non-scientific people, a global dynamic system analysis as a synthesis of the details. All the different points should be joined together in a synergetic system as was done 20 years ago, leading to a fantastic impact on the Club of Rome in 1972. Mistakes after that time could be corrected with modern data and computer analysis.

V. Goldsmith: Much of the coastal zone retreat (e.g. Houston) can be alleviated (not solved, just decreased) by better sand management; for example, dredged sand from inlets placed back on beach. In general better management of our coastal resources is something we can do today without further studies.

18. Comments on Session VIII

18.1.   C. DE MEYER'S PAPER

18.1.1. *C. Kogbe:* If it is realised that what is needed is a supplementary water supply to supplement the discharge of the Chavi River into Lake Chad during drought years, the cost of the Water Transfer Project from Zaire to the Chad basin will be considerably reduced. More field data is therefore required before more realistic costing can be feasible.

18.2.   C. OLLIER'S PAPER

18.2.1. *A. Issar:* You spoke about the problem of recharge flow estimated to take a few million years. Did you consider the velocity under a pressure head which is much quicker than the Dárcy type of flow?
Ollier: No. The estimate is by Habermehl (1980).

18.2.2. *R.W. Fairbridge:* Speaking as a geological engineering consultant to the Snowy Mountains Hydroelectric Scheme some 40 years ago, I recall the discussions at that time about the greening of the irrigated semi-arid interior. Ollier's warning about salinization should be seriously considered in the various Lake Chad schemes.
Ollier: I quite agree. In all irrigation schemes it is necessary not only to add water but to remove it to prevent water table rise. The second factor may be the most expensive but is usually left out of cost estimates.

18.3.   P. KAPLIN'S PAPER

18.3.1. *C. Baeteman:* How do you explain the fact that the "new" beaches remain there and are not eroded again?
Kaplin: Because we place material, the quantity and particle size of which correspond to natural conditions (wave climate).

18.3.2. *V. Goldsmith:* What was the depth of placing course sediment on the sea floor and what was the significant wave height?
Kaplin: Depth about 4 m; waves 2-3 m.

18.3.3. *G. Wells:* According to your data, the Caspian Sea level has risen 1.2 m in the last decade. Beginning in 1981, observations from the NASA Space Shuttle and Landsat satellites show no water communicated from the Caspian into the Kara Bogaz Gol, which serves as a shallow evaporation basin for Caspian overflow. Has a artificial barrier been placed between the Caspian and the Kara Bogaz to halt the decline in sea level caused by reservoirs and irrigation diversions along the Volga?
Kaplin: Yes, a dam was built but without scientific rationale. It was a mistake and now it will be destroyed.

18.4.   PANEL DISCUSSION

E. Custodio: Large-scale surface water schemes compared to groundwater development in local aquifers is a classical issue, most relevant to developing areas and countries. Large-scale projects exhibit very difficult financial problems, long delays for implementation, presently acute political and social problems and long periods of under-use, adding to the present debt of these areas. If there are aquifers with large enough usable resources, even if deep drawdowns are needed, use of their reserves is generally advisable for development to cope with persistent drought, adapting development to reality. Investments in groundwater are more gradual and users want to know how to use water efficiently and economically. Good examples can be found of economical failures related to expensive schemes by comparison with successful projects using groundwater and simple local water resources, in which owners profitably use apparently much more expensive water, whilst others use grant-aided water, and yet lose money.

C.A. Kogbe: It is obvious from discussions at this workshop that there is a growing gap between theoretical knowledge and basic field knowledge. It can be misleading to depend on the results of studies which are not based on recent and reliable field data. The objective of the workshop is to provide expert scientific and technical advice on some current global problems. It is the politicians who will have to take the financial decisions

H.N. Le Houerou: Among solutions one should not forget "Population Control Policy" because in many arid countries this is the overriding factor for drought mitigation.

A. Issar: I think that plans to divert rivers from the Congo Basin and Niger Basin to Chad are not practical. We have heard that a cubic metre of water will cost about 0.7 $ U.S. As it seems unlikely that local agriculture will pay more than 0.1 $, who is going to pay back about half a million dollars per year?

B. Denness: Formal contributions under the heading of Engineering Solutions have been concerned with proposals for activities to mitigate climate change *after* it has occurred. We might instead (or as well) address the possibility of circumventing at last the greenhouse component of potential climate change.

The point has already been raised that emphasis on nuclear power in favour of fossil fuel consumption could help avert greenhouse problems: I would prefer to consider the full range of non-fossil fuels instead of nuclear alone. Among other options are some already available and completely tested. We need not think only of hydro-electric power, biomass conversion, wind energy and so on - all of which exhibit various temporal or geographic limitations: there is also endothermic energy, requiring only a heat pump to convert atmospheric (or other) heat into useful energy for internal space heating and hot water, i.e. the main domestic and commercial requirement in the energy-wasteful temperate zones. This technology is in effect an inside-out refrigerator.

A coefficient of performance (the ratio of energy recovered to energy input) of more than 4 has already been achieved in long-term prototypes of whole buildings on the Isle of Wight in the UK by Alan Ridett. Converted into the large scale this could be sufficient. For example. to provide (via the consequent saving in input electricity and hence reduction in current fossil fuel burning for energy generation all the reduction of CO_2 emissions stipulated for the UK by the EEC up to the year 2005 - applied only to domestic housing at no significant cost beyond normal installations in new housing or repair of older buildings and by applying no restraints at all on industry. An additional benefit arises with respect to the greenhouse inasmuch as endothermic buildings act directly on the atmosphere and the greenhouse by *reducing* air temperature. They also operate in the dark (at night) and even at sub-zero temperatures - additional advantages, in temperate zones, beyond purely solar systems.

One might also think of applying this internationally patented technology to desalinisation plants. In that case it should be possible to collect the energy required to drive the heat pumps from solar power: after all, the problem of drought in populated areas is generally closely associated with interminable sunshine. The cost of desalinated water should drop considerably. The designer of the original space and water heating endothermic system (Alan Ridett) would doubtless advise on such an application if required.

An alternative means of achieving increased energy efficiency is through better insulation to prevent energy loss. Unfortunately this normally also prevents moisture dissipation and causes condensation, thereby discouraging its use. However, this problem has been solved by Professor Alex Hardy of Newcastle University, UK who has invented a cheap, non-energy-consuming dehumidifier.

R.W. Fairbridge: Many coastal investments, e.g. for a motel in a tourist area, are built on a very short-term return basis. If the structure is destroyed after 10 or 15 years, the original investment is long-since paid off.

A. Issar: I suggest that we recommend a standard measure of the cost of water, referred always to the cost per m^3.

C.A. Kogbe: It is not possible to be precise in costing different types of projects. We could classify projects into 2 categories: relatively cheap - small scale projects, small dams; and capital-intensive projects - water export and big dams.

A.J. Roebert: Guiraud has referred to boreholes; are they the same sort of boreholes that are intended in the Chad-area?

R. Guiraud: No: in Chad there are continuous aquifers at larger depths, e.g. Cap Verde, where the capacity of wells is 3 m^3/hr mainly in fissures at 50-60 m depth.

I should also like to give a brief historical summary of the rural groundwater supply in West and Central Africa. In 1973, in order to fight drought, we proposed to drill boreholes in the fault zones of the basement and of the old sedimentary basins. We dug 6 wells in 1973, about 40 in 1974, and about 10,000 since then!

As for the underground dams, this technique was recommended by the International Hydrologic Program of Unesco at the end of the seventies. Since 1979 only a few dams have been built, except in the Cape Verde Archipelago, although several hundred could be built in the Sahel each year. Recently, there has been an acceleration of damming, and a strong demand from the local population. I suggest that we encourage that tendency. *We must advertise small dams.*

Some samples of costs:
- for sand dams in N Cameroon (Mandara Mountains) the average cost is about 1000 US $ for wall cemented, but one can easily find good sites there, there is a cement factory not far away and the local population works for no salary;
- basic sand dams in SW Chad, built with gabions, cost about 20,000 US $; when complete and furnished with a well or a fountain the cost will be about 100,000 US $, but this installation will probably provide drinking water for a big village (_ 3,000 inhabitants);
- subsurface dams in Southern Air (N.Niger), using boreholes and cement injection, cost about 200,000 US $ for an expected yield of 50 m^3/hr.

19. Comments on L.V. Da Cunha's presentation of the Scientific Affairs Division of NATO

M. de Boodt: The input of Nato to the CCMS programs is not very clear to me. Is it correct that, although Nato is not putting in funds, it may "member" countries. How does this work in practice?

Da Cunha: The representatives of the different countries decide together if proposed studies are of general interest or not. If the interest is general the decision is taken to run a pilot study. Development of the study is open to participation of all countries wishing to do so and results of the studies are accessible to all.

A. Issar: Can non-NATO members participate in NATO research programs?

Da Cunha: Yes, in Advance Research Workshops and in Advanced Study Institutes but not in fellowships or in collaborative research grants.

A.J. Roebert: How are fellowships for example divided between the several NATO-countries?
Da Cunha: Originally there was a rigid proportional system. Now there is a preference for the less developed countries.

MEETING REPORT

RHODES W. FAIRBRIDGE
Columbia University; and
NASA-GISS, 2880 Broadway
New York, NY 10025

1. What to Do About Rising Sea Level?

In a recent U.S. National Academy of Sciences report (ed. Dean, 1986), it emerged that basically, for a rising sea-level scenario, there were three options:
(a) Retreat (the "Pilkey solution"); voluntary action would be reinforced by legislative incentives and prohibitions;
(b) Barrage and dike-building (the "Dutch solution"); this is very costly, but inevitable where national need is established (e.g., Venice, Amsterdam, London);
(c) Geohydrologic Management (proposed by Newman and Fairbridge, 1986; and widely noted in the press and TV, e.g. *N.Y. Times*, April 11, 1986; *The Economist*, April 12, 1986); management involves controlled interruption of the river discharge from land areas to the ocean. This last question has recently been studied by a NATO-ARW (March 1-7, 1989), and "Advanced Research Workshop" held at Fuerteventura, Spain, which will be reported here.

Rising sea level is certainly among the great life-threatening hazards facing mankind. More that 100,000 people have been drowned and many billions of dollars in material losses have been suffered in this category during the last decade. During the same decade a second, but far greater life-threatening phenomenon has killed several million people and condemned perhaps another ten million or more children to permanent mental retardation due to famine-induced dietary deficiency.

It occurred to the writer (jointly with the late Prof. Walter Newman) that these two problems could be addressed simultaneously, by the disarmingly simple solution of transferring water from the rainfall-excess regions of the world to subsurface storage in the semi-arid regions where they could be tapped during periods of drought (Newman & Fairbridge, 1986). Improved river flood control would also be provided. A tentative, exploratory proposal was sent to the National Science Foundation (Washington, D.C.) and it was successful. One persistent problem is this: to whom does one submit a highly interdisciplinary proposal

691

R. Paepe et al. (eds.), Greenhouse Effect, Sea Level and Drought, 691–706.

of this sort? Is it Earth Science? Oceanography? Meteorology? It is not really any of the above. It was actually funded by the "Critical Engineering Systems" program, and a global study was on its way. To our collaborating scientists it did indeed seem to be feasible.

A key follow-up was to organize an international conference or workshop to review the various options, environmental problems and social aspects. Accordingly an executive committee was selected and a proposal submitted for a NATO-Advanced Study Workshop. Funds were granted and the NSF of both Belgium and the U.S. provided generous supplementary support.

The site selected for the meeting was on Fuerteventura, in the Canary Islands, an integral province of Spain, which is a member of NATO and a newly elected partner in the EEC. Fuerteventura was chosen because it is semi-arid (less than 100 mm, or 4 inches of rain annually) and is the site of an energetic program for reactivated (traditional) water conservation, small dam construction and desalination plants that support a dramatically developing tourist industry.

The writer was invited to be President of the meeting. The rest of the Executive Committee were chosen from Belgium, Holland and France because these are the countries most directly interested in both the sea-level question and in the drought problem (which is especially critical in the French-speaking Sahel belt of Africa). The Executive Chairman was Roland Paepe (of the Belgian Geological Survey and the director of the Earth Technology Institute of the Free University of Brussels). His committee consisted of: Saskia Jelgersma (Geological Survey of the Netherlands, and President of the INQUA Shorelines Commission), Hugues Faure (past-President of INQUA, University of Marseille-Luminy), and Christian de Meyer (director of Haecon S.A., Harbour & Engineering Consultants, Gent, Belgium).

Another 45 delegates attended, including Spanish representatives who helped immeasurably with local arrangements, press and TV coverage, and field trips. Delegates came mostly from NATO countries, including the U.S. and Canada, but several by special invitation cam from Australia, Sweden, China and USSR (this last a most agreeable indicator of the new "glasnost" policy and much appreciated by all). The International Association of Hydrogeologists was represented by its secretary, Dr. E. Romijn; the International Union for Quaternary Research was represented by its President, Dr. Nat Rutter, and by the Chairman of the Chinese Committee for its next meeting Dr. Liu Tungshen; and by members or chairmen of its various commissions - shorelines, paleohydrology, paleoclimates, Holocene, neotectonics. The ICSU-IUGS International Geological Correlation Projects 200 (sea-level chronology) and 252 (deserts) were also represented. The interest of NATO's Science Division was shown by the presence of its director, Dr. Luis Da Cunha who has recently edited an important book on "Drought".

The program for discussions was planned to focus primarily on the two main areas: sea level rise (measurement and prediction) and drought (prediction and engineering solutions). Geographically, sea level was seen of course as a global problem, whereas the drought questions were oriented mainly towards the African sector. Inevitably there were also constant references to the problem of climate change and the role of the greenhouse gases (CO_2, CH_4, etc.). Noted with great satisfaction was an announcement in the press during the meeting that the European Community has just agreed to a total ban on CFC's.

1.1. THE GREENHOUSE TRIGGER

It is probably the greenhouse effect that has played the most important role in bringing to world attention the sea-level problem. While, over the last 45 years, the writer has devoted much of his professional life to the latter question, it is only during the last few years that it has come into the spotlight. Prior to that, most research funding proposals were summarily rejected as "lacking intrinsic scientific merit".

At the meeting, problems concerning the greenhouse effect, atmospheric ozone and global hydrology were critically reviewed in light of complex climatic feedback mechanisms by Michael Rampino (in a paper by Rampino, Volk & Hoffert). As a physical phenomenon, there is no question whatever that the greenhouse effect is real, a "fact of life", and that the progressive rise of CO_2 (and the other gases involved) in the atmosphere is accurately recorded by independent observers. The role of CO_2 in the terrestrial climate has been known since 1896 when it was enunciated by the distinguished Swedish scientist Svante Arrhenius. And there is no doubt that mankind is upsetting the natural cycle of carbon and CO_2 by the accelerating destruction of the earth's forests (thus decreasing nature's way of reducing CO_2), and the accelerating combustion of CO_2 generating fossil fuels (coal, oil and natural gas). The key U.S. publications on the overall problems concerned are edited by Nierenberg et al. (for the U.S. National Academy of Sciences, 1983), by McCracken & Luther (for U.S. Dept. of Energy, 1985), and by Titus (for the U.S. Environmental Protection Agency, 1986).

Faced by the above-mentioned incontrovertible facts, it required only a small "leap of faith" for the scientific community to point to two there incontrovertible facts which are commonly taken to be logical corollaries to the greenhouse data. These facts are (a) that a statistical average of tide-gauge data from around the world has indicated that sea level was rising, (at least during the first part of the 20th century; Fairbridge & Krebs, 1962; Gornitz et al., 1982, 1986; Barnett, 1984; and others). And (b) that an appreciable number of mid-latitude glaciers were in retreat, i.e., melting, during the same period (Thomarinsson, 1940; Ahlman, 1953; Meier, 1984). The writer long ago observed that in tropical seas, the tips of corals which do not grow above low-tide level, showed a recent upward growth (Fairbridge, 1947).

That "leap of faith", mentioned above, was made by many in the scientific community and some alarming projections were forecast for the 21st century, with global sea level in one scenario rising 3.5 m (nearly 12 feet) above its present level. Several federal and private agencies in Washington have laid out many millions of dollars in symposia and investigations; as to the publications, their actual weight it is said runs to some 100 kg. A popular volume (edited by Barth & Titus, 1984) summarizes the key points. Some of the NATO meeting's papers and discussions, reported below, may serve to disclose that serious gaps still exist between substantive knowledge of these "leaps of faith".

The nature of all the feedback mechanisms in the atmosphere-hydrosphere of planet Earth is still imperfectly known, but from studies of paleontology, sedimentology and geochemistry it seems likely that the mean temperature of the planet has generally remained at about $18+/-3°C$ for the best part of 4 billion years. Otherwise, many organisms would have gone extinct and it is a fact that no major phylum in this long history has even gone extinct; this is called "the law of Biological Continuity", a key axiom in the fundamental laws of the planet Earth (Fairbridge, 1980).

It has been suggested that the planet tends to fluctuate from a "greenhouse state" to an "icehouse state". Although perhaps a little overgeneralized, this is a useful concept. The key to the maintenance of the long-term equilibrium is feedback, which exists in many forms.

Modelling experiments have been run by a number of specialists to test the contemporary CO_2 rose. The GISS/GCM results, with various feedbacks, suggests a mean mid-latitude warming of 0.6 to 1.5°C over the next 40 years (Hansen et al., 1987, 1988).

For a doubling of atmospheric CO_2 a global rise of 5°C might be possible; this a mid-latitude mean value, where the equatorial rise might be minimal in contrast to a very large warming in polar latitudes, which could be likely to affect ice melting. The latter, however, is a complex problem. High latitude glacier ice is mostly at stable temperatures, far below a potential melting point; Antarctic ice averages about -17°C, so that even a large atmospheric warm-up would have little immediate effect. Secondly, there is the so-called "Simpson Effect" (see Fairbridge, 1961, p. 116) which is based on the observation that a warming of surface seawater causes increased evaporation and the initial result (for some decades) is likely to be an increase of snowfall, thus building up the glacier, although later on there will be rapid melting.

Two feedback mechanisms tend to amplify a warming trend. The warmer global climate today leads to an equator-pole thermal gradient that is flatter than in a cold cycle (Fig. 1, from Fairbridge, 1964) and accordingly the atmospheric pressure gradients will be reduced, so that upwelling rates fall. As a result there is less nutrient delivered to the oceanic phyto-plankton, the principal marine sink of CO_2. The present flux is about 30 Gt (i.e., billion tons), which is about half the total CO_2 taken into the ocean by organisms, the rest being recycled. This is what Roger Revelle (of La Jolla) has called the "biological pump" mechanism. The warmer water also retains less CO_2, so the combined effect is net flux to the atmosphere, amplifying any existing rise. In the same sense, a high-latitude ocean warming will decrease sea ice coverage, resulting in a dramatic fall of albedo so that more solar radiation is absorbed by the ocean, warming it further. On land, however, increasing high-latitude vegetation and growth of forests can help reduce CO_2. In the intertidal zone and swamplands, in contrast, a rise in methane (CH_4) production will have the opposite effect.

It should be appreciated that there has been a warming trend of the climate since the Little Ice Age. A similar rise occurred in the Viking-Norman time about 1,000 years ago. Interestingly, our distinguished friend Lorius and his team in Grenoble have discovered, in air bubbles in an ice core collected in Antarctica by a Russian drilling group, that the CO_2 level in the 11th century rose at about the same rate as it did in the 20th century. Although there is undoubtedly a major input today of fossil-fuel-related CO_2, we do not know the rate of marine metabolic withdrawal. Much of the temperature rise in the 20th century may be related thus to post-Little Ice Age warming.

Decreases in marine productivity may be predicted not only from warming surface temperatures, but also from declining ozone (through increasing UV penetration) as well as from declining dimethyl sulphide (DMS, produced by marine phytoplankton). The DMS supplies cloud condensation nuclei in the marine atmosphere, and this could play a positive feedback role in the potential aridification of mid-latitudes.

The current outlook is not all one of total confusion. An international program, JGOFS (Joint Global Ocean Flux Study) plans to try and unravel some of the marine "biological pump" problems. In higher latitudes the physical oceanographic questions become more important, including for example the rates at which sea-ice buildup causes cold, dense water to trigger the circulation of deep, bottom-water; this is included in the WOCE (World Ocean Circulation Experiment).

Yet another group, IGBP (International Geosphere-Biosphere Program) hopes to bring the various modellers, measurers and theoreticians together in a decade-long series of collective analyses of global change. Hugues Faure, reporting to our conference, is chairman of a special committee of INQUA (represented at the meeting by its President, Nat Rutter) to endeavour to bring about some input from the "Quaternarists", who mainly consist of geologists, botanists, paleontologists and glaciologists. Up to the present time, however, there has been a deplorable lack of interdisciplinary understanding. In a recent report on the British national committee for IGBP at the Royal Society of London, Len Morgan remarked: "As you know, it is all highly uncoordinated", (New Scientist, p. 34).

As remarked above, it was not the intention of our meeting to spend much time on the greenhouse debate. The subject was only drawn in because it has played and continues to play a major role in triggering global studies of sea level and also in refocusing ideas on drought.

Notwithstanding widespread differences of opinion as to mechanisms, there seems to be a total consensus in the scientific community that for health reasons alone, and for the health of our trees and our general environment, urgent remedial action is needed in regard to man-made pollutants.

An expert introduction to the sea level question was provided by Saskia Jelgersma who is exceptionally well qualified to speak on it in her double role as a president of the INQUA Shorelines Commission and from a lifelong study of Dutch coastal stratigraphy for the Netherlands Geological Survey. It was followed by lively discussions. With nearly 60% of that country below sea level, it is significant the Dutch spend 6% of their GNP every year on coastal engineering works, which incidentally is larger in proportion than the U.S. spends on its military budget (N.Y. Times, 1986; Feb. 18).

There is a very clear consensus that, as of this date (1989), in most parts of the world there is so much geological structural deformation (neotectonics), and other forms of subsidence (due to compaction, natural and man-made de-watering), that other factors in sea-level change tend to be over-shadowed. Those factors include volumetric expansion ("steric change") of the seawater during the first part of this century, in parallel to a small global warming (Gornitz et al., 1982, 1987; Wigley and Raper, 1987). It was also matched by a retreat of many (but nor all, by any means) of the small mid-latitude glaciers, and thus arguably reflected by a eustatic component (Meier, 1984). However, there is opposing evidence from a mass balance study of Greenland snowfall, which shows an accumulating trend, and may therefore have a contrary eustatic effect.

1.2. KEY POINT ABOUT MODERN SEA LEVEL

(a) The almost worldwide reports of coastal erosion, gathered and documented by Eric Bird (1985) and his coastal commission of the International Geographical Union, can be in part explained in terms of lithology and inherited beach materials. The erosion os in soft, unconsolidated sedimentary materials that are in general characteristic of geologically subsiding and quasi-stable prograding shores. It may be remarked that hard rock coasts are for the most part emerging, i.e. in regions of geological uplift. Because of long-term geological processes (sediment loading, isostasy) the sedimentary coasts in general have been subsiding on and off for 90% of geologic time. Alternating erosion and progradation are thus the "norm" for some 4 billion years.

(b) The world's tide-gauges are customarily set in place at the request of commercial shipping interests, and because sea ports are, for economic reasons, usually located at the mouths of rivers, on estuaries and deltas that are in geologically subsiding parts of the earth's crust, local sea level almost always seems to be rising. All speakers emphasized that this is a RELATIVE rise. Accordingly, there is no generally accepted statistical analysis of world mean sea level that can be accepted today as indicative of any suspected greenhouse warming. At the meeting, two speakers presented detailed analyses of tide-gauge data to demonstrate that neotectonic up's and down's of the earth's crust have (so far) conspired to mask any secular trends of climatic origin. Paolo Pirazzoli, former leader of the now-completed IGCP-200, gave a talk that served to update and reinforce the work that has already been summarized in the *Journal of Coastal Research* (special paper, 1986).

Victor Goldsmith, speaking on the same theme, but specifically detailing the heavily populated region of the Mediterranean and Middle East, brought out the diversified tectonic character of the area (Aubrey & Goldsmith, 1988). He developed another very important point: intra-annual variances often relate to wind systems that are liable to be out-of-phase from one sector to another. By the same token, we may postulate that secular climate change may lead to change in prevailing wind patterns, and thus to a *perceived* sea-level change.

(c) Certain long-established tide-gauge records are available from glacioisostatically rising crustal areas, as in Scandinavia. Inasmuch as the rates of crustal uplift there has been very precisely measured by repeated geodetic surveys, the residual values reflect mainly the steric and eustatic components. The President drew attention to recent reports, that refer to the Stockholm gauge which covers more than 200 years (Ekman, 1988), and to the Hanko gauge (Finland) with 100 years (Mälkki, 1987). Both suggest very small, secular warming effects in one since the 18th century, but neither shows an acceleration of sea-level rise in the last three decades.

(d) Paolo Pirazzoli's global study of sea-level trends (that emerged from his work with IGBP-200) disclosed that only the central Pacific showed "00" values. The president pointed out that this region is far removed from plate boundaries, volcanism, glacioisostatic disturbance, sedimentary loading, and the major geostrophic currents. He mentioned specifically the example of Truk atoll where a half century of measurements disclosed the expected fluctuations due to El Niño effects, but there was no linear trend, up or down.

(e) Certain regions, notably those adjacent to the great geostrophic currents like the Gulf Stream or the Kuro Shio, show anomalously high rates of local sea-level rise that are not compatible with the local rates of long-term (slow) crustal subsidence. The President (Fairbridge, 1989) mentioned the fact that there is a well-known annual variation in the dynamic slope of the sea surface of such currents, relating to the Coriolis Effect. When the current slows down (in summer), the dynamic tilt decreases, so that at Miami, for example, sea level rises while in the Bahamas it falls. This pattern reverses in winter. In view of the global warming since the Little Ice Age, we would expect the mean velocity of the great geostrophic currents to decrease slightly. Thus on the left-hand side of the gyres, i.e. the mainland coasts, both on the western and eastern sides of the oceans, and in both hemispheres, the tide-gauges should all disclose anomalously high rates of sea level rise. Significantly, perhaps, the Baltic gauges (e.g., Stockholm and Hanko) and the mid-Pacific atolls are little affected by these geostrophic-Coriolis influences. Since the Little Ice Age, over the course of about 300 yr, the Baltic has warmed up by about 2°C, but the mid-Pacific, being equatorial, has probably experienced no secular warming.

(f) The President remarked that for nearly 50 years the INQUA Shorelines Commission and its many regional subcommissions have been systematically mapping the geology of the world's coastal formations and geomorphic features, dating mostly in the time-frames of 0-6000 yr BP (Holocene) and around 100,000-130,000 yr BP (Last Interglacial). Where these formations are horizontal and at uniform heights above sea level over many hundreds of km (as they are, e.g. along the highly stable cratonic coasts of Brazil, West Africa, India, Western Australia, and on numerous mid-ocean islands), it would not be logical to attempt to explain away their tide-gauge anomalies in terms of neotectonics. Short-term variation of oceanographic dynamics, in contrast, are appropriate to the climatic trends of the present century. An extended review of the eustatic question, with a tentative and generalized eustatic curve was presented by the writer (Fairbridge, 1961), but not universally approved, largely because of regional tectonism.

(g) At the meeting, Cliff Ollier (Armidale, NSW, Australia) reviewed the coastal record of Australia, noting that some long-term coastal studies, e.g., in southeastern Australia, show that severe coastal erosion of sandy beaches has resulted because of increases in the frequency and direction of coastal storminess. No sea level change is involved.

Long-term alternations in storminess and relative calm climate cycles have been systematically dated in many parts of the world (e.g. by Curray in Mexico, Moore in Alaska, Fairbridge & Hillaire-Marcel in Arctic Canada, Helle and Alestalo in Finland and by Fairbridge, and Searle & Woods in S.W. Australia).

The periodicities are mostly in the range of 11 to 45 yr, which seem to suggest some sort of astronomic forcing, although much more research is still needed. The climatologic spectral study by Guiot (1988) for Arctic Canada shows that both solar and lunar periodicities are present. The fact that the 45-yr period seems to be as well represented in the southern hemisphere (S.W. Australia) as in Canada certainly implies a planetary phenomenon. The impression left with the writer is that most of the coast erosion of unconsolidated materials from around the world as reported by Bird's IGU commission is directly due to a cyclic change in prevailing wind and swell vectors.

But one thing is perfectly clear: dynamic fluctuations of high frequency over the last few thousand years are part of the natural signal of climatic fluctuations that have had nothing to do with anthropogenic influences. In order to gauge the amplitude of man-made climatic effects we must learn more about the natural "ground-swell".

2. What About the Global Water Availability?

First, what is the global availability and use of fresh water? Some 96.5% of the world's water is salty. Of the rain or snow that falls annually on the land surfaces of the globe, i.e. a mean precipitation of 800 mm, a total of some 119,000 km^3, much is lost to evaporation, evapotranspiration or seepage, but 47,000 km^3 returns to the ocean annually by rivers or submarine springs (the latter about 4%). Over a 20yr interval, the annual discharge varies from about 45,000 to 50,000 km^3, thus a 1% variance. For comparison, a 0.1% part of this discharge (500 km^3) is equal to a 1.2 mm change in sea level. The Amazon alone discharges about 7 km^3 each year (220,000 m^3/sec).

Out of the total world evaporation, 87.5% comes from the ocean. The problem with this abundance of water is its uneven distribution. Some parts of the world (much of Africa, the Middle East, Pakistan, Australia, N.E. Brazil, and parts of the United States) have too little, at least, from time to time. It seems, to this writer, that those times of shortage are critical, and any ways of predicting or anticipating them should be vigorously explored. Advance planning by administrations and world agencies should not be left on an ad hoc basis.

Human management of the hydrologic cycle is already showing that *in principle* the idea of management is perfectly feasible. L'vovich (1978) states that we now manage about 15% of the Earth's river discharge. In the quarter-century 1957-1982, an average of 125 km^3/yr was retained in the 107 major dam storages (Newman & Fairbridge, 1986). About 3000 km^3/yr is used in irrigation. Assuming the annual increase in planned capacity to be a reasonable approximation of downstream discharge reduction, this value would constrain sea-level rise. To the 125 km^3/yr for major dams, we have taken another 125 km^3/yr as a minimum for the retention by the millions of small dams and barrages, and a similar value for the added amount lost to evaporation or retained as soil moisture and seepage, and thus annually lost to river-mouth discharge. Thus a total of 375 km^3/yr represents the annual reduction in fluvial discharge to the world ocean. This value is equivalent to a fall of world sea level of close to 1.00 mm/yr, i.e. an amount of water be deducted from any rise that might otherwise have taken place, if no dams had been built. If there were no steric change or eustatic rise, mean global sea level would be falling at the present time.

The contrary effects may be balancing one another.

Mankind's total use of water is divided into two categories: the water that is "borrowed" and then returned to rivers (e.g., in hydropower and most domestic supplies), and the "irrecoverable" fraction (e.g., lost to evapotranspiration and during irrigation). The irrecoverable fraction, according to L'Vovich (1978), was 270 km^3 (out of 400) in AD 1900, 1600 km^3 (out of 2600) in 1970, and will be 3000 km^3 (out of 6000) in AD 2000. Theoretically, this should cause a fall of global sea level, were it not for progressive deforestation which is increasing runoff.

At the meeting these approximations were discussed at some length. The trouble is: the monitoring of the global hydrologic cycle is still very inadequate, and we clearly need much more data. A meticulous study of available discharge rates of some selected, but representative rivers over an 80-yr span by Yves Tardy (France) disclosed a small but steady increase during the present century. If we assume that over long periods like this the mean solar radiation is constant and that the mean evaporation rate from the world ocean is constant, then the secular change in discharge requires a special explanation. One important watershed parameter is NOT constant: deforestation. This has been constantly rising: indeed, it is accelerating at an alarming rate in the last decade. The fall in world sea level due to dam construction appears to be neatly countered by increased runoff due to deforestation, besides glacio-eustasy and steric expansion.

In the light of its multiple ecological and hydrological effects there is little doubt that tropical deforestation is one of the great contemporary threats to the environment by mankind.

However, in a recent article, Colvinaux (1989) emphasizes the long-term resilience of equatorial forests on a glacial/interglacial time scale. Several centuries, however, may be needed for regrowth. Several of us have observed how the forests of eastern New Guinea, cleared for cattle-raising half a century ago, have still not recovered. Damuth & Fairbridge (1970) showed that semi-arid erosion (and feldspar preservation in sediments) replaced the humid, chemical weathering of the interglacial phase. The writer has long asserted that "MAN IS AN ICE-AGE AGENT" (Fairbridge, 1976).

The great tropical equatorial rain forests (Central America, Amazon, Congo/Zaire, and Southeast Asia) are essentially a series of isolated endogenetic climate and hydrologic systems. They are not like the boreal and temperate forests which are sustained by onshore winds carrying moisture evaporated from the world ocean, a hydrologic transfer that is balanced by evapotranspiration and return runoff. As emphasized by many observers and outlined at the meeting by Marcel Leroux (Lyon, France), the low-latitude forests develop local climates with convective cells that diurnally return as precipitation most of the moisture evaporated. The act of cutting down extensive segments of this forest interrupts a critical part of the hydrologic cycle and can change a high rainfall area into a semi-arid one. In the high latitudes, in contrast, the usual forest-management economy commonly has little effect on global CO_2 balance because the lumbered areas are usually in strips that are immediately replanted (e.g., in Finland, eastern Canada, South Korea, and USSR). This low-damage scenario is not true, however, for high-relief areas (e.g., western United States) where even limited clear-cutting may have catastrophic effects on soil erosion and runoff acceleration.

If there are some fundamental answers to be obtained from this brief consideration of water availability, they must include the following statements: Yes, global water is abundant; BUT, it is not naturally distributed evenly; Yes, it can be managed and it does have an effect on sea level. Most importantly, environmental damage from uncontrolled exploitation can be catastrophic, not only locally, but globally. In other words, the timber-feller and ranch-clearers of the western United States, Brazil and elsewhere have been guilty, not wilfully, but guilty of ignorance of the fundamental laws of nature and earth science.

3. What About the Drought Question?

First of all, one man's "drought" is merely a "dry spell" to another. In 1988 Mrs. Thatcher declared a drought emergency for parts of Britain. In March, 1989, Mayor Koch declared a drought emergency for New York City. Both examples reflect the *relative* meaning of the term, inasmuch as both areas lie in the climatic zone of the humid westerlies, and neither can be classified strictly as an under-developed region. These water shortages can hardly be compared with the droughts in sub-Saharan Africa (notably Sahel-Ethiopia) with tens of millions dying of starvation or suffering irreversible malnutrition. Both questions were addressed at the conference.

Geologists and archaeologists have proved conclusively that during the last few thousand years extraordinary shifts of climate belts have taken place, most dramatically along the northern and southern margins of the great desert belts like the Sahara and Kalahari, and those of Rajasthan, Central Australia and southwestern United States. In the 1950's French archaeologists (notably Balout) discovered that the present Sahara desert was a relatively humid savanna during the Neolithic period, and extensive parts were still well-watered in the days of Imperial Rome. The widespread Neolithic lakes of the Sahara were discussed already by Faure (1966), and their recent exploration reviewed for the meeting by Petit-Maire (1986).

At the meeting, Roland Paepe (chairman of the executive committee, and director of E.T.I, Free University of Brussels) emphasized the global importance of these climatic shifts. Not only are they apparent on the two well-known glacial/interglacial-type cyclicities (400 Ka and 100 Ka) but even on the scale of 10 Ka and shorter, as shown by Paepe's own researches in Belgium, Iraq and Greece. While the larger cycles are most clearly demonstrated in the deep-ocean sediments, the shorter ones (because of deep-sea bioturbation) are difficult to separate and Paepe drew attention to the unique value of paleosols and river deposits in this respect.

Ancient lake deposits in the Mediterranean area also provide long-term evidence. As discussed by George Kukla (LGDO, New York), the Macedonian cores obtained by Van der Hammen, span no less than 0.9 million years with wet-dry cyclicity of the order of 20,000 years, clearly the Milankovitch precession cycle that controls terrestrial seasonality. Farther inland in central Europe, central Asia and northern China, the loess cycles (dry/humid alternation) number no less than 44 over a 2.5 million year interval (Kukla, 1988). The Chinese delegate Liu Tungshen spoke of the loess-type dust generated over northern China every winter. It is generated by wind erosion in the Gobi desert and other semi-arid areas of central Asia. Its cyclicity is clearly evidence of strong climatic fluctuations.

In Greece (and elsewhere in the Mediterranean belt) dramatic fluctuations between signs of extreme desiccation of mainly arid periods and evidence of abundant (winter-type) rains in between may be seen in both outcrops and deep borings. Cycles of the order of 500 to 1000 years are demonstrated even during the era of written (human) history. Evidently the political history of these regions needs to be rewritten in terms of climatic-environmental determinism. It is debatable how far aridification has been anthropogenically amplified, but if fundamentally natural in principle, it has shifted progressively through time in a latitudinal sense. Secular changes are thus constantly in progress.

At the decadal to century scale the studies by Sharon Nicholson (Florida) and Peter Lamb (Illinois) illustrate the modern and historical climatologic data bases (Nicholson & Flohn, 1980; Lamb, 1980). Particularly striking is the fact that in Morocco the precipitation fluctuates on a wavelength of several decades (calculated from numerous tree-ring data for a 1000-yr interval by Claudine Till and Joel Guiot (1988). For the Sahel using the longest available rainfall record from Mauritania (Kaedi, 1904-1988), it was shown (by Gaston Demarée, Brussels) that the decadal-type fluctuations in the early part of the century were grossly exceeded by those associated with the drought that began in 1968.

The rise and fall of Morocco precipitation is often exactly out of phase with that of the Sahel as is shown by Lamb and Peppler (1987). Offshore, in the tropical atlantic it is found that sea-surface temperatures are useful monitors, discussed at the meeting by Benno Blumenthal (LGDO, New York). The non-regional, i.e. global importance of the Sahelian drought history has already been shown (by Klaus, 1980) to be closely reflected by the 0-18/0-16 ratios (cold phases favour the heavier isotope).

It has also been noticed recently that the El Niño years of the Pacific are followed also by fluctuations in the Benguela Current off Namibia with warm years matched by high sea levels (Brundrit et al., 1987), by heavy rains in Brasil's Northeast (Hastenrath, 1985) and in the Sahel, but by droughts in Morocco and the Mediterranean belt. Following the droughts usually come extremely heavy rains (Till, 1988). The SST warming and high MSLs seem to propagate from the equator polewards.

The episodicity of periodicity of drought intervals can also be gauged by the supply of dust from desert regions. Louis Berkofsky (Beer Sheva, Israel) has shown experimentally that as aridification proceeds, the roughness index of the soil surface increases accompanied by progressive deflation. Modelling discloses a positive feedback develops that accelerates desertification. Yves Tourre (LGDO, New York; and ORSTOM, Paris) has demonstrated that a positive pressure correlation exists between Darwin, Australia, and the tropical Atlantic, in fact more powerful than the classic Southern Oscillation. He calls it the "PAO" (Pacific-Atlantic Oscillation). A low pressure anomaly and SST warming at Darwin (SO$^+$) is found to precede (leading by 8 to 14 month) the lows and SST warmings in the Tropical Atlantic. The Darwin high pressure (SO$^-$) which often follows the low, is matched by a low in the Eastern Pacific that is from time to time associated with an El Niño. This explains in part the above-mentioned relationship of the ENSO to the rainy periods in the Sahel. An additional teleconnection exists between the strong ENSO years and the Indian droughts (monsoon failures), as shown in a table of monsoons back to AD 1781 (Fairbridge, 1986).

4. Historical and Recent Geological Record of Drought

The meeting addressed itself only to the large-scale phenomenon, and especially to the Sahel drought belt of North Africa, where the ancillary phenomenon of famine or traumatic undernourishment hat attracted a good deal of worldwide attention. In order to adjust to time limitations for the workshop, it was decided that almost no consideration would be given to the admittedly serious water deficiencies in other parts of the globe.

The critical question of what to do about drought on a continent-wide scale has been severely clouded, up to the present day, by the unresolved controversy between (a) the socio-economic arguments and (b) the geological-paleoclimatic arguments. At the meeting, the 20th century acceleration of dry years and environmental deterioration was brought out by several speakers which by itself might support the case for a one-way anthropogenic desertification brought on by several factors such as overpopulation, political mismanagement and so on. The historico-geological record shows, however, that catastrophic droughts are nothing new in the Sahel, and have been recorded in both written history and, earlier, through many thousand years of the geologic record.

At the meeting, Nicole Petit-Maire (Marseille; leader of the IGCP-252 Desert program) illustrated some aspects of the astonishing history of the Saharan Holocene paleoclimate. The former existence of giant freshwater lakes there has been dramatically testified by the discovery of rich fossil deposits, including fresh-water shell beds up to 10 m thick. But equally important is the fact that these lakes dried up completely from time to time, in their late phases accumulating many metres of salt deposits. Those Northwest of Timbuctu in Mali, have supported a continuous commercial salt exploitation for at least 2000 years.

Reporting on the eastern Sahara, Stefan Kröpelin (West Berlin) has show evidence that at times in the mid-Holocene the Sahel belt reached 500-600 km farther North, from about 17° N to 22° N (roughly the Egypt-Sudan border). In contrast, the same boundary shifted some 400 km farther South in the late Pleistocene, at the same time as sea level has fluctuated through more than 130 m.

Comparable changes in the coastal belt of Mauritania and Senegal are referred to in a report submitted by J.P. Barusseau, C. Descamps and P. Giresse (of Perpignan), which correlate in a very significant way with sea-level fluctuations. This belt is a highly stable sector and thus the up's and down's of sea level cannot be explained by neotectonics, but can provide evidence for eustatic control. Early Man camped on the pre-existing beach ridges (sites known as "Sambaqui" in Brazil) exploiting the abundant shell fisheries. Dating the middens has disclosed an occupation and economy chronology. A notable feature that was discovered on the Mauritanian, (i.e., northern) sector is believed by the writer to be a non-local phenomenon. The sudden decline in abundance of *anadara senilis* (the warm-water cockle) at about 2800 BP that brought the Neolithic food supply to a halt, coincided with a 1-2 m fall of sea level and the onset of a new and unfavourably dry climate. The coastal geomorphology also underwent a dramatic change. Up to that date (2800 BP) the long-shore drift was from South to North, evidently dominated by the strong monsoon winds and an expanded intertropical convergence zone (Hastenrath, 1986). After that time the principal shore drift abruptly switched around from North to South, as illustrated by the enormous South-oriented sand spit at Cap Blanc in Mauritania. In comparable latitudes around the world analogous shifts in wind systems document the beginning of a late-Holocene drought-prone "era", e.g., in the Gulf of California (Curray, 1967). In contrast to the coast of Mauritania, on the southern part of the Senegal coast (Soloum delta) it is evident that the warm coastal waters prevailed and a more or less continuous shell-based food economy has continued up till the present day.

The point of this discussion is to bring out the fact that important global controls must be responsible for cultural changes of this type. To try and explain the latter in sociological terms is to ignore the main agency. Those socio-economic problems that have been widely associated with the various aspects of arid climatic cycles such as drought, desertification and specific man-made factors that collectively exacerbate the associated famine and related human catastrophes should always by perceived within the framework of global forcing functions of strictly natural origin. It is simply a cruel hoax, or wilful self-delusion that leads some writers to assume either that there are simple socio-political solutions for drought problems, or even worse, that the situation is hopeless and that there is no solution.

The problems of drought amelioration are multiple, but the fundamental question is immediately answerable from the evidence of the established record: droughts are normal experiences in the semi-arid regions. And this carries the corollary: if we can establish an adequately well-documented record of historico-geological droughts and their dynamic linkage with other long-term climatic time series, then we should be able to predict future drought potentials, at least on some basis of probability.

Several speakers referred to the potential of the rising level atmospheric "greenhouse gases" (CO_2, CH_4, CFC's, etc.). Modelling studies by Hansen et al. (1988), for example, suggest that in the next century Sahelian temperatures may be higher than today, although the rising sea-surface temperatures may well increase the oceanic evaporation so as to augment monsoonal duration and cloud cover which in turn could quite well reduce summer temperatures (although winter temperatures would become higher). The possibility of heavier monsoonal rains brings a message of hope for the Sahel, although the continued role of the natural fluctuating cycles is not likely to be totally eclipsed.

The nature and the predictability of the natural cycles was addressed at the meeting by the President. A paper published by Fairbridge and Shirley (1986) was based upon an over 12,000 year analysis of planetary motions that demonstrate fundamental cycles of 19.857 years and 178.713 years, set up by the outer planets and affecting the angular momentum of the sun. Each cycle varies somewhat from its preceding one. During the last millennium it is known that intervals of reduced solar activity (known as the Oort, Wolf, Spörer, Maunder minima) correspond to times when the C-14 flux, as recorded in tree-rings, is high; these high-flux intervals reflect higher cosmic ray incidence that is normally shielded by the Earth's magnetic field which in turn is modulated by the solar wind, thus reflecting solar activity. Fairbridge & Shirley showed that those low activity intervals on the Sun correspond to the 178-year cycle (varying with each recurrence) and to calculated variance in the Sun's angular momentum and circum-barycentric variance. Thus, with certain reservations, the Sun's radiation potential can be calculated and predicted (forward and backward) on the basis of astronomic ephemerides and formulae of orbital motion and dynamics.

A second series of calculations refer to the orbital motions and gravitational potential of the Moon. These periodicities are found to be commensurable with those of the Sun and Planets. The subject has been reviewed by Fairbridge and Sanders (1986, to which a lengthy bibliography is supplied). Recent research discloses that the El Niño appears to be triggered by internal waves at the oceanic density boundaries, those waves being modulated by lunisolar tides. Again, a predictive potential is now offered. The El Niño is now recognized as not merely a Pacific phenomenon, but spreads to the Atlantic, often with a one-year delay factor. There it relates to warmer waters and heavy rains in the Mediterranean latitudes (e.g., Morocco and the northern Sahara fringe); droughts in the Sahel and the savanna-lands lie south of the subtropical high pressure cells.

5. Practical Ideas About Drought Incidence

Many papers at the meeting introduced ideas and suggestions about the practical treatment of the drought threat, mostly with the Sahel example foremost in mind. These papers were intended rather for "general education" because engineering solutions were not the main thrust of the meeting. Nevertheless, some of the ideas or procedures are worth a few words at this point.

First of all two very disparate approaches became apparent at the meeting:

(a) *Mega-schemes.* Much attention was paid to a major plan for one or multiple large dams along the Ubangi, the most important northern tributary of the Congo/Zaire system. One of our executive committee, Christian de Meyer, has a company, Haecon, that has worked for 10 years on the plans. Another scheme, involving an Italian group, has been reported in the press. Serious environmental problems that may result from the mega-schemes (e.g., Nile, Zambesi, Upper Volta, etc.) were discussed. The denial of sediment to delta coasts and the resultant shore erosion and retreat is particularly important in some cases, but not all; in the Congo/Zaire the proposed scheme would not involve more than 10% of total discharge and year-round flow (analyzed at the meeting by Le Houérou) might help overcome the problem; in any case, much of the sediment catchment is upstream of the gorge and the principal marine sedimentation is one the continental rise (i.e., far beyond the coastal belt).

(b) *Small-scale Strategies.* Considerable attention was paid to various practical ways of capturing and subsurface storage on a scale appropriate to an single (ephemeral) stream bed or a single village. This "small is beautiful" strategy is appealing because it can be achieved without a vast superstructure, can be installed and maintained by the villagers themselves.

6. References

Ahlmann, H.W. (1953) "Glacier variations and climatic variations", Amer. Geogr. Soc., Bowman Mem. Lect. 2

Barnett, T.P. (1984) "The estimation of global sea level change: a problem of uniqueness", J. Geophys. Res. 89(C5), pp. 7980-7988

Barth, M.C., & Titus, J.G. (eds.) (1984) *Greenhouse Effect and Sea Level Rise*, Van Nostrand Reinhold, New York

Brundrit, G.B., De Cuevas, B.A., & Shipley, A.M. (1987) "Long-term sea-level variability in the eastern South Atlantic and a comparison with that of the eastern Pacific", South African J. Mar. Sci. 5, pp. 73-78

Curray, J.R., Emmel, F.J., & Crampton, P.J.S. (1967) "Holocene history of a strand plain, lagoonal coast, Nayarit, Mexico", in Costenares, A.A., et al. (eds.) *Laguna Costeras*, pp. 63-100 (reprinted in Swift, D.J.P. and Palmer, H.D. (1978) *Coastal Sedimentation*, Dowden, Hutchinson & Ross, Stroudsburg)

Damuth, J.E., & Fairbridge, R.W. (1970) "Equatorial Atlantic deep-sea arkosic sands and ice-age aridity in tropical South America", Geol. Soc. Amer. Bull. 81, pp. 189-206

Dean, R.G., et al. (1987) *Responding to Changes in Sea Level: Engineering Implications*, National Academy Press, Washington

Fairbridge, R.W. (1961) "Eustatic changes in sea level", in Ahrens, L.H., et al. (eds.) *Physics and Chemistry*, Pergamon Press, London, 4, pp. 99-185

Fairbridge, R.W. (1964) "The importance of limestone and its Ca/Mg content to palaeoclimatology", Wiley-Interscience, London, pp. 431-530

Fairbridge, R.W. (1980) "Prediction of long-term geologic and climatic changes that might affect the isolation of radioactive waste", in *Underground Disposal of Radioactive Waste*, vol. 2, International Atomic Energy Agency, Vienna, 1 AEA-SM-243/43, pp. 385-405

Fairbridge, R.W. (1986) "Monsoons and paleomonsoons", Episodes 9(3), pp. 143-149

Fairbridge, R.W. (1990) "Solar and Lunar cycles embedded in the El Niño periodicities", Cycles 41(2), pp. 66-73

Fairbridge, R.W., & Krebs, O.A.Jr. (1962) "Sea level and the Southern Oscillation", Geophys. Jour. 6, pp. 532-545

Fairbridge, R.W., & Sanders, J.E. (1987) "The Sun's orbit, A.D. 750-2050: basis for new perspectives on planetary dynamics and earth-moon linkage", in Rampino, M.R., et al. (eds.) *Climate: History, Periodicity, and Predictability*, Van Nostrand Reinhold, New York, pp. 446-471 (with extended Bibliography, pp. 475-541)

Fairbridge, R.W., & Shirley, J.H. (1987) "Prolonged minima and the 179-yr cycle of the solar inertial motion", Solar Physics 110, pp. 191-220

Faure, H. (1966) "Evolution des grand lacs sahariens à l'Holocène", Quaternaria 8, pp. 167-175

Gornitz, V., Lebedeff, S., & Hansen, J. (1982) "Global sea level trend in the last century", Science 215, pp. 1611-1614

Guiot, J. (1987) "Reconstruction of seasonal temperatures in central Canada since A.D. 1700 and detection of the 18.6 - and 22-year signals", Climatic Change 10, pp. 249-268

706

Hansen, J., et al. (1988) "Global climate changes as forecast by Goddard Institute for Space Studies three-dimensional model", Jour. Geophys. Res. 93, pp. 9341-9364

Hastenrath, S. (1985) *Climate and Circulation in the Tropics*, Reidel, Dordrecht

Kukla, G.J. (1987) "Loess stratigraphy in central China", Quaternary Science Reviews 6, pp. 191-219

L'Vovich, M.I. (1979) *World Water Resources and their Future*, Amer. Geophys. Union (translation of Russian text, 1973), Washington

MacCracken, M.C., & Luther, F.M. (1985) *Detecting the Climatic Effects of Increasing Carbon Dioxide*, U.S. Dept. of Energy, DOE/ER-0235, Washington

Meier, M.F. (1984) "Contribution of small glaciers to global sea level", Science 226(468), pp. 1418-1421

Meier, M.F., et al. (1984) *Glaciers, Ice Sheets and Sea Level: Effect of a CO_2-induced Climatic Change*, U.S. Dept. of Energy, DOE/ER/60235-1, Washington

Newman, W.S., & Fairbridge, R.W. (1986) "The management of sea level rise", *Nature* 320, pp. 319-321

Nicholson, S.E., & Flohn, H. (1980) "African environmental and climatic changes and the general atmospheric circulation in Late Pleistocene and Holocene", Climate Change 2, pp. 313-348

Nierenberg, W.A., et al. (1983) *Changing Climate*, National Academy Press, Washington

Petit-Maire, N. (1986) "Paleoclimates in the Sahara of Mali: a multidisciplinary study", Episodes 9(1), pp. 7-16

Thorarinsson, S. (1940) "Present glacier shrinkage and eustatic changes of sea level", Geogr. Annaler, Stockholm, 22, pp. 131-159

Till, C. (1988) "Tree-ring pointer years as proxy records of past climatic events in Morocco since A.D.", ms. in preparation

Titus, J.G. (ed.) (1986) *Effects of Changes in Stratospheric Ozone and Global Climate*, Vol. 4: *Sea Level Rise*, Environmental Protection Agency (and UN Environmental Programme), Washington

Wigley, T.M.L., & Raper, S.C.B. (1987) "Thermal expansion of sea water associated with global warming", *Nature* 330, pp. 127-131

INDEX

712

714